SPRINGER PROCEEDINGS IN PHYSICS 91

Springer
Berlin
Heidelberg
New York
Hong Kong
London
Milan
Paris
Tokyo

SPRINGER PROCEEDINGS IN PHYSICS

79 **Nonlinear Dynamics and Pattern Formation in Semiconductors and Devices**
Editor: F.-J. Niedernostheide

80 **Computer Simulation Studies in Condensed-Matter Physics VIII**
Editors: D.P. Landau, K.K. Mon, and H.-B. Schüttler

81 **Materials and Measurements in Molecular Electronics**
Editors: K. Kajimura and S. Kuroda

82 **Computer Simulation Studies in Condensed-Matter Physics IX**
Editors: D.P. Landau, K.K. Mon, and H.-B. Schüttler

83 **Computer Simulation Studiesin Condensed-Matter Physics X**
Editors: D.P. Landau, K.K. Mon, and H.-B. Schüttler

84 **Computer Simulation Studies in Condensed-Matter Physics XI**
Editors: D.P. Landau and H.-B. Schüttler

85 **Computer Simulation Studies in Condensed-Matter Physics XII**
Editors: D.P. Landau, S.P. Lewis, and H.-B. Schüttler

86 **Computer Simulation Studies in Condensed-Matter Physics XIII**
Editors: D.P. Landau, S.P. Lewis, and H.-B. Schüttler

87 **Proceedings of the 25th International Conference on the Physics of Semiconductors**
Editors: N. Miura and T. Ando

88 **Starburst Galaxies**
Near and Far Editors: L. Tacconi and D. Lutz

89 **Computer Simulation Studies in Condensed-Matter Physics XIV**
Editors: D.P. Landau, S.P. Lewis, and H.-B. Schüttler

90 **Computer Simulation Studies in Condensed-Matter Physics XV**
Editors: D.P. Landau, S.P. Lewis, and H.-B. Schüttler

91 **The Dense Interstellar Medium in Galaxies**
Editors: S. Pfalzner, C. Kramer, C. Straubmeier, A. Heithausen

92 **Beyond the Standard Modell 2003**
Editor: H.V. Klapdor-Kleingrothaus

93 **ISSMGE: Experimental Studies**
Editor: T. Schanz

94 **ISSMGE: Numerical and Theoretical Approaches**
Editor: T. Schanz

95 **Computer Simulation Studies in Condensed-Matter Physics XVI**
Editors:D.P. Landau, S.P. Lewis, and H.-B. Schüttler

96 **Electromagnetics in a Complex World**
Editors: I.M. Pinto, V. Galdi, and L.B. Felsen

Homepage: springeronline.com

Volumes 50–78 are listed at the end of the book.

S. Pfalzner C. Kramer
C. Staubmeier A. Heithausen
(Eds.)

The Dense Interstellar Medium in Galaxies

Proceedings of the
4th Cologne-Bonn-Zermatt-Symposium
"The Dense Interstellar Medium in Galaxies"
Zermatt, 22-26 September, 2003

Springer

Priv.-Doz. Dr. Susanne Pfalzner
Priv.-Doz. Dr. Carsten Kramer
Dr. Christian Straubmeier
Universität Köln
I. Physikalisches Institut
Zülpicher Str. 77
50937 Köln
Germany

Priv.-Doz. Dr. Andreas Heithausen
Universität Bonn
Radioastronomisches Institut
Auf dem Hügel 71
53121 Bonn
Germany

ISSN 0930-8989

ISBN 3-540-21254-X Springer-Verlag Berlin Heidelberg New York

Cataloging-in-Publication Data applied for

Bibliographic information published by Die Deutsche Bibliothek Die Deutsche Bibliothek lists this publication in the Deutsche Nationalbibliografie; detailed bibliographic data is available in the Internet at <http://dnb.ddb.de>

Springer-Verlag is a part of Springer Science+Business Media

springeronline.com

Printed in Germany

Typesetting by the authors/editors
Data conversion: Steingraeber Satztechnik GmbH Heidelberg
Cover concept: eStudio Calamar Steinen
Cover production: *design & production* GmbH, Heidelberg

Printed on acid-free paper 62/3141/ts 5 4 3 2 1 0

Preface

The "4th Cologne-Bonn-Zermatt Symposium: The Dense Interstellar Medium in Galaxies" took place in Zermatt, Switzerland, from September 22nd to 26th, 2003. This symposium continued the series of conferences on the dense interstellar medium which has been organized every 5 years since 1988 by the I. Physikalisches Institut der Unversität zu Köln. Reflecting the by now well established and fruitful collaboration between the MPI für Radioastronomie, the Radioastronomisches Institut der Universität Bonn and the I. Physikalisches Institut der Universität zu Köln in the context of the Sonderforschungsbereich 494 "The Evolution of Interstellar Matter: THz-Spectroscopy in the Laboratory and in Space", the symposium is now organized jointly by these three institutes.

The conference covered a wide range of topics: starting at the largest scales and earliest times in the history of the Universe, it focussed next on galaxy evolution and their structure and dynamics. New and very exciting results on the black hole in the center of the Milky Way and the violent ISM in its immediate vicinity were presented. It continued on the cosmic cycle of star formation and evolution, and reached down to the microphysical processes determining the physical and chemical conditions of the interstellar matter as the important agent for this cosmic cycle. The conference thus brought together a broad spectrum of scientists: well established experts in their respective fields as well as young postdocs and students were confronted with the amazing breadth and complexity of the physics and chemistry of the interstellar medium. This gave a good opportunity for many stimulating discussions, which went far beyond the lecture series and poster presentations and continued into hiking excursions and other, smaller group activities, well supported by the splendid weather and the stimulating surroundings of Zermatt.

The conference had 289 participants from all over the world. The presentations included 16 reviews, 34 invited and 37 contributed talks, as well as 187 posters. The large interest in the conference, resulting in the traditional limit of 200 participants being exhausted already 2 months before the registration deadline, stimulated us to extend the size of the conference. Generous support by the staff of the Zermatterhof and the Burgergemeinde Zermatt made the accommodation of a larger number of participants possible by clever arrangement of the splendid conference center facilities and in particular the perfect logistics and service during the coffee breaks.

The conference has shown the impressive progress in the field over the last 5 years, nicely summarized in the topical reviews in this volume. It has also shown the great expectations that we face ahead of us, given the large new observatory missions and telescope projects and the still rapidly progressing detector and instrument technology. These great opportunities call for a strong effort in the complementary laboratory research and theoretical studies, that will be necessary to fully exploit this unique potential.

With a special session on "Molecular Spectroscopy and Astrophysics" we took the opportunity to honour Gisbert Winnewisser: one of the outstanding scientists who has devoted most of his career to this aspect of combining laboratory and astrophysical research. Gisbert retired in September 2001. His many important and still ongoing contributions to the field are well documented in the invited lectures from this special session, but also in several of the contributed talks and posters. Gisbert has been the driving force in establishing the series of Cologne-Bonn-Zermatt symposia and is certainly the one to be credited most for making them as successful as they are. This conference was the first one that we organized without his immediate involvement and we can only hope that we can meet the expectations towards continuing his work. On behalf of the organizers and also for the whole community we want to say once again: thank You, Gisbert, for having, over the years, given us the great opportunity of these stimulating and splendid conferences.

A personal revue of Gisbert's career, covering as well some interesting historical aspects of the early times of microwave spectroscopy, was delivered by his brother Manfred at the occasion of the conference reception, and is reproduced in a separate booklet delivered to the conference participants together with the proceedings. This booklet also contains an article by Gisbert in honour of Gerhard Herzberg, who died on March 23rd, 1999 at the age of 95 and to whom we devoted the session on astrochemistry; this article is the summary of Gisbert's opening remarks for this session.

The 4th Cologne-Bonn-Zermatt symposium would not have been possible without the help of many people. We thus thank the many members of the three organizing institutes who supported the conference with their activities. We want to mention in particular Andreas Heithausen, Carsten Kramer, Christian Straubmeier, Robert Simon and Friedrich Wyrowski as members of the local organizing committee, who clearly carried the largest burden in preparing the conference. A particular "Thank You" also goes to Steffi Krämer and Marianne Dieckmann as the responsible conference secretaries for their smooth handling of many special wishes (as we all know, scientist come up with innovative approaches; this also holds for conference registration, payment rules and many other issues), and to Frank Krämer for his continuous support with all kinds of expected and unexpected bugs and features of the audio and beamer system in the conference center.

A particular "Thank You" goes to the members of the scientific organizing committee: their advice was essential in composing the exciting program and in selecting the excellent speakers; and to the editors of the proceedings and a number of volunteering proof readers, who helped with their patience and endurance to combine these talks into this book as a valuable source of reviews and up-to-date reports on the various fields and topics.

A special "Merci vielmals" goes to the staff of the Zermatterhof under the direction by J.-P. Lanz and the Burgergemeinde Zermatt under its president Andreas Biner for their excellent support in preparation of and during the conference. And, last not least, to the people of Zermatt, who, as generous hosts to the scientists visiting the conference, are largely responsible for the pleasant atmosphere that contributed to making the conference a success.

Finally, we want to thank the various financial sponsors of the conference, who helped in making the conference possible:

- the Deutsche Forschungsgemeinschaft through the Sonderforschungsbereich 494
- the Universität zu Köln, the Universität Bonn and the MPI für Radioastronomie
- the Internationale Stiftung Hochalpine Forschungsstationen Jungfraujoch und Gornergrat
- the Burgergemeinde Zermatt and the Gornergrat-Monte Rosa-Bahnen.

Köln
December 2003

Andreas Eckart
Ulrich Mebold
Karl Menten
Jürgen Stutzki

Scientific Organizing Committee:

J. Cernicharo
F. Combes
D. Downes
A. Eckart
Y. Fukui
O. Gerhard
M. Gerin
T. Henning
E. Herbst
K. Menten
M. Morris
P. Saraceno
P. Shaver
A. Sternberg
J. Stutzki
P. Strittmatter
E. van Dishoeck
G. White
C. Wilson
H. Yorke

Local Organizing Committee:

A. Eckart
A. Heithausen
C. Kramer
S. Pfalzner
R. Schieder
C. Straubmeier
J. Stutzki
F. Wyrowski

Contents

Part IV Structure and Dynamics of the ISM in the Milky Way

Part V Galactic Center

Part VI New Observatories and Instrumenation

Part VII Magnetic Fields in the ISM

Part I

Gas and Dust at High Redshifts

The First Stars and the Epoch of Reionization

V. Bromm and A. Loeb

Department of Astronomy, Harvard University, Cambridge, MA 02138, U.S.A.

Abstract. We discuss the physical processes that govern the formation of the first stars. These so-called Population III stars are predicted to be predominantly very massive, and to have contributed significantly to the early reionization of the IGM. Such an early reionization epoch is inferred from the recent measurement of the Thomson optical depth by the *WMAP* satellite. We address the importance of heavy elements in bringing about the transition from an early star formation mode dominated by massive stars, to the familiar mode dominated by low mass stars, at later times, and present possible observational probes.

1 Introduction

How did the first stars in the universe form, what were their physical properties, and what was their impact on cosmic history? The first stars, formed at the end of the cosmic dark ages, ionized [36, 10] and metal-enriched [13] the intergalactic medium (IGM) and consequently had important effects on subsequent galaxy formation [2] and on the large-scale polarization anisotropies of the cosmic microwave background [15]. *When did the cosmic dark ages end?* In the context of popular cold dark matter (CDM) models of hierarchical structure formation, the first stars are predicted to have formed in dark matter halos of mass $\sim 10^6 M_\odot$ that collapsed at redshifts $z \simeq 20-30$ [2, 37].

Results from recent numerical simulations of the collapse and fragmentation of primordial clouds suggest that the first stars were predominantly very massive, with typical masses $M_* \geq 100 M_\odot$ [3, 4, 20, 1]. Despite the progress already made, many important questions remain unanswered. An example for an open question is: *How does the primordial initial mass function (IMF) look like?* Having constrained the characteristic mass scale, still leaves undetermined the overall range of stellar masses and the power-law slope which is likely to be a function of mass. In addition, it is presently unknown whether binaries or, more generally, clusters of zero-metallicity stars, can form. *What is the nature of the feedback that the first stars exert on their surroundings?* The first stars are expected to produce copious amounts of UV photons and to possibly explode as energetic hypernovae. These negative feedback effects could suppress star formation in neighboring high-density clumps.

Predicting the properties of the first sources of light is important for the design of upcoming instruments, such as the *James Webb Space Telescope*

(JWST)[1], or the next generation of large (> 10m) ground-based telescopes. The hope is that over the upcoming decade, it will become possible to confront current theoretical predictions about the properties of the first stars with direct observational data.

2 Population III Star Formation

The metal-rich chemistry, magnetohydrodynamics, and radiative transfer involved in present-day star formation is complex, and we still lack a comprehensive theoretical framework that predicts the IMF from first principles (see [17] for a recent review). Star formation in the high redshift universe, on the other hand, poses a theoretically more tractable problem due to a number of simplifying features, such as: (i) the initial absence of heavy metals and therefore of dust; and (ii) the absence of dynamically-significant magnetic fields, in the pristine gas left over from the big bang. The cooling of the primordial gas does then only depend on hydrogen in its atomic and molecular form. Whereas the initial state of the star forming cloud is poorly constrained in the present-day interstellar medium, the corresponding initial conditions for primordial star formation are simple, given by the popular ΛCDM model of cosmological structure formation.

How did the first stars form? A complete answer to this question would entail a theoretical prediction for the Population III IMF, which is rather challenging. Let us start by addressing the simpler problem of estimating the characteristic mass scale of the first stars. This mass scale is observed to be $\sim 1 M_{\odot}$ in the present-day universe. To investigate the collapse and fragmentation of primordial gas, we have carried out numerical simulations, using the smoothed particle hydrodynamics (SPH) method. We have included the chemistry and cooling physics relevant for the evolution of metal-free gas (see [4] for details). Improving on earlier work [3, 4] by initializing our simulation according to the ΛCDM model, we focus here on an isolated overdense region that corresponds to a 3σ−peak: a halo containing a total mass of $10^6 M_{\odot}$, and collapsing at a redshift $z_{\rm vir} \simeq 20$ [7]. In Figure 1 (*left panel*), we show the gas density within the central ~ 25 pc, briefly after the first high-density clump has formed as a result of gravitational runaway collapse.

How massive were the first stars? Star formation typically proceeds from the 'inside-out', through the accretion of gas onto a central hydrostatic core. Whereas the initial mass of the hydrostatic core is very similar for primordial and present-day star formation [23], the accretion process – ultimately responsible for setting the final stellar mass, is expected to be rather different. On dimensional grounds, the accretion rate is simply related to the sound speed cubed over Newton's constant (or equivalently given by the ratio of

[1] See http:// ngst.gsfc.nasa.gov.

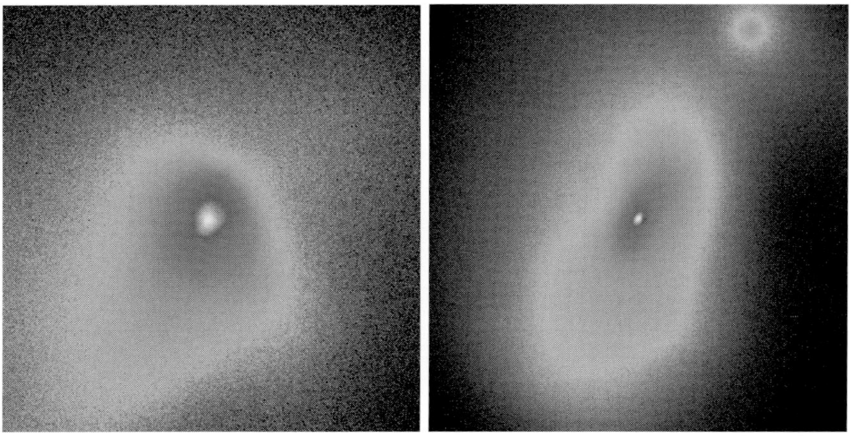

Fig. 1. Collapse and fragmentation of a primordial cloud. Shown is the projected gas density at a redshift $z \simeq 21.5$, briefly after gravitational runaway collapse has commenced in the center of the cloud. *Left:* The coarse-grained morphology in a box with linear physical size of 23.5 pc. *Right:* The fine-grain morphology in a box with linear physical size of 0.5 pc. The central density peak, vigorously gaining mass by accretion, is accompanied by a secondary clump. (Adapted from [7].)

the Jeans mass and the free-fall time): $\dot{M}_{\rm acc} \sim c_s^3/G \propto T^{3/2}$. A simple comparison of the temperatures in present-day star forming regions ($T \sim 10$ K) with those in primordial ones ($T \sim 200-300$ K) already indicates a difference in the accretion rate of more than two orders of magnitude.

Our high-resolution simulation enables us to study the three-dimensional accretion flow around the protostar (see also [24, 27, 31]). We allow the gas to reach densities of 10^{12} cm^{-3} before being incorporated into a central sink particle. At these high densities, three-body reactions [25] have converted the gas into a fully molecular form. In Figure 2, we show how the molecular core grows in mass over the first $\sim 10^4$ yr after its formation. The accretion rate (*left panel*) is initially very high, $\dot{M}_{\rm acc} \sim 0.1 M_\odot$ yr^{-1}, and subsequently declines with time. The mass of the molecular core (*right panel*), taken as an estimator of the proto-stellar mass, grows approximately as: $M_* \sim \int \dot{M}_{\rm acc} {\rm d}t \propto t^{0.75}$ at $t < 10^3$ yr, and $M_* \propto t^{0.4}$ afterwards. A rough upper limit for the final mass of the star is then: $M_*(t = 3 \times 10^6 {\rm yr}) \sim 500 M_\odot$. In deriving this upper bound, we have conservatively assumed that accretion cannot go on for longer than the total lifetime of a very massive star (VMS).

Can a Population III star ever reach this asymptotic mass limit? The answer to this question is not yet known with any certainty, and it depends on whether the accretion from a dust-free envelope is eventually terminated by feedback from the star (e.g., [24, 27, 31, 22]). The standard mechanism by which accretion may be terminated in metal-rich gas, namely radiation pressure on dust grains [35], is evidently not effective for gas with a primordial

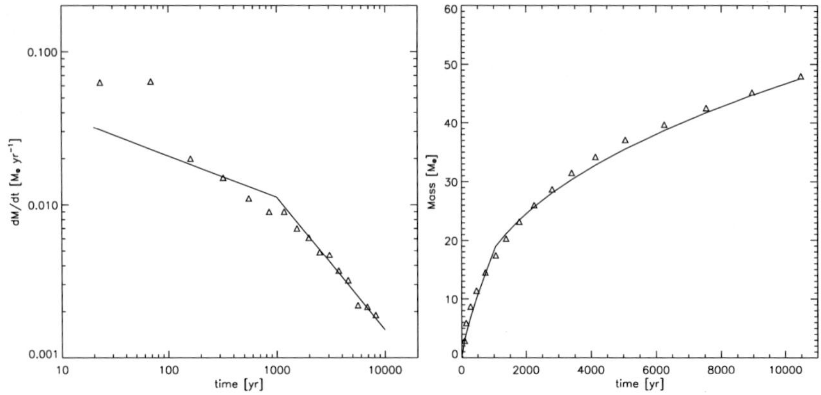

Fig. 2. Accretion onto a primordial protostar. The morphology of this accretion flow is shown in Fig. 1. *Left:* Accretion rate (in $M_\odot$ yr^{-1}) vs. time (in yr) since molecular core formation. *Right:* Mass of the central core (in $M_\odot$) vs. time. *Solid line:* Accretion history approximated as: $M_* \propto t^{0.75}$ at $t < 10^3$ yr, and $M_* \propto t^{0.4}$ afterwards. Using this analytical approximation, we extrapolate that the protostellar mass has grown to $\sim 120 M_\odot$ after $\sim 10^5$ yr, and to $\sim 500 M_\odot$ after $\sim 3 \times 10^6$ yr, the total lifetime of a very massive star. (Adapted from [7].)

composition. Recently, it has been speculated that accretion could instead be turned off through the formation of an H II region [22], or through the radiation pressure exerted by trapped Lyα photons [31]. The termination of the accretion process defines the current unsolved frontier in studies of Population III star formation.

3 The Second Generation of Stars

How and when did the transition take place from the early formation of massive stars to that of low-mass stars at later times? In contrast to the formation mode of massive stars (Population III) at high redshifts, fragmentation is observed to favor stars below a solar mass (Population I and II) in the present-day universe. The transition between these fundamental modes is expected to be mainly driven by the progressive enrichment of the cosmic gas with heavy elements, which enables the gas to cool to lower temperatures. The concept of a 'critical metallicity', $Z_{\rm crit}$, has been used to characterize the transition between Population III and Population II formation modes, where Z denotes the mass fraction contributed by all heavy elements ([21, 5, 29, 30, 18]). These studies have constrained this important parameter to only within a few orders of magnitude, $Z_{\rm crit} \sim 10^{-6} - 10^{-3} Z_\odot$, under the implicit assumption of solar relative abundances of metals. This assumption is likely to be violated by the metal yields of the first SNe at high-redshifts, for which strong deviations from solar abundance ratios are predicted (e.g., [14, 26, 32, 33]).

Recently, we have shown that the transition between the above star formation modes is driven primarily by fine-structure line cooling of singly-ionized carbon or neutral atomic oxygen [6]. Earlier estimates of $Z_{\rm crit}$ which did not explicitly distinguish between different coolants are refined by introducing separate critical abundances for carbon and oxygen, $[{\rm C/H}]_{\rm crit}$ and $[{\rm O/H}]_{\rm crit}$, respectively, where $[{\rm A/H}]= \log_{10}(N_{\rm A}/N_{\rm H})-\log_{10}(N_{\rm A}/N_{\rm H})_\odot$. Since C and O are also the most important coolants throughout most of the cool atomic ISM in present-day galaxies, it is not implausible that these species might be responsible for the global shift in the star formation mode. Under the temperature and density conditions that characterize Population III star formation, the fine-structure lines of O I and C II dominate over all other metal transitions. Cooling due to molecules becomes important only at lower temperatures, and cooling due to dust grains only at higher densities (e.g., [21, 30]). Numerically, the critical C and O abundances are estimated to be: $[{\rm C/H}]_{\rm crit} \simeq -3.5 \pm 0.1$ and $[{\rm O/H}]_{\rm crit} \simeq -3.1 \pm 0.2$.

Even if sufficient C or O atoms are present to further cool the gas, there will be a minimum attainable temperature that is set by the interaction of the atoms with the thermal CMB: $T_{\rm CMB} = 2.7\,{\rm K}(1+z)$ (e.g., [16, 11]). At $z \simeq 15$, this results in a characteristic stellar mass of $M_* \sim 20 M_\odot (n_f/10^4\,{\rm cm}^{-3})^{-1/2}$, where $n_f > 10^4\,{\rm cm}^{-3}$ is the density at which opacity prevents further fragmentation. It is possible that the transition from the high-mass to the low-mass star formation mode was modulated by the CMB temperature and was therefore gradual, involving intermediate-mass ('Population II.5') stars at intermediate redshifts ([18]). This transitional population could give rise to the faint SNe that have been proposed to explain the observed abundance patterns in metal-poor stars ([32, 33]). When and how uniformly the transition in the cosmic star formation mode did take place was governed by the detailed enrichment history of the IGM. This in turn was determined by the hydrodynamical transport and mixing of metals from the first SN explosions (e.g., [19, 8, 28, 34]).

4 Relics from the End of the Dark Ages

It has long been realized that the most metal-poor stars found in our cosmic neighborhood would encode the signature from the first stars within their elemental abundance pattern. For many decades, however, the observational search has failed to discover a truly first-generation star with zero metallicity. Indeed, there seemed to have been an observational lower limit of $[{\rm Fe/H}] \sim -4$ (e.g., [9]). In view of the recent theoretical prediction that most Population III stars were very massive, with associated lifetimes of $\sim 10^6$ yr, the failure to find any 'living' Population III star in the Galaxy is not surprising, as they would all have died a long time ago. Furthermore, theory has predicted

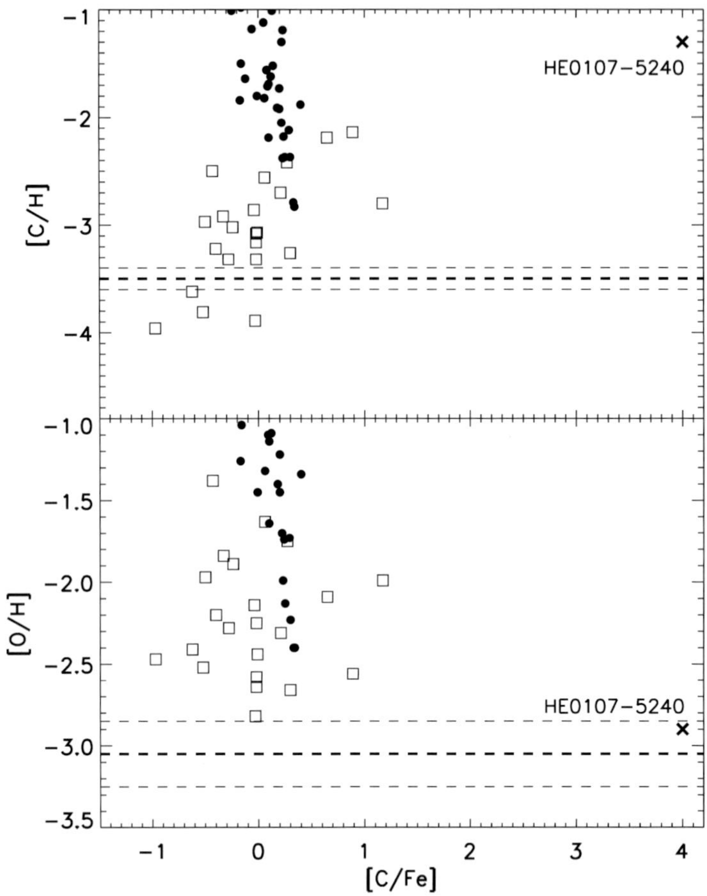

Fig. 3. Observed abundances in low-metallicity Galactic halo stars. For both carbon (*upper panel*) and oxygen (*lower panel*), filled circles correspond to samples of dwarf and subgiant stars, and open squares to a sample of giant stars (see [6] for references). The dashed lines indicate the predicted critical carbon and oxygen abundances. Highlighted (*cross*) is the location of the extremely iron-poor giant star HE0107-5240. (Adapted from [6].)

that star formation out of extremely low-metallicity gas, with $Z \leq Z_{\rm crit} \sim 10^{-3.5} Z_{\odot}$, would be essentially equivalent to that out of truly primordial gas. Again, this theoretical prediction was in accordance with the apparent observed lower-metallicity cutoff.

Recently, however, this simple picture has been challenged by the discovery of the star HE0107-5240 with a mass of $0.8 M_{\odot}$ and an *iron* abundance of $[\mathrm{Fe/H}] = -5.3$ [12]. This finding indicates that at least some low mass stars could have formed out of extremely low-metallicity gas. Does the existence of this star invalidate the theory of a metallicity threshold for enabling low-mass star formation? A possible explanation [33] could lie in the unusually high abundances of carbon and oxygen in HE0107-5240.

In Figure 3, the theoretical C and O thresholds [6] are compared to the observed abundances in metal-poor dwarf and giant stars in the halo of our Galaxy (see [6] for references). As can be seen, all data points lie above the critical O abundance but a few cases lie below the critical C threshold. All of these low mass stars are consistent with the model since the corresponding O abundances lie above the predicted threshold. The sub-critical [C/H] abundances could have either originated in the progenitor cloud or from the mixing of CNO-processed material (with carbon converted into nitrogen) into the stellar atmosphere during the red giant phase. Note that the extremely iron-poor star HE0107-5240 has C and O abundances that both lie above the respective critical levels. The formation of this low mass star ($\sim 0.8 M_{\odot}$) is therefore consistent with the theoretical framework considered by [6].

The lessons from stellar archaeology on the nature of the first stars are likely to increase in importance, since greatly improved, large surveys of metal-poor Galactic halo stars are under way, or are currently being planned.

Acknowledgements

This work has been supported by the NSF (AST 00-71019) and NASA (NAG5-13292).

References

1. Abel, T., Bryan, G. L., & Norman, M. L. 2002, Science, 295, 93
2. Barkana, R., & Loeb, A. 2001, Physics Reports, 349, 125
3. Bromm, V., Coppi, P. S., & Larson, R. B. 1999, ApJ, 527, L5
4. Bromm, V., Coppi, P. S., & Larson, R. B. 2002, ApJ, 564, 23
5. Bromm, V., Ferrara, A., Coppi, P. S., & Larson, R. B. 2001, MNRAS, 328, 969
6. Bromm, V., & Loeb, A. 2003, Nature, 425, 812
7. Bromm, V., & Loeb, A. 2004, New Astronomy, in press (astro-ph/0312456)
8. Bromm, V., Yoshida, N., & Hernquist, L. 2003, ApJ, 596, L135
9. Carr, B. J. 1987, Nature, 326, 829
10. Cen, R. 2003, ApJ, 591, L5
11. Clarke, C. J., & Bromm, V. 2003, MNRAS, 343, 1224
12. Christlieb, N., et al. 2002, Nature, 419, 904
13. Furlanetto, S. R., & Loeb, A. 2003, ApJ, 588, 18
14. Heger, A., & Woosley, S. E. 2002, ApJ, 567, 532
15. Kaplinghat, M., Chu, M., Haiman, Z., Holder, G., Knox, L., & Skordis, C. 2003, ApJ, 583, 24
16. Larson, R. B. 1998, MNRAS, 301, 569
17. Larson, R. B. 2003, Rep. Prog. Phys., 66, 1651

18. Mackey, J., Bromm, V., & Hernquist, L. 2003, ApJ, 586, 1
19. Mori, M., Ferrara, A., & Madau, P. 2002, ApJ, 571, 40
20. Nakamura, F., & Umemura, M. 2001, ApJ, 548, 19
21. Omukai, K. 2000, ApJ, 534, 809
22. Omukai, K., & Inutsuka, S. 2002, MNRAS, 332, 59
23. Omukai, K., & Nishi, R. 1998, ApJ, 508, 141
24. Omukai, K., & Palla, F. 2003, ApJ, 589, 677
25. Palla, F., Salpeter, E. E., & Stahler, S. W. 1983, ApJ, 271, 632
26. Qian, Y.-Z., & Wasserburg, G. J. 2002, ApJ, 567, 515
27. Ripamonti, E., Haardt, F., Ferrara, A., & Colpi, M. 2002, MNRAS, 334, 401
28. Scannapieco, E., Schneider, R., & Ferrara, A. 2003, ApJ, 589, 35
29. Schneider, R., Ferrara, A., Natarajan, P., & Omukai, K. 2002, ApJ, 571, 30
30. Schneider, R., Ferrara, A., Salvaterra, R., Omukai, K., & Bromm, V. 2003, Nature, 422, 869
31. Tan, J. C., & McKee, C. F. 2004, ApJ, in press (astro-ph/0307414)
32. Umeda, H., & Nomoto, K. 2002, ApJ, 565, 385
33. Umeda, H., & Nomoto, K. 2003, Nature, 422, 871
34. Wada, K., & Venkatesan, A. 2003, ApJ, 591, 38
35. Wolfire, M. G., & Cassinelli, J. P. 1987, ApJ, 319, 850
36. Wyithe, J. S. B., & Loeb, A. 2003, ApJ, 588, L69
37. Yoshida, N., Abel, T., Hernquist, L., & Sugiyama, N. 2003, ApJ, 592, 645

HI at High Redshift

N. Kanekar[1] and J.N. Chengalur[2]

[1] Kapteyn Astronomical Institute, University of Groningen, The Netherlands
nissim@astro.rug.nl

[2] National Centre for Radio Astrophysics (TIFR), Pune, India
chengalu@ncra.tifr.res.in

1 Introduction

Damped Lyman-α systems (DLAs), the highest HI column density absorbers seen in QSO spectra, are believed to arise in the precursors of present-day galaxies [1]. Despite their importance in the context of galactic evolution, the typical size and structure of DLAs (as well as physical conditions in them) have long been issues of much controversy. Radio studies of DLAs in the redshifted HI 21cm line provide a new dimension in our understanding of the absorbers. For example, if the QSO behind the DLA is radio-loud, 21cm absorption studies can be used to probe the kinematics, temperature and distribution of the absorbing gas. Conversely, 21cm emission observations allow a direct determination of the HI mass of the absorber, as well as its physical size and velocity structure. In this article, we describe recent progress in such radio studies, primarily based on work with the newly commissioned Giant Metrewave Radio Telescope (GMRT).

2 The Spin Temperature

In the case of a radio-loud background quasar illuminating a DLA, the 21cm optical depth can be combined with the total HI column density (obtained from the Lyman-α profile) to yield the spin temperature T_s of the absorbing gas (e.g. [2]). Earlier 21cm studies have found T_s values in DLAs to be significantly higher than those seen in the Milky Way or local spirals (e.g. [3]); however, the small size of the sample made it very difficult to detect any clear trends. Further, hardly any absorbers of the sample were at low redshifts, $z < 1.6$. We have used the GMRT to carry out a deep search for 21cm absorption in 21 DLAs, with $0.09 \le z \le 3.4$ (see [2] and references therein); this has resulted in the detection of absorption in eight systems (all at $z < 0.6$) and fairly stringent upper limits on the 21cm optical depth in the other cases. The sample of DLAs with 21cm absorption searches now consists of 31 systems, of which 24 have reliable estimates of the spin temperature. The complete sample is presented in [2].

Fig. 1[A] shows a plot of T_s v/s z for the above 24 absorbers. The dotted horizontal line in the figure is at 350 K — more than 80 % of spin temperature measurements in the Milky Way and M31 have values < 350 K [4]. It can be immediately seen that the spin temperatures of the majority of DLAs are considerably higher than those typical of lines of sight through the disks of the Milky Way or nearby spirals. The only exceptions are five absorbers at low z (shown as open squares) and the DLA at $z \sim 2.04$ towards PKS 0458−020, for which $T_s \sim 385 \pm 100$ K; the latter is discussed in [2].

Interestingly enough, *all* low z DLAs with low spin temperatures ($T_s \leq 350$ K) have been identified as arising in luminous $L \sim L_\star$ galaxies (see [2] and references therein). Conversely, $z < 1$ DLAs with $T_s \geq 1000$ K have been found to be associated with dwarf or low surface brightness (LSB) systems. This can be understood as the measured spin temperature is the column density weighted harmonic mean of the spin temperatures of different HI phases along the line of sight; it thus contains information about the *relative fractions of gas in the cold and warm phases*, besides their actual temperatures. It should also be emphasized that T_s is biased towards cold gas; for example, a line of sight with half of the HI at 100 K and the other half at 10^4 K would have $T_s \sim 200$ K. Similarly, T_s is only ~ 1000 K even if 90 % of the HI is at 10^4 K and the rest at ~ 100 K. Higher T_s values are thus to be expected in smaller systems like dwarf galaxies whose low metallicities and pressures are not conducive to the formation of the cold neutral medium (CNM) [5]. Such systems hence have a substantially higher fraction of warm HI (WNM) as compared to normal spirals, and, therefore, a higher T_s on an average line of sight. On the other hand, spirals like the Milky Way have roughly equal fractions of warm and cold HI; the above bias of T_s towards colder temperatures implies that lines of sight through such systems would, on average, yield lower T_s values. It thus appears plausible that T_s can indeed "distinguish" between dwarfs and large spirals, as indicated by the low z DLAs of Fig. 1[A].

It would be very interesting if T_s were indeed an broad indicator of galaxy type as one might then use T_s measurements to verify whether galaxies at high z are small or large systems, in a statistical sense. An obvious first test of the above hypothesis is to attempt to directly measure the fractions of cold and warm HI in specific DLAs to verify that the WNM fraction is indeed higher in systems with a high T_s. This has so far been carried out for two low z DLAs, towards the quasar OI363, where high spectral resolution 21cm studies have succeeded in separating the contributions of the cold and warm phases to the 21cm optical depth [6, 7]. These observations have shown that the WNM fraction is ≥ 70 % in the $z \sim 0.0912$ absorber and ≥ 75 % in the $z \sim 0.2212$ system. Thus, the high T_s values obtained in at least these two DLAs is indeed the result of a higher WNM fraction.

Next, if high T_s values arise because the ISM in small galaxies is inefficiently cooled due to the lack of metals (and hence a lack of radiation pathways), one would expect an anti-correlation to exist between spin temperature and metallicity. We have recently used VLT-UVES to measure the metallicity (i.e. [Zn/H]) in a sample of DLAs with T_s measurements. Fig. 1[B] shows a

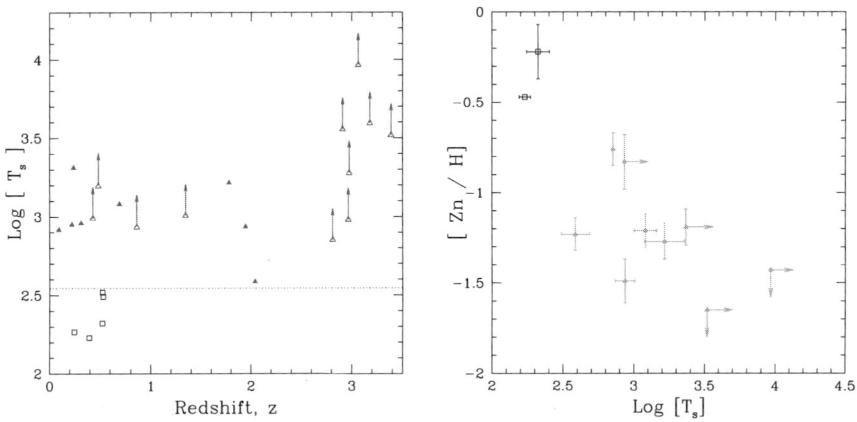

Fig. 1. Left panel [A] The spin temperature T_s as a function of redshift for the 24 DLAs of the sample. Right panel [B] Metallicity ([Zn/H]) plotted as a function of T_s. DLAs identified with luminous galaxies are shown as open squares.

plot of [Zn/H] v/s T_s for the 11 absorbers of the present sample; the predicted anti-correlation is seen to indeed be present, albeit with some spread. Note that this sample contains DLAs from both high and low redshifts.

The above results suggest that the spin temperature is indeed a good indicator of the nature of the absorbing galaxy. Fig. 1[A] then indicates that the majority of damped systems of the present 21cm sample arise in dwarf or LSB galaxies, with only a few systems (at low redshifts) being luminous disks. Further, all seven DLAs of the sample at $z \geq 3$ show $T_s \geq 1000$ K, indicating that these systems contain very small CNM fractions (< 10 % for five of the seven systems; [2]). An independent line of evidence for high temperatures at high z comes from the exceedingly low H_2 fraction observed in the majority of high z DLAs [8]; this can be understood in models in which the HI is at a high temperature, $T > 3000$ K [9].

3 HI 21cm Mapping Studies

While the results of Section 2 suggest that T_s might be used to differentiate between large and small galaxies, it would be even more exciting to actually measure the physical sizes of the absorbers, their kinematics and gas distributions. The most direct way of doing this is through 21cm emission observations; unfortunately, the sensitivity of today's radio telescopes limits such studies to fairly low redshifts. We have recently used the GMRT to make spatially resolved images of the 21cm emission from a $z = 0.009$ DLA, towards the quasar HS1543+5921 [10]; this is the first case of mapping of HI emission from a damped system. We obtain a total HI mass of $\sim 1.4 \times 10^9$

$M_\odot$, considerably smaller than the value of M^*_{HI} determined from blind 21cm emission surveys. It was also possible to measure the rotation curve of the galaxy out to a radius of $\sim 60''$, i.e. a linear distance of ~ 11 kpc; this resulted in a total dynamical mass $M_{dyn} \sim 5 \times 10^9$ $M_\odot$.

Constraints on the size and gas kinematics of DLAs can also be obtained from 21cm absorption studies, if the background radio continuum is extended on scales larger than the telescope beam. We have used the GMRT to obtain such spatially resolved images of 21cm absorption at $z \sim 0.437$ towards the radio galaxy 3C196. The two main components of the 21cm profile were shown to arise from absorption in two arms of a large barred spiral, against the S-W hotspot and the eastern lobe of 3C196, as originally argued by Briggs et al. [11]. HI absorption was detected out to a radius of ~ 70 kpc, far beyond the extent of the optical galaxy (as is common in local spiral disks). Similarly, we have also mapped 21cm absorption from the $z \sim 0.395$ DLA towards PKS 1229−021 [12], showing that the absorber has a physical size larger than ~ 30 kpc. These observations have only now become possible due to the combination of high sensitivity, excellent frequency coverage and long interferometric baselines at the GMRT.

In summary, 21cm studies of DLAs provide an excellent complement to information on these systems obtained from other wavebands. Present results suggest that the spin temperature may be a good indicator of the nature of the absorbing galaxy. The recent commissioning or upgrade of telescopes such as the GMRT, the GBT and the WSRT has resulted in a dramatic improvement in radio spectroscopic instrumentation and make this an exciting time for studies of the evolution of galaxies using redshifted radio lines.

Acknowledgments. We thank Alain Smette and Frank Briggs for allowing us to present results from our unpublished VLT-UVES studies.

References

1. A. M. Wolfe, D. A. Turnshek, H. E. Smith, R. D. Cohen: ApJS **61**, 249 (1986)
2. N. Kanekar, J. N. Chengalur: A&A **399**, 857 (2003)
3. C. L. Carilli, W. M. Lane, A. G. de Bruyn, et al.: AJ **111**, 1830 (1996)
4. R. Braun, R. A. M. Walterbos: ApJ **386**, 120 (1992)
5. M. G. Wolfire, D. Hollenbach, C. F. McKee et al.: ApJ **443**, 152 (1995)
6. W. M. Lane, F. H. Briggs, A. Smette: ApJ **532**, 146 (2000)
7. N. Kanekar, T. Ghosh, J. N. Chengalur: A&A **373**, 94(2001)
8. C. Ledoux, P. Petitjean, R. Srianand: MNRAS, in press (astro-ph/0302582)
9. P. Petitjean, R. Srianand, C. Ledoux: A&A **364**, L26 (2000)
10. J. N. Chengalur, N. Kanekar: A&A **388**, 383(2002)
11. F. H. Briggs, A. G. de Bruyn, R. C. Vermeulen: A&A **373**, 113 (2001)
12. R. L. Brown, R. E. Spencer: ApJ **230**, L1 (1979)

A Spectroscopic Survey of the Submillimeter Galaxy Population: 85 Redshifts Using Keck/LRIS-B

S.C. Chapman[1], A. Blain[2], R. Ivison[3], and I. Smail[1]

[1] Caltech, Pasadena, CA, 91125, USA
[2] Royal Observatory, Edinburgh EH9 3HJ, UK
[3] University of Durham, Durham DH1 3LE, UK

We present spectroscopic identifications for a sample of 85 millimeter and submillimeter galaxies detected with the SCUBA/JCMT and MAMBO/IRAM-30m facilities, all identified with deep 1.4GHz VLA radio observations. The galaxies lie at redshifts $z = 0.5$ to 3.5, we describe their properties, especially the presence of active galactic nuclei (AGN) in the sample, and discuss the connection of the mm/submm galaxies and the formation of spheroidal components of galaxies, which requires knowledge of the timescales of their very luminous activity. For a subset of the galaxies, we show their disturbed and diverse *Hubble Space Telescope (HST)* optical morphologies.

1 Introduction

Millimeter/Submillimeter galaxies – hereafter SMGs (Smail, Ivison & Blain 1997; Barger et al. 1998; Hughes et al. 1998; Eales et al. 1999; Bertoldi et al. 2000) are an important population (Blain et al. 1999, 2002). They are both numerous, with a surface density about 10% that of optically-selected Lyman-break galaxies (LBGs; Steidel et al., 2003), and luminous (possibly because of very high SFRs), with typical luminosities about 10 times greater than LBGs, assuming plausible spectral energy distributions (SEDs). The existing SMGs produce most of the extragalactic mm/submm background radiation intensity, and a significant fraction of the background at far-infrared (FIR) wavelengths (Blain et al. 2002; Smail et al. 2002). They represent a population which is generally difficult to detect at optical wavelengths, with at least the fainter half clearly being missed in current optical cosmological surveys. Unfortunately, the SMG population is notoriously difficult to study! Until now, we have been able to gather almost nothing about their redshifts and morphologies as fundamental observable properties. As a consequence, deriving properties such as total mass and dust temperature, and finding both the fraction of power contributed by AGN and starbursts, and their connection to optically-selected star-forming LBGs, have been the topic of largely idle speculation over the five years since their discovery.

The principal hurdle has always been identification at other wavelengths. SCUBA/MAMBO surveys have large beam sizes (10"–15"), a situation that will remain true for upcoming single-antenna, wide-field instruments: Bolocam, SHARC-II, a bolometer camera on the APEX telescope and SCUBA2. Candidate counterparts to SMGs cannot therefore be identified unambiguously without interferometry to pinpoint their positions. However, mm/submm interferometry is currently an arduous process, with tens of hours of integration required to detect a single object at the OVRO MMA or IRAM PdB interferometers.

As a consequence, the 20-cm radio emission from SMGs has become an important, but *second best* surrogate with which to probe both the energy generation processes and morphology of submm galaxies (Ivison et al. 1998; Smail et al. 2000; Barger, Cowie & Richards 2000; Chapman et al. 2001, 2002a, 2003a; Ivison et al. 2002) The problem remains that in order to use a radio wavelength as a surrogate for submm/FIR emission, we would like a clear physical principle connecting the emission at the different wavelengths. We don't yet, but we do have a strong empirical connection: the far-IR–radio correlation (e.g., Helou et al. 1985), which has an RMS dispersion of only 0.2 dex over a large range in luminosity at low redshifts. Radio identification of SMGs has allowed their optical properties to be explored in detail (Chapman et al. 2003a; Ivison et al. 2002). A large range of optical properties is observed for the radio-selected SCUBA galaxies, with 65% fainter than $I > 23.5$. They have red optical-IR colors with $I - K = 3$ to 6 and $< I - K >= 4.3$, but they are not all extremely red objects (EROs). Deep radio observations allow SMGs to be pre-selected with an efficiency of about 40%.

2 Redshifts for SMGs

We have been able to secure redshifts for 85 radio-identified SMGs through deep Keck/LRIS observations (Chapman et al. 2003b, 2004; Ivison et al. in preparation), and the sample is growing rapidly.[4] The spectrum of the brightest example is shown in Fig. 1a. The redshifts allow, for the first time, accurate calculation of luminosities and dust temperatures for the SCUBA galaxy population. We emphasize that obtaining redshifts is not easy, relying heavily on the superb blue sensitivity of the new LRIS-B mutliobject spectrograph (Steidel et al. 2004), and often have no detectable continuum emission with which to extract the spectrum. In addition the galaxies are hard to identify, being faint, messy, composite objects in optical images. Because of their small radio/optical offsets (~ 0.5") it is often difficult to assess the best position at which to align the slit. Often we designated several slit positions on different masks for each target. Our spectroscopic completeness is 74% over the magnitude range of the sample from $I = 22$ to 27.

[4] Data presented herein were obtained at the W.M. Keck Observatory, which is operated as a scientific partnership between Caltech, the University of California and NASA. The Observatory was made possible by the generous financial support of the W.M. Keck Foundation.

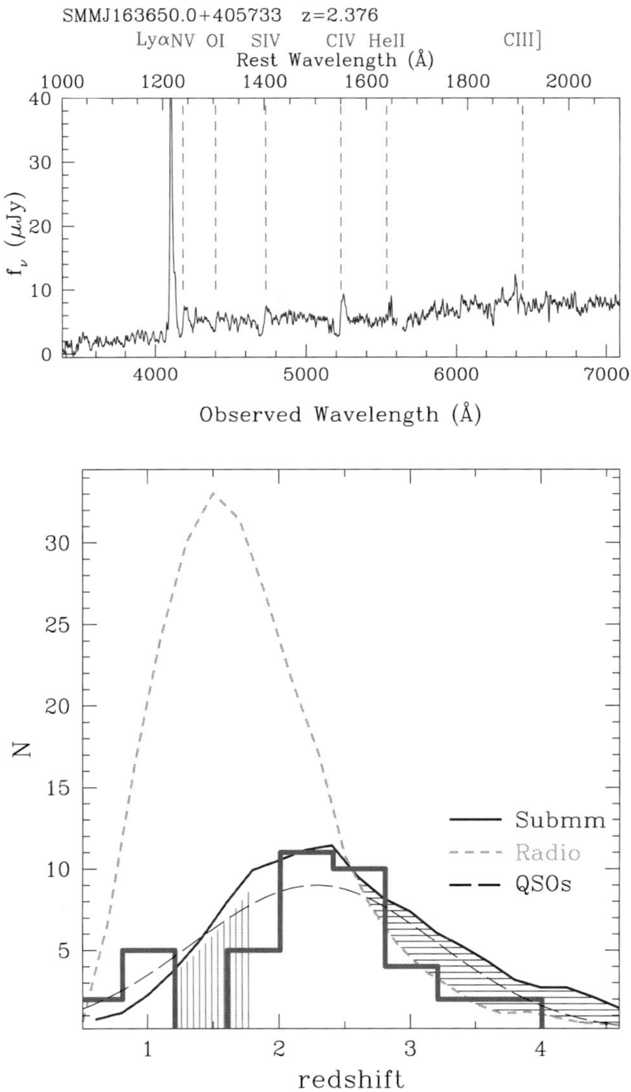

Fig. 1. (a) The rest-UV spectrum of the radio–SCUBA galaxy N2.4, showing P-Cygni absorption line profiles, and a host of stellar and interstellar features. The spectrum is best fit by a ∼5 Myr-old starburst, with an inferred star-formation rate of $1000\,M_\odot\,yr^{-1}$. **(b)** The observed histogram of the redshift distribution for our radio-identified SCUBA galaxies. Curves derived for a model of the radio/SMG populations are overplotted, suggesting that the redshifts of the sources missed in the radio identification process lie at moderate redshifts between the radio and submm model tracks. The redshift distribution of radio selected QSOs (Shaver et al. 1998) are shown with a dashed line, and reveal a very similar evolution to the SMGs.

While the issue of correctly identifying the submm galaxy is concerning, 8 out of 12 CO detections with IRAM-PdB have already been made, realizing an unequivocal confirmation of the redshifts (Neri et al. 2003, Greve et al. in preparation). We also note that beyond the radio positional identification, we are finding the correct *type* of object through optical spectroscopy. The identifications are at high-z, have apparent SFRs of several $100\,\mathrm{M}_{\odot}\,\mathrm{yr}^{-1}$, often show AGN features, and are very rare objects in the LBG distribution. We ask, what else can they be?

We further note that we believe we have spectroscopically identified a sample of blank-field SCUBA galaxy counterpart candidates without radio detections, effectively through trial and error by targeting faint optical sources lying within the SCUBA beam using otherwise redundant spectrograph slits. Observations of these identifications were tried based on our cumulative experience of with the properties of the radio-identified sample: often faint, distorted blue/red composite sources near the SCUBA beam centroid, are found to have Type-II AGN spectra and inferred star formation rates of order $100\,\mathrm{M}_{\odot}\,\mathrm{yr}^{-1}$.

In Fig. 1b, we show the observed redshift distribution and a model (Chapman et al. 2003c) for the radio and submm distributions derived from an evolving far-IR luminosity function. The model is very useful for understanding selection effects, and in particular the bias against the highest-redshift galaxies due to the requirement of a radio selection. This bias has a strong dependence on dust temperature (T_{d}). The sources missed by the radio are expected to fill the region lying between the model radio and submm distributions, overlapping significantly in redshift with the radio-identified sample if their dust temperatures are in the cold to warm regime ($T_{\mathrm{d}} < 35\,\mathrm{K}$).

3 Spheroids in Formation?

We have presented a plausible redshift distribution for submm galaxies, the most important ingredient for addressing whether they could be spheroids in formation. Encouragingly, they lie at the correct redshifts to be proto-ellipticals or the forming bulges of spiral galaxies. (all the stellar population constraints point to most stars being formed at z=2–3). In particular, we have demonstrated that most submm galaxies do not lie at $z > 4$, and thus most are not very high redshift Population-III sources.

However, to assess what type of formation mechanism is at work we must assess the AGN/starburst contribution to the dust heating, the timescales of their luminous phase, and compare their *HST* morphologies. The AGN versus starburst issue is always difficult to address. We have firstly from our restframe UV spectra, the possibility of diagnosing the presence of an AGN through high ionization lines. Indeed, we find signs of CIV, and other lines which cannot easily be excited in a starburst, in approximately 40% of our sources. Our near-IR spectra of SMGs (e.g., Smail et al. 2003) can

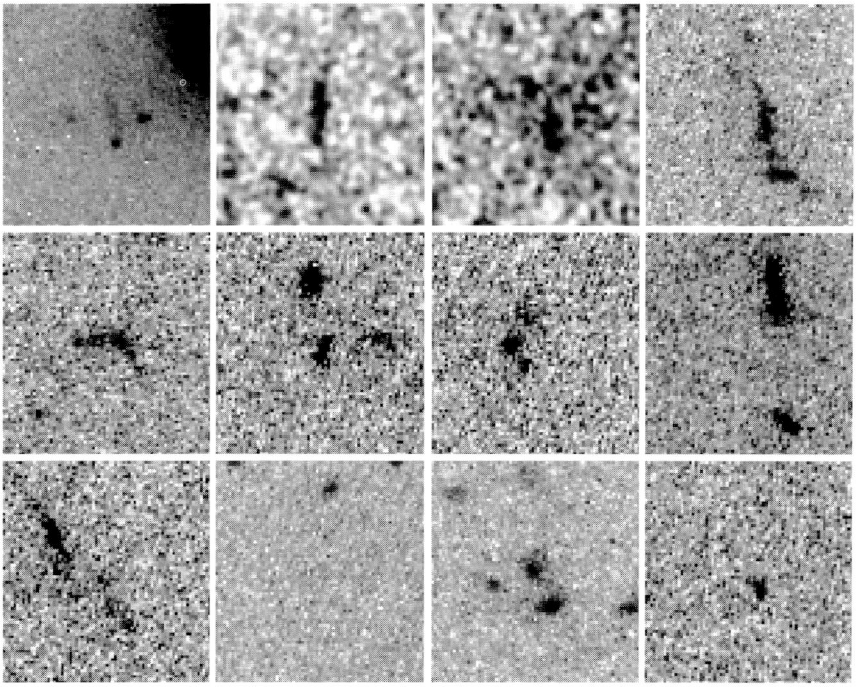

Fig. 2. The morphologies of 8 SCUBA galaxies as studied with *HST*-STIS (Chapman et al. 2003d; 4" images centered on the radio position). All are multi-component, distorted galaxy systems reminiscent of mergers in progress.

reveal OIII/Hβ ratios typical of Seyfert-2 galaxies as well as strong NeV emission. We have evidence in the X-ray from the Chandra Deep Field North that ~90% of the radio-identified SCUBA galaxies have X-ray detections (Alexander et al. 2003, 2004), but only ~35% of these are too bright to be generated by star formation alone (scaling from Nandra et al. 2002 for LBGs). Finally, ~20% of our sources appear to be unusually bright at radio wavelengths, departing significantly from the far-IR–radio distribution for low-redshift starburst galaxies. Together, this suggests that while AGN are prevalent in the sample (as expected from the likely coeval evolution of the BH and bulge required to generate the tight correlation between their masses; Magorrian et al. 1998), but not necessarily dominant in the bolometric energy released by SCUBA galaxies.

The duration of the very luminous activity of the SCUBA galaxies are important: are their timescales 10 Myr or 1 Gyr? If the former, then several bursts would be required to form a massive elliptical galaxy. In the spectra with the highest signal-to-noise ratios, stellar and interstellar features have been observed. These can be fitted using starburst synthesis models, to yield a timescale for the starburst activity visible at ultraviolet (UV) wavelengths, and a relation between the star-formation rates at observed optical and submm wavelengths. For our brightest object, N2.4 (Fig. 1a; Smail et al. 2003) the best fit is for a ~5 Myr-long instantaneous burst of star formation. The implied star formation rate is about $1000\,M_{\odot}\,yr^{-1}$, which is similar to that infered directly from its submm and radio flux densities. While not all sources may contain such a short burst of star-formation dominated energy output, this example is likely to define the paradigm for the nature of a significant fraction of SCUBA galaxies.

The morphologies of a subset of SCUBA galaxies have recently been studied with *HST*-STIS by Chapman et al. (2003d). Fig. 2 shows a montage of several representative examples, typically revealing multi-component, distorted galaxy systems that are reminiscent of mergers in progress. There are no examples of isolated, compact sources, and we conclude generally that the morphologies of SCUBA galaxies are generally consistent with hierarchical galaxy formation scenarios in which the most intense activity occurs when gas-rich, high-redshift galaxies collide and merge. Although they could be the sites of spheroid formation, most spheroids do not form in a monolithic collapse of primordial gas clouds at extreme redshifts.

References

1. Alexander, D., et al., ApJ, 2003, in press (astro-ph/0211267)
2. Alexander, D., Smail, I., Chapman, S., Blain, A., Ivison, R., ApJ, 2004, submitted
3. Barger, A., et al., Nature, 1998, 394, 293
4. Barger, A., Cowie, L., Richards, E., 2000, AJ 119, 2092
5. Bertoldi, F., et al., 2000, A&A, 476, 124
6. Blain, A.W., Smail, I., Ivison, Kneib, J.-P., 1999, MNRAS, 302, 632
7. Blain, A.W., Smail, I., Ivison, Kneib, J.-P., Frayer D.T., 2002, Physics Reports, 369, 111 (astro-ph/0202228)
8. Chapman, S., Richards, E., Lewis, G., Wilson, G., Barger, A., 2001, ApJ, 548, L147
9. Chapman, S.C., Lewis, G.F., Scott, D., Borys, C., Richards, E., 2002a, ApJ, 570, 557
10. Chapman S.C., Barger A., Cowie L., et al., 2003a, ApJ, in press (astro-ph/0211075)
11. Chapman, S., Blain, A., Ivison, R., Smail, I., 2003b, Nature, 422, 695
12. Chapman, S., Helou, G., Lewis, G., Dale, D., ApJ, in press
13. Chapman, S., Windhorst, R., Odewahn, S., Haojing, Y., Conselice, C., ApJ, 2003d, in press
14. Chapman, S., Blain, A., Ivison, R., Smail, I., 2004, ApJ, submitted
15. Eales, S., et al., 1999, ApJ 515, 518
16. Helou, P., et al., 1985, ApJ 440, 35
17. Hughes, D., et al., Nature, 1998, 394, 241
18. Ivison, R., et al., 1998, MNRAS, 298, 583
19. Ivison, R., et al., 2002, MNRAS, 337, 1

20. Magorrian, J. et al., 1998, AJ, 115, 2285
21. Nandra, P., et al. 2002, ApJ, in press (astro-ph/0205215)
22. Neri, R., et al. 2003, A&A, in preparation
23. Smail, I., Ivison, R.J., Blain, A.W., 1997, ApJ 490, L5
24. Smail, I., Ivison, R.J., Blain, A.W., Kneib, J.-P., 2002, MNRAS, 331, 495
25. Smail, I., et al., 2003, MNRAS, submitted
26. Steidel, C., et al., 2003, ApJS, submitted

The Quasar Q0957+561: Lensed CO Emission from a Disk at z∼1.4?

M. Krips[1], R. Neri[2], A. Eckart[1], J. Martín-Pintado[3], P. Planesas[3], and L. Colina[4]

[1] I. Phys. Inst. der Universität zu Köln Zülicher Str. 77, 50937 Köln, Germany krips@ph1.uni-koeln.de
[2] IRAM, 300, rue de la Piscine, Domaine Univ.,F-38406 Saint Martin d'Hères
[3] Observatorio Astronómico Nacional (IGN), Alcalá de Henares, Spain
[4] Instituto de Estructura de la Materia (CSIC), Madrid, Spain

In recent years large efforts have been made to detect molecular gas towards high redshifted objects. Up to now the literature reports on only two cases of CO-detection in quasars at a redshift between 1 and 2 – Q0957+561, a gravitationally lensed system at $z = 1.41$ (Planesas et al. 1999) , and HR10 at $z = 1.44$ (Andreani et al. 2000). According to Planesas et al. (1999), $^{12}CO(2\rightarrow1)$ emission was detected towards both the lensed images of Q0957+561 with the IRAM Plateau de Bure Interferometer (PdBI; Fig. 1). In contrast to the optical spectra of the two images which support the idea that they are images of one and the same object, the CO-spectra were surprisingly different: the southern image (named CO-B) shows a single blueshifted line whereas a double-peaked line profile with a blue- and a redshifted part appears towards the northern image (CO-A). Based on the observations and on simulations with a gravitational lens program, we are tempted to argue that the line profile traces the presence of molecular gas of a disk in the host galaxy around the quasar. We have now new observations with the PdBI providing the necessary sensitivity to corroborate our disk model.

1 Introduction

Since the discovery of Q0957+561, the first confirmed gravitationally lensed quasar (Walsh et al. 1979) at a redshift of $z = 1.4$, several models have been developed to understand the lensing potential of the intervening galaxies: a giant elliptical galaxy (G1) at a redshift of $z = 0.36$ with a surrounding cluster at $z = 0.355$ and probably another group of background galaxies at $z = 0.5$ (e.g. Angonin-Willaime et al. 1994). Planesas et al. (1999, in the following P99) have recently achieved in observing the CO(2-1) line in Q0957+561, but lacked sensitivity and a detailed lensing model to interpret the origin of the molecular emission. To confirm and substantiate this line profile, we have carried out new observations with the IRAM interferometer making Q0957+561 potentially very valuable for understanding the evolution of quasars at redshifts $1 < z < 2$. To further improve on P99's interpretation, we have developed a numerical code incorporating existing lensing models of Q0957+561.

2 Observations

CO(2-1) and (5-4) observations of Q0957+561 were done simultaneously in 1998 and again in 2003 with the IRAM interferometer.

Continuum emission: In the 3.1mm radio continuum, two lensed images of the quasar, labeled A and B (pointlike), and a radio jet C (extended) appear at this wavelength (Fig. 1), in agreement with VLA observations (Harvanek et al. 1997). The positions coincide with the optical ones within the errors. There is no evidence for variability above 10% between May 1998 and April 2003 in all three components. The lack of any clear detection at 1.3mm could also be a sign for resolving out flux in the smaller beam.

Line emission: After subtracting the continuum the integrated CO(2-1) map shows two lensed images labeled CO−A and CO−B, as in P99. The emission centroids of the two images are separated by $\sim 1''$ more than in the optical/radio. The two CO line components show different spectral profiles. A double-peaked profile is visible towards the northern image CO−A whereas only a single (blueshifted) velocity component is detected towards CO−B. No line emission was detected towards CO(5-4).

3 Modelling Q0957+561

We developed a numerical code based on the standard gravitational lens equation to explain the absence of the double-peaked line profile towards CO−B in Q0957+561. The code has been applied on 6 different models of Q0957+561, all based on previous work by Barkana et al. (1999). They can be mainly divided into 2 main groups: one set of models is based on a King profiles, as proposed by Falco et al. (1985– models: FGS, FGSE, FGSE+CL), the second set on a softened power-law distribution, as suggested by Grogin et al. (1996–models: SPLS, SPEMD, SPEMD+CL). Each of these two sets can be subdivided into three subgroups assuming either elliptical or spherical profiles and/or a shear term or a Single Isothermal Sphere for the surrounding cluster. Besides the sizes of the lensed images, the positions, continuum and the blueshifted line intensity ratios, also the optical time delay (Kundic et al. 1997) has been taken into account to distinguish among the models. For simplicity, the respective components were simulated as slightly extended gaussians. The SPEMD+CL and FGSE+CL models reproduce the observed constraints with the lowest χ_r^2 (≤ 3) showing that the contribution of the cluster is important: The models that best explain these observations all require a SIS cluster.

Line emission: Let us first forget about the redshifted CO-A component and concentrate on the blueshifted line. A compact region centered on the nucleus can be directly ruled out by the small line ratio B-blue/A-blue of 0.4 since this would be in contradiction with the measured 3mm continuum ratio A/B of 1.5 originating in a compact region. An extended region of the blueshifted line indeed reproduces the derived small line ratio of 0.4 but it would have a particular shape (very extended in direction of the nuclear jet (∼ 5”) but very thin perpendicular to it). Also a non-nuclear position of the blueshifted emission region with a rather compact shape and closer to the inner tangential caustic originates in a small line ratio.

The difference in the line profiles observed towards CO−A and CO−B can be explained by the location of the redshifted gas component relative to the lens caustic. In the best-fit simulation for the SPEMD+CL model the two caustics are so close together that the blueshifted component of Q0957+561 is still located between them whereas the redshifted line must originate above the northern outer caustic. This calculation used two separated components of the red- and blueshifted line but we can also start with an extended region centered on the nucleus where the north-eastern part corresponds to the redshifted component and the south-western part to blueshifted one. The morphology that reproduces the derived line ratio (A-blue+A-red)/B-Blue∼ 1.1 agrees well with the model of the host galaxy used by Keeton et al. for the Barkana FGSE+CL model being quite close to our best-fit FGSE+CL model.

4 Discussion and Conclusions

What can we conclude about the origin of the redshifted CO(2-1) velocity component? Mainly two groups of hypotheses are possible: it either originates from an independent system (companion etc.) or traces the presence of molecular gas in the rotating disk of the host galaxy in Q0957+561. We have reasons to discard the first class of hypotheses. One of the strongest arguments is that the host galaxy detected in HST observations by Keeton et al. (2000) is oriented in the direction of the two line components in CO−A and is extended on the same scale. Also the integrated line ratio of (CO-A-red+CO-A-blue)/CO-B ≈1.0 is consistent with Keeton et al. (2000). Furthermore, the simulation of the two velocity profiles together supports also an extended disk similar in shape to the used host galaxy models of Keeton et al. 2000 whereas the model for a single blueshifted component would either suggests a very thin but very elongated disk not at all similar to the host galaxy detected in the optical or a quite compact CO region offset from the nucleus. The redshift of this region would then also be significantly different from the one of the host galaxy and the quasar. Thus, the latter two approaches seem to be quite unlikely compared to Keeton et al.'s (2000) results for the host galaxy. Furthermore, the double-peaked line profile at CO−A appears to be symmetric and is centered within the errors at $z_0 = 1.4141$. Based on these arguments, we favour the second hypothesis: the presence of an important reservoir of molecular gas in a disk of the host galaxy surrounding Q0957+561. Based on the lower magnification from the FGSE+CL model, we have estimated an upper limit to the molecular gas mass for each integrated blue and redshifted velocity profile in CO−A and CO−B. Assuming a standard M_{gas} to $L'_{\mathrm{CO}(1-0)}$ ratio ($\simeq 5\,M_\odot(\mathrm{K\,km\,s^{-1}pc^2})^{-1}$, Downes et al. 1993) and R_{21} =CO(2-1)/CO(1-0)≤ 1, we find $M_{\mathrm{gas}} = 2.3\times10^{10}\,M_\odot$ for both the blue and red velocity profiles. The low upper limit on $R_{54} \simeq 1$, the velocity averaged line ratio, characterizes low excitation conditions, and thus agrees with global CO emission from a disk. The agreement between molecular gas masses obtained with the integrated CO luminosities and individual magnification factors, each tracing line emission from half of the quasar host, is a further support for the rotating disk hypothesis.

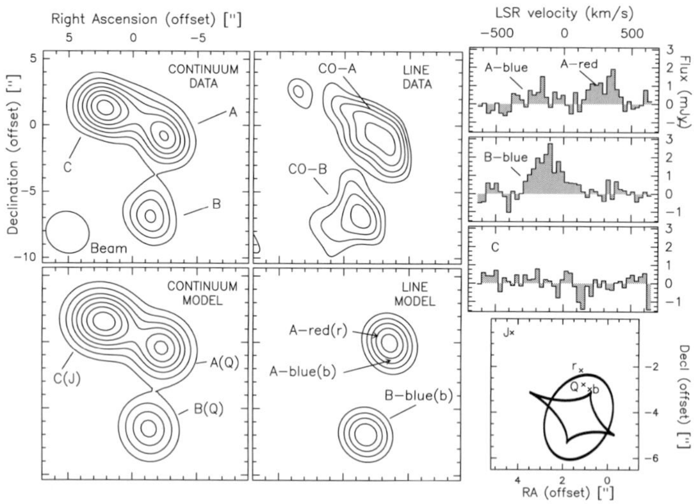

Fig. 1. Upper panel: observed continuum (left) & line emission (middle) plus spectrum (right); lower panel: simulated data with the lens plane (right).

References

1. Angonin-Willaime, M.-C., Soucail, G., Vanderriest, C., 1994, A&A, 291, 411
2. Barkana, R., Lehár, J., Falco, E.E. et al., 1999, ApJ, 520, 479
3. Downes, D., Solomon, P.M. & Radford,E. 1993, ApJ 414, L13
4. Falco, E.E., Gorenstein, M.V., Shapiro, I.I. 1985, ApJ, 289, L1
5. Grogin, N.A., Narayan, R. 1996, ApJ, 464, 92
6. Harvanek, M., Stocke, J.T., Morse, J.A., Rhee, G., 1997, AJ, 114, 2240
7. Keeton, C.R., Falco, E.E., Impey, C.D., et al., 2000, ApJ 542, 74
8. Kundic, T., Turner, E.L., Colley, E.N. et al. 1997,ApJ 531, 635
9. Walsh, D., Carswell, R.F., Weynman, R.J., 1979, Nature, 279, 381

Part II

Molecular Spectroscopy and Astrophysics

This session was held in honour of Prof. G. Winnewisser, who retired in 2001 as director of the I. Physikalische Institut of the Universität zu Köln

Molecular Astrophysics: An Integral Part of Astronomy

E.F. van Dishoeck

Leiden Observatory, P.O. Box 9513, 2300 RA Leiden, The Netherlands
ewine@strw.leidenuniv.nl

1 Introduction

In 1986, Dalgarno wrote a paper with the provocative title 'Is interstellar chemistry useful?'.The purpose of this brief review is to show that this question can now indeed be answered with a resounding 'yes!'. Observations of molecules, both in the gas and in the solid phase, have become an integral part of astronomy and are now routinely used to probe the formation of the earliest stars and galaxies in the distant universe; to determine the mass and kinematics of dense gas in starbursts and AGN's; to make an inventory of molecular clouds and star-forming sites throughout our Galaxy and nearby galaxies; to study the detailed physics, chemistry and dynamics of star- and planet-forming regions in our solar neighborhood; to investigate the fate of dying stars and their mass return to the interstellar medium; and to search for complex organic, perhaps even prebiotic, molecules.

Molecular astrophysics is an inherently interdisciplinary field, and requires an integrated approach combining astronomical observations and instrumentation with laboratory studies of basic molecular processes and astrochemical models. Gisbert Winnewisser has been well aware of the importance of this 'golden triangle' throughout his career, and has successfully built up expertise in all three areas at the University of Cologne. In this paper, a few examples of the importance of molecules and applications throughout astronomy will be given.

2 Molecules in the Distant Universe

At the time of the first Zermatt meeting in 1988, CO observations were limited to our own galaxy and nearby galaxies with redshifts $z < 0.2$. At the second Zermatt meeting five years later, CO had just been identified in the galaxy IRAS F10214+4724 at z=2.29 (Brown & vanden Bout 1992), and a few years later the record was raised to z=4.69 (BR 1202-0725, Ohta et al. 1996, Omont et al. 1996). Thanks to the advent of bolometer arrays on large single-dish telescopes, dusty galaxies have now been detected throughout the universe (see review by Bertoldi, this volume) and CO and dust have even been seen in the host galaxy of the most distant quasar known at $z = 6.42$, SDSS J1148+5251 (Bertoldi et al. 2003, Walter et al. 2003). These exciting data confirm that molecular astrophysics started shortly after the Big Bang, and that dust and molecules are prime tracers of the mass, temperature, dynamics and star formation activity of the earliest galaxies.

Molecules, in particular H_2, also play a key role in the formation of the very first stars in the universe (e.g. Galli & Palla 1997, Abel et al. 2002). The cooling of the gas through H_2 rotational lines allows dense clumps of cold gas to form, which may be the sites for primordial star formation. Astrochemistry itself started at the era of recombination, around $z \approx 3000$, with the formation of the first molecules HeH^+ and He_2^+ (e.g. Stancil et al. 1998).

3 Molecules in our Galaxy

Extensive CO surveys with a variety of telescopes have revealed the presence, mass and kinematics of numerous molecular clouds throughout our Galaxy (e.g. Dame et al. 2001). These clouds are exposed to many sources of energy which can heat, compress and disrupt them. The increased temperatures and densities result in enhanced excitation of the molecules and stronger far-infrared radiation. Thus, analysis of the radiation emitted by molecules and dust can be used to trace the physical conditions and importance of various energetic processes. The central role of molecules as diagnostics of both local (e.g. stellar winds, shocks, UV radiation) and global (e.g. cosmic rays) processes has been nicely illustrated by Winnewisser & Ungerechts (1984) (see Fig. 1).

In addition to their well-known role as physical diagnostics, there has been renewed interest in using molecules as chemical diagnostics. Indeed, the detection of planets around other stars and the possibility of life elsewehere in the universe provide strong motivation to chart the inventories of molecular abundances at the various stages of star- and planet formation. More than 120 different species have been detected in molecular clouds (not counting isotopic or isomeric species), with abundances ranging from 10^{-4} to 10^{-11} with respect to H_2. This large dynamic range in abundances, coupled with observed variations from region to region, illustrates the potential of molecules as chemical tracers.

4 Some Recent Highlights

4.1 H_2O: Probing Temperature and Gas-Grain Interactions

The ISO, SWAS and ODIN satellites have opened up the possibility to directly observe thermal radiation from H_2O (see reviews by Melnick and Hjalmarson, this volume). Although H_2O is only a trace molecule in cold molecular clouds with abundances as low as $10^{-8} - 10^{-9}$, it becomes the dominant oxygen-bearing molecule in warm regions where the dust temperature is high

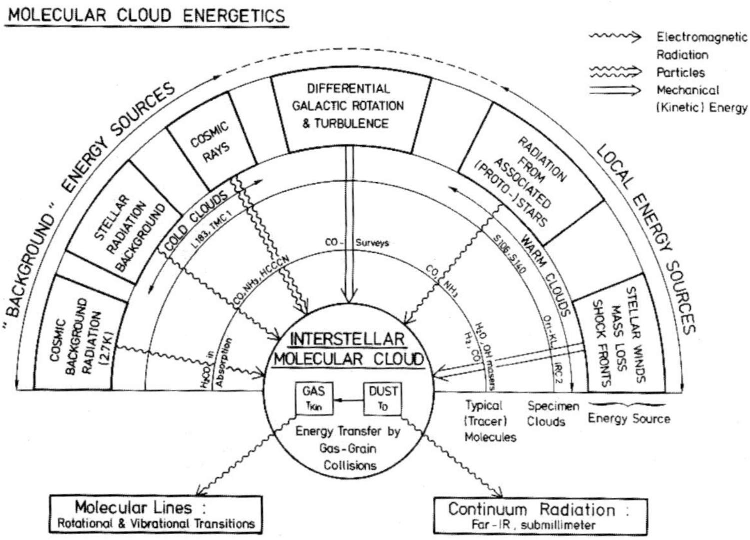

Fig. 1. Schematic representation of the energetics of molecular clouds and use of molecules as diagnostics (from Winnewisser & Ungerechts 1984).

enough to evaporate ices and in high-temperature shocks. This large variation in abundance makes water a unique probe of the physical structure of the region, and of the fundamental chemical interactions between the gas and the grains. Moreover, its level of deuteration provides an important record of the temperature history of the cloud and the conditions during grain surface formation. Water can also play an active role in the energy balance of the gas, both as a coolant and as a heating agent, although its importance with respect to other coolants such as CO has to be quantified.

4.2 Large Carbonaceous Molecules: The Organic Inventory

Laboratory spectroscopy of pure rotational transitions of long carbon-chain molecules and small rings has made enormous progress since the last Zermatt meeting and is now well ahead of observations (see Thaddeus, this volume). These molecules, which are traditionally observed in cold clouds like TMC-1, serve as diagnostic of the atomic carbon content of a cloud, and thus presumably its "youth" (e.g. Suzuki et al. 1992). Two additional avenues are being pursued in the laboratory. The first is optical spectroscopy of carbon chains in an effort to identify any of the diffuse interstellar bands (see Maier, this volume). The second is far-infrared spectroscopy by the Cologne and other groups to measure their low-frequency bending modes.

Large saturated molecules like CH_3OCH_3, $HCOOCH_3$ and C_2H_5CN are known to be abundant in massive star-forming regions, and form the building blocks for even more complex organic species, some of which may be of prebiotic importance. Recent detections include interstellar antifreeze (ethylene glycol, $HOCH_2CH_2OH$; Hollis et al. 2002) and the simplest sugar (glycolaldehyde, CH_2OHCHO; Hollis et al. 2000). The first detection of an interstellar amino acid, glycine, has recently been claimed (Kuan et al. 2003), although its identification is still under debate (Menten, this conference).

Yet another class of complex organic molecules is formed by the polycyclic aromatic hydrocarbons (PAHs), which reveal their presence through a distinct set of emission bands seen in the mid-infrared spectra of virtually all objects. Although no unique identification is possible in this case, they are among the more abundant species containing up to a few % of the cosmic carbon (see Tielens, this volume). Indeed, PAHs have become such an integral part of astronomy that they are routinely cited as redshift indicators of distant obscured galaxies and as diagnostics of starburst versus AGN activity (e.g. Genzel et al. 1998).

4.3 H_3^+ and Related Species: Ionization Tracers

One of the simplest interstellar molecules, H_3^+, also tells a distinct story (see Oka, this volume). Long predicted to be a cornerstone of interstellar chemistry networks, the ion was detected only eight years ago through infrared spectroscopy. Its abundance is directly proportional to the cosmic-ray ionization rate ζ, a key parameter in the chemistry and physics of molecular clouds. Recent observations of H_3^+ in diffuse clouds indicate values of $\zeta \approx 10^{-15}$ s^{-1} (McCall et al. 2003), which are more than an order of magnitude higher than those found for dense clouds (e.g. van der Tak & van Dishoeck 2000). If such high values of ζ were widespread throughout the diffuse ISM, it would have significant consequences for our understanding of the two-phase medium (Wolfire et al. 2003). Abundances of simple molecules like HD and OH are also proportional to ζ and comprehensive models which explain all three species need to be developed.

Chemically related molecules such as HCO^+ and N_2H^+, produced by reactions of CO and N_2 with H_3^+, have traditionally been used as tracers of the ionization fraction of dense clouds (e.g., Williams et al. 1998). The inferred values of $e/H_2 \approx 10^{-7}$ imply that the clouds are weakly coupled to the magnetic field and that the ambipolar diffusion timescale is significantly larger than the free-fall time scale.

4.4 Deuterated Molecules: Temperature History and Freeze-Out

Enormous deuterium fractionation ratios have long been recognized as a striking characteristic of low-temperature ion-molecule chemistry in dense clouds.

Observed ratios of DCN/HCN and DCO/HCO^+ are often in excess of 0.01, three orders of magnitude more than the cosmic [D]/[H] ratio. These enhancements are driven by the isotopic exchange reaction of $H_3^+ + HD \rightarrow H_2D^+ + H_2$, which proceeds primarily in the forward direction at temperatures below ~50 K. H_2D^+ has finally been detected in dense clouds through its lowest ortho transition at 372 GHz, with abundances that approach or even exceed those of H_3^+ (Stark et al. 1999, Caselli et al. 2003). Moreover, a wealth of doubly- and even triply-deuterated species has been found (see Ceccarelli, this volume). Not only do these detections indicate very low temperatures, but they also require significant freeze out of CO (up to 99%) to further enhance the H_3^+ and H_2D^+ abundances (Roberts et al. 2003).

5 Molecules During Star- and Planet Formation

In the last few years, our qualitative and quantitative understanding of the chemistry of protostellar environments has improved dramatically, and it provides a good example of the use of molecules as tracers of certain evolutionary stages (Table 1; see van Dishoeck 2003 for a recent review).In the cold ($\leq$10 K) pre-stellar cores the time scales for freeze-out are so short that most molecules stick onto the grains. Strong N_2H^+ emission and high deuterium fractionation ratios are chemical characteristics of this phase (see §4.4; Bergin, this volume). Once the protostar starts to heat its surroundings (10–100 K), the ices evaporate in a sequence according to their sublimation temperatures. This is reflected both in the gas/ice ratios seen in infrared spectra, and in the abundance and excitation of molecules observed at millimeter wavelengths. Also, the evaporated molecules can participate in a rapid high-temperature gas-phase chemistry leading to even more complex organic molecules for a limited period of time, the so-called 'hot core' phase (see §4.2). Indeed, complex molecules are now the primary diagnostics of this very early stage of massive star formation. At even higher temperatures ($\geq$230 K), most of the oxygen is driven into water and yet a different chemistry ensues.

Progress in this field is due to at least three developments. First, the physical structure of the circumstellar envelope —temperature and density as a function of radius— can now be accurately constrained from bolometer array continuum data and radiative transfer modeling of the spectral energy distributions (e.g. Jørgensen et al. 2002, Young et al. 2003). Second, molecular line radiative transfer codes have been developed and tested, which, in conjunction with the inferred physical structure, constrain the chemical abundances throughout the envelope (e.g. Hogerheijde & van der Tak 2000). By simultaneously fitting lines of the same molecule with different excitation conditions, chemical gradients can be determined on scales which are an order of magnitude smaller than the primary beam of the observations (e.g. Schöier et al. 2002, Maret et al. 2003). Third, thanks to the increased sensitivity of submillimeter and infrared facilities, not only high mass sources but also low-mass

Table 1. Chemical characteristics of young stellar objects

Component	Chemical characteristics	Submillimeter diagnostics	Infrared diagnostics
Surrounding cloud	Low-T chemistry	Ions, long-chains (HC_5N, ...)	Simple ices (H_2O, CO)
Cold envelope	Low-T chemistry, Heavy freeze-out	Simple species (N_2H^+, deuterated mol.)	Ices (H_2O, CO_2, CH_3OH)
Inner warm envelope	Evaporation	High T_{ex} (CH_3OH, H_2CO)	High gas/solid, High T_{ex}, Heated ices (C_2H_2, H_2O, CO_2)
Hot core	High-T chemistry	Complex organics (CH_3OCH_3, CH_3CN, vib. exc. mol.), high HCN	Hydrides (H_2O)
Outflow: Direct impact	Shock chemistry, Sputtering	Si- and S-species (SiO, SO)	Atomic lines, Hydrides ([S I], H_2O, H_2 high-J)
Outflow: Turbulent entrainment	Evaporation	Morphology (CH_3OH, HCO^+, HCN)	Hydrides (H_2 low-J)
PDR, Compact H II regions	Photodissociation, Photoionization	Ions, radicals (CN/HCN, CO^+)	Ionic lines, PAHs, H_2 ([NeII], [CII])
Disks	Photodissociation, Freeze-out	Radicals, simple + ??? mol. (CN, HCO^+, ???)	CO gas, PAHs, Ices

objects can now be probed. Indeed, the first evidence for complex organic molecules around a solar-mass protostar has just been reported (Cazaux et al. 2003).

Whether or not these complex molecules become part of future planetary systems is a question that cannot yet be addressed observationally. Protoplanetary disks are two orders of magnitude smaller in size and mass than the envelopes, and are therefore much more difficult to observe. Submillimeter detections of gas-phase molecules other than CO in disks have become possible only recently (e.g. Dutrey et al. 1997, Kastner et al. 1997, van Zadelhoff et al. 2001, Qi et al. 2003). Currently, these observations are only sensitive to the outer disk at radii greater than 50 AU. Several simple molecules such as HCO^+, HCN, CN and H_2CO have been detected, indicating the importance of ion-molecule reactions and photodissociation in disks. Freeze-out in the colder parts is evidenced by the detection of strong solid CO infrared absorption in at least one edge-on disk (Thi et al. 2002). Strong infrared emission from gas-phase CO can be used to test the temperature structure of the inner and outer disk (Brittain et al. 2003, Najita et al. 2003, Blake & Boogert 2003). Chemical modeling indicates that most of the active gas-phase chemistry occurs in an intermediate layer of the disk just below the surface, which is warm enough to prevent freeze-out of the molecules and sufficiently shielded from ultraviolet radiation to prevent rapid photodissociation (e.g. Aikawa et al. 2002). Eventually, chemistry may be the best diagnostic of the importance of radial and vertical mixing processes in planet-forming zones.

6 Concluding Remarks

The preceding sections, and in fact this entire book, illustrate how molecular astrophysics has become a natural part of astronomy. This will only be strengthened in the coming decade, with exciting facilities like SIRTF, SOFIA, ASTRO-F, Herschel, ALMA, and JWST on the horizon. Young and enthousiastic astrochemists, such as those trained at the University of Cologne in the inspiring environment created by Gisbert Winnewisser, will be essential to make the best use of these outstanding opportunities.

References

1. T. Abel, G.L. Bryan, M.L. Norman: Science **295**, 93 (2002)
2. Y. Aikawa, G.J. van Zadelhoff, E.F. van Dishoeck, E. Herbst: A&A **386**, 622 (2002)
3. F. Bertoldi, C.L. Carilli, P. Cox et al.: A&A **406**, L55 (2003)
4. G.A. Blake, A.C.A. Boogert: Ap. J., submitted (2003)
5. S.D. Brittain, T.W. Rettig, T. Simon et al.: Ap. J. **588**, 535 (2003)
6. R.L. Brown, P.A. vanden Bout: Ap. J. **397**, L19 (1992)
7. P. Caselli, F.F.S. van der Tak, C. Ceccarelli, A. Bacmann: A&A **403**, L37 (2003)
8. S. Cazaux, A.G.G.M. Tielens, C. Ceccarelli et al.: Ap. J. **593**, L51
9. A. Dalgarno: Q. J. R. Astr. Soc. **27**, 83 (1986)
10. T.M. Dame, D. Hartmann, P. Thaddeus: Ap. J. **547**, 792 (2001)
11. A. Dutrey, S. Guilloteau, M. Guélin: A&A **317**, L55 (1997)
12. D. Galli, F. Palla: A&A **335**, 403 (1998)
13. R. Genzel, D. Lutz, E. Sturm et al.: Ap. J. **498**, 579 (1998)
14. M.R. Hogerheijde, F.F.S. van der Tak: A&A **362**, 697 (2000)
15. J.M. Hollis, F.J. Lovas, P.R. Jewell, L.H. Coudert: Ap. J. **571**, L59 (2002)
16. J.M. Hollis, F.J. Lovas, P.R. Jewell: Ap. J. **540**, L107 (2000)
17. J.K. Jørgensen, F.L. Schöier, E.F. van Dishoeck: A&A **389**, 908 (2002)
18. J.H. Kastner, B. Zuckerman, D.A. Weintraub, T. Forveille: Science **277**, 67 (1997)
19. Y.-J. Kuan, S.B. Charnley, H.-C. Huang et al.: Ap. J. **593**, 848 (2003)
20. S. Maret, C. Ceccarelli, E. Caux et al.: A&A, in press (astro-ph/0310536)
21. B.J. McCall, A.J. Huneycutt, R.J. Saykally et al.: Nature **422**, 500 (2003)
22. J. Najita, J.S. Carr, R.D. Mathieu: Ap. J. **589**, 931 (2003)
23. K. Ohta, T. Yamada, K. Nakanishi et al.: Nature **382**, 426 (1996)
24. A. Omont, P. Petitjean, S. Guilloteau et al.: Nature **382**, 428 (1996)
25. C. Qi, J.E. Kessler, D.W. Koerner et al.: Ap. J. **597**, 986
26. H. Roberts, E. Herbst, T.J. Millar: Ap. J. **591**, L41 (2003)
27. F.L. Schöier, J.K. Jørgensen, E.F. van Dishoeck, G.A. Blake: A&A **390**, 1001 (2002)
28. P. Stancil, S. Lepp, A. Dalgarno: Ap. J. **509**, 1 (1998)
29. R. Stark, F.F.S. van der Tak, E.F. van Dishoeck: Ap. J. **521**, L67 (1999)
30. H. Suzuki, S. Yamamoto, M. Ohishi et al.: Ap. J. **392**, 551 (1992)
31. W.F. Thi, K.M. Pontoppidan, E.F. van Dishoeck et al.: A&A **394**, L27 (2002)

32. F.F.S. van der Tak, E.F. van Dishoeck: A&A **358**, L79 (2000)
33. E.F. van Dishoeck. In *Chemistry as a Diagnostic of Star Formation*, ed. C. Curry, M. Fich (NRC Press, Canada, 2003), in press
34. G.-J. van Zadelhoff, E.F. van Dishoeck, W.-F. Thi, G.A. Blake: A&A **377**, 566 (2001)
35. F. Walter, F. Bertoldi, C. Carilli et al.: Nature **424**, 406 (2003)
36. J.P. Williams, E.A. Bergin, P. Caselli et al.: Ap. J. **503**, 689
37. G. Winnewisser, H. Ungerechts: Excitation conditions in molecular clouds. In *Galactic and Extragalactic Infrared Spectroscopy*, 16'th ESLAB symposium (Reidel, Dordrecht 1984) pp 177–191
38. M.G. Wolfire, C.F. McKee, D. Hollenbach, A.G.G.M. Tielens: Ap. J. **587**, 278 (2003)
39. C.H. Young, Y.L. Shirley, N.J. Evans, J.M.C. Rawlings: Ap. J. Suppl. **145**, 111 (2003)

The Ubiquitous H_3^+

T. Oka

Department of Astronomy and Astrophysics and Department of Chemistry, The Enrico Fermi Institute, The University of Chicago t-oka@uchicago.edu

1 Ubiquitous H_3^+

Although it took many years from the discovery of the laboratory spectrum of H_3^+ (Oka 1980) and its first astronomical search (Oka 1981) to its detection in interstellar space (Geballe & Oka 1996), subsequent observations have revealed ubiquity of this fundamental molecular ion in a wide variety of astronomical objects. H_3^+ with column densities on the order of $\sim 10^{14}$ cm^{-2} has been detected not only in many dense clouds (McCall et al. 1998[a], 1999, Geballe et al. 2002, Kulesa et al. 2002), where its presence had been anticipated from chemical model calculations (Herbst & Klemperer 1973, Watson, 1973), but also in many diffuse clouds (McCall et al. 1998[b], Geballe et al. 1999, McCall et al. 2002) where its abundance was not anticipated and introduced an enigma in the chemistry of the diffuse interstellar medium. Many astronomers, physicists and chemists are currently attempting to understand this problem (Oka 2000, McCall & Oka, 2000).

In this contribution I would like to summarize our H_3^+ observations toward stars with a wide range of extinctions (Sect. 2), special characteristics of the chemistry of H_3^+ (Sect. 3), H_3^+ toward the Galactic center (Sect. 4), H_3^+ in metastable rotational level (Sect. 5) and some recent highlights (Sect. 6).

2 From Dense to Diffuse Clouds

The initial detections of interstellar H_3^+ (Geballe & Oka 1996) were toward young stellar objects W33A and AFGL 2136 that are deeply embedded in molecular clouds and have high magnitude of extinction ($A_v \sim 100$). Subsequently, H_3^+ have been found in many dense clouds and the observed column densities are approximately proportional to the extinction.

It was a big surprise when nearly equal column density of H_3^+ was discovered in the diffuse interstellar medium toward a faint visible star Cygnus OB2 No. 12 whose extinction is much smaller ($A_v \sim 10$, V=11.98) (McCall et al. 1998b, Geballe et al. 1999). It had been thought that the number density of H_3^+ in diffuse clouds was two to three orders of magnitude lower than that in dense clouds since the H_3^+ destruction mechanism in the former (dissociative recombination with an electron) is much more rapid than that in the latter (proton hop reaction from H_3^+ to CO) as discussed in the next section.

Subsequently, however, H_3^+ has been observed toward even brighter stars with lower extinction such as the classic translucent sightline HD183143 with $A_v \sim 4.0$ and V=6.86 (McCall et al. 2002). The recent observations of H_3^+ toward the bright star ζ Per with $A_v \sim 1$ and V=2.88 is particularly noteworthy (McCall et al. 2003). This universal abundance of H_3^+ in sightlines with a wide range of extinction makes it a general astrophysical probe to study a wide variety of objects.

The observed H_3^+ column densities, $N(H_3^+)$, in many dense and diffuse clouds are plotted against their extinction A_v in Fig. 1.

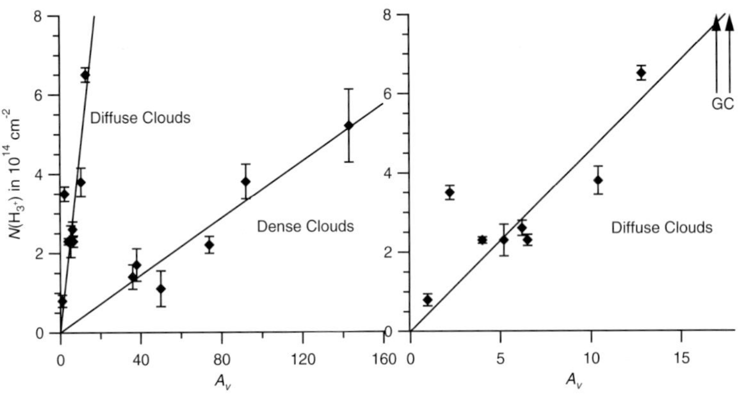

Fig. 1. Observed H_3^+ column density versus A_v for dense and diffuse clouds. $N(H_3^+)$ toward GC are higher than what the slope indicates by a factor of 3$\sim$ 5

We note two surprises in this figure. The first is the huge difference between the slopes of $N(H_3^+)$ versus A_v for dense and diffuse clouds. This is clearly due to the very different chemistry and physics of H_3^+ in the two categories of clouds as discussed in the next section. We are now attempting to fill this singularity of slopes by observing high A_v translucent clouds such as those toward the Stephenson objects (Rawlings et al. 2000).

The second surprise is the extremely high H_3^+ column densities observed toward bright infrared sources GCS 3-2 and GC IRS3 near the Galactic center. These column densities are off from the slope of other diffuse clouds by a factor of $\sim$3. If we include H_3^+ in metastable levels discussed in Sect. 5, this factor is more likely $\sim$ 5. Based on the H_3^+ chemistry discussed in the next section, this indicates that the sightlines cross long paths of low density clouds. This has implications on metastable H_3^+ so far observed only toward the Galactic center as discussed in Sect. 5.

3 Special Characteristics of the H_3^+ Chemistry

H_3^+ is produced via ionization of H_2 to H_2^+ by the ubiquitous cosmic ray or local X-ray, followed by the efficient Langevin reaction, $H_2^+ + H_2 \rightarrow H_3^+ + H$.

Since the latter process ($\sim$ 1/day) is much faster than the former ($\sim 1/10^{7-9}$ years), the production rate is given by $\zeta n(H_2)$ where ζ ($10^{-17} \sim 10^{-15} s^{-1}$) is the H_2 ionization rate. H_3^+ is destroyed by the proton-hop reaction, $H_3^+ + X \rightarrow HX^+ + H_2$, with the rate $k_X n(H_3^+)n(X)$. Equating the production and destruction rates, we have the steady state H_3^+ number density of

$$N(H_3^+) = (\zeta/k_X)[n(H_2)/n(X)] \tag{1}$$

In dense clouds X=CO is the main destroyer and

$$N(H_3^+) = (\zeta/k_{CO})[n(H_2)/n(CO)] \tag{2}$$

$\sim 10^{-4} cm^{-3}$, where the canonical values of $\zeta = 3 \times 10^{-17}$ s^{-1}, $k_{CO} = 2 \times 10^{-9}$ cm^3 s^{-1} and $n(CO)/n(H_2) = 1.5 \times 10^{-4}$ are used.

In diffuse clouds, $X=e^-$ is the main destroyer and

$$N(H_3^+) = (\zeta/k_e)[n(H_2)/n(C^+)] \tag{3}$$

$\sim 10^{-6} cm^{-3}$ where $k_e = 2 \times 10^{-7} cm^3 s^{-1}$ is used.

Note that $n(H_3^+)$ is constant as long as the clouds are typically dense or diffuse. Note also that $n(H_3^+)$ in diffuse clouds is lower than in dense clouds by two orders of magnitude because the dissociative recombination rate k_e is that much faster than the proton hop reaction rate k_{CO}. These situations are schematically summarized in Fig. 2.

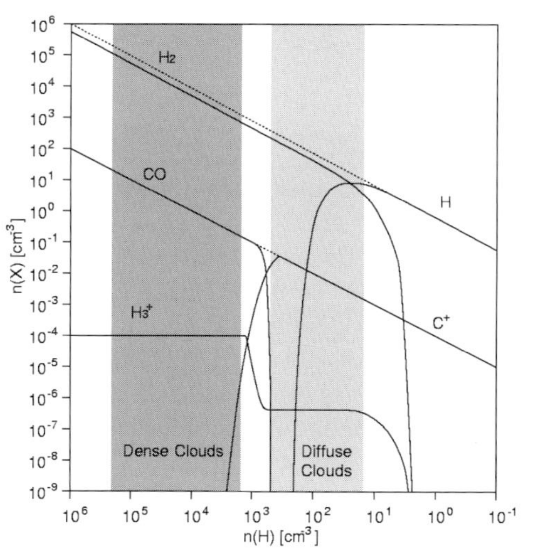

Fig. 2. Number densities of main chemical species n(X) versus cloud density n(H). Note that the H_2 and CO densities scale with the cloud density while that of H_3^+ is constant for typical dense and diffuse clouds. The ratio of the H_3^+ density in dense and diffuse clouds has been controversial. This figure shows results based on canonical values of ζ and k_e where $n(H_3^+)$ in dense cloud is higher than in diffuse cloud by two to tree orders of magnitude.

Since comparable H_3^+ column densities have been seen in dense and diffuse clouds, we need to assume a path in diffuse clouds which is 100 times longer than in dense clouds if we are to accept the above calculations. This contradicts to other astronomical evidences. An easy way out of this dilemma would be to assume smaller recombination rate k_e for which both laboratory data and theory have been highly controversial (Oka 2003). Recently McCall et al. (2003) measured k_e to be equal to the canonical value and claimed 40 times higher ζ in ζ Per than in dense clouds. This is highly controversial and further studies will be needed. Anyhow, observed values of $N(H_3^+)$ should provide crucial information on the ionization of clouds.

4 H_3^+ Toward the Galactic Center

The sightlines toward infrared sources near the Galactic center are the treasure house of H_3^+. They cross several mostly diffuse clouds including local clouds near the solar-system, clouds in the intervening 3kpc and 4kpc spiral arms, and clouds near the Galactic center in the expanding molecular ring and the circumnuclear ring. Luckily, the different radial velocities of the clouds separate their H_3^+ spectrum. Examples of the observed H_3^+ and CO spectra are given in Fig. 3 (Goto et al. 2002).

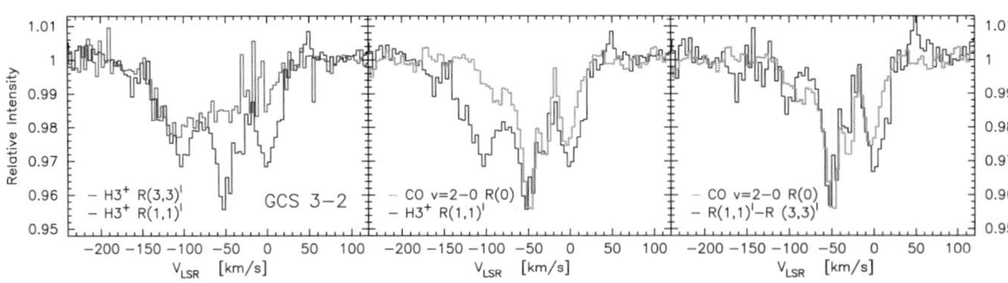

Fig. 3. H_3^+ and CO spectra observed toward the Galactic center

While the peak positions of the H_3^+ and CO absorption agree for each cloud component, their relative intensities differ greatly from cloud to cloud. For example, a strong H_3^+ absorption is noted in the expanding molecular ring at v∼110 km/s, while almost no CO absorption is observed indicating the low density of the area. This difference in the H_3^+ and CO column densities resulting from the different chemical behavior of H_3^+ and CO shown in Fig. 2, make them nice complementary probes. H_3^+ gives a radial dimension of the cloud while CO gives the total amount of molecules on relatively dense clouds.

There are many infrared sources toward the Galactic center. We have observed quintuplets (Nagata et al. 1990) and some NHS stars (Nagata et al. 1993) using the Phoenix Spectrometer at Gemini South.

5 H_3^+ in Metastable Rotational Level

The most exciting observation toward the Galactic center has been our discovery of abundant H_3^+ in the $(J,K)=(3,3)$ metastable rotational level which is 361 K above the lowest (1,1) level. The $R(3,3)^l$ line from the (3,3) level is strongly observed while the $R(2,2)^l$ line from the lower (2,2) level, only 151 K above the lowest level, is undetectable, demonstrating high temperature of the clouds and a very non-thermal rotational distribution of H_3^+. The metastability of the (3,3) level predicted from theory (Pan and Oka 1986) is clearly demonstrated by observations!

The intensity of the metastable $R(3,3)^l$ line is comparable to the ordinary $R(1,1)^l$ line demonstrating the comparable population in the 2 levels. This newly discovered population of H_3^+ amplifies the deviation of $N(H_3^+)$ from the slope of Fig. 1 and make the mystery of H_3^+ overabundance even more pronounced! A theoretical calculation shows that such distribution is only possible in a high temperature (> 200K) low density ($n < 50$) cm^{-3} medium. If we subtract the metastable $R(3,3)^l$ spectrum from the $R(1,1)^l$ spectrum, we obtain a velocity profile nearly identical to that of CO as shown in the right in Fig. 3. This suggests that the right figure give H_3^+ and CO in the ordinary clouds while the subtracted part which contain nearly equal amount of H_3^+ in the (3,3) and (1,1) level result from novel high temperature low density clouds which may not have previously been detected by any other means.

6 H_3^+ in Other Objects

The observed ubiquity of H_3^+ in interstellar clouds suggests that it is also observable in many other objects where molecules and ionization abound. The intense 3.7μm H_3^+ emission spectrum from, Jupiter (Oka & Geballe 1990), Saturn (Geballe et al. 1993) and Uranus (Trafton et al. 1993) has become a general tool to study planetary ionospheres (Connerney & Satoh 2000). Recently, Brittain and Rettig (2002) reported a detection of the H_3^+ emission from the Herbig AeBe star HD141569 and speculated that it is from a protoplanet in the preplanetary disk. The detection remains controversial (Oka 2002).

The high abundance of H_3^+ toward the galactic center suggests abundance of H_3^+ toward obscured active Galactic nuclei (AGN). A detection has been reported toward IRAS 08572 + 3815 with z = 0.05821 (Geballe 2001). With large diameter telescopes, observations of objects such as NGC 7172 (L=8.2) NGC 7479 (L=9.8) and UGC 5101 (L=10.1) are realistic. With low metallicity, Magellanic clouds may also be interesting objects to try H_3^+.

Planetary nebulae and proto-planetary nebulae are also interesting targets. The metastable H_3^+ spectrum may play a great role in such objects.

References

1. Brittain, S.D. & Rettig, T.W. 2002, Nature, 418, 57
2. Connerney, J. E. P. & Satoh, T. 2000, T. Phil. Trans. R. Soc. Lond. A358, 2359
3. Geballe, T. R. 2001, Astro-Ph 10109166
4. Geballe, T. R., Jagod, M.-F. & Oka, T. 1993, ApJ, 408, L109
5. Geballe, T. R., & Oka, T. 1996, Nature, 384, 334
6. Geballe, T. R., McCall, B. J., Hinkle, K. H. & Oka, T. 1999, ApJ, 510, 251
7. Geballe, T. R., McCall, B. J., & Oka, T. 2002, SFCHEM 2002,*Chemistry as a Diagnostic of Star Formation, Waterloo 2002*
8. Goto, M., McCall, B.J., Geballe, T.R., Usuda, T., Kobayashi, N., Terada, H., & Oka, T. 2002 PASJ 54, 951
9. Herbst, E. & Klemperer, W. 1973, ApJ, 185, 505
10. Kulesa, C. A., Black, J. H., Brittain, S. D., & Rettig, T. W. 2002, SFCHEM 2002, *Chemistry as a Diagnostic of Star Formation, Waterloo 2002*
11. McCall, B. J., Geballe, T. R., Hinkle, K. H. & Oka T. 1998a, Science, 279, 1910
12. McCall, B. J., Hinkle, K H., Geballe, T. R. & Oka T. 1998b, Faraday Discuss. 109, 267
13. McCall, B. J., Geballe, T. R., Hinkle, K. H. & Oka, T. 1999, ApJ 522, 338
14. McCall, B. J., Hinkle, K. H., Geballe, T. R., Moriarty-Schieven, G. H., Evans, N. J. II, Kawagichi, K., Takano, S., Smith, V. V. & Oka, T. 2002, ApJ, 567, 391
15. McCall, B. J., et al. 2003, Nature, 422, 500
16. McCall, B. J. & Oka, T. 2000, Science, 287, 1941
17. Nagata, T., Woodward, C. E., Shure, M., Pipher, J. L. & Okuda, H. 1990, ApJ, 351, 83
18. Nagata, T., Hyland, A. R., Straw, S. M., Sato, S. & Kawara, K. 1993, ApJ, 406, 501
19. Oka, T. 1980, Phys. Rev. Lett. 45, 531
20. Oka, T. 1981, Phil. Trans. R. Soc. Lond. A303. 543
21. Oka, T. 2000, Phil. Trans. R. Roc. Lond. A358, 2363
22. Oka, T. 2002, Nature, 418, 31
23. Oka, T. 2003 in *Dissociative Recombination of Molecular Ions with Electrons*, ed. S. L. Guberman, Kluwer Academic/Plenum Publishers, New York, .209-220
24. Oka, T. & Geballe, T. R. 1990, ApJ, 351, L53
25. Pan, F.-S. & Oka, T. 1986, ApJ, 305, 518
26. Rawlings, M. G., Adamson, A J. & Whittet, D. C. B. 2000, ApJS, 131, 531
27. Trafton, L. M., Gaballe, T. R., Miller, S., Tennyson, J. & Ballester, G. E. 1993, ApJ, 455, 761
28. Watson, W. D. 1973, ApJ, 183, L17

New Microwave and Optical Molecules

P. Thaddeus, M.C. McCarthy, and S. Thorwirth

Harvard Smithsonian Center for Astrophysics,60 Garden Street, Cambridge, MA 02138, U.S.A. pthaddeus@cfa.harvard.edu

This article summarizes some recent laboratory spectroscopic activities in our spectroscopy group at the Harvard-Smithsonian Center for Astrophysics. There are currently two main efforts here in laboratory astrophysics, one in the radio band, the other in the optical. The radio work has succeeded beyond expectations, in the last 7 years yielding the 105 new molecules in Fig. 1. Essentially all are plausible constituents of the interstellar gas or circumstellar shells, and 8, as Fig. 1 indicates, have in fact already been detected with large radio telescopes on the basis of the very precise frequencies which our spectroscopy provides. The application of Fourier-transform microwave (FTM) spectroscopy to molecules produced by a gas discharge in a supersonic beam has been crucial to many of these discoveries, which are now occurring almost weekly.

The optical spectroscopy is largely motivated by the long-standing challenge of the diffuse interstellar bands (DIBs), widely recognized as the outstanding unsolved problem in astrophysical spectroscopy. Here and in other laboratories progress on this thorny problem has been slow and frustrating. Several years ago, Maier and colleagues on the basis of elegant laboratory spectroscopy with mass selection tentatively identified a number of weak DIBs with C_7^- and similar carbon chains anions [1], but recent high quality astronomical observations by McCall et al. [2] have failed to support these identifications. In this laboratory, C. D. Ball [3] discovered several years ago a molecular band in exact coincidence (to a few parts in 10^4) with the most intense and best known DIB, λ4430, but it has not yet been possible to identify the carrier of the laboratory band, and there is as yet no certainty that the laboratory and astronomical bands are produced by the same molecule. For want of space we will not go into the details of our optical work on the diffuse interstellar bands, and the many false leads which have been pursued, but will concentrate instead on the radio work, citing a few examples of recent findings.

Almost all the molecules in Fig. 1 are entirely new reactive species, not previously identified in either the gas phase or in inert matrices. They are of astronomical interest because they are generally similar in structure and composition to molecules which have already been found in space (one of the reasons of course that they were sought in the laboratory). Here as in the 130-odd molecules so far identified in the interstellar gas and in circumstellar shells, carbon chains with alternating single and triple bonds or successive double bonds are the dominant structural theme. One dimensional carbon

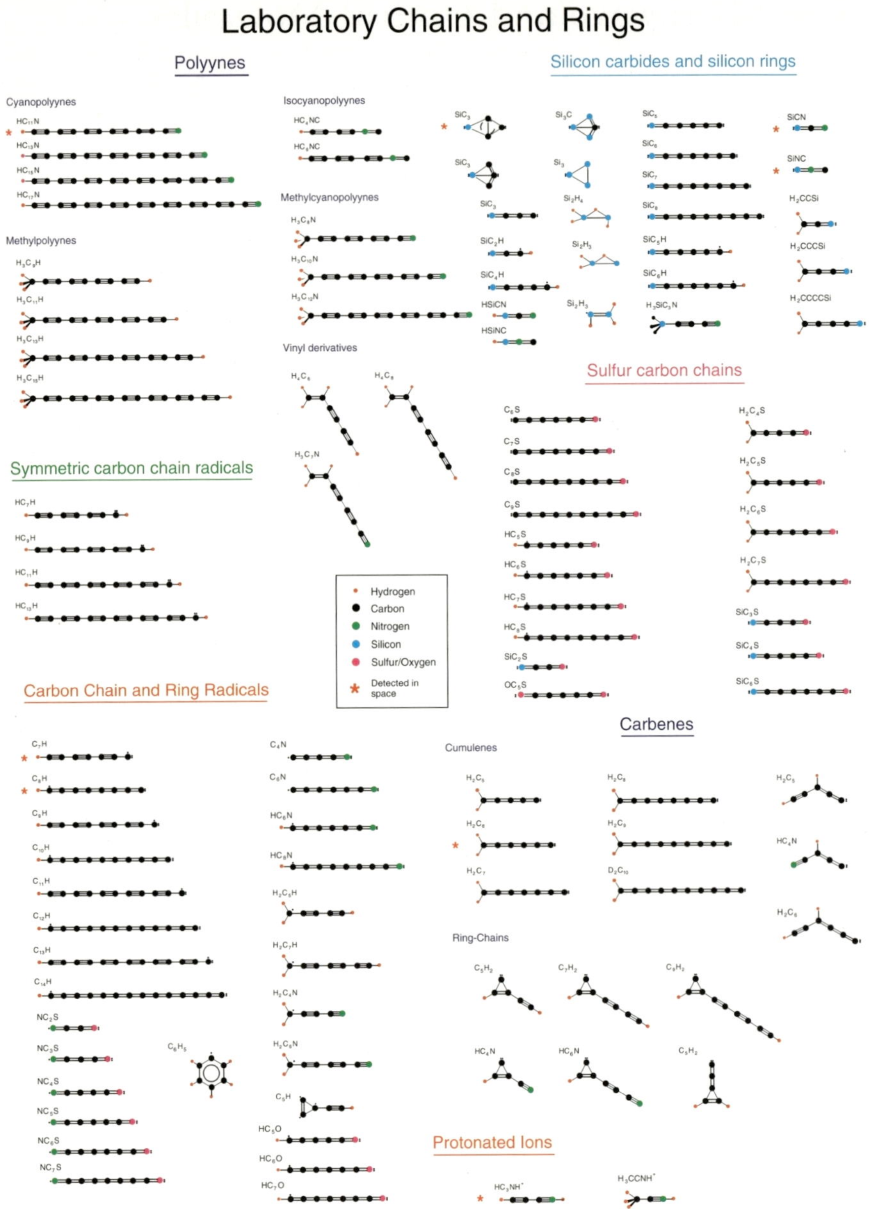

Fig. 1. The 105 molecules discovered here during the past seven years. Asterisks (*) indicate those molecules already detected in space, all but one (HC_3NH^+) on the basis of the present laboratory work.

of this kind constitutes an extremely stable lattice, but one somewhat less stable than the two dimensional lattice of graphite with its strong conjugated bonds. For that reason carbon chains longer than acetylene tend to be highly unstable to explosive polymerization on earth — and acetylene itself is explosive at certain temperatures and pressures. In space however, chain-chain collisions almost never occur, so once formed, chains are highly stable molecules. It may help in the formation of chains that every atom is on the surface, subject to reaction, unlike the situation in grains and large three dimensional molecules.

In our experiments the molecules in Fig. 1 are produced by a gas discharge through an organic vapor, typically acetylene or diacetylene for the chains, silane and acetylene for the silicon molecules, and benzene or toluene for phenyl, in each case heavily diluted in an inert buffer gas, argon or neon.

Most are radio molecules, detected via rotational transitions by FTM spectroscopy in a supersonic molecular beam, and some of these in addition — those generally with the stronger lines — have been detected as well in the millimeter-wave band with a classical free space spectrometer. Four of the molecules in Fig. 1 are non-polar symmetric carbon chains, and have been detected by cavity ringdown laser spectroscopy in the visual.

In the radio band, the laboratory astrophysics is now generally ahead of the radio astronomy, in the sense that many molecules too large to detect in space with present instrumentation are now readily detected in the laboratory. Good examples are the cyanopolyynes HC_nN. The largest interstellar molecule is $HC_{11}N$, with a molecular weight of 147 amu, but that was detected with a large radio telescope only after integrations of tens of hours per line. In contrast, the laboratory lines of $HC_{11}N$ were found with our FTM spectrometer in integrations of only a few minutes, and were ultimately produced in such strength that $HC_{13}N$, $HC_{15}N$, and $HC_{17}N$ could be identified. The decrement in intensity with increasing chain length for the astronomical lines in this sequence is sufficiently shallow to suggest that with better instrumentation, or better sources, these longer cyanopolyynes eventually will be found.

Space does not permit here a detailed description of the laboratory techniques employed for the detections in Fig. 1, or much discussion of specific molecules. A few examples will suffice.

1 The Phenyl Radical

The problem of detecting aromatic carbon molecules in the interstellar gas has long been a challenge to astronomers, and to radio astronomers especially. Since closed shell hydrocarbon rings are generally non-polar by symmetry, soon after the first wave of molecular discoveries in 1968–70 an effort was made to detect organic rings in which the cyclic symmetry was broken by inclusion of the hetero atoms N, O, etc. — well known stable molecules for example like pyridine, pyrrole, and furan. Astronomical radio searches for such stable rings then and over the many years since have unfortunately been almost uniformly disappointing – for reasons poorly understood, since many rings once formed in space are probably quite resistant to photodissociation and other destructive processes. Recently a sharp, weak infrared line at 14.8 μm was detected by the Infrared Space Observatory (ISO) toward a proto-planetary nebula which Cernicharo et al. [4] attribute to the ν_4 bending mode of benzene. The identification is plausible, but because molecular identifications based on a single line in fairly crowded spectra are notoriously unreliable, confirming observations are clearly required.

The existence of non-polar molecules like benzene in space has long been inferred from the detection of closely related polar species obtained by replacing H-atoms with polar groups like CN or CH_3, or by removing H-atoms to form polar radicals like CCH, or by protonation to form positive ions like HCO^+ [5]. We recently detected the phenyl radical, obtained by removing one H from benzene, in a benzene-argon discharge, both with our FTM spectrometer at centimeter wavelengths and with our free space millimeter-wave instrument [6]. The millimeter wave lines turned out to be surprisingly strong. Phenyl is a fairly polar ring with a calculated dipole moment of 0.9 D and a large number of rotational transitions throughout the radio band, each split by roughly 5 MHz into two components of nearly equal intensity by spin-doubling from the unpaired electron. In the centimeter wave band, hfs is also well resolved in the low-J transitions, but it collapses to less than 1 km s^{-1} in equivalent radial velocity in the high-J transitions in the millimeter wave band, and can be neglected there for most practical purposes. We have now measured over 70 transitions, many of which are indicated in the energy level diagram in Fig. 2, and determined with these the rotational constants and the leading centrifugal distortion and spin-doubling terms. Accurate rest frequencies can now be calculated across the radio band, and from the standpoint of radio astronomy the laboratory astrophysics of phenyl can be considered complete — at least for the time being.

There is no evidence of phenyl in several published spectral line surveys of standard Galactic sources, but the column density limits imposed are not particularly tight. The preliminary limits on phenyl are about 10 times higher toward the Orion Nebula, IRC+10216, and Sgr B2 than the column densities determined for c-C_3H_2, largely owing to the fourfold smaller dipole moment of phenyl, and its roughly tenfold larger rotational partition function. There is also no evidence of phenyl in CRL 618, the proto-planetary nebula where Cernicharo et al. have found the possible infrared band of benzene. These are however quite preliminary conclusions, and a dedicated search for phenyl with large radio telescopes like the IRAM 30-m or the new NRAO 100-m is now clearly called for.

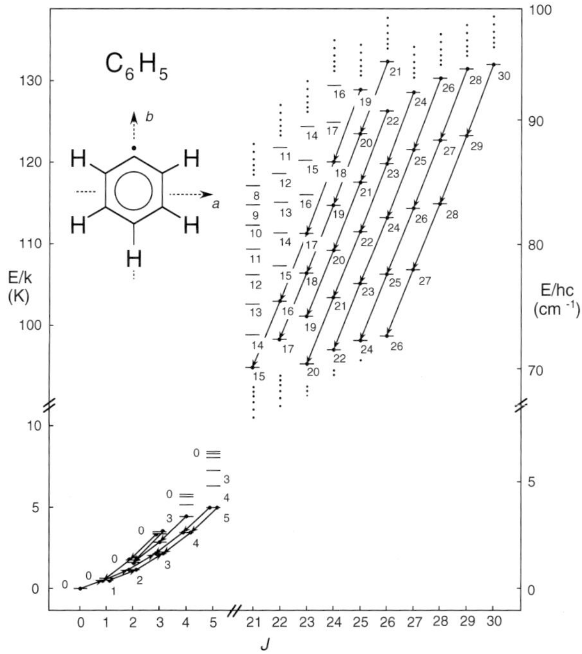

Fig. 2. Geometrical structure of phenyl and part of its rotational energy level diagram showing all the transitions measured by FTM spectroscopy and about 25% of those measured by millimeter-wave spectroscopy.

2 Vinyl Derivatives

The highly polar carbene ring c-C_3H_2 is one of the most widely distributed astronomical molecules [7], and the functional group c-C_3H obtained from it by removing one of its H-atoms can be attached to various carbon chains to produce interesting candidates for laboratory study and astronomical identification. A number of such molecules have now been identified in the laboratory, and are shown in Fig. 1. Following the astronomical discovery of vinyl cyanide in 1975 by Gardner and Winnewisser [8] it was clear that the vinyl group H_2CCH might serve a similar function. Ten years later Kroto et al. [9, 10] succeeded in the chemical synthesis of 1-cyanobut-3-ene-1-yne (**1**), the molecule obtained by substituting a linear CCCN chain for the CN in vinyl cyanide (Fig. 3), and they obtained its laboratory microwave spectrum. With our FTM spectrometer we have now succeeded in detecting the next larger member (**2**) of this sequence with C_5N as the attached chain, as well as the vinyl acetylenes (**3**) and (**4**); all are shown in Fig. 3. With fairly good rest frequencies in hand, the entire set of vinyl derivatives now invites astronomical detection. Dipole moments are expected to be substantial for those with a

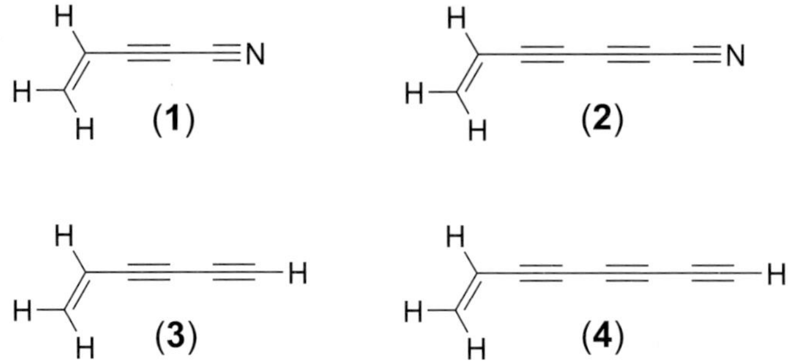

Fig. 3. Vinyl derivatives observed by FTM spectroscopy.

terminal CN: 5–6 D, and smaller but still appreciable for the vinyl acetylenes: 0.7–1.0 D.

The new vinyl molecules (**2**), (**3**) and (**4**) in Fig. 3 have not yet been detected with our millimeter-wave spectrometer, so their centrifugal distortion is not yet determined as precisely as was done for phenyl, for which reliable frequencies can now be calculated into the far infrared. Frequencies are quite adequate however for an unambiguous search in nearly all astronomical sources. In the band 100-150 GHz, it is estimated that rest frequencies of the most interesting astronomical transitions are accurate to roughly $1\,\mathrm{km\,s^{-1}}$ in equivalent radial velocity, or about the line width in a cold, sharp line source like TMC-1.

3 Exotic Silicon Molecules

Considered on earth a highly non-volatile element, and entirely absent from the terrestrial atmosphere, silicon in the interstellar gas is only slightly more depleted than C, N, and O, and in the circumstellar shell of IRC+10216 a surprisingly rich gas phase silicon chemistry has been discovered. Of the 130 gas phase astronomical molecules, 10 contain silicon, and with the wealth of laboratory data now coming in, many more almost certainly will be found. Over the past five years we have detected 25 new silicon molecules with our FTM spectrometer; most are of astronomical interest, and for nearly all rotational rest frequencies are now in hand for an unambiguous astronomical search. Three have already been identified toward IRC+10216: the silicon carbide ring SiC_3, and the linear isomeric pair SiCN and SiNC.

One of the most interesting of these new silicon molecules is Si_3 (Fig. 4), the smallest silicon cluster with a dipole moment ($\sim$0.3 D, according to an ab initio calculation [11]). Very strong lines of this molecule were discovered accidentally among the products of a silane (SiH_4) discharge during an attempt to find new silicon hydrides [12, 13]. The lines persisted in a discharge through fully deuterated silane, indicating to our surprise that we were observing a molecule without hydrogen. Given the high purity of our gas sample, this molecule could contain only silicon and neon, but a van der Waals complex with the neon buffer gas was quickly ruled out by using argon as the buffer. Detection of the isotopic species with silicon-29 (4.7% in natural abundance) and silicon-30 (3.0%) at precisely the calculated isotopic shifts established conclusively that we were observing Si_3 — and by the standard technique of isotopic substitution yielded the precise molecular geometry shown in Fig. 4.

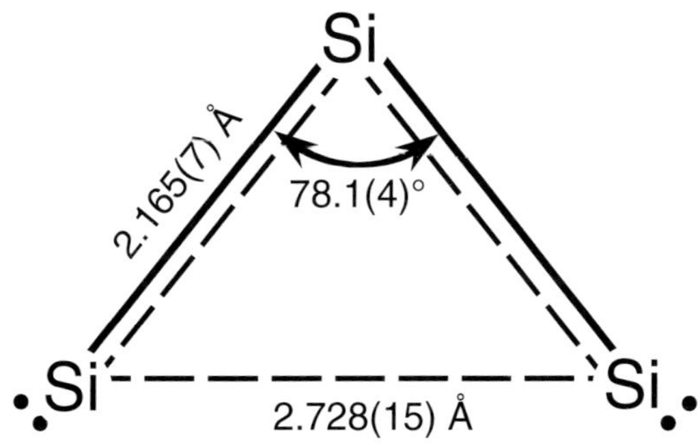

Fig. 4. Geometrical structure of Si_3.

Si_3 is a fairly nonrigid molecule; its bending potential surface is apparently extremely shallow, the bent isomer we observe being only about 9.5 kcal/mol lower in energy than linear Si_3, even though the two differ in bond angle by more than 100°. As a result the centrifugal distortion expansion converges only slowly, and with five distortion constants (the five quartic terms) we have only been able to fit the 21 measured centimeter-wave transitions of Si_3 to 66 kHz — more than an order of magnitude larger than the 2 kHz measurement uncertainty. Additional centrifugal distortion terms improve the fit only slightly. The upshot is that predicted frequencies much beyond the range of measurement (8–42 GHz), and specifically at millimeter wavelengths where it is most likely that Si_3 could be detected, must be treated with caution. At the top of the 100-150 GHz band, we estimate that uncertainties may be as large as $20\,\mathrm{km\,s^{-1}}$ in equivalent radial velocity. That is sufficiently small to allow an unambiguous search in a wide line source like IRC+10216, but large enough to motivate a laboratory search for the millimeter-wave spectrum of Si_3 to better determine its centrifugal distortion.

In recent spectral line surveys of IRC+10216 [14, 15] there is no evidence for Si_3, but that is hardly surprising, because of the small dipole moment of this molecule, and because lines of comparably massive silicon molecules (e.g., SiC_3) have astronomical lines only in the 5–10 mK range — at the sensitivity limit of the present surveys. A dedicated astronomical search is worth undertaking for several reasons: because reactions analogous to those that produce the fairly abundant molecules SiC_2, C_3, and C_4 in space may also work for Si_3, and because several of the recently discovered silicon molecules were found only after a sensitivity in line intensity of a few mK was achieved by long integrations.

References

1. M. Tulej, D. A. Kirkwood, M. Pachkov, J. P. Maier: Astrophys. J. 506, L69 (1998)
2. B. J. McCall, J. Thorburn, L. M. Hobbs, T. Oka, D. G. York: Astrophys. J. 559, L49 (2001)
3. C. D. Ball, M. C. McCarthy, P. Thaddeus: Astrophys. J. 528, L61 (2000)
4. J. Cernicharo, A. M. Heras, A. G. G. M. Tielens, J. R. Pardo, F. Herpin, M. Guélin, L. B. F. M. Waters: Astrophys. J. 546, L123 (2001)
5. E. Herbst, S. Green, P. Thaddeus, W. Klemperer: Astrophys. J. 215, 503 (1977)
6. R. J. McMahon, M. C. McCarthy, C. A. Gottlieb, J. B. Dudek, J. F. Stanton, P. Thaddeus: Astrophys. J. 590, L61 (2003)
7. S. C. Madden, W. M. Irvine, H. E. Matthews, P. Friberg, D. A. Swade: Astron. J. 97, 1403 (1989)
8. F. F. Gardner, G. Winnewisser: Astrophys. J. 195, L127 (1975)
9. H. W. Kroto, D. McNaughton, L. T. Little, N. Matthews: Mon. Not. R. Astron. Soc. 213, 753 (1985)
10. J. August, H. W. Kroto, D. McNaughton, K. Phillips, D. R. M. Walton: J. Mol. Spectrosc. 130, 424 (1988)
11. I. Vasiliev, S. Öğüt, J. R. Chelikowsky: Phys. Rev. Lett. 78, 4805 (1997)
12. M. C. McCarthy, P. Thaddeus: Phys. Rev. Lett. 90, 213003 (2003)
13. M. C. McCarthy, P. Thaddeus: Astrophys. J. 592, L91 (2003)
14. K. Kawaguchi, Y. Kasai, S.-I. Ishikawa, N. Kaifu: Publ. Astron. Soc. Japan 47, 853 (1995)
15. J. Cernicharo, M. Guélin, C. Kahane: Astron. Astrophys. Suppl. Ser. 142, 181 (2000)

Two Problems of Interstellar Chemistry

W. Klemperer

Department of Chemistry, Harvard University, 12 Oxford Street, Cambridge, MA 02138, USA klemperer@chemistry.harvard.edu

I speculate upon two topics of interstellar chemistry. These are first, the ubiquitous abundance of CH^+in diffuse clouds; second, the rich organic chemistry of dense molecular clouds. The problem of the first topic is that CH^+ is relatively difficult to make but very easy to destroy[1]. The second problem is that the cosmic abundance of oxygen is more than twice that of carbon, thus how can this rich, reducing chemistry be produced.

In diffuse clouds, CH^+ is observed in absorption by the visible X $^1\Sigma$ - A $^1\Pi$ transition. Only population in its lowest state, J=0, v=0 , of the $^1\Sigma$ ground state is detected. We may note this appears to be entirely different than observations in planetary nebula such as NGC 7027 where a rich rotational emission spectrum is observed in the infrared[2] and modeled[3] in terms of a hot dense medium. Here we examine whether a cold diffuse, atomic medium may support J=0, v=0 , $^1\Sigma$ CH.

In such a region carbon is primarily C^+ and we assume that hydrogen is atomic. The formation of CH^+ at low temperatures by the radiative association of C^+ and H has a rate constant at 10 K estimated[4,5] to be 5×10^{-16} cm^3 sec^{-1}. This radiative association reaction is dominated by resonances. The abundance of CH^+ is 10^{-8} n_H or equivalently 3×10^{-4} $[C^+]$. Thus it is required that the rate of destruction of the CH^+ balance this slow rate of formation. This is the general problem. The dominant reactions which destroy CH^+ are

(1)	$CH^+ + H = C^+ + H_2$	exothermic by 3100 cm^{-1} (0.393 eV)
(2)	$CH^+ + e = C + H$	very exothermic
(3)	$CH^+ + H_2 = CH_2^+ + H$	exothermic

Reaction (1) and reaction (3) have been studied by Ferguson et al[6] at 300 K. Reaction (1) proceeds at 1/3 the (Langevin) collision frequency, while reaction (3) occurs at essentially collision frequency. Reaction (2) has been measured in merged beams[7] and has a rate constant of 10^{-7} with relatively warm reactants. In effect we ask whether J=0 CH^+ behaves differently kinetically than the 300K material generally studied[6].

We first examine the dissociative attachment, reaction (2). It is generally agreed that the state of CH responsible for the reaction is the repulsive E $^2\Pi$ correlating asymptotically with C ^1d + H ^2s. The angular momentum of the products of the reactive system is then non-zero. The angular momentum of the reactants are that of the electron since diffuse cloud CH^+ is J=0. At low temperatures the electron is an s-wave. Thus the reactants at very low temperatures have zero angular momentum. In this manner the rate of dissociative attachment may be considerably smaller at very low temperatures.

Consider next the reaction (1) the hydrogen atom abstraction. The reactants at very low temperature will be in a state of zero orbital angular momentum. This may result in a reaction rate lower than collision frequency. The stringency of the high $[CH^+]/[C^+] = 3 \times 10^{-4}$ abundance ratio coupled with inefficient radiative association reaction will require that the fraction of collisions leading to H atom abstraction from CH^+ be 1/1000.

In view of the observations of Ferguson and discussions of Armentrout, it is difficult to be optimistic about the reality of such a low destruction efficiency. In addition to destruction by atomic hydrogen the destruction by oxygen is almost certain.

We mention this apparent failure since CH^+ is ubiquitous in diffuse interstellar clouds. There is the assumption that CH^+ exists in the primarily atomic region of diffuse clouds. The question we raise is whether the J=0 state of CH^+ is uniquely unreactive with cold electrons and cold atomic hydrogen.

We turn to the second question, that of the abundance of reduced carbon species in molecular clouds. The problem arises from the cosmic abundance ratio of O/C=2.3. The problem is most easily described by means of a specific kinetic formation scheme, that of cosmic ray induced ion molecule chemistry. Here the scheme[8] is volume ionization of H_2 and He by 100 Mev and higher energy cosmic rays producing H_2^+ and He^+ The subsequent secondary reactions with H_2 destroys H_2^+ producing H_3^+ which now has been detected in both dense molecular clouds as well as diffuse clouds[9]. H_3^+ is an important protonating agent and was indirectly inferred from the high abundance of protonated carbon monoxide, HCO^+[10,11,12]. He^+ does not react with H_2 but does with CO producing C^+. The enhancement of carbon ion production over direct cosmic ray dissociative ionization of CO is at least the He/C cosmic abundance ratio of 500.

Since C^+ plays a dominant role in producing the rich organic chemistry of the interstellar medium it is important that it not destroyed by reaction with oxygen containing molecules[13] such as H_2O or O_2. Since reaction of C^+ with these species is fast, they must be rare in the interstellar gas. This is indeed the case as has been amply demonstrated by the Submillimeter Wave Astronomy Satellite (SWAS) observations. H_2O has been shown to be less than 1% as abundant as CO and O_2 is not observed. The likely repository of the oxygen excess is in refractory grains,. The seeming problem of the adequacy of the metals budget disappears if in addition to the accepted refractory repositories of Mg, Si, Ca and Fe, sulfur in the form of H_2SO_4 is included[14]. We have not included in Table 1. minor species forming refractory oxides such as Na, Al, Ca, Ni.

Some indication of the plausibility of this suggestion is found in the different molecular abundances of cold and heated regions. The valuable list of Walmsley[15] shows an increase of 3 to 4 orders of magnitude of H_2S in the

Table 1. Depletion of Oxygen

Cosmic Abundance		Oxygen in Solid Form	
C	100		
O	235		
Mg	10.6	$MgCO_3H_2O$	42
Si	9.9	$Si(OH)_4$	40
S	5.1	$H_2SO_4H_2O$	25
Fe	8.9	Fe_2O_3	13

hot core of Orion from that in cold molecular clouds. Attendant with this increase is the decrease in HCO^+. Heating of grains produces H_2O followed by the proton transfer reaction $HCO^+ + H_2O = CO + H_3O^+$. The reduction of H_2SO_4 to H_2S by hot hydrogen would be in accord with the suggestion that the grains are a repository for sulfuric acid.

The suggestion we make is that the kinetic stability of the rich, reducing chemistry of carbon is a consequence of having the gas phase O/C abundance ratio near 1. Effectively both carbon and oxygen is in the form of CO in the gas phase of cold molecular clouds.

It is an especial pleasure to acknowledge thirty years of numerous and most enjoyable discussions with Professor Gisbert Winnewisser.

Acknowledgment

This work was supported by the Chemistry Division of the National Science Foundation.

References

1. S.R.Federman, D.E.Welty,J.A.Cardelli: Ap.J. **481**, 795 (1997)
2. J.Cernicharo, X.-W.Liu, et al: Ap. J. **483**, L65-L68 (1997)
3. M.Yan, S.R.Federman, A.Dalgarno, and J.E.Bjorkman: Ap. J. **515**, 640 (1999)
4. G. Barinvos and M.van Hemert (presentation at Telluride workshop on motion of multiple potential surfaces July 2003)
5. M.M.Graff, J.T.Moseley and E.T.Roueff: Ap. J. **515**, 640 (1983)
6. W.Federer, H.Villinger, F.Howorka, W.Lindinger, P.Tosi, D.Bassi, and E. Ferguson: Phys. Rev. Lett, **52**, 2084 (1984)
7. Z.Amitay, D.Zajfman, P.Forck, U.Hechtfischer, B.Seidel, M.Grieser, D.Habs, R.Repnow, D.Schwalm, and A.Wolf: Phys.Rev. A **54**, 4032 (1996)
8. E.Herbst and W. Klemperer: Ap. J. **185**,505 (1973)
9. B.J.McCall, T.R.Geballe, K.H.Hinkle, T.Oka: Astrophys..J. **522**,322(1999)
10. L.E. Snyder and D. Buell: Nature **227**,862 (1970)
11. W. Klemperer: Nature **227**,1230 (1970)

12. R.C.Woods, T.A.Dixon, R.J.Saykally, P.G.Szanto: Phys.Rev. Lett. **29**,1269 (1975)
13. http://cfa-www.harvard.edu/swas/ SWAS is designed to survey O_2, and H_2O in the interstellar gas clouds.
14. F.Scappini, C.Cecchi-Pestellini, H.Smith, W.Klemperer, and A.Dalgarno: Mon. Not. R Astron.Soc **341**,657 (2003)
15. C.M.Walmsley: J.Chem.Soc. Faraday Trans. **89**,2119 (1993)

Electronic Spectroscopy of Carbon Chains and Their Relevance to Astrophysics

J.P. Maier

Department of Chemistry, University of Basel, Klingelbergstrasse 80, CH-4056 Basel, Switzerland

Our research activity of the past decade has focused on the measurement of electronic spectra of selected carbon chains and their ions in the gas phase. Part of this has been concerned with the development of the appropriate techniques, part in their application to obtain the spectra and finally the comparison with astronomical observations. The relevance of carbon chains to the latter arises from two considerations. The identification of numerous polar chains in dark interstellar clouds [1] leads to the question which of these and related species will have strong electronic transitions in the wavelength region where diffuse interstellar bands (DIBs) are observed [2]. Secondly, the expected physico-chemical properties of carbon chains has lead to the suggestion that they are potential DIB carriers [3]. The purpose of our studies was to obtain a few such spectra in the gas phase to test this hypothesis directly.

There are three problems associated with the measurement of the electronic spectra of transient species such as the carbon chains. These are the production of the species in a sufficient concentration, the availabilty of a sensitive method to detect the absorption features sought and the knowledge where the strong electronic transitions will be. The latter was the first task solved and allows us now to predict which types and sizes of carbon chains, their ions and simple derivatives including hydrogen or nitrogen will have their strong electronic transitions in the DIB region (4000–9000 Å).

The understanding which carbon chains should be selected for the gas phase studies has been acquired by the development of a technique which enables us to measure the electronic absorption spectra of mass-selected species in neon matrices. Through the systematic study of a number of homologous series such as the bare carbon chains, C_{2n} or C_{2n+1} , or cations HC_nH^+, or anions C_n^-, or derivatives C_nH, we located the strong electronic transitions and could consequently predict where they would absorb in the gas phase to within 100 cm^{-1} [4]. Thus for example the species C_6^-, C_6H, HC_6H^+, HC_5N^+, are all isoelectronic and show strong absorption bands between 5000–6000 Å, right at the heart of astronomical DIB data. Because species such as C_6H are known to be present in the dark clouds as result of mm-wave detections [5],

This article is dedicated to Gisbert Winnewisser for his inspirational contributions to laboratory and astronomical spectroscopies.

the obvious question was then if they could be responsible for some of the DIBs.

In the gas phase it has proven possible to prepare the carbon chain-like species in supersonic expansions combined with an electrical discharge. Depending on the conditions of this "synthesizer", neutral chains, cations or anions can be produced in concentrations sufficient for detection. Thus for example the linear chains $C_{2n}H$ with n=3–5 can be generated by passing a mixture of few percent acetylene in a rare gas through a pulsed nozzle while maintaining a DC discharge across it [6]. The electronic transitions of the species are then measured by a number of methods; e.g., detection of the absorption directly using pulsed or cw cavity ring down techniques or multiphoton excitation-ionisation methods for the neutrals or anions. Through the analysis of the rotational profiles on the bands, the spectral feature at a chosen temperature can then be predicted. This is an important aspect for comparison with astronomical data, because the rotational profiles change with temperature and can lead to the apparent maximum of the band shifting an Å between 10 and 100 K. Clearly a good understanding of the spectroscopic constants for the species is a requisite for comparison with astronomical data.

As example we can consider the comparison of the laboratory spectrum of C_6H, measured at 20 K, with absorption features through diffuse clouds towards reddened stars [7]. With the current astronomical detection limits for the equivalent width (EW) in the optical region, C_6H is not identifiable in the diffuse medium by its electronic spectrum. This can be put on a quantitative scale with the knowledge of the oscillator strength f of the transition. The latter values are mainly available from theoretical calculations, or by an educated guess based on isoelectronic molecules and scaling the values with the length of the chain. The data for C_6H are:

$$\mathrm{C_6H} \quad \lambda_{0-0} = 5265.8\ \text{Å} \quad f_{0-0} = 0.06 \quad \mathrm{EW} = 0.03\ \text{Å} \quad N \leq 2 \times 10^{12}\ \mathrm{cm^{-2}}$$

and lead to the upper limit of the column density N(C_6H) in the diffuse clouds of less than about 10^{12} cm^{-2}. A number of similar values have been deduced for the isolectronic species C_6^-, HC_6H^+, HC_5N^+ by a first time possible comparison of laboratory and astronomical data [7]. The upper limits of the abundances are of the same order of magnitude ($N \leq 10^{12}$), the main uncertainties coming from the values of the oscillator strengths.

To folllow this up further, the search for the electronic spectra of the three neutral bare carbon chains, C_3, C_4 and C_5 in diffuse clouds was undertaken in the period June 2000 – July 2003 using the Canadian-French-Hawaii telescope on Mauna Kea. The electronic spectrum of C_3 has been known for a long time, whereas those of C_4 and C_5 were obtained recently in our laboratory. In 2000 the detection of C_3 was successful and lead to the column density of 1.6×10^{12} cm^{-2} and a temperature of around 80 K was inferred from the rotational line intensities [8]. On the other hand C_4 and C_5 could not be detected in spite of very high signal to noise attained [9]. This leads to the column density values $N(C_4) \leq 5 \times 10^{11}$ cm^{-2} and $N(C_5) \leq 1 \times 10^{11}$ cm^{-2} [10]. According to chemical model predictions [11], which obtained correct density for C_3, the species C_4 and C_5 should have been detected.

These observations lead to two important conclusions as result of this ten year program to obtain the electronic spectra of carbon chains in the gas phase. The first is that such chains, and related species comprising say a handful or two or carbon atoms can not be responsible for the strong DIBs. Although these molecules are undoubtedly present in the diffuse clouds, they will contribute only to the weakest DIBs, with equivalent widths in the mÅ order of magnitude. The second concerns the goal of laboratory studies for the future.

Concerning the first conclusion, we can consider C_6H as a prototype system. It has been detected in dark clouds by rotational spectroscopy and the typical column densities lie in the $10^{12} - 10^{13}$ cm^{-2} range [12]. A number of chemically and structurally related molecules, such as C_2H, C_3H_2, have been observed also in diffuse regions [13], usually with an abundance some factor up to ten lower than in the dense clouds. It is then reasonable to assume that C_6H is likely to have a column density about $10^{11} - 10^{12}$ cm^{-2} in diffuse clouds. We can subsequently predict using the spectroscopic data given above for C_6H that the equivalent width of the origin band absorption would be mÅ in magnitude, which is hardly detectable. Thus although such carbon chains are probably present in the diffuse clouds, as is the case for C_2H, C_3H_2, they can not be responsible for the prominent DIBs which have equivalent widths around 0.1 Å [2]. This would only be possible either if they had electronic transitions with oscillator strength two orders of magnitude larger than for C_6H, which we know from our studies on a number of carbon chain series comprising up to around ten or so carbon atoms is not the case, or they would have to have column densities in the 10^{14} cm^{-2} range, which is unlikely because only a few diatomics, or H_3^+ [14] attain such abundances in diffuse clouds. As corollary to this, polar chains such as C_6H will be easier to identify in the diffuse medium via their transitions in the mm, or THz, region rather than in the visible by their electronic absorptions.

The second conclusion as results of this comparision addresses the question "where to now" as far the carbon chains are concerned. This aspect is pointed out in relation to the following table.

C_3	EW $= 10^{-4}$ Å	$f = 10^{-2}$	$N_R(C_3) = 10^{10}$ cm^{-2}
C_{21}	EW $\approx 10^{-1}$ Å	$f \approx 10$	$N_R(C_{21}) \approx 10^{11}$ cm^{-2}

In this the detection of C_3 in diffuse clouds is the basis of the argument. Given is the typical oscillator strength of an individual rotational transition of C_3 detected, the corresponding equivalent width of the line in the astronomical spectra and the typical column density inferred for a single rotational level (the total column density being 1.6×10^{12} cm^{-2}). In the line below is considered a long carbon chain, say C_{21}, with a much stronger electronic transition, namely with an oscillator strength say around 10, and thus a factor of 10^3 stronger than that detected for C_3.

This corresponds to the actual situation in the electronic spectra of these carbon chains; whereas the detected transition of C_3 at 4027 Å is of $^1\Pi - {}^1\Sigma$ symmetry, the much stronger $^1\Sigma - {}^1\Sigma$ system with oscillator strength of about unity lies in the VUV, near 1700 Å. As the length of the chain grows, the $^1\Sigma - {}^1\Sigma$ transition shifts by regular increments to the red and by the time sizes larger than C_{15} are reached the band system lies above 4000 Å [15], the astrophysically relevant region for DIBs. Because the oscillator strength is increasing with the length of the chain, for C_{21} it may well be around 10. Thus if we ask the question as to the column density that would be required for C_{21} in order to have an absorption with an EW ≈ 0.1 Å, that of a strong DIB, then 10^{11} cm^{-2} would be sufficient. This is not a prohibitively large amount and therefore such longer chains remain potentially relevant candidates for the stronger DIBs and the laboratory measurement of their electronic transitions in the gas phase is an appealing and challenging proposition.

The bare carbon chains with odd number of carbon atoms appear to fall in a rather special category of homologous series because it is their lowest electronic transition which carries the large oscillator strength. In fact there appears to be considerable selectivity as to the type and size of the carbon chains which should be considered "astrophysically relevant". This is exactly an aspect that has been often pointed out; some selectivity has to be present among the molecules to be consistent with the observation of a limited number of strong DIBs in the optical range. Thus the criteria required are that the species has an oscillator strength of unity or more in the 4000–9000 Å DIB range and that the molecules in question should be chemically sensible. To illustrate this aspect, we refer to the simplified molecular orbital diagram shown below.

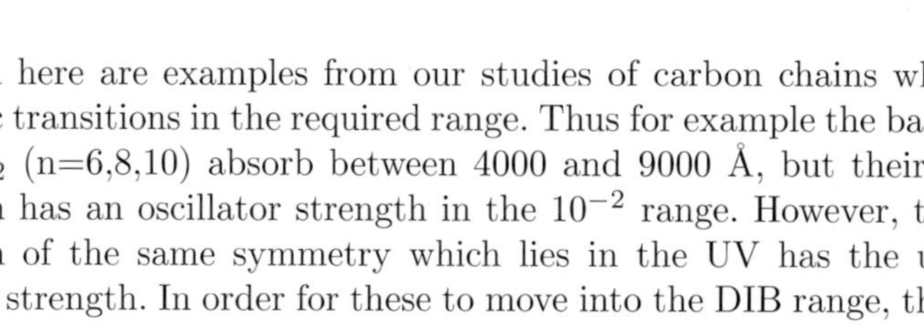

Taken here are examples from our studies of carbon chains which have electronic transitions in the required range. Thus for example the bare carbon chains C_2 (n=6,8,10) absorb between 4000 and 9000 Å, but their $^3\Sigma - {}^3\Sigma$ transition has an oscillator strength in the 10^{-2} range. However, their next transition of the same symmetry which lies in the UV has the unity like oscillator strength. In order for these to move into the DIB range, the system would have to comprise 20 carbon atoms or more, but these linear forms appear to terminate around C_{14}. This is in contrast to the C_{2n+1} series where the linear forms as large as C_{21} have actually been experimentally observed [15]. The $C_{2n}H$ species also show this pattern. As was discussed above for C_6H the first $^2\Pi - ^2\Pi$ transition does lie in the 5000–6000 Å region, but the oscillator strength is small, around 0.03. The stronger transition of the same symmetry lies in the UV and the chains would have to have be very long for this to shift into the visible region.

The reason for the behaviour that the lowest electronic transition has a relatively low oscillator strength whereas the higher energy one has unity of more values is result of configuration interaction, and has been rationalized theoretically long time ago [16]. As a rule this will take place in the systems where the two excitations above have comparable energy; one is within the filled MO manifold, and the other is from the HOMO to LUMO. These are the example in the middle and to the right.

On the other hand the system to the left, is the "classical" electronic transition of closed shell molecules involving HOMO–LUMO excitation. Here the lowest transition carries the large oscillator strength, as is exemplified by the C_{2n+1} series and also the polyacetylenes, $HC_{2n}H$. These strong transitions have been measured by us in the laboratory in the gas phase for systems as large as $HC_{26}H$ [17], which has its origin band of the $^1\Sigma - ^1\Sigma$ transition around 3800 Å with several units of oscillator strength. However, only for sizes above $HC_{40}H$ will the absorptions lie above 4000 Å. On the other hand the polyacetylenes with odd number of carbon atoms, $HC_{2n+1}H$, belong to example shown in the middle of the diagram; their $^3\Sigma - ^3\Sigma$ transition lies in the visible for sizes n=2–5, but the oscillator strength is too small to account for strong DIBs according to the arguments presented above. In order for their UV transition of this symmetry to move above 4000 Å the system will again have to comprise 40 or more atoms. Thus there is a natural selectivity as more favoured geometries may limit the number of possibilities.

The take home lesson of the above analysis is that there are some rather special carbon chain systems which make them attractive candidates for laboratory characterization and a subsequent comparison with DIB data. Prominent among these are the odd carbon chains C_{15}–C_{31} and the isolectronic $C_{2n+1}H^+$, or related $C_{2n}O$, species.

Support by the Swiss National Science Foundation (project 20-100019) is gratefully acknowledged.

References

1. G. Winnewisser: Naturwissenschaften **62**, 200 (1975)
2. G.H. Herbig: ARA & A **33**, 19 (1995)
3. A.E. Douglas: Nature **269**, 130 (1977)
4. J.P. Maier: J. Chem. Phys. **102**, 3462 (1998)

5. M.C. McCarthy & P. Thaddeus: Chem. Soc. Rev. **30**, 177 (2002)
6. H. Linnartz, T. Motylewski & J.P. Maier: J. Chem. Phys. **109**, 3819 (1998)
7. T. Motylewski, H. Linnartz, O. Vaizert, J.P. Maier, G.A. Galazutdinov, F.A. Musaev, J. Krelowski, G.A.H. Walker & D.A. Bohlender: ApJ. **531**, 312 (2000)
8. J.P. Maier, N.M. Lakin, G.A.H. Walker & D.A. Bohlender: ApJ. **553**, 267 (2001)
9. J.P. Maier, G.A.H. Walker& D.A. Bohlender: ApJ. **566**, 332 (2002)
10. J.P. Maier, G.A.H. Walker& D.A. Bohlender: ApJ. (2004), in press
11. E. Roueff, P. Felenbok, J.H. Black & C. Gry: A & A **384**, 629 (2002)
12. B.E. Turner: ApJ. **542**, 837 (2000)
13. R. Lucas & H.S. Liszt: A & A **358**, 1069 (2000)
14. T. Oka, J.A. Thornburn, B.J. McCall, S.D. Friedman, L.M. Hobbs, P. Sonnentrucker, D.E. Welty & D.G. York: ApJ. **582**, 823 (2003)
15. M. Wyss, M. Grutter & J.P. Maier: Chem. Phys. Lett. **304**, 35 (1999)
16. M.J.S. Dewar & H.C. Longuet-Higgins: Proc. Phys. Soc. London, Sect. A **67**, 795 (1954)
17. T. Pino, H. Ding, F. Güthe & J.P. Maier: J. Chem. Phys. **114**, 2208 (2001)

Highlights from the *SWAS* Mission

G.J. Melnick

Harvard-Smithsonian Center for Astrophysics, 60 Garden Street, MS 66, Cambridge, MA (USA) 02138 gmelnick@cfa.harvard.edu

1 Introduction

The *Submillimeter Wave Astronomy Satellite* (*SWAS*) is a mission primarily dedicated to the study of star formation, molecular cloud composition, and the structure of clouds. To carry out these investigations, *SWAS* was designed to observe $H_2^{16}O$, $H_2^{18}O$, O_2, CI, and ^{13}CO in transitions that are either difficult or impossible to detect from within the Earth's atmosphere. *SWAS* is a complete radio telescope in space, capable of observing the transitions listed in Table 1 with a spectral resolution, $\nu/\Delta\nu$, of 3×10^5 ($\Leftrightarrow$1 km s^{-1}) [1]. At the time of launch, the primary mission goals were to answer three important questions:

1. Where is all of the oxygen in the dense interstellar medium (and, in particular, are H_2O and O_2 major reservoirs of oxygen)?
2. Are H_2O and O_2 significant gas coolants?
3. What is the large-scale distribution of CI and warm ^{13}CO in molecular clouds?

Because of the importance of these questions, a number of pre-*SWAS* ground-based and airborne observations had been undertaken that sought to overcome the problems imposed by atmospheric interference, primarily by focusing on emission (and absorption) due to isotopomers of H_2O and O_2 which occur at frequencies where the atmosphere is partially transparent (e.g., [2, 3, 4]). Though limited observing time restricted these efforts to relatively few lines-of-sight, these measurements succeeded in establishing sufficient discrepancies with theory that the need for further, more sensitive observations became clear.

During its 3-year baseline mission, *SWAS* has succeeded in answering these questions by observing the most abundant isotopomers, increasing the number of sources studied, and reducing the flux limit of these measurements. To date, *SWAS* had conducted pointed observations toward more than 200 sources and along more than 5000 lines-of-sight, most toward dense ($n(H_2)>10^3$ cm^{-3}) molecular clouds throughout our Galaxy. In the process, new and important questions have been raised which – thanks to the continued flawless performance of the observatory and the much better than expected radiometric stability of the instrument – *SWAS* is addressing in an extended mission. *SWAS* has also contributed to solar system science by

Table 1. Spectral Lines Observed by *SWAS*

Species	Transition	Energy Above Ground State (E/k)	Frequency (GHz)	Wavelength (μm)	Critical Density (cm^{-3})
O_2	$3,3-1,2$	26 K	487.249	615.276	10^3
CI	$^3P_1 - {}^3P_0$ †	24 K	492.161	609.134	10^4
$H_2^{18}O$	$1_{10}-1_{01}$ †	26 K	547.676	547.390	10^9 ‡
^{13}CO	$J=5-4$	79 K	550.926	544.161	2×10^5
$H_2^{16}O$	$1_{10}-1_{01}$ †	27 K	556.936	538.289	10^9 ‡

† Ground-state transition.

‡ The critical density for $H_2^{16}O$ will likely be less than this value by a factor of 10^2-10^4 due to significant radiation trapping in this line. The critical density for $H_2^{18}O$ could be reduced by a factor of 1.5–50 due to the same effect.

detecting water vapor emission from several comets as well as in the atmospheres of Jupiter, Saturn, and Mars. Page limitations prevent a comprehensive review of the diverse topics being researched using *SWAS*. Instead, three observational projects will be summarized: (1) the understanding of interstellar chemistry to have emerged from *SWAS* observations of H_2O and O_2; (2) *SWAS* observations of comets orbiting another star; and, (3) *SWAS* observations of Mars. For a more in-depth review of *SWAS* discoveries, the interested reader is directed to the 20 August 2000 issue of *Ap. J. Letters*, which was dedicated to early *SWAS* results, as well as a number of more recent papers in the *Astrophysical Journal, Nature, and Icarus.*

2 H_2O and O_2 in the Interstellar Medium

One of the primary science goals of *SWAS* is to determine the abundance of water vapor in various components of the interstellar medium by observing the lowest rotational transitions of ortho-$H_2^{16}O$ and its isotopomer ortho-$H_2^{18}O$. The study of water vapor has several motivations.

Measurements of the water vapor abundance provide crucial constraints on astrochemical models that seek to elucidate the gas-phase chemical reactions and gas-grain interactions that determine the composition of the interstellar medium. In particular, chemical models (e.g., [5, 6]) predict that at temperatures $\gtrsim 300$ K, water vapor will account for most of the gas-phase oxygen that is not bound as CO, as a result of the neutral-neutral reactions:

$$H_2 + O \rightarrow OH + H \tag{1}$$

$$H_2 + OH \rightarrow H_2O + H \tag{2}$$

At temperatures less than 300 K, however, these reactions are negligibly slow because they possess significant activation energy barriers; gas-phase water is then produced solely by means of a cosmic-ray driven ion-neutral chemistry. These ion-neutral reactions proceed more slowly than the neutral-neutral reactions – requiring approximately 10^6 years to reach chemical equilibrium in well-shielded gas with $n(H_2) \sim 10^5$ cm^{-3} – and result in a maximum H_2O abundance more than 100 times lower than in warmer gas.

To date *SWAS* has studied more than 200 sources (and more than 5000 lines-of-sight) and has detected water vapor towards a diverse sample of astronomical objects. These objects include both dense and diffuse interstellar gas clouds, circumstellar envelopes, planetary atmospheres, and comets. As described below, the water abundances inferred from these *SWAS* observations vary by more than a factor of 10^5 and confirm some aspects of previous theoretical predictions but also turn up some surprises. In particular:

- In warm ($T \gtrsim 300$ K), dense ($n(H_2) \gtrsim 10^3$ cm^{-3}) gas, the gaseous water abundance is 10^{-5} to $>10^{-4}$ relative to H_2, in agreement with warm-cloud chemical models. Within this warm gas, H_2O is a significant repository of oxygen as well as a major gas coolant – in accord with model predictions.
- In cold ($T \lesssim 30$ K), dense clouds the abundance of gaseous H_2O is ~ 100 to 1000 times below the predictions of cold-cloud gas-phase chemical models (see Fig. 1) and the O_2 abundance is at least 100 times below predictions. As a result, gaseous H_2O and O_2 are neither significant reservoirs of oxygen in cold gas nor are they important gas coolants in such clouds.
- In cold, but diffuse ($n(H_2) < 10^3$ cm^{-3}), gas there is limited evidence that the H_2O abundance is higher than in the cold dense gas, but O_2 remains undetected.

Since the observations appear to agree with theory for warm gas, the remaining challenge has been to reconcile discrepancies in the chemical models for cold clouds. Successful models should at least meet the following observational constraints:

1) The models should produce $[O_2/H_2] < 10^{-6}$ [7].
2) The models should be able to account for low water abundances, i.e., $10^{-9} < [H_2O/H_2] < 7 \times 10^{-8}$, observed toward star forming regions ($n(H_2) \geq 10^5$ cm^{-3}, T_{gas} = 20–50 K, $T_{dust} \leq 50$ K, $A_V > 10$) and starless cores ($n(H_2) \geq 10^4$ cm^{-3}, $T_{gas} = 10$ K, $T_{dust} < 15$ K, $A_V > 10$) (see [8, 9, 10, 11]).
3) The models should simultaneously reproduce the higher water abundance, i.e., $[H_2O/H_2] \sim 10^{-6}$, in the low density gas along the line-of-sight to Sgr B2 ($n(H_2) \leq 10^4$ cm^{-3}, $T_{gas} \sim$ 10–30 K) (see [12]).
4) The models should account for a water-ice abundance, $[H_2O_{ice}/H_2]$, of about 10^{-4} (e.g., [13, 14]) in gas with $n(H_2) \sim 10^3$ - 10^6 cm^{-3}, and $T_{gas} \sim$ 10–50 K.

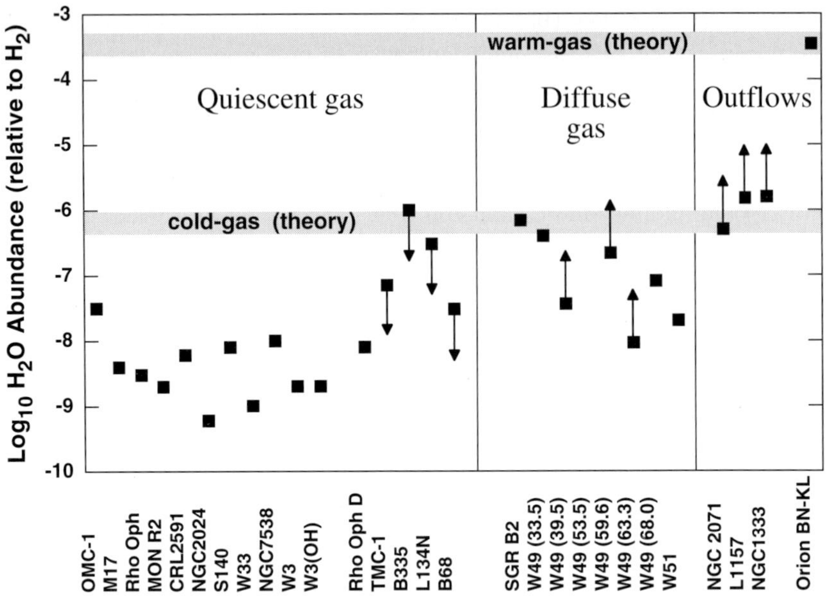

Fig. 1. *SWAS* H_2O abundance results obtained toward a sample of sources. The equilibrium H_2O abundances predicted by pure gas-phase chemical models are shown as grey bands for both cold- and warm-cloud conditions (see text).

Several ideas have been advanced to explain the *SWAS* results. However, the most appealing assumes that – toward cold clouds – gaseous H_2O exists only within a relatively narrow range of visual magnitudes near the cloud surface (see Fig. 2). Closer to the surface than an A_V of a few, both H_2O and O_2 are photo-destroyed by the ambient galactic UV field. Deeper into the cloud than an A_V of $\sim$4-8 (depending on H_2 density and UV intensity) H_2O, along with many other species, depletes onto grains. The preferential depletion of several of the more abundant oxygen-bearing species serves to increase the gas-phase C/O ratio deep within the cloud, further suppressing the gas-phase production of H_2O and O_2 (e.g., [15]). This picture is attractive since it naturally accounts for the low water vapor abundance inferred toward high column density lines-of-sight, the relatively higher water vapor abundances inferred toward lower column density lines-of-sight, and the high water-ice abundance derived toward many clouds.

3 Water Around Carbon-Rich Stars: A Probe of Extra-solar Cometary Systems

The discovery of extra-solar planetary systems has been one of the most exciting developments of the last decade. Jupiter-sized planets have now been

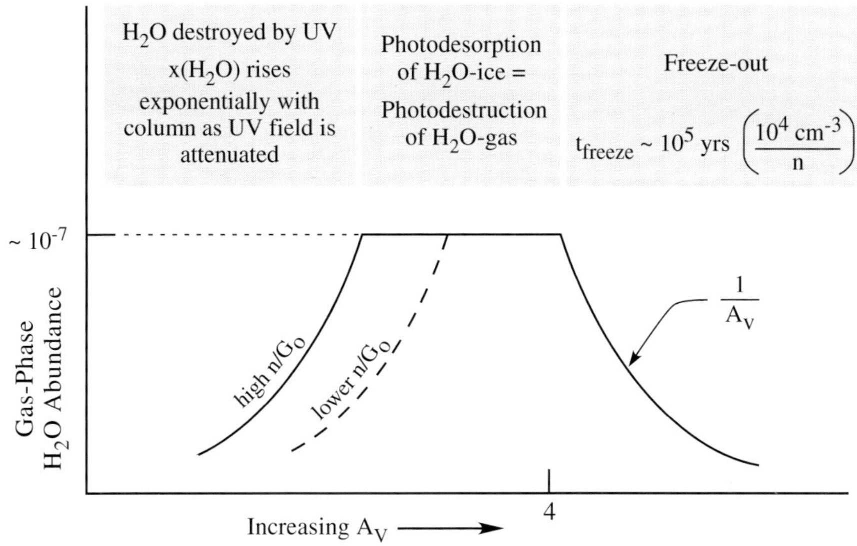

Fig. 2. Schematic depiction of water vapor zone within molecular clouds.

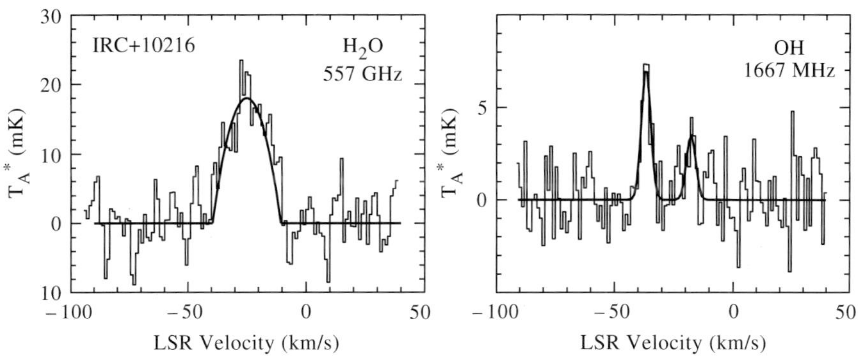

Fig. 3. Spectra of water (from *SWAS*) and its photodissociation product OH (from Arecibo) obtained toward IRC+10216.

detected around more than 100 stars by means of the tiny reflex motions that they induce in the stars they orbit. However, even the most sensitive Doppler observations of other stars are incapable of detecting analogs of the smaller constituents of our solar system, such as terrestrial-sized planets, asteroids and comets. Furthermore, the study of extra-solar planetary systems via periodic Doppler shifts in stellar spectra provides no information whatsoever about the chemical composition of the planets thereby discovered.

SWAS observations have demonstrated a complementary approach that probes the presence of icy bodies in orbit around stars other than the Sun.

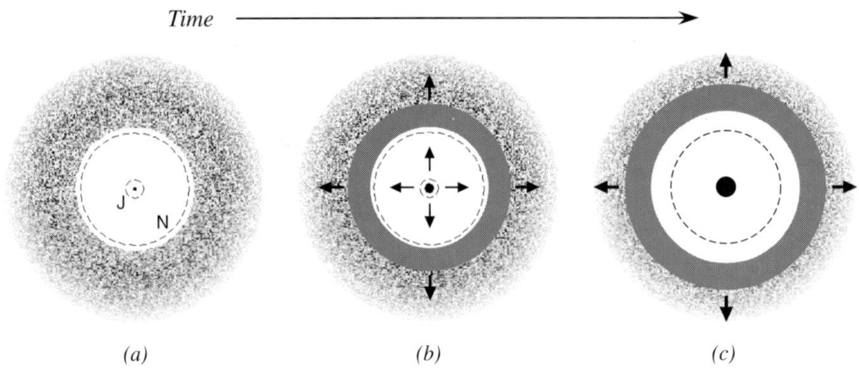

Fig. 4. Schematic drawing of IRC+10216 as it evolves from its main sequence to its asymptotic giant branch phase accompanied by an increase in both its size and luminosity, causing a wave of evaporation to propagate through the orbiting icy bodies. (*a*) During its main sequence phase, the luminosity of IRC+10216 is insufficient to vaporize icy bodies orbiting more distant than a few AU – for scale, distances equal to the orbits of Jupiter (J) and Neptune (N) are shown as dashed circles; (*b*) with the hydrogen and helium depleted within its core, IRC+10216 begins to expand, its luminosity increases, and a zone of vaporization (solid grey) begins to move outward through the population of icy bodies, depositing water vapor into the stellar outflow; (*c*) IRC+10216's diameter has increased to a size roughly equivalent to the orbit of Jupiter, its luminosity has increased greatly, and the zone of vaporization expands (after [16]).

One of the most surprising results obtained with *SWAS* has been the discovery of water vapor in the carbon-rich outflow that surrounds the AGB star IRC+10216 (a.k.a. CW Leonis) [16]. The discovery spectrum is shown in Fig. 3 (*left panel*). The presence of water vapor is entirely unexpected in environments where the abundance of carbon exceeds that of oxygen; astrochemical models for such environments are unanimous in predicting that almost all the oxygen nuclei should be bound as CO, with a small fraction in SiO and almost none remaining to form water vapor. The *SWAS* identification of water vapor is corroborated by our recent detection of OH using the Arecibo radiotelescope ([17]; Fig. 3, *right panel*), OH being the photodissociation product of H_2O that is produced once the outflowing gas encounters the interstellar radiation field. A thorough consideration (see [16]) of the possible origins of the water vapor detected in the IRC+10216 outflow – ranging from shock chemistry in the inner outflow [18] to photochemistry in the outer outflow (e.g. [19]) – has produced only one plausible theory for its origin: the H_2O is produced by the vaporization of orbiting icy bodies located at a distance $\sim$75 to 300 AU from the star.

The vaporization of orbiting icy objects is controlled by the luminosity evolution of the star. The stellar evolution is characterized by a rise in luminosity, modulated by periodic stellar thermal pulsations. As illustrated in

Fig. 4, as the luminosity of the star increases, a zone of evaporation moves outwards through the orbiting material. The inner edge of the evaporation zone is defined by the distance within which the largest icy bodies have already been entirely vaporized, and the outer edge is defined by the distance at which the icy bodies are warmed above the sublimation temperature of water-ice. By the time the convection of carbon has made the photospheric C/O ratio greater than 1, it is found that for distances up to 75 AU even icy bodies as large as Pluto will have been completely vaporized [20]. Thus, in the evaporation scenario, the release of water vapor currently observed occurs beyond about 75 AU. The evaporation is a continuous process in which the ongoing vaporization of icy bodies replenishes the water vapor as it flows through the outflow and is ultimately photodissociated by the interstellar ultraviolet radiation field. New icy bodies continuously replenish the sublimated objects because the evaporation zone steadily moves outwards as the stellar luminosity increases. Detailed theoretical calculations of the ice vaporization process (see [20]) indicate that the mass of icy bodies required to account for the measured water vapor abundance in IRC+10216 is ~ 10 Earth masses, comparable to the mass inferred to have been originally present in the solar system's Kuiper Belt [21].

4 Atmospheric Temperature and Water Vapor on Mars

During the 2001 opposition, *SWAS* observed Martian dust storms and provided unique measurements of changes in the upper atmospheric temperature profile caused by these storms. In particular, simultaneous *SWAS* line observations of the ground state transitions of H_2O and $H_2^{18}O$, and the ^{13}CO $5 \rightarrow 4$ transition, allow a determination of atmospheric temperature and the water column from the surface well into the upper atmosphere (>90 km). Also, when operating in a chop-nod mode for sensitive continuum measurements, *SWAS* 490 GHz data were used to determine the global surface temperature. For comparison, the Thermal Emission Spectrometer (TES) aboard Mars Global Surveyor has been used for temperature determinations between altitudes of ~ 40 and 55 km, depending on the instrument look angle. While TES provides high-spatial and -temporal resolution of the atmosphere, only *SWAS* had the capability to observe the whole atmospheric column to 90 km.

Data were obtained over the period 23 May to 13 September 2001, bracketing the closest approach date of 23 June 2001. The highlights were measurements of the atmosphere and surface during the onset and progression of a planet-encircling dust storm which enveloped Mars from late June through September. *SWAS* spectroscopic results show that after the storm began, the low to middle atmosphere ($\sim$5-45 km) experienced a rapid temperature increase of up to 40 K while the physical surface temperature dropped by up to 20 K on the visible disk. The effect of the storm on the global temperature structure at altitudes above ~ 70 km appears to be minimal. *SWAS* spectral line observations of $H_2^{18}O$ also show that the total column of water in the atmosphere stays essentially constant, but that it is redistributed upward as the atmosphere is turbulently mixed by the storm. The *SWAS* data provide the only measure of the strong response of the upper atmosphere to the storm.

References

1. G. J. Melnick et al.: ApJ Letters **539**, L77 (2000).
2. P. Wannier et al.: ApJ, **377**, 171 (1991).
3. F. Combes et al.: A&A, **248**, 607 (1991).
4. J. Zmuidzinas et al.: in *Airborne Astronomy Symposium on the Galactic Ecosystem*, edited by M. R. Haas, J. A. Davidson, and E. F. Erickson, ASP Conference Series, **73**, 33 (1995).
5. M. Elitzur, T. de Jong: A&A, **67**, 323 (1978).
6. D. A. Neufeld, S. Lepp, G. J. Melnick: ApJ Suppl., **100**, 132 (1995).
7. P. F. Goldsmith et al.: ApJ Letters **539**, L123 (2000).
8. R. L. Snell et al.: ApJ Letters **539**, L93 (2000).
9. R. L. Snell et al.: ApJ Letters **539**, L97 (2000).
10. R. L. Snell et al.: ApJ Letters **539**, L101 (2000).
11. M. L. N. Ashby et al.: ApJ Letters **539**, L119 (2000).
12. D. A. Neufeld et al.: ApJ Letters **539**, L111 (2000).
13. A. G. G. M. Tielens et al.: ApJ, **381**, 181 (1991).
14. W. A. Schutte: in *Laboratory Astrophysics and Space Research*, edited by P. Ehrenfreund & H. Kochan, Dordrecht: Kluwer, 69 (1999).
15. E. A. Bergin et al.: ApJ Letters **539**, L129 (2000).
16. G. J. Melnick et al.: Nature, **412**, 160 (2001).
17. K. E. Ford et al.: ApJ, **589**, 430 (2003).
18. K. Willacy, I. Cherchneff: A&A, **330**, 676 (1998).
19. T. J. Millar, E. Herbst, R. P. A. Bettens: MNRAS, **316**, 195 (2000).
20. K. E. Ford, D. A. Neufeld: ApJ, **557**, L113 (2001).
21. D. Jewitt, J. Luu: in *Protostars and Planets IV*, edited by V. Mannings, A. Boss & S. Russell, Univ. of Arizona Press, Tucson, Arizona, 1201 (2000).
22. M. D. Smith et al.: Icarus, **157**, 259 (2002).

Perspective from a Younger Generation – The Astro-spectroscopy of Gisbert Winnewisser

K.M. Menten

Max-Planck-Institut für Radioastronomie, Auf dem Hügel 69, D-53121 Bonn, Germany kmenten@mpifr-bonn.mpg.de

Gisbert Winnewisser's astronomical career was practically coextensive with the whole development of molecular radio astronomy. Here I would like to pick out a few of his many contributions, which I, personally, find particularly interesting and put them in the context of newer results.

1 Beginnings – When Molecular Radio Astronomy Was Still *Radio* Astronomy

1.1 Introduction

The late 1960s and early 70s were an era of great excitement in all branches of radio astronomy. In spectral line research, 12 years after the detection of the first line, the 21 cm HI hyperfine structure (hfs) transition (Ewen & Purcell 1951), finally the first molecule was found, the hydroxyl (OH) radical (Weinreb et al. 1963).

The OH Λ-doublet transitions, as well as many other lines at m-, cm- and(!) mm-wavelengths from a variety of molecules, had been identified as possible astronomical targets in a visionary article by Charles Townes (1957)[1].

Shortly after the detection of OH, Hoglund & Mezger (1965) discovered the first radio recombination line, H109α at 5009 MHz. *Molecular* radio astronomy started in earnest in 1968/9, with the discoveries of ammonia (NH_3; Cheung et al. 1968) and water vapor (H_2O; Cheung et al. 1969) by a group around Townes[2], as well as the first complex organic species, formaldehyde (H_2CO; Snyder et al. 1969), breaking loose a real gold rush that has yet to stop. Even today, approximately 2–4 "new" interstellar or circumstellar[3] molecules are found each year, totaling 126 species (not counting isotopomers)[4].

[1] see also Shklovsky 1952.

[2] Both lines were on Townes' (1957) list!

[3] Circumstellar molecules exist in the envelopes of evolved giant stars. Many species are detected there that are also found in interstellar clouds, whereas some others, have so far *only* been detected in a circumstellar environment.

[4] See http://www.ph1.uni-koeln.de/vorhersagen/molecules/main_molecules.html

The early history of "molecule-hunting", interesting us here, is well documented by the review by Snyder (1972; "Molecules in Space"), which gives a very useful introduction to molecular radio astronomy, that is still very much readable today. In the same volume, the review following Snyder's is by Winnewisser, Winnewisser, & Winnewisser and presents an equally excellent overview of the laboratory side of "Millimetre Wave Spectroscopy".

Table 1. The First Interstellar "Radio" Molecules

Name	Species	Transition	Frequency (MHz)	Reference
Hydroxyl	OH	$^2\Pi_{\frac{3}{2}}, J=\frac{3}{2}$	1665.4018(2)[a]	Weinreb et al. 1963
Ammonia	NH_3	(1, 1)	23694.4955(1)[b]	Cheung et al. 1968
Formaldehyde	H_2CO	$1_{11}-1_{10}$	4829.660(1)[b]	Snyder et al. 1969
Water	H_2O	$6_{16}-5_{23}$	22235.08(2)[b]	Cheung et al. 1969
Methyl alcohol	CH_3OHH_2CO	1^+-1^-	834.267(2)	Ball et al. 1970
Formic acid	HCOOH	$1_{11}-1_{10}$	1638.804(1)	Zuckerman et al. 1971
		$2_{11}-2_{12}$	4916.319(3)	*GWi & Churchwell 1975*
Formamide	NH_2CHO	$1_{11}-1_{11}$	1539.58(2)[c]	Palmer et al. 1971
		$2_{11}-2_{12}$	4618.55(2)[c]	Rubin et al. 1971
Cyanoacetylene	HC_3N	$1-0$	9098.6537(3)[c]	Turner 1971
	$H^{13}C_3N$	$1-0$	∼ 9060[d]	*Gardner & GWi 1975a*
	$H^{12+13}C_3N$	$1-0$	∼ 9000[d]	*Churchwell et al. 1977*
Acetaldehyde	CH_3CHO	$1_{10}-1_{11}$	1065.075(5)	Ball et al. 1971
Methylidine	CH	$^2\Pi_{\frac{1}{2}}, J=\frac{1}{2}$	3263.788(10)[e]	Rydbeck et al. 1973
Thioformaldehyde	H_2CS	$2_{11}-2_{12}$	3139.38(3)	Sinclair et al. 1973
Methanimine	CH_2NH	$1_{10}-1_{11}$	5290.31(4)	Godfrey et al. 1973
Methylamine	CH_2NH_2	$2_{02}-1_{10}$	8776.2(1)[f]	Fourikis et al. 1974
Vinyl cyanide	H_2CCHCN	$2_{11}-2_{12}$	1371.8262(1)	*Gardner & GWi 1975b*
Methyl formate	$HCOOCH_3$-*A*	$1_{10}-1_{11}$	1610.2445(7)	Brown et al. 1975
	$HCOOCH_3$-*E*	$1_{10}-1_{11}$	1610.91(10)	Churchwell & GWi 1975
Dimethyl ether	$(CH_3)_2O$	$2_{02}-1_{11}$	9119.668(15)[c]	*GWi & Gardner 1976*
Cyanodiacetylene	HC_5N	$4-3$	10650.650(4)	Avery et al. 1976
		$2-1$	5325.328(2)	*Gardner & GWi 1978*

Only molecules first detected at cm-wavelengths are listed. For early detections in the mm-range, see Snyder 1972. All molecules were first detected in the transitions listed, except for dimethyl either, which was discovered in its $6_{06}-5_{15}$ line at 90938.10(3) MHz by Snyder et al. 1974. No detections for isotopomers are listed, except for HC_3N.
Contributions coauthored by GWi are in *italics*. In most cases these contain essential laboratory spectroscopy work by GWi.
[a] Apart from the $F=1-1$ line listed, also the $2-2$ line at 1667.3590(2) MHz was reported in the detection paper. [b] Line with hfs, intensity-weighted frequency is given [c] Line with hfs, average of lowest and highest hfs line frequency is given [d] Various lines with differently placed ^{13}C between 8815 and 9100 MHz. [e] Apart from the $F=0-1$ line listed, also the $F=1-1$ and $F=1-0$ lines at 3335.475(10) and 3349.185(10) MHz, resp., were reported in the detection paper (frequencies from Rydbeck et al. 1974). [f] Frequency from Fourikis et al. 1974.

As shown in Table 1[5], many of the early molecular detections were made at low radio frequencies (and toward Sagittarius B2). A reason for the propensity to find *such* lines *there* is given in §1.2

Almost concurrently with cm-wavelength molecular radio astronomy, shorter wavelength exploration started. Driven by new high-frequency receiver technology and, generally, higher line intensities and a much higher "line density" (i.e. lines per GHz) the "millimeter explosion" started with the discoveries of CO by R. Wilson et al. and HCO^+ ("X-ogen") by Buhl & Snyder, both in 1970 (the latter published in 1971), and continues through today, having reached the submillimeter range in the 1980s.

Once the millimeter window was opened, (most) people spontaneously lost interest in the cm-range because:

- Most molecules have the bulk or the entirety of their rotational spectrum at (sub)millimeter wavelengths
- Almost all "simple" (di- or triatomic), light molecules have (sub)mm or far-IR (FIR) rotational spectra
- Many of the complex molecules were (for a long time) *only* found in Sgr B2

Interest remained high in special cm-wavelength lines, including the NH_3 inversion transitions near 24 GHz, a number of prominent maser lines from OH (hfs, 1.7 GHz and higher frequencies), H_2O (22.2 GHz), CH_3OH (25, 12.2, and 6.7 GHz), as well as the CH hfs lines (3.3 GHz). All of these, except for CH_3OH, have all or the bulk of their *rotational* spectrum at submillimeter and far-infrared wavelengths.

1.2 Sagittarius B2

Considering the mm-wavelength data coming in, soon following the radio observations, it became clear that the emission from many low frequency, low-J lines of complex molecules toward Sgr B2 must be (weakly) inverted. This can be understood (at least for linear or near-linear species) from the fact that "when selection rules for collisional transitions in linear molecules are such that dipole collisions do not dominate, the $J = 1-0$ transitions may have population inversion over a wide range of physical conditions (see, for example Goldsmith 1972)" (quoted from Morris et al. 1976). This is borne out by their statistical equilibrium calculations.

However, *direct* proof for this assertion came only very recently, from cm-wave interferometry of the CH_3CHO $1_{11} - 1_{10}$ (1065 MHz) and HC_3N $1-0$

[5] The frequencies are taken from the JPL catalog (Pickett et al. 1998; see http://spec.jpl.nasa.gov/) or the Cologne Database for Molecular Spectroscopy (Müller et al. 2001; see http://www.ph1.uni-koeln.de/vorhersagen/molecules/main_molecules.html). Numbers in parentheses are uncertainties in the last significant digit(s).

(9100 MHz) lines with the Giant Metre Wave Radio Telescope (GMRT; Chengalur & Kanekar 2003) and the Australia Telescope Compact Array (ATCA; Hunt et al. 1999), respectively. The molecular emission is extended over a $2' \times 4'$ region and shows a one-to-one correspondence with the continuum emission, proving that it is inverted. For the CH_3CHO line a typical optical depth of 0.035 is derived.

Apart from the mentioned lines, Gardner & Winnewisser (1975b) concluded that the vinyl cyanide line listed in Table 1 is inverted, as did Winnewisser & Gardner (1976) for the dimethyl ether line and Sinclair et al. (1973) for the thioformaldehyde line. Broten et al. (1976) inferred the same for the 2.663 GHz $J = 1 - 0$ transition of HC_5N.

Once the first complex molecules were found in Sgr B2, this source turned into a Bonanza! The described inversion mechanism together with a relatively strong, extended continuum background "boosted" many otherwise undetectable lines into "observability". As a consequence, the vast majority of interstellar molecules was and still is found in Sgr B2, many of the most complex ones *only* there. All of the molecules listed in Table 1 are found in this source.

To put GWi's Sgr B2 effort into perspective: Several of the lines measured by Gardner and him confirmed "shaky" detections, based on noisy single line observations. The vinyl cyanide work presents the first detection of this molecule in space. The study on different ^{13}C isotopes of HC_3N by Churchwell et al. (1977) indicated chemical fractionation and put a cautionary remark on attempts at determining the carbon isotope ratio in the Galactic center.

1.3 Extended and Compact Organic Molecular Emission in Sgr B2

Using the (low) rotation temperatures and column densities determined for many of the species in Table 1 by Cummins et al. (1986) based on their 2 – 3 mm wavelength line survey it is completely clear that *all* of the lines in that table must be inverted, which greatly helped their discovery at the time.

Why are these "old" Sgr B2 results still highly relevant today? In large beam observations, such as those of Cummins of al. (angular resolution $90''$–$180''$) one observes very low rotation temperatures for many species. However, interferometer observations by Snyder and collaborators with resolutions $< 5''$ indicate $T_{\rm rot} \sim 200$ K for the same species in a very compact region, dubbed the "Large Molecule Heimat" (LMH) by Snyder et al. (1994). This source, which is in the immediate vicinity of the Sgr B2(N) continuum source, is the most prolific source of complex molecules known.

Perhaps most interestingly, in recent years arcsecond resolution interferometry with the BIMA Array has resulted in the detection and imaging of increasingly complex organic species, such as CH_2CHCN, $HCOOCH_3$, and CH_3CH_2CN; all Miao et al. 1995, 1997), NH_2CHO, HNCO, and $HCOOCH_3$ (Kuan & Snyder 1996), CH_3COOH (acetic acid; Mehringer et al. 1997; Remijan et al. 2002), HCOOH (formic acid; Liu et al. 2001), and $(CH_3)_2CO$ (acetone; Snyder et al. 2002); recent single dish detections include CH_2OHCHO (glycolaldehyde, the first sugar; Hollis et al. 2000; 2001); $HOCH_2CH_2OH$ (ethylene glycol: interstellar "antifreeze"; Hollis et al. 2002), and CH_2CHOH (vinyl alcohol: Turner & Apponi 2001).

The elevated complex molecule abundances in a confined, hot, region, such as the LMH, may be explainable by the evaporation of icy grain mantles, during which molecules that have grown there to complexity by continual hydrogenation over a long time are suddenly released into the gas phase by the ignition of a high-mass embedded star[6]. However, such an event cannot possibly explain (for luminosity reasons alone) the unique large-scale distribution of complex molecules in (and maybe around) Sgr B2, which is the most prolific source of organic molecules anywhere and, so far, has only been probed by the low frequency lines described above. *If* these molecules come off grain mantles, other processes, such as shocks created, e.g., by cloud-cloud collisions must be invoked. There actually is evidence for the latter in Sgr B2 (Mehringer & Menten 1997; Sato et al. 2000).

1.4 A Giant Organic Cloud Around the Galactic Center?

There is actually evidence for a giant repository of organic molecules beyond and including Sgr B2, coextensive with the Central Molecular Zone (CMZ, see e.g. Morris & Serabyn 1996), i.e., stretching from $l = +1\overset{\circ}{.}6$ to $-1\overset{\circ}{.}1$ in a $\sim \pm 0^{\circ}3$ wide band around the Galactic center.

The first evidence for extended organic material in the CMZ came from widespread 4.8 GHz H_2CO absorption (Scoville et al. 1974). Given the ubiquity of formaldehyde in molecular clouds, one might dismiss this "as nothing special". Methanol (CH_3OH) on the other hand has usually quite low abundance and is difficult to detect outside hot, dense cloud cores. Nevertheless, Gottlieb et al. (1979) find the emission in the 834 MHz $(1^+ - 1^-)$ line (see Table 1) in emission and *extended* relative to their $40'$(!) beam, concluding it is inverted.

Furthermore, mapping of the HNCO $5_{05} - 4_{04}$ transition (made serendipitously simultaneously with a $C^{18}O$ survey), shows that the emission in this line is extending continuously from $l = -0\overset{\circ}{.}2$ to $+1\overset{\circ}{.}7$ (Dahmen et al. 1997). The possible existence of such a huge mass of organic material in our Galactic center is extremely exciting and its extent, chemistry, and excitation should be studied with suitable tracers.

1.5 Randbemerkung: TMC-1

After Sgr B2, GWi and his collaborators turned their attention to complex molecules in cold dark clouds. In the abstract of a paper by Churchwell,

[6] A review of complex molecule formation involving grain mantle processes has recently been presented by Ehrenfreund & Charnley (2000).

Winnewisser, & Walmsley (1978) titled fetchingly "Molecular observations of a possible proto-solar nebula in a dark cloud in Taurus", these authors reported "a small molecular condensation near the south eastern edge of Heiles' cloud 2" in which they find "strong emission from the $J = 9 - 8$ transition of HC_5N and the $J = 1 - 0$ transition of HC_3N", and go on to "refer to this small cloud as the Taurus Molecular Cloud 1, or TMC 1". It turned out that they started a vogue: As of this writing (2003 October 25) the Astronomical Data System[7] lists a total of 267 abstracts containing TMC 1 in their text.

Churchwell et al. were by no means the first to notice this spectacular source, which lies in the heart of the Taurus molecular cloud complex, and is well known for its high abundance of linear carbon chains, in particular the cyanopolyynes, the longest of which, $HC_{11}N$ has *only* been found there (Bell et al. 1997). Citing them: "Morris[8] et al. (1976) first drew attention to this region by their detection of the $J = 5 - 4$ transition of HC_3N; this was followed by the detection of the $J = 4 - 3$ transition of HC_5N by MacLeod et al. (1978)[sic][9] and the $J = 9 - 8$ transition of HC_7N by Kroto et al. (1977)".

A great amount of work on TMC 1 was done by the group at the Herzberg Institute of Astrophysics, GWi's onetime home, and at the Nobeyama Radio Observatory (NRO), whose 45 m telescope is ideally suited for the 22 – 50 GHz range in which many heavy species at TMC 1's cold (10 K) temperature have their strongest emission. In a single paper, Kaifu et al. (1987) reported seven(!), strong(!) unidentified lines, which, aided by laboratory work of Saito et al. (1987) and Yamamoto et al. (1987), led to the first identifications of CCS and CCCS.

The NRO's TMC 1 work culminated with a complete 8800 – 50000 MHz line survey, described by Ohishi & Kaifu (1998), but unfortunately not (yet) published in its entirety, which contains a total of 404 lines from 38 molecules (not counting isotopomers). Eleven of the molecules were newly identified in the course of the survey. It is a great testimony to the laboratory spectroscopists involved that only a single line remains unidentified!

2 Light Intermezzo – Interstellar Hydrides

When I was a student, an article by G. Winnewisser, Aliev & Yamada (1982) drew my attention. To quote from its abstract: "The great progress molecular astronomy has made in the past decade has focused attention to one special class of molecules: the hydrides. The presently known spectra of the hydrides are reviewed, both interstellar and laboratory. No interstellar metal hydrides

[7] http://adsabs.harvard.edu/

[8] Mark Morris tells me that the first person to point out TMC 1 to him was Nick Scoville.

[9] That paper actually appeared in 1979.

are presently known. A large number of important hydrides exhibit spectra only in the submillimetre and far infrared region. Further technical advances in this area are expected to yield new interstellar detections."

2.1 Metal-Bearing Molecules in the Interstellar Medium

Observations of species containing metals, as well as other refractory species such as SiO, deliver important information on depletion into dust grains. Moreover, when released either from sputtering of the grains or their mantles by violent dynamics, such as protostellar outflows, they can be used as probes of these phenomena (see, e.g. Bachiller 1996).

Today the third sentence quoted above, "No interstellar metal hydrides ..." is, unfortunately, still true, if we use the strict (chemistry) meaning of "metal" (e.g. Na, K, Mg, Al, Fe,...). Four diatomic, non H-bearing metal compounds (NaCl, KCl, AlCl, and AlF and three triatomic ones (MgNC, MGCN, NaCN, and AlNC) have been detected in the dense envelope of the carbon star IRC+10216 (Cernicharo & Guélin 1987; Guélin et al. 1993; Kawaguchi et al. 1993; Ziurys et al. 1995; Turner et al. 1994; Guélin et al. 1996; Ziurys et al. 2002). MgNC, NaCN, AlF, and NaCl were also found, recently, toward the proto-planetary nebula (PN) CRL 2688 (Highberger et al. 2001, 2003; Ziurys et al. 2002; see Petrie 1999 for a discussion of other candidate species for detection). Still, only one metal-containing molecule, FeO, has been detected in the interstellar medium, toward Sgr B2 (Walmsley et al. 2002).

As to metal hydrides: Radio and (sub)millimeter wavelength searches for interstellar NaH (Plambeck & Erickson 1982) and LiH[10] have so far been unsuccessful, as was a search for MgH in IRC+10216's envelope (Avery et al. 1994). Optical wavelength absorption from CaH, AlH, FeH, and MgH has been found in cool stars and the Sun, MgH even in the integrated light of external galaxies (see, e.g., Spinrad & Wood 1965)

2.2 Light Hydrides – Present Status

The history of interstellar hydrides begins with the detection by Dunham (1937) of three optical absorption lines at 3957.7, 4232.6, and 4300.3 Å from diffuse interstellar clouds, which subsequently were found to arise from CH (4300.3 Å; McKellar 1940; suggested by Swings & Rosenfeld 1937) and CH^+ (the former two; Douglas & Herzberg 1941).

It has been thought for a long time that hydrides might play an important rôle in cooling the highest density regions of interstellar clouds. Hydrides have, firstly, large level spacings, allowing them to emit a relatively large

[10] The ground state $J = 1-0$ transition of LiH is near 444 GHz, an atmospherically very unfavorable frequency. A tentative detection of this line was made in the redshifted ($z = 0.68$) dense absorbing cloud toward the B0218+357 gravitational lens system (Combes & Wiklind 1998).

amount of energy per quantum. Secondly, their large dipole moments require high critical densities for their excitation. Thus, their rotational levels are only excited in high density regions with $n > 10^6$ cm^{-3}, making them effective coolants in this regime, where even rare CO isotopes are so optically thick that CO cooling becomes negligible compared to that of (the hydride) H_2O.

This regime was excluded in the classical study by Goldsmith and Langer (1978) on "Molecular Cooling and Thermal Balance in Dense Molecular Clouds", but was re-addressed by Neufeld et al. (1995), who found that for elevated temperatures, $T = 100$ K, H_2O cooling dominates and the hydrides' contribution is comparable to that *of all other molecules combined*, i.e. $\approx 30\%$ of the former.

Recent detections of hydride rotational transitions were made with the Caltech Submillimeter Observatory (CSO), namely H_3O^+ [$J_K = 1_1^- - 2_1^+, 3_2^+ - 2_2^-, 3_0^+ - 2_0^-$/ 307, 365, 396 GHz, Wootten et al. 1991; Phillips et al. 1992], SiH ($^2\Pi_{\frac{1}{2}}, J = \frac{3}{2} - \frac{1}{2}$/625, 628 GHz; Schilke et al. 2001)[11], HCl ($J = 1 - 0$/626 GHz); Blake et al. 1985), H_2O, $H_2^{18}O$ ($1_{10} - 1_{01}$/ 557, 548 GHz; Melnick et al. 2000), HDO (the $1_{01} - 0_{00}$/ 465 GHz and $1_{11} - 0_{00}$ 894 GHz ground-state transitions; Schulz et al. 1991, Pardo et al. 2001)[12], and NH_2 ($J = \frac{3}{2} - \frac{3}{2}, \frac{1}{2} - \frac{1}{2}, \frac{3}{2} - \frac{1}{2}$/462, 459, 461 GHz); van Dishoeck et al. 1993). H_2D^+ ($1_{10} - 1_{11}$/372 GHz), the deuterated isotope of the key H_3^+ molecule was discovered with the James-Clerck-Maxwell telescope (Stark et al. 1999)[13]. Observations with the Long Wavelength Spectrometer aboard the Infrared Space Observatory led to the discoveries of HF ($J = 2-1$/2.47 THz; Neufeld et al. 1997), HD ($J = 1-0$/2.68 THz and $2-1$/5.36 THz; Polehampton et al. 2002), plus various H_2O, OH, and CH rotational and some H_3O^+ inversion transitions (many toward Sgr B2; Goicoechea & Cernicharo 2001 and these proceedings).

A preliminary conclusion is that (for the only two sources observed, Orion-KL and Sgr B2) the measured HCl and HF abundances indicate that only a few percent of all chlorine is in these molecules, suggesting significant depletion. If this should hold true for other hydrides as well, their cooling contribution could be much smaller than discussed above. More sensitive observations of more transitions toward more lines of sight are needed. These will be possible with the Herschel Space Observatory[14] and the Stratospheric Observatory For Infrared Astronomy (SOFIA[15]).

---

[11] Tentative detection

[12] A number of higher excitation cm and mm-wavelength H_2O, $H_2^{18}O$, and HDO transitions have also been found.

[13] H_3^+ itself was finally detected, after many arduous attempts, in absorption at near-infrared wavelengths by Geballe & Oka (1996), benefitting from a monumental body of laboratory work, mainly gathered by Oka and collaborators (see these proceedings).

[14] http://astro.estec.esa.nl/SA-general/Projects/First/first.html

[15] http://sofia.arc.nasa.gov/

Under GWi's direction the KOSMA Spectroscopy Group has produced definitive submillimeter spectroscopy of NH_2, C, ^{13}C, NH, SH, SD, and CF.

3 The Quest for High Spectroscopic Precision – "Deutsche Gründlichkeit" Gone Loose or Urgent Necessity?

Over the years, the KOSMA lab, with GWi as the driving force, has produced "definitive" spectroscopy (i.e. frequencies to a few times 10 kHz rms up to 1 THz or higher) for a large number of astrophysically important molecules, including the first useful data on many *rare isotopomers* and various *vibrationally excited states*.

These are[16]: HCS^+, $HC^{14+15}N$, $^{12+13}CO$, $C^{17+18}O$, $H_2^{12+13}C^{16+18}O$, ground-state + vibrationally excited $H^{12+13}C_3^{14+15}N$, C_2H_5OCH3, CCS, CCC, ground-state + vibrationally excited HNC, CH2NH, CH_3CCH, CCH, $^{32+33}SO_2$, $S^{16+17}O_2$, $C^{14+15}N$, CH_2CCHCN, CCCO, CH_3C_4H, NH_2D (hfs), HC_5N, HNCO, and the seven simple species with submillimeter spectra listed above[17].

The first astronomical detections of $C^{13}CH$ and $HC^{15}N$ were made by the Cologne group with the KOSMA telescope (Saleck et al. 1992, 1993) and of $^{13}C^{17}O$ with the Swedish-ESO-Submillimeter Telescope (Bensch et al. 2001).

Among others, the Cologne spectroscopy effort is characterized by, first, comprehensive presentations of the *complete* spectrum of a given species, in general up to higher (several THz) frequencies than many other listings, and, second, significantly higher accuracy data for many "common" molecules, such as CO and its isotopomers, than available before.

Here we argue, using two examples, that both high frequency *accuracy* and *comprehensiveness* are necessary for progress in various different areas.

3.1 Accuracy

Ambipolar diffusion – ion/neutral slip – is an important mechanism in the star-formation process by which magnetic fields initially supporting clouds against collapse slip relative to the neutral gas, leading to the formation of dense cloud cores (see, e.g. Shu et al. 1987). Theory predicts velocity differences < 1 km s^{-1}, corresponding to 1 MHz at 300 GHz, between ionic and neutral species (Roberge et al. 1995). In particular, at submillimeter wavelengths present frequency uncertainties for many potentially suitable lines are much too high to draw any reliable conclusions.

[16] For references, see http://www.ph1.uni-koeln.de/vorhersagen/catalog/catdir.html

[17] Note the tenacity with which cyanoacetylene was pursued!; cf. §2

3.2 Comprehensiveness and Line Confusion

In a recent paper, Kuan et al. (2003) claim to have detected the simplest amino acid, glycine (H_2NCH_2COOH) in three warm, dense interstellar clouds, including Sgr B2. If true, this would be the conclusion of a long quest (see, e.g., Snyder et al. 1983; Berulis, Winnewisser, Krasnov, & Sorochenko 1985; Combes et al. 1996; Ceccarelli et al. 2000).

Any such "identification" must rely on a large number of lines. Even so (or maybe *because of this*) the *confusion problem* is a formidable one. Inspecting any figure presented by Kuan et al. we find comparably many glycine as unidentified ("U") lines. For a wider view, see the 218 – 263 GHz line survey by Nummelin et al. (1998, 2000), which, for Sgr B2(N), is at or near the confusion limit throughout that band, even at their relatively moderate sensitivity (compared to the interferometric data by Snyder and his collaborators).

In the high density ($n \sim 10^7$ cm^{-3}) Sgr B2 "Large Molecule Heimat" (see §1.2) level populations are in Local Thermodynamic Equilibrium characterized by rotation temperatures, $T_{\rm rot}$, around 200 K (Liu et al. 2001). At this temperature, the lower vibrationally excited states of a large number of complex molecules are populated, giving rise to a plethora of emission lines, probably with comparable intensities to the putative glycine lines. The width of the LMH's lines, typically $\sim 6 - 10$ km s^{-1}, adds additional complication. Given the gigantic partition functions, many of the lines are weak.

In my opinion, it is completely impossible to conclusively identify *very* complex molecules in a "line jungle", such the one found in the Sgr B2 Large Molecule Heimat[18], if one has not a much better knowledge of, a.o., vibrationally excited lines of many more species than is available today. This means that much more (tedious and little-appreciated) lab work is needed to make progress here: A big job for the Cologne group, which has already measured the spectra (although not necessarily those of the vibrationally excited states) of a number of extant molecules.

Apology

This contribution concerned itself mostly with (some aspects of) GWi's early astronomical career. Among many others, I didn't discuss such interesting and important topics as

- NH_3 hfs anomalies (Stutzki et al. 1982, 1984; Stutzki & Winnewisser 1985a,b,c)

[18] High spatial resolution interferometry as employed by L. Snyder and collaborators helps significantly in situations as complicated as the LMH: Individual molecules or groups of molecules in general have slightly different spatial distributions, so that imaging adds an additional dimension to a confident identification.

- his pushing the Cologne Acousto-Optical Spectrometers and receivers to become world-class, frontline instrumentation (see, e.g., Schieder et al. 1989; Klumb et al. 1995; Graf et al. 1998, 2003; see also Gary Melnick's SWAS contribution in these proceedings)
- any of all the nice KOSMA/Gornergrat results
- any of his non-astronomical spectroscopy; little pre-Cologne spectroscopy.

Acknowledgement: I would like to thank Malcolm Walmsley and Lew Snyder for their encouragement.

References

1. Avery, L. W., Bell, M. B., Cunningham, C. T., Feldman, P. A., Hayward, R. H., McLeod, J. M., Matthews, H. E., Wade, J. D.: ApJ, 426, 737 (1994)
2. Avery, L. W., Broten, N. W., MacLeod, J. M., Oka, T., Kroto, H. W.: ApJ, 205, L173 (1976)
3. Bachiller, R.: ARA& 34, 111 (1996)
4. Ball, J. A., Gottlieb, C. A., Lilley, A. E., Radford, H. E.: ApJ, 162, L203 (1970)
5. Ball, J. A., Gottlieb, C. A., Lilley, A. E., Radford, H. E.: IAU Circ. No. 2350 (1971)
6. Bell, M. B., Feldman, P. A., Travers, M. J., McCarthy, M. C., Gottlieb, C. A., Thaddeus, P.: ApJ, 483, L61 (1997)
7. Bensch, F., Pak, I., Wouterloot, J. G. A., Klapper, G., Winnewisser, G.: ApJ, 562, L185 (2001)
8. Berulis, I. I., Winnewisser, G., Krasnov, V. V., Sorochenko, R. L.: Soviet Astronomy Letters, 11, 251 (1985)
9. Blake, G. A., Keene, J., Phillips, T. G.: ApJ, 295, 501 (1985)
10. Broten, N. W., MacLeod, J. M., Oka, T., Avery, L. W., Brooks, J. W., McGee, R. X., Newton, L. M.: ApJ, 209, L143 (1976)
11. Brown, R. D., Crofts, J. G., Godfrey, P. D., Gardner, F. F., Robinson, B. J., Whiteoak, J. B.: ApJ, 197, L29 (1975)
12. Buhl, D., Snyder, L. E.: Nature, 232, 161 (1971)
13. Ceccarelli, C., Loinard, L., Castets, A., Faure, A., Lefloch, B.: A&A, 362, 1122 (2000)
14. Cernicharo, J., Guélin, M.: A&A, 183, L10 (1987)
15. Chengalur, J. N., Kanekar, N.: A&A, 403, L43 (2003)
16. Cheung, A. C., Rank, D. M., Townes, C. H., Thornton, D. D., Welch, W. J.: Phys. Rev. Letters, 21, 1701 (1968)
17. Cheung, A. C., Rank, D. M., Townes, C. H., Thornton, D. D., Welch, W. J.: Nature, 221, 626 (1969)
18. Churchwell, E., Walmsley, C. M., Winnewisser, G.: A&A, 54, 925 (1977)
19. Churchwell, E., Winnewisser, G.: A&A, 45, 229 (1975)
20. Churchwell, E., Winnewisser, G., Walmsley, C. M.: A&A, 67, 139 (1978)
21. Combes, F., Wiklind, T.: A&A, 334, L81 (1998)
22. Combes, F., Q-Rieu, N., Wlodarczak, G.: A&A, 308, 618 (1996)
23. Cummins, S. E., Thaddeus, P., Linke, R. A.: ApJS, 60, 819 (1986)
24. Dahmen, G. et al. 1997, A&AS, 126, 197 (1997)

25. Douglas, A. E., Herzberg, G.L: ApJ, 94, 381 (1941)
26. Dunham, T. PASP, 49, 26 (1937)
27. Ehrenfreund, P. , Charnley, S. B.:ARA&A, 38, 427 (2000)
28. Ewen, H. I., Purcell, E. M.: Nature, 168, 356 (1951)
29. Fourikis, N., Takagi, K., Morimoto, M.: ApJ, 191, L139 (1974)
30. Gardner, F. F. Winnewisser, G.: ApJ, 195, L127 (1975a)
31. Gardner, F. F. Winnewisser, G.: ApJ, 197, L73 (1975b)
32. Gardner, F. F. Winnewisser, G.: MNRAS, 185, 57P (1978)
33. Geballe, T. R., Oka, T.: Nature, 384, 334 (1996)
34. Godfrey, P. D., Brown, R. D., Robinson, B. J., & Sinclair, M. W.: Ap. Lett., 13, 119 (1973)
35. Goicoechea, J. R., Cernicharo, J.: ApJ, 554, L213 (2001)
36. Goldsmith, P. F.: ApJ, 176, 597 (1972)
37. Goldsmith, P. F. Langer, W. D.: ApJ, 222, 881 (1978)
eds. M. A. Gordon L. E. Snyder (New York, NY: John Wiley and Sons 1973) p 181
38. Gottlieb, C. A., Ball, J. A., Gottlieb, E. W., Dickinson, D. F.: ApJ, 227, 422 (1979)
39. Graf, U. U., Haas, S., Honingh, C. E., Jacobs, K., Schieder, R., Stutzki, J.: Proc. SPIE, 3357, 159 (1998)
40. Graf, U. U. et al.: Proc. SPIE, 4855, 322 (2003)
41. Guélin, M., Lucas, R., Cernicharo, J.: A&A, 280, L19 (1993)
42. Highberger, J. L., Savage, C., Bieging, J. H., Ziurys, L. M.: ApJ, 562, 790 (2001)
43. Highberger, J. L., Thomson, K. J., Young, P. A., Arnett, D., Ziurys, L. M.: ApJ, 593, 393 (2003)
44. Hoglund, B., Mezger, P. G.: Science, 150, 339 (1965)
45. Hollis, J. M., Lovas, F. J., Jewell, P. R.: ApJ, 540, L107 (2000)
46. Hollis, J. M., Vogel, S. N., Snyder, L. E., Jewell, P. R., Lovas, F. J.: ApJ, 554, L81 (2001)
47. Hollis, J. M., Lovas, F. J., Jewell, P. R., Coudert, L. H.: ApJ, 571, L59 (2002)
48. Hunt, M. R., Whiteoak, J. B., Cragg, D. M., White, G. L., Jones, P. A.: MNRAS, 302, 1 (1999)
49. Kaifu, N., Suzuki, H., Ohishi, M., Miyaji, T., Ishikawa, S., Kasuga, T., Morimoto, M., Saito, S.: ApJ, 317, L111 (1987)
50. Kawaguchi, K., Kagi, E., Hirano, T., Takano, S., Saito, S.: ApJ, 406, L39 (1993)
51. Klumb, M., Frerick, J., Horn, J., Schieder, R., Winnewisser, G. F.: Proc. SPIE, 2583, 420 (1995)
52. Kuan, Y. Snyder, L. E.: ApJ, 470, 981 (1996)
53. Kuan, Y., Charnley, S. B., Huang, H., Tseng, W., Kisiel, Z.: ApJ, 593, 848 (2003)
54. Liu, S., Mehringer, D. M., Snyder, L. E.: ApJ, 552, 654 (2001)
55. MacLeod, J. M., Avery, L. W., Broten, N. W.: ApJ,, 233, 584 (1979)
56. McKellar, A.: PASP, 52, 187 (1940)
57. Mehringer, D. M., Menten, K. M.: ApJ, 474, 346 (1997)
58. Mehringer, D. M., Snyder, L. E., Miao, Y., Lovas, F. J.: ApJ, 480, L71 (1997)
59. Melnick, G. J. et al.: ApJ, 539, L87 (2000)
60. Miao, Y., Mehringer, D. M., Kuan, Y., Snyder, L. E.: ApJ, 445, L59 (1995)

61. Miao, Y. Snyder, L. E.: ApJ, 480, L67 (1997)
62. Morris, M., Serabyn, E.: ARA&A, 34, 645 (1996)
63. Morris, M., Turner, B. E., Palmer, P., Zuckerman, B.: ApJ, 205, 82 (1976)
64. Müller, H. S. P., Thorwirth, S., Roth, D. A., Winnewisser, G.: 2001, A&A, 370, L49 (2001)
65. Neufeld, D. A., Lepp, S., Melnick, G. J.: ApJS, 100, 132 (1995)
66. Neufeld, D. A., Zmuidzinas, J., Schilke, P., Phillips, T. G.: ApJ, 488, L141 (1997)
67. Nummelin, A., Bergman, P., Hjalmarson, Å., Friberg, P., Irvine, W. M., Millar, T. J., Ohishi, M., Saito, S.: ApJS, 117, 427 (1998)
68. Nummelin, A., Bergman, P., Hjalmarson, Å., Friberg, P., Irvine, W. M., Millar, T. J., Ohishi, M., Saito, S.: ApJS, 128, 213 (2000)
69. Ohishi, M. & Kaifu, N.: Faraday Discussions No. 109 (1998)
70. Palmer, P., Gottlieb, C. A., Rickard, L. J., Zuckerman, B.: BAAS, 3, 499 (1971)
71. Pardo, J. R., Cernicharo, J., Herpin, F., Kawamura, J., Kooi, J., Phillips, T. G.: ApJ, 562, 799 (2001)
72. Petrie, S.: MNRAS, 302, 482 (1999)
73. Phillips, T. G., van Dishoeck, E. F., Keene, J.: ApJ, 399, 533 (1992)
74. Pickett, H. M.., Poynter, R. L., Cohen, E. A., Delitsky, M. L., Pearson, J. C., Müller, H. S. P.: J. Quant. Spectrosc. & Rad. Transfer, 60, 883 (1998).
75. Plambeck, R. L., Erickson, N. R.: ApJ, 262, 606 (1982)
76. Polehampton, E. T., Baluteau, J.-P., Ceccarelli, C., Swinyard, B. M., Caux, E.: A&A, 388, L44 (2002)
77. Remijan, A., Snyder, L. E., Liu, S.-Y., Mehringer, D., Kuan, Y.-J.: ApJ, 576, 264 (2002)
78. Roberge, W. G., Hanany, S., Messinger, D. W.: ApJ, 453, 238 (1995)
79. Rubin, R. H., Swenson, G. W., Benson, R. C., Tigelaar, H. L., Flygare, W. H.: ApJ, 169, L39 (1971)
80. Rydbeck, O. E. H., Elldér, J., Irvine, W. M.: Nature, 246, 466 (1973)
81. Rydbeck, O. E. H., Elldér, J., Irvine, W. M., Sume, A., Hjalmarson, Å.: A&A, 34, 479 (1974)
82. Saito, S., Kawaguchi, K., Yamamoto, S., Ohishi, M., Suzuki, H., Kaifu, N.: ApJL, 317, L115 (1987)
83. Saleck, A. H., Eigler, K., Simon, R., Vowinkel, B., Winnewisser, G.: Astronomische Gesellschaft Abstract Series, 7, 83 (1992)
84. Saleck, A. H., Simon, R., Schneider, N., Winnewisser, G.: ApJ, 414, L133 (1993)
85. Sato, F., Hasegawa, T., Whiteoak, J. B., Miyawaki, R.: ApJ, 535, 857 (2000)
86. Schieder, R., Tolls, V., Winnewisser, G.: Experimental Astronomy, 1, 101 (1989)
87. Schilke, P., Benford, D. J., Hunter, T. R., Lis, D. C., Phillips, T. G.: ApJS, 132, 281 (2001)
88. Schulz, A., Güsten, R., Walmsley, C. M., Serabyn, E. 1991, A&A, 246, L55 (1991)
89. Scoville, N. Z., Solomon, P. M., Jefferts, K. B.: 1974, ApJ, 187, L63 (1974)
90. Shklovsky, I. S.: Astr. Zhur., 29, 144 (1952)
91. Shu, F. H., Adams, F. C., Lizano, S.: 1987, ARA&A, 25, 23 (1987)
92. Sinclair, M. W., Fourikis, N., Ribes, J. C., Robinson, B. J., Brown, R. D., & Godfrey, P. D.: Australian J. Phys., 26, 85 (1973)

93. Snyder, L. E.: *Molecules in Space*. In: *MTP International Review of Science* vol 3, Spectroscopy, ed by D. A. Ramsay (Butterworths, London 1972) pp 193–240
94. Snyder, L. E., Buhl, D., Zuckerman, B. Palmer, P.: Phys. Rev. Lett., 22, 679 (1969)
95. Snyder, L. E., Buhl, D., Schwartz, P. R., Clark, F. O., Johnson, D. R., Lovas, F. J., Giguere, P. T.: ApJ, 191, L79 (1974)
96. Snyder, L. E., Hollis, J. M., Brown, L. W., Buhl, D., Suenram, R. D., Lovas, F. J.: ApJ, 268, 123 (1983)
97. Snyder, L. E., Kuan, Y.-J., Miao, Y.: LNP Vol. 439: The Structure and Content of Molecular Clouds, 187 (1994)
98. Snyder, L. E., Lovas, F. J., Mehringer, D. M., Miao, N. Y., Kuan, Y., Hollis, J. M., Jewell, P. R.: ApJ, 578, 245 (2002)
99. Spinrad, H., Wood, D. B. 1965: ApJ, 141, 109 (1965)
100. Stark, R., van der Tak, F. F. S., van Dishoeck, E. F.: ApJ, 521, L67 245 (1999)
101. Stutzki, J., Ungerechts, H., Winnewisser, G.: A&A, 111, 201 (1982)
102. Stutzki, J., Olberg, M., Winnewisser, G., Jackson, J. M., Barrett, A. H.: A&A, 139, 258 (1984)
103. Stutzki, J., Winnewisser, G.: A&A, 144, 1 (1985a)
104. Stutzki, J., Winnewisser, G.: A&A, 144, 13 (1985b)
105. Stutzki, J., Winnewisser, G.: A&A, 148, 254 (1985c)
106. Swings, P., Rosenfeld, L.: ApJ, 86, 483 (1937)
107. Townes, C. H.: *Microwave and radio-frequency resonance lines of interest to radio astronomy*. In: *IAU Symp. 4: Radio astronomy*, ed. H. C. van de Hulst (Cambridge University Press 1957) pp 92–103
108. Turner, B. E.: ApJ, 163, L35 (1971)
109. Turner, B. E., Apponi, A. J.: ApJ, 561, L207 (2001)
110. Turner, B. E., Steimle, T. C., Meerts, L.: ApJ, 426, L97 (1994)
111. van Dishoeck, E. F., Jansen, D. J., Schilke, P., Phillips, T. G.: ApJ, 416, L83 (1993)
112. Walmsley, C. M., Bachiller, R., Forêts, G. P. d., Schilke, P.: ApJ, 566, L109 (2002)
113. Weinreb, S., Barrett, A. H., Meeks, M. L.: Nature, 200, 829 (1963)
114. Wilson, R. W., Jefferts, K. B., Penzias, A. A.: ApJ, 161, L43 (1970)
115. Winnewisser, G., Aliev, M. R., Yamada, K.: *Interstellar Hydrides*. In: *Scientific Importance of Submillimetre Observations* (ESA: SP-189) pp 23 – 28 (1982)
116. Winnewisser, G. Churchwell, E.: ApJ, 200, L33 (1975)
117. Winnewisser, G. Gardner, F. F.: A&A, 48, 159 (1976)
118. Winnewisser, G., Winnewisser, M., Winnewisser, B. P.: *Millimetre wave spectroscopy*. In: *MTP International Review of Science* vol 3, Spectroscopy, ed by D. A. Ramsay (Butterworths, London 1972) pp 241–296
119. Wootten, A., Turner, B. E., Mangum, J. G., Bogey, M., Boulanger, F., Combes, F., Encrenaz, P. J., Gerin, M.: ApJ, 380, L79 (1991)
120. Yamamoto, S., Saito, S., Kawaguchi, K., Kaifu, N., Suzuki, H.: ApJ, 317, L119 (1987)
121. Ziurys, L. M., Apponi, A. J., Guélin, M., Cernicharo, J.: ApJ, 445, L47 (1995)
122. Ziurys, L. M., Savage, C., Highberger, J. L., Apponi, A. J., Guélin, M., Cernicharo, J.: ApJ, 564, L45 (2002)
123. Zuckerman, B., Ball, J. A., Gottlieb, C. A.: ApJ, 163, L41 (1971)

Gisbert F. Winnewisser

E. Herbst

Departments of Physics, Chemistry, and Astronomy, The Ohio State University, Columbus, Ohio 43210 USA herbst@mps.ohio-state.edu

1 Early Years

Gisbert Fritz Winnewisser was born in 1936 in Karlsruhe, Germany into a family whose roots can be traced back to the Huguenots. This was not an auspicious year for a birth, because the ravages of the Second World War were soon to engulf the world. Most but not all of Gisbert's family survived the war, while Gisbert attended elementary school in the picturesque village of Bad Peterstal, in the Black Forest. Later, the family moved back to Karlsruhe, where Gisbert first attended the Goethe Gymnaisum (1947-1956) and subsequently studied physics at the Technische Hochscule Karlsruhe (1956-1963). After obtaining his Diplom (Masters), he crossed the Atlantic Ocean in 1963 to study molecular spectroscopy at Duke University, in Durham, North Carolina. At that time, Durham was a small, parochial southern town, and it is hard at first to fathom why Gisbert would choose to study there. But, Gisbert was following in his elder brother Manfred's footsteps; Manfred was already at Duke University pursuing postdoctoral studies. Both Manfred and later Gisbert were attracted to this remote location because Walter Gordy, a very famous microwave spectroscopist, was a professor there. A pioneer in the use of ultra-high frequency microwaves, Walter Gordy can indeed be called the originator of millimeter-wave and even submillimeter-wave spectroscopy.

With his strong academic background, it did not take Gisbert very long to earn the Ph.D. degree at Duke; his thesis, dated 1967, is concerned with the rotational-torsional spectrum of the small (but not simple!) internal rotor molecule HSSH, known as disulfane [1]. A figure from his thesis is shown here; in it one can see the structure of the molecule and the potential energy for torsional motion. Notice that the H atoms lie with a dihedral angle near to perpendicular, while there are barriers at both the *cis* and *trans* positions. Motion against this potential leads to torsional energy levels; these are characterized by the quantum number v = 0,1,2,3... Each torsional level is split into four sublevels, characterized either by the quantum number τ = 1,2,3,4 or by the symmetry of the wave function against reflection through both the *cis* and *trans* planes. Rotational transitions occur along with transitions from one torsional sublevel to another, so that the rotational lines are effectively split, although this splitting is very small in HSSH for the ground torsional state. Since the dipole moment lies along the c axis, only c-type transitions are seen; these transitions occur between ortho or para levels and so show a 3:1 intensity alteration. Gisbert has studied HSSH and its isotopomers many times since his Ph.D. period, and enough is known about this molecule to fill a good-sized volume [2]. Most recently, Gisbert studied the related "brethren" molecule HSOH, oxadisulfane. Here both b-type and c-type transitions are seen, and the torsional splittings are so far unexplained [3].

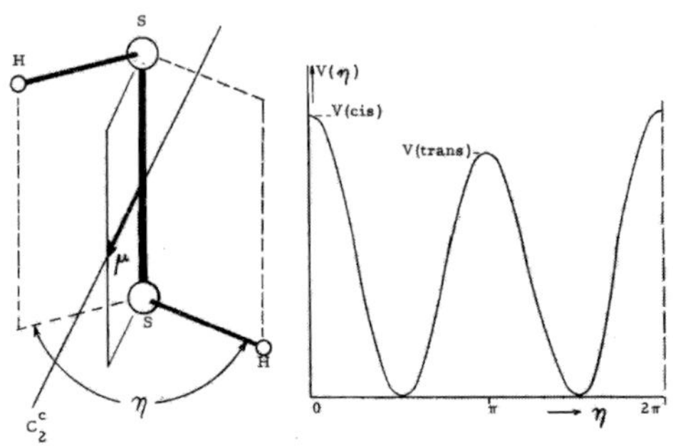

Fig. 1. Structure and potential for torsional motion in HSSH

After finishing his Ph.D., Gisbert remained in Gordy's laboratory at Duke for 1-2 years, and then, with wife Helga and son Carsten, moved to the National Research Council of Canada, in Ottawa in 1968 to accept a postdoctoral fellowship there. At the time, this institute, headed by Gerhard Herzberg, was literally the Mecca of spectroscopy throughout the world. Many well-known spectroscopists spent at least part of their earlier careers there, and Gisbert was no exception. At the time of his arrival, the so-called "glory" period of the NRC [4] was coming to an end, but it must still have been an exciting place for the young spectroscopist to work. While at NRC, Gisbert worked mainly with Cis Costain, although he also associated with Takeshi Oka. Gisbert's papers during this period reported work on the millimeter-wave spectra of HCN, DCN, HCP, DCP, HSSD, and ketene [5], [6]. Gisbert left NRC in 1970 to take up an appointment as a Research Professor in the Department of Astronomy and Astrophysics at the University of British Columbia. The campus, on the outskirts of the city of Vancouver, has one of the most beautiful locales of any academic institution, but Gisbert's appointment was short-lived, and he left after only a year to return to the NRC, this time as a Visiting Scientist.

2 Astronomy Beckons

When Gisbert returned to the NRC, molecular astronomy was just beginning its first period of glory; Townes, Snyder and others had begun to search the heavens for interstellar molecules, and molecules such as NH_3, CO, and H_2CO were found in the interstellar medium. It must have been natural for him to consider going into this field. Eventually, employment was found as a staff member at the Max Planck Institut fuer Radioastronomie in Bonn, Germany, and Gisbert and his family returned to his native country in 1972. Gisbert plunged into molecular astronomy, and soon made a name for himself in this fast-developing field.

Along with colleagues and collaborators, Gisbert found a variety of new molecules in space, the identifications of some of which were based on his own laboratory investigations in microwave and millimeter-wave spectroscopy. Looking at the galactic center source Sgr B2, Gisbert discovered the molecules vinyl cyanide (C_2H_3CN) [7] and formic acid (HCOOH) [8], while somewhat later, he, Snyder, and others found methyl diacetylene (CH_3CCCCH) in Taurus [9]. On the topic of isotopic fractionation, Gisbert was the first to discover singly deuterated ammonia (NH_2D), which was detected in three interstellar clouds [10] and the first to see (in Sgr B2) all three ^{13}C isotopomers of cyanoacetylene (HCCCN), the abundances of which were found to differ from one another [11]. Along with his elder brother Manfred, who had also moved back to Germany, and was now working as a Professor of Physical Chemistry with his wife Brenda at the Justus Liebig Universität in Giessen, Gisbert toiled in the laboratory to identify the carrier of the famous U90.66 line, seen by Snyder & Buhl [12] Although he, Manfred, and the other collaborators succeeded in producing HNC in the laboratory and confirming the interstellar identification [13], they were beaten by a month by a team headed by Saykally and Woods [14].

Although many people find the maps that radioastronomers produce of individual sources somewhat boring, this is not always the case. In particular, along with Ed Churchwell and Malcolm Walmsley, Gisbert was responsible for a most famous map. As reported in a paper entitled "Molecular Observations of a Possible Proto-Solar Nebula in a Dark Cloud in Taurus", these authors looked carefully at a small condensation in Taurus previously labeled Heiles Cloud 2 and partially mapped it in the J=9-8 transition of HC_5N at a frequency of 23.96 GHz [15]. They referred to the region as "the Taurus Molecular Cloud 1, or TMC 1." And so, TMC-1, the subsequently favored site for the detection of large molecules by radioastronomers, was given a name, and the particular part of TMC-1, now known as the CP (cyanopolyyne) peak, mapped out. Their original map is reproduced below as Figure 2. If one looks at a CO ($J = 1 \rightarrow 0$) picture of the Taurus region, it is amazing that the small condensation now known as TMC-1 was ever found; it hardly stands out at all. But it remains the most famous source for large unsaturated molecules, and it is unclear whether another such source in the interstellar medium will ever be found.

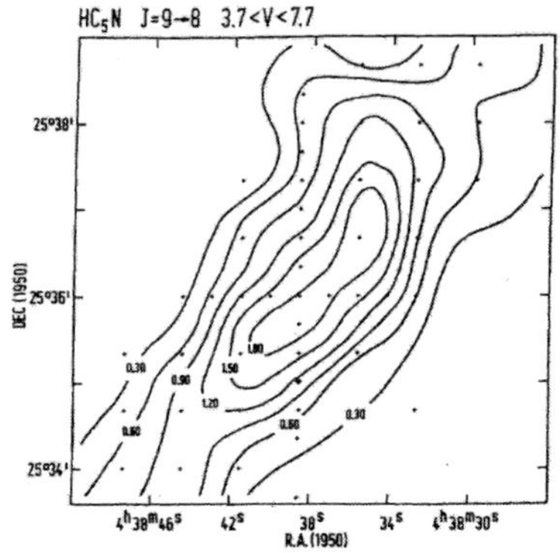

Fig. 2. The CP peak of TMC-1 in the J = 9-8 transition of HC_5N. Permission to reproduce this figure from Churchwell et al. A&A, 67, 139 [1978]) given by A&A.

3 Der Ruf

Gisbert worked at the Max Planck Institut for seven to eight years. During those years, the lack of a strong laboratory program under his control must have bothered him. He applied for a C4 (Full Professor) position at the University of Köln (Cologne), received "the call", and moved there in 1979. He became the Director of the I. Physikalisches Institut which, under the pyramidal German system, gave him the possibility of organizing a large and well-funded effort, similar in scope to the National Research Council in Ottawa. Since Cologne and Bonn are rather close (it takes less than 20 minutes to go by train from the Bonn Hauptbahnhof to the Köln Süd train station), Gisbert did not even have to change his residential address. He was now ready to embark on a career in which both radioastronomy and molecular spectroscopy could be studied.

Within a few years, a broad program had been started in both rotational and vibrational molecular spectroscopy. Some of this effort was related to radioastronomy and some obviously was not. For example, ammonia two-photon studies were accomplished with Doppler-free resolution [16], a thorough study of the infrared spectrum of D_2CO was undertaken [17], and the millimeter-wave rotational spectra of the interstellar molecules methyl diacetylene and the tricarbon monoxide radical were measured [18], [19]. The spectroscopy laboratory seemed to grow and grow. When I arrived as a sabbatical visitor during the year 1988-89, I was amazed at the scope of the institute, and at how many different spectroscopic experiments could be undertaken at the same time. The growth has continued in the 1990's.

Meanwhile, Gisbert had the idea to pursue a new line of work in molecular radioastronomy by the construction of a small radiotelescope, to be used both as a teaching and as a research tool. In 1984, the 3-meter telescope saw first light while atop the roof covering the I. Physikalisches Institut [21], [22].

But Gisbert had a far better site in mind: high atop the Gornergrat, near Zermatt, Switzerland. Accessible by train, this high-altitude and superb site was already being used by an infrared telescope, the 1.5 m TIRGO dish. By 1985, the 3-meter telescope had been moved to its new site atop the southern end of the Kulmhotel, and was ready for millimeter-wave and eventually submillimeter-wave and even THz observations [23].

In addition to running the telescope and continuing to develop a spectroscopy program, Gisbert introduced a large program in receiver technology and development. Acousto-optical spectrometers made in Cologne are now known world-wide. With three areas of effort, the I. Physikalisches Institut has become quite sizeable, with over 100 people and a substantial annual budget. Perhaps 100 Ph.D. students have come through its doors under Gisbert's direction.

3.1 The KOSMA 3-m Telescope

The history of the KOSMA (Köln Observatorium für Submm-Astronomie) telescope is a long and fruitful one, and would be difficult for me to even summarize here. As a small telescope at a superb site, it has been used mainly for CO and CI observations, with an increasing emphasis on high frequencies. Some of the many sources studied are listed in Table 1. In some of the most recent studies, Gisbert's name no longer appears as a co-author. Although the torch has been passed to the new generation, Gisbert has continued to do radioastronomical studies such as his collaboration with the SWAS program. At present, high frequency observations of CI and CO are possible on the KOSMA telescope using an array receiver called SMART (Submillimeter array receiver for two frequencies) which consists of two 2 $\times$ 4 pixel subarrays, one operating in a frequency range around 490 GHz and the other around 810 GHz. The arrangement allows observation of both the $^3P_1 \rightarrow {}^3P_0$ and the $^3P_2 \rightarrow {}^3P_1$ fine-structure transitions of CI at 490 GHz and 809 GHz, respectively [20]. More information can be gleaned from the web site http://www.ph1.uni-koeln.de/gg/.

4 The CH_2 Radical

In recent years, a major part of the spectroscopy in Cologne has revolved around the submillimeter and even THz spectroscopy of small free radicals. The push to higher and higher frequencies has been aided immeasurably by high frequency sources of Russian manufacture [34]. Of these radicals, perhaps the most important to Gisbert is the methylene radical (CH_2), since the spectroscopic study of his species had been originated by Gerhard Herzberg [4]. Below is a portion of the spectrum of CH_2 measured in Cologne at over 900 GHz.

Table 1. Some KOSMA Studies By Year

Year	Source
1984	TMC-1, OMC-1, W51 A, S140 (HCN scans) [21]
1990	nearest cloud (L1457) fragmentation [24]
1991	L1495 survey [25]
1991	fragmentation in California nebula [26]
1992	molecular cloud/HII interface (S 140) [27]
1996	Orion B [28]
1996	Rosette complex [29]
1998	Comet Hyakutake [30]
2000	galactic 5 kpc ring [31]
2002	C I emission (2-1 & 1-0) from star-forming regions [32]
2003	Virgo cluster spiral galaxies [33]

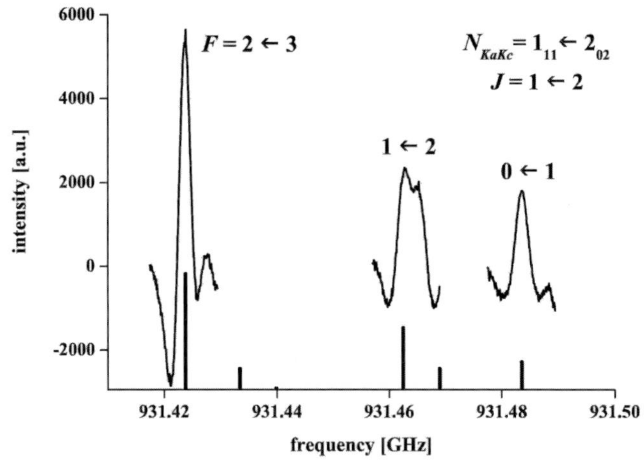

Fig. 3. A portion of the 940 GHz rotational transition of ortho-CH_2. Reproduced from Michael et al. ApJ 596, 1356 [2003] by permission of the AAS.

5 The Zermatt Conferences

One of Gisbert's lasting legacies will be the Zermatt conferences. Counting this gathering, there have now been four Zermatt meetings, all of which have been affairs notable for both the science and the lovely surroundings. I well remember the first one, since I was staying in Cologne at the time, and had the honor of travelling with Gerhard Herzberg and his wife by train from Cologne to Zermatt for the meeting. Perhaps the most unusual of the meetings was the second one, held in September 1993, during which torrential rains came down for days, blocking the valley off by mud slides. The result was that many participants, including the author, had to be evacuated by helicopter. The ride down the narrow valley was unforgettable! Compared with this denouement, the third meeting was a rather tame affair, blessed with nice weather. This last meeting has been, for me, the most interesting scientifically, because of the immense growth of the field of radioastronomy. Volumes representing the major talks at all four meetings have been published [35], [36], [37].

6 The Future

Gisbert is formally retired now from the directorship of the I. Physikalisches Institut in Cologne, but he is still as active as ever, especially in leading the spectroscopy group in a wide variety of fascinating and important experiments. Best wishes for the future, Gisbert, and thank you for all of the exciting times we have spent and will continue to spend together.

References

1. G. F. Winnewisser: Millimeter Wave Rotational Spectrum of HSSH. Ph. D. Thesis, Duke University, Durham, NC (1967)
2. S. Urban, E. Herbst, P. Mittler, G. Winnewisser, K. M. T. Yamada, and M. Winnewisser: J. Mol. Spectrosc. **137**, 327 (1989)
3. G. Winnewisser, F. Lewen, S. Thorwirth, M. Behnke, J. Hahn, J. Gauss, and E. Herbst: Chemistry, A European Journal, in press
4. B. Stoicheff: *Gerhard Herzberg: An Illustrious Life in Science*, (McGill-Queens University Press, Montreal, CA 2003)
5. G. Winnewisser: J. Mol. Spectrosc. **38**, 437 (1971)
6. J. W. C. Johns, J. M. R. Stone, and G. Winnewisser: J. Mol. Spectrosc. **42**, 523 (1972)
7. F. F. Gardner and G. Winnewisser: Astrophys. J. **195**, L27 (1975)
8. G. Winnewisser and E. Churchwell: Astrophys. J. **200**, L33 (1975)
9. C. M. Walmsley, P. R. Jewell, L. E. Snyder, and G. Winnewisser: Astron. Astrophys. **134**, L11 (1984)
10. M. Olberg, M. Bester, G. Rau, T. Pauls, G. Winnewisser, L. E. B. Johansson, and A. Hjalmarson: Astron. Astrophys. **142**, L1 (1985)
11. E. Churchwell, C. M. Walmsley, and G. Winnewisser: Astron. Astrophys. **54**, 925 (1977)
12. L. E. Snyder and D. Buhl: Bull. Amer. Astron. Soc. **3**, 388 (1971)
13. R. A. Creswell, G. Winnewisser, E. F. Pearson, and M. Winnewisser: Sterne und Weltraum **15**, 118 (1976)
14. R. J. Saykally, P. G. Szanto, T. G. Anderson, and R. C. Woods: Astrophys. J. **204**, L143 (1976)
15. E. Churchwell, G. Winnewisser, and C. M. Walmsley: Astron. Astrophys. **67**, 139 (1978)
16. R. Guccione-Gush, H. P. Gush, R. Schieder, K. Yamada, and G. Winnewisser: Phys. Rev. A **23**, 2740 (1981)
17. H.-J. Clar, K. Yamada, G. Winnewisser, F. W. Birss, and H. P. Gush: Can. J. Phys. **62**, 1834 (1984)

18. M. Bester, K. Yamada, G. Winnewisser, W. Joentgen, H.-J. Altenbach, and E. Vogel: Astron. Astrophys. **137**, L20 (1984)
19. W. Klebsch, M. Bester, K. M. T. Yamada, G. Winnewisser, and W. Joentgen: Astron. Astrophys. **152**, L12 (1985)
20. U. U. Graf, S. Heyminck, E. A. Michael et al: SMART: The KOSMA Sub-Millimeter Array Receiver for Two Frequencies. In: *Millimeter and Submillimeter Detectors for Astronomy* ed by T. G. Phillips, J. Zmuidzinas (Proceedings of the SPIE, Vol. 4855, 2003), pp. 322-329
21. T. Pauls and R. Ewald: Observations with the Cologne 3m radio telescope. In: *ESO-IRAM-Onsala Workshop on (Sub)Millimeter Astronomy* (Proceedings (A86-39226 18-90) of the European Southern Observatory, 1985) pp 85-90
22. G. Winnewisser and B. Vowinkel: Sterne und Weltraum **23**, 132 (1984)
23. G. Winnewisser, M. Bester, R. Ewald et al: Astron. Astrophys. **167**, 207 (1986)
24. T. Zimmermann and H. Ungerechts: Astron. Astrophys. **238**, 337 (1990)
25. C. Kramer and G. Winnewisser: Astron. Astrophys. Suppl. **89**, 421 (1991)
26. R. Herbertz, H. Ungerechts, and G. Winnewisser: Astron. Astrophys. **249**, 483 (1991)
27. J. Hernichel, D. Krause, R. Röhrig, J. Stutzki, and G. Winnewisser: Astron. Astrophys. **259**, L77 (1992)
28. C. Kramer, J. Stutzki, and G. Winnewisser: Astron. Astrophys. **307**, 915 (1996)
29. N. Schneider, J. Stutzki, G. Winnewisser, and L. Blitz: Astrophys. J. **468**, L119 (1996)
30. J. G. A. Wouterloot, A. Lingmann, M. Miller, B. Vowinkel, G. Winnewisser, and F. Wyrowski: Plan. Space Sci. **46**, 579 (1998)
31. M. Brill, C. Kramer, and J. Stutzki: CO Multiline Observations near the Galactic 5kpc Ring with KOSMA. In: Astronomische Gesellschaft Abstract Series, Vol. 17, 40 (2000)
32. C. Kramer, B. Mookerjea, H. Jakob, S. Jeyakumar, and J. Stutzki: Mapping of C I emission from Galactic Star Forming Regions. In: *Chemistry as a Diagnostic of Star Formation*, ed by C. L. Curry, M. Fich (NRC Press, Ottawa, Canada 2003), p, 47
33. H. Hafok and J. Stutzki: Astron. Astrophys. **398**, 959 (2003)
34. G. Winnewisser: Vib. Spectrosc. **8**, 241 (1995)
35. G. Winnewisser and J. T. Armstrong (eds): *The Physics and Chemistry of Interstellar Molecular Clouds* (Springer-Verlag, Berlin, 1989)
36. G. Winnewisser and G. Pelz (eds): *The Physics and Chemistry of Interstellar Molecular Clouds* (Springer-Verlag, Berlin, 1993)
37. V. Ossenkopf, J. Stutzki, and G. Winnewisser (eds): *The Physics and Chemistry of the Interstellar Medium* (GCA-Verlag, Herdecke, 1999)

trans-Ethyl Methyl Ether: Laboratory Measurements and Analysis

U. Fuchs[1], G.W. Fuchs[1], G. Winnewisser[1], P. Groner[2], F.C. DeLucia[3], and E. Herbst[3]

[1] I. Physikalisches Institut, Universität zu Köln, Germany, feldt@ph1.uni-koeln.de
[2] Department of Chemistry, University of Missouri-Kansas City, Kansas City, MO, USA, gronerp@umks.edu
[3] Department of Physics, The Ohio State University, Columbus, OH, USA, herbst@mps.ohio-state.edu

High-resolution measurements of rotational-torsional transitions of ethyl methyl ether (EME) were performed in its vibrational-torsional ground state at frequencies up to 350 GHz [1]. The new results correspond to about 1000 new observed transitions including internal rotation splitting of each transition due to torsion of the two methyl groups. A simultaneous fit to experimental accuracy of these and approximately 100 previously assigned transitions (see [2]) allows reliable prediction of additional transition frequencies up to 400 GHz, based on a newly calculated parameter set. The new data set should enable the astrophysical detection of ethyl methyl ether in the interstellar medium. Ethyl methyl ether was tentatively detected in hot cores which are quiescent high-density sources associated with regions of high-mass star formation by Charnley et al. [3].

1 Molecular Structure of *trans*-Ethyl Methyl Ether

Ethyl methyl ether ($C_2H_5OCH_3$) is a nearly-prolate, asymmetric rotor, i.e. the B and C rotational constants are nearly equal while the A rotational constant is significantly larger. The dipole moment lies nearly parallel to the inertial b-axis evoking strong b-type transitions and weak a-type transitions (see Fig. 1).

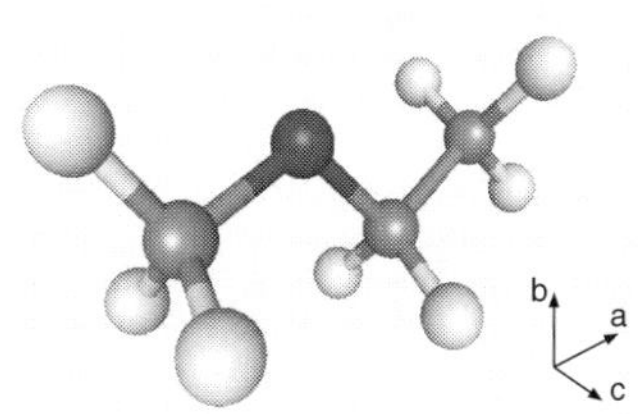

Fig. 1. Molecular structure of *trans*-ethyl methyl ether

Moreover, ethyl methyl ether consists of two methyl groups (CH_3) which are connected to a -CH_2-O- frame by a single bond. Each of the methyl groups are related to different potential barriers and may rotate in respect to the molecular frame. This phenomenon of internal rotation leads to a splitting of the rotational energy levels and hence to a splitting of the rotational transitions.

Table 1. Parameter for ethyl methyl ether including asymmetry and internal rotation, see [1]. The values in parenthesis are the uncertainties.

Rotational Constants		**Distortional Constants**		
constant	value	constant		value
A / MHz	27991.7139(15)	Δ_J	/kHz	0.97355(14)
B / MHz	4159.44314(19)	Δ_{JK}	/kHz	-2.4808(19)
C / MHz	3891.13634(19)	Δ_K	/kHz	73.712(62)
	.	δ_J	/kHz	0.087217(22)
	.	δ_K	/kHz	-0.9368(20)
	.	Φ_J	/Hz	0.000184(35)
	.	Φ_{KJ}	/Hz	-0.089(19)
	.	ϕ_J	/Hz	0.0000549(57)
Internal Rotor Parameters		**Tunneling Parameters**		
parameter	value	parameter		value
ρ_1	0.1282(37)	ϵ_{10}	/ MHz	-0.612(16)
ρ_2	0.16859(59)	ϵ_{01}	/ MHz	-3.082(22)
β_1 / degrees	5.52(17)	[(B-C)/4]$_{10}$	/ MHz	0.0000196(48)
β_2 / degrees	182.962(49)			

2 Theory

The simplified total Hamiltonian including the hindered rotation of the two methyl groups can be written as

$$\begin{aligned} H &= H_R + H_D + H_I + H_{RI} \\ \text{with} \quad H_I &= F\hat{j}^2 + V(\alpha_i) \end{aligned}$$

where H_R is the rigid rotor leading to the rotational constants and H_D the centrifugal distortion. H_I, the internal rotation and H_{RI}, the coupling term between internal and overall rotation give the internal rotor and tunneling parameter as listed in Table 1, while F is the reduced rotational constant of the internal rotation, $\hat{j}$ a reduced angular momentum of the internal rotation, $V(\alpha_i)$ the potential barriers, and α_i are the angles between CC and CH of one of the CH_3 groups and CO and CH of the other CH_3 group (see the Hamiltonian in more detail in [4]).

3 Asymmetry and Internal Rotor Splitting

In case of symmetric tops the two energy levels of K ($\pm$ K) are degenerated, whereas for an asymmetric rotor as ethyl methyl ether, this degeneracy is lifted. The resulting splitting of the EME levels depends on the quantum numbers J and K. Additionally, the internal rotation leads to a further splitting of the energy levels. Fig. 2 displays the dependency of torsional splitting on the quantum number J for the complete $K_a = 1 \leftarrow 0$, K_c = J-1 $\leftarrow$ J subband, covering J between 2 and 60. For this subband, the smallest splitting is found for the lowest J-values for which only a doublet can be resolved but in general this varies for the different subbands.

One rotational transition is depicted in the right picture of Fig. 2 which shows the splitting into four lines, due to internal rotation of the methyl groups showing one smaller and one bigger splitting.

4 Spectrometer

The rotational spectrum of ethyl methyl ether between 130 and 350 GHz was investigated with two special-purpose instruments in order to achieve efficient and highly precise measurements within a broad spectral range. One of them was the FASSST spectrometer located at The Ohio State University, Columbus which provides detailed spectral information over wide spectral regions in very short time (10 GHz/s) [6].

The other spectrometer, the Cologne Terahertz Spectrometer (I. Physikalisches Institut, Köln), can be used to perform highly precise measurements with Doppler as well as Sub-Doppler resolution at frequencies as high as 1.2 THz (see [7] and [8]). A phase-lock-loop (PLL) system provides phase stabilization and thus frequency stabilization of the BWO spectrometer. In the current Doppler-limited experiments with ethyl methyl ether, the accuracy of the line frequencies is estimated to be $\sim$ 0.05 MHz whereas for Sub-Doppler measurements an accuracy of $\sim$ 0.005 MHz can be achieved.

Hence, the combination of these two experiments suits perfectly to analyze the complex spectrum of an asymmetric top like ethyl methyl ether with two internal rotors causing additional small splittings of each transition.

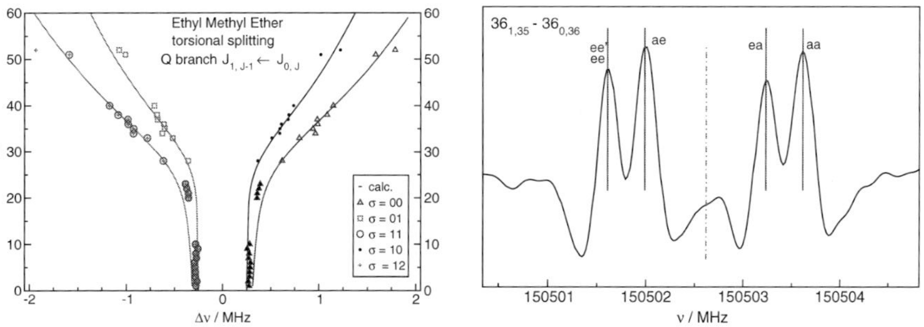

Fig. 2. Left: Example of internal splitting, showing the Q-branch rotational transition $J_{1,J-1} \leftarrow J_{0,J}$. The σ denotes the symmetry properties of the torsional splitting, see [5]; Right: Exemplary spectrum $36_{1,35} \leftarrow 36_{0,36}$. The ee, ee', ae, ea, and aa are symmetry labels of each transition.

5 Astronomy

Dimethyl ether $(CH_3)_2O$ has already been detected in space in hot cores of high density sources [9]. It is therefore likely that ethyl methyl ether ($C_2H_5OCH_3$) is abundant in hot cores as well [3]. A search with the IRAM 30m telescope and SEST dedicated to identify ethyl methyl ether in hot cores such as SgrB2N, NGC6334 and several other sources by Fuchs did not lead to a clear detection so far. This may be due to the dense and complex spectrum typically found in these types of sources. Thus, the amount of lines (assigned and unassigned) and the relative weakness of even the strongest lines of ethyl methyl ether compared to lines belonging to other molecules make it extremely difficult to assign ethyl methyl ether as the carrier of the newly found emission lines in the above mentioned sources.

Acknowledgments

This work was supported by the DFG (SFB 494), the DAAD, and NASA.

Markus Behnke is kindly acknowledge for the preparation of the EME sample for the measurements in Köln.

References

1. U. Fuchs, G. Winnewisser, P. Groner, F. De Lucia, and E. Herbst. *Astrophys. J. Suppl. Ser.*, 144:277–286, 2003.
2. M. Hayashi and K. Kuwada. *J. Mol. Struct.*, 28:147–161, 1975.
3. S. B. Charnley, P. Ehrenfreund, and Y. J. Kuan. *Spectrochim. Acta A*, 57:685–703, 2001.
4. P. Groner. *J. Chem. Phys.*, 107(12):4483–4498, 1997.
5. U. Fuchs. *High Resolution Spectroscopy of Ethyl Methyl Ether and Hydrogen Cyanide Isotopomers.* Cuvillier Verlag, Göttingen, 2003.
6. D. T. Petkie, T. M. Goyette, R. P. A. Bettens, S. P. Belov, S. Albert, P. Helminger, and F. C. DeLucia. *Rev. Sci. Instr.*, 68:1675–1683, 1997.
7. G. Winnewisser. *Vibrational Spectroscopy*, 8:241–253, 1995.
8. F. Lewen, R. Gendriesch, I. Pak, D. G. Paveliev, M. Hepp, R. Schieder, and G. Winnewisser. *Rev. Sci. Instr.*, 69(1):32–39, 1998.
9. M. Ikeda, M. Ohishi, A. Nummelin, J. E. Dickens, P. Bergman, Å. Hjalmarson, and W. M. Irvine. *Astrophys. J.*, 560:792–805, 2001.

The Cologne Database for Molecular Spectroscopy, CDMS

H.S.P. Müller, F. Schlöder, S. Thorwirth, and G. Winnewisser

I. Physikalisches Institut, Universität zu Köln, Zülpicher Str. 77, 50937 Köln, Germany, Email: hspm@ph1.uni-koeln.de

1 Introduction

The unambiguous assignment of atomic or molecular lines in the interstellar or circumstellar medium (ISM or CSM) depends critically on the availability of laboratory data with appropriate accuracy. Above approximately 500 GHz the amount of transitions measured in the laboratory decreases fast. However, with upcoming missions such as the Stratospheric Observatory For Infrared Astronomy (SOFIA), the Herschel Space Observatory, or the Atacama Large Millimeter Array (ALMA) the need for accurate rest frequencies up to at least 2 THz increases considerably. The catalog section of the Cologne Database for Molecular Spectroscopy (CDMS) has been created to provide the astronomical community with line frequencies of atoms and molecules of astronomical interest. The CDMS is available on the internet free of charge *via* `http://www.ph1.uni-koeln.de/vorhersagen/` or *via* the short-cut `http://www.cdms.de/`.

The CDMS has been decribed in detail in Ref. [1]. The present contribution documents new additions and changes in the database and highlights selected aspects.

2 The Catalog

The catalog section of the Cologne Database for Molecular Spectroscopy (CDMS), formerly known as New line predictions section, consists of several parts: *General*, *Entries*, *Search*, *Partition Functions*, and *Archive*.

General

The *General* part provides information on the content of the catalog, namely pure rotational transition frequencies of atomic and molecular species of astronomical interest. The entries are not restricted to species already observed in the ISM or CSM. Hence, the catalog can be used also to identify new molecules in space. The radio frequency and microwave to the far infrared spectral regimes are presently covered. Hence, vibration-rotation transitions are provided only rarely. Nevertheless, this type of data is used occasionally in calculating the transition frequencies. Separate entries are provided not only for each molecular or atomic species, but also for isotopomers and often also for excited vibrational states when appropriate.

A brief desciption of the format of the catalog entries, remarks on the partition function, and some useful equations are also present in this part.

Entries

The *Catalog Directory* lists the six-digit tag, which is the sorting criterion, the name of the species, the number of lines in the entry, the version number, the catalog file, the documentation file, and the date of creation of the entry. The transition frequencies are provided separately in units of cm^{-1} for selected species. Experimental line frequencies and uncertainties are given in the catalog whenever available and are indicated as such. However, many transition have to be calculated based on the experimental data and using appropriate spectroscopic models. The documentation lists the author and version of the entry, sources of the experimental data, special considerations for spin-statistics or fine and hyperfine structure, rotational constants, dipole moment values, values of the partition function at selected temperatures (see below) along with other information characterizing the entry. In addition, we have started to include a flag indicating whether a species has been detected in the ISM or CSM.

Each transition is given in exactly one line that contains the transition frequency, its uncertainty, intensity, the degree of freedom in the rotational partition function, the lower state energy, the upper state degeneracy, coding of the format of the quantum numbers, and finally the upper and lower state quantum numbers. Special considerations as well as details on the format and the units are given in the *General* part and in Ref. [1].

As of October 2003, the catalog contains 200 species, almost twice as many as listed in Ref. [1]. In addition to creating many new entries, several have been updated. Among the new or updated entries are light hydrids such as OH^+ and ND_3, complex molecules such as the aGg' lowest state conformer of ethylene glycol, long carbon chain molecules, such as $HC_{13}N$, many metal containing compounds such as NaCl and NaF, HCN $v_2 = 0$ and 1, CH_3OH and $^{13}CH_3OH$, and the ν_2 band of C_3. While the relevant astronomical, astrophysical, and laboratory spectroscopic literature is scanned constantly for the consideration of new entries, we welcome suggestions in this respect. Some background information is desirable. Also, we welcome suggestions on the need for additional laboratory data. E. g., there is a great need for new or more accurate data of light hydride species, including cationic ones such as CH^+ and for vibrationally excited states of many hot core species such as methyl formate, methanol and its isotopomers, ethanol, ethyl cyanide etc. In addition, the observation of highly deuterated species, in particular in dark clouds, e. g. the recent detection of ND_3 [2, 3], indicates an increased need for data on deuterated isotopic species.

In the near future, we intend to update the entries on CO and CS and to create new entries for light hydride species, such as HD_2^+, some Si containing molecules, isotopomers of HC_3N, excited states of HCN, and many more.

For some molecular species rest frequencies determined in cold molecular clouds have been used to derive spectroscopic parameters. However, the determination of very accurate rest frequencies in the ISM is usually rather difficult. For that matter, secondary standards are commonly employed. Among the species in the CDMS catalog particularly well suited as secondary standards are HCN $v_2 = 0$ and 1, isotopomers of CO, H_2CO, $H_2^{13}CO$, CH_3CCH, SO_2 $v_2 = 0$ and 1, and, in the near future, isotopomers of CS. The relative accuracy is about 10^{-8} for many transitions and over a wide frequency range.

Search

A search form is available to scan for transition frequencies of selected or all species in a specific frequency region. Because of the large number of transition frequencies in the catalog it is strongly recommended to scan in small frequency windows. The output can be in text form or in graphical form. If the latter option is chosen and more than one species has been selected transitions of a certain species can be highlighted. If the text form has been selected the transition frequencies can be sorted according to species, to frequencies, and to intensities. Intensities are calculated at 300 K by default. The search form offers recalculation at selected lower temperatures. In addition, the Einstein A-value can be requested.

Partition Functions

The partition function Q is very important to estimate intensities of molecular lines assuming thermal population. Its values at selected temperatures are given in the documentation. The decadic logarithm of these values has been gathered in a separate file for conversion purposes. Rather frequently no or only very imprecise data are available on low-lying excited vibrational states. Therefore, in most cases only data for the ground vibrational state have been considered in the calculation of Q for a certain species. When excited vibrational states have been taken into account this is mentioned in the documentation. Usually, individual contributions of the vibrational states are given in the documentation, or a link is given to a separate file containing that information.

Spin-statistical weight-ratios have been considered in the calculation of Q in most instances. Note, however, that normally common factors have been devided off in order to keep Q and the upper state degeneracy g_{up} small. Hence, it is strongly advisable to compare Q/g_{up} rather than Q from various databases. Non-trivial spin-statistical weight ratios are given explicitely in the documentation of a given species. Spin-statistical weight-ratios are the prime reason for Q values of different isotopomers to be quite dissimilar on occasion.

Other reasons may be that the number of vibrational states considered differ substantially.

In very cold regions of the ISM it may be important to consider *ortho* and *para* states separately. The energy of the lowest rotational or rotation-hyperfine level is 0 by default. The energy of the lowest level for a different spin-modification has been stated in the documentation throughout. Strictly speaking, one should consider Q values for the individual spin-modifications separately – at least at low temperatures. We intend to provide this information in the near future.

3 Molecules in Space

A new section in the CDMS lists the molecules detected in the ISM or CSM. Tentative detections and questionable detections are marked with a question mark. Because of the existence of the latter two types of molecules we refrain from stating any specific number. For some of the entries background information on the initial detection is given. Since in some cases it is hard to decide whether a report on a detection is reliable we welcome comments on the list.

4 Other Sections in the CDMS

Besides these sections of major interest for the astronomical community, other parts of the CDMS should be mentioned briefly. They are *What's New*, *Fitting Spectra* with information and examples for selected spectroscopy programs, *Cologne spectroscopy data* with selected spectroscopic data related to projects in Cologne, *Links*, and *Contact*.

5 Acknowledgments

The CDMS is supported by the Deutsche Forschungsgemeinschaft (DFG) through grant SFB 494.

References

1. H. S. P. Müller, S. Thorwirth, D. A. Roth, G. Winnewisser: Astron. Astrophys. **370**, L49 (2001)
2. D. C. Lis, E. Roueff, M. Guerin, et al.: Astrophys. J. **571**, L55 (2002)
3. F. F. S. van der Tak, P. Schilke, H. S. P. Müller, et al.: Astron. Astrophys. **388**, L53 (2002)

Spectroscopy of Linear Carbon Chain Radicals – The Case of C_nN Chains

G.W. Fuchs[1], U. Fuchs[1], T.F. Giesen[1], G. Winnewisser[1], F. Wyrowski[2], M.C. McCarthy[3], C.A. Gottlieb[3], J. Kucera[3], and P. Thaddeus[3]

[1] I. Physikalisches Institut, Universität zu Köln, Germany
fuchs@ph1.uni-koeln.de

[2] Max-Planck-Institut für Radioastronomie, Auf dem Hügel 69, D-53121 Bonn

[3] Harvard-Smithsonian Center for Astrophysics, 60 Garden Street, Cambridge, MA 02138, USA, mmccarthy@cfa.harvard.edu

Carbon chains are the dominant architectural scheme of interstellar molecular chemistry. The cyano radicals $C_3N$and C_5N, for example, have been detected in the late type star IRC+10216 and in the molecular cloud TMC-1; C_3N also in IRAS 15194-5115, TMC-2, HCL 2, CRL618, and CRL2688 ([1], Appendix D). Other cyano radicals like C_2N with a much smaller dipole moment have not been observed so far. The astronomical detection of these species depends strongly on the knowledge of their transition frequencies obtained by laboratory work. Longer C_nN chains or isotopomers of C_nN are difficult to observe in the laboratory and very sensitive spectrometers are required to detect their transition frequencies.

1 Laboratory Spectroscopy of C_nN Radicals

Pure rotational spectra of linear C_4N and C_6N, in their $^2\Pi$ electronic ground state and four C_3N isotopomers in their $^2\Sigma$ ground state have been measured by Fourier Transform Microwave Spectroscopy in a supersonic jet at the Harvard-Smithsonian Institution in Cambridge, MA. A detailed description of the experiment is given in McCarthy *et al.* [2]. Briefly, a pulsed supersonic molecular beam of an organic precursor vapor heavily diluted in an inert gas is produced by a commercial solenoid valve. Reactive molecules of many kinds are produced by applying a small electrical discharge prior to a supersonic expansion of the gas into the Fabry-Perot cavity. The molecules are irradiated with a short (1 μs) microwave pulse at frequency ν_0 which polarizes a range of frequencies which may contain one of the resonant transitions of the radicals. The polarized gas coherently emits at its resonant frequencies. This free-induction decay is detected and subsequently Fourier transformed to obtain the spectrum. The molecule emission frequency is the mean value of the Doppler shifted lines which originate from the electromagnetic waves traveling back and forth in the FP-cavity relative to the molecules' flight direction, see Fig. 1.

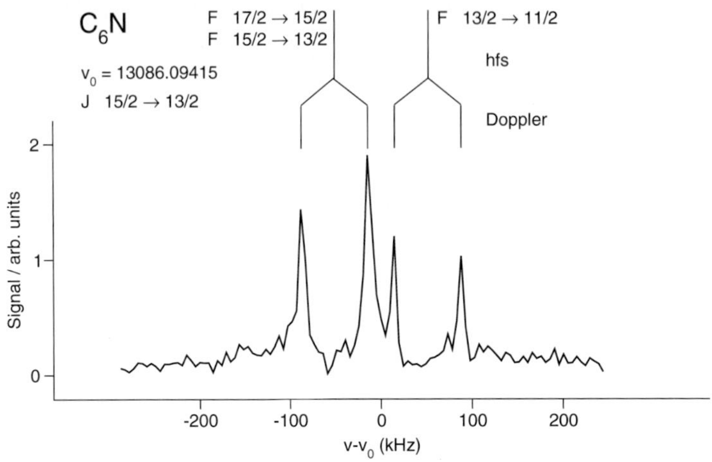

Fig. 1. A J 15/2 → 13/2 transition of C_6N with additional fine structure and the Doppler splitting typical for the used Fourier Transform experiment.

1.1 Measurements on Mono-substituted C_3N Isotopomers

Measurements of ^{13}CCCN, C^{13}CCN, CC^{13}CN, and CCC^{15}N were made between 9.5 - 38.4 GHz. A standard Hamiltonian $\mathbf{H} = \mathbf{H}_{rotation} + \mathbf{H}_{fine} + \mathbf{H}_{hfs}$ to fit the $^2\Sigma$ states has been applied and Herb Pickett's program [3] was used for the analysis of the data. The rotational and leading centrifugal distortion constants were determined to high accuracy by using the microwave data and previously measured millimeter data [4]. The Fermi contact $b(^{14}$N), dipole-dipole $c(^{14}$N), and the nitrogen quadrupole hyperfine coupling constants for ^{13}CCCN, C^{13}CCN, and CC^{13}CN have been determined and the previously published $b(^{13}$C) and $c(^{13}$C) values [4] were stated more precisely. The magnetic hyperfine coupling constants of the ^{13}C isotopic species of C_3N differ from those of the isoelectronic chain C_4H, but are fairly close to those of the isoelectronic C_2H, indicating a rather pure $^2\Sigma$ electronic ground state. The CCC^{15}N b and c magnetic hyperfine constants follow the expected values derived from the values of the ^{14}N species.

During a line survey towards IRC+10216, Cernicharo *et al.* [5] detected three ^{13}C mono-substituted C_3N isotopomers. In the case of ^{13}CCCN, the assignment of two of the astronomical observed lines is not in agreement with the data derived with the new laboratory data set. Nevertheless, this detection supports the claim that also in other sources, like the cold molecular cloud TMC-1, isotopic species of C_3N can be detected.

1.2 Linear C_4N and C_6N Radicals

Two new cyano radicals, linear C_4N and C_6N in their $^2\Pi$ electronic ground state were analyzed. The Ω=1/2 states of C_4N and C_6N were observed between 7 and 22 GHz and their hyperfine structure (hfs) and Λ-type doubling was resolved. J.M. Brown's Hamiltonian for a Hund's case (a) was used for the data analysis [6]. At most, 9 spectroscopic constants were required to reproduce their spectra to a few parts in 10^7. The spectroscopic constants derived for C_4N and C_6N allow the prediction of the most interesting radio lines of these species up to 50 GHz to an uncertainty of 0.30 km s^{-1} or better.

1.3 Longer C_nN Chains

C_nN chains with n=7,9,11,13 have only be detected in matrix measurements in the near IR or visible part of the spectrum [7]. Searches for C_7N in the gas phase have so far been unsuccessful. In the above described experiment the absence of lines at the predicted frequencies suggests that the ground state of C_7N may not be $^2\Sigma$ but $^2\Pi$, for which the dipole moment is calculated to be much smaller.

2 A Search for C_2N Towards IRC+10216

Recently in a theoretical work, Mebel & Kaiser [8] examined the formation of C_2N isomers via neutral-neutral reactions in the interstellar medium and concluded that these species might well be observable in the envelopes of late type stars like IRC+10216 which represent ideal targets for an astronomical search in the infrared as well as in the microwave region. After an unsuccessful attempt to find C_2N in IRC+10216 by Guélin *et al.* [9] and the identification of at least two lines which coincided with transition frequencies of C_2N in 2002 by Fuchs [1], a new search for C_2N was started in July 2003 towards IRC+10216. This extensive search for C_2N has been performed in the frequency range 80-272 GHz, thus covering 8 rotational transition frequencies of C_2N (the J 15/2 - 13/2 transition has not been considered because of blending with HCN), using the IRAM 30m telescope at Pico Velata, Spain. The total on/off time spent on each line ranges from 250 to 585 minutes depending on the performance of the telescope at certain frequencies. With an r.m.s noise of the bands between 9 and 13 mK (effective antenna temperature T_A^*) lines as weak as 0.04 K were found during the observation. Many lines could not be assigned to any known molecule but again some lines have been found which correspond to C_2N transition frequencies. However, analysis of these lines support no evidence so far that C_2N is the carrier, see Fig. 2. An upper limit of the column density could be estimated to be $N_T = 1.6 \cdot 10^{13}\,\mathrm{cm}^{-2}$ with T_{rot}=22 K using the calculated C_2N dipole moment μ= 0.425 D by [10]. Astrochemical models by Millar & Herbst [11] predict the C_2N radical to be

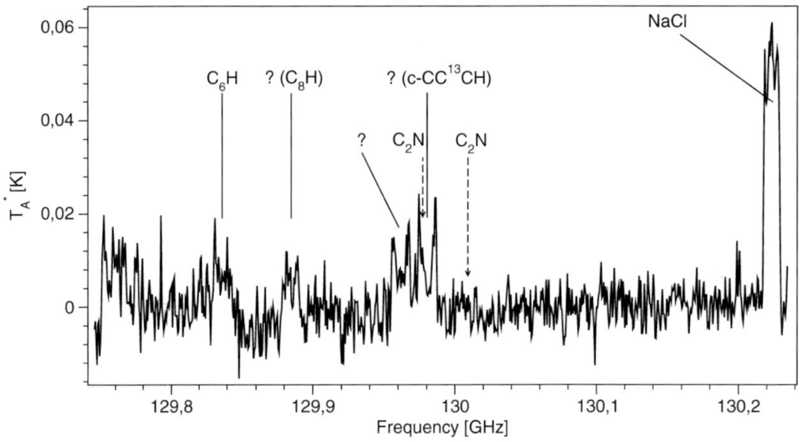

Fig. 2. IRAM observation at 130 GHz towards IRC+10216. At the transition frequencies of C_2N (dashed) which both should have similar intensities only on line is observed which possibly belongs to c-$CC^{13}CH$ as a carrier.

more abundant then the C_4N and C_6N chains in the carbon rich envelope of IRC+10216.

Acknowledgments

The present study was supported by the DFG, the Ministerium für Wissenschaft und Forschung des Landes Nordrhein-Westfalen, the Smithsonian Institution, NSF, NASA and IRAM.

References

1. G. W. Fuchs. *Carbon Chain Molecules: Production and Spectroscopic Detection.* Cuvillier Verlag, Göttingen, 2003.
2. M. C. McCarthy, G. W. Fuchs, J. Kucera, G. Winnewisser, and P. Thaddeus. *J.Chem.Phys.*, 118:3549, 2003.
3. H. Pickett. *J.Mol.Spectrosc*, 148:371, 1991.
4. M. McCarthy, C. Gottlieb, P. Thaddeus, M. Horn, and P. Botschwina. *J.Chem.Phys.*, 103:7820, 1995.
5. J. Cernicharo, M. Guélin, and C. Kahane. *Astron.Astrophys.Suppl.Ser.*, 142:181, 2000.
6. J. Brown and J. Schubert. *J.Mol.Spectrosc*, 95:194, 1982.
7. M. Grutter, M. Wyss, and J. Maier. *J.Chem.Phys.*, 110:1492, 1999.
8. A. M. Mebel and R. I. Kaiser. *Astrophys.J.*, 564:787, 2002.
9. M. Guelin and J. Cernicharo. *Astron.Astrophys.*, 244:L21, 1991.
10. R. Pd and P. Chandra. *J.Chem.Phys.*, 114:1589, 2001.
11. T. Millar, E. Herbst, and R. Bettens. *Mon.Not.R.Astron.Soc.*, 316:195, 2000.

Precise Laboratory Rest Frequency of a Submillimeter-Wave Line of D_2H^+

T. Hirao and T. Amano

Institute for Astrophysics and Planetary Sciences, Ibaraki University, Mito, Ibaraki, 310-8512, JAPAN; yoshi@mcs.ibaraki.ac.jp, amano@mx.ibaraki.ac.jp

A pure rotational transition ($J_{K_aK_c} = 1_{10} - 1_{01}$) of D_2H^+ is measured in the laboratory with a high-sensitivity submillimeter-wave spectrometer. The transition frequency is determined to be 691.660440(19) GHz with 1 σ in parentheses. The molecular constants for the ground state as well as the first excited vibrational states have been revised by refitting all available observed data.

1 Introduction

In these several years, multiply deuterated molecules have been discovered in extreme cold dark clouds. Several new astrochemical models have been developed to explain the high abundances of such species. Among one of recent chemical models, Roberts *et al.*[1] showed that, if the gas phase reactions are dominant in deuterium fractionations, D_2H^+ and D_3^+ play key roles. In view of these developments, a submillimeter-wave observation of one of the low lying rotational transitions of D_2H^+ was made and the result has been published very recently[2]. In this report, we present a summary of the laboratory observation and discuss spectroscopic aspects of the experiments.

2 Spectroscopy of D_2H^+

The first spectroscopic detection of D_2H^+ was made by Lubic and Amano by observing the ν_1 fundamental band in the 3.7 μm region[3]. The measurements were extended to the Coriolis coupled ν_2 and ν_3 fundamental bands by McKellar and his coworkers[4]. Polyansky and McKellar[5] carried out an extensive critical analysis of all the available data including two THz transitions measured by Evenson and his coworkers[6]. In parallel with the experimental developments, high-level calculations of the vibration-rotation energies based on ab initio potential surfaces of H_3^+ and its deuterated species were presented by Tennyson and coworkers[7]. Such theoretical results were often instrumental in assigning the spectra. More recently Fárník et al.[8] were successful in observing the overtone and combination bands in near infrared. The molecular constants derived by Polyansky and McKellar[5] yielded a predicted transition frequency for the $J_{K_aK_c} = 1_{10} - 1_{01}$ transition to be 691.705 GHz.

3 Submillimeter Observation

A submillimeter-wave spectrometer system with Russian-made backward-wave oscillators (BWO) combined with an extended negative glow discharge source for production of D_2H^+ was used[9]. The discharge cell was cooled by flowing liquid nitrogen through the outer jacket, and baseline reduction was carried out by subtraction of signals with the magnetic field on and off.

We first monitored a known line of H_2D^+, the 1_{10}-1_{11} transition at 372.42134(20) GHz[10, 11] to optimize the reaction conditions for the D_2H^+ search. The optimum gas mixing ratio for H_2D^+ was found to be $H_2/D_2/Ar$ = 3/1/17 mTorr, and the optimum discharge current was found to be about 8 mA. The spectra thus obtained is illustrated in Fig. 1(a).

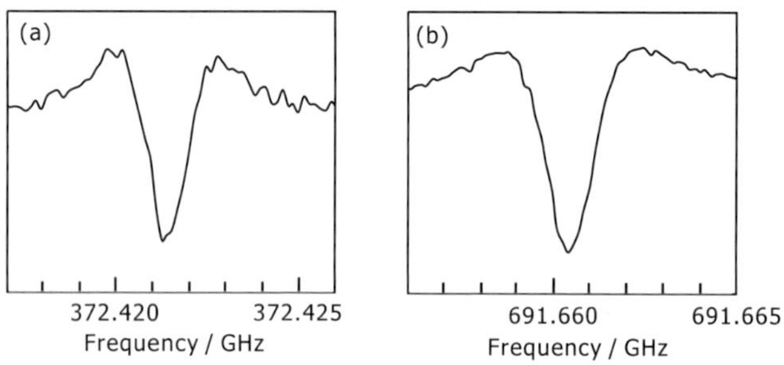

Fig. 1. The observed lines of (a) H_2D^+ $1_{10} - 1_{11}$ at 372.42134 GHz and (b) $D_2H^+ 1_{10} - 1_{01}$ at 691.660440 GHz. Integration time is ~600 s, and the time constant of the lock-in amplifier is set to 1.25 ms.

Using the optimum condition for H_2D^+ as a starting point, the rotational transition of D_2H^+ was searched for within a range of ±90 MHz around the predicted transition frequency, 691.705 GHz[5]. Eventually we found only one line at 691.660440(19) GHz in the search frequency range that was highly magnetic field sensitive. The line profile was fitted to a Gaussian profile with the assumption that the pressure broadening was negligible. The transition frequency was determined by averaging the line centers from five sets of independent upward and downward scans. Figure 1(b) illustrates an example of a section of the spectrum of D_2H^+ under the optimum conditions: $H_2/D_2/Ar$ = 4/2/17 mTorr with the discharge current of 8 mA.

The Doppler shifts of line frequency of ions are noticeable in typical glow discharges due to the discharge electric field. However, the line positions of ions are hardly affected in the extended negative glow discharge, since most ions are present in the negative glow discharge region where the electric field is almost zero[12]. Indeed reversing the polarity of the discharge voltage resulted in no significant change of the transition frequency. Thus we concluded that the uncertainty of the measured line frequency mainly arose from the random measurement uncertainties with no systematic line shift.

The characteristics of this line were carefully examined, as described in Hirao and Amano[2]. Therefore we do not go into detail here. In addition, the line width was carefully measured and was found to be consistent with that expected for D_2H^+, and it provided us with another evidence that the transition was not due to ArH_2D^+ or ArD_2H^+[13].

4 Spectroscopic Analysis

The measured transition frequency turned out to be lower by 45 MHz than the predicted frequency calculated from the molecular constants determined by Polyansky and McKellar[5], and the difference is right at the edge of the 3σ uncertainties of their calculated transition frequency. We carried out a least squares analysis to check if the new line does not upset the fit. The analysis was carried out by including all the available infrared transition wavenumbers rather than using the ground state combination differences. Because the combination differences calculated from the infrared data were limited to only $J = 4$, the centrifugal distortion constants were not determined very well for this light molecule. So the analysis was essentially the same as that by Polyansky and McKellar with only one difference; inclusion of the new submillimeter-wave transition. A set of the revised molecular constants for the ground state thus obtained are listed in Table 1. The results are very reasonable, with the standard deviation about the same as that obtained by Polyansky and McKellar[5], and the 45 MHz difference almost entirely accounted for by a difference in the two rotational constants, $A-C$, with slight readjustments of the centrifugal distortion constants.

Table 1. The molecular constants of D_2H^+ in the ground state(in MHz).[a]

	This work	Ref.[5]
A	1085193.7(25)	1085222.3(114)
B	655612.8(126)	655615.5(123)
C	391847.3(51)	391825.1(99)
Δ_J	168.27(43)	167.71(47)
Δ_{JK}	68.4(27)	69.3(27)
Δ_K	587.4(35)	592.9(40)
δ_J	63.00(16)	63.37(21)
δ_K	35.5(28)	35.3(28)

[a] Values in the parentheses indicate one sigma to the last significant digits.

In conclusion, we have determined the rest frequency of $J_{K_aK_c} = 1_{10}-1_{01}$ of D_2H^+ to be 691.660440(19) GHz. Inclusion of this line frequency led to slight revision of the molecular constants. This submillimeter-wave line will prove to be an important probe into the chemistry and gas-phase depletion in cold dense clouds.

The authors thank A. R. W. McKellar for providing his analysis program. The authors are grateful to the Ministry of Education, Science and Culture of Japan and the JEM/SMILES program for financial support. T. H. acknowledges Japan Society for Promotion of Science (JSPS) Fellowship and Grant-in-Aid for JSPS Fellows.

References

1. H. Roberts, E. Herbst, and T. J. Millar, *Astrophys. J.*, **591**, L41 (2003).
2. T. Hirao and T. Amano, *Astrophys. J.*, **597**, L85 (2003), and references cited therein.
3. K. G. Lubic and T. Amano, *Can. J. Phys.*, **62**, 1886 (1984).
4. S. C. Foster, A. R. W. McKellar, and J. K. G. Watson, *J. Chem. Phys.*, **85**, 664 (1986).
5. O. Polyansky and A. R. W. McKellar, *J. Chem. Phys.*, **92**, 4039 (1990).
6. D. A. Jennings, C. Demuynck, M. Banek, and K. M. Evenson, unpublished.
7. O. Polyansky and J. Tennyson, *J. Chem. Phys.*, **110**, 5056 (1999), and references cited therein.
8. M. Fárník, S. Davis, M. A. Kostin, O. L. Polyansky, J. Tennyson, and D. J. Nesbitt, *J. Chem. Phys.*, **116**, 6146 (2002), and references cited therein.
9. T. Amano and A. Maeda, *J. Mol. Spectrosc.*, **203**, 140 (2000).
10. M. Bogey, C. Demuynck, M. Denis, J. L. Destombes, and L. Lemoine, *Astron. Astrophys.*, **137**, L15 (1984).
11. H. E. Warner, W. T. Conner, R. H. Petrmichl, and R. C. Woods, *J. Chem. Phys.*, **81**, 2514 (1984).
12. F. C. De Lucia, E. Herbst, F. C. Plummer, and G. A. Blake, *J. Chem. Phys.*, **78**, 2312 (1983).
13. M. Bogey, H. Bolvin, C. Demuynck, J. L. Destombes, B. P. van Eijck, *J. Chem. Phys.*, **88**, 4120 (1988).

Part III

Structure and Dynamics of the ISM in External Galaxies

Molecular Gas in Spiral Galaxies

M.D. Thornley

Bucknell University, Department of Physics, Lewisburg, PA 17837
(mthornle@bucknell.edu)

1 Abstract

In this review, I highlight a number of recent surveys of molecular gas in nearby spiral galaxies. Through such surveys, more complete observations of the distribution and kinematics of molecular gas have become available for galaxies with a wider range of properties (e.g., brightness, Hubble type, strength of spiral or bar structure). These studies show the promise of both interferometers and single-dish telescopes in advancing our general understanding of molecular gas in spiral galaxies. In particular, I highlight the contributions of the recent BIMA Survey of Nearby Galaxies (SONG).

2 Introduction

Molecular gas is arguably one of the most important constituents to consider in any study of galaxy evolution. It is the raw material from which stars form, and it responds strongly to dynamical influences which can move large amounts of gas. Spiral galaxies, in particular, provide a variety of environments in which to study the processing of molecular gas. For instance, galactic bars are likely important in fueling central star formation by promoting the flow of gas inward. The star formation resulting from the influx of gas may in turn promote the development of a bulge-like structure, or even the destruction of the bar itself (e.g.,[1],[2],[3]). In addition, the presence of a bar or spiral potential may influence the large-scale pattern of gas, and therefore new star formation (e.g.,[4],[5],[6]).

The symmetry of the H_2 molecule effectively limits the direct detection of molecular material to the small fraction of warm molecular gas typically found in the vicinity of shocks or star formation. Due to its relatively high abundance and low effective excitation, the CO molecule has become a standard proxy for H_2, and its emission (particularly from the J=1-0 transition) has been detected in many galaxies, in some cases out to very large redshift (e.g., z=6.419,[7]). There is still significant debate about the appropriate conversion factor between CO integrated intensities and H_2 surface densities in different environments (see presentations by M. Guelin, F. Walter, and A. Bolatto in this volume; see also, e.g., [8],[9],[10],[11],[12]). However, there is no question that CO observations play a key role in discovering the whereabouts and motions of molecular gas in galaxies outside the Milky Way.

In the last decade, the improved sensitivity of millimeter-wave facilities has enabled studies of the distribution and kinematics of molecular gas in galaxies on smaller spatial scales, at larger radii, and over smaller velocity shifts. In particular, the advent of a number of surveys has significantly advanced the study of molecular gas in nearby galaxies since the previous Zermatt Symposium. We now have a full accounting of the molecular gas in Local Group spirals, as well as a self-consistent accounting of the detailed molecular gas distributions in wider samples of Hubble type, bar contribution, and star formation activity. For the purposes of this presentation, I include techniques such as On The Fly (OTF) mapping (with single-dish telescopes) and mosaicking (for interferometers) as "survey" techniques.

3 Surveys of Local Group Spirals

In the disks of our nearest spiral neighbors, M31 and M33, we can resolve populations and kinematics of Galactic giant molecular cloud (GMC) analogs. Their proximity also requires on the order of 1000 pointings for a typical millimeter telescope or interferometer to cover the angular size subtended by the star-forming disk. Therefore, complete studies of these individual galaxies can rightly be called surveys, and they provide the means to study cloud populations in external galaxies in an unbiased way.

3.1 M31

M31 is the nearest spiral to our own, offering an alternative view of a large, bright spiral galaxy. The large inclination of M31 motivates the study of molecular gas at high angular resolution, in order to separate arm and interarm regions and assess the properties and development of spiral structure. Recent efforts to attain a complete assessment of the molecular gas in M31 have seen a dramatic improvement in spatial resolution, from the first complete CO survey at 1.7 kpc resolution in 1993 (with the CfA 1.2m,[13]) to a higher resolution (200 pc) survey of the southwestern half of the galaxy (with the FCRAO 14m, [14],[15]) and a full survey at better than 90pc resolution (with the IRAM 30m,[16]) in the last few years. We can now assess the properties of the spiral structure and molecular cloud population in M31, and compare it with those of the Milky Way (see presentation by M. Guelin, this volume). Furthermore, the wealth of kinematic information over a wide range of radii can be used to constrain galaxy properties such as the shape and rotation speed of the bulge [17],[18].

3.2 M33

M33 offers a contrasting view of the properties of a nearby, later-type spiral. After a seminal study of the properties of molecular gas in the nuclear region of M33 [19],[20], it has taken a decade to expand on this important work. The recently completed 759-field survey covering the star-forming disk of M33 at 50 pc resolution (with BIMA, [21]) is the first flux-limited sample of distinguishable GMCs in a spiral galaxy, and creates a database of 148 GMCs with which to study the properties of molecular clouds in M33. In turn, it has enabled a higher resolution (20 pc) study of a high-mass sample of giant molecular clouds in M33 [22]. These studies confirm that M33 seems to be devoid of the most massive GMCs seen in the Galaxy, and enable comparisons with other tracers such as HI and Hα which can be used to constrain GMC lifetimes, a gas depletion timescale, and cloud formation mechanisms.

4 Surveys Outside the Local Group

As we leave the Local Group, we return to the more traditional definition of a survey, that of studying many different objects rather than many pointings on a single object. Through surveys, we are rapidly gathering enough information on molecular gas in spiral galaxies to begin making general statements directly from statistical samples.

4.1 Single-Dish Surveys

Much of our basic understanding of the global properties of gas distributions in spirals come from single dish surveys. Single dish telescopes are able to observe a significant fraction of a galaxy in a single beam, and do not suffer the uncertainties of missing large-scale flux inherent in interferometer studies. The groundwork for our understanding of the molecular gas content of spiral galaxies was laid by earlier surveys (e.g.,[23],[24],[25],to name a few), which indicated that molecular gas can be detected over a range of Hubble types, and tends to be more enhanced in the centers of galaxies than in the outskirts. Further, we find that molecular gas and star formation tracers are generally closely associated, though interpretations of the form of this association vary (e.g., [26],[27],[28],[29]).

Recent single-dish surveys (e.g., [30], [31], [32], [33], [34],[35]) highlight the fact that virtually all single-dish millimeter telescopes are actively engaged in surveys of a wider range of molecular gas properties in spirals. In recent years, these and other studies have been used to improve our measures of molecular gas in very late type spirals (Scd-Sm), construct higher resolution central rotation curves than are available from HI, and show the promise of mapping variations in molecular gas excitation and density through observations of higher rotational transitions of CO.

Single-dish telescopes have also been used increasingly in a mapping mode such as On-The-Fly (OTF) mapping. Rather that measuring only the central surface density or the variation of surface density along the major or minor axis, these studies map the detailed variation of the emission from various parts of the galaxy. An excellent example is the fully-sampled mapping of 5 nearby spiral galaxies with the NRAO 12m (now the UASO 12m), including IC 342 and M83 ([36],[37],[38]). These single dish studies discern individual arm and interarm regions, and can trace the relationship of molecular and atomic gas in detail. These studies show smooth connections between the molecular and atomic gas components, which suggests that the gas phase may be determined by local variations in turbulent pressure or dynamical conditions (e.g., [39], [40],[41], [42], [43], [44]).

4.2 Interferometric Surveys

Large single-dish telescopes such as the IRAM 30m and the Nobeyama 45m can achieve moderate angular resolutions of $\sim$10-20$''$, corresponding to $\sim$1-2 kpc at d=20 Mpc. Therefore, achieving kiloparsec or even sub-kiloparsec resolution in a large number of nearby spirals requires the use of interferometers. The observations possible with current millimeter interferometers provide important information for planning future surveys with the Combined Array for Research in Millimeter-wave Astronomy (CARMA) and the Atacama Large Millimeter Array (ALMA).

Circumnuclear Molecular Gas

With significantly higher resolution, interferometric studies are well suited to study central molecular gas properties, and their relationship with central activity. The progress of one of the most recent such surveys, The NUclei of GAlaxies project (NUGA), has been reported at this meeting by S. Garcia-Burillo and L. Tacconi. A study with the NRO and OVRO interferometers of CO emission in 10 barred and 10 unbarred spirals [45],[46] reveals that molecular gas is more concentrated in barred than unbarred spiral galaxies, thus providing the first statistical evidence for bar-induced inflow in transporting gas to the centers of galaxies. However, this study found no correlation between central gas concentration and type of nuclear activity, which may indicate that the resolution of the survey (4$''$) was not sufficient to assess separately the properties of the circumnuclear gas.

Recent dissertation projects by S. Jogee [47] and A. Baker [48], as well as the OVRO MAIN survey (see description in [49],[50]) complement these studies by targeting particular aspects of the interaction of molecular gas with other activities in the centers of galaxies. Jogee compared central molecular gas mass and central star formation activity, and showed that information about gas kinematics as well as the size of the gas reservoir was necessary to determine whether or not a central starburst would form. The Baker study of molecular gas in AGN with broad Hα line emission, as well as the OVRO MAIN multi-line survey, are being used to study central kinematic and heating conditions, and to test which processes are most influential in determining central activity in galaxies.

The Nobeyama Millimeter Array (NMA) has been used recently to target spirals in the Virgo cluster, using the advantage of a common distance to minimize ambiguities when comparing structures between galaxies ([51],[52],[53]), thus being able to compare directly the inferred surface mass densities and effects of interactions on molecular gas distribution and motions.

Gas in Disks

The interferometric surveys described thus far have included CO brightness as a criterion for selection into the study. Such studies may be biased by an unknown factor toward galaxies with more significant, or more highly concentrated, molecular gas masses. The BIMA Survey of Nearby Galaxies, or SONG, is the first large interferometric CO survey to observe galaxies which were not chosen on the basis of central CO brightness [54],[55]. The 44 spirals in SONG were chosen by the following criteria: Hubble type Sa-Sd, $V_{sys} < 2000$ km s^{-1}, $i < 70^o$, $\delta > -20^o$, and $B_T < 11.0$[1]. Observations for SONG effectively mapped a ~10-kpc diameter region around each galaxy center, enabling a more accurate assessment of the properties of gas in the inner disk. CO J=1-0 emission was detected in 41 of the 44 chosen galaxies (within a radius of $120''$), providing a broad sample of galaxies with which to explore the properties of gas and star formation over a range of radii[2].

The BIMA SONG data (see Figure 1) show that the molecular gas distributions in spiral galaxies are very heterogeneous, and even the azimuthally averaged radial profiles show a range of properties with respect to the radial variation of the stellar light. A comparison of the molecular gas distributions in barred (SB/SAB) and unbarred (SA) galaxies in the full SONG sample [56] confirms and strengthens the results of the NMA-OVRO survey: the average central gas surface density of barred spirals is three times higher than that of unbarred galaxies. It also appears that in many spirals the increase in molecular gas surface density occurs over the same range of radii in which the stellar light increases above that of an exponential profile [54],[57]. As this central "excess" occurs commonly in unbarred galaxies as well as barred galaxies, there may be some other mechanism than bar inflow required to explain central gas excesses in all spiral types. The complete coverage of molecular gas in the inner disk has enabled detailed studies of the relationship of molecular gas and star formation, targeting both the validity and formulation of a star formation "law" ([58]) as well as the influence of galactic bars and gas inflow on the placement of star formation sites ([59]). In addition, the wealth of kinematic information available has enabled a systematic study of low-level streaming motions in a sub sample of ~20 SONG galaxies [60].

---

[1] SONG also imposed the criterion $D_{25} < 70'$ to exclude M33, which is the subject of the Engargiola et al. (2003) survey described in §3.2.

[2] The BIMA SONG data are available at the following address: http://nedwww.ipac.caltech.edu/level5/March02/SONG/SONG.html

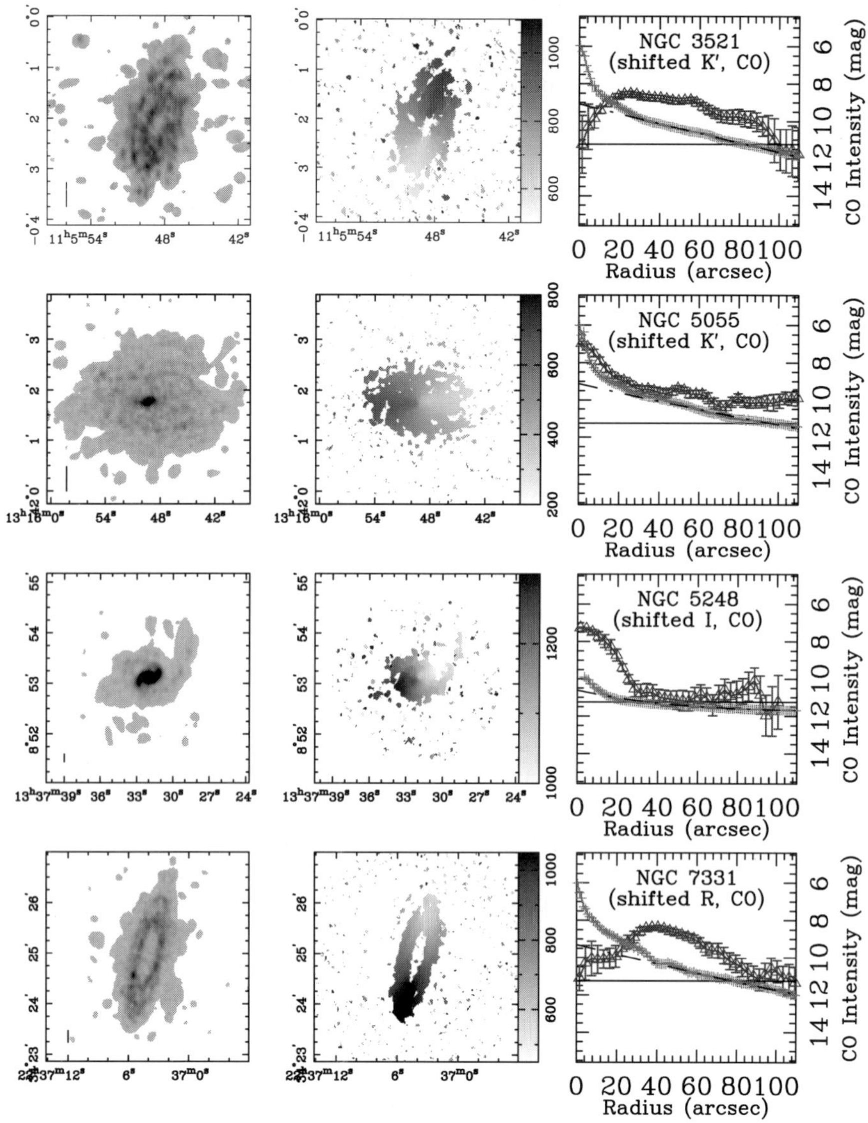

Fig. 1. BIMA SONG data for NGC3521, NGC5055, NGC 5248, and NGC7331 (top to bottom). **(left:)** integrated intensity (I_{CO}, with 1 kpc bar at lower left); **(center:)** velocity field (in km s^{-1}); and **(right:)** azimuthally averaged radial profiles for I_{CO} (triangles) and optical/NIR broadband light (crosses). For the purposes of comparison, I_{CO} has been converted to magnitudes for the radial profile (see [54]) and the optical/NIR profiles have been shifted vertically. An exponential fit to the outer part of each optical/NIR profile has been overplotted as a dot-dash line.

Acknowledgements

I would like to thank my fellow members of the BIMA SONG consortium, particularly T. Wong, M. Regan, and K. Sheth, for interactions that continue to enhance the scientific value of SONG. I would also like to thank R. Magee and C. Spohn-Larkins, two Bucknell University students who contributed to the research presented here.

References

1. D. Friedli & W. Benz 1993, A&A, 301, 649
2. C.A. Norman, J.A. Sellwood, & H. Hasan 1996, ApJ, 462, 114
3. E. Athanassoula 2003, MNRAS, 341, 1179
4. B.G. Elmegreen & D.M. Elmegreen 1990, ApJ, 355, 52
5. A. Toomre & A.J. Kalnajs 1991, in Dynamics of Disk Galaxies, Ed. B. Sundelius,(Göteberg:Göteberg University), 195
6. S.Chakrabarti, G. Laughlin, & F.H. Shu 2003, ApJ, 596, 220
7. F.Bertoldi, et al. 2003, A&A, 409, L47
8. P. Maloney & J.H. Black 1988, ApJ, 325, 389
9. C.D. Wilson 1995, ApJ, 448, L97
10. N. Arimoto, Y. Sofue, & T. Tsujimoto 1996, PASJ, 48, 275
11. T.M. Dame, D. Hartmann, & P. Thaddeus 2001, ApJ, 547, 792
12. A. Boselli, J. Lequeux, & G. Gavazzi 2002, A&A, 384, 33
13. T.M. Dame, E. Koper, F.P. Israel, P. Thaddeus 1993, ApJ, 418, 730
14. L. Loinard, T.M. Dame, E. Koper, J. Lequeux, P. Thaddeus, & J.S. Young 1996, ApJL, 469, L101
15. L. Loinard, T.M. Dame, M.H. Heyer, J. Lequeux, & P. Thaddeus 1999, A&A, 351, 1087
16. N. Neininger, C. Nieten, M. Guelin, et al. 2001, in Galaxies and their Constituents at the Highest Angular Resolutions, ed. R. T. Schilizzi, Proceedings of IAU Symposium 205, 352
17. S. Berman 2001, A&A, 371, 476
18. S. Berman & L. Loinard 2002, MNRAS, 336, 477
19. C.D. Wilson & N. Scoville 1989, ApJ, 347, 743
20. C.D. Wilson & N. Scoville 1990, ApJ, 363, 435
21. G. Engargiola, R. Plambeck, E. Rosolowsky, & L. Blitz 2003, astro-ph/0308388
22. E. Rosolowsky, G. Engargiola, R. Plambeck, L. Blitz 2003, astro-ph/0307322
23. J. Braine, F. Combes, F. Casoli, C. Dupraz, M. Gerin, U. Klein, R. Wielebinski, & N. Brouillet 1993, A&AS, 269, 7
24. L.J. Sage 1993, A&AS, 100, 537
25. J.S. Young et al. 1995, ApJS
26. B.K. Rownd & J.S. Young 1996, AJ, 118, 670
27. R.C. Kennicutt, Jr. 1989, ApJ,344, 685
28. R.C. Kennicutt, Jr. 1998, ApJ, 498,541
29. T. Wong & L. Blitz 2002, ApJ, 569, 157
30. M. Dumke, Ch. Nieten, G. Thuma, R. Wielebinski, & W. Walsh 2001, A&A, 373, 853

31. T.A.D. Paglione et al. 2001, ApJS, 135, 183
32. K. Nishiyama & N. Nakai 2001, PASJ, 53, 713
33. K. Nishiyama, N. Nakai, & N. Kuno 2001, 53, 757
34. T. Böker, U. Lisenfeld, & E. Schinnerer 2003, A&A, 406,87
35. H. Hafok & J. Stutzki 2003, A&A, 398, 959
36. L.P. Crosthwaite, J.L. Turner, R.L. Hurt, D.A. Levine, R.N. Martin, & P.T.P. Ho 2001, AJ, 122, 797
37. .P. Crosthwaite, J.L. Turner, L. Buchholz, P.T.P. Ho, & R.N. Martin 2002, AJ, 123, 1892
38. L.P. Crosthwaite 2002, PASP, 114, 929
39. P. Maloney 1988, ApJ, 334, 761
40. B.G. Elmegreen 1993, ApJ, 411, 170
41. B.G. Elmegreen & A. Parravano 1994, ApJL, 435, L121
42. Y. Sofue, M. Honma, N. Arimoto 1995, A&A, 296, 33
43. M. Honma, Y. Sofue, N. Arimoto 1995, A&A, 304, 1
44. M. Hidaka & Y. Sofue 2002, PASJ, 54, 223
45. K. Sakamoto, S.K. Okamura, S. Ishizuki, & N.Z. Scoville 1999a, ApJS, 124, 403
46. K. Sakamoto, S.K. Okamura, S. Ishizuki, & N.Z. Scoville 1999b, ApJ, 525. 691
47. S. Jogee 1999, Ph.D. thesis, Yale University
48. A.J. Baker 2000, Ph.D. thesis, California Institute of Technology
49. S. Jogee, A.J. Baker, K. Sakamoto, N.Z. Scoville, J.D.P. Kenney 2001, in The Central Kiloparsec of Starbursts and AGN: The La Palma Connection, ed. J. H. Knapen, J. E. Beckman, I. Shlosman, and T. J. Mahoney, (San Francisco: Astronomical Society of the Pacific), 249, 612
50. A.J. Baker, S. Jogee, K. Sakamoto, & N.Z. Scoville 2003, in Active Galactic Nuclei: from Central Engine to Host Galaxy, ed. S. Collin, F. Combes and I. Shlosman,(San Francisco:Astronomical Society of the Pacific), 290, 479.
51. Y. Sofue, J. Koda, H. Nakanishi, S. Onodera, K. Kohno, A. Tomita, & S. Okumura 2003, PASJ, 55, 17
52. Y. Sofue, J. Koda, H. Nakanishi, S. Onodera 2003, PASJ, 55, 59
53. Y. Sofue, J. Koda, H. Nakanishi, M. Hidaka 2003, PASJ, 55, 75
54. M.W. Regan, M.D. Thornley, T.T. Helfer, K. Sheth, T. Wong, S.N. Vogel, L. Blitz, & D.C.-J. Bock 2001, ApJ, 561, 218
55. T.T. Helfer, M.D. Thornley, M.W. Regan, T. Wong, K. Sheth, S.N. Vogel, L. Blitz, & D.C.-J. Bock 2003, ApJS, 145, 259
56. K. Sheth, S.N. Vogel, M.W. Regan, P.J. Teuben, A.I. Harris, M.D. Thornley, & T.T. Helfer 2004, ApJ, submitted.
57. M.D. Thornley, C.J.L. Spohn-Larkins, M.W. Regan, & K. Sheth 2002, BAAS, 201, 1317
58. T. Wong & L. Blitz 2002, ApJ, 569, 157
59. K. Sheth, S.N. Vogel, M.W. Regan, P.J. Teuben, A.I. Harris, & M.D. Thornley 2002, AJ, 124, 2581
60. M.D. Thornley, M.W. Regan, S.N. Vogel, R. Fraga-Encinas, & K. Sheth, in preparation

The Spiral Arms of M 51: Physical Properties of the Gas

E. Schinnerer[1], A. Weiß[2], S. Aalto[3], and N.Z. Scoville[4]

[1] NRAO, Socorro, U.S.A. eschinne@nrao.edu
[2] IRAM, Granada, Spain aweiss@iram.es
[3] OSO, Sweden susanne@oso.chalmers.se
[4] California Institute of Technology, Pasadena, U.S.A. nzs@astro.caltech.edu

We have mapped key molecular line probes (^{12}CO, ^{13}CO, C^{18}O, HCN, HCO+) in two distinct regions in the spiral arms of the Whirlpool galaxy M 51 using the Owens Valley Radio Observatory (OVRO) mm-interferometer and the IRAM 30m single dish telescope. Line Velocity Gradient (LVG) analysis performed at a linear resolution of 210 pc (4.5"; both regions) and 135 pc (2.9"; western region only) suggests physical conditions in the Giant Molecular Cloud complexes (GMCs) of M 51 very similar to those in the Milky Way: cold ($T_{kin} \sim 15$ K) clouds with moderate H_2 density ($n(H_2) < 10^{2.7}$ cm^{-3}). We find indications for a galacto-centric trend with higher kinetic temperature for smaller radii. The data show little evidence for cloud heating by the massive star formation at our resolution of 135 pc. CO(2-1) data at our highest angular resolution of 1.3" reveal the presence of multiple unresolved GMCs, with a tendency to higher velocity dispersion upstream, i.e. the inner side, of the spiral arm.

1 Introduction

The Whirlpool galaxy M 51 is one of the closest (9.6 Mpc; 1" $\sim$ 46.5 pc) almost face-on grand-design spiral galaxies. Most of its molecular gas is found in the spectacular spiral arms (e.g. Scoville & Young 1983; Aalto et al. 1999). Therefore it offers the unique opportunity to study the physical properties of the molecular gas within the spiral arms. Due to its low inclination ($i \sim 20^o$) non-circular motions can be separated from the galactic rotation. The unusually large streaming motions (60 - 150 km s^{-1}) found in M 51 imply a very strong density wave whose general features are velocity discontinuities and streaming across the spiral arms (e.g. Roberts & Stewart 1987). Besides the large number of Giant Molecular Cloud Associations (GMAs) (up to 16 within one spiral arm, Aalto et al. 1999), numerous OB star clusters reside in the spiral arms as is obvious in high resolution HST data (Scoville et al. 2001). Nevertheless, there is no simple correlation between the gas reservoir and the star formation rate which indicates varying star formation efficiencies and thus varying physical condition of individual GMAs. This view is supported by a previous low spatial resolution study of the physical cloud conditions in M 51 (Garcia-Burillo et al. 1993).

2 Observations

The OVRO M 51 mosaic in the CO(1-0) line (Aalto et al. 1999) and the IRAM 30m maps of the CO(1-0) and CO(2-1) lines (Garcia-Burillo et al. 1993) were combined with new OVRO and IRAM 30m data for analysis (see also Fig. 1). Two distinct regions in the spiral arms of M 51 were observed in the CO(2-1), ^{13}CO(1-0), and C^{18}O(1-0) line with the Owens Valley Radio Observatory (OVRO) mm-interferometer. In addition, we obtained IRAM 30m data to apply short spacing corrections (SSC) to avoid missing flux problems. The final short spacing corrected data cubes were binned to two common resolutions of 2.9" (135 pc) and 4.5" (210 pc) for Line Velocity Gradient (LVG) analysis. The line ratios were measured from spectra coming from the same position. Even at 135 pc resolution a few spectra showed hints for multiple line components suggesting blending of GMCs.

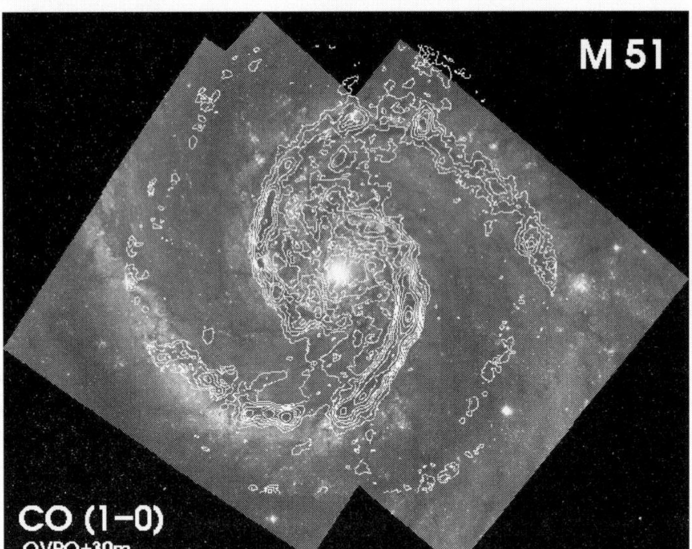

Fig. 1. Integrated ^{12}CO(1-0) line emission (contours) overlaid onto a composite optical HST color image (from Scoville et al. 2001). The OVRO CO data of Aalto et al. (1999) were short spacing corrected (SSC) using the 30m single dish data from Garcia-Burillo et al. (1993). At a current resolution of $\sim$ 150 pc the CO emission is well-correlated with the more extincted regions in the spiral arms.

3 Results from the Combined OVRO and 30m Data

LVG analysis: To constrain the physical conditions of the GMCs we use a isothermal, spherical large velocity gradient model (for details see Weiß et al. 2001). Since $C^{18}O$(1-0) is only marginal detected at the strongest positions we did not include the transition in the analysis. For the models presented here we fixed the ^{12}CO abundance per velocity gradient and the ^{13}CO abundance relative to ^{12}CO to 8×10^{-5} and 30, respectively. The numbers correspond to values found in GMCs in the inner 5 kpc of the Milky Way (see e.g. Langer & Penzias 1993). LVG line intensity ratios were computed for kinetic gas temperatures between 5 and 200 K and H_2 densities between $\log(n(H_2))$ of 2.0 - 6.0. The LVG models suggest that the physical conditions in the GMCs of M 51 are very similar to those in the Milky Way: The bulk of the CO emission arises from cold ($T_{kin} \sim 15$ K) clouds with moderate H_2 density ($n(H_2) < 10^{2.7}$ cm^{-3}) (Fig. 2). The most obvious spatial variation of the physical gas condition is a decrease of the kinematic temperature with increasing galacto-centric radius along the western spiral arm. Temperatures decrease from about 20 K at 1.3 kpc (Pos W1) to 5 K at 2.2 kpc (Pos W9). In the southern region, with almost constant galacto-centric radius, no such gradient is detected. A comparison of the LVG results from star-forming clouds (as traced by Hα) and non star-forming regions suggests that the gas temperature in the star-forming regions is on average lower ($T_{kin} \sim 10$ K) than the gas temperature in non star-forming regions ($T_{kin} \sim 23$ K). The mean density for both environments is about $n(H_2) \approx 10^{2.4}$ cm^{-3}. This finding suggests that the CO emission from the star-forming regions is dominated by cold gas surrounding the HII regions. The impact of SF on the molecular clouds (e.g. heating at the cloud surfaces) is not visible at a linear resolution of 210pc (or even for the higher 135pc resolution data).

High resolution CO(2-1) data: The molecular gas distribution closely follows the dust lanes seen in the optical. The average H_2 volume densities of $\sim 10^{2.5}$ cm^{-3} inferred from the LVG analysis imply an extinction A_V of $\sim 15^{mag}$ (assuming a scale of 15 pc for the molecular gas layer and a screen geometry). This number is quite large compared to the average A_V of 3^{mag} found for the HII regions (Scoville et al. 2001), and suggests that a significant fraction of on-going star formation could be hidden in the dust. We find a weak correlation between the presence of HII regions as traced in their Hα and/or Paα line emission and peaks in the molecular gas distribution: All HII regions within the spiral arms are close to, but usually not coinciding with molecular gas peaks. The internal velocity dispersion is changing dramatically across the western spiral. The observed large difference might explain why - despite the peak in the intensity - no HII regions are observed/formed on the inner side of the spiral arm.

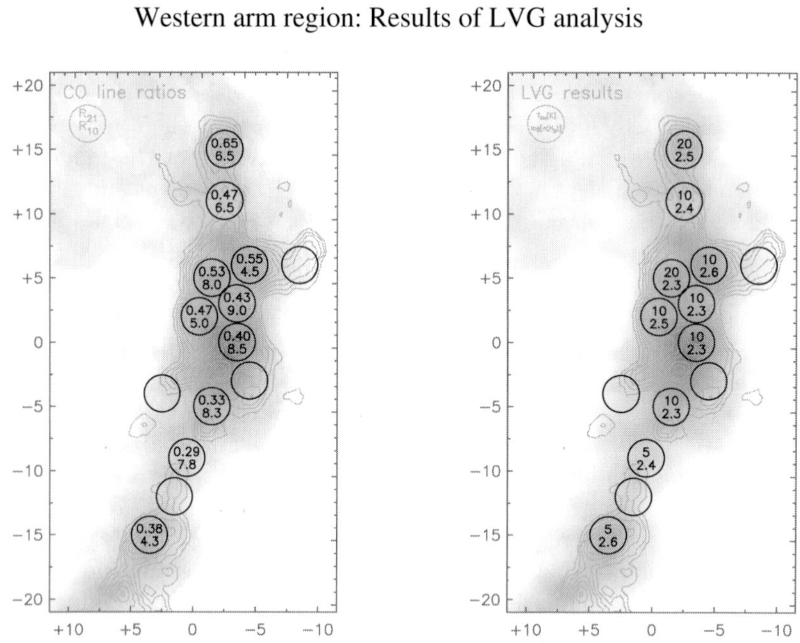

Fig. 2. The integrated CO(2-1) line emission (contours) is shown in both panels overlaid onto the integrated CO(1-0) line emission (grayscale). The circles indicate the positions of the spectra used for the LVG analysis at 135 kpc linear resolution. *Left:* The derived values for the CO line ratios are given: R_{21}=CO(2-1)/CO(1-0) (top number) and R_{10}=CO(1-0)/^{13}CO(1-0) (bottom number). *Right:* The results from the LVG analysis are presented: T_{kin} [K] (top number) and log(n(H_2)) (bottom number).

References

1. Aalto, S., Hüttemeister, S., Scoville, N.Z., Thaddeus, P.: ApJ, **522**, 165 (1999)
2. García-Burillo, S., Guélin, M., Cernicharo, J.: A&A, **274**, 123 (1993)
3. Langer, W.D., Penzias, A.A., ApJ, **408**, 539 (1993)
4. Roberts, W.W., Stewart, G.R., ApJ, **314**, 10 (1987) Scoville et al. 2001
5. Scoville, N.Z., Polletta, M., Ewald, S., Stolovy, S.R., Thompson, R., Rieke, M., AJ, **122**, 3017 (2001)
6. Scoville, N.Z., Young, J.S., ApJ, **265**, 148 (1983)
7. Weiß, A., Neininger, N., Hüttemeister, S., Klein, U., A&A, **365**, 571 (2001)

Molecular Gas in the Andromeda Galaxy

M. Guélin[1], S. Muller[1], C. Nieten[2], N. Neininger[3], H. Ungerechts[1], R. Lucas[1], and R. Wielebinski[2]

[1] IRAM, 300 rue de la Piscine, F-38406 Saint Martin d'Hères, France, guelin@iram.fr
[2] MPIfR, Auf dem Hügel 69, D-53121 Bonn, Germany
[3] Radioastron. Institut der Univ. Bonn, Auf dem Hügel 71, D-53121 Bonn, Germany

1 Molecular Spiral Structure

M 31, the closest large spiral galaxy to our own, is the best object for studying molecular clouds and their relation to the spiral structure. As one of the astronomical objects with the best known distance (0.78 ± 0.02 Mpc), it is also one of the best places where to estimate molecular clouds masses through the Virial Theorem.

With the 30-m telescope, we have mapped the CO(1-0) emission in M 31 (Fig. 1.a) with an unprecedented angular resolution and sensitivity: equivalent HPBW=23″, or 90 pc in the plane of the sky, $\Delta Tmb \simeq 25$ mK per 1 MHz channel (Nieten et al. 2003, see also Neininger et al. 1998). In addition, several arm portions have been mapped in the (2-1) line with the 30-m telescope (angular resolution 11″, or 45 pc) and several cloud complexes in the (1-0) and (2-1) lines with the PdB interferometer (resolution 3-10 pc, Muller et al. 2004, in prep).

One incentive for a study with such a large spatial dynamics is to learn about the formation and destruction timescales of molecular clouds. We have analyzed data cubes with the CLUMPFIND (Williams et al. 1994) and GAUSSCLUMP (Stutzki & Güsten 1990) programs that decompose the signals into compact 3-d structures, which, in the case of the 30-m data, we identify with 'giant clouds' (GCs) or cloud complexes, and, in the case of the interferometer data, we identify with 'individual' clouds.

Fig. 1.a shows a least-squares fit to M 31's brightest CO 'clouds' by two logarithmic spirals. The GC positions were determined from the CO (1-0) map with CLUMPFIND. The best fit yields a pitch angle 7-8° for both arms. Up to galactocentric radii $R = 12$ kpc, most of the bright CO clouds and HII regions lie on the two-armed pattern. A notable exception is the region of the giant stellar association NGC 206, in the bottom part of the map ($x = -40'$, $y = 0$). The CO arms are 2-3 thinner than the HI and radio continuum arms and the CO arm-interarm brightness ratio about ten times higher. This is illustrated in Fig. 1.b-d, which shows the average CO, HI and 20-cm continuum brightness profiles across the two spiral arms in the NW quarter of the galaxy.

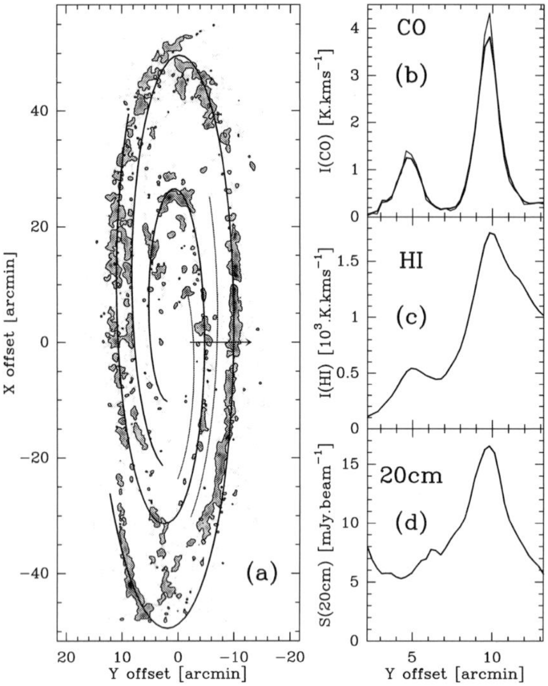

Fig. 1. a) The CO (1-0) velocity-integrated emission in M 31, I(CO), observed with a resolution of 23″ with the IRAM 30-m telescope. The thick lines represent the fitted two-arm spiral pattern. **b)** The CO intensity profile across both arms, calculated by averaging I(CO) along spiral segments parallel to the arms (two such segments are shown on Fig. 1a). The thin curve corresponds to the original map (HPBW 23″) and the thick curve to a map smoothed to HPBW=45″; the abscissa is the galactocentric radius of the spiral segment when it crosses the minor axis. **c)** and **d)** same as (b) for the HI and 20-cm continuum emissions (HPBW= 45″). From Nieten et al. 2003.

At least in the region of interest here ($3 < R < 12$ kpc) the CO velocity-integrated intensity, $I(\mathrm{CO}_{10}) = \int T(\mathrm{CO}_{10})dv$, is statistically a good tracer of the H_2 column density $N(H_2)$, even through the CO-to-H_2 'conversion factor', $\mathbf{X} = \mathrm{N(H_2)}/I(\mathrm{CO}_{10})$, is poorly known and may vary from cloud to cloud. This has been shown by Neininger et al. (1998), who noted the remarkable spatial correlation between the CO clouds and the dust extinction lanes in this region: all dark patches on the near side of M 31 have CO counterparts, i.e. there is no cloud with a large visual extinction undetected in CO. The almost complete lack of CO emission (and of dark patches) in the interarm region of Fig. 1.a thus implies that there are no hidden molecular clouds in this region and that the arm-interarm contrast in H_2 is high.

Since the molecular clouds are mostly confined into narrow lanes following the arm crest, they should be relatively short-lived. As can be seen on Fig. 1.b, the CO intensity profile across the arms is nearly symmetrical. Molecular clouds must form rapidly in the stellar arm potential dip ($<$ few$\times 10^7$ yr) and be destroyed equally fast.

1.1 Molecular Cloud Properties

The well known distance to M 31 and the high sensitivity of the observations allow, in principle, to derive accurate clouds sizes and fluxes. The 30-m telescope beam at 115 GHz is however too large to resolve many cloud complexes, and one has to turn to the higher resolution (2-1) line data (HPBW= 45 pc) to get a less biased representation of the clouds. We note that the (2-1) maps smoothed to 23″ show clouds with sizes and shapes almost identical to those observed in the (1-0) map, so that excitation effects should not significantly affect the sizes.

Fig. 2.a presents the size distribution of the 50 brightest CGs found by GAUSSCLUMP in the CO(2-1) maps. These clouds have galactocentric radii comprised between 5 and 12 kpc. The average size (FWHP in the plane of the sky) is 80 pc. Only $\sim 40\%$ of the clouds are large enough to be resolved in the 1-0 line and many of those are so large that they may well not be gravitationally bound.

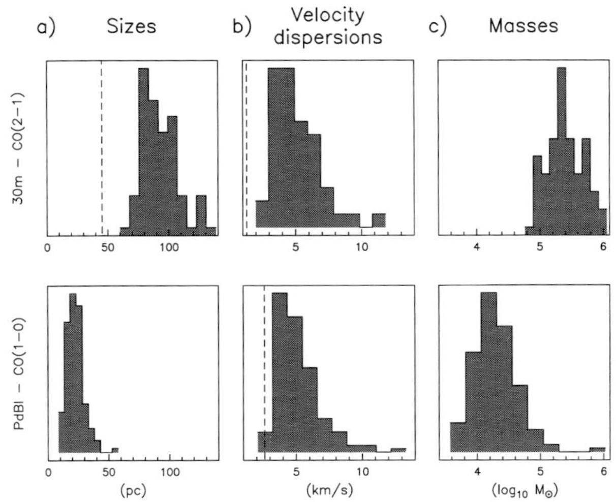

Fig. 2. Histograms of the sizes (FWHP in pc), velocity dispersion (FWHP in kms^{-1}) and masses (calculated from the CO and HI luminosities) for the 50 brightest GCs found by GAUSSCLUMP in the CO(2-1) 30-m telescope maps (upper diagrams –resolution 45 pc) and in the CO(1-0) interferometer maps (lower diagrams –resolution ≤ 10 pc). From Muller 2003.

Fig. 2.b shows the velocity dispersions found by GAUSSCLUMP in the 50 clouds. The average velocity dispersion is 5 kms^{-1}. Fig. 2.c shows the cloud 'total masses', $M_{\mathrm{CO+HI}} = 1.36\,(M_{\mathrm{CO}} + M_{\mathrm{HI}})$, where the molecular masses M_{CO} are estimated from the CO integrated line intensities, assuming $\mathbf{X} = N(\mathrm{H}_2)/I(\mathrm{CO}_{10}) = 2\,10^{20}\ \mathrm{cm}^{-2}\mathrm{K}^{-1}\mathrm{km}^{-1}\mathrm{s}$.

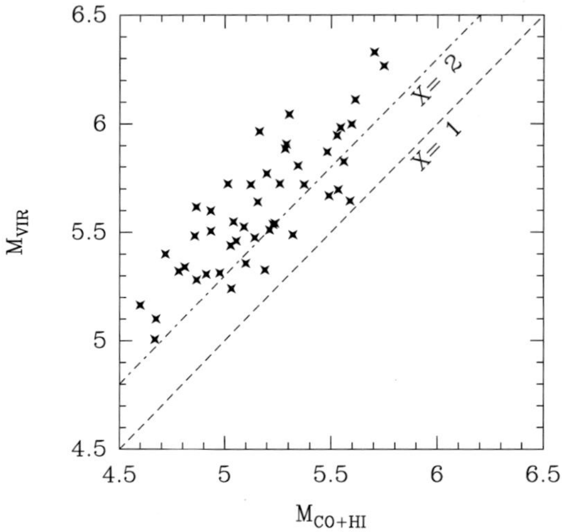

Fig. 3. Comparison of the cloud masses, calculated from the CO and HI luminosities, with those calculated from the Virial Theorem. For this figure, we used a CO-to-H_2 conversion factor **X** = 1 10^{20} $cm^{-2}K^{-1}km^{-1}s$. Adapted from Muller 2003.

The above $M_{\rm CO+HI}$ estimates are compared in Fig. 3 with the masses $M_{Vir} = 190\sqrt{\Delta x \Delta y}\Delta V$ that can be derived with the Virial Theorem by assuming that the clouds are in Virial equilibrium.

The data points in Fig. 3 show a relatively large scatter. This scatter partly reflects variations of **X** from cloud to cloud (and with galactocentric radius) but also that many clouds are not at virial equilibrium. The observation of broad or multiply-peaked line profiles, e.g. caused by streaming motions, and, for the largest cloud complexes, of a galactocentric rotation gradient, are good illustrations of this.

In order to furher investigate the dynamical state of the GCs, we have resolved with the interferometer several complexes into individual clouds. The mass of a complex can then be estimated from the relative motions of its clouds, considered as test particles in the complex gravitational field, without any assumption on the field distribution. We have used several estimators of the dynamical mass that have different sensitivity to runaway or unrelated clouds. This method has allowed to select 3 quiescent cloud complexes whose motions are compatible with Virial equilibrium. The dynamical masses of these complexes yield values of **X** between 1 and 2.6 10^{20} $cm^{-2}K^{-1}km^{-1}s$.

References

1. S. Muller: Caractérisation de la Matière Interstellaire Extragalactique. PhD Thesis, Université Joseph Fourier, Grenoble (2003)
2. N. Neininger, M. Guélin, H. Ungerechts et al.: Nature **395**, 871 (1998)
3. C. Nieten: Untersuchung der CO-Emission in der Anromeda-Galaxie. PhD Thesis, Universität Bonn, Bonn (2001)
4. C. Nieten, N. Neininger, M. Guélin et al: A&A *submitted for publication* (2003)
5. J. Stutzki, R. Güsten: ApJ **356**, 513 (1990)
6. J.P. Williams, E.J. De Geus, L. Blitz: ApJ **428**, 693 (1994)

CO Emission Associated with a Cooling Flow

F. Combes and P. Salomé

LERMA, Observatoire de Paris, 61 Av. de l'Observatoire, 75014, Paris, France
francoise.combes@obspm.fr, philippe.salome@obspm.fr

1 Abstract

The existence of cooling flows in the center of galaxy clusters has always been a puzzle, and in particular the fate of the cooling gas, since the presence of cold gas has never been proven directly. X-ray data from the satellites Chandra and XMM-Newton have constrained the amount of cooling, and it was realized that feedback and heating from a central AGN and its jets could reduce the amount of cold gas. Recently, a few central galaxies of cooling flow clusters have been detected in the CO lines. For the first time, we show IRAM interferometer maps of CO(1-0) and CO(2-1) emission in a cooling flow, showing a clear association of the cold gas (at about 20K) with the cooling flow. This shows that although the AGN provides a feedback heating, the cooling phenomenon does occur, with about the expected rate.

2 Recent Results on Cooling Flows

Cooling flows are suspected to occur in the center of rich galaxy clusters, since the cooling time of the hot gas is shorter than Hubble time. Until recently, large flow rates were expected, of the order of 100 to 1000 $M_{\odot}$/yr. But the fate of the cooled gas has always been a mystery, the gas or the stars formed out of it could not be seen. Attempts to observe this cold gas either in HI (Shostak et al 1983, Dwarakanath et al 1995) or CO (McNamara et al 1994, Braine & Dupraz 1994, O'Dea et al 1994), have always obtained upper limits. Only in NGC1275 (Perseus A) has molecular gas been detected (Lazareff et al 1989), but being associated to merging galaxies, its origin was ambiguous. Recently, CO emission has been detected in 3C31 (RXJ0107+32) and 3C264 (A1367) at the center of cooling flows (Lim et al 2001), and towards 16 central galaxies of clusters (Edge 2001).

One of the main results of the X-ray satellites Chandra and XMM-Newton was to change considerably our view on cooling flows: the deduced cooling rates have been reduced by at least one order of magnitude; the old view of quiet and regular, quasi-spherical cooling has given place to partial and intermittent cooling, perturbed by re-heating processes, feedback due to the central AGN, and associated shocks, jets, sound waves, bubbles... A recent spectacular illustration of this perturbed cooling is the Chandra image of the cooling flow in Perseus, with bubbles, gas streaming up and down from the center, and ripples looking like emitted sound waves (Fabian et al. 2003).

3 Detection of CO Emission

We have undertaken a survey of CO emission in central galaxies of cooling flow clusters (Salomé & Combes 2003a), with the IRAM-30m telescope. The clusters were selected from their previously derived cooling rates, and their observed Hα luminosities. It is indeed expected that the Hα emission is also related to the cooling, either diretly or because gas is photoionised by stars formed in the cooling. Out of 32 galaxies observed, between 6 and 10 detections were obtained (according to whether both CO(1-0) and CO(2-1) detected, or a tentative single one). Masses between 3 10^8 and 4 $10^{10} M_\odot$ are found. The excitation derived from the ratio of the two first rotational lines of the CO molecule is compatible with optically thick cold gas (beam corrected ratio close to 1, within the uncertainties).

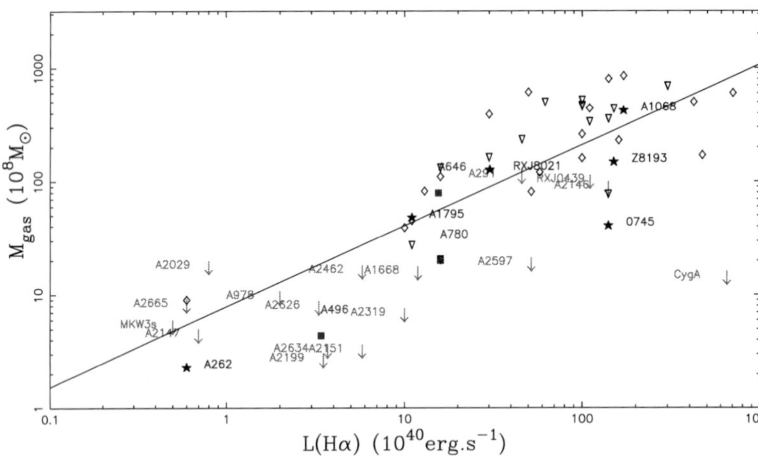

Fig. 1. H_2 gas mass derived from CO emission, as a function of Hα luminosity; results both from Edge (2001) and Salomé & Combes (2003a) have been included.

Figure 1 reveals a correlation between the H_2 mass detected toward cooling flows, and their Hα luminosity. This may be interpreted in terms of common excitation mechanisms, Hα is either shocked gas from cooling flows or gas photo-ionized by the young stars formed out of the cooled gas.

It is also interesting to compare the amount of gas detected to what is expected from the cooling flow. With respect to the cooling rates deduced from Einstein (White et al 1997), the CO gas reaches about 10% of the expected value. We have to note that the region observed with the IRAM beam is only the center of the cooling flow. The cooling region extends up to 10 times its radius, and the cooling mass inside a given radius grows linearly with radius. The H_2 gas detected corresponds to cooling rates during 1 Gyr about 10 times less than the previously computed value.

Our results do not show any clear correlation with the radio flux at 1.4 GHz. However, it is clear that cooling flows help to fuel the central black hole, when there is one. About 71% of cD galaxies with cooling flow have an AGN, while only 23% of cD without cooling flow have one. It is important to try to derive the gas-to-dust ratio in the cooled gas, since dust is expected to be destroyed in the hot gas by sputtering. The cooled gas should be dust depleted. We have derived the dust mass from IRAS data (assuming a dust temperature of the order of 35K), and indeed, the dust-to-gas ratio was small. Cold dust observed at 1mm would yield clearer results, due to the lower temperature sensitivity, in the Rayleigh-Jeans domain.

4 Abell 1795 Cooling Flow

The Abell 1795 cluster (z=0.063) has a conspicuous cooling wake at its center in the form of a long (80kpc) North-South filament (Fabian et al 2001). The cooling time is about 300 Myr, and the cooling rate $\sim$ 100 $M_\odot$/yr within 200kpc (Ettori et al 2002). The filament presents also Hα line emission (Cowie et al 1985), at the velocity of the cluster. The central cD galaxy is moving at V=374km/s with respect to the cluster, and this central oscillation is probably the origin of the wake.

We have mapped CO(1-0) and CO(2-1) emission with the IRAM interferometer with 3.2" and 1.8" beams respectively. The cold gas is clearly coincident with the cooling gas, both in X-ray and Hα, but not with any galaxy (Salomé & Combes 2003b). 3mm continuum emission is detected at the position of the AGN, with a flux of 7mJy corresponding to a synchrotron spectrum with slope -0.98. Figure 2 shows the very good coincidence between CO(2-1) and Hα line emission, and its avoidance of the radio jets. Cold gas may have deflected the expanding radio lobes, alternatively the jet creates a hole (bubble) in the hot gas, which is compressed at the boundaries, cools down and condenses in molecular gas.

The interferometer retrieves only 20% of the 30m flux, which means that there exists an important extended CO emission, in addition to the mapped features. The total H_2 mass is 4.8 10^9 $M_\odot$. The CO kinematics confirm the association of the molecular gas with the cooling flow. The CO velocity is not associated to the central galaxy, but to the cluster (-350km/s), as is the Hα.

5 Conclusions

Out of 32 cooling flow galaxies observed, 6-10 were detected, with H_2 masses up to $10^{10} M_\odot$, increasing the total number to 23 (Edge 2001). The interferometer map of Abell 1795 shows CO clearly associated to the cooling wake, and not in rotation around the cD galaxy. The CO emission is closely associated to Hα, confirming the global CO-Hα correlation found with the single

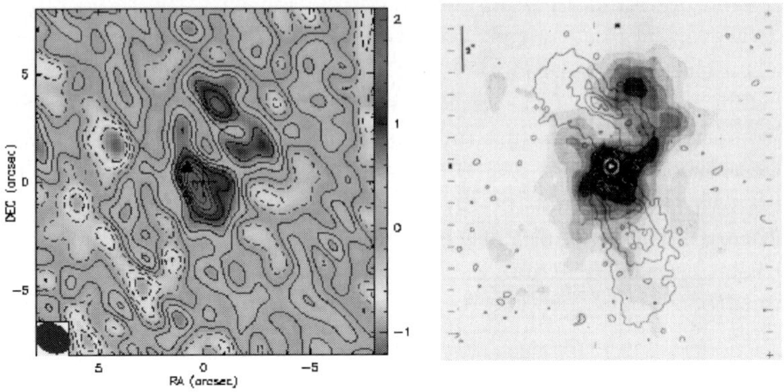

Fig. 2. Left CO(2-1) map obtained with the IRAM interferometer. **Right** Hα +[NII] line emission (grey scale), with 6cm contours from van Breugel et al. (1984).

dish. The H_2 mass found corresponds to what is expected from the cooling rate, 100 $M_\odot$/yr within 200kpc. There is tight coupling between the AGN and the cold gas. The AGN creates cavities in the hot gas. Cooling occurs along the edges of cavities, where CO and Hα are observed.

References

1. Braine J., Dupraz C.: 1994, A&A 283, 407
2. Cowie, L. L., Hu, E. M., Jenkins, E. B., York, D G.: 1983 ApJ, 272, 29C
3. Dwarakanath K.S., Owen F.N., van Gorkom J.H.: 1995, ApJ 442, L1
4. Edge A.C.: 2001, MNRAS 328, 762
5. Ettori, S., Fabian, A. C., Allen, S. W. et al.: 2002 MNRAS 331, 635E
6. Fabian A.C., Sanders J.S., Ettori S. et al: 2001, MNRAS 321, L33
7. Fabian A.C., Sanders J.S., Allen, S. W. et al.: 2003, MNRAS 344, L43
8. Lazareff B., Castets, A., Kim, D.-W., Jura, M.: 1989, ApJ 336, L13
9. Lim J., Leon, S., Combes, F., Dinh-V-Trung: 2001, ApJ 545, L93
10. McNamara B.R., Jaffe W.: 1994, A&A 281, 673
11. O'Dea C.P., Baum S.A., Maloney P.M., et al.: 1994a, ApJ 422, 467
12. Salomé P., Combes F.: 2003a, A&A in press, 2003b, A&A, sub
13. Shostak G.S., van Gorkom J.H., Ekers R.D. et al.: 1983, A&A 119, L3
14. Van Breugel W., Heckman, T., Miley, G.: 1984, ApJ 276, 79
15. White D. A., Jones, C., Forman, W.: 1997, MNRAS 292, 419

Triggered Star Formation on Galactic Scales

J. Palouš[1], S. Ehlerová[1], R. Wünsch[1], G. Tenorio-Tagle[2], S. Silich[2], G.A. Medina-Tanco[3], and C. Muñoz-Tuñon[4]

[1] Astronomical Institute, Academy of Sciences of the Czech Republic, Bocni II 1401, 141 31 Prague 4, Czech Republic, palous@ig.cas.cz
[2] Instituto Nacional de Astrofísica Optica y Electronica, AP 51, 7200 Puebla, México, gtt@inaoep.mx, silich@inaoep.mx
[3] Instituto de Astroñomico e Geofísico, USP, Av. Miguel Stéfano 4200, 04301-904 Sao Paulo, Brazil, gustavo@iagusp.usp.br
[4] Instituto de Astrofísica de Canarias, 38200 La Laguna, Tenerife, Spain, cmt@ll.iac.es

Abstract. Star formation in the turbulent ISM may be triggered by the collapse of an expanding sheet driven by stellar winds, stellar radiation and supernova explosions. The collapse produces clumps, some of which are gravitationally bound. In the low density environment of the galactic disks, clumps have masses in the range $10^3 - 10^6$ $M_\odot$ and they form if 1) the disk gas surface density is high enough, 2) the disk rotation is slow enough and 3) the gas cooling is fast enough. In the high density environment of molecular clouds, masses of fragments are in the range $1 - 100$ $M_\odot$. The mass spectrum can be approximated by a power law with the spectral index close to -2.3. We also introduce a model of the star forming factory, where a compact stellar cluster results from the collapse of an ISM giant cloud.

1 Structures and Processes in the Turbulent Interstellar Medium

The interstellar medium has a turbulent nature: supersonically moving turbulent eddies, sheets and filaments expand and collide forming temporary structures. Large-scale galactic rotation, gravitational forces and star formation are driving mechanisms of the turbulence, continuously straggling to ballance dissipation and cooling.

Supersonic turbulence may be responsible for the formation of superclouds, giant molecular clouds and even the dense cores inside of GMCs. Only some of them are self-gravitating and long lasting. Self-gravitating cores are places of the star formation. New and massive stars release stellar winds and radiation, forming HII regions and subsequently they turn to supernovae driving supersonic motions in the interstellar medium. Recent review on the control of star formation by supersonic turbulence has been published by Mac Low and Klessen [1].

In this paper, we discuss the supersonic expansion of a sheet of cold material. We use the thin shell approximation and 3-dimensional hydrodynamic

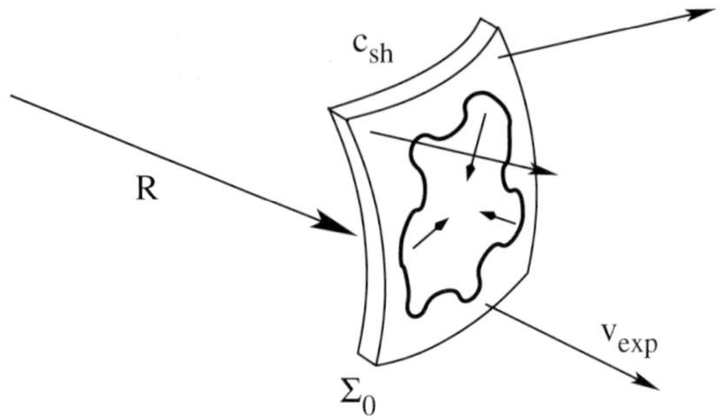

Fig. 1. The expanding sheet. R is the radius of the curvature, v_{exp} is the expansion velocity, Σ_0 is the unperturbed column density and c_{sh} is the speed of sound inside the layer.

simulations to show the formation of gravitationally bound and unbound fragments. We also propose a scenario for the formation of a massive stellar cluster inside a collapsing cloud.

2 The Expanding Sheet

In this section we describe the gravitational fragmentation of sheets expanding into the ISM, why and how the fragmentation happens and what are the necessary conditions. In a thin layer expanding with the supersonic speed into the ambient medium, there is the competition between gravity, internal pressure and stretching. Using the linear stability analysis given in [2], we estimate the time t_b, when the instability begins (see [3]) and develops. In low densities, $n \leq 1$ cm^{-3}, the gravitational instability and the formation of fragments is a rather slow process, $t_b \sim$ a few $\times 10^7$ yr. While in high densities, $n \geq 10^6$ cm^{-3}, the gravitational instability is very fast, $t_b \leq 10^4$ yr.

The linear stability analysis has been extended to quadratic terms in [4]. In [5] and [6] environmental dependences for star formation triggered by an expanding shell collapse are discussed. The conclusion is that the gravitational instability, fragmentation and star formation triggered by the collapse of an expanding shell requires three basic conditions: 1) the disk gas surface density has to surpass a critical value, 2) the disk should not rotate too fast and the shear should not be too large, and 3) the ISM should be able to cool sufficiently fast to decrease the random velocities in the ambient medium to the lower value in the accumulated gas sheet within the instability time (see also [7]).

In Fig. (2) we compare the unperturbed density Σ_0 from the thin shell approximation, [8], to its perturbations from the linear analysis. In the studied case, t_b is about 1 Myr. The linear analysis shows that at the time $t \sim 4 \times t_b$

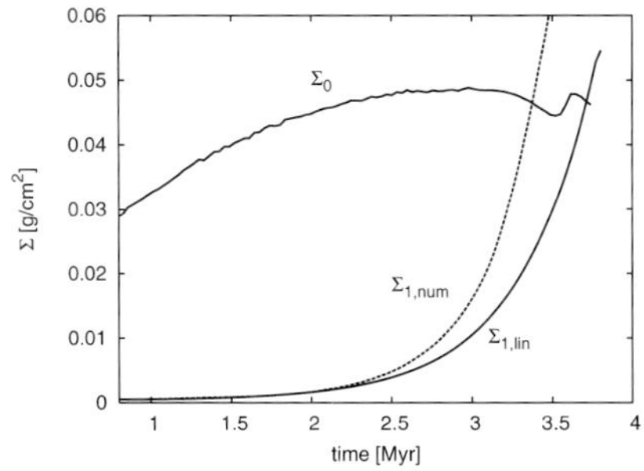

Fig. 2. The time evolution of the surface density Σ_0 based on a thin shell approximation and the evolution of perturbations computed from linear analysis [4] $\Sigma_{1,lin}$ and from 3 dimensional simulations $\Sigma_{1,num}$.

the perturbation $\Sigma_{1,\rm lin}$ is nearly equal to the unperturbed surface density Σ_0. This gives the time when the fastest modes detach from each other, the mass accretion to the most developed fragments decreases and their total mass becomes fixed.

The 3-dimensional simulations using ZEUS code with Šemík, a code that accounts for the self-gravity of the sheet, has been performed. Šemík was developed as a part of the PhD project of R. Wünsch [9]. The results are also given in Fig. (2): perturbations $\Sigma_{1,\rm num}$ based on 3D simulations grow faster reaching the value Σ_0 at an earlier time. This is due to cooling, which causes that the high density fragments have a lower temperature than the average value in the expanding shell. This results into a faster growth-rate of fragments in comparison to the linear case when the shell temperature is taken as a constant.

The mass spectrum of fragments has been discussed in [3] and [10]. We conclude that in a high density environment, where the fragments acquire stellar masses and the collisional time-scales are longer than typical t_b [11], the gravitational fragmentation shapes the mass spectrum causing a power law with a slope in the range $(-2.25, -2.40)$.

3 Star Forming Factory

Star formation appears to proceed in a variety of different modes and without a doubt, major bursts of star formation or superstar clusters (SSC), are among the most striking entities that we know of. With a typical half-light radius of the order of 3 pc and masses in the range of a few times $10^5 \mathrm{M}_\odot$ to several times 10^6 $\mathrm{M}_\odot$, young SSC remind us of the power and beauty that our collection of globular clusters must have displayed at the beginning of the Galaxy formation. Contrary to the above described scenario, SSCs demand of a mechanism that could gather sufficient ISM mass, similar to

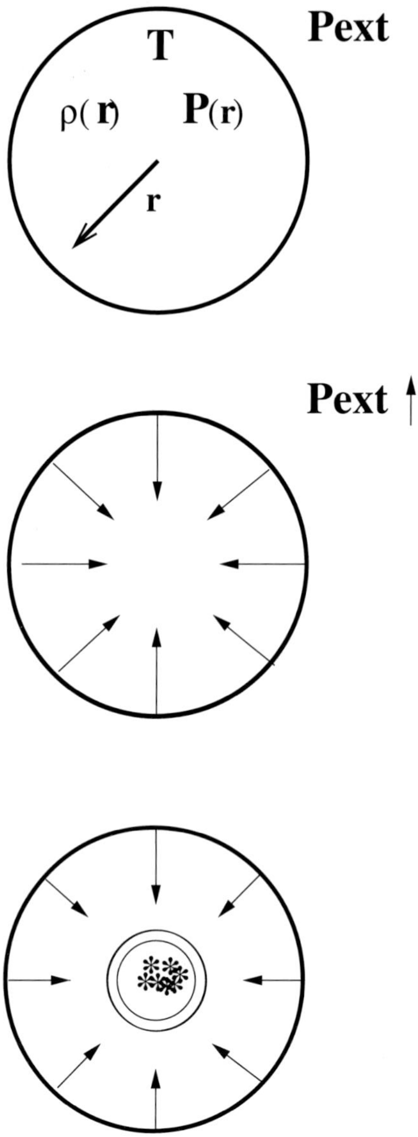

Fig. 3. The star forming factory. The equilibrium cloud (upper panel), the cloud in an isothermal collapse initiated by the increased external pressure (middle panel), the static shell in a collapsing cloud formed by action of the first generation of stars (lower panel).

what one finds in giant molecular clouds, into a small volume (a few pc across) where all the sources will form. The problems to explain such episodes of star formation are indeed enormous: Which stars formed first? and how to cope with the disruptive effects of feedback from massive stars (photoionization and mechanical energy), are also major issues.

The star forming factory ([13]) is a model that addresses many of these issues and also make strong predictions that can be corroborated from an observational point of view. The model assumes (see Fig. 3) that during the isothermal collapse phase (T = 100K) of a massive cloud, which has been described in [12], the density at the center of such a configuration attains sufficiently large values ($n \geq 10^4$ cm^{-3}) as to exceed the Jeans criterion for the formation of a first generation of massive stars (100 $M_\odot$ - 10 $M_\odot$). The impact of these stars, with their winds and supernova explosions, leads initially to a perfect balance where the wind ram pressure is, at a given radius, identical to the infalling cloud ram pressure and thus a standing shell able to process eventually the whole cloud, is then established. The shell presents two layer of different material: the inner section contains the stellar ejecta and the outer layers are filled with shocked cloud matter. We have shown that after a few 10^4 yr the shell becomes gravitationally unstable and fragments. The linear analysis of the standing shell has lead us to the mass distribution of fragments. This presents a maximum at about a solar mass, and a power law distribution for massive sources with a slope of -2.25, very similar to that found in the galaxy and in other galaxies.

The fragments are then assumed to fall towards the center of the configuration as they evolve into stars, adding mass to the forming cluster, while enhancing also their energy output. The situation has been shown to self regulate itself. The equilibrium criteria for a standing shell although now demands that the gravitational force exerted on the standing shell is taken into consideration, has been shown to be in a perfect balance with the enhanced ram pressure of the increasingly more powerful (more massive) wind. This equilibrium can be supported for more than 20 Myr, time during which a massive cloud of up to several times 10^6 $M_\odot$ can be processed at the standing shell. In the factory model everything happens at the same time. As the collapse proceeds, the standing shell fragments, new sources appear at the center of the configuration, all with a well defined distribution of masses. New sources become contaminated by the supernova products of former stellar generations and thus massive clusters are to present a large metallicity spread and a mixture of stellar populations with different ages.

It is clear that massive clusters cannot form overnight, and the factory model strongly addresses this issue. It will take more than 20 Myr to lead to a cluster of 10^6 $M_\odot$, which by them will have many stellar generations resultant from a continuously accelerated SFR.

The factory model makes also strong predictions regarding the UV photon output from the forming cluster and it has shown how such a disruptive mechanism is kept well under control by absorbing all the ionizing radiation within the inner skin of the standing shell. The factory model thus accounts for the feedback of massive stars, it reproduces an IMF very similar to the observed one, and predicts the self-contamination in massive compact clusters.

Acknowledgments. Authors gratefully acknowledge a financial support by the Grant Agency of the Academy of Sciences of the Czech Republic under the grant No. B3003106, a support by the grant project of the Academy of Sciences of the Czech Republic No. K1048102 and the project AV0Z 1003909 and CONACYT-Mexico research grant 36132-E.

References

1. M.-M. Mac Low, R. Klessen: astro-ph/031093 (2003)
2. B. G. Elmegreen: ApJ **427**, 384 (1994)
3. J. Palouš, R. Wünsch, S. Ehlerová: In Recycling intergalactic and interstellar matter, IAU Symp. 217, eds. Pierre-Alain Duc, Jonathan Braine and Elias Brinks, in press (2004)
4. R. Wünsch, J. Palouš: A&A **374**, 746 (2001)
5. S. Ehlerová, J. Palouš: MNRAS **330**, 1022 (2002)
6. B. G. Elmegreen, J. Palouš, S. Ehlerová: MNRAS **334**, 693 (2002)
7. J. Palouš, S. Ehlerová, B. G. Elmegreen: Astrophys. and Space Sci. **281**, 101 (2002)
8. G. S. Bisnovatyi-Kogan, S. A. Silich: Rev. Mod. Phys. **67**, 661 (1995)
9. R. Wünsch: Gravitational instabilities of expanding shells, PhD thesis, Charles University, Prague (2003)
10. J. Palouš, S. Ehlerová, R. Wünsch: Astrophys. and Space Sci. **284**, 873 (2003)
11. B. G. Elmegreen, M. Shadmehri: MNRAS **338**, 231 (2003)
12. R. B. Larson: MNRAS **145**, 271 (1969)
13. G. Tenorio-Tagle, J. Palouš, S. A. Silich, A. G. Medina-Tanco, C. Muñoz-Tuñon: A&A **xxx**, accepted(2003)

CO ($J = 7 \rightarrow 6$) Observations of NGC 253: Exited Molecular Gas in the Nucleus

C.M. Bradford[1,2], G.J. Stacey[1], T. Nikola[1], A.D. Bolatto[3,4], and J.M. Jackson[3]

[1] Cornell Univ., Ithaca, NY,
[2] Caltech, Pasadena, CA – bradford@caltech.edu
[3] Boston Univ., Boston, MA, [4]Univ. of California, Berkeley, CA

1 Abstract

We report observations of the CO $J = 7 \rightarrow 6$ transition toward the starburst nucleus of NGC 253. This is the highest–excitation CO measurement in this source to date, and allows an estimate of the molecular gas excitation conditions. We find that the bulk of the $2\text{–}5 \times 10^7$ $M_\odot$ of molecular gas in the central 180 pc is highly excited. The inferred mass of warm, dense molecular gas is 10–30 times the atomic gas mass as traced through its [C II] and [O I] line emission. This large mass ratio is inconsistent with photodissociation region models where the gas is heated by far-UV starlight. It is also not likely that the gas is heated by shocks in outflows or cloud-cloud collisions given the lack of evidence for substantial heating in fast shocks. We propose two mechanisms for heating the molecular gas: cosmic rays ionization and dissipation of supersonic turbulence, both of which which provide a natural means of uniformly heating the full volume of molecular clouds, and provide a good match to the cooling observed in the CO lines.

2 Mass and Properties of the Molecular Gas

The 809 GHz CO $J = 7 \rightarrow 6$ observations were carried out with our submm Fabry-Perot spectrometer SPIFI [2] on its first run at the JCMT. The data are plotted in Figure 1. We have incorporated all of the published ^{12}CO and ^{13}CO transitions observed toward the NGC 253 nucleus with our data to compare with a Large Velocity Gradient (LVG) escape probability model for the CO excitation. Though there is no unique solution, the run of CO intensity with J places strong constraints on the molecular gas conditions, as illustrated in Figure 2. We have found (see also Bradford et al (2003) [1]):

1) The molecular gas traced with CO is both warm and dense. For temperatures less than 300 K, the density must be greater than $2 - 3 \times 10^4\,\mathrm{cm}^{-3}$. A temperature less than 150 K requires a density greater than $\sim 4 \times 10^4\,\mathrm{cm}^{-3}$, clearly different from the typical Galactic molecular cloud.

2) Large values of the velocity gradient are required to reproduce the high-J lines intensities and the ^{12}CO to ^{13}CO $J{=}3 \rightarrow 2$ ratio. The opacities in the

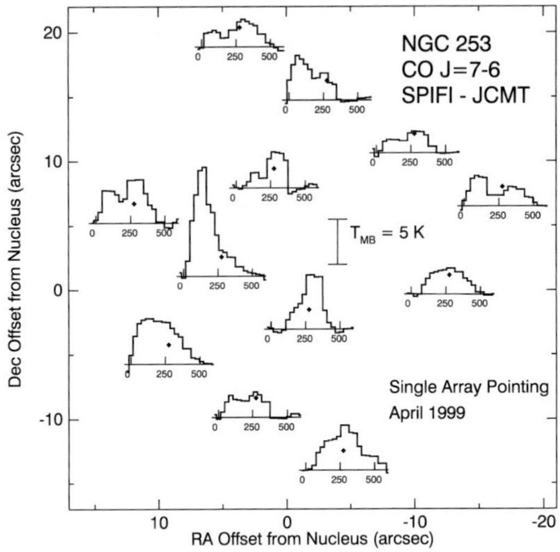

Fig. 1. CO J $= 7 \rightarrow 6$ spectra toward the nucleus of NGC 253. The (0,0)-position is at the IR nucleus at, $R.A = 0^{\rm h}\ 45^{\rm m}\ 06^{\rm s}.0, Dec. = -25^\circ\ 33'40''$ (1950) and the beam size is $11''$.

high-J lines are 2–4. The results require velocity gradients $> 10\ \mathrm{km\,s^{-1}\,pc^{-1}}$, larger than traditionally employed in LVG studies. Even larger velocity gradients ($> 20\ \mathrm{km\,s^{-1}\,pc^{-1}}$) are required if the temperature is constrained to be below 200 K. This is not implausible–the required values are a factor of a few larger than would required required by motion in a self-gravitating cloud at the modeled density, and there are additional sources of confinement in the nuclear ISM: the gravitational potential of the stellar mass density, and the thermal pressure of the hot diffuse ISM P $\sim 7 \times 10^6\ \mathrm{K\,cm^{-3}}$.

3) The CO column density is well-determined: $N_{CO} = 3–7 \times 10^{18}\ \mathrm{cm^{-2}}$, slightly larger but comparable to the larger-beam estimates with only the lower-J lines and their isotopes (Harrision, Henkel & Russel 1999 [4]).

4) Because the J$=7 \rightarrow 6$ transition is not high enough to see the turn-over in CO excitation, the CO analysis does not provide a good upper limit to the gas temperature. We can use the ISO H_2 pure rotational line observations of Rigopoulou et al. (2002) [6] to provide a temperature constraint. The modeled molecular hydrogen column density of $N_{H_2} = N_{CO}/X_{CO} \sim 4–8 \times 10^{22}\ \mathrm{cm^{-2}}$ is consistent with the S(0) data presented by [6] for gas temperatures in the range of 95–120 K. We therefore favor models on the low end of the temperature range allowed by CO analysis, as shown in Figure 2, right side.

5) There is no reason to invoke a large component of cool gas as is found, for example, in the central regions of the Milky Way.

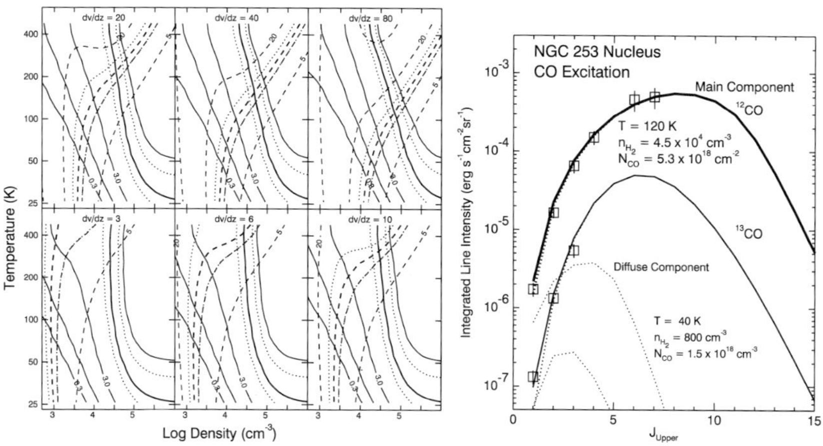

Fig. 2. Left: Contours of two line ratios in the density–temperature plane: ^{12}CO J=7→6 to J=3→2 (solid lines) and ^{12}CO to ^{13}CO J=3→2 (dashed lines). The gas is highly excited and large velocity gradients are required to match the observed ratios (thick curves). Right: The favored model given the temperature constraints of the mid-IR rotational H_2 lines (see also [1]).

6) There is, however, evidence of some UV-shielding with the presence of HCN. Based on the ratio of atomic gas mass [3] to the molecular gas mass, we estimate a clump size of $\sim$2 pc.

3 Heating of the Molecular Gas

The CO intensity vs. J curve measures the luminosity of the molecular gas in the central 15″ of NCG 253. The total flux in the CO lines from the modeled region is 1.5×10^{-11} ergs s^{-1} cm^{-2}. This is 25% of the [C II] 158 μm and [O I] 63 μm intensities from the entire inner 45″ [3], and a substantial 1.4×10^{-4} fraction of the total IR flux of the galaxy. At a distance of 2.5 Mpc, this molecular gas flux corresponds to a luminosity of 2.9×10^6 L$_\odot$, giving a luminosity to mass ratio of $\sim$ 0.1 L$_\odot$/M$_\odot$ for the gas containing CO. Simple arguments outlined in [1] show that traditional heating sources for warm molecular gas in the Galactic Disk – far-UV photons, outflows, and cloud-cloud collisions – are not likely important contributions to this luminosity.

An excellent potential heating source is cosmic rays. The heating rate for cosmic rays is thought to be proportional to their density, which can be scaled from their density in the Milky Way disk according to:

$$\frac{\zeta_{p,\,\mathrm{N253}}}{\zeta_{p\,\mathrm{Gal}}} = \frac{n_{cr\,\mathrm{N253}}}{n_{cr,\,\mathrm{Gal}}} \propto \frac{\psi_{\mathrm{N253}}}{\psi_{\mathrm{Gal}}} \frac{v_{\mathrm{Galdisk}}}{v_{w,\mathrm{N253}}}. \tag{1}$$

Values for the various factors are outlined in [1], the result we derive is $\zeta_{\rm p,\,N253} = 750\,\zeta_{\rm p,\,Gal} = 1.5 - 5.3 \times 10^{-14}{\rm s}^{-1}$ With a ΔQ of 20 eV per ionization, the power deposited per H_2 molecule is then 5–18 $\times 10^{-25}\,{\rm erg\,s^{-1}}$, an excellent match to the measured luminosity per H_2 molecule in all of the CO lines, $7 \times 10^{-25}\,{\rm erg\,s^{-1}}$.

Dissipation of supersonic turbulence is another potential heating source. Mac Low (1999) [5] and references therein have estimated the dissipation rate and shown that it is independent of the presence of a magnetic field. Rearranging their expression, we can write a luminosity to mass ratio as

$$\frac{L}{M} = 0.42 \frac{v_{\rm rms}^3}{\lambda_{\rm d}}, \tag{2}$$

where $v_{\rm rms}$ is the typical turbulent velocity and $\lambda_{\rm d}$ is the turbulent wavelength, or size scale of typical turbulent cells. Our modeled velocity gradient of $40 {\rm km\,s^{-1}\,pc^{-1}}$ provides a relationship between the $\lambda_{\rm d}$ and $v_{\rm rms}$, and we can express Eq. (2) as

$$\frac{L}{M} = 0.42 \frac{dv}{dz} v_{\rm rms}^2 = 7.0 \times 10^{-5} \frac{dv}{dz} v_{\rm rms}^2 \left[\frac{L_\odot}{M_\odot}\right]. \tag{3}$$

Naturally, the size scale of the turbulence ought to be smaller than a single clump, modeled to be 2 pc. If it is 1/10 the clump size, then the corresponding typical turbulent velocity is $0.2 \times 40 \sim 8 km\,s^{-1}$, and we have $L/M \sim 7.5 \times 10^{-2}$ in solar units, comparable to the observed value. We note that this modest velocity is consistent with the lack of evidence for fast shocks in this nucleus. This estimate is subject to many uncertainties, but it demonstrates that with the large velocity gradients observed and inferred, dissipation of turbulence is a good energetic match to the observed molecular luminosity.

If the supersonic turbulence is to persist as a steady-state condition, it must be constantly driven. The high supernova rate in the nuclear starburst is a likely source. Indeed, the luminosity in the molecular gas is a small fraction of the mechanical energy released in the supernovae:

$$L_{\rm mech,\ 253\ nuc.} \sim 0.1 yr^{-1} \times 10^{51} erg \times f_{\rm mech} = f_{\rm mech} \times 8 \times 10^8 {\rm L}_\odot, \tag{4}$$

a factor of 300 larger than the observed molecular luminosity for $f_{\rm mech} = 1$.

References

1. C.M. Bradford et al: Astrophys J. **586**, 891 (2003)
2. C.M. Bradford et al: Applied Optics **41**, 2561 (2002)
3. P. Carral et al: Astrophys. J. **423**, 223 (1994)
4. A. Harrison, C. Henkel,& A. Russell: Mon. Not. Royal Astron. Soc. **303**, 157 (1999)
5. M.M. Mac-Low: Astrophys. J. **524**, 169 (1999)
6. D. Rigopoulou, et al: Astron. & Astrophys. **389**, 374 (2002)

NUclei of GAlaxies (NUGA): The IRAM Survey of Low Luminosity AGN

S. García-Burillo[1], F. Combes[2], A. Eckart[3], L.J. Tacconi[4], L.K. Hunt[5], S. Leon[6], A.J. Baker[4], P. Englmaier[7], F. Boone[8], E. Schinnerer[9], and R. Neri[10]

[1] Observatorio Astronómico Nacional (OAN), Madrid, Spain s.gburillo@oan.es
[2] Observatoire de Paris, LERMA, Paris, France francoise.combes@obspm.fr
[3] Universität zu Köln, Köln, Germany eckart@ph1.uni-koeln.de
[4] MPE, Garching, Germany linda@mpe.mpg.de, ajb@mpe.mpg.de
[5] Instituto di Radioastronomia/CNR,Firenze, Italy hunt@arcetri.astro.it
[6] Instituto de Astrofísica de Andalucía, Granada, Spain stephane@iaa.es
[7] Universität Basel, Binningen, Switzerland ppe@astro.unibas.ch
[8] Bochum University, Bochum, Germany fboone@astro.rurh-uni-bochum.de
[9] NRAO, Socorro, USA eschinne@nrao.edu
[10] IRAM, Grenoble, France neri@iram.fr

The NUGA project (García-Burillo et al [2],[3]) is the first high-resolution ($\sim$0.5″–1″) CO survey of low luminosity AGN including the full sequence of activity types (Seyfert 1–2 and LINERs). Other interferometric surveys had not been focused on the study of AGN thus far (e.g., Helfer et al [4]). NUGA aims to study systematically the different mechanisms for gas fueling of the AGN. There is evidence that the critical spatial scales for AGN feeding are <100 pc. Any causal link between a particular type of instability and the onset of activity must therefore be sought in *secondary* modes embedded in the kpc-scale perturbations (bars or spirals). High-resolution observations are thus paramount to achieve a sharp view of the distribution and kinematics of molecular gas down to the critical scales in AGN.

NUGA has a multi-wavelength approach: the availability of HST and ground-based optical/NIR images for most of the targets allows us to probe the stellar potentials and to study the star formation patterns in NUGA galaxies. Ongoing VLA (HI) and VLBI (radiocontinuum) NUGA-related projects probe the large scale disks and the compact nuclear sources. Altogether, this multi-wavelength information allows us to test the efficiency of AGN feeding mechanisms thanks to a systematic evaluation of the stellar gravitational torques exerted on the gas and, also, with the help of numerical simulations of the gas dynamics for the galaxies in our sample.

1 First Results of NUGA

The first CO NUGA maps made with the Plateau de Bure Interferometer reveal a rich variety of morphologies in the circumnuclear disks of AGN hosts.

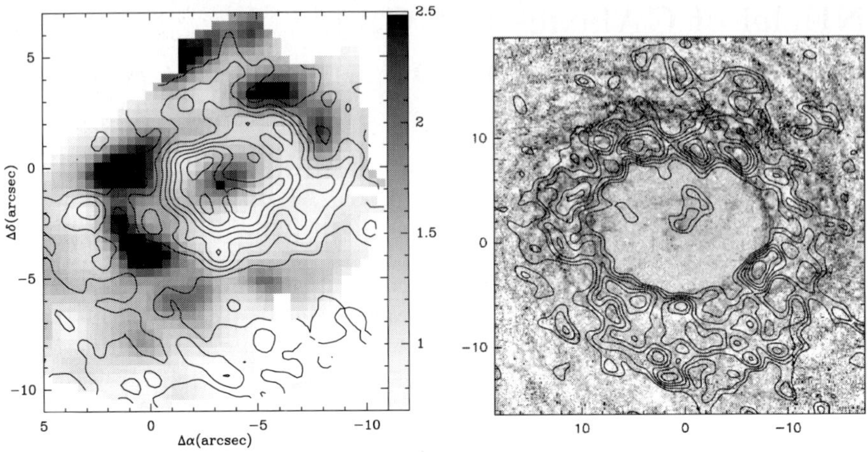

Fig. 1. Left:a) Map of the Toomre Q parameter (grey scale) overlaid on the ^{12}CO(2–1) integrated intensity contours of NGC 4826 from García-Burillo et al [3]. Gas self-gravity seems important in the m=1 modes identified in NGC 4826. **Right:b)** ^{12}CO(1–0) integrated intensity contours in NGC 7217 from Combes et al [1], overlaid on HST V–I color map. Molecular gas forms a contrasted ring. A low-mass non self-gravitating gas unit is situated on the AGN, at the center of the map.

Various gravitational instabilities have been identified at different spatial scales, including m=2, m=1 and m=0 features. Some galaxies host several coexisting perturbations, while others host mainly one type of instability.

The distributions and kinematics of molecular gas in several galaxies reveal the presence of m=1 perturbations, which appear as one-arm spirals and lopsided disks developing from several tens to several hundreds of pc. NUGA observations address for the first time the role of m=1 instabilities in AGN fueling. Paper I (García-Burillo et al. [3]) focused on the counter-rotating LINER NGC 4826: a prototype for m=1 modes. The BCD-maps of NGC 4826 show that the molecular gas is distributed in a 40 pc-radius lopsided disk and two m=1 spirals (Fig. 1). Streaming motions linked with the m=1 perturbations are compatible with the inner modes being *fast* trailing waves. This implies that the lopsided instability may slow down or even temporarily halt gas infall. An independent confirmation that AGN feeding is inhibited at present in NGC 4826 has been obtained from an evaluation of the stellar gravitational torques exerted on the gas.

We have compared the NUGA observations of the ringed LINER NGC7217 with numerical simulations in Paper II (Combes et al. [1]). This galaxy is in direct contrast with what we observe in NGC 4826: in NGC 7217 the molecular gas is distributed in an astonishingly axisymmetric ring (Fig. 1). N-body simulations, including gas and star-formation, monitor the formation and evolution of the ring; results from these simulations led us to conclude that

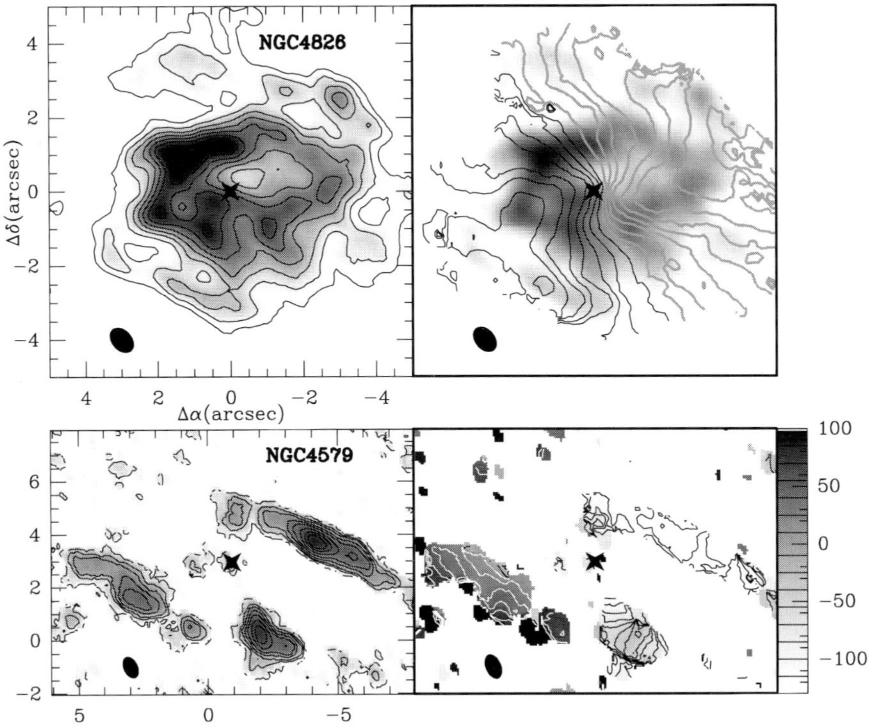

Fig. 2. Top: High-resolution (∼0.5″–1″) new data gives a sharp view of gas kinematics in the lopsided circumnuclear disk of NGC 4826 . The ^{12}CO(2–1) integrated intensity map is shown in *left* panel (2.5-8.5 Jy km s^{-1}) and the overlay with velocity contours in *right* panel (-120 to 135 by 15 km s^{-1}). **Bottom:** The m=1 mode identified in the ^{12}CO(2–1) map (*left*) of NGC 4579 (scale=0.1 to 9 Jy km s^{-1}) is revealed by a strongly asymmetrical velocity gradient detected only on the eastern arm (*right*). Figures taken from García-Burillo et al (2004), in prep.

gas inside the ring, instead of helping AGN feeding, is experiencing an outward flow. Only a GMC-like non self-gravitating unit is linked with the AGN at the center of the map.

2 New Subarcsec Resolution Maps

The first ∼0.5″–1″ resolution maps, incorporating AB-configuration data, have been recently obtained for five galaxies: NGC6951, NGC4826, NGC4579, NGC 2782, and NGC 1961. We briefly report here on new results obtained for two galaxies. The new CO(2–1) data for NGC 4826 (Fig. 2) have provided the first CO-line probe of the kinematics and distribution of a molecular gas disk near an AGN at a spatial resolution of $\leq$ 10 pc. In the new maps, the inner *fast* lopsided instability is fully resolved; a secondary m=2 outer spiral leaves its imprint of *slow* trailing perturbation in the highly perturbed gas kinematics. The combination of m=1+m=2 modes seems to cancel out any gas inflow to the AGN at present. New CO(2–1) A-configuration data of NGC 4579 have fully resolved the asymmetric kinematics of molecular gas (Fig. 2). The new observations have revealed a transverse velocity gradient on the eastern arm which has surprisingly no counterpart on the western arm; this is further evidence for the existence of a lopsided instability.

3 Conclusions

The large variety of circumnuclear disk morphologies found in NUGA galaxies is a challenging result that urges the refinement of current dynamical models. The lack of a clear correlation between activity type and nuclear morphology of the AGN host suggests that the AGN duty cycle in low-L AGN is typically shorter than the time scale needed to build up the described gravitational instabilities at scales of a few ten-to-hundred pc.

Most of the nuclear disk perturbations thus far identified in NUGA targets at scales of ∼a few 100 pc are related to self-gravitating gas instabilities (e.g., in NGC 4826; see Fig. 1). This finding is at odds with initial claims based on HST surveys of AGN spiral hosts which suggested non self-gravitating perturbations are most likely linked with AGN feeding at these scales (Martini & Pogge [5]). However, the new subarcsec NUGA maps have unveiled in some galaxies (NGC 7217 (Combes et al [1]), NGC 6951 (Schinnerer et al 2004, in prep.) and NGC 4579 (García-Burillo et al 2004, in prep.)) the presence of isolated GMC-like gas units (of high Q) closely linked to the AGN source; these low mass gas concentrations, suspicious of being directly responsible of AGN feeding, seem to coexist with the larger-scale self-gravitating gas instabilities. Most remarkably, gas instabilities like the ones analyzed in the two LINERs NGC 4826 and NGC 7217 may have, for quite different reasons, an *inhibiting* role in AGN feeding. Whether this puzzling scenario could be extrapolated to other low-L AGN is being explored for the whole NUGA sample using diagnostic tools similar to the ones developed in Papers I-II.

References

1. Combes, F., García-Burillo, S., Boone, F. et al: A&A, in press (2003)
2. García-Burillo, S., Combes, F., Eckart, A. et al: NUGA. In *Active Galactic Nuclei: from Central Engine to Host Galaxy*, ed by S. Collin, F. Combes, and I. Shlosman (ASP Conference Series, San Francisco 2003) pp 423–426 (2003a)
3. García-Burillo, S., Combes, F., Hunt, L. K. et al: A&A, **407**, 485 (2003b)
4. Helfer, T. T, Thornley, M. D., Regan, M. W. et al: ApJS, **145**, 259 (2003)
5. Martini, P. & Pogge, R. W.: AJ, **118**, 2646 (1999)

The Nearby QSO Host I Zw 1: NIR Probing of Structural Properties and Stellar Populations

J. Scharwächter[1], A. Eckart[1], S. Pfalzner[1], J. Moultaka[1], C. Straubmeier[1], J. Staguhn[2], and E. Schinnerer[3]

[1] I. Physikalisches Institut, Universität zu Köln, Zülpicher Str. 77, 50937 Köln, Germany scharw@ph1.uni-koeln.de
[2] NASA/Goddard Space Flight Center, Greenbelt, MD 20771, USA
[3] National Radio Astronomy Observatory, PO Box 0, Socorro, NM 87801, USA

The likely merger process and the properties of the stellar populations in the I Zw 1 host galaxy are analyzed on the basis of multi-wavelength observations[4] and N-body simulations. The data give a consistent picture of I Zw 1, with properties between those of ultra-luminous infrared galaxies (ULIRGs) and QSOs as displayed by transition objects in the evolutionary sequence of active galaxies.

1 Introduction

The formation and evolution of active galactic nuclei (AGN) and the associated star-formation activity are conditioned by the rate at which gas is supplied from the surrounding host galaxy. Interactions and mergers are identified as one possible mechanism to provide the transport of gas towards the central regions of a galaxy ([1] and references therein). The hypothesis of a merger-driven evolution of active galaxies from ULIRGs to unobscured QSOs is motivated by the spectral energy distributions of template objects showing a sequence of decreasing far-infrared excess and increasing UV brightness [4]. In accordance with the assumption that the evolution is triggered by interactions, the vast majority of ULIRGs show evidence for ongoing merger processes [3]. Merger signs are less clear in the case of objects close to the QSO phase. These transition objects are important sources to confirm the evolutionary sequence. As a nearby representative ($z = 0.0611$, [9]) of a template transition object in the original evolutionary sequence (Fig. 17 in [4]), I Zw 1 is suitable to address the merger properties, the structural composition, and the starburst activity using near-infrared (NIR) and molecular line observations at sub-kpc angular resolution.

[4] With the ISAAC camera at the Very Large Telescope (VLT/UT1) of the European Southern Observatory (ESO), Chile (Paranal), with the interferometer of the Berkeley-Illinois-Maryland Association (BIMA), USA (Hat Creek/California), and with the IRAM Plateau de Bure Interferometer (PdBI), France

2 Observations and Results

2.1 Merger Process

The J-band image of I Zw 1 (Fig. 1), obtained with ISAAC at a median seeing of 0.6" ($\approx$ 0.7 kpc), provides a detailed view of the host galaxy with its two-armed spiral, its likely western companion galaxy, and the northern foreground star. The ongoing merger process with the companion is indicated by the tidal bridge towards the companion and by the tidal tail and the elongated shape of the companion [5]. The parameter space of this interaction is investigated with the help of restricted stellar-dynamical N-body simulations. As suggested by the surface-brightness profile of the companion, the study focuses on companions with less than 10% of the total mass of I Zw 1. On a bound elliptical orbit, even a companion with only 3% of the total mass of I Zw 1 can evoke a two-armed spiral structure similar to the one observed in the J-band image (Fig. 1).

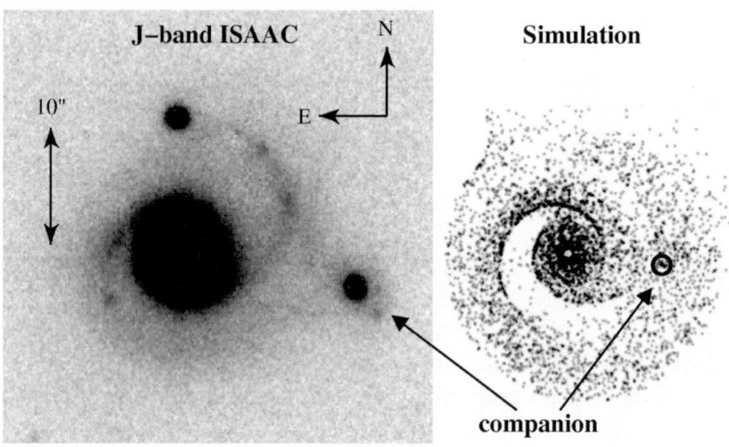

Fig. 1. J-band image of I Zw 1 (left panel), taken with the ISAAC camera at the VLT of ESO, and a snapshot of a restricted N-body simulation for the interaction between I Zw 1 and the companion (right panel). The companion in the simulation has a mass of 3% of the total mass of I Zw 1 and orbits I Zw 1 at an eccentricity of 0.2 and with a pericentric distance of 15.3 kpc.

2.2 Hydrogen Emission and Stellar Absorption Lines

The K-band spectrum of I Zw 1 shown in Fig. 2 was obtained with the 1" long-slit of ISAAC and extracted in a 3" aperture centered on the I Zw 1 nucleus. It is characterized by strong hydrogen emission lines (Pa α, Br γ, Br δ). In agreement with I Zw 1 being classified as a narrow-lined Seyfert 1, the hydrogen lines are narrow with a full width at half maximum of about 1,000 km s^{-1}. The narrow lines with broad wings indicate a mixture of emission from narrow-line regions and from broad-line regions in the surroundings of the AGN. The flux ratio of the Pa α and the Br γ line is about 11, as expected for HII regions. The overall fluxes and line widths are in good agreement with the results from previous K-band spectroscopy of I Zw 1 [8]. There are indications for the ^{12}CO(2-0) bandhead (inset in Fig.2) which is mainly coming from late-type giants and/or super-giants. The preliminary analysis of the depth of this bandhead suggests a stellar contribution of about 20% to the total nuclear NIR flux of I Zw 1. This is in good agreement with the results presented in [8], which are based on the analysis of the ^{12}CO(6-3) overtone bandhead in the H-band spectrum of I Zw 1. By comparing stellar and dynamical masses, it becomes evident that such a high stellar flux contribution can only be produced by a massive starburst (see [8]).

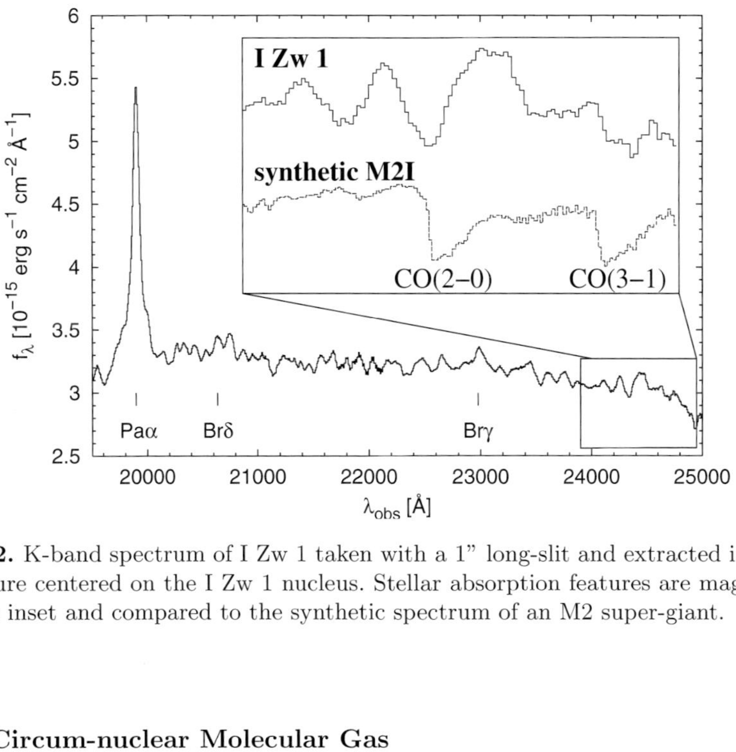

Fig. 2. K-band spectrum of I Zw 1 taken with a 1" long-slit and extracted in a 3" aperture centered on the I Zw 1 nucleus. Stellar absorption features are magnified in the inset and compared to the synthetic spectrum of an M2 super-giant.

2.3 Circum-nuclear Molecular Gas

The circum-nuclear ^{12}CO(1-0) emission in I Zw 1, observed with the BIMA interferometer at an angular resolution of 0.7" ($\approx$ 0.8 kpc), has a ring-like structure with a radius of about 1.5 kpc (see [10]). The comparison of the BIMA ^{12}CO(1-0) measurements with PdBI ^{12}CO(2-1) data indicates that the molecular gas in the disk region is cold or subthermally excited while the gas in the circum-nuclear region is warm and optically thick. The large molecular clouds in the ring-like structure are most probably the sites of the massive starburst activity discussed in [2], [8], and Sect. 2.2.

2.4 Structural Decomposition and Mean Stellar Populations

Mass-to-light (M/L) ratios of the mean stellar populations in the bulge and disk components of the I Zw 1 host are derived by decomposing the J-band surface-brightness profile and the gas rotation curve (see [5], [6], [7]). The disk component and, in particular, the bulge component, show sub-solar M/L ratios. In accordance with the findings of starburst activity, the low M/L ratios indicate young mean stellar populations with an enhanced fraction of hot stars and super-giants.

3 Conclusion

The multi-wavelength data converge to a consistent picture of I Zw 1 as a transition object. A merger-driven evolution is suggested by the J-band image and the N-body simulations showing evidence for a merger process with the companion. The decomposition, together with the large stellar contribution to the nuclear NIR flux, indicates young mean stellar populations and starburst activity which is most likely located in the giant clouds of the circum-nuclear molecular ring.

References

1. J. E. Barnes & L. Hernquist: Dynamics of interacting galaxies. In: *Annual review of astronomy and astrophysics*, vol 30, pp 705–742 (1992)
2. A. Eckart, P. P. van der Werf, R. Hofmann, R., & A. I. Harris: ApJ **424**, 627 (1994)
3. D. B. Sanders: Ap&SS, **266**, 331 (1999)
4. D. B. Sanders, B. T. Soifer, J. H. Elias et al: ApJ, **325**, 74 (1988)
5. J. Scharwächter, A. Eckart, S. Pfalzner et al: A&A **405**, 959 (2003)
6. J. Scharwächter, A. Eckart, S. Pfalzner et al: Astrophysics and Space Science, vol 284, Issue 2, pp 507–510 (2003)
7. J. Scharwächter, A.Eckart & S. Pfalzner: Rotation curve and mean stellar population of the I Zw 1 QSO host. In: *QSO Hosts and Their Environments*, Proceedings of an international workshop, held January 10-12, 2001, in Granada, Spain, ed by Isabel Márquez et al. (Dordrecht: Kluwer Academic / Plenum Publishers) p 289 (2001)
8. E. Schinnerer, A. Eckart & L. J. Tacconi: ApJ **500**, 147 (1998)
9. P. M. Solomon, D. Downes, S. J. E. Radford & J. W. Barrett: ApJ, **478**, 144 (1997)
10. J. Staguhn, E. Schinnerer & A. Eckart, A.: IAU Symposium, **205**, 340 (2001)

The Neutral ISM in Nearby Luminous Compact Blue Galaxies

C.A. Garland[1], D.J. Pisano[2], J.P. Williams[1], R. Guzmán[3], and F.J. Castander[4]

[1] Institute for Astronomy, University of Hawai'i, 2680 Woodlawn Drive, Honolulu, HI 96822 USA garland or jpw @ifa.hawaii.edu
[2] CSIRO Australia Telescope National Facility, P. O. Box 76, Epping NSW 1710, Australia Daniel.Pisano@atnf.csiro.au
[3] Department of Astronomy, University of Florida, 216 Bryant Space Science Center, P.O. Box 112055, Gainesville, FL 32611 USA guzman@astro.ufl.edu
[4] Institut D'Estudis Espacials de Catalunya/CSIC, Edifici Nexus, Gran Capità, 2-4, 08034 Barcelona, Spain fjc@ieec.fcr.es

1 Introduction

Luminous Compact Blue Galaxies (LCBGs) are $\sim$L*, blue, high surface brightness, high metallicity, vigorously starbursting galaxies with an underlying older stellar population [1, 2]. They include a variety of morphological types, such as spiral, polar-ring, interacting/merging and peculiar galaxies. They have optical diameters of a few kpc, but are more luminous and more metal rich than the Blue Compact Dwarf Galaxies widely studied in the nearby Universe, e.g. [3, 4].

When Jangren et al. [5] compared intermediate redshift LCBGs with local normal galaxies, they found that they can be isolated quantitatively on the basis of color, surface brightness, image concentration and asymmetry, with color and surface brightness giving the best leverage for separating LCBGs from normal galaxies. Specifically, LCBGs have B$-$V < 0.6, SBe < 21 mag arcsec^{-2}, and M$_B < -18.5$, assuming H$_0$ = 70 km s^{-1} Mpc^{-1}. LCBGs are quite common at intermediate redshifts, but by z$\sim$0 their number density has decreased by a factor of ten. At z$\sim$1, they have a total star formation rate density equal to that of grand-design spirals at that time, but today they contribute negligibly to the star formation rate density of the Universe [6]. Therefore, LCBGs must undergo dramatic evolution. From studies at intermediate redshift, Koo et al. [7] and Guzmán et al. [1] suggest that some LCBGs may be the progenitors of local low-mass dwarf elliptical galaxies. Alternatively, Phillips et al. [8] and Hammer et al. [9] suggest that others may be more massive disks forming from the center outward to become local L* galaxies.

In order to discriminate between the possible evolutionary scenarios it is essential to measure the dynamical masses of the galaxies: Are they as massive as implied by their high luminosities? It is also essential to measure their gas content for future star formation in order to constrain the amount of fading of their stellar populations. We have undertaken a survey of local LCBGs in H I and CO to address these questions. The H I provides a measure of the dynamical mass, while both H I and CO provide measures of the gas content: H I for long-term star formation and CO for the current burst of star formation.

2 Observations

The current sensitivity of telescopes limits us to detecting CO in LCBGs within ∼70 Mpc. We used the first data release of the Sloan Digital Sky Survey to select our sample of nearby LCBGs, using Jangren et al.'s [5] selection criteria. Out of the ∼million galaxies in the first data release, only ∼100 are LCBGs, and only 16 are within 70 Mpc. To these 16 local LCBGs, we added four more from the literature, for a local LCBG sample of 20 galaxies. We observed this sample in 21 cm H I using the Green Bank Telescope at the National Radio Astronomy Observatory in Green Bank, West Virginia in Winter 2002. The James Clerk Maxwell Telescope on Mauna Kea, Hawai'i was used to observe our sample in CO(J=2−1) in 2002 - 2003. All sources were detected in H I; 13 were detected in CO(J=2−1).

3 Results

3.1 Dynamical Masses

We find that local LCBGs span a wide range of dynamical masses, from 4×10^9 to 1×10^{11} $M_\odot$ (measured within R_{25}). Figure 1 compares the dynamical masses of local LCBGs with intermediate redshift LCBGs and local spiral galaxies of all Hubble types. Many local LCBGs are ∼ten times less massive than local galaxies of similar luminosities, as found for LCBGs at intermediate redshifts [8]. However, others are as massive as local galaxies of similar luminosities.

3.2 Gas Depletion Time Scales

We find our 13 LCBGs detected in CO(J=2−1) have molecular gas masses ranging from 5×10^7 to 2×10^9 $M_\odot$ (assuming a Galactic CO-to-H_2 conversion factor of 1.8×10^{20} cm^{-2} K^{-1} km^{-1} s [10]). Note that these are most likely underestimates of the molecular gas masses since we are using CO(J=2−1). The fraction of molecular to atomic gas mass is small, ranging from 0.03 to 0.3, similar to local late-type spiral galaxies [11].

We estimated star formation rates from available IRAS data, using 60 and 100 μm fluxes as outlined in Kewley et al. [12]. The star formation rates for local LCBGs range from ∼1 to 15 $M_\odot$ year^{-1}. For comparison, local

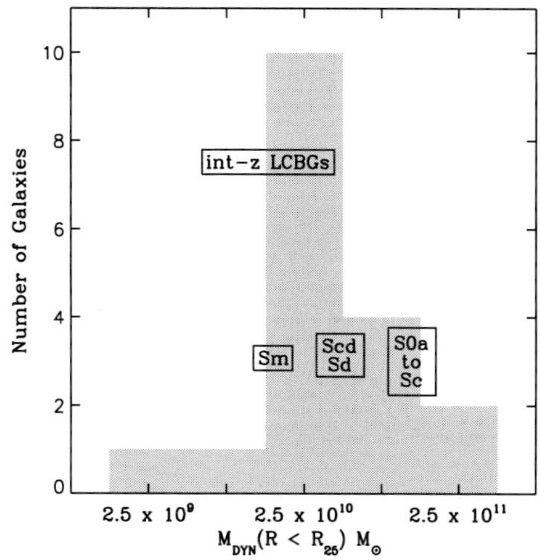

Fig. 1. Dynamical masses (within R_{25}) for local LCBGs, as measured from H I observations. For comparison, the ranges of dynamical masses for intermediate redshift LCBGs [8], and local spiral galaxies [11] are indicated. Note that "Sm" indicates Magellanic or low-luminosity spirals.

spirals of all types have star formation rates $\sim$2 $M_\odot$ year^{-1} [11], so these LCBGs do not have unusually high star formation rates. However, they do have very high *specific* star formation rates–the ratio of star formation rate to dynamical mass (within R_e). As seen in Figure 2, local LCBGs have specific star formation rates from $\sim$3 to 40 times those of local normal spirals [11]. The specific star formation rates of local LCBGs are in the same range as local H II (starbursting) galaxies [13].

We find that the molecular gas in local LCBGs is depleted quite quickly, in 30 to 200 million years. The molecular plus atomic gas is depleted in 30 million to 10 billion years; however, $\sim$80% of the local LCBGs deplete their gas in less than 5 billion years. Therefore, most LCBGs will not be able to sustain their current rates of star formation and will eventually fade.

4 Conclusions

Both in dynamical masses and gas depletion time scales, we find that local LCBGs have a wide range of characteristics and are unlikely to evolve into one galaxy class. They have dynamical masses consistent with a range of galaxy types, such as dwarf ellipticals, Magellanic (low-luminosity) spirals and normal spirals. The majority have atomic plus molecular gas depletion time scales less than five billion years; such galaxies may have masses, sizes

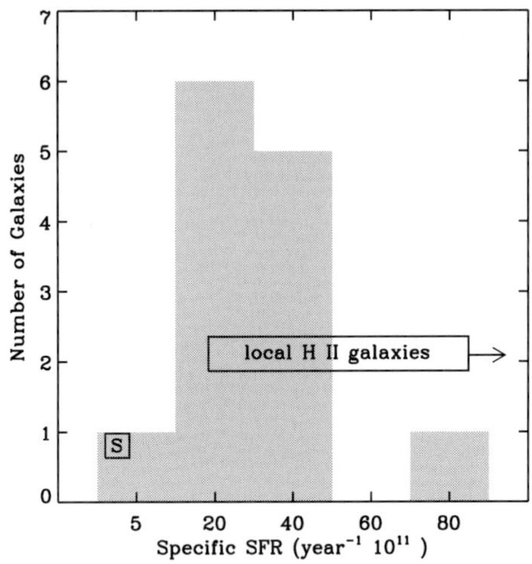

Fig. 2. The *specific* star formation rate (ratio of star formation rate to dynamical mass within Re) for the local sample of LCBGs. Their specific star formation rates are much higher than all types of local spirals (S); they are similar to local H II galaxies.

and faded luminosities and surface brightnesses consistent with the brightest local dwarf ellipticals. A few local LCBGs have longer gas depletion time scales, approaching a Hubble time. These may fade very little, becoming spirals or Magellanic irregulars.

Acknowledgements Support for this work was provided by the NSF through award GSSP02-0001 from the NRAO. Support for conference attendance was provided by the AAS and NSF in the form of an International Travel Grant.

References

1. R. Guzmán, D. C. Koo, S. M. Faber et al: ApJ **460**, L5 (1996)
2. R. Guzmán, A. Jangren, D. C. Koo et al: ApJ **495**, L13 (1998)
3. T. X. Thuan & G. E. Martin: ApJ **247**, 823 (1981)
4. C. L. Taylor, E. Brinks, R. M. Grashuis et al: ApJS **99**, 427 (1995)
5. A. Jangren, M. A. Bershady, C. Conselice et al: AJ submitted (2003)
6. R. O. Marzke, J. J. Geller, J. P. Huchra et al: AJ **108**, 437 (1994)
7. D. C. Koo, M. A. Bershady, G. Wirth et al: ApJ **427**, L9 (1994)
8. A. C. Phillips, R. Guzmán, J. Gallego et al: ApJ **489**, 543 (1997)
9. F. Hammer, N. Gruel, T. Thuan et al: ApJ **550**, 570 (2001)
10. T. M. Dame, D. Hartmann & P. Thaddeus: ApJ **547**, 792 (2001)
11. M. S. Roberts & M. P. Haynes: ARA&A **32**, 115 (1994)
12. L. J. Kewley, M. J. Geller, R. A. Jansen et al: AJ **124**, 3135 (2002)
13. R. Guzmán, J. Gallego, D. C. Koo et al: ApJ, **489**, 559 (1997)

Giant Molecular Clouds and On-Going Star Formation in the LMC and SMC

N. Mizuno, T. Onishi, Y. Fukui, and the NANTEN team

Department of Astrophysics Nagoya University, Japan
norikazu@a.phys.nagoya-u.ac.jp

After completion of the first CO survey of the Large Magellanic Cloud with NANTEN, we are continuing more sensitive CO survey of the both Large and Small Magellanic Clouds. These will help us to obtain a new insight into formation of GMCs and clusters. The new NANTEN results have reinforced our conclusion that the youngest stellar clusters and H II regions show good spatial correlation with GMCs, and that SNRs are only poorly correlated with CO emission. This support the view that cluster formation is very rapid, in a few Myr, and that the GMCs are quickly dissipated by the UV photons from the clusters. In addition, the novel aspects of this study include the identification of starless GMCs, apparently very rare in the Galaxy.

1 Introductions

There is an increasing interest on the formation of rich stellar clusters including the globulars in the last few years. This has been strongly motivated by the magnificent images of young stellar clusters in Mergers, possible counterparts of the Galactic globulars, taken with the Hubble Space Telescope. Our ultimate goal should be comprehensive understanding of the cluster formation mechanism and this is hardly accomplished without the knowledge on the proto-cluster molecular clouds, cold gaseous precursors of the clusters. It is of particular interest to observe the dense molecular gas in the Magellanic Clouds unique to the southern sky, where populous clusters, mini-globulars, are still being formed. These galaxies are located at unrivaled closeness to us of around 50–60 kpc and their metallicities are significantly less than that of the Galaxy. These unique properties of the Clouds make them an ideal target to study formation of rich stellar clusters, and even individual star formation of under low metallicity environments.

Our group has been intensively working on a complete survey for giant molecular clouds in the LMC and SMC by using a 4-m mm and sub-mm telescope NANTEN located in Las Campanas, Chile. This telescope equipped with the most sensitive receiver at 115 GHz, the frequency of CO molecules, has provided the first complete images of the giant molecular clouds of the two galaxies at a spatial resolution of 40 pc (see the papers in the two NANTEN special issues of PASJapan, Vol.51, No. 6, 1999 and Vol.53, No. 6, 2001). The major results of the survey are the discovery of nearly 300 giant molecular clouds (GMCs) in the LMC and 20 of them in the SMC.

2 New NANTEN Results

The new survey was started in 1999 April and has covered $\sim$ 80% of the 36 square-degrees field (Fukui et al. 2001). It is still on going to fully cover the LMC. The survey spends about 4–5 min per a point, resulting an improvement in noise fluctuations by a factor of 3 compared to the first one. The typical rms noise level is 0.16 K in a 0.1 km s^{-1} velocity resolution. The number of the clouds detected has been $\sim$ 270 by the new survey for a present spatial coverage of $\sim$ 80%. We estimate that the new survey is complete down to $1{\times}10^5 M_\odot$, a factor of 3 lower than that of the first one.

Figure 1 shows the map of the new survey for the LMC. This clearly illustrates that the new survey detects weaker CO emission in the peripherals of GMCs as well as smaller clouds not seen in the first survey. This improvement generally supports and reinforces the conclusions reached by Fukui et al. 1999, like the clumpiness of the molecular clouds.

The cloud mass spectrum from 168 GMC is well fitted by a power law index of -1.9 ± 0.1, somewhat steeper than that of the Galactic GMCs, -1.5 (e.g., Solomon et al. 1987). A possible interpretation for this is that the dissipation of GMCs in the LMC is more enhanced due to the stronger UV radiation fields by the high mass stars, as well as due to low interstellar extinction in the LMC.

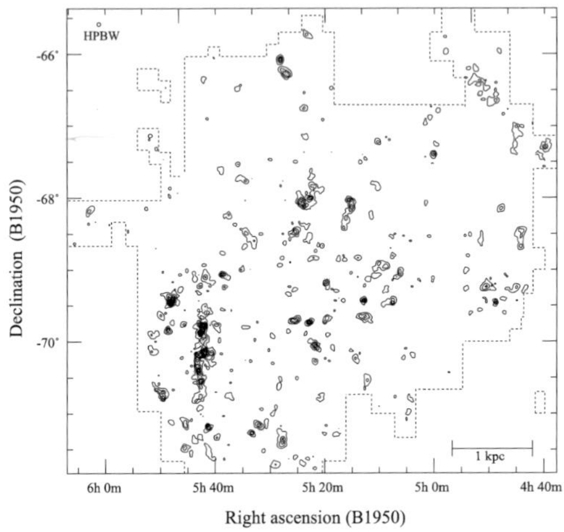

Fig. 1. CO(J = 1–0) velocity-integrated intensity map of the LMC. Contour levels begin at 1.2 K km s^{-1} (= 3 σ noise level), and increase by steps of 2.4 K km s^{-1}. The broken lines indicate the observed area.

Fukui et al. 1999 (also see Yamaguchi et al. 2001a) first demonstrate that the GMCs in the LMC are the sites of populous cluster formation by showing that about one third of the 55 CO clouds are positionally well correlated with the youngest stellar clusters optically identified, SWB0. The age of SWB0 is estimated to be less than 10 Myr (Bica et al. 1996). Even older clusters, SWB1 or later, show little such correlation. This is interpreted as indication of rapid dissipation of GMCs in the LMC. The fact that SNRs either show little correlation with GMCs is also consistent with this interpretation. The mass range of the GMCs is similar between the LMC and the Galaxy. The mass spectrum may be even steeper in the LMC than in the Galaxy. This makes us to abandon a simple idea that a rich cluster forms in a large GMC. The process of mass contraction should be different in forming populous clusters from forming open clusters.

Fukui et al. 1999 note that some of the GMCs in the LMC are not associated with any H II regions or young stellar clusters. This conclusion is not affected by the new data even after the weaker, more extended outer skirts of GMCs are detected. If they are not forming stars at all, the GMCs in the LMC may be considerably different from the Galactic GMCs, since all the Galactic GMCs are believed to be forming massive stars except for one, the Maddalena's cloud which is forming low mass stars only with no H II regions (Maddalena & Thaddeus 1985). This point perhaps needs further detailed quantitative analyses. We should test how inactive these GMCs are in star formation by taking into account the detection limits in various signs of star/cluster formation, like for infrared emission for embedded sources, between the Galactic GMCs and those in the LMC. In this respect, some of the apparently starless GMCs like those in Supergiant shells and in the CO Arc should be of particular interest in our continuing future works.

We should also pay attention to the supergiant shells (SGSs) and their relation to cluster and cluster formation. SGSs, whose diameter is ~ 1000 pc have been considered as formed by the combined action of stellar winds and supernovae due to the OB stars inside the shell (e.g., Tenerio-Tagle & Bodenheimer 1988). As they grow, they sweep up and accumulate interstellar matter by the pressure of the hot gas. Because of their large volume, they may play an important role in formation of molecular clouds and stellar clusters. A statistical study of the effects of SGSs on the formation of stars and molecular clouds was made based on the new sensitive data of the LMC (Yamaguchi et al. 2001b). The surface number and mass densities of the CO clouds are higher by a factor of 1.5–2 at the edge of the SGSs than elsewhere. The number of clusters is increased by a factor of 2 on the side facing to the center of the SGSs with respect to the CO peaks compared with on the side far from the center. These results strongly suggest that cluster formation is triggered by dynamical effects of the SGSs.

The results of a new CO survey of the Small Magellanic Cloud are presented in Mizuno et al. (2001b).

3 Follow-Up Observations for NANTEN

To reveal the properties (density, temperature, size) of dense gas in the GMCs, NANTEN GMCs are mapped into more detail with SEST at several pc resolution, under a collaboration with Onsala Observatory, and their cloud properties will be compared in detail with those of the well known galactic GMCs in order to highlight the difference in the physical characteristics.

Recently, we have also started CI observations with ASTE (Atacama Submillimeter Telescope Experiment, see Kohno et al. 2003; this proceedings). This is the first high resolution map of CI toward the Magellanic Clouds (figure 2).

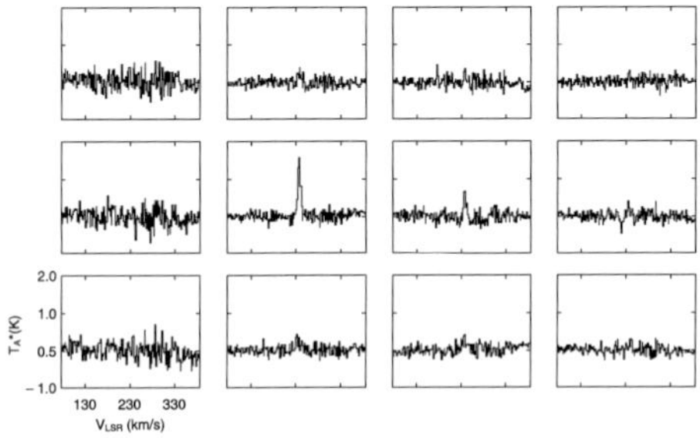

Fig. 2. CI($^3P_1 -^3 P_0$) spectra from the N159 region in the LMC. The grid spacing is 30".

References

1. E. Bica et al. 1996, APJS, **102**, 57 (1996)
2. Y. Fukui, N. Mizuno, et al. 1999, PASJ, **51**, 745 (1999)
3. Y. Fukui, N. Mizuno, et al. PASJ, **53**, L41 (2001)
4. R. J. Maddalena & P. Thaddeus ApJ, **294**, 231 (1985)
5. N. Mizuno, et al. PASJ, **53**, 971 (2001a)
6. N. Mizuno, et al. PASJ, **53**, L45 (2001b)
7. P. M. Solomon etg al. ApJ, **319**, 730 (1987)
8. G. Tenorio-Tagle & P. Bodenheimer ARA&A, **26**, 145 (1988)
9. R. Yamaguchi, N. Mizuno et al. PASJ, **53**, 985 (2001a)
10. R. Yamaguchi, N. Mizuno et al. PASJ, **53**, 959 (2001b)

Molecular Gas in Nearby Dwarf Galaxies: Single Dish and Interferometric Results

A.D. Bolatto[1], A. Leroy[1], J.D. Simon[1], L. Blitz[1], and F. Walter[2]

[1] Department of Astronomy, University of California at Berkeley
bolatto@astro.berkeley.edu
[2] National Radio Astronomy Observatory

1 Introduction: The MIDGET Survey

During the past two years we have undertaken a systematic millimeter–wave survey of molecular gas in a sample of 150 nearby, northern dwarf galaxies with IRAS emission. We have called this survey MIDGET (Millimeter Interferometry of Dwarf Galaxies). MIDGET has two parts: a single–dish search for CO done at the UASO Kitt Peak 12m telescope (Leroy et al., in preparation), and an interferometric follow-up of the galaxies with strong CO emission carried out at the Berkeley–Illinois–Maryland Array (BIMA). The single–dish survey targeted the centers of northern galaxies that are nearby ($V_{LSR} \leq 1000$ km s^{-1}), of small mass (HI linewidths $W_{20} \leq 200$ km s^{-1}), compact in size, and detected by IRAS. The single–dish observations were very successful, finding 41 new CO emitters and more than doubling the number of previously known CO sources within the defined sample. The typical 1σ sensitivity attained was ~ 0.2 K km s^{-1}. Carbon monoxide emission was detected in galaxies with $12 + \log([O/H]) \sim 7.9 - 8.5$, a metallicity regime similar to that of the Magellanic Clouds. We are using these data to address several questions related to molecular cloud and star formation in these objects. This paper concentrates on the issue of how CO traces molecular gas as a function of heavy element abundance Z.

2 Does the X_{CO} Factor Depend Strongly on Metallicity?

Measuring the amount of molecular hydrogen present using CO observations requires assuming a CO–to–H_2 conversion factor $X_{CO} = N(H_2)/\int I(CO)\,dv$ (e.g., Sanders, Solomon, & Scoville 1984). This factor will be a product of abundance, excitation, and cloud structure averaged over a large area. The theoretical expectation is that the X_{CO} factor will depend on the local properties of the interstellar medium (ISM) such as volume density, temperature, and metallicity. The dependence of X_{CO} on metallicity Z remains a point of contention. Some authors find X_{CO} strongly increasing for decreasing Z (e.g., Wilson 1995; Israel 1997), while other, recent work finds no apparent trend (e.g., Rosolowsky et al. 2003).

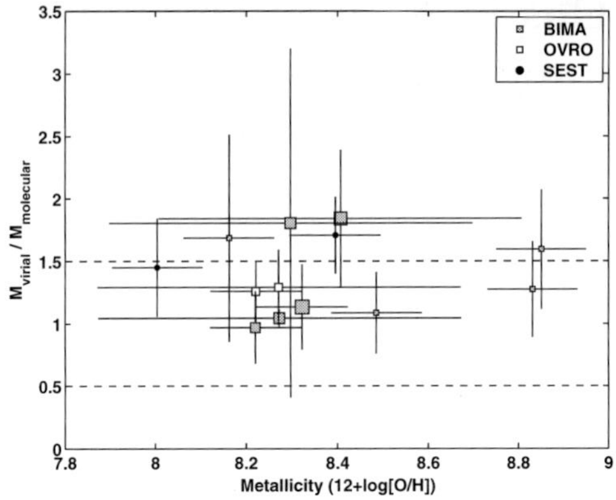

Fig. 1. Virial CO–to–H_2 conversion factor as a function of metallicity, normalized to the Galactic value. Each point represents the average of the ratio of virial to molecular mass for all the GMCs analyzed in one galaxy, with the molecular mass computed using the integrated CO intensity and the Galactic X_{CO} factor. Symbol color identifies the telescopes used for the measurements, while size is proportional to the spatial resolution attained in the observations. The data include clouds in M 33, the SMC (N83/N84), the LMC (N159/N160), IC 10, NGC 2976, NGC 3077, NGC 4214, NGC 4449, and NGC 4605.

We have used the interferometric data to measure the properties of individual molecular clouds (size, velocity dispersion, CO luminosity) in dwarf galaxies where we can accurately derive virial masses. For these measurements we have developed a robust algorithm called `CLOUDALYZE`, which uses moments of the intensity distribution computed over connected regions of the datacube as a function of a threshold intensity (Bolatto et al. 2003). These moments, adequately corrected to account for angular resolution and signal–to–noise effects, are then used to compute the size and velocity dispersion of a cloud, thus measuring its virial mass. In designing this experiment we have explicitly tried to avoid problems common in some extragalactic X_{CO} determinations found in the literature. In particular, we deconvolved the beam from our cloud size estimates and we avoided unresolved clouds. Another potentially complicating effect is cloud blending in velocity space. In poor signal–to-noise data, emission from two or more clouds may be confused and assigned to only one entity. Because the velocity dispersion is squared to compute the virial mass, velocity cloud blending can introduce large errors that overestimate virial masses.

Figure 1 shows the ratio of virial mass to molecular mass traced by CO using the Galactic conversion factor ($X_{CO} = 2\times10^{20}$ cm^{-1}/K km s^{-1}) found by using `CLOUDALYZE` on several extragalactic GMCs. These include observations performed by BIMA and OVRO, as well as SEST observations of regions in the LMC and SMC. This plot shows that resolved extragalactic GMCs have an approximately Galactic CO–to–H_2 conversion factor regardless of metallicity. The offset from $M_{virial}/M_{mol} = 1$ is probably due to the algorithm, which has not been scaled to produce $M_{virial} = M_{mol}$ in Galactic GMCs (e.g., $M_{virial}/M_{mol} = 1.5$ in the Rosette GMC).

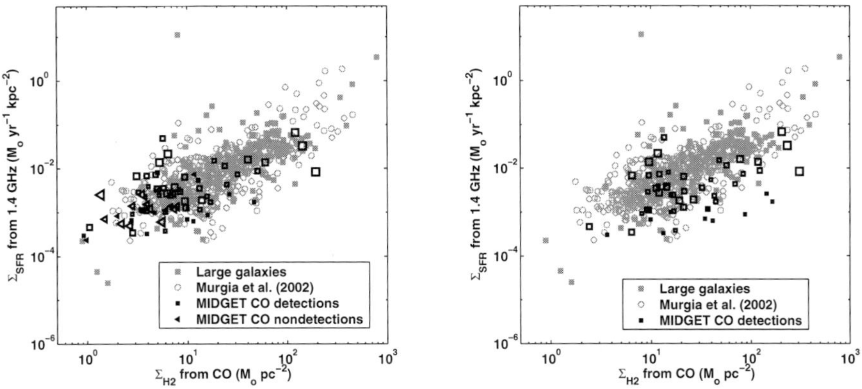

Fig. 2. (a) Surface density of star formation rate versus surface density of molecular gas, for the central pointings of galaxies in the FCRAO catalog (gray symbols), the sample from Murgia et al. (2002) which includes all FCRAO pointings (gray circles), and the MIDGET galaxies (black symbols). Triangles indicate CO upper limits. The SFR was obtained from the 1.4 GHz radio continuum NVSS images, which have a resolution matched to that of the CO observations. The column density of molecular gas was obtained from the CO using the Galactic conversion factor. **(b)** Same plot as (a), but the MIDGET points now show the effects of applying a metallicity correction to X_{CO} such as the one by Wilson (1995). The symbol sizes for the MIDGET points are proportional to metallicity.

Is there any evidence for a constant CO–to–H_2 conversion factor apart from the virial arguments made above? Using NVSS radio continuum images and following Murgia et al. (2002) we can readily obtain the star formation rate (SFR) in the central regions of these galaxies with a beam size matched to the single–dish CO measurement. Figure 2a shows the surface density of SFR against that of molecular gas. In this diagram, dwarf galaxies occupy the same locus as more massive galaxies. If the Galactic X_{CO} factor used to obtain Σ_{H_2} were systematically underestimating the amount of H_2 available for star formation in dwarf galaxies then their points would be displaced to the left of those of bigger galaxies and they are not. In fact, attempts to correct X_{CO} for metallicity effects (using the metallicities inferred from M_B; Richer & McCall 1995) such as that in Figure 2b clearly destroy the agreement and appear to overestimate the surface density of molecular gas for a given star formation rate.

It is possible that the radio continuum emission from dwarf galaxies might be less bright for a given star formation rate — perhaps as a result of lower magnetic field strength in these galaxies. This effect could collude with a metallicity-dependent X_{CO} factor to keep dwarf galaxies on the same locus as the large galaxies despite the application of an incorrect X_{CO} factor. We have looked into this possibility. We find that small galaxies with $V_{rot} \sim 70 - 110$ km s^{-1} fall in the same radio continuum to far infrared correlation as the larger galaxies: in this regard the radio continuum seems to be as good a tracer of SFR as the far infrared for these objects. For this alternative explanation to work, there needs to be almost perfect cancellation between the increase in X_{CO} and the reduction in the SFR measured by the radio continuum.

3 Conclusions

Using virial mass arguments, we showed that the X_{CO} factor measured in resolved extragalactic GMCs is not strongly dependent on metallicity. Using the correlation between radio continuum and molecular gas, we argued that a Galactic CO–to–H_2 conversion factor appears to explain the observed SFR in small galaxies as well as in large galaxies. Thus, at least for the molecular gas relevant to star formation processes, X_{CO} in these metal–poor environments appears to be approximately Galactic regardless of metallicity.

A few cautionary statements are in order. First, and perhaps more importantly, this result does not mean that $X_{CO} = 1$ in every conceivable environment. We are only able to obtain virial mass estimates for very special objects, the brightest extragalactic GMCs. This result thus may have little bearing on the CO–to–H_2 conversion factor over an entire galaxy, except perhaps that molecular gas with high X_{CO} does not appear to be important for star formation. Second, the physics relating the FIR and radio continuum emission to the star formation rate are poorly understood. Therefore it is possible that other effects come into play that cancel an increase in X_{CO}. This cancellation, however, needs to be almost perfect. Finally, to properly make quantitative arguments such as those in Figure 2b it is desirable to have direct metallicity measurements on a large sample of dwarf galaxies where CO is observed. We have used a rather indirect estimator of metallicity with a large intrinsic scatter (M_B), because no other data are available.

References

1. Bolatto, A. D., Leroy, A., Israel, F. P., & Jackson, J. M. 2003, ApJ, 595, 167
2. Israel, F. P. 1997, A&A, 328, 471
3. Murgia, M., Crapsi, A., Moscadelli, L., & Gregorini, L. 2002, A&A, 385, 412
4. Richer, M. G. & McCall, M. L. 1995, ApJ, 445, 642
5. Rosolowsky, E., Plambeck, D., Engargiola, G., & Blitz, L. 2003, ApJ, in press
6. Sanders, D. B., Solomon, P. M., & Scoville, N. Z. 1984, ApJ, 276, 182
7. Wilson, C. D. 1995, ApJ, 448, L97.

Molecular Gas in Dwarf Galaxies

F. Walter

NRAO, P.O. Box 0, Socorro, NM 87801 fwalter@nrao.edu

1 Introduction

Dwarf galaxies are now widely acknowledged to be the building blocks of today's massive galaxies. In a sense, today's dwarf galaxies can be considered to be the survivors of an initially much richer galaxy population – in the local universe, dwarf galaxies remain the most numerous type of galaxies. With metallicities of around one–tenth solar, dwarf galaxies are believed to be the closest counterparts to the early star–forming fragments. In–depth studies of nearby dwarf galaxies are therefore key to understanding the conditions in and evolution of galaxies in the early universe. Studies focusing on the molecular gas phase of the ISM of these low–metallicity objects are particularly important since it is the molecular gas out of which stars form.

Studies of the molecular gas phase in dwarf galaxies have proven to be a challenge since dwarf galaxies are much smaller and fainter than typical massive spiral galaxies. Molecular gas in galaxies is typically traced by the carbon–monoxide (CO) molecule which emits radiation in its ground state at 115 GHz (2.7 mm). Taylor et al. (1998) noted that no dwarf galaxy with metallicities less than about one–tenth solar has been detected in CO emission. Even today this finding still holds, despite quite a number of deep searches. This is especially surprising since quite a number of low–metallicity dwarfs show violent ongoing star formation which, according to our current understanding, requires the presence of molecular gas. One suggestion in the literature to resolve this puzzle is that the conversion factor X_{CO} to calculate H_2 column densities from CO luminosities increases as a function of decreasing metallicity; i.e. that the same amount of H_2 emits 'less' CO luminosity in a low–metallicity environment. This argument was invoked by a number of authors (e.g., Wilson 1995, Israel 1997, Arimoto et al. 1996) in studies which derived the conversion factor using a number of different techniques.

2 The Crux with X_{CO}

One way to derive the conversion factor X_{CO} in nearby galaxies is to resolve individual molecular clouds. Once the size and the line–width is determined for an individual cloud, and assuming that the clouds are in virial equilibrium, the total molecular gas mass can be derived using the virial theorem ($M_{H2} \sim 210 \times r \times \Delta v^2$). This virial mass can then directly be compared to the H_2 mass derived from the CO luminosity by assuming a Galactic conversion factor ($X_{CO} = 2.3 \times 10^{20} cm^{-2} (K\,km\,s^{-1})^{-1}$; e.g., Strong et al. 1988). The ratio of these two numbers give X_{CO} in the cloud under study (in units of the Galactic X_{CO}).

One problem with this technique is that individual clouds (with typical sizes of a few to tens of parsecs) need to be resolved both spatially and in velocity space. In extragalactic systems such as the dwarf galaxies discussed here, achieving arcsecond resolution is absolutely critical – this can only be achieved using millimeter interferometers. The results presented here were all obtained with the Caltech OVRO millimeter interferometer, located near Bishop, CA in the Owens Valley. From 1999–2002 about ten dwarf galaxies were observed at high velocity resolution and sensitivity at OVRO. Another dwarf galaxy CO survey was undertaken at the BIMA interferometer by Bolatto et al. (see the contribution by Bolatto et al. in these proceedings). Some of the highlights of the OVRO survey are summarized in the following.

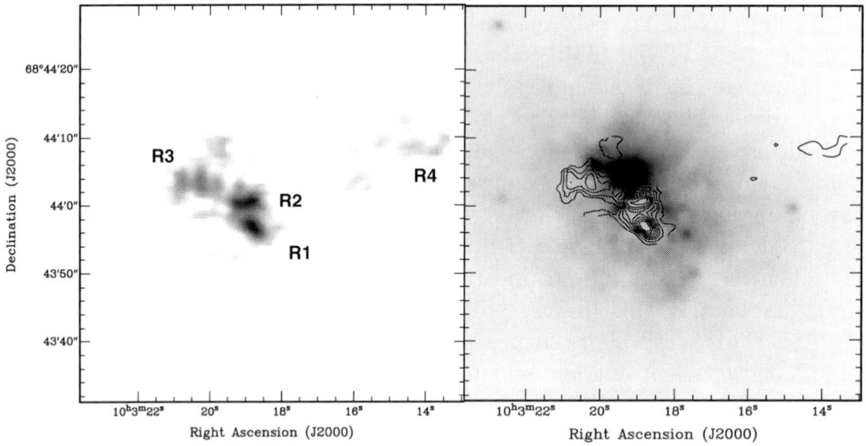

Fig. 1. *Left:* CO emission in NGC 3077 as seen by OVRO at 2.4″ × 1.9″ resolution (~ 30 pc). R1–R4 label four major molecular complexes in NGC 3077 which actually break up into 16 clouds (see Walter et al. 2002a for details). *Right:* The CO contours are plotted over an optical B band image. CO emission is mostly restricted to the center and is associated with a dust absorption feature seen in the optical.

3 NGC 3077

Together with the prototypical starburst galaxy M 82 and the spiral galaxy M 81, NGC 3077 forms the famous M 81 triplet of interacting galaxies at a distance of ~ 3.7 Mpc (e.g., Yun et al. 1994). Our OVRO observations of CO emission in this galaxy reveal the presence of 16 molecular complexes near the center of NGC 3077 (total mass: $\sim 1.6 \times 10^6\,M_\odot$, Walter et al. 2002a, see also Meier et al. 2001). An integrated CO intensity map is shown in Fig. 1 (left), indicating 4 global regions of molecular gas emission – the CO contours are overlaid on an optical image in Fig. 1 (right). A virial mass analysis of the individual complexes yields a *lower* X_{CO} factor in NGC 3077 than the Galactic value – this is a surprising result for a dwarf galaxy. The lower conversion factor can most likely be explained by extreme excitation conditions of the molecular gas. Low X_{CO} factors have indeed been found in other starburst environments such as, e.g., M 82 (Weiss et al. 1999, Walter et al. 2002b). Including high–resolution HI data in the analysis, the total gas content in the center of NGC 3077 is displaced from the stellar component of NGC 3077 – this implies that not only the gas at large galactocentric radii is affected by the interaction within the triplet but also the center. We speculate that the starburst activity of NGC 3077 was triggered by this redistribution of gas in the center. Globally, CO emission in NGC 3077 is found where one would expect the molecular gas to be present, i.e. near the centre. We will see below that this is not necessarily the case for other irregular dwarf galaxies.

4 NGC 4214

NGC 4214 is a nearby dwarf galaxy currently experiencing high levels of massive star formation (D$\sim$ 4.1 Mpc, metallicity: about 1/4 solar, Kobulnicky & Skillman 1996). Using OVRO, we detected three regions of molecular emission in the northwest, southeast, and center of NGC 4214 that are in very different and distinct evolutionary stages (total molecular mass: $\sim 5.1 \times 10^6$ $M_\odot$). These differences are apparent most dramatically when the CO morphologies are compared to HST imaging: massive star formation has not started yet in the northwestern region; the well-known starburst in the center is the most evolved, while star formation in the southeastern complex has begun more recently (Fig. 2, Walter et al. 2001). No clear correlation was found between the peaks of HI, CO, and the sites of ongoing star formation. This emphasizes the irregular nature of dwarf galaxies. In total, we cataloged 14 molecular clouds in NGC 4214. A virial mass analysis of the individual clouds show that the conversion factor in NGC 4214 is compatible with the Galactic X_{CO} factor. Our observations show that CO emission is present in regions where no star formation is going on (and where future star formation may occur) and that the onset of star formation can potentially completely disrupt the parental molecular cloud.

5 IC 10

IC 10 is one of the gas-rich dwarf irregular galaxies that inhabit the outskirts of the Local Group of galaxies. It is at a distance of only 0.82 Mpc, and

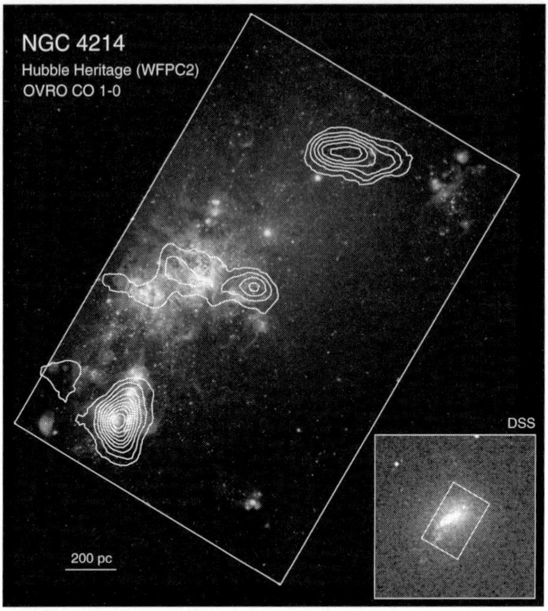

Fig. 2. NGC 4214: CO emission in contours are superposed on a multiwavelength optical image created by the Hubble Heritage Team from HST WFPC2 data (see Walter et al. 2001 for details).

characteristic of dwarf irregulars, it has a low metallicity (12 + log (O/H) = 8.16, Lequeux et al. 1979), making it the lowest–metallicity galaxy of the dwarf galaxies presented in this contribution. 14 pointings with OVRO were needed to cover most of the extent of the galaxy. Figure 3 shows the CO contours overlaid on an Hα image of IC 10. No clear correlation between the sites of ongoing star formation and molecular gas is found. Strikingly, molecular gas is found at large galactocentric radii, far outside the main optical body of IC 10. This implies that stars can form and enrich the ISM at these large radii. We note that the energy input of the most massive stars that form at large radii may have created the impressive HI structures seen in the outskirts of IC 10 (Wilcots & Miller 1998). Due to the proximity of IC 10 we reach unprecedented resolutions of the molecular gas phase of $\sim$ 12 pc and 0.6 km s^{-1}. We note that the velocity resolution is particularly important because of the Δv^2 dependence of the virial mass. In total we identified 18 well resolved (both spatially and in velocity space) CO clouds and derive an average conversion factor of X_{CO} = 1.1 $\pm$ 0.4 times the Galactic factor for IC 10. This result was unexpected given earlier work by Wilson et al. (1995) but can be explained by the higher sensitivity and resolution of our new data. With the new conversion factor the total molecular gas mass in IC 10 is only 1.3×10^6 M$_\odot$, or only 1/20 of the total mass in HI (Wilcots & Miller 1998)

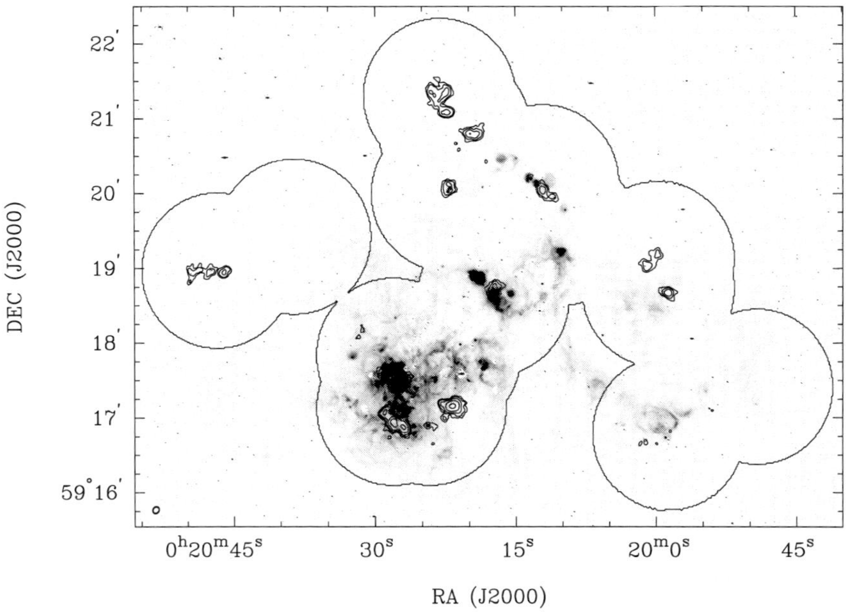

Fig. 3. Molecular gas emission in IC 10 is plotted in contours over an Hα image. The circles indicate the position of the 14 OVRO pointings needed to cover most of the extent of IC10. The beamsize of the observations ($\sim 3''$, 12 pc) is shown in the lower left corner. No clear correlation is found between the molecular gas emission and ongoing star formation in IC 10.

6 Concluding Remarks

Enormous progress has been made in the last few years in the studies of the molecular medium in dwarf galaxies, most notably due to the 2 surveys performed at the OVRO and BIMA millimeter interferometers. These surveys increased the number of high–quality observations of dwarf galaxies by more than an order of magnitude. Our observations show that it is very difficult to predict where the molecular gas is situated in dwarf galaxies and that molecular gas can be present out to very large galactocentric radii. The recent high–resolution studies have unambiguously shown that the conversion factor X_{CO} can *not* be a strong function of metallicity. Even for metallicities of about one–fifth solar we consistently obtain a Galactic X_{CO}. However, one should keep in mind that this result is obviously only applicable for the regions within molecular clouds where CO can actually be detected (i.e. using the method described here no conclusion can be reached regarding molecular gas that may potentially be devoid of CO). Future instruments such as ALMA will be instrumental to detect molecular gas in more metal–poor systems and to study other tracers of the molecular gas phase in the low–metallicity ISM.

Acknowledgments

I would like to thank the organizers for inviting me to this stimulating meeting. It is a pleasure to thank my colleagues Axel Weiss, Alberto Bolatto, Chris Taylor and Nick Scoville.

References

1. Arimoto, N., Sofue, Y., & Tsujimoto, T. 1996, PASJ, 48, 275
2. Israel, F. P. 1997, A&A, 328, 471
3. Kobulnicky, H. A. & Skillman, E. D. 1996, ApJ, 471, 211
4. Lequeux, J., Rayo, J. F., Serrano, A., Peimbert, M., & Torres-Peimbert, S. 1979, A&A, 80, 155
5. Meier, D. S., Turner, J. L., & Beck, S. C. 2001, AJ, 122, 1770
6. Strong, A. W. et al. 1988, A&A, 207, 1
7. Taylor, C. L., Kobulnicky, H. A., & Skillman, E. D. 1998, AJ, 116, 2746
8. Walter, F., Taylor, C. L., Hüttemeister, S., Scoville, N., & McIntyre, V. 2001, AJ, 121, 727
9. Walter, F., Weiss, A., Martin, C., & Scoville, N. 2002a, AJ, 123, 225
10. Walter, F., Weiss, A., & Scoville, N. 2002b, ApJ Letters, 580, L21
11. Weiß, A., Neininger, N., Hüttemeister, S., & Klein, U. 2001, A&A, 365, 571
12. Wilcots, E. M. & Miller, B. W. 1998, AJ, 116, 2363
13. Wilson, C. D. 1995, ApJ Letters, 448, L97
14. Yun, M. S., Ho, P. T. P., & Lo, K. Y. 1994, Nature, 372, 530

Detection of Molecular Gas in the Interarm Regions of M83

S.D. Lord[1], M. Yang[1], A.C.A. Boogert[1], and R.P.J. Tilanus[2]

[1] California Institute of Technology, Pasadena, CA lord@ipac.caltech.edu
[2] Joint Astronomy Center, HI

1 Observations and Comparisons with Other Tracers

We have mapped the eastern half of the nearby gas-rich starburst galaxy M83 in the CO(2-1) transition with the Caltech Submillimeter Observatory (CSO). CO emission was detected at all 135 positions in the mapped area using a $32''$ beam (FWHM). Throughout the interarm regions the CO emission was detected at a level consistently greater than 3σ. This is one of the few galaxies for which interarm molecular emission has been measured. We have used the bar, arm, and interarm CO intensity results in conjunction with other tracers of the ISM in the disk of M83 to draw a more complete set of conclusions about the properties of the interstellar gas and its heating.

At this resolution, the CO(2-1) arm/interarm emission ratio ranges between 2 and 3. A similar arm/interarm emission ratio was obtained in a series of CO(3-2) observations conducted at $24''$ resolution (not discussed further here). By comparing our CO(2-1) single-dish flux with CO(1-0) OVRO interferometer flux (Rand et al. 1999), we confirm that only 5% of the flux was resolved by the 3.5 x $7''$ (84 x 168 pc) synthesized beam of OVRO (assuming $I_{CO(2-1)}/I_{CO(1-0)} = 1$). This fact, combined with our detection of widespread interarm emission suggests that a spatially extended CO component is prevalent in the disk of M83. In Figure 1 we display the integrated CO(2-1) flux map. We compared the CO emission with maps of HI 21-cm, Hβ, 20-cm non-thermal radio emission and blue light, all smoothed to a lower resolution similar to the CO resolution. The non-thermal radio emission was computed by Deutsch & Allen (1993) by subtracting a thermal component of the radio emission (as estimated from their Hβ observations) from their observed 20-cm VLA image.

Outside the nuclear region, HI emission follows the same general pattern as the CO, but along the Eastern arm the HI emission is displaced 5–$10''$ downstream from the CO arm. This effect may be seen in Figure 2, which traces the radial centroids of the various emission distributions across the spiral arm. We note that the CO peak in this galaxy is itself located downstream from the arm's inner edge dust lane, a configuration quite distinct from that of other grand design spirals such as M51 (see e.g., Lord & Kenney 1991.) The further downstream displacement of HI is likely the result of accumulated photodissociation of the molecular hydrogen, as discussed in Tilanus and Allen (1993). While Hβ emission and non-thermal emission in

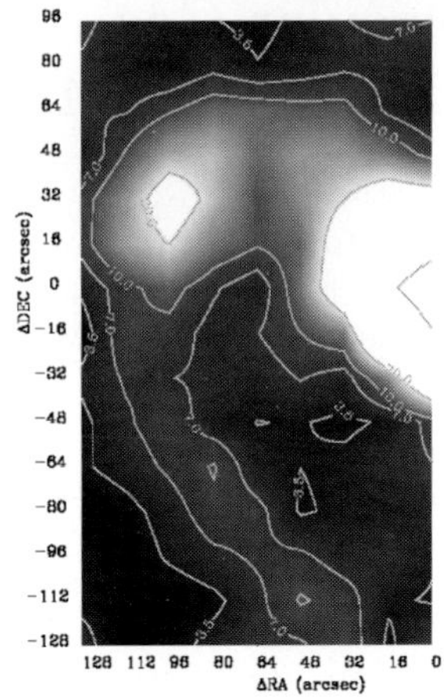

Fig. 1. M83 CO(2-1) Eastern Arm and Eastern Bar flux map. Contour levels are 3.5, 7, 10, 20, 90 K km/s (1 σ =1.1 K km/s)

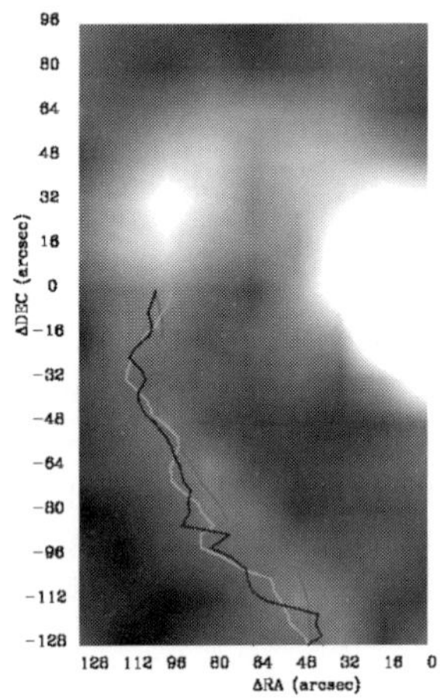

Fig. 2. The red, green, and blue lines overlaid on the CO emission image represent the centroid locations of the radial distributions of CO, HI, and Hβ emission, respectively. The displacement of the HI and Hβ downstream from the CO is evident.

general coincided well with CO emission, the non-thermal radio emission actually showed a higher correlation with the CO distribution than did the Hβ emission. This effect may be seen in the ratio maps of Figures 3 and 4, and the correlation plots of Figure 5.

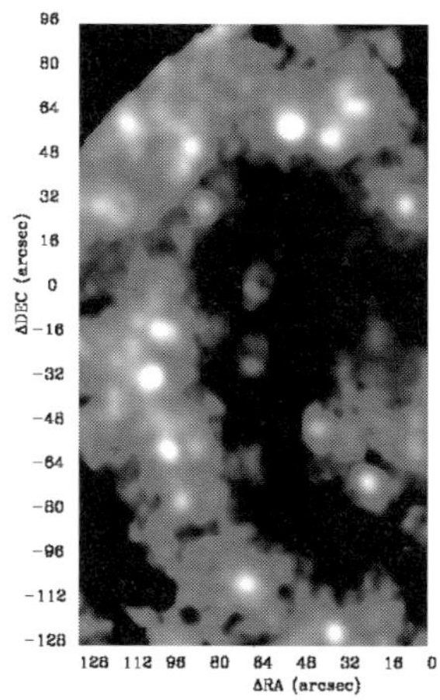

Fig. 3. Ratio map of Hβ/CO

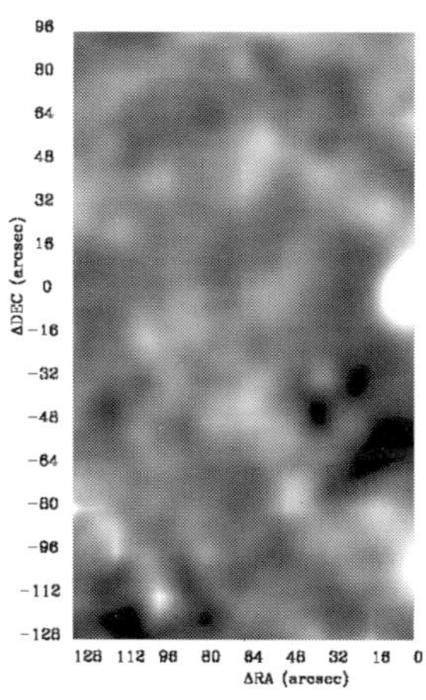

Fig. 4. Ratio map of Non-thermal emission/CO emission

2 Implication for ISM Heating

Both low-energy cosmic rays and UV radiation from young O and B stars can heat the ISM and enhance observed CO emission beyond what is expected from molecular gas of nominal temperature and density N(H_2)=$I_{co} \times X$, where X goes as T_0/T. In M83 the $\frac{I_{H\beta}}{I_{CO}}$ emission ratios are strongly peaked along the spiral arm, suggesting a nonlinear relation between the star formation rate and molecular gas density. UV radiation from HII regions cannot travel far without being absorbed and scattered, and thus UV heating will chiefly affect CO brightness within the vicinity of the spiral arm. By contrast, the $\frac{I_{non-thermal}}{I_{CO}}$ emission ratios maintain a relatively constant value through-

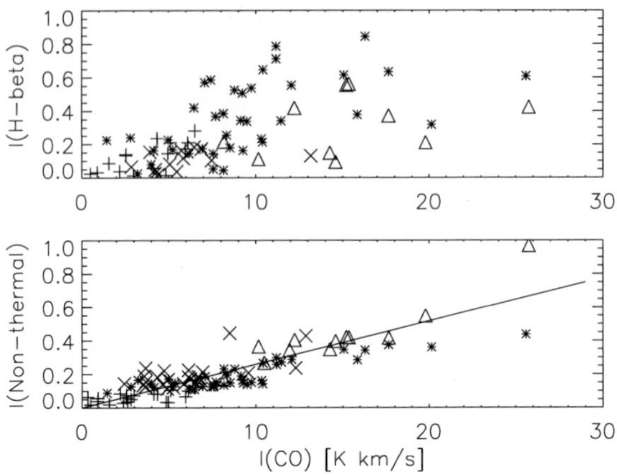

Fig. 5. Correlation plots showing H(beta) plotted against CO (upper panel) and Non-thermal continuum against CO (lower panel). Bar positions are marked with triangles, Eastern arm positions with asterisks, interarm positions with X's and outer disk positions (East of the Eastern arm) with croses. For this plot the Hβ and non-thermal maps were smoothed exactly to the CO resolution.

out the arm and interarm regions (as evidenced by the uniformity of Figure 4). It also follows that non-thermal radio emission is as widespread as CO emission, which might be expected from M83's exceptionally high supernova rate. The importance of cosmic ray heating for M83 was first suggested by Deutsch and Allen (1993).

CO and Hβ, are compared in Figure 5 (upper panel). A correlation is present but with large scatter at the higher intensities. In Figure 5(lower panel), we see a tight linear correlation between the non-thermal intensity and the CO intensity that is maintained in the bar, interarm and arm regions. We argue that in the disk of M83, low-energy cosmic rays are ubiquitous, and that their heating effects provide an significant contribution to the CO emission brightness in both the arms and interarms. As an important caveat, we note there exists strong observational evidence (e.g.: Young & Scoville 1991; Rand et al. 1999) to suggest that molecular gas column density remains the primary factor in determining the overall CO emission.

References

1. E. W. Deutsch & R. J. Allen: AJ **106**, 1812 (1993)
2. R. J. Rand, S. D. Lord, & J. L. Higdon **513**, 720 (1999)
3. S. D. Lord, & J. D. Kenney **381**, 139 (1991)
4. R. P. J. Tilanis & R. J. Allen: AA **274**, 707 (1993)
5. J. S. Young & N. Z. Scoville: AARA **29**, 581 (1991)

A Masing Molecular Cloud in the Central Parsecs of Mrk 348

A.B. Peck[1], C. Henkel[2], and J.S. Ulvestad[3]

[1] Harvard Smithsonian CfA, SMA Project, P.O. Box 824, Hilo, HI 96720 USA apeck@cfa.harvard.edu
[2] Max-Planck-Institut für Radioastronomie, Auf dem Hügel 69, 53121 Bonn, Germany p220hen@mpifr-bonn.mpg.de
[3] NRAO, P.O. Box 0, Socorro, NM 87801 USA julvesta@nrao.edu

We report new observations of the H_2O megamaser in the Seyfert 2 galaxy Mrk 348. The line is redshifted by ~130 km s^{-1} with respect to the systemic velocity, is extremely broad, with a FWHM of 130 km s^{-1}, and has no detectable high velocity components within 1500 km s^{-1} on either side of the observed line. The unusual line profile led us to suspect that this source, like NGC 1052 [1], might belong to a class of megamaser galaxies in which the amplified emission is the result of an interaction between the radio jet and an encroaching molecular cloud, rather than occurring in a circumnuclear disk. Our initial VLBA observations show that the maser emission emanates entirely from a region $\leq$0.25 pc in extent, located toward a continuum component thought to be associated with the receding jet [6]. The very high linewidth occurring on such small spatial scales and the rapid variability indicate that the H_2O emission is more likely to arise from a shocked region at the interface between the energetic jet material and the molecular gas in the cloud where the jet is boring through, than simply as the result of amplification by molecular clouds along the line of sight to the continuum jet. The orientation of the radio jets close to the plane of the sky also results in shocks with the preferred orientation for strong masers from our vantage point. Single-dish monitoring with the Effelsberg 100m telescope showed that the line and continuum emission "flared" on very similar timescales. The close temporal correlation between this activity in the maser emission and the continuum flare further suggest that the masing region and the continuum hotspots are nearly equidistant from the central engine and may be different manifestations of the same dynamical events. The study of this newly discovered type of H_2O megamaser can provide detailed information about the conditions in the ISM in the central 1-10 pc of active galactic nuclei.

1 Introduction

H_2O megamasers are best known as a means to probe the accretion disks in Seyfert galaxies. In the most famous source, NGC 4258, a thin, slightly warped, nearly edge-on disk orbits in Keplerian rotation around a central mass of 4×10^7 $M_\odot$ (e.g. [5]). VLBI studies have been used to determine the size and shape of this warped molecular disk as traced by the maser spots. A few other sources show evidence of a toroidal structure, but the distribution of maser spots is not as well understood. There is now evidence, however, for a distinct class of H_2O megamaser. In these sources the amplified emission is the result of an interaction between the radio jet and an encroaching molecular cloud, rather than occurring in a circumnuclear disk. The only known sources in this class were NGC 1068 [3] and the Circinus galaxy [4], which appear to have both a circumnuclear disk and maser emission arising along the edges of an ionization cone or outflow, and NGC 1052, in which the masers appear to arise along the jet and have a full width at half maximum (FWHM) ~90 km s^{-1} [1]. We have recently identified the fourth such source, Mrk 348, a Seyfert 2 galaxy with a low inclination angle and an exceptionally bright and highly variable nuclear radio source [2]. Observations of Mrk 348 using Global VLBI at 4.8 GHz [7] show two faint continuum hotspots on either side of a much brighter central peak. The relative intensities of the two outer components are quite similar, indicating that relativistic beaming effects are probably minimal and the jet axis should be close to the plane of the sky. Ground-based observations [8] show evidence of an ionization cone with a half-angle of ~45°, which also suggests a jet axis fairly close to the plane of the sky.

The initial detection of the flaring maser in Mrk 348 using the Effelsberg 100m telescope took place in 2000 March [2]. The H_2O maser line in Mrk 348 is extremely broad, with a FWHM of ~130 km s^{-1}, though in many of the monitoring epochs the emission appears to consist of 2 lines which can be tentatively fit with a broad Gaussian function at ~4609 km s^{-1} with FWHM~100 km s^{-1} and a narrower one at ~4678 km s^{-1} with FWHM~60 km s^{-1}. The amplitudes of each component vary significantly on very short timescales. There are no detectable high velocity components within 1500 km s^{-1} on either side of the strong emission line [2].

Figure 1a shows 5 spectra which were chosen to emphasize the significant variations in the line profiles over our first 16 months of observations. The first profile is the discovery spectrum, where both components had a flux of around 30 mJy. In the second profile, taken within a week of our VLBA observations, both components verged on undetectability and then, in the third profile, the higher velocity component flared to twice the intensity of the year previous, and the lower velocity component disappeared altogether. This change was reversed in June, shown in the fourth profile, when the lower velocity component was present, though the intensity was low, and the higher velocity component vanished. Within two weeks, both components were once again nearly equal. Figure 1b shows the variation in line and continuum flux density with respect to time. Comparison shows that the continuum flux varies loosely with the maser flux. The close temporal correlation between the flaring activity in the maser emission and the continuum flare suggest that

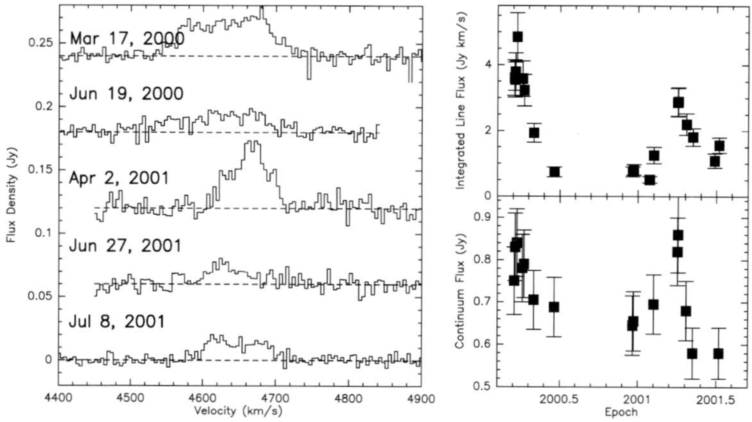

Fig. 1. Left) 5 H_2O spectra from the Effelsberg 100-m telescope chosen to illustrate the variations in the profiles seen on timescales of days or months. **Top right)** Integrated line flux density vs. time; **Bottom right)** continuum flux density vs. time measured with the Effelsberg 100-m telescope. The variations in line and continuum intensity can be seen to vary concurrently.

the masing region and the continuum hotspots are nearly equidistant from the central engine and may be different manifestations of the same dynamical events. VLBA continuum observations indicate that the center of activity in Mrk 348 is likely to be located between the two strongest central components [6]. We hypothesize that at an epoch prior to 1998.75, jet components were ejected simultaneously in the approaching and receding jets, leading to both the continuum flare and the new flaring activity of the maser.

Figure 2a shows three integrated line profiles toward Mrk 348 taken using the VLBA. The maser emission is clearly seen to lie along the line of sight to the fainter northern continuum component, rather than the brightest region of the continuum source which is thought to contain the core. The Gaussian fit to the line shown in the upper left profile has an amplitude of 14±2 mJy and an integrated flux of 2.11±0.34 Jy/beam/km s^{-1}, indicating that all of the flux measured in the Effelsberg May 2 observation has been recovered. The FWHM is 139±11 km s^{-1} centered on 4641±2.2 km s^{-1}, consistent with the single-dish measurements and redshifted by 134 km s^{-1} with respect to the systemic velocity.

In order to produce the maser emission detected in this source, the gas in the molecular cloud within the central parsecs of Mrk 348 must have a pre-shock density ranging from around 10^6 cm^{-3} to a few times 10^9 cm^{-3}. An expanding bow-shock being driven into this cloud by the AGN jet has a velocity between 135 km s^{-1} and 0.5c in the direction of jet propagation, and between 35 km s^{-1} and 300 km s^{-1} at various points along the oblique edges (Figure 2b). This shock generates a region of very high temperature, ($\leq 10^5$ K), which dissociates the molecular gas and to some extent shatters

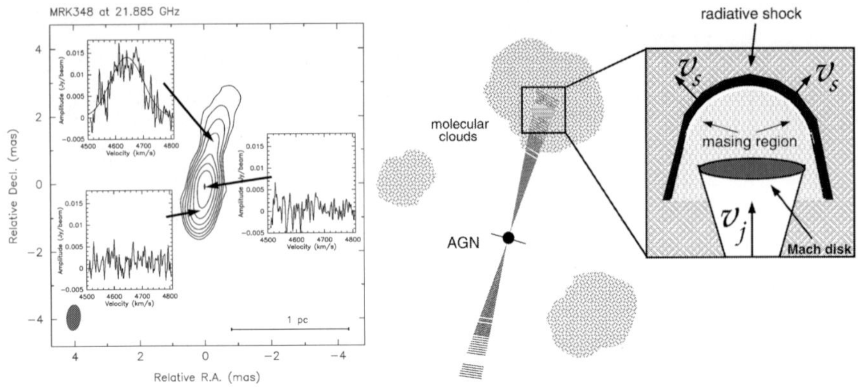

Fig. 2. Left)H_2O line profiles toward Mrk 348, superimposed on the continuum map made from 20 line-free channels extracted from the low frequency end of the observed frequency range. The RMS noise in the line profiles is ~4 mJy beam^{-1} channel^{-1}. The continuum image is naturally weighted and the lowest contour is 5 mJy. The RMS noise in the continuum map is less than 1 mJy beam^{-1}. **Right)**A cartoon model of the expanding bubble caused by the jet material impacting the molecular cloud. The maser emission arises within the region surrounded by the radiative shock.

the dust grains expected to be present and/or evaporates their icy mantles. Immediately following this shock, H_2 begins forming on the surviving dust grains when the temperature has dropped to ~1000 K, and this in turn provides sufficient heating to stabilize the temperature at ~400 K, with gas densities of ~10^8 cm^{-3} and ultimately an H_2O abundance as high as 10^{-5}. Ammonia is not found because the levels of UV radiation remain too high, but the detection of other molecular species will help to test this scenario.

References

1. Claussen, M. J., Diamond, P. J., Braatz, J. A., *et. al.* 1998, ApJL, 500, 129
2. Falcke, H., Henkel, C., Peck, A. B., *et. al.* 2000, A&A, 358, L17
3. Gallimore, J. F., Baum, S. A., O'Dea, C. P., *et. al.* 1996, ApJ, 462, 740
4. Greenhill, L. J., Moran, J. M., Booth, R. S., *et. al* 2001, in *IAU Symposium 205* eds. R. Schilizzi, S. Vogel, F. Paresce, M. Elvis, 334
5. Miyoshi, M., Moran, J., Herrnstein, J., *et. al.* 1995, Nature 373, 127
6. Peck, A. B., Henkel, C., Ulvestad, J. S., *et. al.*, 2003, ApJ, 590, 149
7. Roy A. L., Ulvestad J. S., Wilson A. S., *et. al.* 1999, in *Perspectives on Radio Astronomy: Science with Large Antenna Arrays*, [Dwingeloo: NFRA], Ed. M.P. van Haarlem, p. 173
8. Simpson, C., Mulchaey, J. S., Wilson, A. S., *et. al.* 1996, ApJ, 457, L19

The Spectral Line Survey of NGC 253

S. Martín[1], R. Mauersberger[1], J. Martín-Pintado[2], C. Henkel[3], and S. García-Burillo[4]

[1] IRAM, Avda. Divina Pastora 7 NC, E-18012 Granada, Spain martin@iram.es
[2] IRAM, Avda. Divina Pastora 7 NC, E-18012 Granada, Spain mauers@iram.es
[3] DAMIR, IEM, CSIC, Serrano 121, E-28006 Madrid, Spain martin@damir.iem.csic.es
[4] MPIfR, Auf dem Hügel 69, D-53121 Bonn, Germany p220hen@mpifr-bonn.mpg.de
[5] OAN, C/ Alfonso XII 3, 28014 Madrid, Spain s.gburillo@oan.es

We present the first 2 mm spectral line survey of an extragalactic source, the nucleus of the starburst galaxy NGC 253. The survey covers the range from 129 to 172 GHz. We detected emission from 7 species not previously observed outside the Milky Way. One of the most outstanding result is the detection of NO with a relatively large fractional abundance of a few 10^{-7}. We discuss the origin of the chemistry in the nucleus of NGC 253 by comparing the measured abundances with those in prototypical galactic sources. We propose that the molecular abundances in NGC 253 are consistent with a chemistry dominated by large scale shocks.

1 The Survey

We present a complete unbiased spectral line survey at the 2 mm atmospheric window towards the nuclear region of the starburst galaxy NGC 253. At a distance of 2.5 Mpc this spiral galaxy is one of the nearest prototypical starburst galaxies. Its nuclear region is the most prolific extragalactic source of molecular emission. In fact, most of the 24 molecular species so far detected outside the Galaxy have been observed towards this source.

Observations were carried out at the IRAM 30 m telescope. Two SIS receivers operating at 2 mm were tuned to adjacent frequencies allowing us to cover an instantaneous 2 GHz band with a resolution of 4 MHz (~ 8 km s^{-1}). We used the wobbling secondary mirror with a symmetrical beam throw of $4'$ in azimuth and a switching frequency of 0.5 Hz. We integrated three hours on each frequency setup to achieve a rms noise of $\sim$4–6 mK.

Fig. 1 shows the full spectral scan. We used the line catalogues of Lovas (1992)[8] and Pickett et al. (1998)[16] to identify the features in the scan. Out of the 21 molecules and isotopes (Table 1), seven are detected for the first time in the extragalactic interstellar medium. Table 2 shows the physical parameters derived from the newly detected molecules assuming LTE and optically thin emission. To calculate the fractional abundances, we have assumed the H_2 column density of $1.7\,10^{22}$cm^{-2} (Mauersberger et al. 2003[9]).

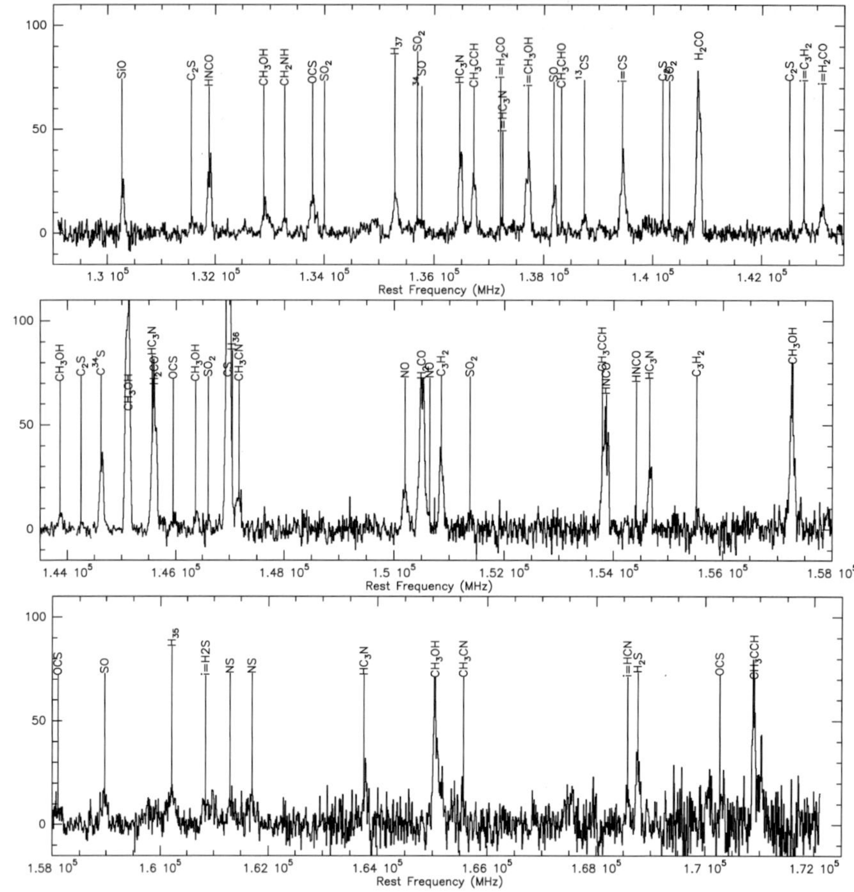

Fig. 1. Line survey towards the nuclear region of NGC 253.

2 Molecular Emission from the Nucleus of NGC 253

The chemistry of the molecular material associated with a starburst is expected to be dominated by two different processes: the photodissociation of the gas by the UV field from OB associations and the processing of grains by strong high-velocity shocks waves in the molecular clouds. Previous studies of species such as NH_3 (Mauersberger et al. 2003 [9]) and SiO (García-Burillo et al. 2000 [1]) in the nuclear region of NGC 253 point out the presence of strong shocks as the most likely explanation for the observed emission.

In this context, this line survey and the newly detected species will allow us to establish the mechanism leading the chemical complexity of the nuclear region of NGC 253. With this aim, in Table 3 we compare the fractional abundances obtained in this work for the new molecules detected and those observed in prototypical Galactic molecular clouds. We have selected two dark

Table 1. Molecules and isotopes identified in the survey

CS	^{34}CS	^{13}CS	**NO**
NS	SO	^{34}**SO**	SiO
C_2S	**H_2S**	OCS	**SO_2**
H_2CO	HNCO	C_3H_2	**CH_2NH**
HC_3N	CH_3CN	CH_3OH	CH_3CCH
CH_3CHO			

In bold, molecular species detected for the first time outside the Milky Way

Table 2. Physical parameters derived for the newly detected species

Molecule	N cm^{-2}	T_{rot} K	[X]/[H_2]
C_2S	$2\,10^{13}$	27 (12)	$1\,10^{-9}$
CH_2NH	$9\,10^{13}$	15^a	$5\,10^{-9}$
CH_3CHO	$2\,10^{13}$	12 (8)	$1\,10^{-9}$
H_2S	$8\,10^{13}$	15^a	$5\,10^{-9}$
NO	$5\,10^{15}$	7 (2)	$3\,10^{-7}$
NS	$5\,10^{13}$	8 (1)	$3\,10^{-9}$
SO_2	$7\,10^{13}$	14 (9)	$4\,10^{-9}$

[a] Parameter fixed

Table 3. Fractional abundances compared to Galactic molecular clouds.

Source	[SO_2]/[H_2] 10^{-8}	[NS]/[H_2] 10^{-9}	[NO]/[H_2] 10^{-8}	[C_2S]/[H_2] 10^{-10}	[CH_3CHO]/[H_2] 10^{-10}	[H_2S]/[H_2] 10^{-8}
NGC 253	0.4	3	30	10	10	0.5
Dark Clouds						
L134N	0.4 [a]	0.2–0.6 [b]	20 [c]	16 [h]		0.08 [a]
TMC1	< 0.1 [a]	0.7–1.2 [b]	2.7 [c]	55 [h]	60 [j]	
Hot Cores						
Orion H.C.	9.4 [d]	0.4 [e]	30 [c]		6 [j]	500 [k]
Sgr B2(N)	3 [f]	10 [f]	20 [f]	2.4 [i]	0.8 [j]	0.02 [l]
Sgr B2(M)	40 [f]	0.03 [f]	30 [f]	3.9 [i]	2 [f]	8 [l]
PDRs						
Orion Bar	0.01 [g]		0.2 [g]			

[a] Ohishi et al. 1992 [15]; [b] McGonagle et al. 1994 [10]; [c] Gerin et al. 1992 [3], 1993 [4]; [d] Charnley, 1997 [2]; [e] McGonagle et al. 1997 [11]; [f] Nummelin et al. 2001 [14]; [g] Jansen et al. 1995 [6]; [h] Suzuki et al. 1992 [17]; [i] Kuan & Snyder 1994 [7]; [j] Ikeda et al. 2001 [5]; [k] Minh et al. 1990 [12]. [l] Minh et al. 1991 [13].

clouds to illustrate the ion-molecule chemistry associated to quiescent gas, three hot cores to illustrate grain and shock chemistry associated to massive star-forming regions and one photodissociation region (PDR) to illustrate the chemistry dominated by the UV field from OB stars in the main sequence.

Emission arising from PDR dominated chemistry can be ruled out from this comparison. The NO abundance in NGC253 is a factor of 100 larger than in PDRs. Abundances of molecules such as NS and NO seem to be similar to that found in hot cores, but the low excitation temperature derived (Table 2) is much lower than the expected in these kind of objects ($T > 70$ K). SO_2 and H_2S show large abundances in hot cores associated to star-forming regions. Such high abundances are not found in the nuclear region of NGC 253, indicating a different origin. However, the abundance of these molecules are still higher than what we find in dark molecular clouds. The temperature derived from SO_2 is also higher than typical kinetic temperature in dark clouds

(10 K). In hot cores the sulfur chemistry is driven by the evaporation H_2S from grain-mantles. SO_2 is then rapidly formed from H_2S when T$< 200\,K$. A similar chemistry is found in molecular clouds processed by shock waves, where H_2S is injected into the gas phase by shocks or produced as a result of the hydrogenation of the atomic sulfur in the hot postshock material (Charnley 1997 [2]). CH_3CHO is found both in dark and hot molecular clouds. Ikeda et al.2001 [5] find low rotational temperatures (10–40 K) towards a number of star-forming regions, as what we find towards NGC 253.

We propose the emission of these molecules arises from a warm shocked component in the nuclear region of NGC 253. It is surprising the presence of C_2S, observed to be much more abundant in dark clouds than in hot cores (Suzuki et al. 1992 [17]). Nevertheless, the temperature we infer seems to exclude a dark cloud origin. Interferometric maps of the Sgr B2 (Kuan & Snyder 1994 [7]) show an emission of C_2S which supports the idea of an emission in NGC 253 steming from a warm component similar to the Sgr B2 envelope.

Acknowledgements. J.M.-P. has been partially supported by the MCyT under grant mumbers ESP2002-01627 AND AYA2002-10113-E.

References

1. S. García-Burillo, J. Martín-Pintado, A. Fuente, R. Neri: ApJ, **355**, 499 (2000)
2. S.B. Charnley: ApJ, **481**, 396 (1997)
3. M. Gerin, Y. Viala, F. Pauzat, Y. Ellinger: A&A, **266**, 463 (1992)
4. M. Gerin, Y. Viala, F. Casoli: ApJ, **268**, 212 (1993)
5. M. Ikeda, M. Ohishi, A. Nummelin et al.: ApJ, **560**, 792 (2001)
6. D.J. Jansen, M. Spaans, M.R. Hogerheijde, E.F. van Dishoeck: ApJ, **303**, 541 (1995)
7. Y.-J. Kuan, L.E. Snyder: ApJ, **94**, 651 (1994)
8. F.J. Lovas: J. Phys. Chem. Ref. Data, **21**, 181 (1992)
9. R. Mauersberger, C. Henkel, A. Weiß, A.B. Peck, Y. Hagiwara: A&A, **403**, 561 (2003)
10. D. McGonagle, W.M. Irvine, M. Ohishi: ApJ, **422**, 621 (1994)
11. D. McGonagle, W.M. Irvine: ApJ, **477**, 711 (1997)
12. Y.C. Minh, L.M. Ziurys, W.M. Irvine, D. McGonagle: ApJ, **360**, 136 (1990)
13. Y.C. Minh, J.M.C. Rawlings, A. Bennett, P.D. Brown, S.B. Charnley: Ap&SS, **87**, 585 (1991)
14. A. Nummelin, P. Bergman, ÅHjalmarson et al.: ApJS, **128**, 213 (2001)
15. M. Ohishi, W.M. Irvine, N. Kaifu: IAU Symp. **150**, 171 (1992)
16. H.M. Pickett, R.L. Poynter, E.A. Cohen et al.: J. Quant. Spectrosc. & Rad. Transfer, **60**, 883 (1998)
17. H. Suzuki, S. Yamamoto, M. Ohishi et al.: ApJ, **392**, 551 (1992)

Starburst at the Molecular Superbubble in M82

S. Matsushita[1], R. Kawabe[2], K. Kohno[3], H. Matsumoto[4], T.G. Tsuru[4], and B. Vila-Vilaró[5]

[1] Institute of Astronomy and Astrophysics, Academia Sinica, P.O. Box 23-141, Taipei 106, Taiwan, R.O.C. satoki@asiaa.sinica.edu.tw
[2] National Astronomical Observatory of Japan, 2-21-1 Osawa, Mitaka, Tokyo 181-8588, Japan kawabe@nro.nao.ac.jp
[3] Institute of Astronomy, University of Tokyo, 2-21-1 Osawa, Mitaka, Tokyo 181-0015, Japan kkohno@ioa.s.u-tokyo.ac.jp
[4] Department of Physics, Faculty of Science, Kyoto University, Sakyo-ku, Kyoto 606-8502, Japan matumoto@cr.scphys.kyoto-u.ac.jp, tsuru@cr.scphys.kyoto-u.ac.jp
[5] Steward Observatory, University of Arizona, Tucson, AZ 85721 bvila@as.arizona.edu

Abstract. We present high spatial resolution 100 GHz continuum emission observations with the Nobeyama Millimeter Array (NMA) toward an expanding molecular superbubble in the central region of M82. The 100 GHz continuum image, which is dominated by free-free emission, revealed that the strongest peaks are concentrated at the inner edge of the superbubble along the galactic disk. Star formation at these peaks are an order of magnitude more active in terms of ionizing flux than those from the most massive star forming regions in our Galaxy. At this region, high velocity ionized gas, masers, and diffuse hard X-ray emission are also concentrated. These observations suggest that past strong starbursts produced plasma and the superbubble expansion, and induced the present starburst regions. These results provide the first clear evidence of self-propagating starburst in external galaxies.

1 Introduction

M82 is one of the nearest starburst galaxies (3.25 Mpc [1]), and therefore well suited for a case study about starburst phenomena. Soft X-ray, optical recombination lines, dust, and molecular gas observations show kpc-scale outflows from the central region, and these observations suggest that the outflows are the result of numerous supernova explosions, namely, starburst. Detailed information about the starburst region itself is, however, still not clear, because the central region of M82 is deeply obscured by dust. In this paper, we report the nature of the present starburst regions in M82 based on our 100 GHz continuum emission and molecular gas observations, which have a small effect of dust obscuration.

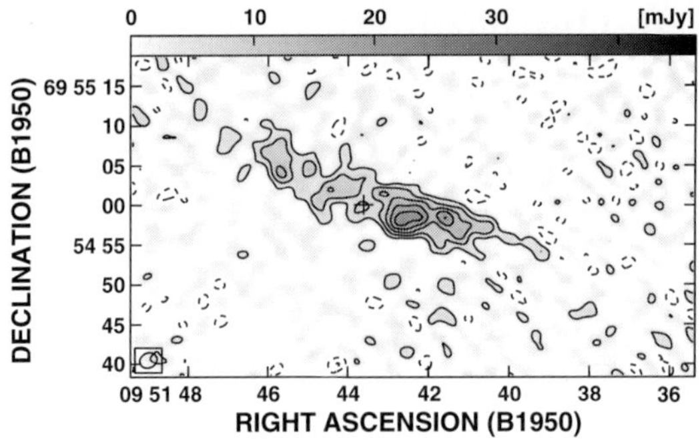

Fig. 1. Uniform *uv* weighting 100 GHz continuum image of the central region of M82. The contour levels of the map are $-4, -2, 2, 4, 6, 8$, and 10σ, where $1\sigma = 2.0$ mJy beam^{-1}. The plus mark indicates the position of the galactic nucleus [3]

2 Observations and Results

We obtained aperture synthesis 100 GHz continuum image toward the central region of M82 with the Nobeyama Millimeter Array (NMA) during 1997 December – 1998 April. Three configurations of six 10 m antennas, equipped with tunerless SIS receivers, were used for the observations. As back-end, we used an XF-type spectro-correlator Ultra Wide Band Correlator (UWBC [2]), which has a 1 GHz bandwidth and a 6 GHz intermediate frequency. The band-pass calibration was done with 3C273, and 0923+392 was observed every 10 minutes as a phase and amplitude calibrator. The flux scale of 0923+392 was determined by comparisons with Mars and Uranus, and the uncertainty is estimated to be $\sim$ 20%. We CLEANed the data with uniform *uv* weighting, and the resultant synthesized beams are 2.3" $\times$ 1.9" (36 pc $\times$ 30 pc).

The 100 GHz continuum map shows several peaks along the galactic disk, but the strongest peaks are located south-west side of the nucleus (Fig. 1). Note that our low spatial resolution and high signal-to-noise ratio map (natural *uv* weighting map; not shown in this paper) is consistent with the previously published data [4, 5]. We overlaid the ^{12}CO(1–0) molecular superbubble image [6] on the 100 GHz continuum map (Fig. 2a), and it clearly shows that the strongest peaks in the 100 GHz continuum image are located at the inner edge of the molecular superbubble along the galactic disk of M82. Since the 100 GHz continuum emission in M82 is dominated by free-free emission [4], this distribution strongly suggests that the ongoing starburst regions in M82 are concentrated at the inner edge of the molecular superbubble.

Since the 100 GHz continuum emission is mostly free-free emission, its flux is directly related to star formation activity like Hα emission at optical wavelength. Using the 100 GHz continuum emission flux (i.e., free-free emission flux), S_{T}, we can estimate the production rates of Lyman continuum photons, N_{Ly}, as

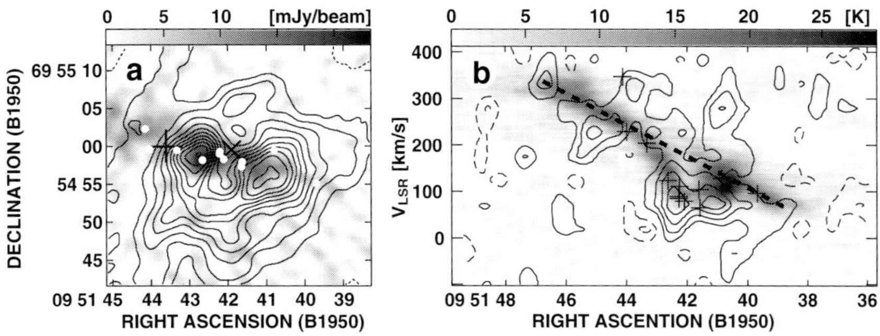

Fig. 2. (**a**) 100 GHz continuum emission image (grayscale) overlaid on the ^{12}CO(1–0) superbubble map (contours [6]). The plus and cross marks indicate the positions of the nucleus [3] and the 2.2μm secondary peak [7], respectively. Contours for the ^{12}CO map are -10, 10, 20, $\cdots$, 110σ, where 1σ is 8.9 mJy beam^{-1}. White filled circles indicate the positions of OH and H_2O masers [8, 9, 10]. (**b**) PV diagram of ^{12}CO (grayscale) and H41α (contours [5]) emissions along the galactic disk of M82. The thick dashed line shows rigid rotation velocity. The plus marks indicate the positions and velocities of the OH and H_2O masers

$$\left(\frac{N_{\rm Ly}}{\rm s^{-1}}\right) \sim 1.3\times10^{51}\left(\frac{T_{\rm e}}{10^4\ \rm K}\right)^{-0.45}\left(\frac{\nu}{100\ \rm GHz}\right)^{0.1}\left(\frac{D}{3.25\ \rm Mpc}\right)^{2}\left(\frac{S_{\rm T}}{1\ \rm mJy}\right),$$

where $T_{\rm e}$ and ν are the electron temperature and the observation frequency, respectively [11]. In this calculation, $T_{\rm e}$ of 5,000–10,000 K was used [5]. We substituted peak flux densities averaged over the beam size (30–40 pc), and derived $N_{\rm Ly}$ at the continuum peaks inside the superbubble. We find that $N_{\rm Ly}$ in each starburst region is as high as $\sim (2-4)\times10^{52}$ photons s^{-1}. These $N_{\rm Ly}$ are an order of magnitude higher than those of active star forming regions in our Galaxy: The giant HII region NGC 3603 has $N_{\rm Ly}$ of 1.1×10^{51} photons s^{-1} [12]. The HII region complex G49.5$-$0.4 in one of the most luminous star forming regions W51 has $N_{\rm Ly}$ of 8.9×10^{50} photons s^{-1} [13]. The diameters of these regions are about 50 pc and 30 pc, respectively [13], which are almost the same linear scale as our observations. This comparison suggests that the ongoing starburst regions in M82 are at least 20 times more active in terms of ionizing flux than the active star forming regions in our Galaxy.

3 Self-propagating Starburst

We compared the position-velocity (PV) diagram of molecular gas with that of H41α line [5] and found that the deviation from the rigid rotation of the recombination line is larger than that of the molecular gas (Fig. 2b). Similar deviation difference is also seen in the comparison between the molecular gas and the [NeII] line [14]. Since the deviation is an evidence of the superbubble expansion, this velocity difference suggests that the ionized gas is expanding faster than the neutral gas. We compiled all the observed OH and H_2O masers toward M82 [8, 9, 10] and found that the masers are obviously concentrated at the 100 GHz continuum emitting regions, namely, the intense starburst regions (Fig. 2a). In addition, we overplotted the masers on the PV diagram and found that the masers are strongly concentrated at the superbubble with good velocity coincidences (Fig. 2b). The masers in M82 are not so luminous as megamasers around AGNs, but much more similar to the Galactic sources [9, 10]. Furthermore, interferometric SiO line observation toward the superbubble suggests that the shocked regions may locate the inner wall of the superbubble [15]. Recent Chandra X-ray Observatory observation, on the other hand, found a diffuse hard X-ray source with an extent of $7.2'' \times 5.4''$. This component could be due to hot thermal plasma, and this plasma seems to be overpressurized [16]. We found that the peak of this diffuse hard X-ray component is also located inside the superbubble. These results suggest that ionized gas inside the bubble expands faster than the molecular gas, and creates the shocked and starburst regions at the inner wall of the superbubble.

Based on our observations and other previously published results, we can see a hierarchal structure in the superbubble, namely, from the center, a starbursted star cluster, overpressurized hot thermal plasma, high velocity ionized gas, shocked molecular gas, the present starburst regions, and the expanding molecular superbubble. This structure and the physical properties of these objects/mediums suggest that the past starburst produced the high-velocity gas, swept surrounding gas, produced the superbubble, and the compression of surrounding gas induced new starburst regions. This result is the first clear evidence of self-propagating starburst in an external galaxy.

References

1. A. Sandage, G.A. Tammann: Astrophys. J. **196**, 313 (1975)
2. S.K. Okumura, et al: Publ. Astron. Soc. Japan **52**, 393 (2000)
3. D.F. Lester, J.S. Carr, M. Joy, N. Gaffney: Astrophys. J. **352**, 544 (1990)
4. J.E. Carlstrom, P.P. Kromberg: Astrophys. J. **366**, 422 (1991)
5. E.R. Seaquist, et al: Astrophys. J. **465**, 691 (1996)
6. S. Matsushita, et al: Astrophys. J. **545**, L107 (2000)
7. R.D. Dietz, et al: Astron. J. **91**, 758 (1986)
8. L. Weliachew, et al: Astron. & Astrophys. **137**, 335 (1984)
9. A. Baudry, N. Brouillet: Astron. & Astrophys. **316**, 188 (1996)
10. E.R. Seaquist, D.T. Frayer, D.A. Frail: Astrophys. J. **487**, L131 (1997)
11. R.H. Rubin: Astrophys. J. **154**, 391 (1968)
12. L.F. Smith, P. Biermann, P.G. Mezger: Astron. & Astrophys. **66**, 65 (1978)
13. S. Okumura, et al: Astrophys. J. **543**, 799 (2000)
14. K.A. Wills, et al: Mon. Not. R. Astron. Soc. **309**, 395 (1999)
15. S. García-Burillo, et al: Astrophys. J. **563**, L27 (2001)
16. R.E. Griffiths, et al: Science, 290, 1325 (2000)

Gas Dynamics in the Non-barred Seyfert Galaxy NGC 4501

S. Onodera[1], J. Koda[2], Y. Sofue[1], and K. Kohno[1]

[1] Institute of Astronomy, The University of Tokyo, Tokyo
sonodera@ioa.s.u-tokyo.ac.jp
[2] California Institute for Technology, Pasadena

We report high resolution interferometer observations of the central 5 kpc region of NGC 4501. The observations were carried out in CO line using the Nobeyama Millimeter Array (NMA). The CO major features are (1) a nuclear concentration which consists of double peaks, (2) spiral arms. The feature (1) has a low star-forming efficiency, which might be due to low gas-to-dynamical mass ratio. The central double peaks are located on the root of optical spiral arms in a *HST* image. The spiral arms are extending out from the nuclear region, and they show systematic non-circular motions. They seem to be associated with the *HST* spiral dust lanes and may be linked to the central double peaks. We calculated gas cloud orbits in a stellar spiral potential, and could explain the observed CO spiral arms and non-circular motion in the arms. In a stellar spiral potential, two mechanisms of angular momentum transfer are possible – galactic spiral shocks in orbit-crowding regions, and gravitational torques exerted by the stellar spirals. We found that the galactic shock is dominant. These mechanisms lead to gas inflow which possibly explains the central-condensed double peaks in NGC 4501.

1 Introduction

Radial inflow of the interstellar gas in disk galaxies fuels active galactic nuclei, central starburst, and leads to a flattening of the metallicity gradient by gas mixing. Stellar bars have long been considered as the most viable candidate for transferring mass from large to small scales (e.g. [9]). In addition, recent statistics in observations report the circumstantial evidence of radial gas inflow by bars; the degree of gas condensation to the central kiloparsec is higher in barred galaxies than in unbarred ones [8]. On the other hand, these kinds of central gas condensations are common both in barred and non-barred CO-luminous galaxies [7]. Therefore, the gas should be transported to the centers even without a bar. Thus, it is crucial to study gas-inflow mechanisms, other than bar-induced ones. Here we study gas dynamics in spiral arms, which has not been thoroughly investigated compared to bar-driven gas dynamics.

In order to investigate the gas dynamics and possible inflow mechanisms in spirals, we observed the spiral galaxy NGC4501, which shows a high central gas concentration [8] but has no dintinct bar.

2 Observations and Results

Observational data of ^{12}CO(J=1-0) emission in NGC 4501 were obtained with the NMA in the course of a CO-line survey of Virgo spirals [10]. They were mapped using CLEAN procedure with natural ($\sim 5''$) and uniform ($\sim 2''$) weightings. The low-resolution map and the velocity field are shown in Fig. 1 (left, middle). We found kpc-scale spiral arms and strong central condensation ($r \sim 5''$) which resolves into double peaks with a $2''$ (150 pc) beam. The molecular spiral arms are associated with dust lanes. The velocity field on the spiral arms shows non-circular motion about $\sim$ 50km s^{-1} along the galactic minor axis. The S-wise isovelocity distortions around $r \sim 38''$ (the ellipse in Fig. 1) and Z-wise slight distortions in $r < 10''$ can be attributed to a spiral density wave (Sect. 4). Overlaid on an *HST* image (Fig. 1 right), it is shown that the double peaks inhabit the root of spiral dust lanes penetrating into the nucleus. The gas-to-dynamical mass ratio within $r = 5''$ is only 3.5% . The peak surface brightness in a $\sim 2''$ beam is 290 K km s^{-1}, which corresponds to $\Sigma_{\rm gas} = 620 M_\odot {\rm pc}^{-2}$.

3 Central Double Peaks and Nuclear Spirals

The peak surface brightness above expects a star formation rate of $1.3 \times 10^3 M_\odot {\rm Gyr}^{-1} {\rm pc}^{-2}$ with Schmidt law by [2]. However, the actual star formation rate estimated from Hα and far-infrared surface brightness [8] is one or

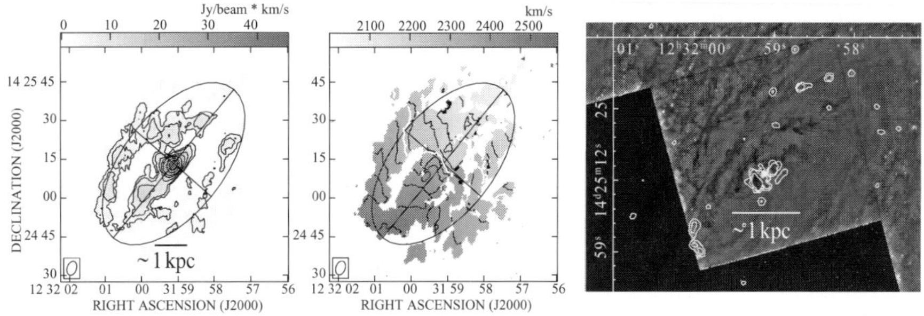

Fig. 1. (*Left, Middle*) Observed CO intensity map and velocity field. The ellipse represents a projected circle with diameter $75''$ inclined with position angle (140°) and inclination (58°) of NGC 4501. (*Right*) High-resolution CO map overlaid on *HST* *V*-band image.

two orders of magnitude lower than it. This low star-forming efficiency in the double peaks can even be an upper limit, since diffuser components may be missing in our interferometer observation. The low gas-to-dynamical mass ratio, the fast rigid rotation, and the large velocity dispersion of clouds in the nuclear region, suggest that the gas clouds in central region are gravitationally stable. It is not common to find "twin peaks" in non-barred galaxies although they are often associated with barred galaxies (e.g. [3]). Their physical condition is significantly different from that in barred galaxies, which regularly accompany with nuclear star formation. This difference may arise from the difference of gas-to-dynamical mass ratio which indicates the effectiveness of self gravity for gas. The twin-peaks in barred galaxies are often explained as a consequence of bar-induced gas dynamics. However, a different explanation is necessary in the case of the non-barred galaxy NGC 4501. The nuclear spiral dust lanes have been recently found in many galaxies, and they are thought to be the location of shock fronts or turbulence, and therefore, the sites of angular momentum dissipation (e.g. [5]). Centrally-condensed double peaks could be consequences of angular momentum transfer and in their accreting period.

4 Gas Dynamics in NGC 4501: Gaseous Orbits in a Spiral Potential

NGC 4501 has predominant spiral arms rather than a bar [1], and we calculate gas cloud orbits in a spiral potential to understand the observed gas distributions and dynamics. We modified a damped-orbit model by [11]. This model gives closed gas orbits in a barred potential by solving the equations of motion which include a damping-force term to emulate the collisional nature of the gas. It has been applied to galaxies NGC 5005 [6] and NGC 3079 [4] and successfully explains the observed gas dynamics in barred galaxies. In this model, the first-order equations of motion in the frame rotating with the potential $\Phi(R, \psi)$ are

$$\ddot{R}_1 + \left[\frac{d^2\Phi_0}{dR^2} - \Omega^2\right]_{R_0} R_1 - 2R_0\Omega_0\dot{\psi}_1 + 2\lambda\dot{R}_1 = -\left[\frac{\partial\Phi_1}{\partial R}\right]_{R_0}, \tag{1}$$

$$\ddot{\psi}_1 + 2\Omega_0\frac{\dot{R}_1}{R_0} = -\frac{1}{R_0^2}\left[\frac{\partial\Phi_1}{\partial\psi}\right]_{R_0}. \tag{2}$$

The term $+2\lambda\dot{R}_1$ in the left-hand side of (1) represents the effects of gas viscosity. We calculated gas orbits in a spiral potential instead of a straight bar. We set the model potential as follows:

$$\Phi(R, \psi) = \Phi_0(R) + \Phi_p(R)\cos 2[\psi + \cot\theta\log(R)], \tag{3}$$

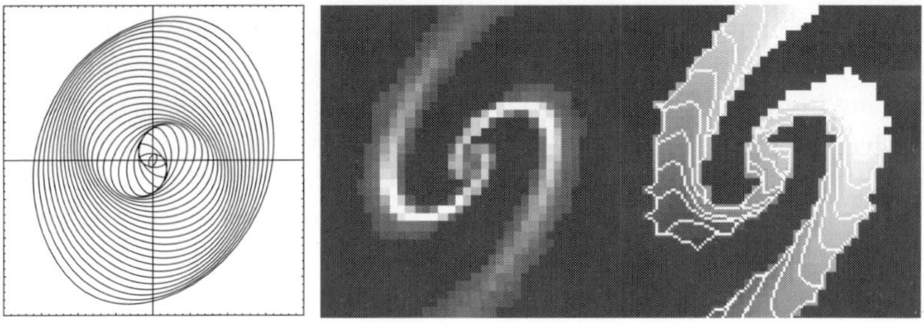

Fig. 2. (*Left*) Particle orbits in the fixed spiral potential. (*Middle, Right*) Gas distributions and velocity fields derived with our model.

where $\Phi_0(R)$ is the Plummer potential of the axisymmetric disk, and the second term is the first-order perturbation by a two-armed spiral with a pitch angle θ. The resultant orbits are shown in Fig. 2 (left). Spiral-shaped orbit-crowding regions are generated. If we calculate the gas density from the staying time of clouds ($\propto 1/\mathrm{v}$), the observed gas distribution can be well reproduced (Fig. 2 middle). The observed non-circular motion along the molecular spiral arms are reproduced as well (Fig. 2 right). The large S-wise distortions along the spiral arms and the small Z-wise distortions in the nuclear region of the model velocity field well explain those in the observations. Gas dynamics in NGC 4501 can be understood as being controlled by the stellar spiral potential.

We estimated the angular momentum loss of the gas in the spiral potential. There are two possible mechanisms in angular momentum transfer – galactic spiral shocks and gravitational torques. Both of them are dependent on the pitch angle of stellar spiral arms. We found the galactic shocks dominate gravitational torques by an order of magnitude for the case of NGC 4501. It is possible that the centrally-condensed double peaks in NGC 4501 are generated as a result of the angular momentum transfer due to shocks.

References

1. Jungwiert, B., Combes, F.,& Axon, D. J. : A&AS, **125**, 479 (1997)
2. Kennicutt, R. C. : ApJ, **498**, 541 (1998)
3. Kenney, J. D. P., Wilson, C. D., Scoville, N. Z., Devereux, N. A., & Young, J. S. : ApJL, **395**, L79 (1992)
4. Koda, J., Sofue, Y., Kohno, K., Nakanishi, H., Onodera, S., Okumura, S. K., & Irwin, J. A. : ApJ, **573**, 105 (2002)
5. Martini, P., Regan, M. W., Mulchaey, J. S., & Pogge, R. W. : ApJ, **589**, 774, (2003)
6. Sakamoto, K., Baker, A. J., & Scoville, N. Z. : ApJ, **533**, 149 (2000)
7. Sakamoto, K., Okumura, S. K., Ishizuki, S., & Scoville, N. Z. : ApJS, **124**, 403 (1999)

8. Sakamoto, K., Okumura, S. K., Ishizuki, S., & Scoville, N. Z. : ApJ, **525**, 691 (1999)
9. Shlosman, I., Frank, J., & Begelman, M. C. 1989, Nature, **338**, 45
10. Sofue, Y., Koda, J., Nakanishi, H., Onodera, S., Kohno, K., Tomita, A., & Okumura, S. K. : PASJ, **55**, 17 (2003)
11. Wada, K. : PASJ, **46**, 165 (1994)

The Merger QSO 3C 48 and Its Host Galaxy in the Near Infrared

J. Zuther[1], A. Eckart[1], J. Scharwächter[1], M. Krips[1], S. Pfalzner[1] and C. Straubmeier[1]

1. Physikalisches Institut, Universität zu Köln, Zülpicher Str. 77, 50937 Köln
zuther@ph1.uni-koeln.de

In this contribution we present new near-infrared (NIR) data on the quasar 3C 48 and its host galaxy, obtained with ISAAC at the Very Large Telescope (European Southern Observatory, Chile). The NIR images and spectra reveal a reddening of several magnitudes caused by extinction due to molecular material and dust within the host galaxy. For the first time we clearly identify the highly reddened potential second nucleus 3C 48A about $1''$ northeast of the quasar position in the NIR. Its reddening can be accounted for by warm dust, heated by star formation or an interaction of the 3C 48 radio jet with the interstellar medium, or both. The NIR colors and the CO(6-3) absorption feature both give a stellar contribution of about 30% to the QSO-dominated light. These results will contribute to the question of how the nuclear activity and the apparent merger process are influencing the host galaxy properties and they will improve existing models.

1 Introduction

A key goal in understanding the connection between nuclear activity (accretion onto a super massive black hole) and host galaxy structure (e.g. morphology, stellar populations, etc.) is to find out about the physical conditions of quasars and the galaxy hosts they reside in.

Because of its large angular size and its brightness at the redshift of $z \approx 0.37$ [1], 3C 48 is an ideal target for high-resolution imaging and spectroscopy with large telescopes. The near-infrared (NIR) is sensitive for the mass dominating stellar populations and at the same time it is less affected by extinction than the optical. The radio-loud galaxy shows an excess far-infrared (FIR) emission, $L_{\mathrm{FIR}} = 5 \times 10^{12} L_{\odot}$ [2], and a one-sided radio jet extending about $0.5''$ to the north, fanning out to the east to about $1''$ [3]. The quasar is rich in molecular gas, $M_{\mathrm{H}_2} \sim 2.7 \times 10^{10} M_{\odot}$, as CO(1-0) observations show [4, 5]. There is also morphological evidence for a recent merger event. Such are a possible double nucleus and a tidal tail extending several arcseconds to the northwest [6, 7, 8]. Yet it is still unclear, whether the brightness peak about $1''$ northeast (NE) of the QSO is really the nucleus of the companion galaxy, or a region of interaction of the radio jet with the interstellar medium of the host galaxy inducing star formation [8].

2 Near Infrared Properties of 3C 48

J, H, and Ks band imaging as well as H and K band spectroscopy were carried out using the NIR camera-spectrometer ISAAC at the Very Large Telescope (VLT, ESO, Chile) [9].

2.1 $JHKs$ Photometry

The images at 1.2, 1.65, and 2.2 μm (J, H, and Ks) show a morphology quite similar to the optical one [6]. A tidal tail like structure extends to the northwest of the quasar (Fig. 1). Another feature in the southeast of the quasar was previously thought to be a counter tidal tail [8]. Our two-color analysis of the host galaxy showed that this feature is most probably not associated with 3C 48. Instead, with colors typical for low-redshift galaxies, it seems to be a foreground galaxy (Fig. 2, aperture no. 12).

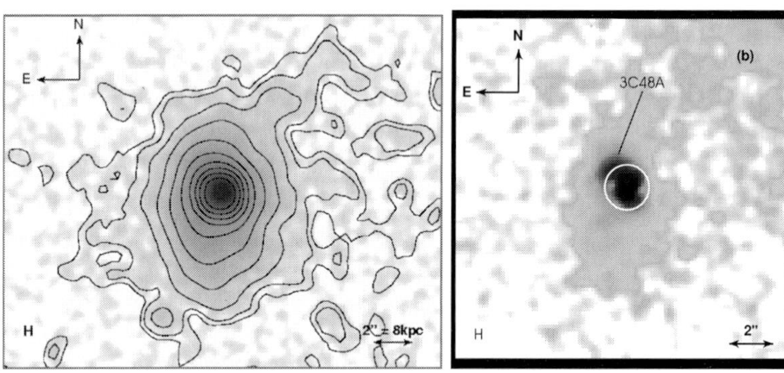

Fig. 1. ISAAC H band image of 3C 48. *Left:* Original image, still being dominated by the QSO. *Right:* Core-subtracted image, emphasizing the excess brightness of 3C 48A. Both taken from [9].

The two-color diagram in Fig. 2 shows that all colors of the host galaxy are reddened with respect to an old (10 Gyr) single starburst population (star). The nucleus itself is also heavily reddened (aperture no. 5; obscuration by circum nuclear material). The grey curve presents the color evolution of a single starburst population from zero (lower left end) to 10 Gyr. Young starbursts therefore have quite blue colors. It has been shown that there is a lot of star formation going on in the host [8]. Starbursts have quite blue NIR colors (cf. evolutionary curve in Fig. 2). Therefore a huge extinction seems to be at work (all data points are redder than the 10 Gyr color point).

In order to study the nuclear region, especially 3C 48A, the nuclear contribution has to be removed. This was done by subtracting a nearby star, scaled to the peak flux of the QSO. The result is presented in the right panel

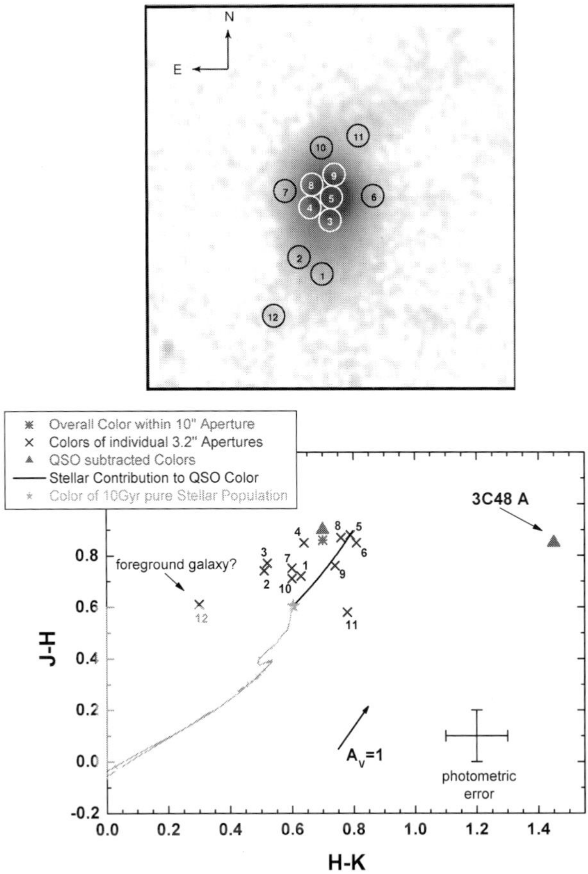

Fig. 2. *Top panel:* H band image of 3C 48 showing the loci of the 3.2″ apertures for measuring colors. *Bottom panel:* NIR two color diagram. The symbols are explained in the figure. The light solid line represents an evolutionary model for colors of a single, passively evolving starburst [12]. Evolution proceeds from the lower left corner at 0 Gyr to the upper right corner at 20 Gyr (10 Gyr are marked with a $\star$). The dark solid line represents the stellar contribution to the QSO light when increasing the aperture. The reddening vector for 1 mag of visual extinction is also shown. Both panels taken from [9].

of Fig. 1, showing an excess brightness about 1″ northeast of the QSO. The colors of this region are strongly reddened due to the presence of warm/hot dust (only in the $H - K$ direction). Furthermore there appears to be an alignment of the radio jet with 3C 48A in the NIR. However, multi-particle simulations by [13] show, that the A component could still be the nucleus of the companion galaxy.

2.2 *HK* Spectroscopy

Emission line spectroscopy allows to probe the interstellar medium (ISM) in the central region of the galaxy, e.g. via the recombination lines Paβ and Paγ. In our case we find a line ratio of about 1.6($\pm$30%). This is a value typical for the ISM of active galaxies being excitated by the central-engine continuum emission ($T \sim 10$ K and $n \sim 10$ cm^{-3}).

Absorption lines can be used to trace the stellar content. The depth of the CO(6-3) absorption line is a measure for a population of cool old stars [10]. In our case the depth of the line measured only as an upper limit corresponds to a stellar contribution of about 30%. This is consistent with the results from photometry. For details see [9].

3 3C 48 Revisited

Our NIR imaging and spectroscopy result in a first detection of 3C 48A in J, H, and Ks. The host galaxy, but especially 3C 48A, are strongly reddened. Both scenarios, a second nucleus or an interaction of the ISM with the radio jet, are still possible to explain the nature of 3C 48A. The presence of significant extinction is an important ingredient for future studies of the stellar populations of the host galaxy. Multi-particle simulations by [13] can explain the missing counter tidal tail as the result of a projection effect, consistent with the optical [8] and our NIR data. The merger morphology, its high FIR luminosity, and its large content of molecular gas put 3C 48 in an evolutionary scheme envisioned by [11]. There it fits nicely as a transition objects between an ultra-luminous infrared galaxy (ULIRG) phase and a pure QSO phase, being still dominated by its FIR excess but also displaying QSO features like its luminosity and rather flat spectral energy distribution.

References

1. T. A. Boroson & J. B. Oke: ApJ **281**, 535 (1984)
2. G. Neugebauer and B. T. Soifer: ApJL **295**, L27 (1985)
3. P. N. Wilkinson, A. K. Tzioumis, J. M. Benson, et al.: Nature **352**, 313 (1991)
4. N. Z. Scoville, S. Padin, D. B. Sanders, B. T. Soifer, & M. S. Yun: ApJ **415**, L75 (1993)
5. J. E. Wink, S. Guilloteau, & T. L. Wilson: A&A **322**, 427 (1997)
6. A. Stockton & S. E. Ridgway: AJ **102**, 488 (1991)
7. E. T. Chatzichristou, C. Vanderriest, & W. Jaffe: A&A **343**, 407 (1999)
8. G. Canalizo and A. Stockton: ApJ **528**, 201 (2000)
9. J. Zuther et al.: submitted to A&A (2003)
10. L. Origlia, A. F. M. Moorwood,& E. Oliva: A&A **280**, 536 (1993)
11. D. B. Sanders, B. T. Soifer, J. H. Elias, et al.: ApJ **325**, 74 (1988)
12. J. B. Hutchings & S. G. Neff: AJ, **113**, 550 (1997)
13. J. Scharwächter et al.: astro-ph/0310740

Part IV

Structure and Dynamics of the ISM in the Milky Way

New Views of Neutral Hydrogen in the Milky Way

N.M. McClure-Griffiths

Australia Telescope National Facility, CSIRO, PO Box 76, Epping NSW 1710, Australia naomi.mcclure-griffiths@csiro.au

1 Introduction

Since the first detection of neutral hydrogen (H I) by Ewen and Purcell in 1951 [4], H I has been a valuable tool for understanding the structure and dynamics of the Milky Way. The first analysis of H I in the Milky Way was also published in 1951 [14] and surveys of the entire Galaxy were immediately undertaken. In the intervening years the atlas of H I in the Milky Way has been continually updated, culminating most recently with the one arcminute resolution International Galactic Plane Survey.

For studying the interstellar medium (ISM), H I is unrivalled in its ability to trace distribution and kinematics across the entire Galaxy. The ubiquity of H I, combined with the fact that it is easily observed in both emission and absorption, make it an excellent tracer of the warm and cold H I and simultaneously a probe of the physical state of the ISM.

2 New Surveys of H I in the Galaxy

Over the past decade surveys of the CO, dust and H-α emission in the Milky Way have improved markedly. Meanwhile, despite the obvious advantages of observing H I, large-scale high resolution surveys of H I were technically infeasible. Fortunately, recent developments in single dish and interferometer data combination have resulted in a resurgence in Galactic H I studies. In 1995 the Dominion Radio Astrophysical Observatory began the Canadian Galactic Plane Survey (CGPS)[16], an arcminute resolution survey of H I in the third Galactic quadrant. This project led to an expanded international effort via the Southern and VLA Galactic Plane surveys to fully image H I in the Galactic disk at arcminute resolution [12, 15]. Together these surveys form the International Galactic Plane Survey (IGPS). The IGPS covers $\sim 80\%$ of the stellar disk of the Galaxy at an angular resolution of ~ 1 arcminute, an rms brightness temperature sensitivity of ~ 1 K, and a velocity resolution of ~ 1 km s^{-1}.

The IGPS is partially motivated by the desire to understand the evolutionary cycle of the ISM. With its high angular resolution and large-scale images, the IGPS offers us, for the first time, the ability to simultaneously observe the entire life-cycle of the ISM. Some of the key questions that we hope to answer with the survey are: How does the HI distribute itself? What are the dynamics of the HI throughout the Galaxy? What are the temperatures of the warm and cold HI? What is the nature of the coldest, most compact HI clouds as seen in HI self-absorption? What is the spatial power spectrum of density and velocity fluctuations? How does the ISM transition from deterministic structures (SNRs, shells, etc.) to turbulence? What are the variations in scale height and midpoint? What are the dominant causes of structure in the Milky Way?

The IGPS probably will not dramatically change our view of the very large-scale properties of the disk, such as characteristics of the warp and flare or the radial and vertical extents of the disk. These have, in fact, changed little since they were first measured in the 1950's. What the IGPS does provide, however, is a revolutionary new view of H I in the Galaxy. The combination of the new high resolution data produced by this survey and developments in data visualization techniques, have allowed us to take H I studies of the ISM and Galactic structure beyond the traditional single plane longitude-velocity diagrams. Figures 1 & 2 are examples of SGPS images, showing the detail and variety of structures in the new H I data. The CGPS has produced a stunning collection of similar vistas, each containing a wealth of information about the gas distribution. Where the IGPS excels most is in its ability to resolve individual small-scale H I features, while maintaining the large-scale context of the original low-resolution surveys. For example, we can now detect variations in the scale height of the disk on the scale of tens of parsecs. Whereas traditional measurements of the scale height have reflected variations in the gravitational potential, IGPS measurements of the scale height reveal the effects of both the gravitational potential and individual dynamic objects like H I shells. Here I review just two of the major new areas opened up to us by the resolution IGPS. First, I focus on the detection — and importance — of individual parsec scale "clouds" on the edges of kiloparsec scale shells and chimneys. Second, I discuss the emerging science from parsec and sub-parsec scale cold clouds observed in H I self-absorption.

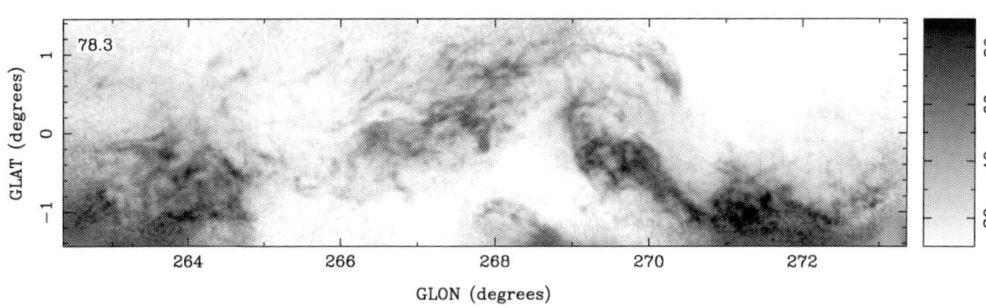

Fig. 1. Grey-scale H I image of $10^\circ \times 2^\circ$ region from the Southern Galactic Plane [9]. The resolution of this image is $\sim$ 2 arcmin. We observe structure on all size scales from the resolution limit to $\sim 10^\circ$.

2.1 ISM Evolution

One of the most fundamental hurdles to understanding the structure and evolution of the Milky Way is our knowledge of the evolutionary cycle of matter in the interstellar medium (ISM). We know that the Galactic ecosystem involves a constant interaction between stars, which inject matter and energy into the ISM and star formation, which removes mass from the ISM. We do not know, however, how the ISM cools, transitioning through its various phases to reach star formation. What factors govern how quickly matter cools and condenses? What role do magnetic fields play in this process? The Milky Way offers a unique laboratory to answer these questions. Studies of the Galactic ISM allow us to look at the detailed interaction of the phases with sensitivity and resolution that is unattainable in external galaxies. H I is a particularly powerful tracer of the ISM evolutionary life cycle. H I emission can be used to explore large shells and bubbles, created by stellar winds and supernovae. Meanwhile, both H I emission and absorption can be used to probe cold atomic gas which lines the edges of these shells and may precede the formation of molecular clouds.

With the IGPS we have begun to resolve the complete life-cycle of H I in the ISM. Figure 2 is an image of a portion of a very large chimney, GSH 277+00+36 discovered in the Southern portion of the IGPS [10]. The image is of H I emission at a single velocity and combines data from the Australia Telescope Compact Array and the Parkes Radio Telescope for sensitivity to all angular scales from 3 arcminutes (6pc at 6.5 kpc) to 10 degrees [11]. GSH 277+00+36 is more than 600 pc in diameter and extends more than 1 kpc out of both sides of the Galactic Plane. The image is fascinating in its level of detail. Along the shell walls there are parsec scale features, which we believe were created by local Rayleigh-Taylor instabilities at the interface between the tenuous shell interior and the dense walls. H I linewidths along the majority of the shell walls indicate standard warm neutral medium temperatures on the order of 5000 K. The small-scale features along the walls, however, have very narrow linewidths, suggesting gas kinetic temperatures as low as 40 - 150 K. Some of these small-scale features even have associated molecular gas, as shown in contours in Fig. 2. Here we are able to resolve, for the first time, the transition from the hot-ionized gas presumably filling the interior of this shell, to the warm neutral medium of the outer walls, to the cold neutral medium along the small-scale instability created features, and finally molecular gas, all in the one object!

2.2 Cold H I

Cold H I can be observed in a number of ways. Historically, much of what we know about the cold state of H I in the Milky Way is derived from H I absorption measurements toward continuum sources [2]. An alternate way to study the cold neutral medium is through observations of H I self-absorption.

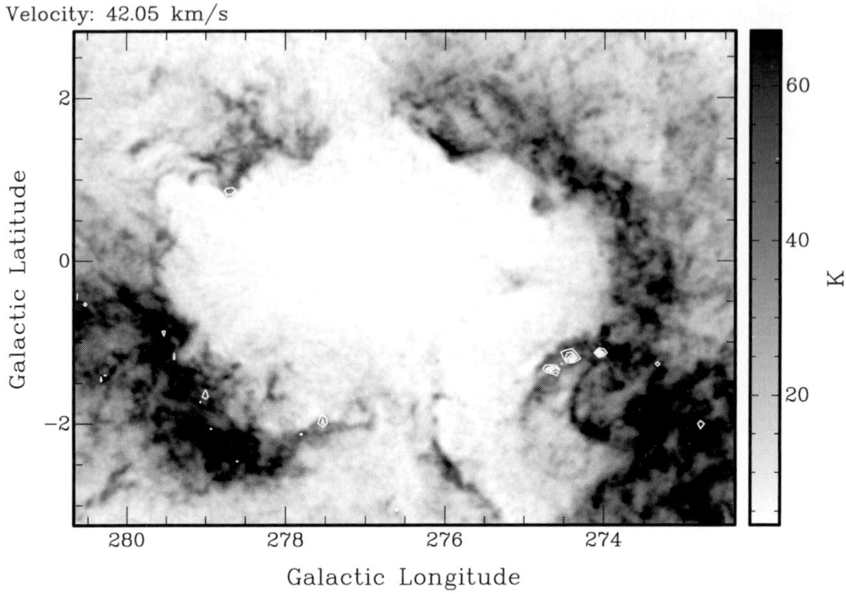

Fig. 2. Grey-scale H I image of GSH 277+00+36 at $v = +42$ km s^{-1}[11], overlaid with ^{12}CO emission contours from the NANTEN CO Southern Milky Way Survey [13]. The white contours show the positions of the positions of CO emitting molecular clouds. These agree well with the positions of cold, small-scale H I structure.

H I self-absorption occurs when cold foreground H I absorbs the emission from warmer background H I at the same velocity. H I self-absorption, as a phenomenon, has been studied for many years [6, 1]. However, the new high resolution surveys of large areas are revealing populations of small cold clouds, many of which would not have been resolved by previous surveys [5]. We hope that the IGPS data will allow us to answer some basic questions about cold H I in the Milky Way. For instance, how much of the H I mass is tied up in cold clouds? How cold is this H I? What is the structure of these H I self-absorption clouds? What is the relationship between cold H I and molecular gas. As the IGPS progresses we hope to be able to answer these questions. Initial results from SGPS indicate that much of the H I self-absorption clouds have temperatures less than 40 K and some may be as cold as 10 - 15 K [3]. In the context of this meeting, which focuses on the dense ISM, the connection of cold H I to molecular gas is particularly pertinent. Comparing H I self-absorption features from the CGPS with ^{12}CO emission Gibson et al. (2000) found that the majority of H I self-absorption features showed no ^{12}CO emission. This finding immediately calls into question a commonly held belief that cold H I is found as a skin on molecular clouds. However, new evidence from the Boston University ^{13}CO Galactic Ring Survey suggests that this may be an opacity and that there is a much better correspondence between ^{13}CO and H I self-absorption [8].

Though much of the observed H I self-absorption is dominated by small, cold clouds, there are several examples of very large-scale cold clouds. One of the most striking examples of large-scale H I self-absorption to come from the IGPS is the so-called "Dark Arc' [7]. The Dark Arc was discovered in the Canadian Galactic Plane Survey as a several degree long feature observed in relief against the bright background H I. Knee & Brunt (2001) show that the arc is in fact a large, cold H I supershell, which is observed in emission away from the Galactic plane and in absorption in the plane, where the H I background is much brighter. The arc is remarkable in its size, about 2 kpc and its temperature, about 10 K.

3 Conclusions

The IGPS is presenting us with a new view of H I in the Milky Way. Earlier Galactic H I surveys have given a complete picture of the large-scale structure of H I in the Galaxy. The IGPS is now allowing us to examine the small scale structure of H I throughout the Milky Way, while maintaining the context of the large-scale emission. Early results from the IGPS show us the power of the survey in studying the full evolutionary cycle of the ISM. We can now resolve the emergence of parsec-scale cold features on the edges of kiloparsec scale H I shells. In addition, we are detecting large populations of cold clouds with H I self-absorption. Many of these features are often small-scale and would have been missed in lower resolution surveys.

References

1. Baker, P. L. & Burton, W. B. 1979, A&AS, 35, 129
2. Dickey, J. M., Terzian, Y. & Salpeter, E. E. 1978, ApJS, 36, 77
3. Dickey, J. M., McClure-Griffiths, N. M., Gaensler, B. M., Green, A. J. 2002, ApJ, 585, 801
4. Ewen, H. I. & Purcell, E. M. 1951, Nature, 168, 356
5. Gibson, S. J., Taylor, A. R., Higgs, L. A., & Dewdney, P. E. 2000, ApJ, 540, 841
6. Knapp, G. 1974, AJ, 79, 527
7. Knee, L. B. G. & Brunt, C. M. 2001, Nature, 412, 318
8. Jackson, J. M., Bania, T. M., Simon, R., Kolpak, M., Clemens, D. P., & Heyer, M. 2002, ApJ, 566, L81
9. McClure-Griffiths, N. M. 2001, PhD Thesis, University of Minnesota
10. McClure-Griffiths, N. M., Dickey, J. M., Gaensler, B. M., Green, A. J., Haynes, R. F., & Wieringa, M. H. 2000, AJ, 119, 2828
11. McClure-Griffiths, N. M., Dickey, J. M., Gaensler, B. M., Green, A. J. 2003, ApJ, 594,833

12. McClure-Griffiths, N. M., Green, A. J., Dickey, J. M., Gaensler, B. M., Haynes, R. F., & Wieringa, M. H. 2001, ApJ, 551, 394
13. Matsunaga, K., Mizuno, N., Moriguchi, Y., Onishi, T., Mizuno, A., & Fukui, Y. 2001, PASJ, 53, 1003
14. Muller, C. A. & Oort, J. H. 1951, Nature, 168, 357
15. Taylor, A. R., Stil, J. M., Dickey, J. M., McClure-Griffiths, N. M., Martin, P. G., Rothwell, T., Lockman, F. J. 2002, ASP Conf. Ser. 276: Seeing Through the Dust: The Detection of HI and the Exploration of the ISM in Galaxies, 68
16. Taylor, A. R., et al. 2003, AJ, 125, 3145

The BU-FCRAO Galactic Ring Survey

R. Simon[1,2], J.M. Jackson[1], R. Shah[1], J. Rathborne[1], T.M. Bania[1], D.P. Clemens[1], and M.H. Heyer[3]

[1] Institute for Astrophysical Research, Boston University
[2] I. Physikalisches Institut, Universität zu Köln simonr@ph1.uni-koeln.de jackson@bu-ast.bu.edu, ronak@bu.edu, rathborn@bu.edu, bania@inanna.bu.edu, clemens@bu.edu
[3] Five College Radio Astronomy Observatory, University of Massachusetts, Amherst heyer@astro.umass.edu

Using the SEQUOIA multi-pixel array receiver on the Five College Radio Astronomy Observatory (FCRAO) 14 m telescope, we are conducting a new molecular line survey of the inner Galaxy. The Boston University (BU)-FCRAO Galactic Ring Survey (*GRS*) will map ^{13}CO (1–0) emission from $l = 15°$ to $52°$ and $b = -1°$ to $1°$. Compared with previous molecular line surveys of the inner Galaxy, the *GRS* offers excellent sensitivity, better spectral resolution, the same or better angular resolution, Nyquist sampling, and the use of ^{13}CO (1–0), a better column density tracer than ^{12}CO.

We have found that 21 cm H I self-absorption is well correlated with *GRS* emission in nearby clouds and use its presence or absence to resolve the near/far kinematic distance ambiguity. These distance measurements to clouds and their embedded young stellar objects allow us to establish their masses, sizes, luminosities, and Galactic distribution.

We have also studied infrared dark clouds revealed by the *MSX* survey. They may be the initial condensations that will ultimately become massive clusters. The results are presented in detail by Simon et al. in this volume.

1 Introduction

The radial distribution of molecular gas in the Milky Way shows a distinct peak about halfway between the Sun and the Galactic Center. Although this so-called "5 kpc molecular ring" dominates the Milky Way's star formation activity, it has remained largely unexplored since it is challenging to image at good angular resolution such a large expanse of sky and to estimate distances to Galactic molecular clouds and star forming regions. Thus, a number of fundamental questions about Galactic structure and cloud and star formation in the ring remain unanswered: Do the properties of molecular clouds in the ring differ from clouds elsewhere in the Galaxy? What are the spatial distribution, luminosity function, and initial mass function of young stars both inside and outside the ring? Is star formation in the ring more efficient? What is the detailed structure of the ring?

Table 1. Molecular Cloud Survey Comparison

	CfA	*UMSB*	Bell Labs	*GRS*
Angular Resolution	480″	46″	100″	46″
Grid	450″	180″	180″	22″
Sensitivity	0.1 - 0.4 K	0.4	0.12 K	0.4 K
Spectral Resolution	0.65 km s^{-1}	1.0 km s^{-1}	0.68 km s^{-1}	0.21 km s^{-1}
Isotope	^{12}CO	^{12}CO	^{13}CO	^{13}CO

Recent advances in mm-wave technology allow us to conduct a new Galactic survey, the *GRS*, which exploits the focal plane array receiver SEQUOIA on the FCRAO 14 meter telescope.

2 Comparison of the GRS with Other Surveys

SEQUOIA's 32 pixels and high sensitivity enable us to efficiently survey the Galactic plane in ^{13}CO (1–0). Although being fainter than ^{12}CO, the ^{13}CO line is optically thinner, and hence a better tracer of column density. Its narrower linewidths allow a cleaner separation of velocity components and it suffers very little from self-absorption, which often complicates the analysis of ^{12}CO emission.

Compared to previous CO surveys, the *GRS* offers: (1) the same or better angular resolution, (2) better spectral resolution, (3) better angular sampling, and (4) the use of ^{13}CO. Fig. 1 and Table 1 compare the *GRS* to previous molecular line surveys: the CfA survey [3], the UMass-Stony Brook survey [2], and the Bell Labs survey [7]. Fig. 1 clearly reveals the great improvement afforded by observing ^{13}CO with the SEQUOIA array. The *GRS* detects new structures and cloud cores completely missed by the older surveys.

As of mid 2003, the *GRS* has mapped about 45 deg^2, a total of over 1,200,000 positions. Eventually, the *GRS* will image the Galactic plane from $l = 15°$ to $52°$ and $b = -1°$ to $1°$ on a $22''$ fully sampled grid.

3 Resolving the Kinematic Distance Ambiguity

Many key physical properties, such as mass, size, and luminosity, of the *GRS* clouds and their associated young stellar objects can only be deduced if their distances are known. The *GRS* radial velocities provide the key to good kinematic distance estimates for molecular clouds in the inner Galaxy. Since the Milky Way rotates differentially, but to a good accuracy has a flat and well known rotation curve, a cloud's radial velocity determines its Galactocentric radius, which, in the inner Galaxy, corresponds to two heliocentric distances: the "near" and "far" kinematic distances.

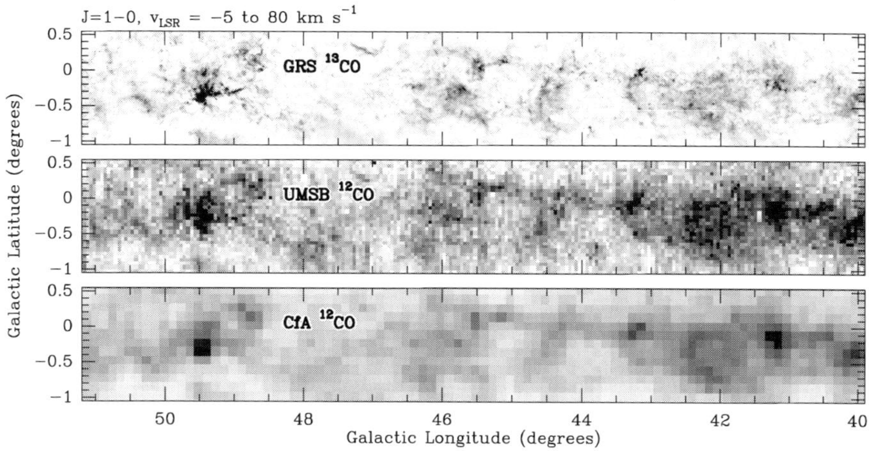

Fig. 1. A comparison of the Galactic Ring Survey (top) with previous CO surveys (UMass-Stony Brook in the middle, CfA at the bottom). All datasets show emission integrated over a velocity range of $V_{LSR} = -10$ to 80 km s^{-1}. The *GRS* clearly reveals new structures.

We resolve the kinematic distance ambiguity for *GRS* clouds by comparing our ^{13}CO data to 21 cm H I data from the Boston University-Arecibo Observatory Galactic H I Survey [6] using an automated procedure. We have, for the first time, found in many clouds a clear correspondence between H I self-absorption and ^{13}CO emission, both in morphology and in velocity [5]. Since hydrogen is ubiquitous throughout the Galaxy, there is an ample amount of warm H I at all velocities at the far distance against which a cold molecular foreground cloud can be seen in silhouette. If the same cloud were placed at the far distance, there would be little or no background H I, and the self-absorption feature would be absent. We suggest that the presence or absence of H I self-absorption at the LSR velocity of the *GRS* molecular cloud establishes whether the cloud lies at the near or far kinematic distance [5].

An example is given in Fig. 3, which shows ^{13}CO and H I spectra toward $(l, b) = (45.33, +0.12)$. According to the presence or absence of H I self-absorption at the velocity of *GRS* emission, we assign the near or far kinematic distances to the clouds. The prominent ^{13}CO emission peak at $V_{LSR} = 25$ km s^{-1} corresponds to deep H I self-absorption. This particular cloud also shows prominent visual extinction, which independently confirms that it is indeed nearby. We will apply this technique throughout the survey region to assign distances to *GRS* clouds and their associated young stellar clusters identified from infrared surveys such as *MSX*, *2MASS*, and (in the future) *SIRTF*.

A particularly interesting new class of clouds are the so-called infrared dark clouds identified by the *ISO* [4] and *MSX* [1] satellites. IR dark clouds absorb a significant infrared Galactic plane background and hence lie on the

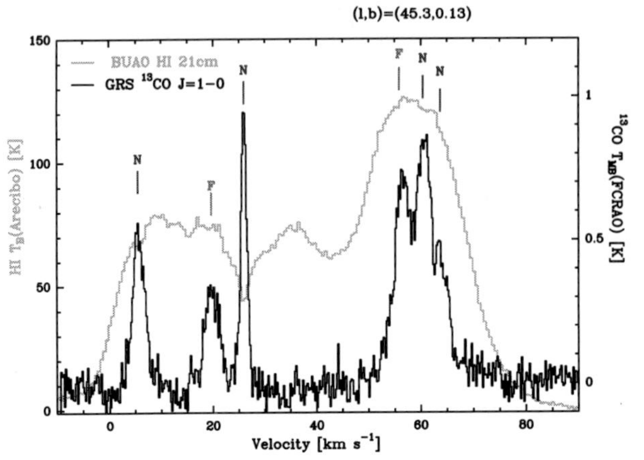

Fig. 2. H I and *GRS* ^{13}CO spectra. Note that many ^{13}CO features correspond to H I self-absorption dips. These are identified with clouds at the near kinematic distance. Clouds with no H I self-absorption are identified as far.

near side of the Galaxy. We have found that these dark clouds are strong sources of ^{13}CO emission. With *GRS* molecular line velocities, we can derive the first good distance estimates for a large number (~ 300) of IR dark clouds (see Simon et al., this volume).

4 Summary

The chief advantages of the *GRS* are its good angular sampling and the use of the ^{13}CO line. Having mapped over 45 square degrees of the Galactic plane, the *GRS* already provides a dataset useful for cloud structure [8] and distance studies [5]. *GRS* and H I data resolve the kinematic distance ambiguity and we have used *GRS* velocities to determine the properties of infrared dark clouds. They are likely to be clouds in the very earliest stages in the formation of massive star clusters. Future *GRS* efforts will include determining Galactic structure and a systematic study of the distances to infrared sources. *GRS* data are available to the community at the website www.bu.edu/grs.

References

1. S.J. Carey, F.O. Clark, M.P. Egan, S.D. Price, R.F. Shipman, T.A. Kuchar: ApJ **494**, L199 (1998)
2. D.P. Clemens, D.B. Sanders, N.Z. Scoville: ApJ **327**, 139 (1988)
3. T. Dame et al.: ApJ **322**, 706 (1987)
4. P. Hennebelle, M. Perault, D. Teyssier, S. Ganesh: A&A **365**, 598 (2001)
5. J. M. Jackson, T. M. Bania, R. Simon, M. Kolpak, D.P. Clemens, M.H. Heyer: ApJ **566**, L81 (2002)
6. T.A. Kuchar, T.M. Bania: ApJ **436**, 117 (1994)
7. Y. Lee, A.A. Stark, H.G. Kim, D.S. Moon: ApJS **136**, 137 (2001)
8. R. Simon, J.M. Jackson, D.P. Clemens, T.M. Bania, M.H. Heyer: ApJ **551**, 747 (2001)

New View of Molecular Gas Distribution of the Southern Sky: CO Surveys with NANTEN

T. Onishi, N. Mizuno, Y. Fukui, and the NANTEN group

Department of Astrophysics, Nagoya University, Chikusa-ku, Nagoya 464-8602, Japan ohnishi@a.phys.nagoya-u.ac.jp

We have been carrying out a CO $J = 1-0$ large-scale survey of molecular gas along the galacti plane with NANTEN radio telescope at Las Campanas observatory, Chile since 1996. More than 1,100,000 spectra have been obtained at a $4'$ grid spacing for $|b| \leq 5°$ and at a $8'$ grid spacing for $10° > |b| > 5°$, respectively. In addition to having obtained a considerably improved better-sampled dataset for the CO clouds in the galactic disk, we have revealed new features outside the disk including molecular supershells, some peculiar off-plane structures in the galactic center region, and distant molecular clouds located in the galactic warp. I will briefly review some of the notable topics of the NANTEN CO galactic plane survey.

1 Introduction

The 4-m radio telescope of Nagoya University equipped with the highest sensitivity SIS receiver at 115 GHz allows to cover a large area within a reasonable time at an angular resolution high enough to resolve dense cores in nearby (within 1kpc) dark clouds and also to resolve distant (up to 30kpc from the sun) giant molecular clouds. In 1996, the 4-m telescope "NANTEN" was installed at Las Campanas observatory in Chile under mutual collaboration with the Carnegie institution of Washington, and we started a CO galactic plane survey of the southern sky.

Figure 1 shows the latest CO total intensity map of the southern galactic plane obtained with the NANTEN telescope. This map consists of more than 1,100,000 spectra. The observing grid spacings are $4'$ between 5 degrees from the galactic plane and $8'$ for the area above 5 degrees in the galactic latitude with a $2.6'$ beam. In the longitudinal direction almost 200 degrees, i.e. $l =$ $220°$ to $60°$ have been covered. The velocity resolution and coverage are 0.65 km s^{-1} and $\sim$500 km s^{-1}, respectively. Typical rms noise fluctuations are $\sim$ 0.4K at a velocity resolution of 0.65 km s^{-1} In this contribution, I would like to present three recent topical results obtained through the NANTEN Galactic plane CO survey. A fuller account of this survey will be published in the third NANTEN special issue of PASJapan in 2004.

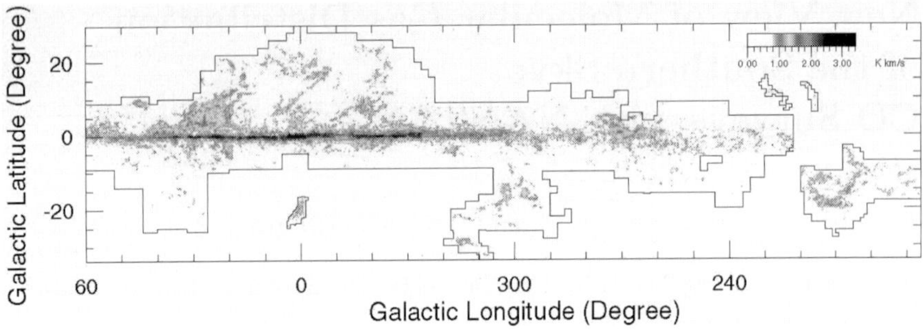

Fig. 1. ^{12}CO total intensity map obtained with the NANTEN telescope.

2 CO Shells and Supershells

Supershells are large cavities created by multiple supernovae, having a size greater than $\sim$ 100 pc in the interstellar space mostly identified in the HI emission in the past. Some of the supershells exhibit expansion at a velocity greater than 10 km s^{-1}, and their kinetic energies are more than 10^{52} erg. So far, more than a few hundred HI shells including galactic worms have been identified in the galaxy (e.g., [6, 8]). On the other hand, the number of CO supershells, large shell structure identified in CO emission, are quite few because of the lack of systematic CO searches out of the plane to date.

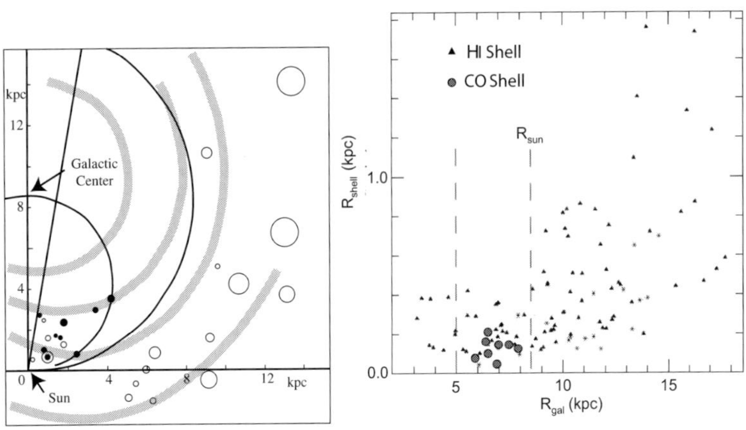

Fig. 2. (Right) Distribution of the CO shells (filled circles, [10]) and the HI shells (open circles, [12]) in the 4th quadrant. The gray thick lines indicate the location of spiral arms. (Left) Galacto-centric radius vs. size distribution of the CO and HI shells.

Recently, we discovered a CO supershell in the Carina region, named "Carina flare", from sensitive observations with NANTEN [2], demonstrating that there actually exist molecular clouds associated with supershells. Subsequently, we started a systematic search for CO supershells with NANTEN. Matsunaga et al.[10] completed the search for CO supershells in the 4th galactic quadrant between 300° and 350° in the galactic longitude. Eight CO supershell candidates have been additionally identified by them. In Figure 2, we show the face-on view of the distribution of CO [10] and HI [12] supershells. It is notable that the coincidence between the CO and HI shells is fairly poor. Only one shell is detected in both CO and HI.

We summarize the characteristics of CO shells along with a comparison with those of HI shells: (1) CO shells are located along the galactic arms, but many HI shells are located in the inter-arm regions (see Figure 2a). (2) CO shells are smaller in size than most of the HI shells (see Figure 2b). (3) The dynamical timescales ($=R_{shell}$ / V_{exp}) of CO shells ($(0.2 - 1) \times 10^7$ yr) are smaller than those of HI shells ($(0.5 - 3) \times 10^7$ yr). These three characteristics are interpreted in terms of evolution of supershells as follows. It is likely that a supershell is formed in an OB association that is located in the spiral arm. In the spiral arm, the shell accumulates the ambient gas as it expands, and molecular clouds are formed in the shell. In this phase, the supershell tends to be observable as a CO shell rather than a HI shell because the cavity of the shell is embedded in and contaminated by the HI gas. After a typical crossing time, a few $\times 10^7$ yr, the shell should gradually move to the inter-arm region due to the galactic rotation, and the galactic shear may distort the shell. Then the expanding shell cannot collect the ambient gas efficiently any longer. Separation between the CO clouds in the shell becomes sparse, and the CO clouds themselves may dissipate due to the expansion or by forming stars. Thus, the shell-like feature of CO clouds is not significant in this phase. In the inter-arm region, the HI contamination becomes less significant, and the shell becomes more obvious in HI than in CO. In this scenario, the HI and CO shell corresponds to different evolutionary stages of a single shell, and this may explain the cause for the poor coincidence between the CO and HI shells.

3 CO Clouds in the Galactic Center Region

The galactic center is a unique region because of the high temperature of molecular gas and high-pressure environment compared with the galactic disk region. Previous CO observations were concentrated into a small area especially in the coverage of the galactic latitude, $b < 0.7°$ except for Bitran et al.[2] whose coverage exceeds up to $b = 2.0°$. It is, however, known that neutral gas traced by HI is extended up to $b \sim 5°$. Thus we have covered the area of $|b| < 5°$ at $4'$ grid spacing with an rms noise of 0.35K at a velocity resolution of 0.65 km s^{-1}. In addition to this, we made more sensitive observations toward HI clouds whose angular distance from the galactic center is greater than several degrees with an rms noise of 0.18K.

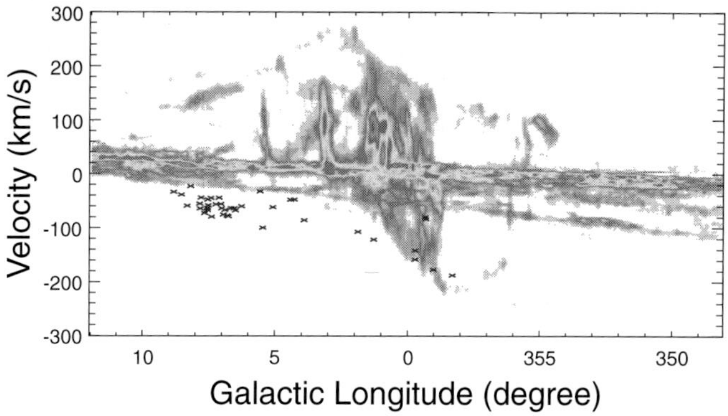

Fig. 3. ^{12}CO L-V diagram of the galactic center region. The crosses indicate the locations of the newly identified clouds at $|b| > 1°$.

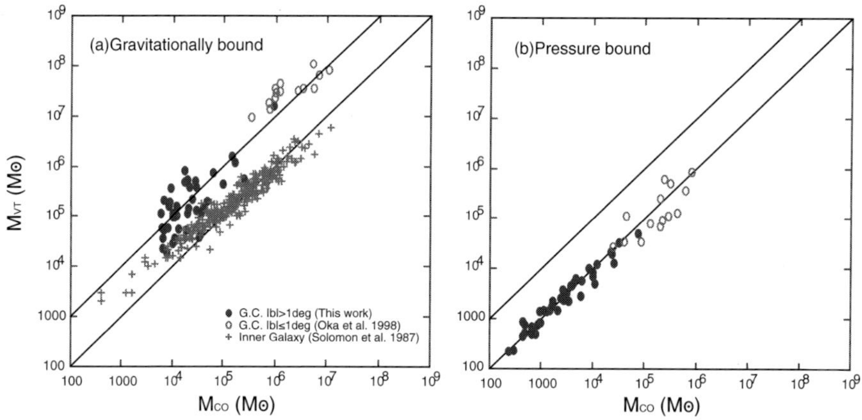

Fig. 4. (Right) M_{CO} vs. M_{vir} plot of the high latitude clouds in the galactic center region (blue filled circle), the galactic center clouds in Oka et al.[13](red open circle), and molecular clouds in the inner disk (red cross, [14]). M_{vir} is calculated by using the canonical X-factor. (Left) M_{CO} vs. M_{vir} plot by using virial theorem incorporating the external pressure term ($p/k = 1 \times 10^5$ K cm^{-3} for the present data and $p/k = 3 \times 10^5$ K cm^{-3} for Oka et al.[13], respectively).

The L-V diagram is shown in Figure 3. In addition to the previously known features, we have newly identified a few peculiar vertical features around $l \sim 355°$ and $\sim 5°$. These vertical features in the L-V diagram indicate unusually violent motions in the cloud and may be related with past energetic events at the galactic center or due to supernova explosions. More detailed discussion of these vertical features will be presented elsewhere (Yamamoto et al. in preparation).

We have further identified 42 smaller molecular clouds whose distribution in the L-V space is indicated also in Figure 3. The location of these clouds appears to coincide with the HI emission in the higher latitude regions where $|b| > 1°$ (van der Kruit, 1970). It is important to note that the radial velocities of the higher latitude clouds are located in so-called "forbidden region". Figure 4(a) shows the distribution of the high latitude clouds in a diagram of the virial mass vs. molecular mass calculated from CO luminosity using a canonical CO luminosity-molecular column density conversion factor, so-called X-factor, $\sim (2-3) \times 10^{20}$ cm^{-2} / K km s^{-1} (e.g., [14, 1]. This plot suggests that the molecular clouds are not bound by their own self-gravity but by the high ambient pressure in the galactic center region. Oka et al.[13] estimated the ambient pressure and X-factor for the molecular clouds in the central part where $|l| < 2.5°$ and $|b| < 1°$ by using virial theorem incorporating the external pressure term. We applied the same fitting method as Oka et al.[13] to the present sample of the high latitude clouds, then the ambient pressure and X-factor are estimated to be $p/k = 1 \times 10^5$ K cm^{-3} and 0.6×10^{20} cm^{-2} / K km s^{-1}, respectively. The values of the ambient pressure and X-factor of the present sample are in between those of the inner galaxy and the galactic center.

4 Molecular Clouds in the Galactic Warp Region

We also surveyed CO clouds in the outer extreme region of the galaxy, the galactic warp. The physical condition in the warp region is characterized by low HI density, weak stellar gravitation field, less stellar UV flux, and perhaps, the less cosmic ray flux, and the warp region is a unique site to study the formation process of molecular clouds and stars under such an exotic environmental conditions. Several efforts have been made to search for CO clouds in the outer galaxy (e.g., [11, 16]), but they were limited in spatial resolution or biased toward IRAS sources.

We covered an area of 60 square degrees from $l = 252°$ to $266°$ and have identified 70 molecular clouds whose kinematic distances are greater than 14kpc from the galactic center. In order to drive the distances, we used the galactic rotation model with R_0 = 8.5kpc and Θ = 220km s^{-1}. Most of the 70 clouds were previously not known. The locations of the CO clouds in the R-z plane are plotted in Figure 5. The distribution of the CO clouds shows good correlation with the warped HI emission. The mass of the clouds ranging from $7 \times 10^2 M_\odot$ to $7.5 \times 10^4 M_\odot$ is significantly smaller than those of giant molecular clouds in the inner disk, but the power index of cloud mass spectrum -1.67 ± 0.16 for cloud mass $> 3000 M_\odot$ is similar to the canonical value in the galaxy (e.g., [9]). The X-factor derived from the 70 warp CO clouds is 5.4×10^{20} cm^{-2} / K km s^{-1}. This value is about 3 times larger than that in the inner disk.

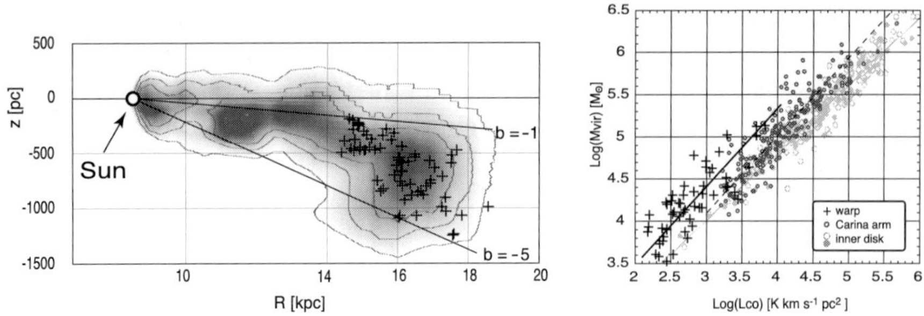

Fig. 5. (Right) Distribution of CO clouds in the warp region superposed on the HI gray-scale map [7] in the galacto-centric radius (R) vs. distance from the galactic plane (z) coordinate. (Left) CO luminosity vs. M_{vir} plot of the warp clouds (cross), Carina-arm clouds (open circle, Matsunaga et al. private communication), and inner disk clouds (gray open and filled circles, [14]).

5 Summary

The NANTEN CO survey has provided a new view of the molecular distribution in the southern sky, from the galactic center to the outer edge of the galaxy. Its larger coverage in the galactic latitude than the previous surveys and high sensitivity has allowed us to identify previously unknown weak features such as CO supershells. Detailed searches for weak CO emission toward the galactic center and the warp regions have revealed physical properties of CO clouds in peculiar environments in the galaxy.

NANTEN telescope is to be moved to Atacama area in 2004 with an upgraded main reflector having three times better surface accuracy. This project, NANTEN2, is motivated to explore the large-scale molecular distribution for the first time in the sub-mm region. The telescope will be equipped with the SMART array receiver[5] developed by KOSMA for mid-J CO and CI observations. This project is a collaboration between the NANTEN group (Nagoya University and Osaka Pref. University, Japan), the KOSMA group (University of Cologne and University of Bonn), the SRAO (Seoul, Korea), and the University of Chile (Chile).

Acknowledgements: The NANTEN project is based on the mutual agreement between Nagoya University and the Carnegie Institution of Washington. We acknowledge that this project was able to be realized by contributions from many Japanese public donators and companies. This work was financially supported by the Grants-in-Aid for the Science Research by the Ministry of Education, Science, Sports, and Culture.

References

1. D. L. Bertsch, T. M. Dame, C. E. Fichtel, S. D. Hunter, P. Sreekumar, J. G. Stacy, P. Thaddeus: ApJ **416**, 587 (1993)
2. M. Bitran, H. Alvarez, L. Bronfman, J. May, P. Thaddeus: A&AS **125**,99 (1997)
3. T. M. Dame, D. Hartmann, P. Thaddeus: ApJ **547**, 792 (2001)
4. Y. Fukui, T. Onishi, R. Abe, A. Kawamura, K. Tachihara, R. Yamaguchi, A. Mizuno, H. Ogawa: PASJ **51**, 751 (1999)
5. U. U. Graf, et al.: Millimeter and Submillimeter Detectors for Astronomy. Edited by Phillips, Thomas G.; Zmuidzinas, Jonas. Proceedings of the SPIE **4855**, 322 (2003)
6. C. Heiles: APJS **229**, 533 (1979)
7. F. J. Kerr, P. F. Bowers, P. D. Jackson, M. Kerr: A&AS **66**, 373 (1986)
8. B. -C. Koo , C. Heiles, W. T. Reach: ApJ **390**, 108 (1992)
9. C. Kramer, J. Stutzki, R. Rohrig, U. Corneliussen: A&AS **329**, 249 (1998)
10. K. Matsunaga, N. Mizuno, Y. Moriguchi, T. Onishi, A. Mizuno, Y. Fukui: PASJ **53**, 1003 (2001)
11. J. May, H. Alvarez, L. Bronfman: A&A **327**, 325 (1997)
12. N. M. McClure-Griffiths, J. M. Dickey, B. M. Gaensler, A. J. Green: ApJ **578**, 176 (2002)
13. T. Oka, T. Hasegawa, M. Hayashi, T. Handa, S. Sakamoto: ApJ **493**, 730 (1998)
14. P. M. Solomon, A. R. Rivolo, J. Barrett, A. Yahil: ApJ **319**, 730 (1987)
15. P. C. van der Kruit: A&A **4**, 462 (1970)
16. J. G. A. Wouterloot, J. Brand, W. B. Burton, & K. K. Kwee: A&A **230**, 21 (1990)

OB Star Formation Across the Galaxy

L. Bronfman[1], L.Å. Nyman[2], J. May[1], S. Faúndez[1], and C. López[1]

[1] Universidad de Chile leo@das.uchile.cl
[2] European Southern Observatory lnyman@eso.org

We present preliminary results of a new CS(2-1) survey of OB star forming regions and provide some statistics of their distribution in the Galaxy. The complete scenario of massive star formation for one particular GMC, at a distance of 1.6 kpc, is revealed through 1.2 mm continuum observations for spatial scales from 1 pc to 50 pc.

1 A Deep CS(2-1) Survey of Massive Star Forming Regions in the Milky Way; Completing the Sample

We have completed a new CS(2-1) survey of UC HII region candidates in the whole galactic plane, using the 15 m Swedish-ESO Submillimeter Telescope (SEST) and the Onsala Space Observatory (OSO) Telescope (Figure 1). A total of 715 IRAS point-like sources, with FIR colors of UC H II regions [1], were observed. Of these sources, 576 had been already observed but not detected in our original BNM CS(2-1) survey [3], because of poor sensitivity or narrow velocity coverage at the time. The remaining 139 sources were chosen to improve the galactic latitude coverage of the survey, which now includes from $b = -4^o$ to $b = 4^o$ all over the Galaxy. The new observations, 3 times more sensitive and with a full galactic velocity coverage, are ment mostly to detect sources far away from the Sun.

Completion of the present survey allows a better tracing of the galactic spiral structure and of the properties of massive star formation in the outer galaxy. Most important, it provides the community with the largest and most homogeneus observational sample of massive star forming regions in the Milky Way(cf. Bontemps et al., this volume). Such sample is presently of importance for testing and comissioning the new southern telescopes ASTE (Atacama Submillimeter Telescope Experiment, Japan) and APEX (Atacama Pathfinder Experiment, ESO-MPI-OSO), and will also serve as ancillary data for infrared space telescopes like SIRTF.

1.1 CS Observations Toward OB Star Forming Regions

One of the best specific tracers of dense gas in the ISM is the CS molecule. Its J=(2-1) rotational line, at a frequency of 99 GHz, has an excitation density threshold of 10^4 - 10^5 cm^{-3} and is ubiquitous toward regions of massive

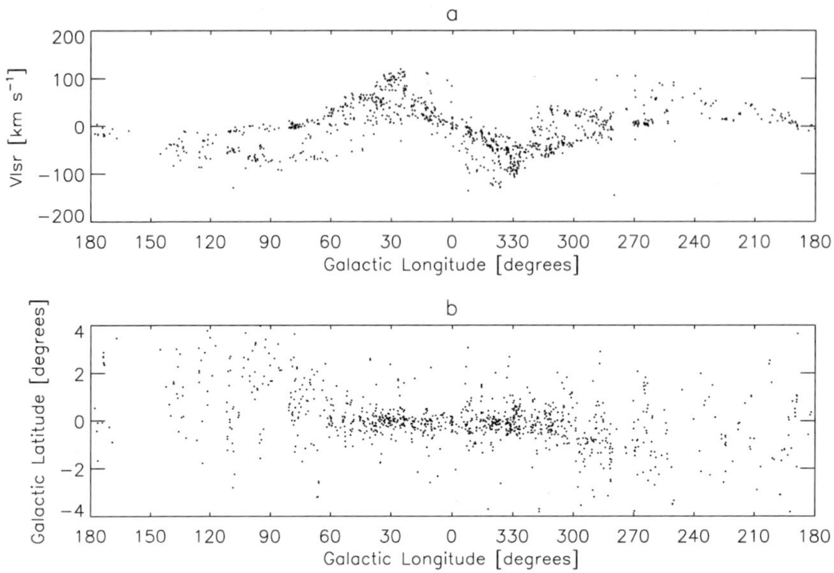

Fig. 1. Distribution of massive star forming regions in the whole Galaxy: (a) Longitude-velocity; and (b) Longitude-latitude.

star formation but seldom elsewhere in the galactic disk - with the exception of the galactic center region. A survey of the CS(2-1) line toward IRAS point-like sources with FIR colors of UC H II regions in the galactic plane, selected according to the criterion by Wood & Churchwell [1], was presented by Bronfman, Nyman, and May 1996 [3]. Nearly in all cases the CS(2-1) profiles, obtained with the SEST and OSO telescopes, are singular; the dense gas cores are associated one-to-one with the heated dust which produces the FIR emission detected by IRAS as a point-like source. More extended maps of 50 sources show that they are resolved at a resolution of 40 - 50 arcsec, generally not larger than 150 arcsec, and that the peak CS temperature is normally at the central IRAS position.

With 851 sources detected out of a total of 1427 observed, the BNM CS(2-1) survey is the most complete observational sample of massive star forming regions in the Milky Way published so far [1]. The usefulness of such sample is two-fold: on one hand, regions of massive star formation are the best tracers of galactic spiral structure and, on the other, within such a large sample we find the most interesting regions where to further study the elusive early stages of the formation of massive stars, like massive circumstellar disks and outflows. For example, the CS(2-1) BNM survey provided the main observing list for testing and comissioning of the SIMBA 1 mm bolometer at SEST, and is expected to play a similar rol for the new southern telescopes ASTE and APEX.

The 576 non-detected sources that are not low luminosity nearby objects, may (a) have been left out of the velocity coverage used in the original survey, that was limited to a longitude dependent range of 120 km s^{-1}, or (b) correspond to medium luminosity sources far from the Sun. Improving the completeness of the CS(2-1) sample with better sensitivity and velocity coverage was then crucial to trace the Milky Way structure far from the Sun as well as to derive the average physical properties of young massive stars and clusters at large galactocentri radii, and in particular to examine whether the IMF is different than for the inner Galaxy.

Completing the CS Survey

We observed with SEST, in the year 2000, 54 sources in the galactic center region, from $l = 346^o$ to $l = 14^o$ and from $b = -4^o$ to $b = 4^o$, detecting 21 of them - mostly falling in cathegory (a), i.e., were left out of the original velocity coverage. In year 2001 we reobserved with the OSO 20 m telescope 382 sources between $l = 14^o$ and $l = 232^o$, 346 undetected in the original sample, and 36 previously unobserved sources at latitudes from $b = -2^o$ to $b = -4^o$ and from $b = 2^o$ to $b = 4^o$, in the I galactic quadrant, where the outer galactic disk is severely warped. We detected a total of 138 sources, with more than half of them in the outer galaxy, an important improvement in our statistics for large galactocentric radii. We have now observed with SEST 279 sources, between $l = 232^o$ and $l = 346^o$, either not detected in the original CS(2-1) survey or at latitudes from $b = -2^o$ to $b = -4^o$ and from $b = 2^o$ to $b = 4^o$. The CS(2-1) survey has been finally completed with homogeneous sensitivity, full velocity coverage, and latitude coverage from $b = -4^o$ to $b = 4^o$ for the whole Galaxy.

Table 1. Complete Survey Statistics

Solar Circ.	IQ	IIQ	IIIQ	IVQ	Total
Within	334	0	0	386	720
Outside	73	140	117	88	418

Table 2. Old Survey Statistics

Solar Circ.	IQ	IIQ	IIIQ	IVQ	Total
Within	272	0	0	296	568
Outside	43	85	93	62	283

The total number of new sources detected in CS(2-1) is of 287 out of a total of 715 observed ones. An important fraction (47 %) of these sources are outside the solar circle. The complete source statistics per galactic quadrant including all the data is presented in Table 1. For comparison, in Table 2 we present the statistics for the original 1996 survey alone. The sources are divided in the table between those within and those outside the solar circle for the I and IV galactic quadrants. Our new survey has increased the original sample by a factor of 27% within the solar circle, and by 48% outside. A more full analysis of the data is in preparation and will be presented elsewhere.

2 The Complete Scenario for High-Mass Star Formation in a Giant Molecular Cloud

We have traced with the SIMBA Bolometer at SEST, in the 1.2 mm continuum emission, the distribution of dust, as well as the high mass star forming clusters and their possible connections, for the first time through one entire giant molecular cloud. Our goal is to determine the nature of the filamentary emission seen in the infrared by MSX, at scales of several tens of parsecs, and in preliminary SIMBA maps of the brightest UC HII regions within the cloud [2]. Are these filaments seen in emission at 8 microns because of hot dust caused by shocks, or they are seen because they harbour on-going star formation (like the Orion filament), or both? With SIMBA it is possible to detect the star forming cores within the filaments; further observations with e.g. ATCA will eventually give an indication of the ages of the embedded star forming clusters.

2.1 Massive Star Formation in GMCs

Massive stars (M> $8M_{\odot}$) are known to be formed in the dense cores of large molecular clouds. The gas in GMCs consist of filaments, sheets, and partial shells in which the active star forming dense cloud cores are embedded. Whether this large scale morphology is a consequence of the formation of massive stars or reflect the initial conditions for the formation of the dense cores is not yet well established.

Most studies of massive star formation have been carried out toward isolated regions, at scales of a few parsecs or less. There is no prescription, however, for the entire scenario of massive star formation within a GMC, about 50 pc in size, of the type found in spiral arms. Is it possible to trace observationally the process of massive star formation through a sequence of evolutionary stages within one GMC?

Many regions of massive star formation are associated with filaments of dust, and some of them are connected by these filaments at scales of several parsecs. We have carried out with SIMBA an extensive 1.2 mm continuum survey of 150 such regions in the southern Galaxy, selected from our CS(2-1) catalog of UC H II regions. In our ongoing analysis of the SIMBA data [2] we detect massive star formation along dust filaments in an important fraction of the images. The maps obtained, 10 arcmin in size, sample a spatial scale of about 4 pc in the best case, at a distance of 1.5 kpc, however, and cannot reveal the large scale distribution of the dusty filaments and the possible connections between individual regions of massive star formation within the GMCs.

2.2 Mapping a Full GMC in the 1.2 mm Continuum

We have now mapped in the 1.2 mm continuum with SIMBA one of the nearest GMCs, containing several separated clusters of massive star formation detected by IRAS and in CS(2-1). The cloud, GMC345.5+1.0, is in the near side of the Carina Spiral Arm, at a kinematic distance of 1.6 kpc, and well above the galactic equator so as to present very little confusion along the line of sight. It has an approximate size of 50 pc and subtends roughly 2 x 2 degrees in the sky (Figure 2). The area mapped was chosen to include all the MSX filamentary features, and to contain all the IRAS-CS sources in one continuous map.

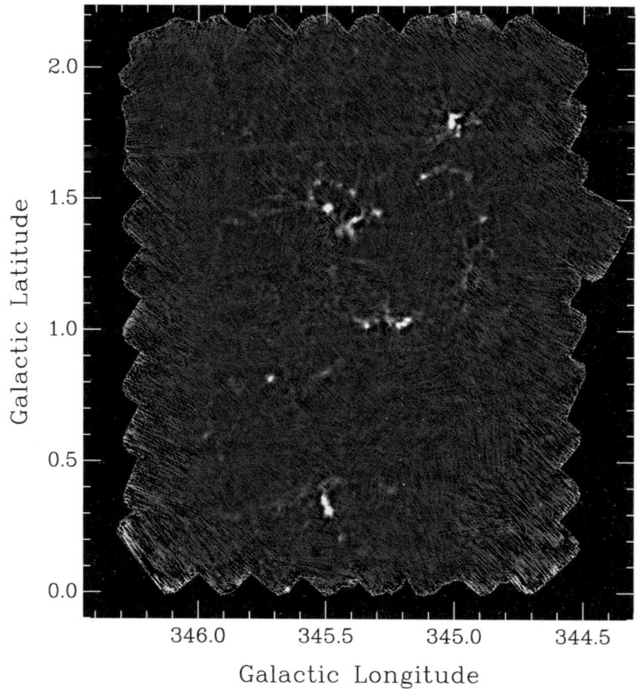

Fig. 2. 1.2 mm continuum map of GMC345.5+1.0 at 1.6 kpc

The selected cloud has a complex filamentary structure, as seen in 1.2 mm continuum, revealed in detail by our preliminar reduction of the SIMBA image. Such filamentary strucure is very similar to that seen by MSX at 8 microns. We have obtained for the first time a complete view of massive star formation in a giant molecular cloud, 50 pc in size, at a spatial resolution of 0.2 pc, unprecedent for such a large region. We have both resolved the sub-parsec structure at the scale of UC H II regions, and characterized the hot dust filaments at the scale of several parsecs. It would appear that massive star formation could occur preferentially along filaments in giant molecular clouds. It seems possible to trace in the 1.2 mm continuum large scale connections between the massive star forming regions through the filamentary structure of the dust within the GMC.

More full analysis of these data is necessary to determine the nature of the filamentary emission seen in MSX and in the SIMBA map. Are these filaments seen in emission at 8 microns because of hot dust caused by shocks, or a are they associated with on-going star formation? The SIMBA observations of this complete GMC should enable us also to ascertain the relation between clusters of star formation at different evolutionary stages, including the earliest ones.

Acknowlegements
We thank S. Casassus for useful discussions and F. Azagra for help with data reduction. We acknowledge support by FONDECYT Grant 1010431 and by FONDAP Center for Astrophysics 15010003.

CII, CI, and CO in the Massive Star Forming Region W3 Main

C. Kramer[1], H. Jakob[1], B. Mookerjea[1], N. Schneider[2], M. Brüll[1], and J. Stutzki[1]

[1] Universität zu Köln, I. Physikalisches Institut, Zülpicher Strasse 77, 50937 Köln kramer@ph1.uni-koeln.de, jakob, bhaswati, bruell, stutzki

[2] Observatoire de Bordeaux, Universite de Bordeaux 1, BP 89, 33270 Floirac, France schneider@obs.u-bordeaux1.fr

1 Introduction

We used the KOSMA 3m telescope to map the core $7' \times 5'$ of the Galactic massive star forming region W3 Main in the two fine structure lines of atomic carbon and four mid-J transitions of CO and ^{13}CO. In combination with a map of singly ionized carbon (Howe et al. 1991), and ISO/LWS data at the center position, these data sets allow to study in detail the physical structure of the photon dominated cloud interface regions (PDRs) where the occurance of carbon changes from CII to CI to CO. The physical and chemical processes in these layers are governed by FUV radiation. Depending on the induced chemistry, the critical densities, and energy levels of their transitions, the lines of [CII], [CI], and CO serve as probes to trace the different temperature and density regimes.

W3 Main is a well known site of high-mass star formation (e.g. Tieftrunk et al. 1998). It is associated with the W3 giant molecular cloud complex within the Perseus arm at a distance of 2.3 kpc (Georgelin & Georgelin 1976, cf. Imai et al. 2000). W3 Main comprises several HII regions of different sizes and evolutionary stages, some of them hypercompact, possibly identifying the earliest stages of high-mass star formation (Tieftrunk et al. 1997).

2 Modelling the Spatial Structure of the FUV Field

Figure 1 shows the positions of the 12 OB stars found in W3 Main (Tieftrunk et al. 1998). The most luminous source is IRS 2, identified as O5 star. The luminosity of IRS 5 at the map center position corresponds to a cluster of ~ 7 B0-type stars (Claussen et al. 1994). We have used the spectral types of the exciting OB stars to calculate the spatial large-scale variation of the far-UV field heating the molecular cloud. Figure 1 shows the resulting FUV distribution relative to that of [CII] emission (Howe et al. 1991).

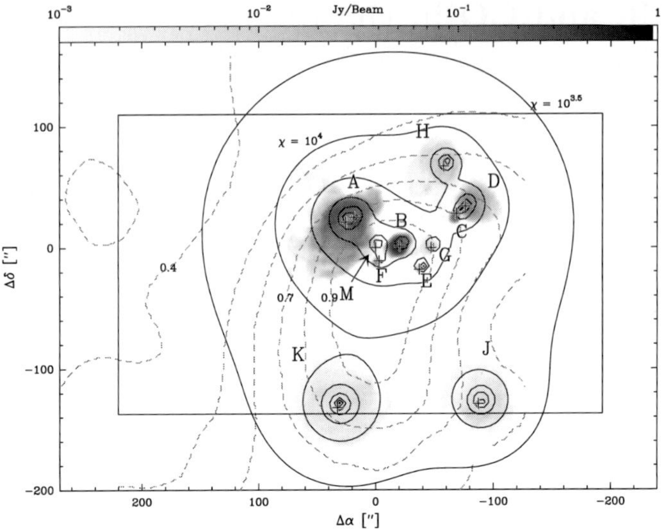

Fig. 1. Overlay of UV field modelled from the spectral types of the OB stars known in W3 Main, [CII] emission, and 6 cm radio continuum emission. **Grey-scale:** Image of the VLA 6 cm radio continuum (Tieftrunk et al. 1997). The box denotes the region mapped with KOSMA in CO 3–2, 4–3, 7–6, [CI] 1–0, 2–1, and in ^{13}CO 8–7. **Solid contours:** Contours of the modelled spatial structure of the FUV field derived from the spectral types of the embedded OB stars (positions are marked with a +). The UV flux χ varies between $10^{3.5}$ at the cloud edge and 10^{6} in the immediate vicinity of IRS 2. Attenuation by the cloud density structure has yet not been taken into account. **Dashed contours:** extended [CII] emission observed by Howe et al. (1991) with the KAO at 80″ resolution.

3 IRS 5 in W3 Main

3.1 KOSMA and ISO/LWS Data

Figure 2 shows representative spectra at the center position IRS 5. ^{12}CO 3–2 peak temperatures indicate an excitation temperature of at least 50 K. A lower limit of the [CI] excitation temperature of 122 K has been derived from the [CI] 2-1/1-0 line ratio assuming a calibration error of 20%. Column densities per 75″ beam (0.84 pc) and the relative abundances of the three major gas phase species CO, CI, and CII, are derived assuming LTE. At IRS 5, the CII:CI:CO column density ratio is 18:8:74, i.e. most of the gas-phase carbon is in CO, ∼ 20% is in CII, and only less than 10% is in atomic carbon. The total H_2 column density derived from ^{13}CO 3–2 observations is $4\,10^{22}\,\mathrm{cm}^{-2}$ ($A_V \sim 40$ mag) within the 75″ beam which agrees well with the result of Tieftrunk et al. (1998) derived from C^{18}O observations.

We combined the KOSMA data at IRS 5 with ISO/LWS data from the ISO data archive taken at the same position. Several fine structure lines of [OIII],

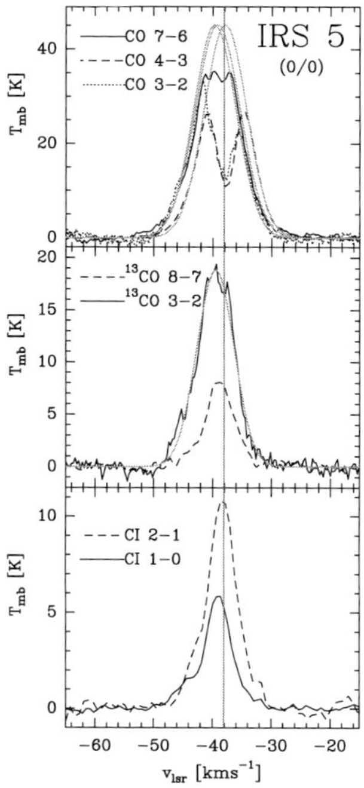

Fig. 2. Spectra at one representative position: IRS 5 in W3 Main. All spectra are on a common spatial resolution of 75″ (0.84 pc) and are plotted on the $T_{\rm mb}$ scale. ^{13}CO and CI lines peak at the velocity of a dip in the ^{12}CO spectra, indicating that the dip is due to a colder foreground component at about 38 km s^{-1} (denoted by a vertical line). To estimate the line integrated ^{12}CO intensity of the background source, we fitted a Gaussian to the high and low velocity parts of the ^{12}CO spectra. The fit results are shown.

[NIII], and [NII] were detected tracing the ionized medium, as well as lines of [CII], [OI], and four high-J rotational transitions of CO tracing the neutral medium at the PDR cloud surfaces. Figure 3 shows the observed CO and ^{13}CO fluxes together with the results of a homogeneous, non-LTE radiation transfer model. The detected line strengths indicate kinetic temperatures of more than 140 K and densities of more than $3\,10^4$ cm^{-3}.

3.2 Detailed PDR Modelling – First Results

We have started to determine the physical structure of the PDR by detailed modelling using the FUV field previously derived. Here, we briefly describe first results obtained from modelling the emission at IRS 5 using the spherical Cologne PDR model (Störzer et al. 1996, 2000). Its basic free parameters are the clump surface density, its total mass, and the impinging FUV field. We kept the FUV field fixed at 10^5 times the Draine field as derived from the contribution of all embedded OB stars. First results indicate a common solution for most ratios (see Figure 4). Future modelling will include the absolute line intensities and explore the possible importance of clumps of different masses (clump mass spectrum), and of pre-shielding by an embedding H_2 interclump medium.

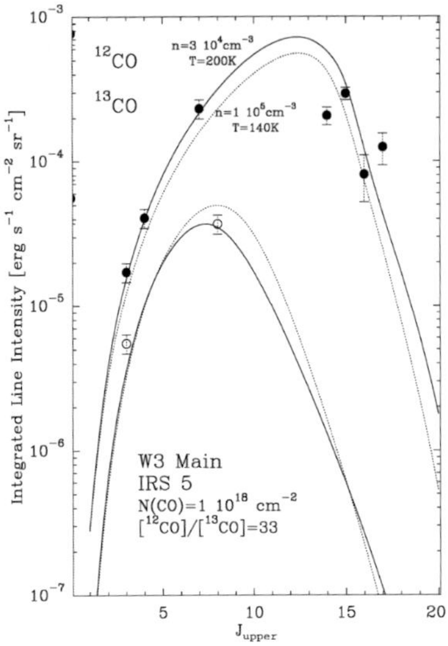

Fig. 3. Integrated CO intensities as a function of upper rotational quantum number J for IRS 5 in W3 Main and CO cooling curves produced by a non-LTE radiative transfer model. Rotational transitions upto $J = 8$ were observed with KOSMA, the high-J transitions upto $J = 17$ were observed with ISO/LWS.

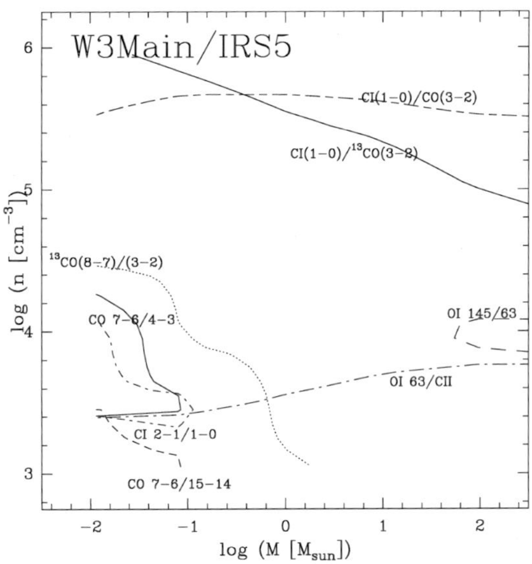

Fig. 4. Results from the PDR model for an external FUV field of $\chi = 10^5\,\chi_0$. Lines denote the observed line ratios at IRS 5 as a function of clump mass and clump surface density. The [CI]/CO and [CI]/^{13}CO flux ratios indicate high volume densities of more than $10^6\,\mathrm{cm}^{-3}$. The [OI 145μm/OI 63μm] ratio is larger than expected, probably due to optically thick [OI] 63μm emission and self-absorption due to the colder foreground material also responsible for the dips in the low- and mid-J CO lines. All other five observed ratios indicate a common solution of about $n \approx 10^4\,\mathrm{cm}^{-3}$ and a clump mass of less than $\approx 0.1\,\mathrm{M_{sun}}$ in this preliminary analysis.

High-Angular Resolution HC_3N and CS Observations of the Dense Core in the Cirrus Cloud MCLD123.5+24.9

C. Böttner[1], A. Heithausen[1], and F. Walter[2]

[1] Radioastronomisches Institut der Universität Bonn, Auf dem Hügel 71, 53121 Bonn cboettne@astro.uni-bonn.de, heith@astro.uni-bonn.de

[2] National Radio Astronomy Observatory, P.O. Box O, Socorro, NM, 87801, USA fwalter@nrao.edu

Abstract: We report on high-angular resolution observations of a dense core in the cirrus cloud MCLD123.5+24.9 obtained with the PdB and the OVRO interferometers. Our maps show substructures down to the scale of the beam ($\sim$ 1000 AU). The chosen molecules CS (2-1) and HC_3N (10-9) trace different regions in the core. This can be explained by time-dependent chemical evolution of the cloud and therefore provides constraints on the timescales of the fragmentation and core formation processes. Our data demonstrate that the chemical evolution plays the decisive role in our attempt to interpret observational data of different gas density tracers.

1 Introduction

Galactic cirrus clouds are typically complexes of diffuse gas and dust most easily seen at high galactic latitudes. On large scales they are dominated by turbulent motions, however, on small scales dense cores have been found and there is evidence that these cores are gravitationally bound (Heithausen et al. 2002). In one core, even the spectral signature of infall motion was discovered (Heithausen 1999).

Core formation constitutes the very first step which leads to star-formation and is still rather poorly understood. The knowledge of how cores form and how they evolve is important for understanding the minimal physical conditions needed for stars to form. Studying the evolution of dense cores in cirrus clouds, which live in a rather quiescent environment and are mainly heated by the interstellar radiation field, may prove to be crucial to interpret observations of star-forming regions correctly.

One of the best studied cirrus cloud cores so far is the one in MCLD 123.5+24.9, located in the Polaris Flare, a huge cirrus cloud complex in the direction of the north celestial pole. Its distance is between 130 and 240 pc (cf. Heithausen et al. 1993), and here we adopt a value of 150 pc. The temperature of the dust was determined to 13 K (Bernard et al. 1999) and the kinetic temperatures of the gas between 6 and 15 K (e.g. Großmann & Heithausen 1992).

Here we present new high-angular resolution HC_3N and CS observations of this core which allow a much closer insight into the fragmentation and core formation processes.

2 Observations

The observations of the HC_3N (10-9) transition at 90.98 GHz were conducted between May and November 2002 with the IRAM Plateau de Bure (PdB) interferometer. The standard CLIC reduction procedure was applied. Phase stability was checked with frequent observations of the quasars 1928+738 and 0716+714 and the amplitude scale was derived from measurements of MWC349 and CRL618.

The CS (2-1) line observations were carried out using the Caltech OVRO millimeter interferometer from January to April 2001. In total, 5 tracks were spent on source in the 'C' and 'L' configurations. The nearby source J1803+784 was used for phase calibration. The data for each array were edited and calibrated separately with the MMA and MIRIAD packages. The uv-data were inspected and bad data points due to either interference or shadowing between telescopes were removed, after which the data were calibrated.

We then corrected the data for zero spacing with our dataset obtained with the IRAM 30m telescope. The effective synthesized beam FWHM of $\approx 6'' \times 8''$ is similar in both observations (but with different position angles of PA $= -72°$ and PA $= -13.5°$, respectively) equivalent to 900 AU $\times$ 1200 AU at the distance of 150 pc.

3 Results

Our maps (see Fig. 1) show substructures down to the scale of the beam and the cores are clearly elongated. Most interesting is the complete misalignment of the HC_3N and the CS clumps, which was already suggested by the single dish data. However, the HC_3N clump was not resolved at all by the single dish observations and also the CS structures were barely resolved. The HC_3N clump resides at the southern edge of the core, is elongated and divides in substructure, whereas the CS emission is much more extended and the maximum, where the spectral signature of infall motion can be seen (cf. Heithausen 1999), lies more towards the north of the core. Nevertheless, there is still CS emission seen along with the HC_3N, even a small local maximum at the position of the HC_3N peak.

Heithausen et al. (2002) presented a dust continuum emission map of the core obtained with the MAMBO array at the IRAM 30m telescope with a resolution of $11''$, hence roughly comparable to our resolution. The dust shows again a somewhat different behavior than the gas components. The

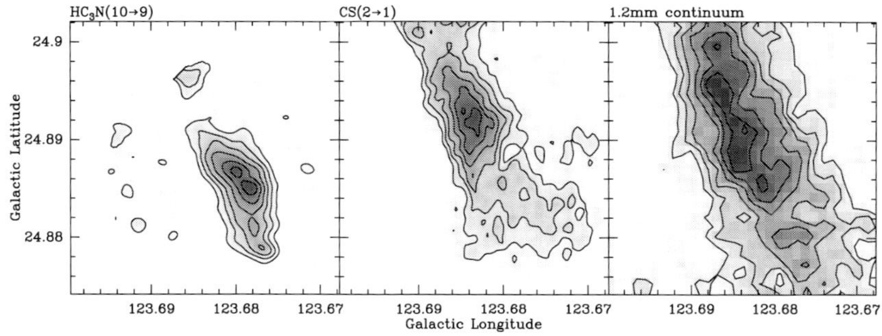

Fig. 1. Maps of the core in MCLD123.5+24.9 in HC_3N (10-9), CS (2-1) and the dust continuum at 1.2mm. The most striking feature is the displacement of the two molecular clumps, which are both embedded in the dust core.

filamentary structure of the whole core can be seen much better and only one elongated dense core, nevertheless with a lot of substructure down to the scale of the beam, is visible. This dust core corresponds well with the CS emission but is even more extended and clearly includes also the HC_3N clump. Therefore we can conclude that we are looking at one cirrus cloud core with a lot of substructure due to fragmentation and strong small scale variations in the molecular abundances.

4 Discussion

The most likely explanation for this appearance is that it is caused by strong abundance variations of the observed molecules due to time dependent chemical evolution. There are several chemical models in the literature which can account for the observed behavior. One that matches our derived properties of the core quite good is the one of Markwick et al. (2000), constructed for TMC-1, a low-mass star forming region. The initial conditions in this model are a temperature of 10 K, a hydrogen number density $n(H_2) = 2 \times 10^4 cm^{-3}$, a cosmic ray ionization rate of $1 \times 10^{-17} s^{-1}$, and the UMIST Database for Astrochemistry 1995 reaction set (Millar et al. 1995). This model predicts a time delay of HC_3N compared to CS in reaching a measurable high abundance of $(3-10) \times 10^5$ years (see Fig. 2). By this time the CS, which attained its highest abundance very fast after only 10^3 years, may already start to deplete due to freezing out on dust grains.

This result is further supported by a time-dependent chemical model of core formation by Taylor et al. (1998), which basically comes to the same results with somewhat different initial conditions. Their main conclusion is, that CS can be seen in more or less the whole core, whereas the HC_3N on the other hand can only be observed in longer-lasting clumps. This behavior is essentially a result of the cold (~ 10 K) gas chemistry where also neutral-neutral reactions play a surprisingly important role.

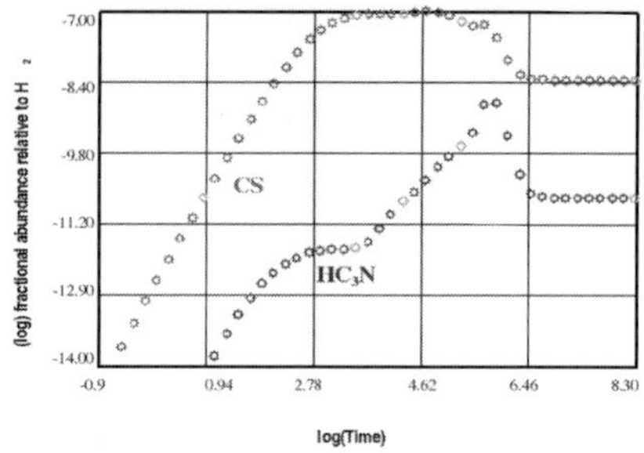

Fig. 2. Abundances of CS and HC_3N relative to H_2 with time in the chemical model by Markwick et al. (2000). HC_3N needs about 3×10^5 years more than CS to reach a measurable high abundance. (from www.astrochemistry.net)

5 Conclusions

We observed a core in the cirrus cloud MCLD123.5+24.9 with high-angular resolution in the molecules HC_3N and CS. Our data show a spatial offset of the maxima of both molecules and substructures down to the scale of 1000 AU. The misalignment can be explained by the interplay of time-dependent chemical evolution and fragmentation. The HC_3N traces only the older part of the core, whereas the CS is much closer tracing the density structure of the whole core as seen also in the dust continuum. This provides us with constraints on the timescales of the core formation process of a few 10^5 years, the time HC_3N needs to reach a measurable high abundance.

References

1. J.P. Bernard, A. Abergel, I. Ristorcelli, et al.: A&A **347**, 640 (1999)
2. V. Großmann & A. Heithausen: A&A **264**, 195 (1992)
3. A. Heithausen, et al: A&A **268**, 265 (1993)
4. A. Heithausen: A&A **349**, L53 (1999)
5. A. Heithausen, F. Bertoldi, & F. Bensch: A&A **383**, 591 (2002)
6. A.J. Markwick, T.J. Millar, & S.B. Charnley: ApJ **535**, 256 (2000)
7. T.J. Millar, P.R.A. Farquhar, & K. Willacy: A&AS **121**, 139 (1995)
8. S.D. Taylor, O. Morata, & D.A. Williams: A&A **336**, 309 (1998)

The KOSMA Multi-line CO Survey of Clouds in the Galactic Molecular Ring

M. Brüll[1], C. Kramer[1], V. Ossenkopf[1,2], J. Stutzki[1], R. Simon[1], and F. Bensch[3]

[1] KOSMA, I.Physikalisches Institut, Universität zu Köln, Zülpicher Strasse 77, 50937 Köln, Germany lastname@ph1.uni-koeln.de

[2] SRON National Institute for Space Research, Postbus 800, 9700 AV Groningen, the Netherlands

[3] Harvard-Smithsonian Center for Astrophysics, Cambridge, MA, U.S.A. fbensch@cfa.harvard.edu

1 Introduction

The Milky Way is known to be a spiral barred galaxy with a prominent ring between 4 and 7 kpc Galactocentric radius (Scoville & Solomon 1975; Clemens et al. 1988). Ring resonances occur frequently in barred spiral disk galaxies by gas accumulation under the continuous influence of gravitational torques from the bar pattern. As density enhancements they are often sites of increased star formation (Buta 1999). In an alternative view, this part of our Galaxy can be separated into two spiral arms, the Sagittarius and the Scutum(Cohen et al.1980). The Galactic Molecular Ring (GMR) shows up prominently in CO J=1-0 (Dame et al. 2001) as well as in C I, C II, and N II in the COBE survey (Fixsen et al.1999). Due to its high star formation rate, most of the Galactic giant H II regions, maser sources, and SN remnants are observed towards the ring and most of the Galaxy's FIR luminosity is emitted here (Burton et al. 1976, Bloemen et al. 1990).

2 The KOSMA Survey

To study the physical conditions of the molecular gas and their variation within the GMR, we have started a low-J multi-line CO survey of clouds in the GMR with the KOSMA 3m telescope on Gornergrat, Switzerland. This survey complements the Boston University ^{13}CO 1-0 GMR-survey at the FCRAO (Simon et al. 2001; Jackson et al. 2002). At present, we have mapped two fields, a 0.375 deg^2 field at l=45-46 and a 1 deg^2 field at l=30-31 on a fully-sampled grid in CO 3-2, 2-1 and ^{13}CO 2-1 with a velocity resolution of 0.3 km/s. One CO 3-2 channel map corresponding to gas with the velocity of the Scutum arm is shown in Figure 1. The star forming activity is associated with dense and warm molecular clouds traced by the CO 3-2 transition. Both regions contain extended foreground clouds. The GMR material in the G30-31 field is kinematically separated into two regions in the Sagittarius and

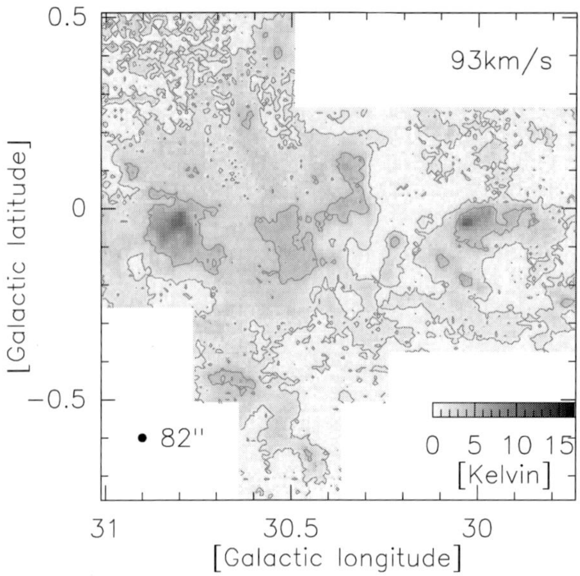

Fig. 1. One CO 3-2 channel map of the G30-31 field. The center velocity of 93 km/s corresponds to emission in the Scutum spiral arm. Contours are at 3σ and 10 σ. The field contains the prominent cloud W43, seen at l=30.8, b=-0.1.

Scutum spiral arms, whereas the GMR material in the G45-46 field is only located in the Sagittarius spiral arm.

The data set reveals many different clouds at various Galactic radii. We have started to analyse their properties using radiation transfer models. Here, we are concentrating on studying the observed emission structure.

3 Structure-Analysis

3.1 Cloud Structure in Different Tracers

In the G45-46 field (not displayed here), the foreground source is more fragmented in CO 3-2 than the GMR source which is dominated by large components. To quantify this observation, we used the Δ-variance analysis (Stutzki et al. 1998; Bensch et al. 2001; Ossenkopf et al. 2002) measuring the relative amount of angular structure σ^2(L) of a map depending on the structure size L. This method allows to separate the scaling behaviour of the cloud structure from observational noise and the telescope beam influence (Bensch et al.2001). Figure 2 shows a Δ-variance spectrum of the G45-46 field. The different degree of fragmentation visible in both G45-46 clouds is reflected by different slopes d_Δ in the Δ-variance spectrum (Fig.2). The result for the foreground cloud is nearly equal for the three tracers used for this analysis: d_Δ(CO3-2)=0.65(0.07), d_Δ(CO2-1)=0.63(0.07) & d_Δ(^{13}CO1-0)=0.59(0.05). For the GMR cloud we find: d_Δ(CO2-1)=0.79(0.07), d_Δ(CO3-2)=0.88(0.03) & d_Δ(^{13}CO1-0)=0.98(0.08). Bensch et al. (2001) found for different clouds a

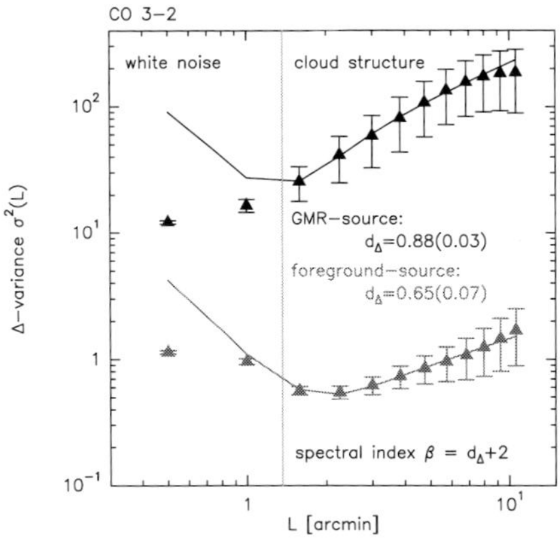

Fig. 2. Δ-variance spectrum of the CO 3-2 integrated maps of the G45-46 field as a function of lag-size L. The upper and lower points correspond to the GMR-source (40-80 km/s) and the foreground source (20-30 km/s), respectively. The data are fitted by a power law with radiometric noise and an approximate beam profile. The vertical lines gives the beam-size of 82″. At smaller lags the analytical fit does not work any more. The power law fit is best perfect in the significant range above the beam-size.

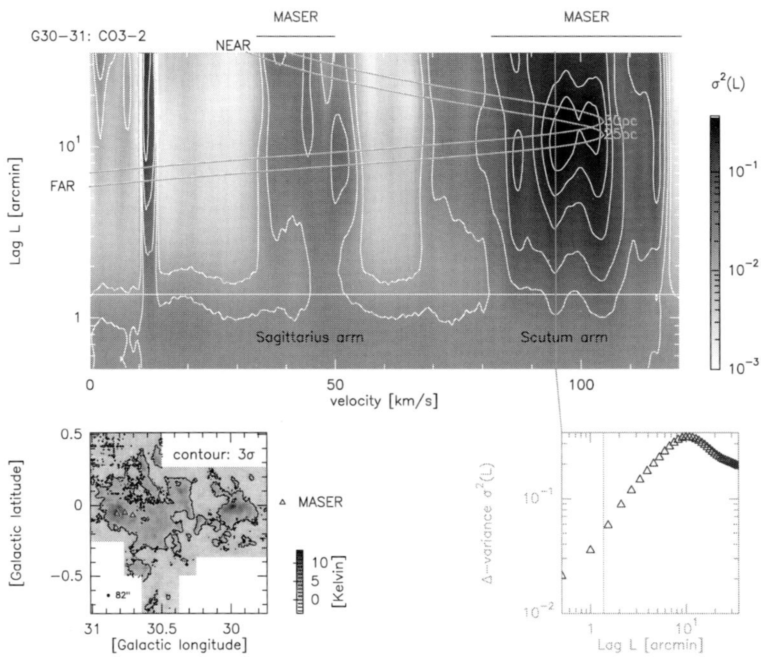

Fig. 3. Upper panel: The Δ-variance applied to velocity-channel maps of the G30-31 field in CO 3-2. In the plot of lag-size L versus LSR-velocity, the Δ-variance σ^2(L) is given in grey scales. The horizontal line corresponds to the beam-size. The contours are 2.5, 5, 10, 20, 40, 60 and 80 % of the maximal σ^2(L). The overlayed curves show the scala of 25 and 30 pc assigned by the rotation curve of Wouterloot & Brand (1989); it further illustrates the near/far distance ambiguity. On top the velocity-spread of OH-, H_2O- and CH_3OH-maser lines are shown. **Lower left panel:** One exemplary channelmap at ~95 km/s. **Lower right panel:** Δ-variance spectrum at this velocity. The vertical line corresponds to the beam-size.

slope of $0.5 \leq d_{\Delta} \leq 1.3$. Our results are in the range of their finding but allows for the first time the comparison of different tracers in the same maps. For all tracers, the slope for the GMR source is higher than for the foreground source, which means that the GMR source is dominated by larger structures. This difference is less obvious in CO 2-1, more pronounced in CO 3-2 and strongest in ^{13}CO 1-0. We explain this finding with widely extended CO 2-1 emission, similar in both clouds. The different slopes in the other tracers then tell us that these, more likely optically thinner lines, reveal different structures for both clouds behind the surface, traced by the CO 2-1 line.

3.2 Cloud Sizes

Figure 3 shows the application of the Δ-variance analysis to channel maps of the G30-31 field. In many velocity channels the Δ-variance spectrum shows a maximum indicating the maximal angular scale up to which the cloud structure is self similar. At the tangent point ($\sim$105 km/s) we find a typical angular length scale of this maximum of $\sim 10'$, corresponding to 25-30 pc (Fig.3). This scale can be assigned to the size of the dominant structures in the map. We find the same scale size for all tracers, used in this analysis: CO 3-2, CO 2-1 and ^{13}CO 2-1. We call this scale a typical structure or cloud size. Under this assumption, we can solve the distance ambiguity.

Many velocity channels also show a significant amount of structure at sizes larger than the maximum described above. They occur in the velocity channels where OH-, H_2O- and CH_3OH-maser lines were found. Their velocities correspond to the Sagittarius and Scutum spiral arms. We interpret this as the signature of extended diffuse material in the spiral arms. There is no material seen in the interarm regions.

References

1. F. Bensch, J. Stutzki, & V. Ossenkopf A&A **366** 636 (2001)
2. J. B. G. M. Bloemen, E. R. Deul, & P. Thaddeus A&A **233** 437 (1990)
3. W. B. Burton ARA&A **14** 275 (1976)
4. R. Buta: AP&SS **269**, 79 (1999)
5. D. P. Clemens, D. B. Sanders, & N. Z. Scoville: ApJ **327**, 139 (1988)
6. R. S. Cohen, et al. ApJ **239**, L53 (1980)
7. T. M. Dame, D. Hartmann, & P. Thaddeus,: ApJ **547**, 792 (2001)
8. D. J. Fixsen, C. L. Bennett, & J. C. Mather ApJ **526** 207 (1999)
9. J. M. Jackson et al. ApJ **566** L81 (2002)
10. V. Ossenkopf, M. Krips, & J. Stutzki A&A submitted (2002)
11. N. Z. Scoville & P. M. Solomon: ApJ **199**, L105 (1975)
12. R. Simon et al. ApJ **551** 747 (2001)
13. J. Stutzki et al. A&A **336** 697 (1998)
14. J. G. A. Wouterloot, & J. Brand A&AS **80** 149 (1989)

The Distribution of MSX Dark Clouds in the Inner Milky Way

R. Simon[1,2], J.M. Jackson[2], T.M. Bania[2], D.P. Clemens[2], and M.H. Heyer[3]

[1] I. Physikalisches Institut, Universität zu Köln simonr@ph1.uni-koeln.de
[2] Institute for Astrophysical Research, Boston University
jackson@bu-ast.bu.edu, bania@inanna.bu.edu, clemens@bu.edu
[3] Five College Radio Astronomy Observatory, University of Massachusetts, Amherst heyer@astro.umass.edu

Recent high resolution surveys with the *ISO* and *MSX* satellites have revealed a large number of Galactic clouds with significant extinction in the mid-infrared. The infrared dark clouds are characterized by their high masses, high column densities, and low temperatures. Little is known, however, about their origin and distribution in the Galaxy. We use *MSX* mid-IR data to catalog IR dark cloud and core candidates in the first and fourth Galactic quadrants. The distribution of IR dark clouds is concentrated around the Galactic mid-plane and traces prominent bright star forming regions. Selected high contrast IR dark clouds in the first Galactic quadrant are correlated with ^{13}CO emission from the Galactic Ring Survey to spectroscopically determine their distances. The majority of IR dark clouds is found towards the 5 kpc Galactic Ring. We suggest that the condensations in the most massive IR dark clouds represent high mass proto-clusters, or OB-associations in the making.

1 GRS and MSX Data Sets

The Galactic Ring Survey (*GRS*) is a collaboration between Boston University and Five College Radio Astronomy Observatory (FCRAO) designed to probe the star-forming and quiescent molecular gas comprising the Galactic Ring. For details on the *GRS*, see Simon et al. (this volume).
The Midcourse Space Experiment (*MSX*) surveyed the entire Galactic plane within $|b| < 5°$ in four infrared bands centered on 8.3, 12.1, 14.7, and 21.3 μm [1]. Observational parameters of the two surveys are summarized in Table 1.

Table 1. Survey parameters for the *GRS* and *MSX* surveys.

Survey	Wavelength	Coverage	Ang. res./Grid
GRS	^{13}CO(1-0) 2.6mm	$l = [15°, 54°]$, $b = [-1°, 1°]$	$46''/22''$
MSX	8μm, 12μm, 15μm, 21μm	l: whole plane, $b = [-5°, 5°]$	$20''/6''$

2 Infrared Dark Clouds: Identification and Distribution

Many high contrast IR dark clouds are readily identified by visual inspection of the *MSX* images (Fig. 1). To systematically identify dark clouds, we determine the depths (contrasts) of the IR absorption features with respect to their local backgrounds from the *MSX* data using median filtering. For this, we use the Band A images, since they have the highest sensitivity and the brightest background due to PAH emission in this Band.

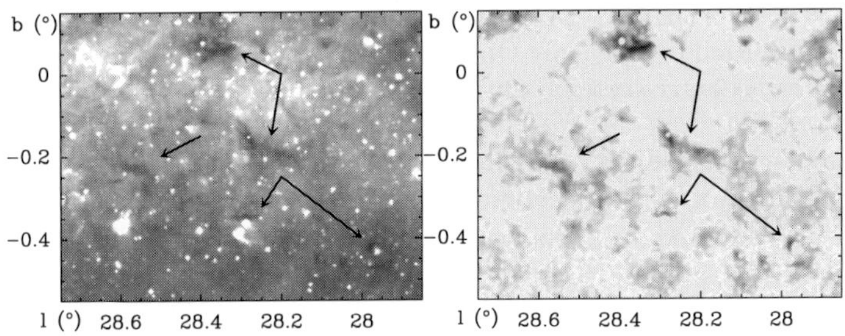

Fig. 1. Left: *MSX* Band A image towards $l = 28.3°$. A number of IR dark clouds can easily be identified against the bright background of the Galactic plane. Right: Dark cloud contrast image obtained from the *MSX* data and the background model. The minimum and maximum contrasts plotted are 0.05 and 0.65. The *MSX* dark clouds show up as high contrast objects (indicated by arrows).

The final cloud identification was done using objects within the closed 10% contrast contour. Our cloud catalog contains ~12,000 IR dark clouds. Their distribution with Galactic longitude and latitude is shown in Fig. 2.

3 The Galactic Ring Survey: A Key to Cloud Distances

With the BU-FCRAO Galactic Ring Survey it is possible to spectroscopically determine distances to the IR dark clouds and to derive physical parameters, such as sizes and masses, of the dark cores and their parental molecular clouds. Based on morphological correlation of IR extinction and *GRS* ^{13}CO emission in velocity channel maps, we assign radial velocities to the IR dark clouds throughout the first Galactic quadrant. Assuming a flat rotation curve with $(D_{\rm GC}, V_0)$=(8.5 kpc, 220 km s^{-1}) and that the clouds are at the near kinematic distance, we solve the distance ambiguity and determine their location in the Galaxy.

3.1 Results

We selected a total of 375 dark clouds in the *GRS* coverage of the first Galactic quadrant with contrasts higher than 20% and sizes larger than 1′. Of

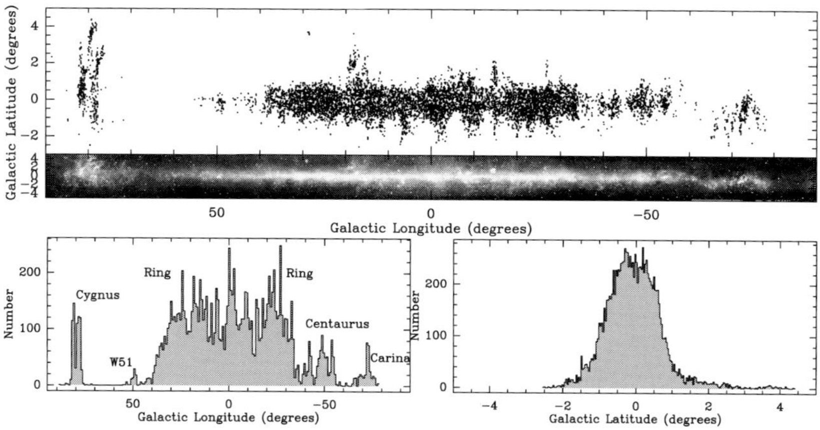

Fig. 2. Top: Galactic distribution of the *MSX* dark clouds. The *MSX* Band A emission is shown in the lower panel. Bottom: Number distributions of IR dark clouds with Galactic Longitude (left) and Latitude (right).

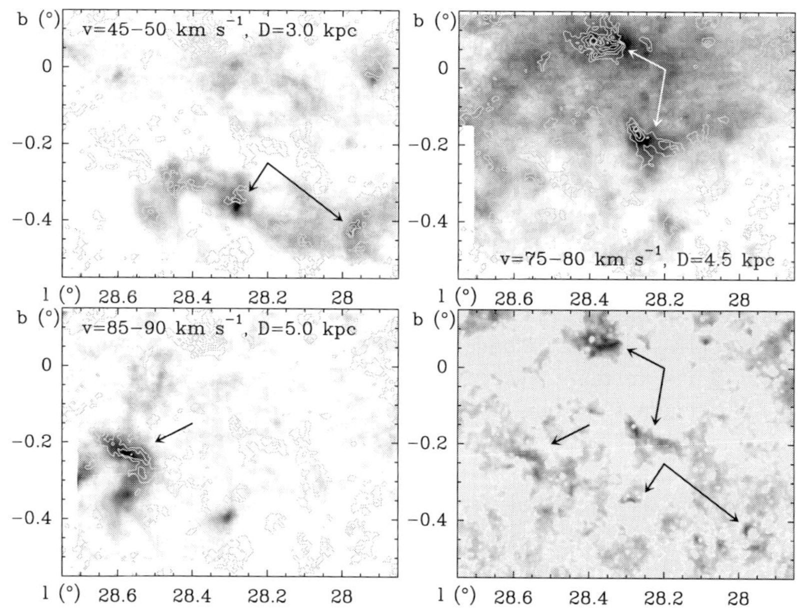

Fig. 3. *GRS* ^{13}CO(1-0) channel maps (grey scale) and dark cloud contrast contours towards the region around l=28.3°. Kinematic distances are assigned using morphological matches (indicated by arrows) of ^{13}CO emission with mid-IR extinction. The lower right panel shows the dark cloud contrast image from Fig. 1.

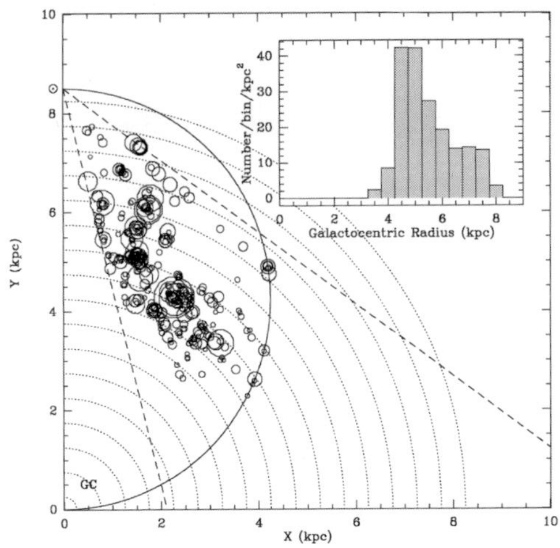

Fig. 4. Face-on distribution of 318 IR dark clouds in the first Galactic quadrant with kinematic distances determined from GRS ^{13}CO data. The sizes of the symbols represent different contrast ranges. The histogram shows the number distribution of IR dark clouds with Galactocentric radius, normalized by the area of the corresponding face-on segment.

these, 318 have clean morphological matches with *GRS* molecular line emission in distinct velocity channels. The remaining dark clouds have either low contrast, only weak *GRS* emission counterparts or molecular line emission in more than one velocity channel, or no *GRS* emission at any velocity. The latter are often found close to bright H II regions and are probably holes in the *MSX* emission.

High contrast IR dark clouds almost always have bright, extended ^{13}CO emission. Thus, in many cases, they are condensations inside larger Giant Molecular Clouds. Core masses are typically a few 100 to a few 1000 $M_{\odot}$. We therefore suggest that the condensations in the most massive IR dark clouds represent high mass proto-clusters, or OB-associations in the making.

This work was supported by the National Science Foundation via grants AST-9800334 and AST-0098562 and NASA LTSA grant NAG5-10808. Processing of the *MSX* data was funded by the Ballistic Missile Defense Organization with additional support from NASA Office of Space Science. For more information and available data, visit the *GRS* web page at www.bu.edu/GRS. The *MSX* data are available at irsa.ipac.caltech.edu/applications/MSX/.

References

1. Price, S. D., Egan, M. P., Carey, S. J., Mizuno, D. R., & Kuchar, T. A.: AJ **121**, 2819 (2001)

The Carbon Content in the Galactic CygnusX/DR21 Star Forming Region

H. Jakob[1], R. Simon[1], C. Kramer[1], B. Mookerjea[1], N. Schneider[2], S. Bontemps[2], and J. Stutzki[1]

[1] KOSMA, I.Physikalisches Institut, Universitaet zu Koeln, Germany
[2] Observatoire de Bordeaux, France

1 Introduction

Observations of Carbon bearing species are among the most important diagnostic probes of ongoing star formation. CO is a surrogate for H_2 and is found in the vicinity of star formation sites. There, [CI] emission is thought to outline the dense molecular cores and extend into the lower density regions, where the impinging interstellar UV radiation field plays a critical role for the dissociation and ionization processes. Emission of ionized carbon ([CII]) is found to be even more extended than [CI] and is linking up with the ionized medium. These different tracers emphasize the importance of multi-wavelength studies to draw a coherent picture of the processes driving and driven by high mass star formation. Until now, large scale surveys were only done with low resolution, such as the COBE full sky survey, or were biased to a few selected bright sources (e.g. Yamamoto et al. 2001, Schneider et al. 2003). A broader basis of unbiased, high-resolution observations of [CI], CO, and [CII] may play a key role to probe the material processed by UV radiation.

We here present a 2 deg^2 large-scale map of ^{13}CO 2-1 (see Figure 1) in the Cygnus X region and follow-up ($12' \times 14'$) maps of the DR21 region in the two [CI] fine structure lines and the ^{12}CO 7-6 rotational transition. Additional KOSMA observations of ^{12}CO 6-5, ^{13}CO 6-5, and ^{12}CO 3-2 are included for comparison. We compare parts of the KOSMA data with a [CII] map taken with KAO in 1994, ^{12}CO 9-8, and ^{13}CO 9-8 observations (Boreiko & Betz 1991, also KAO) and with publicly available ISO-LWS spectra at 7 distinct positions in the vicinity of DR21.

2 The Observations: CI vs. CO

The DR21 HII region/molecular cloud core has long been subject of detailed studies, which have led to a comprehensive view of the region. Measuring several parsecs in diameter at a distance of 1.5 kpc (Bontemps et al., in prep.), it harbors probably one of the most extended bipolar outflow in the Galaxy. The region contains a compact HII region at the center and two extended lobes of shocked H_2 emission running along an axis from NE to

SW. The molecular material found in the lobes was presumably ripped off a dense molecular ridge that is aligned N-S, and the wind originating from DR21 is about to break through the cavity (Lane et al. 1990).

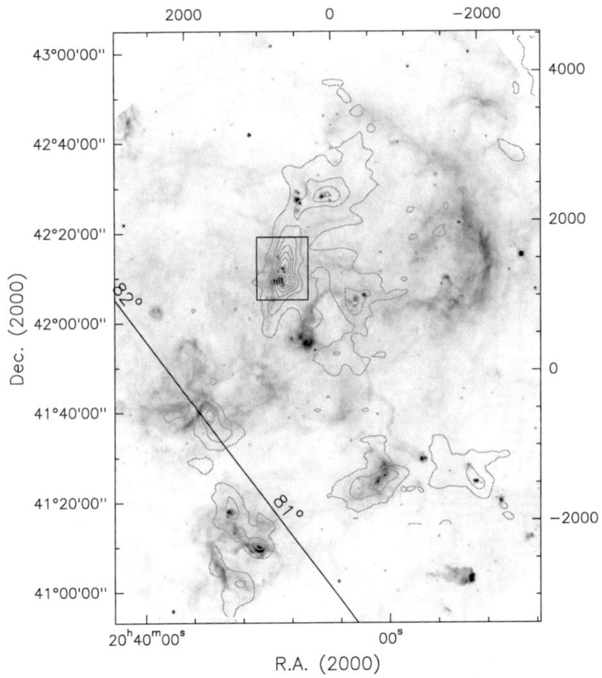

Fig. 1. KOSMA ^{13}CO 2-1 velocity integrated intensity between −6 and 1 kms^{-1} (contours) on top of an 8 μm MSX image. The DR21 region mapped in both CI lines and CO 7-6 is marked by a box (cf. Figure 2).

Besides ^{13}CO 2-1 (Figure 1), we have gathered observational data of both [CI] lines that are assumed to be optically thin. We find, that the ratio of the integrated intensity is almost constant at about 0.9 in the mapped region (compare with Figure 2). This indicates that the [CI] lines have their origin in a warm environment. The emission roughly follows the shape of the low-J ^{13}CO maps. This clearly is in contrast to the CO 7-6 map, which shows strong emission toward the DR21 core and the outflow lobes, but not along the ridge in the vicinity of DR21(OH) and FIR1 (cf. Figure 3).

3 The ISO LWS Sample

Seven ISO-LWS positions (see Figure 3) cover the core (#2), the east and west lobe of the DR21 outflow (#1 and #3) and two positions in between (DR21 East and West). The two nearby sources DR21(OH) and DR21 FIR1 lie inside the dust lane to the north.

At position #2, the continuum peak corresponds to $L_{44-197\mu m} = 4 \times 10^4\ L_\odot$. This result is consistent with earlier results (Harvey et al. 1977), when

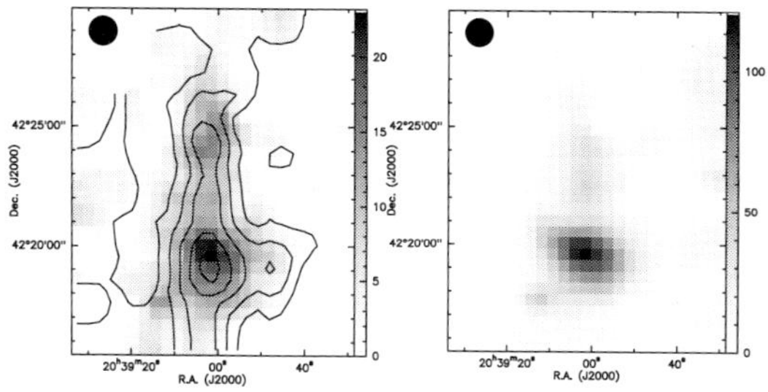

Fig. 2. KOSMA observations of [CI] $^3P_1{-}^3P_0$ (left, in contours), [CI] $^3P_1{-}^3P_0$ (greyscale) and CO 7-6 (right) of the DR21/DR21(OH) region.

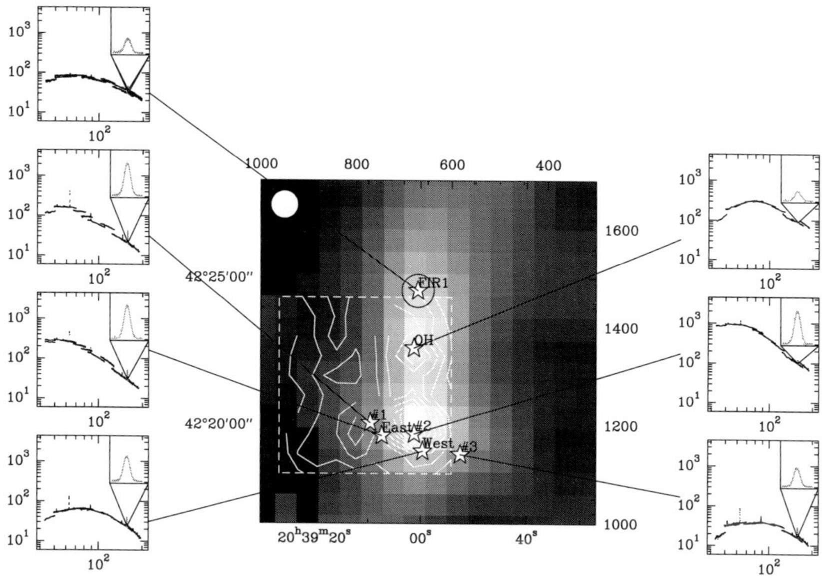

Fig. 3. Overlay of KOSMA ^{13}CO 3-2 (gray) and KAO [CII] data (in contours). The boxes show the ISO/LWS-FIR continuum and lines (e.g. high-J CO, [CII], [OI] and [OIII]) with a closeup view on the (unresolved) [CII] line at 158μm on the right. Circles indicate the KOSMA and ISO beamwidths.

using the old distance estimate of 3 kpc instead of 1.5 kpc. All major cooling lines are most pronounced, but the line cooling efficiency ($L_{\rm Lines}/L_{\rm Cont}$) is much lower than towards the eastern and western lobes (up to 1% at position #1). Especially to the east, the 63μm [OI] line intensity shows a strong east-west asymmetry and extends beyond the continuum source.

4 Molecular Gas: Physical Conditions

We compare the observed ^{12}CO and ^{13}CO line intensities for position #2 with a set of escape probability models at different CO column and H_2 volume densities, as well as kinetic temperatures. Two models are found to match the KOSMA and ISO line intensities (see Figure 4). However, $n(H_2)$ varies by a factor of 10 between the two solutions, whereas N(CO) and T_{kin} are better constrained. We therefore included a ^{12}CO and ^{13}CO 9-8 observation of comparable beam width (Boreiko & Betz 1991). It is compatible with the 10^4 cm^{-3} model, but since ^{12}CO 9-8 is not fitting at all, we presume a calibration error for both lines. More valuable ^{13}CO observations with high-J are planed with KOSMA.

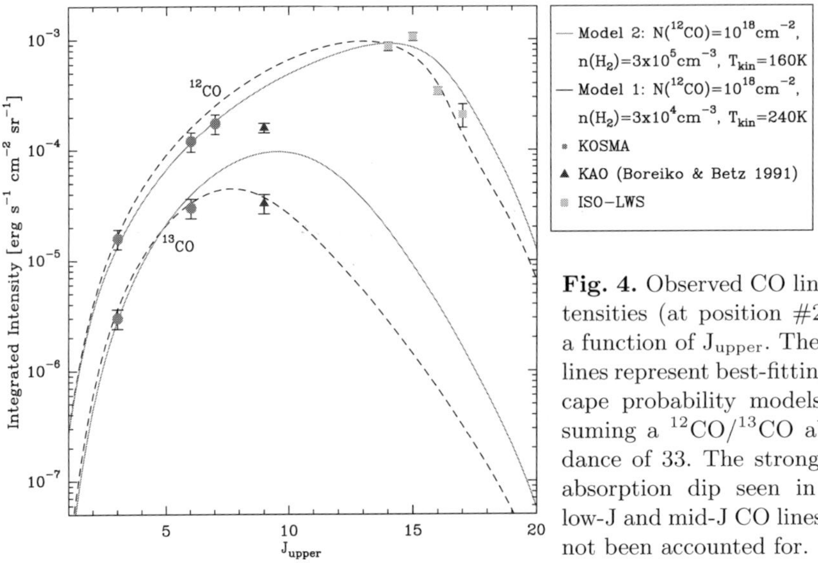

Fig. 4. Observed CO line intensities (at position #2) as a function of J_{upper}. The two lines represent best-fitting escape probability models assuming a $^{12}CO/^{13}CO$ abundance of 33. The strong self absorption dip seen in the low-J and mid-J CO lines has not been accounted for.

5 Perspectives

With this study, we aim at extending our observations of the emission due to the major coolants of PDRs in Galactic star forming regions (e.g. S106 Schneider et al. 2003; W3 Kramer et al., in prep.). These data constitute the basis for the application of radiative transfer and chemical PDR models in order to 1. constrain the physical conditions in regions of high mass star formation and 2. show whether all important physical processes are taken into account in the current models to reproduce the observed line intensities and ratios. Also, it allows to derive a consistent picture of the photon dominated regions in the vicinity of massive star formation. This serves as a preparatory work for future missions like SOFIA and Herschel, which will provide a wealth of new information including velocity resolved [CII] data.

Part V

Galactic Center

Probing the Galactic Black Hole and Its Environment with the Orbits of Stars Experiencing Their Closest Approaches

A.M. Ghez

Department of Physics & Astronomy, UCLA, Los Angeles 90095-1562
ghez@astro.ucla.edu

1 Overview

The proximity of our Galaxy's center presents an opportunity to build a case for a supermassive black hole and to study the black hole's environment and its effects thereon with much higher spatial resolution than can be brought to bear on any other galaxy. In this talk, I reviewed the progress that has been made based on data obtained at the W. M. Keck 10 m telescope; a complementary experiment carried out at the NTT and VLT is covered by the talk presented by Schoedel.

After almost a decade of astrometry from diffraction-limited speckle imaging, we have moved the case for a supermassive black hole at the Galactic Center from a possibility to a certainty, thanks to our recent ability to determine the orbits of individual stars, which confines the central dark mass of 4 million times the mass of the sun to within 90 AU (1 AU = the Earth-Sun distance), or equivalently, 1,000 Schwarzchild radii (see §2). With the advent of adaptive optics, we have significantly expanded our studies of the Galaxy's central black hole through the addition of diffraction-limited spectroscopy and deep imaging at wavelengths other than 2.2 microns. Spectroscopy has revealed that the stars orbiting in such close proximity are apparently massive and young; the origin of these stars is difficulty to explain, given the strong tidal forces, and may provide key insight into the growth of the central black hole (see §3). Thermal infrared imaging (3.8 microns) has led to the direct detection of plasma associated with the central black hole. This source is variable on timescales as short as 40 min, implying that the emission arises quite close to the black hole, within 5 AU, or 80 Schwarzchild radii and providing a new, constantly accessible, window into the physical conditions of the plasma in close proximity to the central black hole (see §4).

2 Stellar Proper Motions: The Case for a Supermassive Black Hole

Significant progress on mapping out the distribution of dark matter at the center of the Milky Way has been made recently with diffraction-limited near-infrared studies of the central stellar cluster. The first phase of these experiments yielded proper motion velocities (Eckart & Genzel 1997; Ghez et al. 1998), which suggested that $(2.6 \pm 0.6) \times 10^6 M_{\odot}$ of dark matter is confined to within 0.015 pc. This increased the implied minimum dark matter density by 3 orders of magnitude to $10^{12} M_{\odot}/pc^3$ and eliminated a cluster of dark objects as a possible explanation of the Galaxy's central dark mass concentration (Maoz et al. 1998), but still left the fermion ball hypothesis (e.g., Tsiklauri & Viollier 1998) as an alternative to a single supermassive black hole. The velocity dispersion measurements also localized the dark matter's centroid to within 100 mas and at a position consistent with the nominal location of the unusual radio source Sgr A* (Ghez et al. 1998), whose emission is posited to arise from accretion onto a central supermassive black hole (e.g., Lo et al. 1985). The detection of acceleration for three stars – S0-1, S0-2, and S0-4 – originating from the same center, localized the dark mass to within 30 mas, increased the dark matter's minimum density to $10^{13} M_{\odot}/pc^3$, and thereby further strengthened both the case for a supermassive black hole and its association with Sgr A* (Ghez et al. 2000; Eckart et al. 2002).

Deviations from linear motions also initiated a new phase for these proper motion experiments, that of direct orbital studies. By making a number of assumptions, including fixing the central mass to the value obtained from the velocity dispersion analysis and its location to that inferred for Sgr A* by Menten et al. (1997), Ghez et al. (2000) and Eckart et al. (2002) obtained first crude orbital solutions; these experiments revealed that orbital periods for S0-2 and S0-1 could be as short as 15 and 35 years, respectively. With more of the orbit being traced, more precise orbital analyses have been carried out for S0-2 by Schödel et al. (2002), who dropped the mass assumption, and by Ghez et al. (2003) and Eisenhauer et al. (2003), who dropped both the mass and center of attraction assumptions and added radial velocity measurements. Most recently, we've carried an analysis in which the orbits for multiple (8) stars are derived *simultaneously* so that they jointly constrain the central dark object's properties: its mass, its position and, for the first time using orbits, its motion on the plane of the sky (Ghez et al. 2004b). This analysis pinpoints the Galaxy's central dark mass to within 1.5 milli-arcsec (mas) and limits its proper motion to 0.8 ± 0.7 mas y^{-1} with respect to central stellar cluster. This localization of the central dark mass is consistent with our derivation of the position of the radio source Sgr A* in the infrared reference frame (± 7 mas), but with an uncertainty that is a factor 5 times smaller, which greatly facilitates searches for near infrared counterparts to the central black hole. The estimated central dark mass from orbital motions is $(4.0 \pm 0.3) \times 10^6 (\frac{R_o}{8kpc})^3 M_{\odot}$; this is a more direct measure of mass than those obtained from velocity dispersion measurements, which are as much as a factor of two smaller, and makes, for the first time, the Galactic center's distance, which adds an additional 19% uncertainty in the estimated mass, the limiting source of uncertainty in the absolute mass. For stars in this

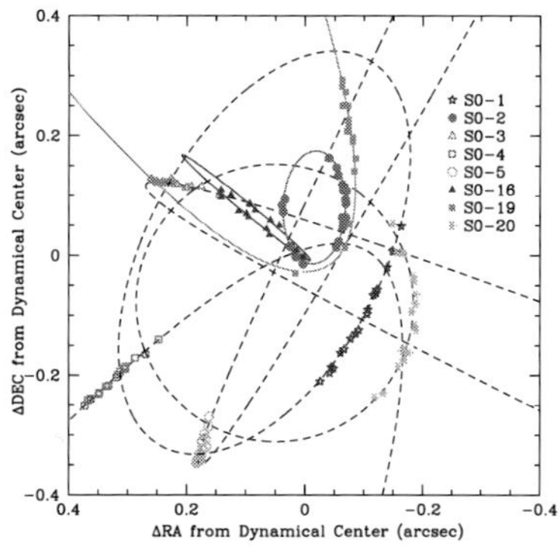

Fig. 1. Astrometric positions and solutions from simultaneous orbital fits for 8 stars that show significant deviation from linear motion. Three stars, thus far, have sufficient curvature to constrain the properties of the central dark object (stars with solid curves). An additional 5 stars have orbital parameters that can be determined in this simultaneous fit (stars with dashed curves).

sample, the closest approach is achieved by S0-16, which confines the mass to within a radius of a mere 90 AU (= 0.0004 pc = 1,000 R_{sh}) and increases the inferred dark mass density by four orders of magnitude compared to earlier analyses based on velocity and acceleration vectors, making the Milky Way the strongest existing case for a supermassive black hole at the center of any normal type galaxy.

3 Stellar Spectroscopy: A Paradox of Youth

With the introduction of an adaptive-optics-fed spectrometer on the W. M. Keck 10 m telescope, we have obtained the first detection of spectral absorption lines in one of the high-velocity stars in the vicinity of the Galaxy's central supermassive black hole (Ghez et al. 2003). Both Br γ (2.1661 μm) and He I (2.1126 μm) are seen in absorption in S0-2 with equivalent widths (2.8 $\pm$ 0.3 Å & 1.7 $\pm$ 0.4 Å) and an inferred stellar rotational velocity (220$\pm$40 km/s) that are consistent with that of an O8-B0 dwarf, which suggests that it is a massive ($\sim$15 $M_\odot$), young (<10 Myr) main sequence star. Less direct measurements of other stars within the central $1'' \times 1''$, which are known collectively as the Sgr A* stellar cluster, imply that these stars might be similarly young; specifically, their similar 2 μm luminosities and the lack of CO absorption in spectra of individual stars (Genzel et al. 1997; Gezari et al. 2002) or in integrated spectra of the Sgr A* stellar cluster (Eckart, Ott, & Genzel 1999; Figer et al. 2000) lead one to conclude that they, like S0-2, have hot photospheres consistent with massive young stars.

While the presence of young stars in close proximity to the Galactic black hole has long been recognized as a problem in the context of the young He I emission line stars (Sanders 1992, Morris 1993), this problem is much worse for the Sgr A* cluster stars, whose distances from the black hole are an order of magnitude smaller. At S0-2's apoapse distance of 0.01 pc, inferred from the orbital solutions, the Roche density is 10^{14} cm^{-3}, whereas the maximum density determined for even the nearby circumnuclear disk, located at radii of $\sim$ 1-3 pc, is only about 10^5 cm^{-3} (e.g., Jackson et al. 1993). Furthermore, at present, the region over which S0-2 is currently orbiting contains only a very low-density plasma, as evidenced by weak Br-γ line emission (Figer et al. 2000; Gezari et al. 2002). It is therefore challenging to form these stars given the tremendous tidal forces presented by the black hole. Several ideas proposed to account for the apparently young He I emission line stars may be applicable to the Sgr A* cluster stars; 1) that the early-type stars are the result of stellar mergers in this dense environment (Lee 1994; Genzel et al. 2003), 2) that they are exotic objects masquerading as normal stars (Morris 1993; Alexander & Morris 2003), 3) that they were formed by a cataclysmic compression of an already dense cloud which wandered into this region as a result of sudden accretion onto the supermassive black hole (Morris 1993; Levin & Belbedodov 2003), and 4) that they were formed elsewhere as part of a massive cluster, but migrated inwards rapidly by dynamical friction Gerhard 2001; Kim & Morris 2003; Portegies-Zwart McMillian & Gerhard 2003; Hansen & Milosavljevic 2003). None of these possibilities is altogether satisfactory, leaving the stars in the Sgr A* cluster as a paradox of apparent youth in the vicinity of a supermassive black hole.

The well determined orbital parameters for these 8 Sgr A* cluster stars provides new constraints on how these apparently massive, young (<10Myr) stars formed in a region that seems to be hostile to star formation. While the distributions of eccentricities, after correcting for selection effects, and the directions of the angular momentum vectors are consistent with those of an isotropic system, the distribution of the apoapse directions is not, at the 2σ level. Furthermore, there appears to be an inner cutoff in the distribution of apoapse distances for stars brighter than $K \sim 15.5$, with a minimum value of $\sim$1800 AU. Nonetheless, the apoapse distances are remarkably small in light of the spectroscopic identification of the majority of these stars as having hot photospheres consistent with massive, young stars (<10 Myr). Unfortunately, the existing theories that produce young stars, or old stars that look young, in close proximity to a central supermassive black hole are all somewhat problematic. However, if the bias in the distribution of apoapse directions is real, this would favor a formation scenario in which the Sgr A* cluster stars are young stars that have been tidally stripped from a dense inspiraling cluster near the cluster's periapse passage (<$\sim$1800 AU) and subsequently scattered inward. Understanding the apparent youth of stars in the Sgr A* cluster, as well as the more distant He I emission line stars, has now become one of the major outstanding issues in the study of the Galactic Center.

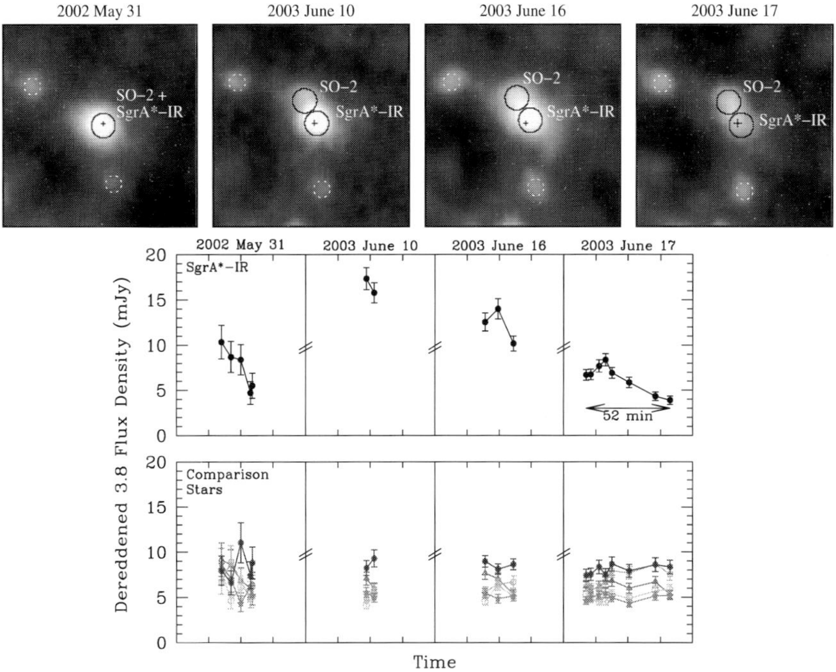

Fig. 2. *(Top)* A ~0.6″ × 0.6″ region of the L'(3.8 μm) Galactic center images obtained with the adaptive optics systems on the W. M. Keck II 10-meter telescope. In each image, a cross denotes the dynamically determined position of the central black hole and its uncertainties. An L' source, SgrA*-IR, is coincident with this position in all images; while in 2002 SgrA*-IR is blended with the stellar source S0-2, S0-2's significant orbital motion makes it well resolved from SgrA*-IR in the 2003 images. *(Bottom)* Dereddened 3.8 μm flux densities for SgrA*-IR and 5 nearby stellar sources of similar brightness. A panel of 80 minutes for each of the four nights of observations is shown. The comparison sources are S0-1 (circles), S0-3 (triangles), S0-4 (squares), S0-6 (asterisks), and S0-11 (stars). While the stellar sources show no significant variation, SgrA*-IR has varied by a factor of 4 over the course of a week, 2003 June 10-17, and a factor of 2 within 40 minutes on the night of 2003 June 17.

4 Infrared Detection of Sgr A*

The advent of Adaptive Optics has also permitted the detection of a variable point source, imaged at L'(3.8 μm) with the W. M. Keck II 10-meter telescope, that is coincident to within 18 mas (1 σ) of the Galaxy's central supermassive black hole and the unique radio source Sgr A* (Ghez et al. 2004a). While in 2002 this source (SgrA*-IR) was confused with the stellar source S0-2, in 2003 these two sources are now separated by 87 mas allowing the new source's properties to be determined directly. On four separate nights, its observed L' magnitude ranges from 12.2 to 13.8, which corresponds to a dereddened flux density of 4 - 17 mJy; no other source in this region shows such large variations in flux density - a factor of 4 over a week and a factor of 2 over 40 min. In addition, it has a K-L' color greater than 2.1, which is at least 1 mag redder than any other source detected at L' in its vicinity. Based on this source's coincidence with the Galaxy's dynamical center, its lack of motion, its variability, and its red color, we conclude that it is associated with the central supermassive black hole. The short timescale for the 3.8 μm flux density variations implies that the emission arises quite close to the black hole, within 5 AU, or 80 R_s. We suggest that both the variable 3.8 μm emission and the X-ray flares arise from the same underlying physical process, possibly the acceleration of a small populations of electrons to ultra-relativistic energies. In contrast to the X-ray flares which are only detectable $\sim$2% of the time (e.g., Baganoff et al. 2001; Goldwurm et al. 2003), the 3.8 μm emission provides a new, constantly accessible, window into the physical conditions of the plasma in close proximity to the central black hole.

References

1. Baganoff, F. K. et al. Nature, **413**, 45 (2001)
2. Eckart, A., & Genzel, R. MNRAS, **284**, 576 (1997)
3. Eckart, A., Ott, T., Genzel, R., A& A, **352**, L22 (1999)
4. Eckart, A., Genzel, R., Ott, T., & Schödel, R., MNRAS, **331**, 917 (2002)
5. Eisenhauer, F., Schödel, R., Genzel, R., Ott, T., Tecza, M., Abuter, R., Eckart, E., & Alexander, T., ApJ, **597**, L121 (2003)
6. Figer, D. F. et al., ApJ, **533**, L49 (2000)
7. Ghez, A. M. et al., ApJ, **586**, L127 (2003)
8. Ghez, A. M., Klein, B. C., Morris, M., & Becklin, E. E., ApJ, **509**, 678 (1998)
9. Ghez, A. M., Morris, M., Becklin, E. E., Tanner, A., & Kremenek, T., Nature, **407**, 349 (2000)
10. Ghez, A. M. et al., ApJ, submitted (astro-ph/0306130) (2004b)
11. Ghez, A. M. et al., ApJLett, in press (astro-ph/0309076) (2004a)
12. Goldwurm, A., Brion, E., Goldoni, P., Ferrando, P., Daigne, F., Dechourchelle, A., Warwick, R. S., Predehl, P., ApJ, 584, 751 (2003)
13. Jackson, J. M., Geis, N., Genzel, R., Harris, A. I., Madden, S., Poglitsch, A., Stacey, G. J., Townes, C. H., ApJ, **402**, 173 (1993)
14. Lo, K. Y., Backer, D. C., Ekers, R. D., Kellermann, K. I., Reid, M., & Moran, J. M. Nature, **315**, 124 (1985)
15. Maoz, E., ApJ, **494**, L181 (1998)
16. Menten, K. M., Reid, M. J., Eckart, A., & Genzel, R., ApJ, **475**, L111 (1997)
17. Morris, M., ApJ, **408**, 496 (1993)
18. Sanders, R. H., Nature, **359**, 131 (1992)
19. Schödel, R. et al. Nature, **419**, 694 (2002)
20. Tsiklauri, D., & Viollier, R. D. ApJ, **500**, 591 (1998)

The Galactic Center: Breakthroughs with VLT/NACO

R. Schödel[1], R. Genzel[2,3], and A. Eckart[1]

[1] Universität zu Köln, I.Physikalisches Institut, Zülpicher Str. 77, 50837 Köln, Germany rainer@ph1.uni-koeln.de
[2] Max-Planck-Institut für extraterrestrische Physik, Giessenbachstr., 85748 Garching, Germany
[3] Department of Physics, Le Conte Hall, University of California, Berkeley, California 94720, USA

1 Introduction

The galactic centre (GC) serves as an exemplary model for a quiescent galaxy nucleus. It can be observed at a level of detail that is not achievable in any external galaxy. Recent observations of the orbit of the star S2 have have provided new, highly significant evidence that the central non-thermal radio source Sagittarius A* (Sgr A*) is indeed a super-massive black hole [15, 10]. The ESO VLT was equipped with the new adaptive optics (AO) module NAOS and the NIR camera CONICA (NACO) at the end of 2001. The unique NIR wavefront sensor, in combination with the advantages of the ESO Paranal site (excellent seeing, GC passes close to Zenith), make NACO at the VLT the ideal instrument for studying the extremely dense GC stellar cluster and the immediate environment of the super-massive black hole Sgr A*. Observations with NACO/VLT have contributed decisive new insights about the GC already after less than two years of operations.

2 Central Stellar Cluster and Cusp Around Sgr A*

The luminosity of the stellar cluster in the central 0.5 pc of the Milky Way is dominated completely by a few dozen bright stars (K$\sim 9-12$). These stars are hot, blue, HeI emission line stars and AGB stars, and are witnesses of recent episodes of star formation (e.g. see [5, 4, 7, 12]). Therefore, in order to determine reliably the density profile of the GC star cluster, one cannot use the surface luminosity, but has to analyse the source counts of the faint stellar component. This requires deep, high-resolution NIR images. We have used NACO/VLT K- and H-band images in order to determine the crowding and incompleteness corrected surface number density of stars in the central 0.5 pc [9].

The surface number density of the cluster (see Figure 1) does not level off near the centre as in an isothermal cluster, but keeps increasing toward Sgr A*, i.e. there is a cusp of stars around the super-massive black hole

Sgr A*, consistent with theoretical expectations. The stellar density distribution can be described by a broken power-law of the form

$$\rho_\star(R) = 1.2 \times 10^6 \left(\frac{R}{10''}\right)^{-\alpha} \quad [\mathrm{M}_\odot\ \mathrm{pc}^{-3}]\,, \tag{1}$$

with $\alpha = 2.0 \pm 0.1$ at $R \geq 10''$, and $\alpha = 1.4 \pm 0.1$ at $R < 10''$ [9]. With these parameters the stellar density of the cusp is $3 \times 10^7\, M_\odot\ \mathrm{pc}^{-3}$ at $R = 1''$, and $7 \times 10^8\, M_\odot\ \mathrm{pc}^{-3}$ at $R = 0.1''$ [9].

The stellar cusp is centred on Sgr A* [$(\Delta\mathrm{RA}, \Delta\mathrm{Dec}) = (+0.09", -0.15")$], with an uncertainty of $\pm 0.2"$. The observations thus resolve the 17 year old puzzle of why Sgr A* is offset from the 2μm emission peak [1]. The offset is caused by the bright stars in the IRS16 complex, and is not a property of the majority of the faint stars in the overall nuclear cluster.

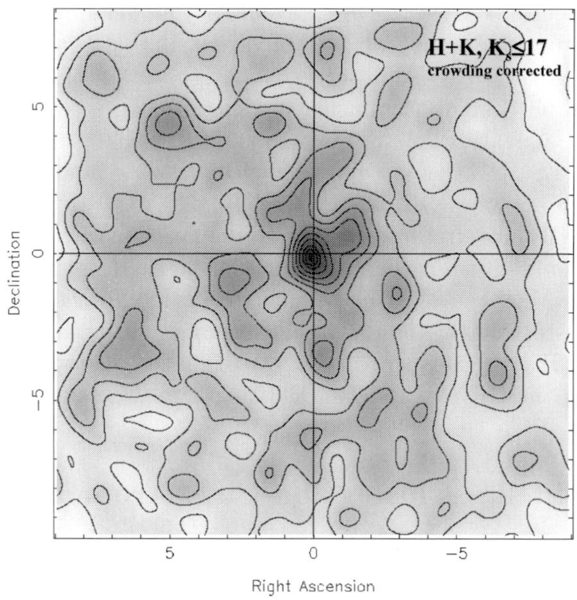

Fig. 1. Two-dimensional map of the smoothed surface density of the GC stellar cluster [9, 17]**.** Two-dimensional map of the smoothed, incompleteness and crowding corrected surface density of stars in the central 0.5 pc around Sgr A*. Contours are 10, 20,...,90, 95, 99 and 99.9% of the peak surface density.

3 Mass, Location, and Distance of Sgr A*

We combined new NACO/VLT AO NIR imaging data from the epochs 2002 and 2003 with NIR speckle imaging data, obtained with the instrument SHARP at the ESO NTT telescope in the epochs 1992 to 2001, in order

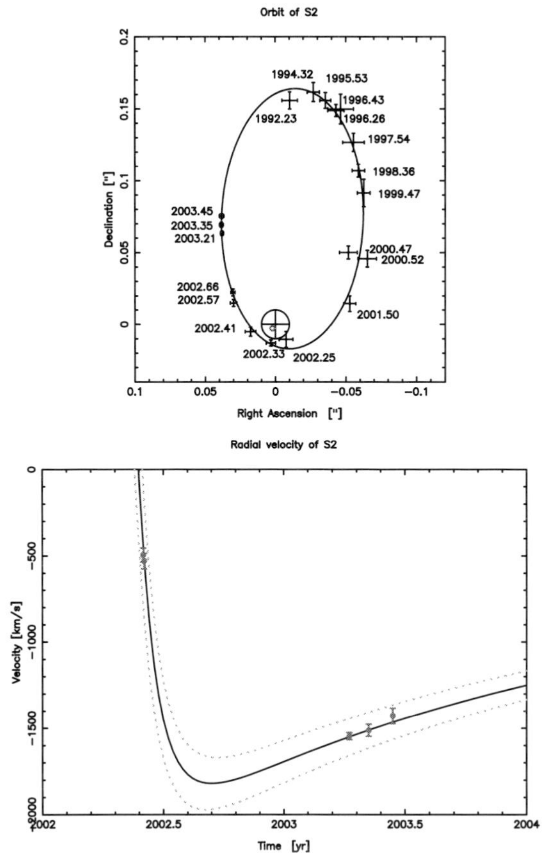

Fig. 2. Orbit and line-of-sight velocities of the star S2. The left panel shows the measured positions of the star S2 at each observing epoch and the best-fit orbit. The right panel shows the line-of-sight velocity of S2 vs. epoch and the spectroscopically determined velocities.

to measure the proper motions of the stars near Sgr A*[16, 17]. We detected significant acceleration for six stars, located in projection at < 20 mpc from Sgr A*, and determined Keplerian orbits for them. All orbits show that the stars move in the gravitational potential of a 3 to 4×10^6 $M_\odot$ point mass [16, 17].

The best-determined orbit is the one of the star S2 (Figure 2). More than 2/3 of its orbit have been observed. In combination with spectroscopic measurements of its line-of-sight velocity, we used this orbit to determine simultaneously the mass and location of Sgr A* as well as its distance from Earth [6]. The mass of Sgr A* is $3.6 \pm 0.6 \times 10^6$ $M_\odot$. Its position with respect to the surrounding stars is known with an uncertainty of < 2 mas and is consistent with the nominal radio position of Sgr A*. The distance to the GC was found to be 7.9±0.4 kpc. The result of this first geometric determination of this fundamental step on the cosmic distance scale is in good agreement with the value of 8.0±0.5 kpc, that was derived from a combination of various indirect measurements [14]. The orbit of S2 and several other stars around Sgr A* have been independently observed and determined by the group from UCLA, using the Keck telescope. See the article by A. Ghez in this volume and also [10].

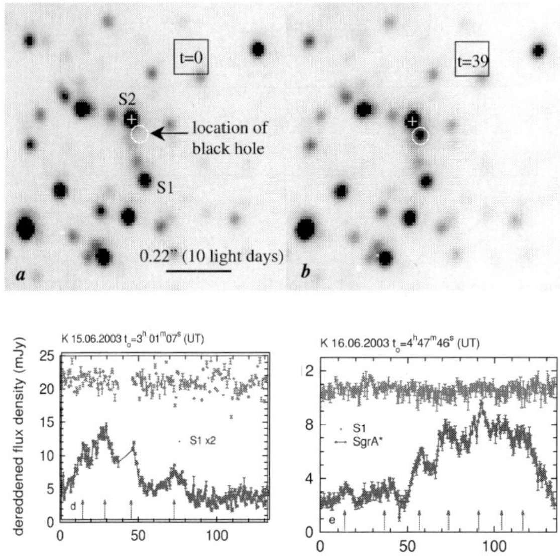

Fig. 3. Observation of NIR flares from Sgr A*. Upper panels: Raw AO images (60 s total exposure time) of an area $\sim 1'' \times 1''$ around Sgr A*, observed on May 9, 2003 UT. The left image was taken at the beginning of the observations, the right image about 40 min later. The position of the flaring source position is offset -1.4 ± 3.0 mas in R.A. and -0.2 ± 3.0 in Dec. from the dynamical position of Sgr A* [6]. The star S2 is marked by a cross, the position of Sgr A* is indicated by a white circle. Lower panels: Two Sgr A* K-band flares observed on June 15 and 16, 2003, with NACO/VLT. The lower light curves show the time-dependent variability of the emission from Sgr A*. Vertical arrows mark the sub-peaks that indicate a periodicity of about 17 min. For comparison, the upper, constant light curves in both plots show the emission from the non-variable star S1.

4 NIR Emission from Sgr A*

The search for a counterpart of Sgr A* was one of the major objectives since the beginning of this GC NIR observations. Surprisingly, the efforts stayed without unambiguous results and only allowed to estimate upper limits for

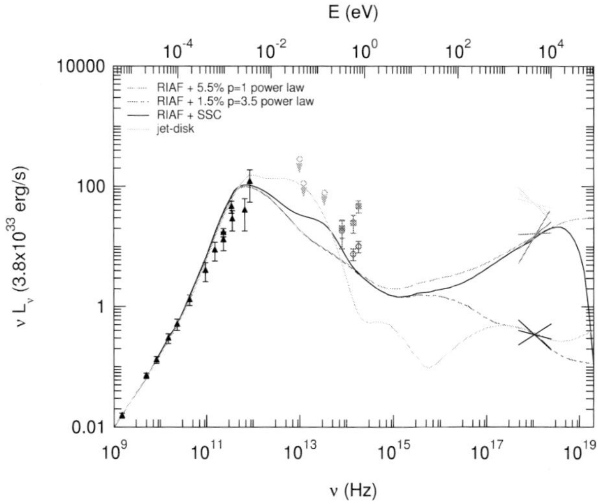

Fig. 4. Spectral energy distribution of the emission from Sgr A*. This plot shows the extinction and absorption corrected luminosities νL_ν, with the observed flux density $S_\nu = L_\nu/4\pi D^2$, where $D = 7.94$pc is the GC distance. All error bars are $\pm 1\sigma$. The data for the plot are taken from [11, 19, 9, 2, 3] and [13] (For a more extensive description of the data see [9]). The thick green solid curve is the jet-starved disk model by [18]. The red long dash-short dash curve is a radiative inefficient accretion flow (RIAF) model of the quiescent emission, where in addition to the thermal electron population of [18] 1.5% of the electrons have a non-thermal power-law energy spectrum of exponent $p = -3.5$. The black thin solid curve is a RIAF model of the flares with 5.5% of the electrons in a power-law of $p = -1$. The long-dash thick curve is a RIAF flare model of the flares with a synchrotron-self Compton component (for the models see Yuan, Quataert, & Narayan, 2003, astro-ph/0304125). This Figure corresponds to Figure 3 in [8]

the emission of Sgr A*. However, in May 2003, we could finally observe the first powerful flare (a factor of ~ 5 above the quiescent level) from Sgr A* in the H-band with NACO/VLT (Fig. 3, upper panels)). In subsequent observations and retrospective data analysis we detected flaring and quiescent emission from within < 10 mas of Sgr A* in the H- (1.7μm), K-2.2μm, and L-bands (3.8μm) [8]. The short rise-and-decay time scales (Fig. 3, lower panels) of ≤ 5 min indicate that the emission came from a region of less than ten Schwarzschild radii diameter (for a 3.6×10^6 $M_\odot$ black hole). The duration of the flares was between 10 and 85 min FWZM, with variability factors (excess emission above quiescent level) between about one (L-band) and five (H-band). From our data, we estimated a flare frequency of two to six per day. The flares seem to have similar properties to the X-ray flares [2, 13]. The NIR regime is critical for constraining models of Sgr A* accretion and emission. The current observations of quiescent and flaring emission are largely compatible with models where the emission is produced by a thermal population of electrons with a small percentage of them in a non-thermal power-law distribution (see Fig. 4 and its caption). Future simultaneous multi-wavelength observations will be essential for understanding the quiescent and flaring state accretion and emission mechanisms of Sgr A*.

The most striking feature of the flares is a $\sim$ 17 min quasi-periodicity found in two independent K-band flares that happened more than 24 hours apart on June 15 and June 16, 2003, respectively. Of the various periodic phenomena near a black hole (orbital motion, Lense-Thirring or orbital node precession, acoustic modes of thin accretion disks) the last stable orbit (LSO) has the shortest characteristic time scale: 27 min for a 3.6×10^6 $M_\odot$ Schwarzschild black hole. Hence, the even shorter observed period of 17 min, intrinsic to the flares, could only be explained by Doppler boosting of the radiation of material near the LSO if the Sgr A* has a spin parameter $a = J/(GM_{BH})/c = 0.52(\pm 0.1, \pm 0.08, \pm 0.08)$ (where G is the gravitational constant, M_{BH} the mass of the black hole, J its spin, and c the speed of light; the errors in brackets include the influence of the uncertainties in M_{BH}, the flare modulation frequency, and the distance to the GC). The observed spin parameter corresponds to half the maximum value for a Kerr black hole [8]. Should upcoming observations confirm this result, then Sgr A* is the first black hole where this fundamental parameter is accessible to observations.

References

1. D. A. Allen, & R. H: Sanders: Nature, **319**, 191 (1986).
2. F. K. Baganoff, M. W. Bautz, W. N. Brandt, et al.: Nature, **413**, 349 (2001)
3. F. K. Baganoff, Y. Maeda, M. Morris, et al.: Astrophys. J. **591**, 891 (2003)
4. Blum, R.D., Sellgren, K., & DePoy, D.L. 1996, Astrophys. J., **470**, 864 (1996)
5. Eckart, A., Gezel, R., Hofmann, R., Sams, B.J., & Tacconi-Gaarman, L.E., Astrophys. J., **445**, L23 (1995)
6. F. Eisenhauer, R. Schödel, R. Genzel, et al.: Astrophys. J., in press (2003)
7. R. Genzel, N. Thatte, A. Krabbe, H. Kroker, & L. E. Tacconi-Garman, Astrophys.J., **472**, 153 (1996)
8. R. Genzel, R. Schödel, T. Ott, T. Alexander, F. Lacombe, D. Rouan, & B. Aschenbach: Nature, **425**, 934 (2003)
9. R. Genzel, R. Hofmann, M. Lehnert, T. Ott, R. Schödel, T. Alexander, A. Sternberg, R. Lenzen, F. Lacombe, D. Rouan, M. Zoccali, & A. Renzini: Astrophys. J., **594**, 812
10. A. Ghez, G. Duchêne, K. Matthews, S. D. Hornstein, A. Tanner, J. Larkin, M. Morris, E. E. Becklin, S. Salim, K. Kremenek, D. Thompson, B. T. Soifer, G. Neugebauer, & I. McLean: 2003, Astrophys. J., **586**, L127 (2003)
11. S. Markoff, H. Falcke, F. Yuan, & P. L. Biermann: Astron. and Astrophys., **379**, L13 (2001)
12. F. Najarro, A. Krabbe, R. Genzel, D. Lutz, R. P. Kudritzki, & D. J. Hillier: Astron. & Astrophys., **325**, 700 (1997)
13. D. Porquet, P. Predehl, B. Aschenbach, et al.: Astron. & Astrophys., **407**, L17 (2003)

14. M. J. Reid: Annu. Rev. Astron. Astrophys., **31**, 345 (1993)
15. R. Schödel, T. Ott, R. Genzel, et al.: Nature, **419**, 694 (2002)
16. R. Schödel, T. Ott, R. Genzel, et al.: Astrophys. J., **596**, 1015 (2003)
17. R. Schödel: High-resolution Near-Infrared Imaging Observations of the Galactic Centre. PhD Thesis, Ludwig-Maximilians-Universität, München (2003)
18. F. Yuan, S. Markoff, & H. Falcke: Astron. & Astrophys., **383**, 854 (2002)
19. J.-H. Zhao, K. H. Young, R. M. Herrnstein, et al.: Astrophys. J., **586**, L29 (2003)

Galactic Center Molecular Clouds

R. Güsten[1] and S.D. Philipp[2]

[1] MPI für Radioastronomie, Auf dem Hügel 69, 53121 Bonn, Germany
rguesten@mpifr-bonn.mpg.de
[2] MPI für Radioastronomie, Auf dem Hügel 69, 53121 Bonn, Germany
sphilipp@mpifr-bonn.mpg.de

1 Introduction

At a distance of 8 kpc only, the center of the Galaxy provides a unique testbed for studies about the physics of the interstellar medium and star formation in the nuclei of galaxies. Since it shares many properties of other, more spectacular nuclei, a detailed investigation of the Galactic Center is a necessary step to gain better understanding of the physical processes governing galactic nuclei in general.

We review the characteristics of the central molecular cloud layer, the physical properties of these clouds, and their interplay with the magnetic field.

2 The Galactic Center Molecular Clouds

2.1 Large-Scale Distribution: Overview

The Galactic Center is characterized by a large concentration of dominantly molecular gas ($\sim 10^8 M_\odot$), associated with the innermost few hundred parsec of the Galaxy (the size of the molecular layer is 450 x 50 pc). Despite its exceptionally high gas surface densities ($\geq$100 $M_\odot pc^{-2}$), the Galactic Center is highly depleted, with only a small fraction of matter left in the interstellar gas phase, $\mu \sim 0.01$. The high degree of astration is reflected in the enhanced chemical enrichment of the gas with ejecta from evolved stars (Table 2).

The spatial distribution of the molecular layer about the nucleus is asymmetric ($\sim$2/3 displaced towards positive longitudes), with the gas column densities peaking towards the Sgr B2 complex (see [1] for an early data base). Much of the mass is concentrated in cloud complexes as large as 50-70 pc.

Already back in the 80ies, molecular excitation studies have revealed the exceptional physical properties of these clouds: compared with clouds in the Galactic disk, GC clouds are denser, warmer, more turbulent and more massive. More recent surveys of the warm CO and also the atomic fine structure lines extended well into the submm-regime ([25][16][17] with AST/RO, [29] with CHAMP at the CSO; see Table 1), thus better constraining the thermodynamics of the gas. Similarly, technology has made possible to develop large

Table 1. Compilation of some of the more recent mm/submm line and continuum surveys towards the Galactic Center. The area surveyed and the angular resolution of the observations is given.

Tracer	Surveyed area [deg] Longitude	Latitude	Beam	Reference
^{13}CO(1-0)	-0.8 to 1.7	-0.35 to 0.35	17"	Oka et al. 1998
^{12}CO(1-0)	-1.5 to 3.47	-0.6 to 0.6	16"	Oka et al. 1998
^{13}CO(2-1)	-0.23 to 0.27	-0.17 to 0.20	12"	Philipp et al. 2004
^{12}CO(2-1)	-6.0 to 6.0	-2.0 to 2.0	540"	Sawada et al. 2001
^{12}CO(4-3)	-0.23 to 0.32	-0.17 to 0.20	16"	Philipp et al. 2004
	-1.3 to 2.0	-0.3 to 0.2	103"	Martin et al. 2003
^{12}CO(7-6)	-1.3 to 2.0	-0.3 to 0.2	58"	Martin et al. 2003
CS(1-0)	-1.1 to 1.7	-0.3 to 0.2	34"	Tsuboi et al. 1999
CS(2-1)	-0.23 to 0.20	-0.17 to 0.20	25"	Philipp et al. 2004
CS(3-2)	-0.23 to 0.20	-0.17 to 0.20	17"	Philipp et al. 2004
CI(^{3}P$_1$-^{3}P$_0$)	-0.10 to 0.27	-0.08 to 0.17	15"	Philipp et al. 2004
			103"	Martin et al. 2003
C 1200 μm	-0.5 to 0.8	-0.2 to 0.2	11"	Zylka et al. 1998
C 850 μm	-1.0 to 1.7	-0.3 to 0.2	15"	Pierce-Price et al. 2000
C 450 μm	-1.0 to 1.7	-0.3 to 0.2	8"	Pierce-Price et al. 2000

submm bolometer arrays, which allowed mapping the large-scale distribution of the warm dust (and more recently, its polarisation) with ~10" resolution across the central layer ([46][31]).

2.2 Physical Characteristics of the GC Cloud Population

The average characteristics of the molecular clouds in the Galactic Center differ significantly from those in the outer Galaxy: elevated gas temperatures ($<T_{kin}>\sim$50-70 K), high densities ($<n> \sim 10^4$ cm^{-3}) and strong turbulence, usually found in compact star forming dense cloud cores only, are pervasive and likely a consequence of tidal friction in the steep gravitational potential of the central bulge: clouds must be more massive in order to survive against the strong tidal forces. Dissipation of the high supersonic internal velocities (V ~15-30 km s^{-1}), comparable to the intercloud velocity dispersion, is a likely source for the heating of the molecular gas. Evidence for a pervasive magnetic field component with milliGauss field strength throughout the Galactic Center region is discussed in Sect.3.

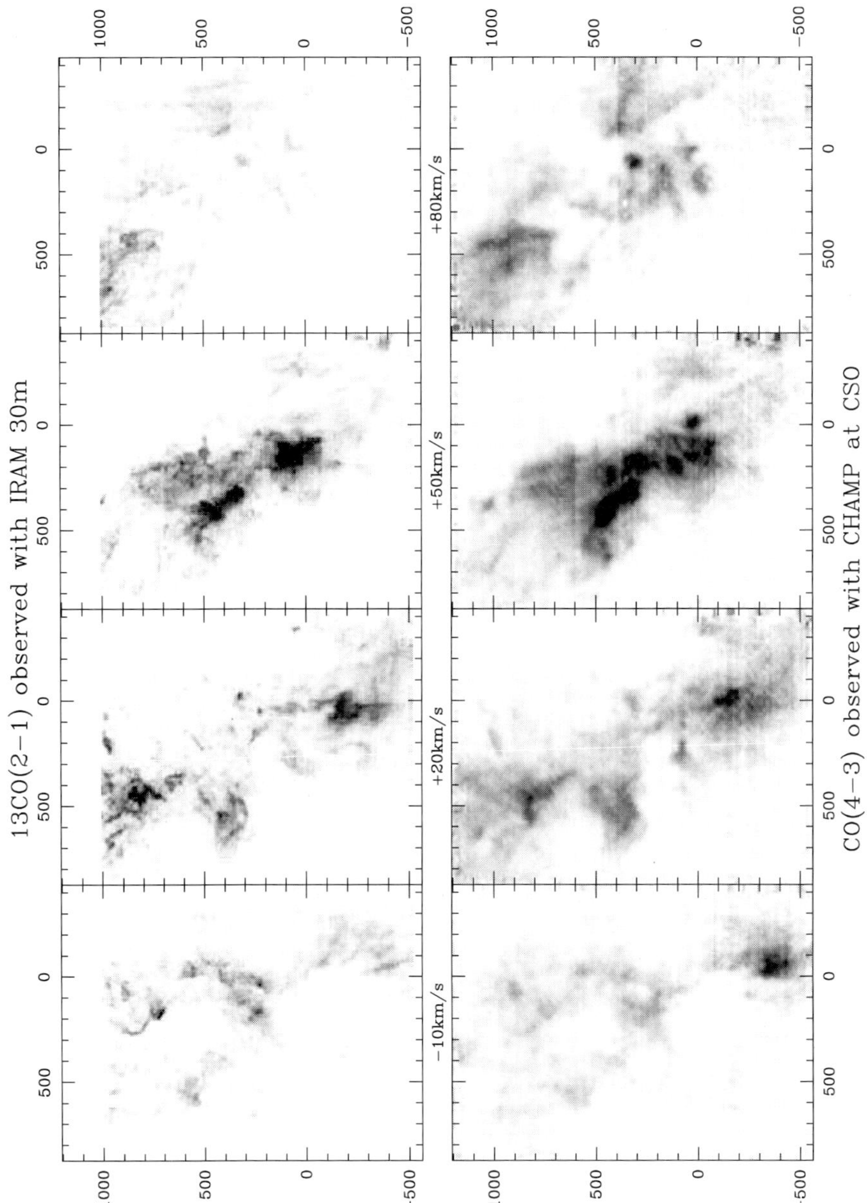

Fig. 1. Selected channels maps of ^{13}CO(J=2-1) and CO(J=4-3) velocity-integrated brightness temperatures, revealing the tremendous morphological fine-structure visible in high-resolution observations [29]. Offsets in arcsec rel. to Sgr A*. The data have been sampled into 5 km/s wide velocity bins. The observations were performed with the IRAM 30m telescope (11") and with CHAMP at the CSO (15" resolution).

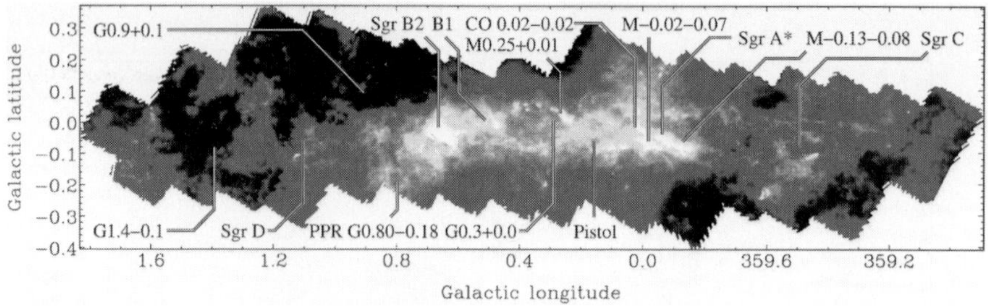

Fig. 2. Submm dust emission at 850 μm as measured with the JCMT/SCUBA array with 15" angular resolution [31]. Prominent GC cloud complexes are labelled.

Statistical Properties: Velocity Dispersion and Mass Spectrum

Based of the recent high-resolution line surveys of CS(1-0) [40] and CO(1-0) [26] (Table 1), a structural decomposition [39] and statistical analysis of cloud properties was performed [28][19]. The velocity linewidth - size dependence of GC clouds basically follow the same relation as disk clouds, $\triangle V \propto R^{\alpha}$, with $\alpha \sim$0.4-0.5, but they are about five times more turbulent (Fig. 3, left panel). The mass spectrum ($M \geq 10^4\ M_{\odot}$) is well represented by a power law, $dN/dM \propto M^{-\gamma}$ with γ=1.6$\pm$0.1, very similar again to indices derived for Galactic disk cloud complexes. Primarily due to the large velocity dispersion, virial masses exceed *observed* masses (derived from the line intensities, assuming LTE) by an order of magnitude (Fig.3, right panel). The discussion about what stabilizes these clouds is ongoing and controversial. External pressure from a hot plasma is limited observationally to $P_{pla} \sim 10^{-9.2}$ erg/cm^3, and falls short by an order of magnitude to the required pressure ($P_{turb} \sim 10^{-8}$ erg/cm^3). A strong magnetic field could bind the cloud: assuming equilibrium between turbulent and magnetic pressure, $B^2/8\pi$, stabilizing intra-cloud fields of $\sim$0.5 mG are required (such field strengths are observationally not excluded, see Sect.3).

Temperatures and Densities

A number of studies with density-sensitive molecular species (like CS [1], H_2CO [11], HC_3N [42]) have shown that average gas densities of order 10^4 cm^{-3} are characteristic for GC clouds. Detailed excitation studies suggest a clumpy morphology with higher density clumps ($\sim 10^5$ cm^{-3}) embedded in a lower density interclump medium ($\sim 10^{3.7}$ cm^{-3}) [42]. High densities of order $10^4 \cdot (75\ \mathrm{pc}/R_{gc})^{1.8}$ [cm^{-3}] have been argued necessary for a cloud at galactocentric distance R_{gc} to survive the tidal stresses in the steep Galactic Center potential.

High kinetic temperatures of the gas have first been deduced from the metastable transitions of ammonia NH_3 [10][20][14]. Warm gas with elaborated bulk temperatures ($\sim$50 K) has been detected towards all the clouds

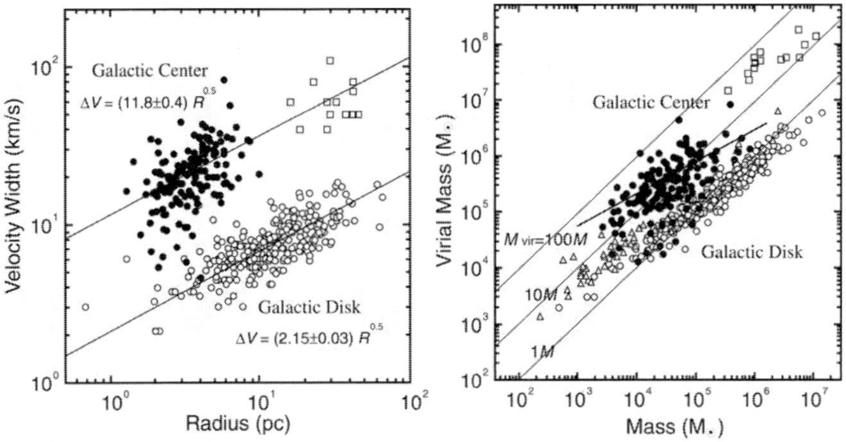

Fig. 3. Properties of the Galactic Center clouds [19], derived from structural decomposition of the CS(1-0)[40] and CO(1-0)[26] survey data. The intracloud velocity dispersion and the virial mass of the clumps are displayed, in comparison with findings for normal Galactic disk clouds (open circles).

surveyed. The new large-scale AST/RO maps in the submm CO lines confirm these findings [16][17]. Soon however, it was noticed that towards all line-of-sights there is evidence in the excitation studies for co-existing gas phases covering a much wider range of temperatures. This is best reflected in the H_2 S(0) to S(5) pure-rotational line studies performed with the SWS on board of ISO [33], covering excitation energies above ground up to a few 1000 K (see their Fig.5). Their dedicated excitation analysis re-assures the earlier findings: warm temperatures are pervasive in galactic center clouds and not necessarily related to star formation activity, and there is a wide hierarchy of temperatures: while temperatures traced by the lower S-transitions measure T_{32} ~150 K, the high-J transitions yield T_{76} ~600 K but derive from only ~1% of the gas column density residing at 150 K.

Such pervasive high gas temperatures have stimulated from the very beginning a lively discussion about the underlying heating mechanisms. Direct heating by collisional dissipation of energy via dust particles has soon been eliminated because of the uniformly low temperature of the dust in the Galactic Center ($T_d = 21 \pm 2$ K $\ll T_{gas}$ [31]). So the quest is up for a heating mechanisms that acts directly on the gas particles. A number of scenarios have been proposed that potentially could deliver the energy required to compensate for the cooling rate of the gas (the combined H_2 and CO cooling rate is estimated to $\Lambda \sim 6 \times 10^{-22}$ erg $s^{-1}cm^{-3}$ [33]): magnetic viscous heating, though requiring mG fields [10], small scale dissipation of supersonic turbulence [43][12], large(r) scale J- and C-shocks as well as UV-heating in exposed photo-dissociation layers. Most likely, all of these processes are on stage:

- towards M-0.96+0.13, [33] find evidence for large-scale shock heating in their H_2-data,
- the molecular gas associated with the Arched Filaments (north of SgrA) and with G0.18-0.04 is known to be exposed to a strong UV field from the Quintuplet [30] and Arches clusters, respectively (see below),
- for the majority of clouds, however, dissipation of supersonic turbulence seems the most attractive heating process. Following Black [2] the heating rate can be estimated as

$$\Gamma \sim 3.5 \cdot 10^{28}\ \mathrm{v_t^3} \cdot \mathrm{n(H_2)} \cdot (1\ \mathrm{pc}/\ell) \qquad [\mathrm{erg\ s^{-1}cm^{-3}}]. \qquad (1)$$

With $\ell \sim$5 pc, $v_t \sim$15 km s^{-1}, n(H_2) $\sim 10^3$cm^{-3} (as typical for GC clouds), we calculate $\Gamma \sim 5{\cdot}10^{-22}$ erg s^{-1}cm^{-3}, comparable to the cooling rate of the gas.

Star Formation in the Galactic Center

In view of these dramatically different initial conditions, star formation proceeds at a surprisingly normal rate (better said, the formation of massive stars, because we have limited insight into the IMF). For a normal IMF, the overall star formation rate across the central layer is $\Phi \sim$0.3-0.6 $M_\odot$/yr [13], about 10% of the galactic star formation rate, with an efficiency Φ/M_g $\sim 5\times10^{-9}$yr^{-1} that compares well with the average galactic rate (Table 2). Details are beyond the scope of this review (see e.g. [21][7]), but there is strong evidence that the peculiar boundary conditions of the GC clouds do affect the dominant mode of star formation: the tidal forces, the large turbulence and, likely, the strong magnetic fields are expected to favor the formation of massive stars. And in fact, the most spectacular young stellar clusters in the Galaxy (the Quintuplet cluster illuminating G0.18-0.04 [6], and the Arches cluster near the thermal Arched Filaments [8], the Central cluster) are all found in the central few 10 parsecs. The Arches cluster compares in age ($2.5 \cdot 10^6$yr) and content (>150 O-stars, $4{\cdot}10^{51}$ N_{lyc} s^{-1}) only to R136 in 30 Dor.

3 The Galactic Center Magnetosphere

In the Galactic disk, energy densities associated with the interstellar magnetic field compare with those derived for e.g. the kinetic energies of the ISM (generally, clouds are observed in approximate energy equipartition). For the enhanced energy densities of the gas in the GC molecular layer, equipartition fields of order mG are estimated. Observationally though the data base is inconclusive. While locally the dominant impact of magnetic fields on the physical processes in the central gas layer is supported by a large number of observations, we have no good handle on the absolute field strength. This

Table 2. Global properties of the central gas layer and cloud characteristics.

	Central 500 pc	**Galactic Disk**
Masses and Densities		
Stars $[M_\star]$	$10^{9.8} M_\odot$	– –
Gas, atomic	$10^{6.4} M_\odot$	$\sim 10^9 M_\odot$
... molecular	$10^{7.9} M_\odot$	$\sim 10^9 M_\odot$
Gas fraction $\mu = M_g/M_\star$	~ 0.01	$\sim 0.05 - 0.10$
Fract. abundance $[HI]/[H_2]$	~ 0.05	~ 2
Gas density $<n>_v$	$100\ cm^{-3}$	$1 - 2\ cm^{-3}$
Gas surface density σ	$\geq 100 M_\odot pc^{-2}$	$\sim 5 M_\odot pc^{-2}$
Star Formation Rate		
Rate Φ	$0.3 - 0.6 M_\odot yr^{-1}$	$\sim 5.5 M_\odot yr^{-1}$
Efficiency Φ/M_g	$5 \cdot 10^{-9} yr^{-1}$	$10^{-9} - 10^{-8} yr^{-1}$

	Center Clouds	**Disk Clouds**
Cloud Characteristics		
Mass spectrum: $dN/dM \propto M^{-\gamma}$	$\gamma \sim 1.6$ ($\geq 10^4\ M_\odot$)	1.6 - 1.7
Vel. dispersion $[kms^{-1}]$	15 – 30	≤ 5
Vel.dispersion-Size: ΔV [km/s]	$12 \cdot R^{0.5}$	$2 \cdot R^{0.5}$
Mean Gas Density $[cm^{-3}]$	$\sim 10^{4.0}$	$\sim 10^{2.5}$
Temperatures [K]: Gas	bulk: 50 – 70 wide range of coex. hot phases	~ 15
: Dust	21 ± 2	
Magnetic Field [mG]	order mG	≤ 0.1
Isotopic Abundances		
$[^{12}C]/[^{13}C]$	~ 25	~ 75
$[^{16}O]/[^{18}O]$	~ 200	~ 400
$[^{14}N]/[^{15}N]$	~ 1000	~ 400

stems from the general difficulties in observing magnetic fields: while we have good tools to derive the field *morphology*, the only direct access to the field *strength* is via difficult-to-perform line polarization measurements.

3.1 Magnetic Fields in the Diffuse GC ISM

The most striking manifestation of the magnetic field in the diffuse ISM (or more precise: in the intercloud medium) are the Non-thermal Filaments (NTFs), evident in radio continuum images of the Galactic Center [44]. These structures - unique to the Galactic Center - are observed isolated or in bundles of parallel strands, they run (within 20-30 deg) orthogonal to the Galactic plane for a few tens of parsecs, while their transverse dimension is only a fraction of a parsec. Their radio spectra with strong linear polarization suggest a non-thermal (synchrotron) nature of the underlying emission process. Most, if not all, of the NTFs are interacting with molecular clouds, but reveal little, if any, distortions. The best example is where the "+25km/s cloud", associated with G0.18-0.04, interacts with the Radio Arc [36]. From the filament's rigidity, a magnetic field strength (in the NTF) of a few mG have been estimated by equating magnetic pressure with the ram and/or intracloud turbulent pressure of the associated molecular cloud [36] [37].

A plethora of theoretical models has been suggested to describe the nature of these filaments, and their relationship with the overall GC field structure (e.g., [21] [22]). Common to most of these models is that the NTFs, independent of their very origin, are manifests of a large-scale poloidal field component. Whether the NTFs are local illuminations of a volume-pervasive (static) strong central dipole field, or whether they represent local (maybe, dynamic) field enhancements of an otherwise much weaker field, is much of a controversy.

3.2 Magnetic Fields in the GC Molecular Clouds

The great advancement of submm/FIR technologies (with large arrays available now for extended mapping) has made possible to perform polarimetry of the thermal emission from magnetically aligned dust grains in the dense Galactic Center clouds, so providing the field orientation $B_{\perp}$ (vertical to the line-of-sight). Fig.4 is a composite of results from polarisation measurements performed over the last couple of years (from [4]).

At first sight, there is little evidence for a poloidal component of the magnetic field associated with the dense clouds: instead, the orientation is predominantly parallel to the galactic plane (following the main axis of the cloud complexes), more suggestive of a toroidal configuration. The most amazing discrepancy - in view of their obvious physical interaction - is noticed towards G0.18-0.04, where the field in the cloud is right perpendicular to the field traced by the NTFs (Fig.4, right panel).

The strong degree of polarization and the rather uniform field orientation within a given cloud complex (as sampled with sub-parsec resolution) suggests a scenario, in which gravity controls magnetic forces, and differential rotation has sheared the field to its azimuthal configuration [23]. Chuss et al. [4], based on higher resolution data from the Hertz polarimeter on the CSO,

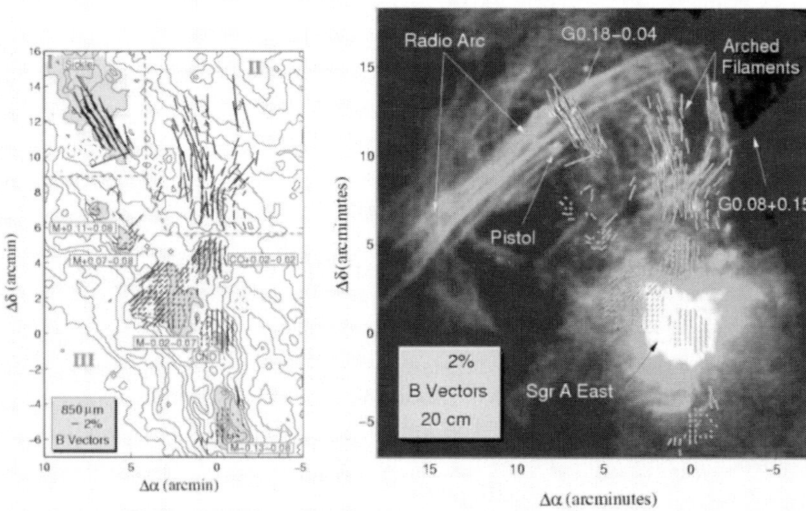

Fig. 4. Magnetic field directions as inferred from polarization measurements of magnetically aligned dust grains (from [4]), superposed on the 850μm dust emission [31]. Polarimetry data are from Dotson et al. [5] [region I at 60μm, region II at 100μm] and from Novak et al. [23] [region III at 350μm]. Offsets refer to Sgr A*.

report that the field orientation is in fact coupled to the column density of material: in the dense clouds, gravity dominates and the field is stretched along the long axis of the cloud, whereas in the cloud envelope the field is bending back to its (original ?) poloidal configuration, perpendicular to the plane. These results, which clearly have to be followed-up and should be complemented by line polarization measurements [9], that take advantage of the velocity information along the line-of-sight, seem to provide the clue to understand the apparently contradictory findings of co-existing poloidal and toroidal field configurations in the center of our Galaxy.

Assuming equipartition conditions in these transition layers, with obvious uncertainties, field strength of order mG are deduced [4]. However, as long as we have no direct determination (means via Zeeman splitting), the actual strength of the magnetic field must remain uncertain. Except for the circumnuclear disk (the structure orbiting about the central mass at a few parsec distance only), for which [15][32] report $B_{||}$ ~2-3 mG, no relevant detection has been reported [fields derived from measuremens of OH masers [45] compare with findings in the Galactic disk, but because they heavily reflect local excitation conditions, they shall not be used to constrain the field in the molecular clouds]. An extended OH absorption survey [41] has yielded upper limits only of ~0.3 mG (3σ, 13 positions) to any coherent l-o-s field component. While the failure to detect any $B_{||}$ component towards, e.g., G0.18-0.04 (where we have a most convincing, though indirect, case of a mG field) could still be explained by its special field geometry (with no field component projected towards the observer), statistically these negative results seem to question the existence of pervasive strong magnetic field in the GC clouds.

4 Perspectives – The Next Steps

Technological progress has provided us with the ability for wide-field mapping with critically high spatial resolution. A unique data base has been collected, that greatly improved our understanding of the excitation conditions in the central cloud layer, but still fundamental questions about the exceptional physics remain to be answered (about the thermodynamics of the gas, its heating processes, the role of the magnetic fields – to mention a few). Our review should have affirmed that (1) understanding the GC is a mandatory (intermediate) step in order to adequately model extragalactic nuclei in general, and (2) that much of the insight derives from "micro"-physics that will only be possible to resolve in our galactic center due to its close proximity.

By the nature of the underlying physics (the main cooling lines of the cold ISM are in the submm/FIR regime), we do expect critical answers from the new submm and FIR facilities, currently under development. APEX and ASTE, ALMA on mid terms, will soon explore the submm atmospheric windows to the technological limits, with the unique airborne (SOFIA) and space FIR missions (Herschel) already on the horizon.

References

1. J. Bally, A. Stark, R. Wilson, C. Henkel: Ap.J. Suppl. **65**, 13 (1985)
2. J. Black: in: *Interstellar Processes*, ed. D. Hollenbach et al., pp 731 (1987)
3. D. Chuss: Mapping Magnetic Fields in the Cold Dust at the Galactic Center. Thesis, Northwestern University, Illinois (2002)
4. D. Chuss, J. Davidson, J. Dotson, C. Dowell, R. Hildebrand, G. Novak, J. Vaillancourt: Ap.J. **599**, 1116 (2003)
5. J. Dotson, J. Davidson, C. Dowell et al.: Ap.J. Suppl. **128**, 335 (2000)
6. D. Figer, I. McLean, M. Morris: Ap.J. **514**, 202 (1999)
7. D. Figer, M. Morris: Starburst Clusters in Galactic Nuclei. In: *Modes of Star Formation*, ASP Conf. Series, ed. by E. Grevel & W. Brandner. (2001)
8. D. Figer, F. Najarro, D. Gilmore, M. Morris, S. Kim, E. Serabyn, I. McLean, A. Gilbert, J. Graham, J. Larkin, H. Teplitz: Ap.J. **581**, 258 (2002)
9. J. Greaves, W. Holland, W. Dent: Ap.J. **578**, 224 (2002)
10. R. Güsten, C. Walmsley, T. Pauls: A&A **103**, 197 (1981)
11. R. Güsten, C. Henkel: A&A **125**, 136 (1983)
12. R. Güsten, C. Walmsley, H. Ungerechts, E. Churchwell: A&A **142**, 381 (1985)
13. R. Güsten: Gas and Dust in the Inner Few Degrees of the Galaxy. In: *The Center of the Galaxy*, IAU Symp. 136, ed by M.Morris. (1989) pp 89

14. S. Hüttemeister, T. Wilson, T. Bania, J. Martin-Pintado: A&A **280**, 255 (1993)
15. N. Killeen, K. Lo, Crutcher, R.: Ap.J. **385**, 585 (1992)
16. S. Kim, C. Martin, A. Stark, A. Lane: Ap.J. **580**, 896 (2002)
17. C. Martin, W. Walsh, K. Xiao, A. Lane, C. Walker, A. Stark: Ap.J. submitted (2003), available on http://cfa-www.harvard.edu/ASTRO/
18. P. Mezger, W. Duschl, R. Zylka: A&A Rev. **7**, pp 289 (1996)
19. A. Miyazaki, M.Tsuboi: Ap.J. **536**, 357 (2000)
20. M. Morris, N. Polish, B. Zuckermann, N. Kaifu: Astron.J. **88**, 1228 (1983)
21. M. Morris, E. Serabyn: Ann. Rev. A&A **34**, 645 (1996)
22. M. Morris: Magnetic Phenomena in Galactic Nuclei. In: *The Central Regions of the Galaxy and Galaxies*, IAU Symp. 184, ed. by Y.Sofue. (1998) pp 331–340
23. G. Novak, J. Dotson, C. Dowell, R. Hildebrand, T. Renbarger, D. Schleuning: Ap.J. **529**, 241 (2000)
24. G. Novak, D. Chuss, T. Renbarger, G. Griffin, M. Newcomb, J. Peterson, R. Loewenstein, D. Pernic, J. Dotson: Ap.J. **583**, L83 (2003)
25. R. Ojha, A. Stark, H. Hsieh, A. Lane, R. Chamberlin, T. Bania, A. Bolatto, J. Jackson: Ap.J. **548**, 253 (2001)
26. T. Oka, T. Hasegawa, M. Hayashi, T. Handa et al.: Ap.J. **493**, 730 (1998)
27. T. Oka, T. Hasegawa, F. Sato, M. Tsuboi et al.: Ap.J. Suppl. **118**, 455 (1998)
28. T. Oka, T. Hasegawa, F. Sato, M. Tsuboi et al.: Ap.J. **562**, 348 (2001)
29. S. Philipp, R. Güsten, D. Lis, H. Ungerechts, R. Zylka et al.: in prep. (2004)
30. S. Philipp, R. Güsten, K. Uchida: in prep. (2004)
31. D. Pierce-Price, J. Richer, J. Greaves et al.: Ap.J. **545**, L121 (2000)
32. R. Plante, K. Lo, R. Crutcher: Ap.J. **445**, L113 (1995)
33. N. Rodriguez-Fernandez, J. Martin-Pintado, A. Fuente, P. de Vicente, T. Wilson, S. Hüttemeister: A&A **365**, 174 (1987)
34. T. Sawada, T. Hasegawa, T. Handa, J. Morino, T. Oka, R. Booth, L. Bronfman, M. Hayasi, A. Castellanos, L. Nyman et al.: Ap.J. Suppl. **136**, 189 (2001)
35. E. Serabyn, R. Güsten: A&A **184**, 133 (1987)
36. E. Serabyn, R. Güsten: A&A **242**, 376 (1991)
37. E. Serabyn, M. Morris: Ap.J. Letters **424**, L91 (1994)
38. E. Serabyn, M. Morris: Nature **382**, 602 (1996)
39. J. Stutzki, R. Güsten: Ap.J. **356**, 513 (1990)
40. M. Tsuboi, T. Handa, Ukita, N.: Ap.J. Suppl. **120**, 1 (1999)
41. K. Uchida, R. Güsten: A&A **298**, 473 (1995)
42. C. Walmsley, R. Güsten, P. Angerhofer et al.: A&A **155**, 129 (1986)
43. T. Wilson, K. Ruf, C. Walmsley, R. Martin et al.: A&A **115**, 185 (1982)
44. F. Yusef-Zadeh, M. Morris, D. Chance: Nature **310**, 557 (1984)
45. F. Yusef-Zadeh, D. Roberts, W. Goss, D. Frail et al.: Ap.J. **466**, L25 (1996)
46. R. Zylka, S. Philipp, W. Duschl, P. Mezger, et al.: in: *The Central Regions of the Galaxy and Galaxies*, IAU Symp. 184, ed. by Y.Sofue. (1998) pp 291

The Line-of-Sight Distribution of Water in SgrB2

C. Comito[1], P. Schilke[1], M. Gerin[2], T.G. Phillips[3], J. Zmuidzinas[3], and D.C. Lis[3]

[1] Max-Planck-Institut für Radioastronomie, Auf dem Hügel 69, D-53121 Bonn, Germany ccomito@mpifr-bonn.mpg.de
[2] Laboratoire de Radioastronomie Millimétrique, Observatoire de Paris and Ecole Normale Supérieure, 24 rue Lhomond, F-75231 Paris, CEDEX 05, France
[3] California Institute of Technology, Downs Laboratory of Physics 320-47, Pasadena, CA 91125, USA

1 H_2O in SgrB2

The SgrB2 cloud is one of the most massive star forming regions in our Galaxy. It has many unique characteristics, among them an exceptional chemistry. Several species (FeO, [21]; NH_2, [20]; HF, [15]; C_3, [7]) have, in spite of searches elsewhere, been detected only toward this source. One possible explanation for this enigmatic chemistry is the existence of a layer of hot gas, which is thought to be produced by a shock. Since this layer has the same velocity as the ambient gas in SgrB2, it has been proven difficult to assess its importance for the chemistry of many species, with a few exceptions. In particular its importance for the water chemistry has been a matter of recent debate ([3]; [17]). While its temperature and density are well known (cf. [9]), its column density and width have remained elusive. In this work (illustrated in detail in [5]) we have been able to determine its importance for water chemistry and give constraints for column density and spatial width, by modeling the HDO and $H_2^{18}O$ emission and absorption.

Water is known to be a fundamental ingredient of the interstellar medium. An ubiquitous tracer of shock-heated gas, it dramatically influences the chemistry in shocked regions (cf. [14]; [1]). In general, it acts as a a major coolant in star-forming clouds ([2]). Although direct observation of non-masing water lines from ground-based observatories is made extremely difficult by atmospheric absorption, H_2O abundances can be estimated via observations of isotopomers, such as $H_2^{18}O$ (e.g. [19]; [10] and [11]; [6]), and, in recent years, by the availibility of satellites such as ISO (e.g. [4]) and notably SWAS (cf. [16]), but also by airborne observations of $H_2^{18}O$ (e.g. [23]). However, the deuterated counterpart of water, HDO, presents many features observable from ground in the cm, mm and submm wavelength atmospheric windows (cf. [8]; [11] and [12]; [18]), and is also used as a tracer of water abundance under the assumption that both the deuterated and non-deuterated species be spatially coexistent, and that their abundance ratio remain constant throughout the studied region.

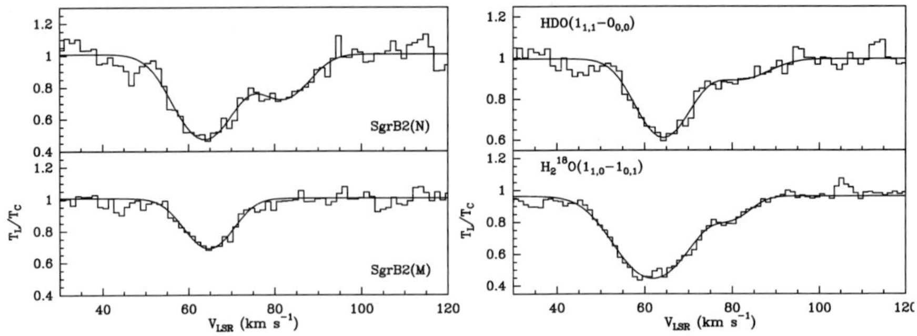

Fig. 1. *Left panel:* the 894-GHz HDO $(1_{1,1} - 0_{0,0})$ transition, observed in absorption against the continuum emission of SgrB2(M) and of SgrB2(N). *Right panel:* comparison between the 894-GHz HDO$(1_{1,1} - 0_{0,0})$ and the 547.7-GHz o-$H_2^{18}O(1_{1,0} - 1_{0,1})$ absorption features observed towards SgrB2. The y-scale gives the line-to-continuum (T_L/T_C) ratio.

We report the detection of the 894-GHz ground-state $1_{1,1} - 0_{0,0}$ transition of HDO *in absorption* against the background continuum sources SgrB2(M) and SgrB2(N). The radial velocity of the observed HDO features suggests that the absorption is directly connected to the SgrB2 complex. The fact that, i), we observed a ground-state transition, and ii), we observed it in absorption, allowed us to estimate, indipendently of assumption on the physical state of the gas, the HDO column density, which we found to be of 1.2×10^{13} cm^{-2} towards SgrB2(M), and of 2.4×10^{13} and 1.2×10^{13} cm^{-2} towards the 65- and the 81-km s^{-1} velocity components of SgrB2(N) respectively. Moreover, using the $H_2^{18}O$ SWAS data ([16]), we were able to estimate a lower limit of 5×10^{-4} for the [HDO]/[H_2O] ratio towards the SgrB2 complex.

The above mentioned results hold under the assumption that all the absorbing water lies in the warm envelope. [3] propose that H_2O might instead be located in the hot gas layer that surrounds the whole SgrB2 complex, implying that gas-phase water in the warm envelope does not contribute at all to the observed absorption features. In fact, while this hypothesis is reasonable for H_2O (the gas-phase production of water is triggered by high temperatures), it would not be applicable to HDO (the same high temperatures inhibit its enhancement). We therefore think it possible that the gas observed in HDO and that seen in H_2O absorption be located in distinct regions.

2 Modeling the Distribution of Water

SgrB2 is often adopted as a template region for the study of sites of massive star formation. For this purpose, it is important to evaluate to what extent the warm envelope and the hot layer *separately* contribute to the observed HDO and H_2O absorption. In particular, in order to better understand the nature of the hot layer and possibly the mechanism that created it, it is vital to learn more about its chemical and physical characteristics, and especially about its water content. A precise determination of the column density of water would set a tight constraint on the physical models that attempt to identify the heating mechanism responsible for the high gas temperatures. Unfortunately, since both gas components show roughly the same radial velocity, it is not possible to separate, by purely observational means, the ground-state water absorption produced in the warm envelope from that produced in the hot layer. However, a number of HDO and $H_2^{18}O$ transitions have been observed towards the SgrB2 cores with a variety of instruments.

Our 893-GHz HDO data, together with the published HDO and $H_2^{18}O$ observations carried out at mm and submm wavelengths, provide strong observational constraints and enable us to model, in a self-consistent manner, the HDO and H_2O abundance in all three components of the SgrB2 cloud. In detail:

a) the observed *emission* lines of p-$H_2^{18}O$ at 203 GHz (SgrB2(N), [6]), and of HDO at 143, 226, 241 GHz (SgrB2(M) and (N), [11]) constrain the HDO and water abundance and the H_2 density in the hot-core-type components;
b) the 893-GHz HDO *absorption* feature observed towards SgrB2(M) and N, together with the [HDO]/[H_2O] ratio estimated at point *a)*, help us set the HDO and H_2O abundance in the warm envelope;
c) finally, having estimated the column density of water in the warm envelope, it is possible to assess whether or not the quantity of non-deuterated water in the warm envelope can be responsible for the $H_2^{18}O$ absorption observed at 548 GHz ([16], [23]).

We use the radiative transfer code described by [24], with a few modifications (see [5] for details), to reproduce the intensities observed for the features. The model consists of two spherically symmetric components representing the hot core and the warm envelope (see § 1). Most of the spectral line data are reproduced within errors of $\sim 30\%$, with two exceptions:

1) The 226-GHz HDO emission line observed towards SgrB2(M) is a factor of 2 weaker, and the 143-GHz HDO line in SgrB2(N) is 4 times stronger than predicted by the model. Such discrepancies are likely to be due to the intrinsic chemical differences between the two hot cores (cf. [13] and references therein), which has not been taken into account in our model.
2) Most interestingly, the predicted ground-state o-$H_2^{18}O$ absorption at 548 GHz is much shallower than observed ([16], cf. Fig. 1): *the abundance of non-deuterated water in the warm envelope is not sufficient to produce the absorption feature observed at 548 GHz.*

This is a very important result: our predictions confirm, at least from a qualitative point of view, the hypothesis of [3] that the most important contribution to the observed $H_2^{18}O$ absorption come from the foreground hot gas layer. Having predicted the contribution of the warm envelope to the $H_2^{18}O$ absorption, it is possible to estimate the spatial width and H_2O abundance necessary for the hot layer to produce the observed feature, given the observational constraints mentioned in section § 1. We use the radiative transport equation to calculate the total intensity emerging from the hot layer, given the background emission produced by hot cores and warm envelope. The data are best reproduced when the o-$H_2^{18}O$ column density in the hot layer is $\sim 10^{14}\,cm^{-2}$. Assuming a $[^{16}O]/[^{18}O]$ ratio of 261±20 ([22]) and an ortho/para ratio of 3, we can estimate the H_2O column density in the hot gas component to be of order $\sim 3.5 \times 10^{16}$ cm^{-2}. If all the atomic oxygen not locked in CO is bound in gas-phase water, then the H_2O abundance in this region is about 5×10^{-4} cm^{-3}, and the spatial width of the hot layer is only 0.02 pc.

References

1. Bergin, E.A., Melnick, G.J., Neufeld, D.A. 1998, ApJ, 499, 777
2. Ceccarelli, C., Hollenbach, D.J., Tielens, A.G.G.M. 1996, ApJ, 471, 400
3. Ceccarelli, C., Baluteau, J.-P., Walmsley, C.M., et al. 2002, A&A, 383, 603
4. Cernicharo, J., Lim, T., Cox, P. et al. 1997, A&A, 323, 25
5. Comito, C., Schilke, P., Gerin, M., et al. 2003, A&A, 402, 635
6. Gensheimer, P.D., Mauersberger, R., Wilson, T.L. 1996, A&A, 314, 281
7. Giesen, T.F., Van Orden, A.O., Cruzan, J.D., et al. 2001, ApJ, 551, 181
8. Henkel, C., Mauersberger, R., Wilson, T.L., et al. 1987, A&A, 182, 299
9. Hüttemeister, S., Wilson, T.L., Mauersberger, R., et al. 1995, A&A, 294, 667
10. Jacq, T., Jewell, P.R., Henkel, C., et al. 1988, A&A, 199, 5
11. Jacq, T., Walmsley, C.M., Henkel, C., et al. 1990, A&A, 228, 447
12. Jacq, T., Baudry, A., Walmsley, C.M., et al. 1999, A&A, 347, 957
13. Miao, Y., Snyder, L.E. 1997, ApJ, 480, L67
14. Neufeld, D.A., Melnick, G.J. 1987, ApJ, 322, 266
15. Neufeld, D. A., Zmuidzinas, J., Schilke, P., et al. 1997 ApJ, 488, 141
16. Neufeld, D.A., Ashby, M.L.N., Bergin, N.R. et al. 2000, ApJ, 539, 111
17. Neufeld, D. A., Bergin, E. A., Melnick, G. J., et al. 2003, ApJ, 590, 882
18. Pardo, J. R., Cernicharo, J., Herpin, F., et al. 2001, ApJ, 562, 799
19. Phillips, T.G., Scoville, N.Z., Kwan, J., et al. 1978, ApJ, 222, 59
20. van Dishoeck, E.F., Jansen, D.J., Schilke, P., et al. 1993, 1993 ApJ, 416, 83
21. Walmsley, C.M., Bachiller, R., Pineau des Forêts, G., et al. 2002 ApJ, 566, 109
22. Whiteoak, J.B., Gardner, F.F. 1981, MNRAS, 197, 39
23. Zmuidzinas, J., Blake, G.A., Carlstrom, J., et al. 1995b, *Proceedings of the Airborne Astronomy Symposium on the Galactic Ecosystem: From Gas to Stars to Dust*, ed. M.R. Haas, J.A. Davidson and E.F. Erickson, San Francisco: ASP
24. Zmuidzinas, J., Blake, G.A., Carlstrom, J., et al. 1995, ApJ, 447, L125

Results from the AST/RO Survey of the Galactic Center Region

A.A. Stark[1], C.L. Martin[1], W.M. Walsh[1], K. Xiao[1], A.P. Lane[1], C.K. Walker[2], and J. Stutzki[3]

[1] Smithsonian Astrophysical Observatory, 60 Garden St. MS 12, Cambridge, MA 02138 USA aas, cmartin, wwalsh, kxiao, alane@cfa.harvard.edu
[2] Steward Observatory, University of Arizona, Tucson, AZ 85721 USA cwalker@as.arizona.edu
[3] Physikalische Institut I, Universität zu Köln, Zülpicher Straße 77, Köln, D-50937 Germany stutzki@ph1.uni-koeln.de

We have used the Antarctic Submillimeter Telescope and Remote Observatory (AST/RO), a 1.7m diameter single-dish submillimeter-wave telescope at the geographic South Pole, to determine the physical state of gas in the Galactic Center region and assess its stability. indexsubmillimeter We present an analysis based on data obtained as part of an ongoing AST/RO key project: the large-scale mapping of the dominant cooling lines of the molecular interstellar medium in the Milky Way [1, 2, 3, 4]. These data are released for general use.

1 Introduction

The distribution of molecular gas in the Galaxy is known from extensive and on-going surveys in CO and ^{13}CO $J = 1 \rightarrow 0$ and $J = 2 \rightarrow 1$; these are spectral lines which indicate the presence of molecular gas. These lines alone do not, however, determine the excitation temperature, density, or cooling rate of that gas. Observations of C I and the mid-J lines of CO and ^{13}CO provide the missing information, showing a more complete picture of the thermodynamic state of the molecular gas, highlighting the active regions, and looking into the dense cores. AST/RO can measure the dominant cooling lines of molecular material in he interstellar medium: the $^3P_1 \rightarrow {}^3P_0$ (492 GHz) and $^3P_2 \rightarrow {}^3P_1$ (809 GHz) fine-structure lines of atomic carbon (C I) nd the $J = 4 \rightarrow 3$ (461 GHz) and $J = 7 \rightarrow 6$ (807 GHz) rotational lines of carbon monoxide (CO). These measurements can then be modeled using the large velocity gradient (LVG) approximation, and the gas temperature and density thereby determined. Since the low-J states of CO are in local thermodynamic equilibrium (LTE) in almost all molecular gas, measurements of mid-J states are critical to achieving a model solution of the radiative transfer by breaking the degeneracy between beam filling factor and excitation temperature.

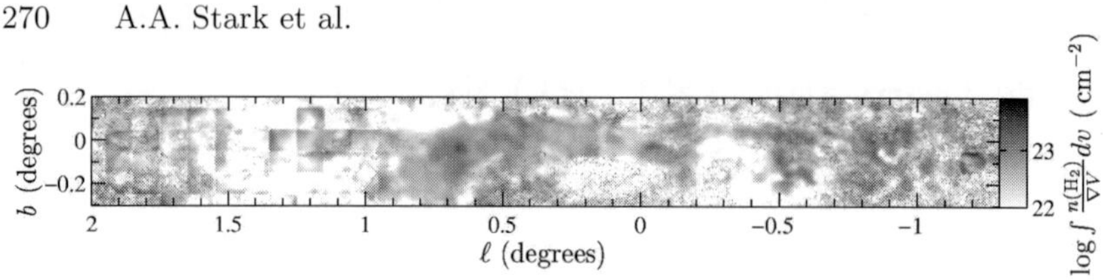

Fig. 1. Greyscale representation of molecular column density in the Galactic Center Region, from an LVG model using AST/RO survey data [1].

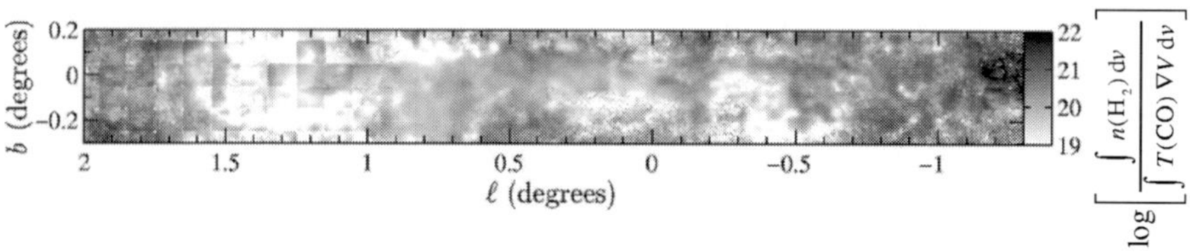

Fig. 2. Greyscale representation of the ratio of molecular column density (as derived from an LVG model, cf. Fig. 1) to the integrated brightness of the CO $J = 1 \rightarrow 0$ line [7]. If both these quantities were accurate measures of molecular column density, then the ratio would be a constant; the ratio actually varies by an order of magnitude either way around the mean.

2 AST/RO Survey Parameters

AST/RO has been operational in the submillimeter-wave atmospheric windows since 1995 [5, 6]. Essential to AST/RO's capabilities is its location at Amundsen-Scott South Pole Station, an exceptionally cold, dry site which has unique logistical opportunities and challenges. Among the key AST/RO projects is mapping of the Galactic Center Region. ky coverage as of 2002 is $-1^\circ\!.3 < \ell < 2^\circ$, $-0^\circ\!.3 < b < 0^\circ\!.2$ with $0'\!.5$ spacing, resulting in spectra of three transitions at 24,000 positions on the sky [2, 1]. The data are available on the AST/RO website[4] for general use.

The C I emission has a spatial extent similar to the low-J CO emission, but is more diffuse [3, 4]. The CO $J = 4 \rightarrow 3$ emission is also found to be essentially coextensive with lower-J transitions of CO, indicating that even the $J = 4$ state is in LTE at most places; in contrast, the CO $J = 7 \rightarrow 6$ emission is spatially confined to far smaller regions [2]. Applying an LVG model to these data [1] together with ata from the Bell Labs 7m [7, 8] yields maps of gas density and temperature as a function of position and velocity for the entire region. indextemperature Kinetic temperature is found to decrease from relatively high values (> 70 K) at cloud edges to lower values (< 50 K) in the interiors. Typical pressures in the Galactic Center gas are $n(\mathrm{H_2}) \cdot T_{kinetic} \sim 10^{5.2}\,\mathrm{K\,cm^{-3}}$.

An estimate of the molecular column density indexdensity can be obtained by integrating the LVG-derived density over all velocities [1], as shown in Fig-

[4] `http://cfa-www.harvard.edu/ASTRO`

ure 1. An alternative estimate can be made utilizing the common assumption that the column density is proportional to the brightness of the $J = 1 \rightarrow 0$ CO line [9, 10]. Both of these estimates can be carried out on each line of sight, and their ratio plotted as a function of position, as shown in Figure 2. If the methodologies were mutually consistent, then the ratio should not vary with position. In fact the ratio varies *by two orders of magnitude.* Both methods incorporate rather ill-determined multiplicative constants, which can arbitrarily be adjusted to bring the two methods into agreement over some of the map but not all of it. The discrepancies are caused by variations in the excitation and optical depth of the CO lines, suggesting that the LVG method should be the more accurate method when submillimeter data are available.

3 Galactic Center Gas Stability

The dynamics of the Galactic Center region depend critically on whether the interstellar medium is self-gravitating or not [11]; given our estimate for the density of the molecular gas, we can now determine its stability against coagulation into self-gravitating lumps. Binney et al. [12] have suggested that gas in the Galactic Center will be found mostly on closed orbits, because gas thrown into a stellar system will tend to shock and violently damp until it settles onto closed, parallel streamlines. In the central bar of the Milky Way, the closed orbits fall into two families: the elongated x_1 orbits running parallel to the outer parts of the bar about 1 or 2 kpc from the center, and the roughly circular x_2 orbits located near the inner Lindblad resonance (ILR) of the bar at ~ 350 and further inwards [13]. Gas falling into the system from outside first settles onto the x_1 orbits. As it continues to lose energy, it slides downwards through x_1 orbits of decreasing size until it gets to the inner x_1 orbits, which are self-intersecting, where it transitions onto the outer x_2 orbits [13, 11] near the ILR—there it will tend to accumulate, because inwards of the ILR the net torque on the gas by the bar reverses sign, so that gas inside the ILR is pushed outward while gas outside the ILR is pushed inward [14]. Gas therefore accumulates on the outer x_2 orbits near the ILR, growing more and more dense until it becomes self-gravitating at a density [14] of

$$n(\mathrm{H}_2) > \frac{0.3\kappa^2}{m_p G} \approx 10^{3.4}\,\mathrm{cm}^{-3}\,,$$

where $\kappa \approx 900\,\mathrm{km\,s^{-1}\,kpc^{-1}}$ is the epicyclic frequency [13]. Gas on x_2 orbits appears in the data on a strip about $50\,\mathrm{km\,s^{-1}}$ wide extending from $(\ell, v) = (-1°, -80\,\mathrm{km\,s^{-1}})$ to $(+1°, +80\,\mathrm{km\,s^{-1}})$. Our LVG model shows that the gas in this strip is near the critical density, indicating that it is only marginally stable against gravitational coagulation into one or two giant clouds [14]. his suggests a relaxation oscillator mechanism for starbursts in the Milky Way, where inflowing gas accumulates in a ring at 350 pc radius until the critical density is reached, and the resulting instability leads to the sudden deposition of a Giant Molecular Cloud onto the Galactic Center.

This work was supported in part by US NSF grant OPP-0126090. We thank our colleagues R. A. Chamberlin, K. Jacobs, J. Kooi, and G. A. Wright for their contributions to AST/RO.

References

1. C. L. Martin, W. M. Walsh, K. Xiao, A. P. Lane, C. K. Walker, and A. A. Stark. *ApJS*, in press `astro-ph/0211025` (2004)
2. S. Kim, C. L. Martin, A. A. Stark, and A. P. Lane. *ApJ*, **580**:896 (2002)
3. R. Ojha, A. A. Stark, H. H. Hsieh, A. P. Lane, R. A. Chamberlin, T. M. Bania, A. D. Bolatto, J. M. Jackson, and G. A. Wright. *ApJ*, **548**:253 (2001)
4. B. Mookerjea, C. L. Martin, J. Stutzki, A. A. Stark, A. P. Lane, W. M. Walsh, and K. Xiao. In this volume. (2003)
5. A. A. Stark, J. Bally, S. P. Balm, T. M. Bania, A. D. Bolatto, R. A. Chamberlin, G. Engargiola, M. Huang, J. G. Ingalls, K. Jacobs, J. M. Jackson, J. W. Kooi, A. P. Lane, K.-Y. Lo, R. D. Marks, C. L. Martin, D. Mumma, R. Ojha, R. Schieder, J. Staguhn, J. Stutzki, C. K. Walker, R. W. Wilson, G. A. Wright, X. Zhang, P. Zimmermann, and R. Zimmermann. *PASP*, **113**:567 (2001)
6. A. A. Stark. AST/RO: A small submillimeter telescope at the South Pole. In T. D. Oswalt, editor, *The Future of Small Telescopes in the New Millennium, Volume II—The Telescopes We Use*, pages 269–284. Kluwer Academic Publishers `astro-ph/0110429` (2003)
7. J. Bally, A. A. Stark, R. W. Wilson, and C. Henkel. Galactic Center molecular clouds. I. Spatial and spatial velocity maps. *ApJS*, **65**:13 (1987)
8. J. Bally, A. A. Stark, R. W. Wilson, and C. Henkel. *ApJ*, **324**:223 (1988)
9. H. S. Liszt. *Comments on Astrophysics*, **10**:137 (1984)
10. D. B. Sanders, P. M. Solomon, and N. Z. Scoville. *ApJ*, **276**:182 (1984)
11. A. Jenkins and J. Binney. *MNRAS*, **270**:703 (1994)
12. J. Binney, O. E. Gerhard, A. A. Stark, J. Bally, and K. I. Uchida. *MNRAS*, **252**:210 (1991)
13. N. Bissantz, P. Englmaier, and O. Gerhard. *MNRAS*, **340**:949 (2003)
14. B. G. Elmegreen. *ApJ*, **425**:L73 (1994)

SPIFI Imaging of the Galactic Center

G.J. Stacey[1], T. Nikola[1], C.M. Bradford[1,2], L. Hall[1], A.D. Bolatto[3,4], J.M. Jackson[3], M.L. Savage[5], and J.A. Davidson[5]

[1] Cornell University, Space Sciences Bldg, Ithaca, NY 14843 gjs12@cornell.edu
[2] Caltech
[3] Boston University
[4] UC Berkeley
[5] USRA

1 Introduction

The center of the Galaxy provides a unique opportunity to study many of the fascinating phenomena assocated with galactic nuclei at relatively high spatial resolution. Of particular interest is the circumnuclear disk (CND) that circulates about Sgr A* with a sharp inner edge at 1.5 pc. Discovered in its far-IR continuum emission, the ring has since been investigated in a wide variety long wavelength spectral line and continuum probes. This work has shown that the ring is warm, dense, and turbulent. The total mass of the ring out to 4 pc radius is about $10^4 M_\odot$. Within the ring lies the "mini-spiral", a group of gas and dust streamers. The ring/mini-spiral morphology is well illustrated by the high resolution KAO images at 31 and 38 μm that clearly illustrate all of these features [1]. The warm dust traced by these images is heated by the far-UV radiation from the central star cluster.

To gain new insights into the molecular gas excitation and gas dynamics within the ring, we have undertaken high spatial resolution imaging of the inner regions in the CO(7→6) rotational transition at 371.651 um and the $^3P_2 \rightarrow ^3P_1$ [CI] fine structure at 370.415 μm using the South Pole Imaging Fabry Perot Interferometer (SPIFI) on the 15 m JCMT on Mauna Kea [2].

1.1 Observations and Results

SPIFI is an imaging Fabry-Perot interferometer designed for use in the 350 and 450 μm telluric windows available on Mauna Kea and the 200, 350 and 450 μm windows available to the AST/RO telescope at the South Pole. SPIFI employs a 5 × 5 array of silicon bolometers from GSFC held at 60 mK using an adiabatic demagnetization refrigerator. The bolometers are fed using $1.4\lambda/D$ (7" on JCMT) Winston cones arranged on a $1.4\lambda/D$ square grid (35"×35" footprint). SPIFI has been deployed on the JCMT 6 times since its first light in April 1999, but unfortunately, only our runs in April 1999, and March 2001 had sufficient submillimeter weather to pursue high quality science. On our first run, we had 12 operating pixels arranged in a 4×4 array and mapped the inner $2'$ of the galaxy (9 pointings, or about 150 spectra including field rotation) in the CO(7→6) line at 70 km/s velocity resolution [3]. The measured beamsize was 10".

Figure 1 displays the velocity integrated CO(7→6) map we obtained superposed on our 38 μm continuum map (5" beam). The far-IR radiation arises from dust heated by far-UV starlight from the central cluster. This radiation is largely deposited in the photodissociated surfaces of the molecular ring. The far-IR emission therefore traces the photodissociation regions (PDRs). The good correspondence between the far-IR and CO(7→6) maps, particularly at the southwestern portion of the CNR then suggests that much of the warm CO resides in the PDRs as well. There is also CO(7→6) emission from the northern arm and streamers strongly suggest that warm molecular gas is flowing into the cavity. In contrast, there is no evidence for CO(7→6) emission from the east-west bar. When superposed on a 5$''$ HCN(1→0) map [4], it is clear that the CO(7→6) emission lies interior to the HCN(1→0) emission consistent with a PDR scenario: HCN must lie further away from the ionization source since it is easier to photo-ionize than CO.

We have modeled the observed run of CO line brightness with J in an escape probability, large velocity gradient approximation [3]. Figure 1 displays the best fit model, obtained for a velocity gradient, $\delta v = 300$km/s/pc, (which corresponds to a gradient of 100 km/s in our 11$''$ beam), a gas temperature of 240 K, density of 7.1×10^4 cm^{-3}, and CO column density of 5.9×10^{18} cm^{-2}. The beam filling factor is 20%. The optical depth of the lower J lines is small, but in the mid-J levels, $\tau \sim 1.5$. The total mass of warm CO in the inner 1.5 pc is about 2000 to 3000 M$_\odot$, or about 20 to 30% of the ring mass.

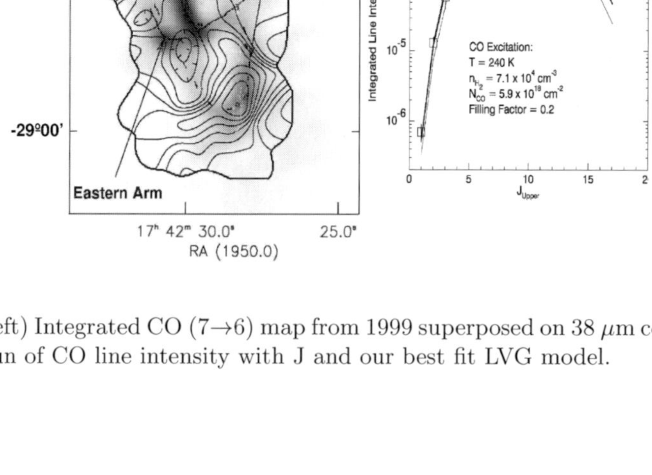

Fig. 1. (left) Integrated CO (7→6) map from 1999 superposed on 38 μm continuum. (right) Run of CO line intensity with J and our best fit LVG model.

Surprisingly, the total cooling in the CO lines is comparable to that of the far-IR lines. Standard PDR models can not reproduce such strong CO line emission, and even clumpy models under predict the CO line emission by a factor of 25 [5]. However, the observed large molecular line widths in the ring suggest another form of gas heating: molecular shocks through cloud-cloud collisions. Indeed, the large CO to far-IR line ratio is similar to that observed in the Orion BN-KL shock. C shock models [6] can reproduce the CO excitation quite well with $n \sim 3 \times 10^4$ cm^{-3}, $v_{shock} \sim 10$ to 20 km/s , and B $\sim$ 0.5 to 0.3 milli Gauss. Such a shock would also dominate the [OI] line emission as well. The orbital decay time for shock dissipation of energy is about 10^5 years, which is similar to the orbital timescales. Therefore, the CNR and streamers are likely transient features formed by infalling clouds.

During our April 2001 run, we had several days of excellent weather during which we obtained a much larger, better sampled map. At this time, SPIFI had its full 5×5 array (21 operating pixels) and we obtain 43 pointings, or over 900 independent spectra. We also had a greatly improve beam profile, $\sim$ $7.8''$ FWHM, and lowered the velocity resolution so that our spectral scans included both the CO(7$\rightarrow$6) and 370 μm [CI] lines. The typical integration time was 12 minutes per pointing.

The CO(7$\rightarrow$6) line is strongly enhanced on the far-IR ring, and rapidly fades further away from the central cluster (Figure 2). In contrast, the 370 μm [CI] line remains uniformly bright in regions at least $1'$ (3 pc) exterior to the far-IR ring. This is expected. The CO(7$\rightarrow$6) line is far more density sensitive than the [CI] line, and there will likely be significant contamination in the [CI] spectrum from the foreground +20 and +50 km/s clouds that is smeared by our rather modest velocity resolution. The large radial extent of the [CI] line suggests that the [CI] emitting gas is not heated by the far-UV starlight from the central cluster since most of that energy is deposited in (and traced by) the far-IR ring. Additional sources of heating may be embedded early type stars in the outer molecular ring. Indeed, close inspection of the 31 and 38 μm images show several localized point sources in the outer ring. The ubiquity of the [CI] emission was previously demonstrated by the mapping in the lower J, 610 μm [CI] line using the CSO [7]. The ratio of observed line brightness indicates the extended CO is quite warm: $T \sim 100$ K.

Figure 2 displays the velocity integrated CO(7$\rightarrow$6) emission superposed on our far-IR map. It is clear there is excellent correspondence between the tracers over much of the CND. Also apparent are strong clumpy features, and bright emission from the Northern Arm. The CO emission from the arm is offset to the east with respect to the far-IR arm, which is what one would expect in a PDR scenario where the source of heating lies to the west (the central cluster). The east-west bar is not detectable in CO, and in fact there are quite deep (7:1) holes in the emission at both the Eastern Arm and the position of the IRS 16 cluster. Also apparent is strong CO emission from the water maser source to the NE of the ring.

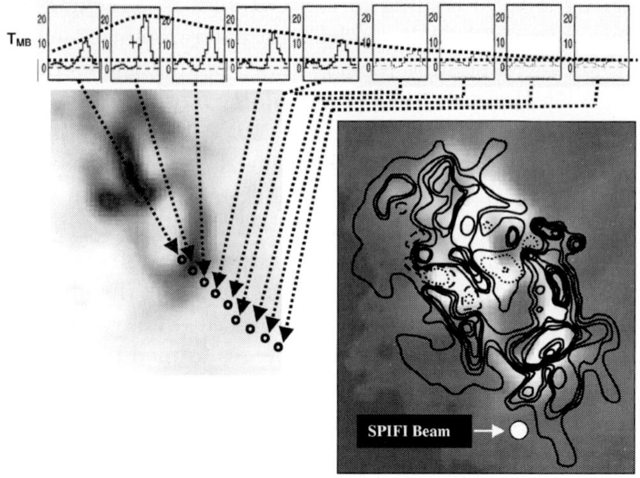

Fig. 2. (left) Cuts in CO (7→6) & [CI] 370 μm across the SW ring. (right) Integrated CO (7→6) from 2001 on 38 μm continuum map.

The total gas mass in the Northern Arm as traced by the CO(7→6) line is of the order 5 to 50 $M_{\odot}$ or roughly 10% of that traced in the [OI] 63 μm line and far-IR continuum [8] [1]. This mass can be better constrained by observing other mid, and high J CO lines, but the Northern Arm will be difficult to detect in the lower J lines due to small optical depth, and smearing with foreground clouds.

Our immediate future plans with SPIFI are to deploy it on the AST/RO telescope at the South Pole for a winter-over in 2004. SPIFI will be used in the 200 and 230 μm windows to observe the [NII] 205 μm and the CO(13→12) and (11→10) lines from the Galactic Center and external galaxies.

This work was supported by NASA grants NAGW-3925 and NAG5-4503, NSF grants OPP-0085182, and subcontract to NSF grant OPP-8920223.

References

1. H. Latvakoski, G.J. Stacey, T.L. Hayward, & G.E. Gull: Ap.J. **511**, 761 (1999)
2. C.M. Bradford, G.J. Stacey, M.R. Swain, et al. Appl.Opt. **41**, 2561 (2002)
3. C.M. Bradford PhD Thesis, Cornell University (2001)
4. R. Güsten, R. Genzel, M.C.H. Wright, D.T. Jaffe, J. Stutzki & A.I. Harris: Ap.J. **318**, 124 (1987)
5. M.G. Burton, D.J. Hollenbach,& A.G.G.M. Tielens: ApJ, **365**, 620 (1990)
6. B.T. Draine, W.G. Roberge, & A. Dalgarno: Ap.J. **264**, 485 (1983)
7. E. Serabyn, J. Keene, D.C. Lis, & T.G. Phillips: Ap.J. **424**, L95 (1994)
8. J.M. Jackson, N. Geis, R. Genzel, A.I. Harris, S. Madden, A. Poglitsch, G.J. Stacey, & C.H. Townes: Ap.J. **402**, 173 (1993)

350 μm Galactic Center Dust Observations with SHARC II

J. Staguhn[1], D. Benford[1], M. Morris[2], and K. Uchida[3]

[1] NASA/GSFC, Code 685, Greenbelt, MD 20771, USA
staguhn@stars.gsfc.nasa.gov

[2] Division of Astronomy & Astrophysics, UCLA, Los Angeles, CA 90095-1562, USA

[3] Dept. of Astronomy, Cornell University, Ithaca, NY 14853-6801, USA

We present first and preliminary submillimeter continuum images of the Galactic Center region obtained with the new Caltech Submillimeter Observatory facility camera SHARC II. The instrument allows 350 μm observations with unprecedented sensitivity and instantaneous spatial coverage. The stability of the SHARC II detectors combined with the large number of available pixels allows a non-differential scanning mode that does not require the observation of a reference off-source position. Here we present large-scale 350 μm dust continuum images from the Sgr A and Sgr C regions, as well as the detection of sub-mm dust continuum emission from the IRAS 25μm source AFGL5376. This source is produced in a large-scale shock that extends well above the Galactic plane.

1 Introduction

Studies of the interstellar medium in the nuclei of Galaxies can be used to derive physical parameters for galaxy evolutionary models. The presence of stronger magnetic fields than in the Galactic disk and strong shear due to differential rotation allow efficient studies of their effects on the interstellar medium and on star formation activity. Observations of the submillimeter continuum emission with SHARC II [1] play an important role in the determination of dust temperatures and opacity indices in Galactic Center molecular clouds. The heating of the dust is typically dominated by the interstellar UV field, and hence is attributable to relatively recent star formation. Other heating mechanisms such as interaction of molecular clouds with large-scale shocks are not well explored. We therefore targeted a large-scale shock region with no apparent star formation activity. Furthermore, submillimeter dust continuum data can be used to identify target positions for polarimetric follow-up observations aimed at obtaining magnetic field directions in molecular clouds [4].

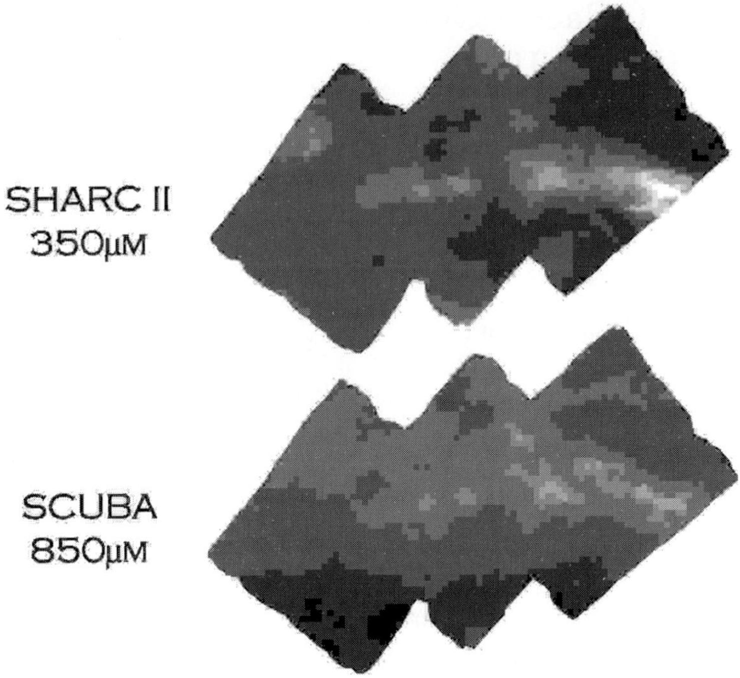

Fig. 1. *top:* SHARC II 350 μm image of the Sgr region. *bottom:* SCUBA 850 μm image of the same region (from [5]).

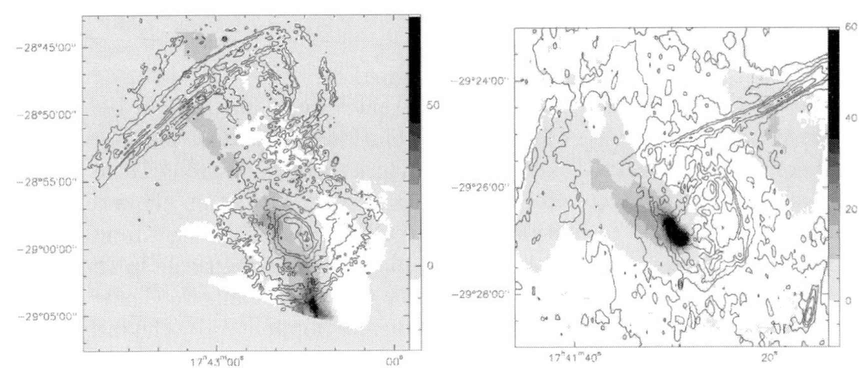

Fig. 2. *left:* SHARC II 350 μm image of the Sgr A region (grey scale) with the 20 cm radio image from [7] superimposed in contours. The most prominent features in the radio image are Sgr A, the Galactic Center Thermal Arches, and the Nonthermal Radio Arc. *right:* SHARC II 350 μm observations of the region around Sgr C (grey scale). Superimposed is the 1.5 GHz radio image from [3]. Note the anti-correlation between the radio emission and the dust continuum towards the NW of the Sgr C HII region.

2 Sgr A and Sgr C

We observed an area of about $20' \times 20'$ at 350 μm which covers the main dust features around Sgr A. Figure 1 shows the SHARC II image and, for comparison, the corresponding area as observed by SCUBA at 850 μm [5]. Figure 2 (left) shows the SHARC II image with an overlay of the 20 cm image from [7]. Figure 2 (right) shows the region around Sgr C at 350 μm. Superimposed on the image are the 1.5 GHz radio contours from [3] which show the Sgr C HII region and the associated nonthermal filaments. An anti-correlation of the dust emission and radio emission is evident north of the HII region.

3 AFGL5376

The IRAS source AFGL5376 is probably the most prominent concentration of warm dust near the Galactic Center which clearly shows no evidence for nearby star formation. This curious structure rises almost a full degree (150 pc) perpendicular to the Galactic plane (Figure 3). Its strong 25 μm emission indicates the presence of warm ($T \sim$100 K) dust which is clearly associated with high velocity ^{12}CO(2-1) emission (Figure 4 left, from [6]). The molecular line emission traces gas that surrounds the warm dust emission peak. Uchida et al. [6] show that the kinematics of the gas as well as the position of AFGL5376 provide strong evidence that it is associated with a large scale shock that coincides with the western edge of the radio continuum Galactic

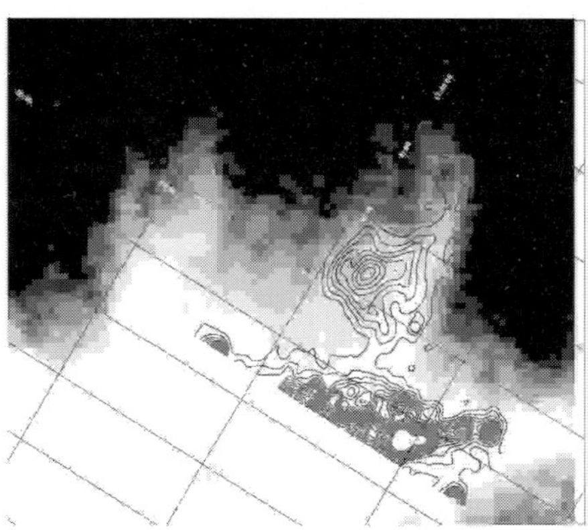

Fig. 3. Large scale Nobeyama 10 GHz image [2] (grey scale) of the Galactic center with IRAS 25 μm contours overlayed. The Galactic Center Lobes East and West can be seen as vertical structures in the grey-scale representation of the Nobeyama data. AFGL5376 is the prominent 25 μm feature at center right in the image, to the immediate East of the radio emission from the Galactic Center Lobe West. The Galactic plane is horizontal, and is indicated by the highest 25 μm contours.

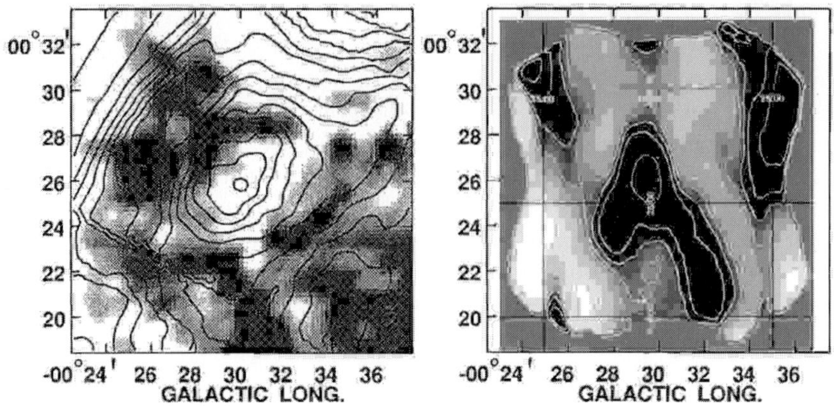

Fig. 4. *left:* AFGL5376 in $^{12}CO(2\text{-}1)$ emission (grey scale, from [6]) with an overlay of the IRAS 25 μm emission (contours) *right:* SHARC II 350 μm continuum image of the region (grey scale and contours). Note that the grey scale representation of the right image shows emission light, whereas it is dark in the left image.

Center Lobe. However, the mass of the warm dust in AFGL5376 is only 10^{-2} $M_\odot$, and until recently it was unclear whether larger amounts of colder dust are associated with this source. In August 2003 we obtained a snapshot of this region at 350 μm with SHARC II in order to address this question. The observations (Fig. 4, right) show strong 350 μm continuum emission which, like the molecular line emission, surrounds the 25 μm peak. A void in the emission stretching perpendicular to the Galactic plane is clearly visible in the SHARC II image. Its position coincides with one of the two shock fronts which are described in [6]. A determination of the spatial extent, column densities and temperatures of the dust and the nature of its association with the Galactic Center Lobe and the associated large scale shock is under way. This will enable us to study the energetics involved in the shock generation and will be important for understanding the relationship between the kinematics and magnetic fields in the Galactic center region.

References

1. C.D. Dowell, et al., SPIE, **4855**, 73 (2003)
2. T. Handa, Y. Sofue, N. Nakai, H. Hirabayashi, M. Inoue, PASJ, **39**, 709 (1987)
3. H. S. Liszt, R.W. Spiker, ApJS, **98**, 259 (1995)
4. G. Novak, et al., ApJ, **583L**, 83 (2003)
5. D. Pierce-Price et al., **545L**, 121 (2000)
6. K.I. Uchida, M.R. Morris, E. Serabyn, J. Bally, APJ, **421**, 505 (1994)
7. F. Yusef-Zadeh, M. Morris, D. Chance, Nature, **310**, 557 (1984)

The Hot Interstellar Medium of the Galactic Center: Observations with Chandra

M. Morris[1], C. Howard[1], M. Muno[1], F.K. Baganoff[2], S. Park[3], E. Feigelson[3], G. Garmire[3], and N. Brandt[3]

[1] Univ. of California, Los Angeles, Box 951562, Los Angeles, CA 90095-1562, USA morris@astro.ucla.edu
[2] Center for Space Research, MIT, Cambridge, MA 02139, USA
[3] Penn. State Univ., 525 Davey Laboratory, University Park, PA 16802, USA

Abstract. Deep observations of the Galactic center with the ACIS instrument on the Chandra X-ray Observatory reveal a wealth of detail in the spatial distribution and the spectrum of the hot interstellar gas. Here we describe a few prominent characteristics of the diffuse emission from this region. The diffuse X-ray emission apparently occupying most of the volume within one parsec of the center appears to be partially density bounded on the outside by the dense circumnuclear disk located between 1 and $\sim$5 pc. Further out, on a scale of $\sim$10 pc, the diffuse emission is dominated by bipolar X-ray lobes which appear to be centered on the central black hole and oriented perpendicular to the Galactic plane. Within the bipolar lobes are several blobs of X-ray emitting gas which display a rough point reflection symmetry about the center and which are thus consistent with a succession of collimated ejections of hot gas from the immediate vicinity of the black hole. Nonthermal emission from a few places around the rim of the northern lobe of this structure indicates that highly relativistic particles are present in the otherwise apparently thermally emitting lobes.

1 Observations and Background

On a number of occasions since 1999, the Chandra X-ray Observatory has been used to observe a single 17' x 17' field centered on the Galactic center, with the detector array oriented at various rotation angles. These observations have revealed the X-ray counterpart to the Galactic black hole, and have led to the discovery of its flaring activity (Baganoff et al. 2001; 2003). They have also strongly supported the supernova remnant hypothesis for SgrA East (Maeda et al. 2001), which encloses the Galactic center in projection, and they have revealed over 2300 point X-ray sources (Muno et al. 2003), most of which are part of the Galactic center population. The diffuse X-ray line emission extracted from these observations is described by Park et al. (2004).

2 SgrA* and Its Surroundings

The X-ray counterpart to the Galactic black hole has shown a steady-state quiescent X-ray intensity (0.6 $L_\odot$, absorption corrected, 2-10 keV) between its discovery in 10/99 and the most recent observation in 6/03. The ~1" size (10^5 Schwarzchild radii) corresponds well with the Bondi radius for a 4 x 10^6 $M_\odot$ black hole (Baganoff et al. 2003). During more than 600 ksec of observations with Chandra, we have noted a number of X-ray flares having durations from 30 min to a few hours, intensities from 5 to 50 times that of the quiescent level, and a frequency of a little less than once per day (Baganoff et al. 2001, & in preparation). Flaring and variability are now observed in the near-infrared as well (Ghez et al. 2003; Genzel et al. 2003). The flares are ascribable by various authors to instabilities or inhomogeneities in the accretion flow, or to shocks in an outflow structure (*i.e.*, a jet nozzle). The streak located ~1 pc to the southwest of SgrA* is the possible jet from SgrA*

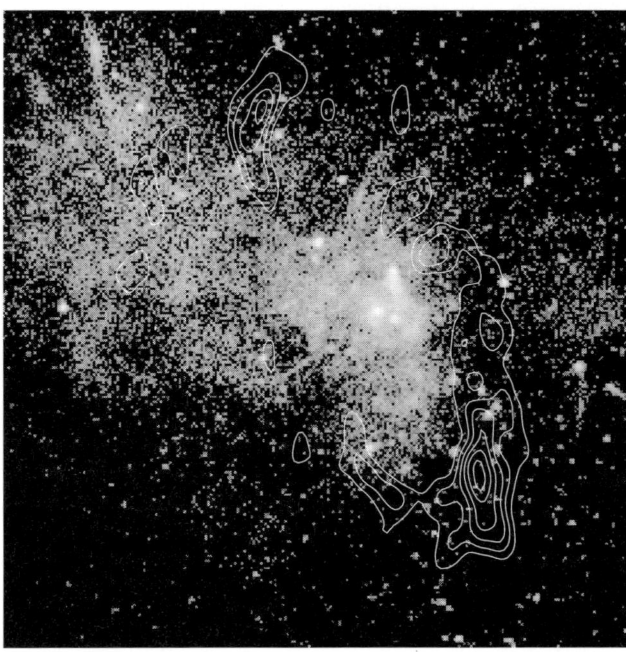

Fig. 1. Unsmoothed image of the vicinity of SgrA* [4 × 4 pc]integrated over the energy range from 2 to 8 keV. The full width at half maximum intensity of the point spread function is slightly larger than the 0.5" pixels. Note that SgrA* – the leftmost of the bright triplet of sources near the center – is noticeably extended with respect to the obvious point sources in the vicinity. The superimposed contours show HCN J=1-0 emission from the circumnuclear disk (CND) [measured with OVRO, M. Christopher, personal communication] superimposed on the unsmoothed X-ray image.

(Baganoff, in preparation). The absence of an oppositely directed counterpart can be ascribed to absorption by the near side of the tilted northern side of the circumnuclear disk (top right in fig. 1).

Interaction of the hot gas with the CND: Note the discontinuity in the X-ray intensity at the inner boundary of the CND, especially along the Western boundary. Spectra measured toward the CND and toward its interior indicate that the discontinuity occurs because the hot, emitting gas is density-bounded along this edge, rather than being caused by partial local absorption of a uniform X-ray background (Howard et al. in preparation). We hypothesize that most of the volume interior to the CND is filled with this hot gas. One possibility for the presence of such hot gas in the inner parsec is that the blast wave from the near-background supernova remnant SgrA East has recently passed through this volume, and is just now impinging on the Northern side of the CND (*c.f.*, Maeda et al. 2001). If so, the blast wave may have fueled a burst of activity as it passed over the Galactic black hole sometime within the last 1000 years. Alternatively, the low-density gas occupying most of the interior of the CND could be heated by dynamical interaction with abundant stars and stellar winds.

The Bipolar Lobes

The most prominent of the diffuse, extended structures in the X-ray image are two thermally emitting lobes aligned along a line passing through SgrA* and oriented perpendicular to the Galactic plane. These bipolar lobes are apparently filled structures having, in places, a discernible sharp edge. Thermal fits to the "blobs" along the central axis of the bipolar lobes give temperatures of $\sim$2 keV, with considerable uncertainty because of the difficulty of finding an appropriate local background to subtract. No radial trend in gas temperature is evident. The bipolar X-ray lobes have counterparts at low radio frequencies (90 cm; Nord et al. 2003). The steep-spectrum, low-frequency radio emission can be interpreted as synchrotron emission from a population of energetic electrons that are closely related to those responsible for the presumably thermal emission in the X-ray lobes. The total amount of mass involved in the X-ray lobes is quite small, about 1 $M_\odot$ per blob.

A compelling interpretation for the bipolar lobes straddling SgrA* is that they result from energetic mass ejections from the immediate environment of SgrA*, presumably from an accretion disk. The 1 - 2 arcminute radial separations of the blobs correspond to time intervals of (2500 - 5000 years) x (1000 $km\ s^{-1}/V$), where V is the unknown outflow velocity of the X-ray emitting material. This characteristic time presumably reflects a time scale in the ejection process.

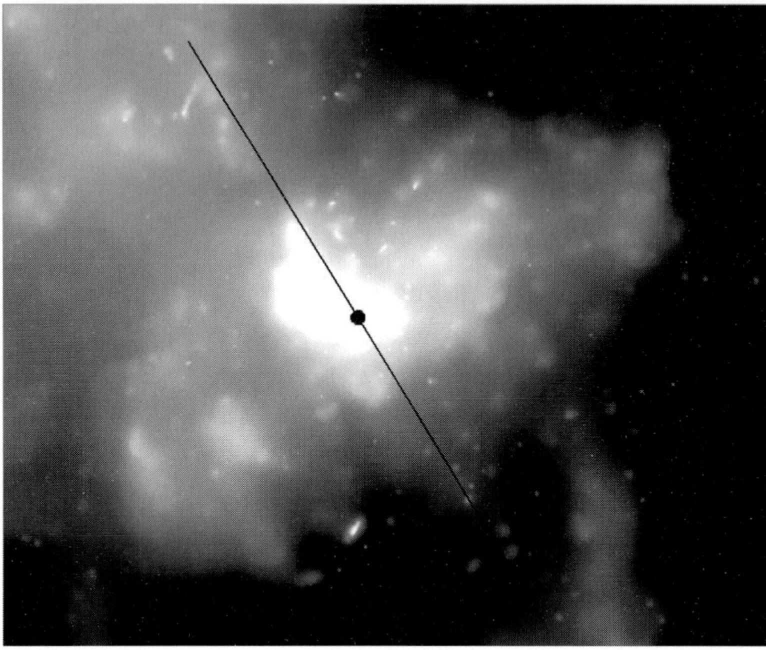

Fig. 2. Broad-band (3.3 - 4.7 keV) X-ray emission from a 35 x 28 pc region centered on SgrA*. The image has been adaptively smoothed, and all point sources identified by Muno et al. (2003) have been removed. The true Galactic plane and the position of SgrA* are indicated in black. The central region enveloped by SgrA East is saturated in order to bring out the dimmer bipolar features.

References

1. F.K. Baganoff, M.W. Bautz, W.N. Brandt, G. Chartas, E.D. Feigelson, G.P. Garmire, Y. Maeda, M. Morris, G.R. Ricker, L.K. Townsley & F. Walter: Nature **413**, 45 (2001)
2. F.K. Baganoff, Y. Maeda, M. Morris, M.W. Bautz, W.N. Brandt, W. Cui, J.P. Doty, E.D. Feigelson, G.P. Garmire, S.H. Pravdo, G.R. Ricker & L.K. Townsley: ApJ **590** (2003)
3. R. Genzel, R. Schüdel, T. Ott, A. Eckart, T. Alexander, F. Lacombe, D. Rouan, B. Aschenbach: Nature **425**, 934 (2003)
4. A.M. Ghez, S.A. Wright, K. Matthews, D. Thompson, D. Le Mignant, A. Tanner, S.D. Hornstein, M. Morris, E.E. Becklin, & B.T. Soifer: ApJL, in press (2003)
5. Y. Maeda, F.K. Baganoff, E.D. Feigelson, M. Morris, M.W. Bautz, W.N. Brandt, D.N. Burrows, J.P. Doty, G.P. Garmire, S.H. Pravdo, G.R. Ricker, & L.K. Townsley: ApJ **570**, 671 (2001)
6. M.P. Muno, F.K. Baganoff, M.W. Bautz, W.N. Brandt, P.S. Broos, E.D. Feigelson, G.P. Garmire, M. Morris, G.R. Ricker & L.K. Townsley: ApJ **589**, 225 (2003)

7. M.E. Nord, C.L. Brogan, S.D. Hyman, T.J.W. Lazio, N.E. Kassim, T.N. LaRosa, K. Anantharamaiah & N. Duric: Astron.. Nachr./AN 324, No. S1: "The Central 300 Parsecs of the Milky Way," eds: A. Cotera, H. Falcke, T.R. Geballe, S. Markoff, in press (2003)
8. S. Park, F.K. Baganoff, M. Morris, Y. Maeda, M.P. Muno, C. Howard, M.W. Bautz & G.P. Garmire: ApJ, in press (2004)

Non-Keplerian Orbits for the Star S2 Around the Black Hole at the Galactic Center

N. Mouawad, A. Eckart, S. Pfalzner, J. Moultaka, and C. Straubmeier

I. Physikalisches Institut, Universität zu Köln, Zülpicher Str.77, 50937 Köln, Germany nelly@ph1.uni-koeln.de

1 Introduction

Ten years of high resolution imaging at the Galactic Center (Sharp/NTT, KECK and VLT), allowed Schoedel et al. (2002, 2003) and Ghez et al. (2003) to determine Keplerian orbits for few stars in the central inner arcsec. Of those stars, the star S2 which shows the shortest orbital period has the best sampled orbit. It orbits in the region of the extended stellar cusp around SgrA* (Genzel et al .2003). Taking this into account, it is of importance to study the evolution of stellar dynamics in a non-point-mass potential. We find that the stellar orbit of S2 can be fitted well using combinations of point mass and extended cusp mass potentials. Assuming that the M/L(2μm) of the cusp is different from that of the outer stellar cluster, we allow high cusp masses.

2 A Realistic Potential

The density profile can be estimated from number density counts. The surface density (fig.1a) is fitted very well by a superposition of 6 Plummer models[1] having different densities and different core radii. The innermost region (1"×1"), representing the stellar cusp around SgrA*, can be fitted separately by a single Plummer model of core radius $R_{mid} = 15.5^{+4.7}_{-1.7}$ mpc. To the density distribution, corresponds a mass distribution[2] which is scaled and fitted to masses derived from dynamical mass estimators (fig.1b). If the M/L ratio fo the inner cusp equals that outer cluster, the total mass of the inner cusp results in $8540 M_\odot$.

3 Fitting Orbits

In order to fit our orbits, we use a fourth order Hermite integrator derived from the one used in high-accuracy N-body simulations (Aarseth 1999,

[1] $\rho(r)=\rho(0)[1+(r/R)^2]^{-\alpha/2}$, ($\alpha$=5, R is the core radius, $\rho(0)$ is the central density)
[2] $M(r)=(M_{tot}\times r^3/R^3)/(1+r^2/R^2)^{3/2}$; M_{tot} is the total mass in [8540 $M_\odot$, 8820 $M_\odot$]

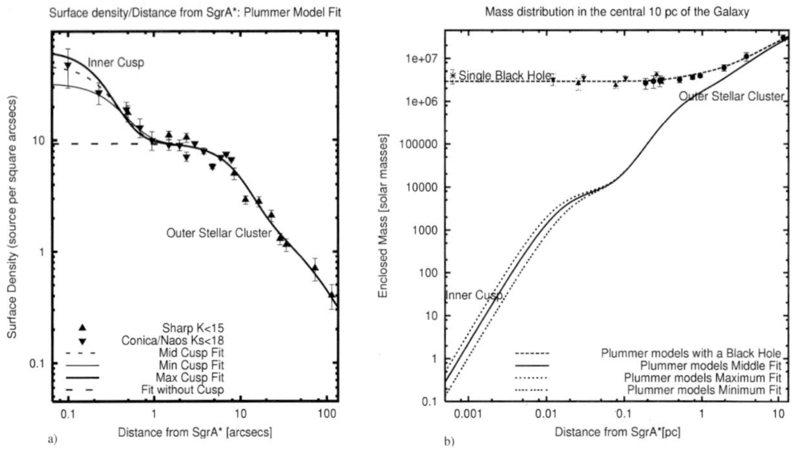

Fig. 1. a): Surface density of stars as function of distance from number counts computed in annuli. The downward triangles represent the CONICA/NAOS data for $K_s \leq 18$, the upward are SHARP/NTT data for $K_s \leq 15$ stars scaled up by a factor of 5 in order to match the fainter CONICA/NAOS counts (Genzel 2000, 2003). b): Mass distribution in the central 10 pc of the Galaxy. The long dashed line is a fit to the measured enclosed mass against distance (see Genzel et al. 2000).That results from a combination of the Plummer model fit to the stellar cluster plus a 2.9×10^6 $M_\odot$ black hole; The same model without a point mass is represented by the 3 other curves, they correspond to acceptable models within the error bars.

Spurzem 1999, for the first introduction of the Hermite scheme see Makino & Aarseth 1992). Adapting the mid-value of our model, we computed, for an S2 like orbit (Schödel et al. 2002), the trajectory of a star through the extended mass and around the BH. The Hermite scheme allows a fourth order accurate integration based on only two time steps. For that it requires the analytic computation of the time derivative of the gravitational force; therefore Plummer model superpositions as they are used here are very convenient. Further studies with more general density distributions should be undertaken, because they may influence the precise value of the periastron shift. It should also be noted that this integration is purely classical so any relativistic periastron shifts are not taken into account (Rubilar & Eckart 2001). We search for solutions over a range of 6 total mass M_{tot} values chosen between $2.6*10^6 M_\odot$ and $4.8*10^6 M_\odot$. For each value of M_{tot}, we allowed the fraction of extended mass to vary between 0% to 50%. We let the position of the central mass vary between 6 different positions within the errobars (0.035mpc) around the nominal radio position of SgrA*; we choose one central position, 4 extreme positions, and one offset position determine as best fitting position by Schoedel (2003, private communication). This results in 6 tables or grids, each corresponding to different central mass positions. The resolution of our search is increased or decreased depending on each case (see

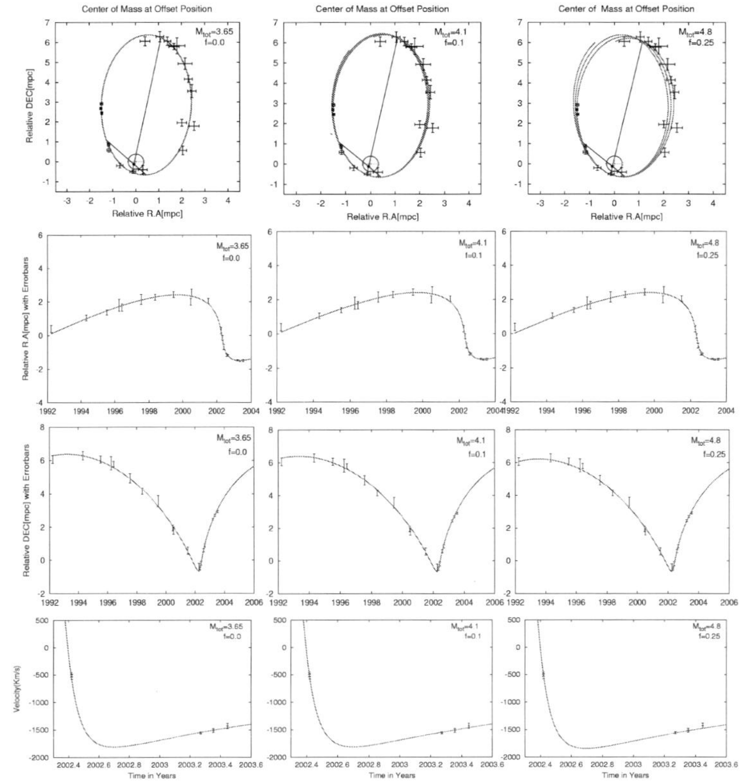

Fig. 2. Example of 3 fits corresponding to: (left) a Keplerian orbit with $3.65\times10_6\mathrm{M}_\odot$ point mass (best fitting orbit given by Schoedel, private communication); (middle) $3.69\times10_6\mathrm{M}_\odot$ BH + $0.41\times10_6\mathrm{M}_\odot$ extended component, and (right) $3.65\times10_6\mathrm{M}_\odot$ + $1.15\times10_6\mathrm{M}_\odot$ extended component. Here the central mass is at the offset position (0.082mpc east and 0.112mpc south from the nominal radio position of SgrA*).

tab.1). χ^2 values for positions and for velocities are separately estimated; a fitting orbit fits both in positions and in velocities. We vary simultaneously the 6 initial stellar coordinates in phase space, which is equivalent to varying the orbital elements. This leads to a number of N^6 orbital runs for each fitting procedure (N = step per parameter); The orbit having the smallest χ^2 is chosen to be the best fitting orbit.

4 Results, Interpretation and Analysis

The only cases where the S2 orbit fits the data with $\chi^2 \leq 2$ are the ones where the position of SgrA* is at the center position or at the offset position

Table 1. a):Values of χ^2. M is the total mass: point mass + cusp component, F is the fraction of the cusp component over the total mass. Dark gray to very faint gray corresponds to decreasingly well fitting orbits. White boxes for non fitting orbits. The 4 best cases we obtained for $1.1 \leq \chi^2 \leq 1.2$. b): Left side the M/L(2μm) estimated for the best fitting cases M/L(est), and on the right side the corresponding mag(k), and calculated M/L(2μm) when possible for the different spectral types, M/L(cal).

M \ F	0	0.05	0.1	0.15	0.2	0.25	0.3	0.35	0.4	0.45	0.5	0.55
6.8									2.28	1.28	1.37	2.92
6.1							3.35	1.67	1.21	1.32	2.9	
5.4						2.04	1.18	1.1	1.86			
4.8	6.41	5.16	3.07	1.91	1.39	1.17	1.52	2.35				
4.1	1.76	1.35	1.1	1.22	1.78	4.36						
3.65	1.1	1.27	1.72	2.24	4.11							
3.3	1.93	2.28	4.22									
3	3.26	5.22	6.9									
2.6	4.86	7.41										

a)

$M_T(M_\odot)$	Cusp	M/L(est)
3.65	0%	2
3.65	5%	4.27
4.1	10%	9.6
4.1	15%	14.5
4.8	25%	26.9
5.4	30%	37.9
5.4	35%	44.3
6.1	40%	57.1
6.8	45%	71.7

M/L(cal)	mag(k)	$M(_{M_{sol}})$	Sp
1.92	17	1.4	F5V
2.73	17.7	1.05	G0V
5.02	18.7	0.79	K0V
5.36	18.9	0.67	K5V
9.23	20.1	0.4	M2V
11.76	20.1	0.51	M0V
16.57	20.91	0.33	M3V
38.48	22.31	0.21	M5V
87.55	23.81	0.12	M7V

b)

(fig.2). For these, we study the extreme cases (core radius R=13.8mpc and R=20.2mpc). Independently of the core radius value, we find that for higher total masses the mass percentage of the extended cusp is bigger.

Our best fitting cases correspond to the offset position (given by Schoedel at 0.082mpc east and 0.112mpc south from the nominal radio position of SgrA*). For this position, we also investigate orbits with total masses higher than $4.8 \times 10^6 M_\odot$ and up to $6.8 \times 10^6 M_\odot$. This results in large cusp masses. However, the total masses are in disagreement with the dynamical mass estimates at large radii (tab.1a).

The exact composition of the cusp is still unknown (Genzel et al. 2003). Especially the existence of the $\sim 20 M_\odot$ S-stars is unexplained. Our methode gives M/L ratios estimated (tab.1b) for each cusp via the normalization to M/L = 2 of the outer stellar cluster (Genzel et al. 2000). For low mass cusps the data is in agreement with main sequence stars, but a model of >10 percent cusp mass gives M0V to M2V stars having 0.4 to 0.51 solar masses. Low masses stars could be a product of destructive collisions (Marc Freitag, private communication). These are faint stars with mag(k)=17 for the brightest; with $A_k = 3$ not yet detectable with observations.

References

1. Eckart et al., Mon.Not.R.Soc., **331**, 917-934(2002).
2. Genzel, R. et al, : ApJ, **549**,Issue 2, 812-832 (2003).
3. Ghez, A. et al., Nature, **407**, 349-351 (2000).
4. Mouawad, N. et al., in preparation (2004).
5. Spurzem, R., Jl. Comp. Appl. Math,**109**, 407-432.
6. Schödel, R. et al.,Nature,**419**, 694-696 (2002).

The Galactic Center Interstellar Medium as Seen by ISO

N.J. Rodriguez-Fernandez[1] and J. Martin-Pintado[2]

[1] LERMA (UMR 8112), Observatoire de Paris / CNRS, 61 Av. de l'Observatoire, 75014 Paris, France nemesio.rodriguez@obspm.fr
[2] DAMIR, IEM, CSIC, Serrano 121, Madrid, Spain

This paper deals with the heating and the ionization of the interstellar medium (ISM) in the 500 central pc of the Milky Way (hereafter Galactic center, GC). We review the results of *Infrared Space Observatory* (ISO) observations of a sample of GC molecular clouds located far from thermal radiocontinuum or far-infrared sources. For the first time, we have been able to study in detail the dust continuum spectra from 40 to 190 μm founding a warm (30-40 K) dust component in addition to the well known 15-20 K component. Fine-structure lines observations have revealed the presence of diffuse ionized gas associated with the molecular clouds. The effective temperature of the ionizing radiation is higher than 33000 K. ISO has also allow us to measure the fraction of warm ($\sim$ 150 K) H_2 in the GC clouds, which is on average of 30 %. The observations of the warm (a few 100 K) neutral gas are compatible with a Photon Dominated Region (PDR) scenario.

1 Dust

We have been able to study the full spectrum of the dust continuum emission from 40 to 190 μm. Two gray bodies are needed to model the spectra: a cold component with similar temperature for all the sources of 15-18 K and a warmer component with a temperature of 26 K to 39 K depending on the source (see Rodriguez-Fernandez et al. 2004). The cold dust is a well known component of the GC ISM from previous far-infrared and sub-millimeter studies. On the contrary, this is the first time we observe the warm dust component. As shown in Fig. 1, the warm component temperature and luminosity are well correlated with the 20 μm intensity in the images taken by *Midcourse Space Experiment* (MSX).

2 Ionized Gas

We have detected fine structure lines of ionized species as [C II] 158 μm and [Si I] 34 μm in all the sources. In most of them we have also detected lines like [N II] 122 μm, [Ne II] 13 μm or [S III] 34 μm. In 11 of the 18 sources we have even

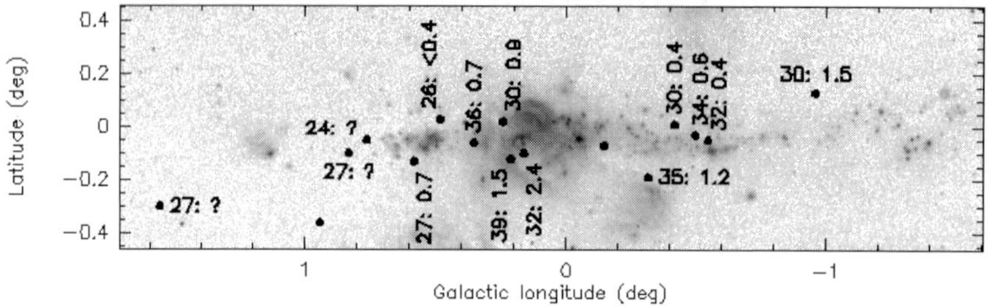

Fig. 1. Sources positions (black dots) overlaid on the 20 μm image by MSX. For each source numbers separated by a colon represent the temperature of the warm dust component in K and the [O III] 88 to [N II] 122 μm flux ratio

detected the [O III] 88 μm (with excitational potential of 35 eV). (Rodriguez-Fernandez et al. 2001b; Rodriguez-Fernandez & Martin-Pintado 2004). The lines observed with the LWS Fabry-Perot (velocity resolution of $\sim 30\,\mathrm{km\,s^{-1}}$) are very broad, with linewidths up to 150 $\mathrm{km\,s^{-1}}$. Left panel of Fig. 2 shows the [N II] spectra overplotted on the ^{13}CO(1-0) data of Rodriguez-Fernandez et al. (2001a). Taking into account the rather different spectral resolutions, for most of the sources the Fabry-Perot lines profiles are similar to those of the ^{13}CO(1-0) lines. [S III] and [O III] lines ratios imply that the electron densities are lower than a few 10 $\mathrm{cm^{-3}}$. Therefore, line profiles and electron densities are consistent with the ionized gas arising in the low density envelopes of the molecular clouds. The good correlation between the the [C II] 158 μm and the [N II] 122 μm fluxes (right panel of Fig. 2) implies that part of the [C II] flux also arises the ionized gas component (Heiles 1994).

Figure 1 shows one of the lines ratios sensitive to the effective temperature of the ionizing radiation that can be derived for most of the sources ([O III] 88/[N II] 122) over the 20 μm MSX image. The higher lines ratios are measured in those sources with the higher dust temperature (Section 1) which are those located in the surrounding of the 20 μm emission features. The association of warm dust and ionized gas imply heating by stars radiation. The ionizing sources are probably located in the vicinity of the 20 μm peaks but the radiation reach large distances due to the inhomogeneity of the medium, i.e., the general scenario is probably similar to that found by Rodriguez-Fernandez et al. (2001b) in the Radio Arc region. CLOUDY models show that the effective temperature of the radiation is higher than 33000 K (Rodriguez-Fernandez & Martin-Pintado 2004).

3 Warm Neutral Gas

Observations of H_2 pure-rotational lines have allow us to measure the total column density of warm gas in the GC clouds. The column density of warm

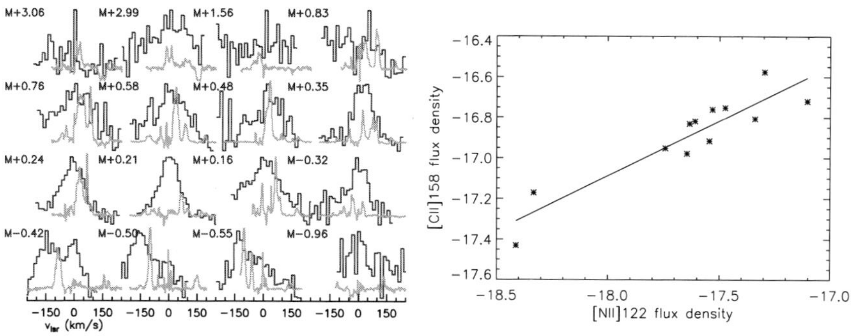

Fig. 2. Left panel: comparison of the [N II] 122 μm and the ^{13}CO(1-0) line profiles. Right panel: [C II] 158 μm versus [N II] 122 μm fluxes and least squares fit.

($\sim$ 150 K) H_2 is $\sim 10^{22}$ cm^{-2}. In addition there is a $\sim$ 500 K gas component with a column density lower than 1% of that at 150 K. Comparing with the H_2 column density derived from CO we see that the fraction of warm H_2 is on average of 30 % but it is as high as $\sim$ 100% for some of the sources. The H_2 line intensities can be reproduced both by shocks or PDRs models. In particular, the observed temperature gradient is in perfect agreement with both model predictions and ISO observations of well-known PDRs (Rodriguez-Fernandez et al. 2001a)

In addition, we have observed the warm (a few 100 K) neutral medium by means of fine structure lines of species like O, C^+ or Si^+, which are the main cooling lines in these medium. The lines to continuum ratio ($\sim$ 0.3%) is typical of PDRs. Indeed, the discovery of ionized gas associated with the molecular clouds implies that PDRs are expected in the interface between the ionized and the molecular gas. From the PDRs diagnostic diagrams of Wolfire et al. (1996) we estimate a density of 10^3 cm^{-3} and an incident field 10^3 times higher than the local interstellar radiation field.

We have also found that the ratio of warm to cold H_2 increases with an increasing fine-structure lines to far-infrared continuum ratio. This fact points to a common origin for the warm H_2 and the fine structure lines. (Rodriguez-Fernandez et al. 2004).

4 Discussion and Conclusions

The heating of the GC clouds is a long-standing problem. The discrepancy of the dust and gas temperatures and the apparently lack of ionized gas associated to the molecular clouds are usually invoked to rule out radiative heating mechanisms. However, the picture arising from the ISO observations is a complex medium where the molecular clouds are irradiated from the exterior by hot radiation arising from relatively hot and distant sources. The surfaces of the clouds are ionized and there is a warm (150 K) neutral gas component in the interface between the ionized and the molecular gas. The 30-40 K dust component should be associated with the 150 K gas component and both heated in PDRs. It is noteworthy that the discrepancy of gas and dust temperatures only rules out gas heating by collisions with hot dust but it does not rule out *any* radiative heating mechanisms as photoelectric effect on the dust grains in a PDR. Indeed a gas temperature of 150 K and a dust temperature of $\sim$ 35 K as measured in the GC clouds it is exactly what it is expected in a 10^3 cm^{-3} and G_0=10^3 PDR (Hollenbach et al. 1991).

On the other hand, the extended distribution and high abundances of fragile molecules when iluminated by ultra-violet radiation (see for instance Martin-Pintado et al. 2001) and the large line-widths are the best probe of the presence of turbulence or low velocity shocks in the GC clouds. Shocks, turbulence and PDRs probably coexist in the GC. Unfortunately, it is still not clear how much gas is heated in PDRs and how much is heated by other mechanisms. In any case, the new ISO data show that the importance of radiation in the heating of the GC clouds should be revisited and that a significant fraction of the warm gas in the GC clouds arise in standard PDRs.

Acknowledgments: The authors acknowledge the collaboration of A. Fuente, P. de Vicente, T.L. Wilson and S. Huttemeister in some of the work presented in this contribution. N.J.R.-F. acknowledges fruitful discussions with J.R. Goicoechea and M. Morris during the conference and support by a Marie Curie Fellowship of the European Community program"Improving Human Research Potential and the Socio-economic Knowledge base" under contract number HPMF-CT-2002-01677. J. M-P acknowledges *Ministerio de Ciencia y Tecnologia* grants ESP2002-01627 and AYA2002-10113-E.

References

1. C. Heiles , ApJ **436**, 720 (1994)
2. D. J. Hollenbach, T. Takahashi, A. G. G. M. Tielens ApJ, **377**, 192 (1991)
3. J. Martin-Pintado, J.R. Rizzo, P. de Vicente, N.J. Rodriguez-Fernandez, A. Fuente, P. Planesas: ApJ **548**, L65 (2001)
4. N.J. Rodriguez-Fernandez, J. Martin-Pintado: (2004, submitted)
5. N.J. Rodriguez-Fernandez, J. Martin-Pintado, A. Fuente, T.L. Wilson, S. Huttemeister: A&A **365**, 174 (2001a)
6. N.J. Rodriguez-Fernandez, J. Martin-Pintado, P. de Vicente: A&A **377**, 631 (2001b)
7. N.J. Rodriguez-Fernandez, J. Martin-Pintado, et al.: (2004, in prep.)
8. M.G. Wolfire, A.G.G.M. Tielens, D. Hollenbach: ApJ **358**, 116 (1990)

Dust Embedded Sources at the Galactic Center – 2 to 4 Micron Spectroscopy in the Central Parsec

J. Moultaka[1], A. Eckart[1], T. Viehmann[1], N. Mouawad[1], C. Straubmeier[1], and T. Ott[2]

[1] I-Physikalisches Institut, University of Cologne Zülpicher str. 77 50937 Cologne, Germany. moultaka@ph1.uni-koeln.de

[2] Max Planck Institut für extraterrestrische Physik, Giessenbach, 85748 Garching, Germany.

1 Introduction

The Galactic Center cavity is about 3 pc in diameter at a distance of 8 kpc. It is surrounded by a circumnuclear ring of dense gas and dust showing clumpy extinction [6]. The visual extinction towards prominent sources within the central stellar cluster range between 29 and 37 magnitudes [8]. In addition to luminous supergiant stars (as IRS7 and IRS9), a population of dusty sources associated with bright dust emission can be found in the Galactic Center stellar cluster as well as young and massive He stars (the IRS16 complex).

By means of MIR L-band low resolution (R=600) spectroscopy obtained at the ESO VLT unit telescope UT1 with ISAAC instrument as well as K-band previously published data [7], we have derived a detailed picture of the ISM in the half central parsec of the Galaxy.

2 Optical Depth Towards the Central Parsec

The optical depth spectra of the central parsec bright sources have been derived after performing simultaneous fits of the K- and L-band spectra with a reddened Blackbody ($3 \leq A_K \leq 4$).

The resulting spectra reveal four absorption features:

- *The water ice absorption feature at* $3.0\mu m$ observed in molecular clouds,
- *the* $3.4\mu m$ *and* $3.48\mu m$ *features* due to aliphatic hydrocarbons which are characterized by their CH_2 and CH_3 stretching modes. These features have been observed toward many Galactic sources [4] and are known to arise in the diffuse ISM.
- *a non-identified feature at* $3.3\mu m$ (visible in fig. 1(b)) also observed in the Galactic Center by [4].

The mean optical depth spectra of the isolated $3.0\mu m$ and $3.4\mu m$ absorption features are shown respectively in figs. 1(a) and 1(b) respectively.

The shape of the 3.0μm feature is consistent with a 40K ice temperature and a maximum mantle thickness of 0.85μm assuming the simple model of core-mantle grains [1] and the hypothesis of a variable mantle thickness [9].

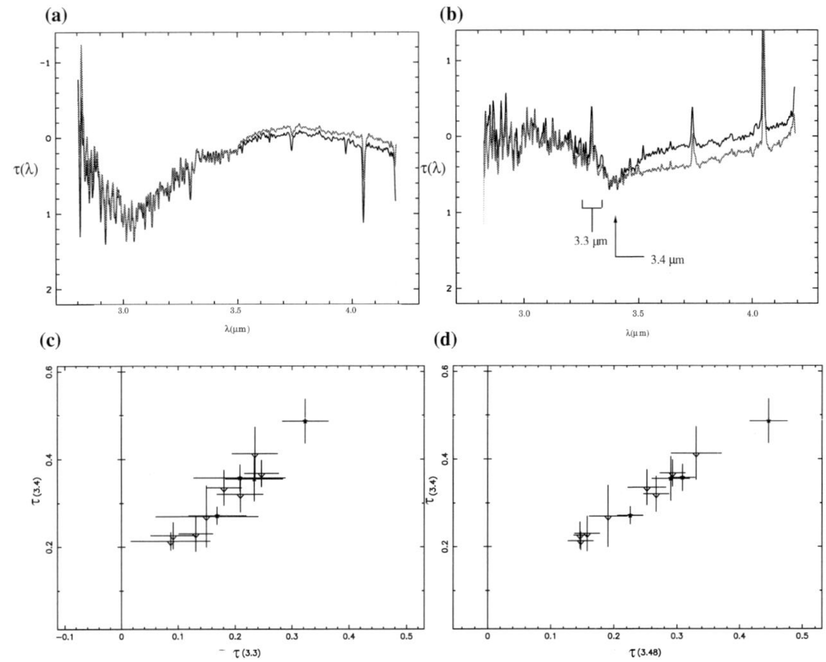

Fig. 1. Figures (a) and (b) show the mean optical depth spectra of the 3.0μm and the 3.4μm isolated absorption features for the He stars of the IRS16 complex (in gray) and the dusty sources and late type stars (in black). In figure (b) one can distinguish the 3.3μm unknown feature. Figures (c) and (d) show the correlations of the optical depths at the 3.3μm, 3.4μm and 3.48μm features.

The optical depths at the different absorption features have been measured for the most luminous sources of the central half parsec. These measurements reveal a high correlation (fig. 1(c)) between the optical depths at 3.3μm and 3.4μm (or 3.48μm) suggesting that the 3.3μm feature originates from the diffuse ISM. They reveal as well a trend of correlation between the 3.3μm feature and the water ice feature suggesting that its origin may be directly associated with the individual sources or the dense ISM in the central parsec.

A significant offset from the origin of the coordinate system is detected in the correlations between the 3.4μm and the 3.48μm (fig. 1(d)) and the 3.4μm versus 3.3μm (fig. 1(c)). It suggests that approximately 50% in τ (and therefore in the corresponding column density) of the 3.4μm line absorption is due to the diffuse ISM along the line of sight to the Galactic Center.

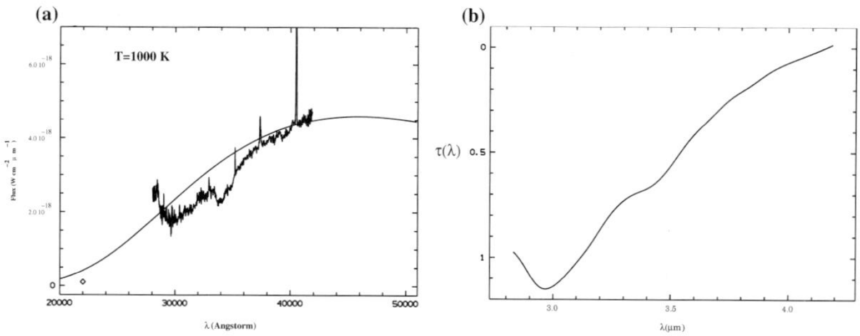

Fig. 2. (a) The spectrum of the IRS 13N region fitted by a blackbody of 1000K (b) The optical depth of the derived L-band extinction along the line of sight.

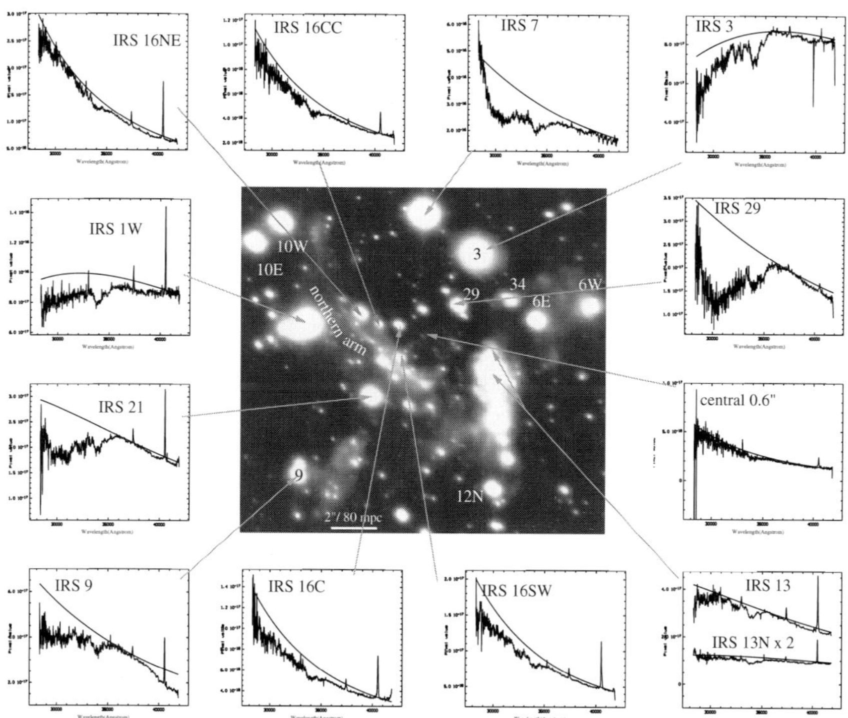

Fig. 3. The corrected spectra for extinction along the line of sight of the most luminous sources of the Galactic central parsec. Blackbody spectra of temperatures equal to the ones obtained for each object from the fitting procedure (see section 1) are also plotted. The good agreement between the spectra and the blackbody continua shows that the optical absorption spectrum of figure 2(b) is well consistent with the values of the K-band absorption used in the fitting procedure (section 1). The L-band image is obtained with ISAAC at the VLT.

The temperatures obtained for each source by the fitting procedure are consistent with the known nature of the sources, this result implies that the K-band absorption magnitude toward the Galactic Center doesn't vary of more than $\Delta A_K = \pm 0.5$ mag.

3 Calibration of the Extinction Along the Line of Sight

The slit positions adopted for our observations run through a late-type star located well off the mini-spiral emission. The K-band spectrum of this star shows clear $2.3\mu m$ CO absorption bandheads and no excess emission at wavelegnths of $3.0\mu m$ or longer are distinguished in its L-band spectrum. Therefore, we assumed that the emission is mostly affected by the line of sight extinction along the 8 kpc towards the Galactic Center and that the L-band spectrum of this star can be represented by a 3600K blackbody spectrum.

Dividing the spectrum of the previous late-type star by the blackbody continuum provides a measurement of the wavelength dependent L-band absorption due to the ISM along the line of sight. Figure 2(b) shows the optical depth of this absorption.

The corrected spectra for extinction along the line of sight of the bright sources are then derived by means of this spectrum and shown in figure 3.

The integrated spectrum of a region located 0.5" north of IRS 13 complex (fig. 2(a)) shows that the IR excess found in this region [5] is due to the contribution of warm dust T=500-1000K. The IR excess is possibly due to the presence of very young objects in this region.

References

1. Bohren, C.F. & Huffman, D.R. asls. book (1983)
2. Butchart, I. McFadzean, A. D., Whittet, D. C. B. et al: A&A **154** (1986)
3. Duley, W. W. & Williams, D. A., Natur **311**, 685 (1984)
4. Chiar, J.E., Adamson, A. J., Pendleton, Y. J. etal: ApJ **570**, 198 (2002)
5. Eckart, A., Moultaka, J. et al: accepted to ApJ (2003)
6. Guesten, R., Genzel, R., Wright, M. C. H. et al: ApJ **318**, 124 (1987)
7. Ott, T., Eckart, A., Genzel, R., ApJ **523**, 248 (1999)
8. Scoville, N.Z., Stolovy, S.R., Rieke, M. et al: accepted to ApJ (9/1/03 issue)
9. Smith, R. G., Sellgren, K. & Brooke, T. Y., MNRAS, 263, 749 (1993)

[CI] 492 GHz Study of the Galactic Center

B. Mookerjea[1], C.L. Martin[2], J. Stutzki[1], A.A. Stark[2], A.P. Lane[2], W.M. Walsh[2], and K. Xiao[2]

[1] KOSMA, I. Physikalisches Institut, Universität zu Köln, Zülpicher Strasse 77, D 50937, Köln,Germany bhaswati,stutzki@ph1.uni-koeln.de

[2] Harvard Smithsonian Center for Astrophysics, Cambridge, USA cmartin,aas,alane,wwalsh,kxiao@cfa.harvard.edu

1 Introduction

Molecular clouds consist mostly of clumps, the surfaces of which are illuminated by UV radiation resulting in Photon Dominated Regions (PDR's), which constitute a large fraction of the molecular material (and the neutral ISM). In the PDRs the dominant carbon bearing species changes from C^+ to C^0 to CO. Neutral carbon has two fine structure transitions at 492 GHz ($^3P_1 \rightarrow {}^3P_0$) and 809 GHz ($^3P_2 \rightarrow {}^3P_1$). The upper state of the transition at 492 GHz is only 23 K above ground and has a critical density of 1000 cm^{-3} for collisions with H_2 molecules. This implies that [CI] is easily excited and also the line is easily detectable even when emitted by moderate density interstellar gas exposed only to a radiation field equal to the mean interstellar radiation field in the solar neighborhood. We present results of preliminary analysis of a large scale 492 GHz [CI] survey of the Galactic center. We present first detection of a spatially extended narrow absorption feature, only weakly seen in ^{12}CO and ^{13}CO, that is due to cold foreground material along lines of sight in the direction of Sgr C.

2 The [CI] Data

Figure 1 shows the map of the integrated [CI] intensity as observed with the AST/RO telescope by Martin et al. 2003 (and A. A. Stark et al. in this

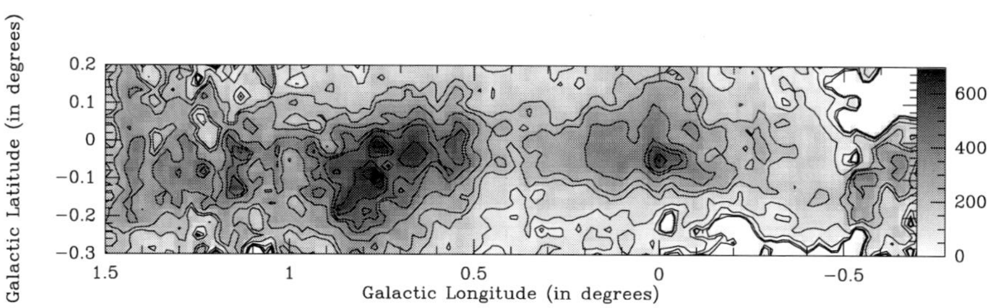

Fig. 1. Integrated intensity map of [CI] at 492 GHz emission from the Galactic center

volume). The [CI] data has been analyzed in combination with the Bell Labs (7m) ^{12}CO(1-0), ^{12}CO (1-0) and CS(2-1) data (Stark et al. 1989 & Bally et al. 1987) and ^{12}CO (4-3) data taken with AST/RO (Martin et al. 2003). For direct comparison, all datasets have been smoothed to the same angular resolution of $2'$ and gridded on a $1'$ grid.

3 Line Intensity Ratios for Selected Emission Features

We have selected 6 emission features corresponding to the peak intensities in the regions chosen based on the position-velocity diagram. Figure 2 shows the spectra corresponding to these emission features and Table 1 list positions and line intensity ratios calculated at each position.

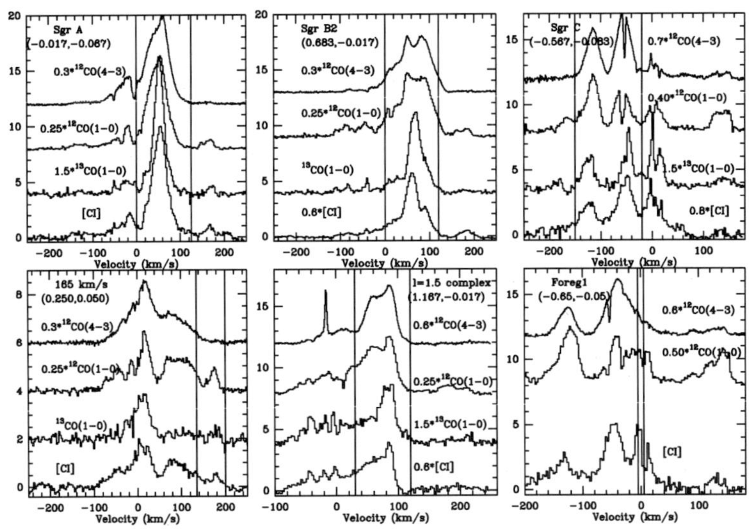

Fig. 2. Observed spectra for selected emission features in the Galactic center. Spectra have been drawn with offsets for better visibility. The vertical lines correspond to the velocity ranges over which the integrated intensities are calculated.

The [CI]/CO(1-0) is fairly constant for most regions, except for Sgr C and the foreground emission feature. The CO(1-0)/CO(4-3) ratio is, within limits of the error, identical for Sgr A, B2 and C and the value, $\sim 1.3\pm0.2$) , reflects optical thickness of the emission. At the other three positions the values deviate substantially. This is in contrast to the observed CO(1-0)/CO(2-1) ratio of ~ 1 for the entire Galactic center (Sawada et al. 2001). Optically thin ^{12}CO emission is evident in the 165 km/s arm, l=1.5 complex and the Foreground emission feature, where the CO(1-0)/CO(4-3) differs substantially from unity, and the ^{13}CO/^{12}CO(1-0) ratio are low (between 0.06 and 0.13).

Table 1. Line intensity ratios for selected emission features in the Galactic Center.

Feature	(l,b)	$\frac{[CI]}{CO(1-0)}$	$\frac{[CI]}{^{13}CO(1-0)}$	$\frac{[CI]}{CO(4-3)}$	$\frac{CO(1-0)}{CO(4-3)}$	$\frac{^{13}CO(1-0)}{CO(4-3)}$	$\frac{^{13}CO(1-0)}{CO(1-0)}$
Sgr A	-0.017,-0.067	0.26	1.80	0.30	1.14	0.17	0.15
Sgr B2	0.683,-0.017	0.27	1.62	0.37	1.35	0.23	0.17
Sgr C	-0.567,-0.083	0.61	4.61	0.88	1.44	0.19	0.13
165 km s^{-1}	0.250,0.050	0.24	4.15	1.32	5.53	0.32	0.06
l= 1.5	1.167,-0.017	0.28	2.07	0.70	2.51	0.34	0.13
Fground1	-0.650,-0.050	0.82	...	1.86	2.26	...	...

Preliminary estimates using calculations based on Escape Probability method (Stutzki & Winnewisser 1985) as well as detailed models for PDRs (Störzer et al. 1996) suggest that the density at Sgr A (and Sgr B2) is $\sim 10^{4.5}$cm^{-3}, while the density at Sgr C is $\sim 10^{3.5}$cm^{-3}.

4 Detection of Extended Foreground Absorption

The [CI] observations (Figure 3) presented here confirm the large spatial extent (at least $35' \times 35'$) of a spectrally narrow absorption feature extending from v_{lsr} =5 to 15 km s^{-1} toward the broad line emission from Sgr C. This feature was first detected at a few positions by Staguhn et al. (1997). The absorption feature clearly shows two velocity components centered at 8 km/s and 12.5 km/s with velocity widths of 4.5 and 3.5 km/s respectively. The relative strengths of the two components vary with position. For quantitative analysis of the absorption feature we spatially average $5' \times 5'$ regions to improve the S/N ratio.

We use a method similar to that outlined by Staguhn et al. (1997) to derive the excitation temperature and column density of the absorbing foreground cloud. Using the geometry explained in Figure 5 of Staguhn et al. (1997) and the constraint that the line is seen in absorption and the beam filling factor of the foreground cloud (α_f) has to be between 0 and 1 we arrive at the following expression for the excitation temperature ($T_{ex,f}$) for the foreground cloud:

$$J_\nu(T_{ex,f}) \leq T_{mb,bg} + J_\nu(2.7\ K) + T_{absorb} \tag{1}$$

where $J_\nu(T) \equiv (h\nu/k)(e^{h\nu/kT} - 1)^{-1}$ is the Rayleigh Jean's function, T_{absorb} is the (negative) antenna temperature corresponding to the absorption feature and $T_{mb,bg}$ is the observed brightness temperature of the background emission.

The upper limits of $T_{ex,f}$ calculated as above for the two absorption components in [CI] range between 3 to 10 K. Using these $T_{ex,f}$ the lower limits for the [C I] column densities calculated at all positions for both components are equal to a few times 10^{17} cm^{-2}.

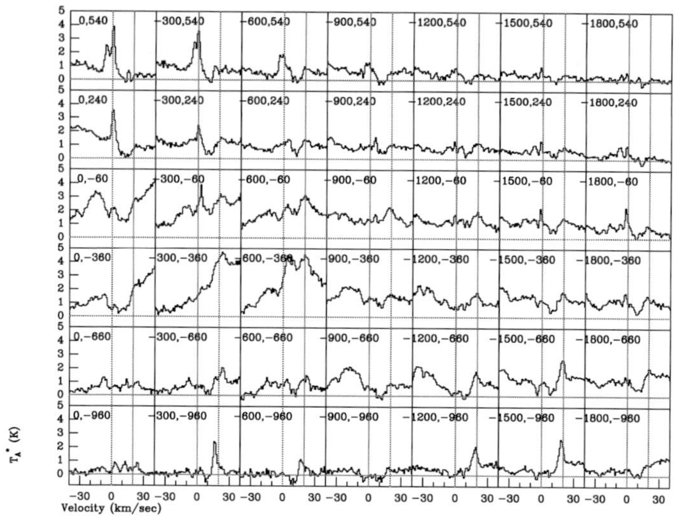

Fig. 3. [CI] spectra showing the extent of the absorption dip. Each position corresponds to a $5' \times 5'$ average centered at the position (offsets in arcseconds) indicated in the panels. Vertical lines are drawn to provide guidance to identify the absorption feature

The ^{12}CO(1-0) data for the same region shows only one absorption component centered around 12.1 km/s and with a width of 3.5 km/s. Using an analysis similar to that for the [CI] data we arrive at excitation temperatures between 4 to 15 K and CO column densities between 10^{15} to 10^{16} cm^{-2}. Both components of the absorption feature have marginally detected at a few positions in the ^{13}CO(1-0) data and preliminary analysis provides an excitation temperature of 3.5 K and a ^{13}CO column density of $\sim 10^{15}$ cm^{-2}.

The reason for the apparent discrepancy between the column densities of CI and CO and the usual relative abundance of these species is not understood. While the ^{12}CO is likely to be optically thick, this is less true of ^{13}CO. Possible explanations for the underabundance of CO and its isotopomers include dissociation or freezing out of CO in the foreground cloud.

This work was supported by the DFG within grant SFB-494 and in part by U.S. NSF grant OPP0126090.

The Galactic Center in the IR – VLT Infrared Observations of the Galactic Center

T. Viehmann, A. Eckart, J. Moultaka, and C. Straubmeier

1. Physikalisches Institut, University of Cologne viehmann@ph1.uni-koeln.de

The ISAAC system at the ESO VLT (Very Large Telescope) UT1 telescope has been used to image the central stellar cluster of the Milky Way in the mid-infrared L- and M-band (3.3 μm and 4.6 μm), with a field of view of 72.6"×72.6".These obsevations were part of a program to monitor the MIR flux density at the position of Sgr A* and were carried out simultaneously to CHANDRA observations. They have resulted in upper limits for the flux density of Sgr A* during an outburst of the associated X-ray source.

Additionally, the NAOS/CONICA adaptive optics assisted imager/spectrometer at the VLT UT4 telescope has been used to obtain diffraction limited H-, K_s- and L'-band (1.6 μm, 2.1 μm and 3.8 μm) images of the central cluster, with a field of view of 17.8"×17.8". These images offer unprecedented sharpness and sensitivity and have resulted in the detection of a small cluster of sources with a strong infrared excess, which might consist of young stars.

Upper Limits for the MIR Flux Density of Sgr A*

On May 29th 2002, the CHANDRA X-ray mission detected an outburst of the X-ray source at the position of Sgr A*, which lasted from about 05:40 to 08:20 UTC, with the peak occurring at approximately 06:10. Unfortunately, the VLT infrared imaging times only cover intervals immediately before and after the burst, however, spectroscopic data acquired during the burst, with the slit positioned on Sgr A*, is available. No mid-infrared burst activity was recorded during these times. The de-reddened upper flux density limits that could be attributed to a variable source are of the order of 10 mJy for the imaging observations in typical seeing conditions of 0.7". The spectroscopic exposures averaged seeing variations over longer integration times. Additionally, the source structure was averaged across the spectrometer slit, and the observations were also taken through the collimator optics of the spectrograph, resulting in a lower resolution of about one arcsecond. The spectroscopic exposures thus result in higher flux density limits of the order of 100 mJy.

These upper limits make models in which the X-ray flares are explained by an increased accretion rate or an increase in electron temperature (i.e. local heating in the accretion flow) less preferable, since large infrared flux density

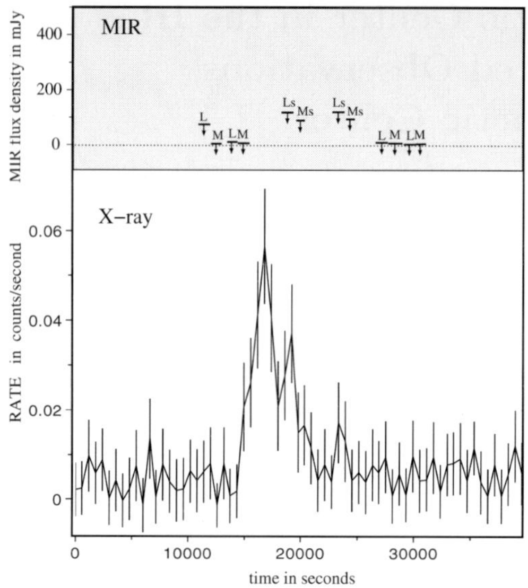

Fig. 1. Sgr A* X-ray burst (May 29th 2002) monitored simultaneously in the mid-infrared L- and M-band with ISAAC at the VLT. Flux density limits from spectroscopic measurements are labeled with an 's'.

variations were not observed. Flares due to (non-thermal) acceleration of electrons through shocks, however, remain consistent with these results [3].

IRS13N – A Cluster of Young Stars?

Close comparison of the NAOS/CONICA images has revealed a previously unknown, small cluster (0.13 ly diameter) of weak (m_K >14), compact (less than 0.09", ~740 AU) sources exhibiting a strong L-band excess (see fig. 3). This complex of sources, called IRS 13N, dominates the flux density of a weak northern extension of a dust continuum emitting region associated with the IRS 13 cluster, which contains the hottest of the Galactic Center He-stars. Spectral data of the region (obtained with ISAAC) shows that the L-band excess is due to emission of warm (T>500 K) dust. The nature of the IRS 13N sources is currently unclear; the following discussion mentions some possibilities.

IRS 13N might be a small association of luminous supergiants or young stars. However, the high L-band excess would require an exceptionally large foreground extinction, exceeding the average value towards the Galactic Center by several tens of magnitudes, and making the objects in question extremely bright (e.g. as the supergiant IRS 7). Moreover, the IRS 13N sources appear too blue in $H-K$ and too red in $K-L$ to be dominated by extinction only. This explanation is therefore rather unsatisfactory.

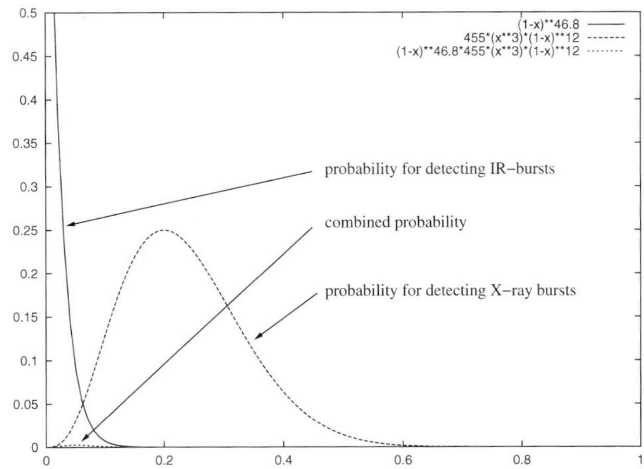

Fig. 2. Probability of detecting a 10 ks flare in the ISAAC and CHANDRA observations for the corresponding duty cycles. The low combined probability suggests that if flaring at IR and X-ray wavelengths is coupled, a flare occurred during the IR monitoring and must have been lower than the detection limit.

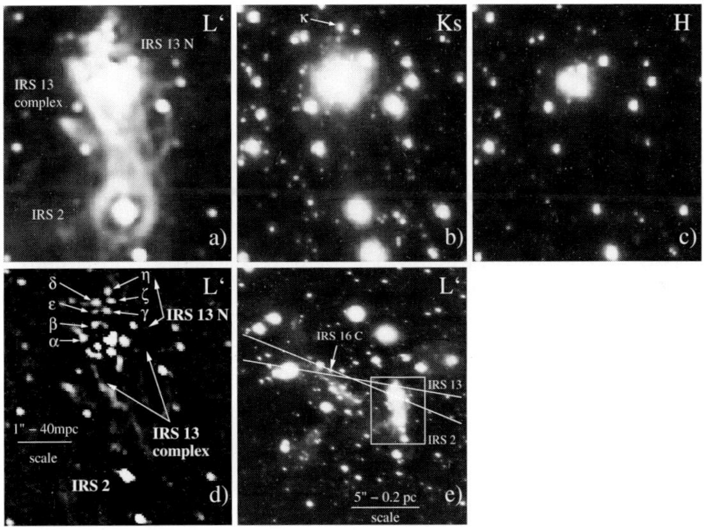

Fig. 3. H-, K_s- and L'-band images showing the newly-discovered IRS 13N sources. Panel (d) was high pass filtered to show the individual sources. Panel (e) is an overview of the region [1].

IRS 13N could consist of main sequence B or giant A stars heating the ambient gas and dust by radiation and stellar winds as they pass through the mini-spiral. The very red intrinsic colors are then due to large amounts of warm dust in their immediate environment. These sources would then be lower luminosity analogues of the 'bow shock sources' [2]. In this case, however, similar objects should be found at other locations along the mini-spiral, since it passes through regions of high stellar density.

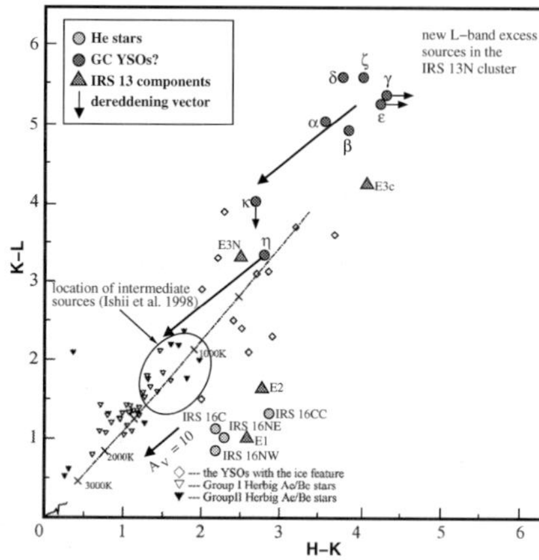

Fig. 4. HKL two-color diagram comparing the location of YSOs and Herbig Ae/Be stars (group I and II) with the IRS 13N sources and Galactic Center He-stars [1].

The most intriguing scenario is that IRS 13N is a cluster of young stars. Their colors, flux densities and luminosities ($\sim 10^3 L_{sol}$) are consistent with those of Herbig Ae/Be (group II) stars or young stellar objects reddened by 30 magnitudes of visual extinction (see fig. 4). In this case, the L-band excess would be caused by warm dust located in their circumstellar disks or envelopes. As the minimum age of such objects (0.1 Myr) is only ten times larger than the orbital timescale for the combined motion of the entire IRS 13 complex and the mini-spiral gas ($\sim 10^4$yr), these sources may have formed well within the central few light years, possibly even within the mini-spiral.

References

1. A. Eckart et al.: New MIR Excess Sources north of the IRS 13 Complex. In: *Proceedings of the Galactic Center Workshop 2002 – The central 300 parsecs of the Milky Way*, Astron. Nachr. 324, S1 Supplement, pp 521–526
2. Angelle M. Tanner, A. M. Ghez, M. Morris, E. E. Becklin: Resolving The Northern Arm Sources at the Galactic Center. In: *Proceedings of the Galactic Center Workshop 2002 – The central 300 parsecs of the Milky Way*, Astron. Nachr. 324, S1 Supplement, pp 597–603
3. S. Markoff, H. Falcke, F. Yuan, P. L. Biermann: The Nature of the 10 kilosecond X-ray flare in Sgr A*, A&A **379**, L13–L16 (2001)

Part VI

New Observatories and Instrumenation

Imaging and Spectroscopy with the Next Generation of Optical/Near-IR Telescopes

T. Herbst

Max-Planck-Institut für Astronomie, Königstuhl 17, 69117 Heidelberg, Germany
herbst@mpia.de

In this paper, I hope to do a little crystal ball gazing into the future of imaging and spectroscopy with ground-based optical and near infrared telescopes. Predicting the future is tricky at best, but I hope to convince the reader that the way forward lies not just in the development of instrumentation, nor in the development of telescopes, but rather in the effective development of both telescopes and instrumentation together.

Any self-respecting telescope or instrument builder must, of course, begin with the science. Specifically, he or she should ask what astronomers really want. The answer to this question has not really changed much over the years: astronomers want an instrument-telescope combination with great sensitivity, large field of view, excellent spatial resolution, wide wavelength coverage, and a variety of available observing modes.

In other words, a machine that does everything.

There is a wonderful expression for this type of device in the German language, namely *Eier-legende-woll-milch-sau*, or "egg-laying, wool-bearing, milk-giving, pig." This simple stereotype is not entirely appropriate, however. I believe that there is a good type of *Eier-legende-woll-milch-sau*, that is, an instrument-telescope combination that offers a variety of options to the observer, allowing multiple science programs, as well as rapid reaction to changing observing conditions. On the other hand, there definitely is a bad type of *Eier-legende-woll-milch-sau*. This is a device that attempts to do so much and is so complex that it is impossible to build, afford, and maintain.

1 A Taxonomy of Current and Future Instrumentation

Trying to see the way forward to a time of more effective telescopes and instrumentation requires a clear sense of where we are today. Therefore, I did a quick survey of the current state of observatories and facilities. There are approximately fifteen operating 4 m class telescopes in the world supporting about 100 different facility instruments. Interestingly, there are now also about fifteen, 8 m class telescopes in operation, with about 50 different facility instruments.

With so many different telescope and instrumentation approaches, extrapolation to the future requires some sort of order and classification. We can establish a taxonomy of telescope-instrumentation by imagining a "capability phase space," (figure 1), with axes corresponding to wavelength, angles, and signal. Improving the capability of an instrument corresponds to increasing its volume in this space. On the wavelength axis, an increase corresponds to greater wavelength coverage. Along the angle axis, pushing down the lower limit corresponds to improved spatial resolution on the sky, while pushing back the upper limit relates directly to an increase in the field of view of the instrument. The extent along this axis for any instrument with proper spatial sampling corresponds to the number of pixels in the detector. The third, signal axis also has limits which the instrument builder wishes to push back. The lower limit corresponds to the noise background in the system, while the upper limit represents the dynamic range of the device. As with the angle axis, the extent along the signal axis relates to a measured quantity, namely the data.

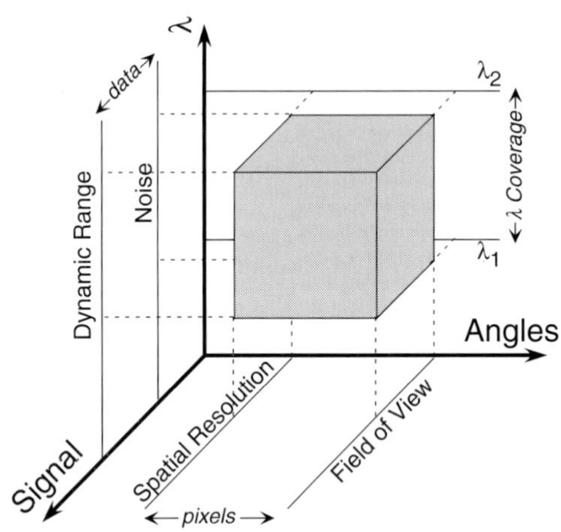

Fig. 1. Capability phase space. See text for discussion

One aspect of instrument design not represented in Figure 1 is the variety of modes offered to the user. Equation 1, below, shows another way of looking at the capability of instrumentation. The "Information Content" (IC) of an instrument is simply the sum of a number of logarithmic quantities related to the wavelength coverage, angular coverage, signal, and observing modes. This formulation, originally suggested by Steve Beckwith, corresponds directly to the number of bits of information contained in a particular observing result. The Information Content is:

$$IC = \log_2\left(\frac{\lambda_{max}}{\lambda_{min}}\right) + \log_2(pixels) + \log_2\left(\frac{Well}{RON}\right) + \log_2\left(\frac{R_{max}}{R_{min}}\right), \quad (1)$$

where $Well$ and RON are the detector well depth and readout noise, respectively, and R is $\frac{\lambda}{\Delta\lambda}$, the spectral resolution.

One possible prediction of the future of imaging and spectroscopy comes from an examination of the evolution of Information Content with time. Figure 2 shows the Information Content of various facility instruments located at 4 and 8 m class telescopes. Most 4 m telescopes were completed in the late '70s and early '80s, while most 8 m class telescopes were completed at the turn of the millenium. Therefore, the horizontal axis in Figure 2 could just as well have been time instead of telescope diameter.

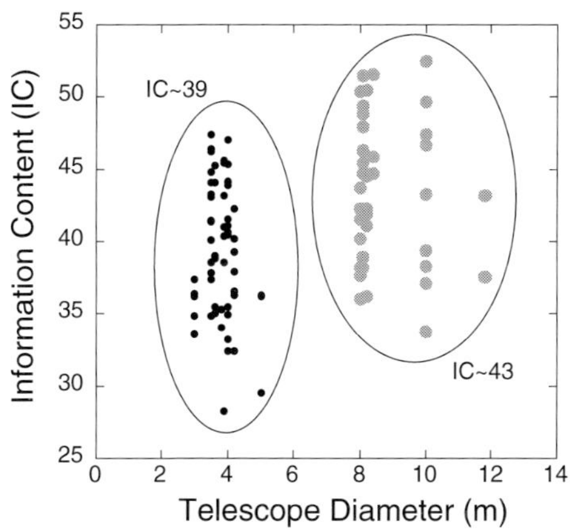

Fig. 2. Growth in Information Content of instrumentation.

The growth in average Information Content from 39 to 43 may not initially appear to be rapid, but recall that IC is a logarithmic quantity. Therefore, the 8 m class instruments provide 10 to 20 times more information content than their 4 m class cousins. The future direction is also clear from Figure 2, whether the horizontal axis is diameter or time. A simple extrapolation to 30 to 100 m class facilities leads to an Information Content value thousands of times greater than the current state-of-the-art.

A caveat regarding ever-increasing Information Content of instruments is in order, however. Those familiar with the second-generation instrumentation plans for large telescopes, such as the VLT or Gemini, will recognize that these instruments are not necessarily more complex or information-rich. Rather, a number of the second-generation efforts are tightly focused on particular science programs, such as high dynamic range imaging, or integral field spectroscopy. Nevertheless, the facility instrumentation for future 30 to 100 m class telescopes will inevitably be more complex than that of current 8 m class telescopes. The responsibility of the telescope/instrument builder is to grow that capability responsibly, in other words, to produce capable instrumentation that avoids the "bad *Eier-legende-woll-milch-sau* syndrome."

2 Case Study: Improving Point-Source Sensitivity

Safely growing the Information Content of instrumentation is a somewhat science-dependent problem. In this section, I will discuss some technological solutions for going forward in the context of a very important astronomical goal: detecting and studying point sources. This is, by necessity, not an exhaustive list of strategies. Rather, it shows a number of approaches in different areas that can lead us to improved point source sensitivity for imagers and spectrometers.

Equation 2 shows the signal to noise ratio (SNR) for point source detection in the background limit. The SNR is a combination of factors related to the telescope as well as the instrument. Making improvements in the future will clearly involve improving both. The signal to noise ratio is:

$$SNR \sim \sqrt{\frac{A_{tel} T \Delta\lambda}{A_{img} B}}, \tag{2}$$

where A_{tel} is the collecting area of the telescope, T is the system throughput, $\Delta\lambda$ is the bandwidth of the observation, A_{img} is the area of the point spread function, and B is the background flux.

2.1 Background B

I will tackle these various components in reverse order, starting with the background factor, B. A combination of of background sources limit our ability to observe at visible and near-infrared wavelengths. Shortward of 1 μm, observations are hampered by scattered light, as well as the inability of the photons to penetrate cosmic dust. At wavelengths longward of 2.2 microns, the thermal background of the telescope and site dominate the spectrum. Clearly, the J and H photometric bands would be ideal for high sensitivity observations, were it not for the presence of numerous bright OH lines.

There are several successful approaches to minimizing this OH backround problem. Unfortunately, such techniques are restricted exclusively to spectroscopic observations. In my view, one of the great unsolved problems of ground-based instrumentation is how to avoid OH contamination for panoramic, imaging observations. Although there have been a number of suggestions in the past, none appears very workable, and the way forward to an effective OH-suppressed wide-field imager is not clear.

There may well be an effective strategy for suppressing the thermal background, however. The amount of background photons, and hence the background noise, relates directly to the temperature of the telescope, atmosphere, and other associated optics. Choosing an extremely cold site, such as Antarctica, will inevitably improve the situation. In particular, one Antarctic site, called Dome C, has attracted a great deal of interest and attention in recent years. This site is high (~3,500 m), cold (220 K during winter/nighttime), and dry, (precipitable water vapor ranging below 200 microns).

There is an added bonus to placing a telescope at Dome C. Recent measurements have shown that the vertical turbulence profile above the Antarctic continent may be very suitable for high angular resolution observations. Both balloon and SODAR measurements show that the turbulence above Dome C appears to be concentrated in the lower 100 meters of the atmosphere. Such a localized turbulence profile lends itself very well to improving the next aspect of point source sensitivity, namely reducing the size of the delivered point spread function (PSF).

2.2 Image Point Spread Function Size A_{img}

Reducing the size of the telescope PSF means adaptive optics. There has been considerable development and discussion in recent years about various types of adaptive optics strategies, including extreme adaptive optics (XAO), multi-conjugated adaptive optics (MCAO), and various other schemes. One particular type of adaptive optics that has not received a great deal of attention, but which may prove to be critically important for both future Extremely Large Telescopes (ELTs) and for taking advantage of Antarctica, is the technique of so-called "ground layer adaptive optics," or GLAO.

Figure 3 illustrates the principle of ground layer adaptive optics. The light from a number of different stars, widely separated on the sky, passes through different patches of the upper layer turbulence, but through the same patch of air in the ground layer. This has three major implications. First, if we can remove the wavefront aberrations introduced in this ground layer, we should be able achieve partial correction over a very wide field of view. Second, we should be able to use natural reference guide stars spread over a fairly wide sky angle. Third, the greater the concentration of turbulence in the ground layer, the greater the level of correction we should achieve. This has direct and significant implications for a site such as Antarctica's Dome C. Partial correction, with typical image sizes of 0.1 arcsecond, may be achievable in the near infrared over a *one-degree* field of view. Instrument builders take note: sampling such a field would require 72,000×72,000 pixels.

2.3 Observation Bandwidth $\Delta\lambda$

The opacity of the Earth's atmosphere limits the improvements we can achieve by increasing the bandwith or wavelength coverage of our observations. Going to a high, dry site, such as Antarctic Dome C, would produce gains in the water absorption features between several of the near-infrared transmission windows. Large gains in bandwith will continue to require a space observatory, however.

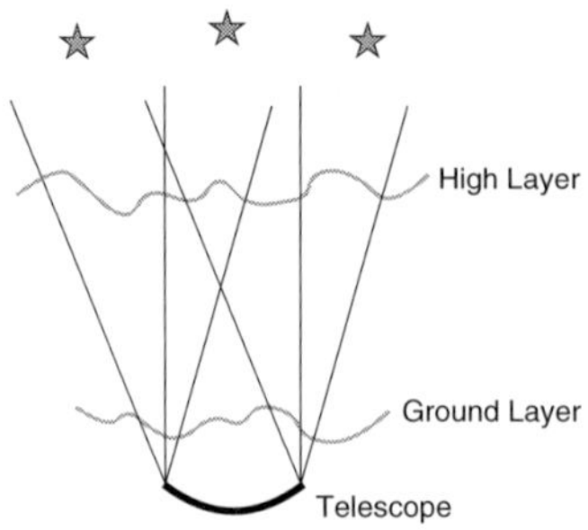

Fig. 3. The principle of Ground-Layer Adaptive Optics (GLAO)

2.4 System Throughput T

The next component in our attempt to gain improvement in point source sensitivity is the throughput, T. As with the bandwidth, we do not have much leverage, since the maximum value of T is one. There are, however numerous efforts underway to improve the transmission of existing and future instrumentation. For example, innovative multilayer filter design, better dispersers (including immersion gratings, volume phase holographic gratings, etc.), and high performance anti-reflection coatings (including solgels), are currently being implemented in ground-based telescope instrumentation. Another interesting recent development is in high performance, durable mirrors. These mirror coatings, consisting of layers of silver, aluminum, and nickel-chrome, exhibit high reflectivity down to the atmospheric cutoff at 320 nm. A further advantage for future extremely large telescopes is the durability of these coatings. Accelerated environmental tests indicate that such mirrors may last for decades without recoating.

2.5 Telescope Collecting Area A_{tel}

The final element in our equation for point source sensitivity is the size of the telescope itself. Unlike some of the other components, huge gains can be achieved if a workable large telescope can be built. There are a number of ongoing and very serious engineering efforts aimed at achieving 30 to 100 m size telescopes within the next decade or two. Prominent among these are Magellan 20 and the Thirty Meter Telescope, or TMT, which represents a fusion of the Giant Segmented Mirror Telescope, CELT, and the Canadian VLOT effort. Within Europe, a consortium led by Sweden is actively investigating a 50 m telescope design called Euro 50. Finally, there is the ambitious OWL effort, led by ESO, which foresees a 100 m ground-based telescope in operation by 2020.

Figure 4a emphasizes how technology has allowed enormous growth in telescope collecting area, and hence, man's ability to do science. If the 30 to 100 m class telescopes succeed, they will represent a huge increase in the cumulative collecting area available to astronomers. Perhaps a better measure of the importance of a certain telescope is a quantity I call "impact": the impact of a given telescope is simply the ratio of its collecting area to the then-current cumulative collecting area of all telescopes available to astronomers (see figure 4b). The impact measures not only the size of the telescope, but also it's timeliness. A 10 m telescope like Keck 1, commissioned in the early 1990's, has a significantly greater impact than an essentially identical telescope, GTC, which should achieve first light in 2005.

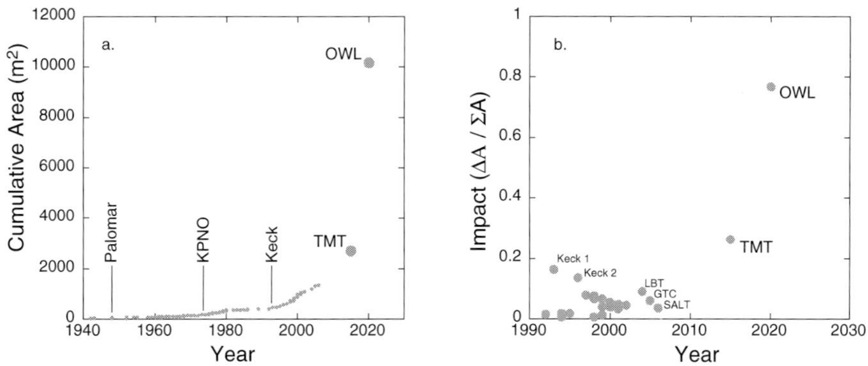

Fig. 4. Growth in the cumulative collecting area (a) and "impact" of telescopes (b). Most of the current 8 meter telescopes (VLT, Gemini, Subaru, etc) lie in the cluster between 1998 and 2002.

By this measure, the TMT and, particularly, the OWL project will have an enormous impact on our ability to do science. In fact, if OWL is completed by 2020, 80 percent of the cumulative collecting area available to earthbound astronomers will be concentrated in this one facility. The last time that a single telescope had such enormous impact was the first refractor of Galileo himself.

3 Extremely Large Telescopes and Their Instrumentation

Needless to say, there are a number of challenges to building and operating such enormous telescopes. There is a great deal of ongoing work in such areas as segment fabrication, active and adaptive optics, integrated modeling of telescopes and their environment, site selection, and last, but certainly not least, instrumentation.

The earlier analysis of Information Content or complexity clearly indicates that future ELT instruments will be a challenge. They will necessarily be much more complicated than current state-of-the-art instruments, and their success will depend on a systems approach. Specifically, it is essential that the design and limitations of the telescope be included in the instrument design, and vice-versa. For such ambitious projects, the "build it and they will come," approach to instrumentation will simply not work. Because of their large size, there will also likely be fewer available facility instruments, but future planners must avoid the "bad *Eier-legende-woll-milch-sau* syndrome."

Cost is an additional daunting aspect of future ELT imagers and spectrographs. It is a sobering fact that 8 m class instruments cost approximately 10 times as much as their 4 m class cousins. Nowadays, instrumentation can represent up to 20 or 30 percent of the total telescope capital investment, and it is not clear how these figures will scale in the 30 to 100 m world. Current instrument plans for Extremely Large Telescopes manifest what I call the "10-10-10 effect": the instruments are tens of cubic meters, tens of tons, and tens of millions of dollars. Clearly, we need to break the cost cycle, in order to build the kinds of instrumentation that astronomers want.

One potential future ELT instrument serves as an interesting example of the way designers are thinking. The MOMSI spectrograph, being developed by a consortium led by the U.K. ATC, takes a practical approach to sampling the huge field of view available at an ELT. The MOMSI spectrograph uses a so-called "smart focal plane," with multiple deployable integral field units sampling up to 32 different subfields. MOMSI contains eight fully cryogenic spectrographs in a dewar approximately 2.5 meters in diameter and 6 m long. With a 70 cm vacuum entrance window, this device represents a significant technological challenge.

There are some interesting stops along the road to Extremely Large Telescope instrumentation. For example, the Large Binocular Telescope (LBT), operating in beam-combination mode, presents many of the challenges that future ELT instrument builders will face. Teams in both Arizona and Europe are working on instruments that take advantage of the combined synthetic aperture that the LBT will provide. One of these instruments, called LINC-NIRVANA, is effectively a 23 m ELT infrared imager. Further details of this instrument appear in the paper of Straubmeier *et al.*, in these proceedings.

4 Conclusion

The road toward imagers and spectrographs on future near infrared and optical telescopes is by no means clear. Nevertheless, the challenges presented, and the potential scientific harvest, make such a road well worth travelling.

LINC-NIRVANA – The Interferometric Near-Infrared Imaging Camera for the Large Binocular Telescope

C. Straubmeier[1], T. Bertram[2], A. Eckart[2], and T. Herbst[3]

[1] I. Physikalisches Institut, University of Cologne, Zülpicher Straße 77, 50937 Cologne, Germany cstraubm@ph1.uni-koeln.de
[2] I. Physikalisches Institut, University of Cologne
[3] Max-Planck-Institute for Astronomy, Königstuhl 17, 69117 Heidelberg

In this article we present the basic principle of operation and the current status of LINC-NIRVANA, the upcoming interferometric near-infrared imaging camera for the Large Binocular Telescope (LBT). Using the exceptional optical design of the LBT for homothetic mapping at a large scale twin-telescope, LINC-NIRVANA will provide an unique combination of high angular resolution ($\approx 9\ mas$), wide field of view ($\approx 100\ arcsec^2$ - both values given for J-band) and large collecting area ($\approx 100\ m^2$).

1 The Large Binocular Telescope

Unlike all other large scale optical interferometers like VLTI (cf. von der Lühe et al. 1997 [1], Glindemann & Lévêque 2000 [2]) or Keck I & II (cf. Booth et al. 1999 [3], Swanson et al. 1997 [4]), the Large Binocular Telescope (LBT) follows the optical design of a Fizeau Interferometer (cf. Herbst et al. 2001 [5]). With its two 8.4 m primary mirrors sharing a common mount, the length of the uv-projected baseline of the LBT remains constant in time while tracking a celestial source, thus avoiding technically complicated optical delay lines. Furthermore, the time constant relation of the pupil diameters of the individual telescopes to the uv-projected length of the interferometric baseline provides the easiest opto-mechanical configuration for homothetic mapping (see Section 1.1).

After the ground works on top of Mt. Graham (3192 m, Arizona, USA) started in spring 1996, the enclosure of the LBT was finished in Fall 2002 (see Fig. 1). Since January 2003 the electro-mechanical structure of the twin-telescope is being erected on the central peer (see Fig. 2). First light of the LBT with one primary mirror is scheduled for Summer 2004 and interferometric operations are planned to start about one year later.

1.1 Homothetic Mapping

So far all major instruments on astronomical interferometers like MIDI (Graser et al. 1998 [7]), AMBER (Hofmann et al. 1998 [8]) or the Keck In-

Fig. 1. Since Summer 2002 the voluminous enclosure (housing about 40,000 m^3 and weighting more than 2,000 metric tons) of the Large Binocular Telescope is in full motorised operation.

Fig. 2. Since January 2003 the mechanical structure of the telecospe is being erected within the enclosure. Fig. 2 shows the status in July 2003 with the instrument platform already installed.

terferometer (Colavita et al. 1998 [9]) are based on Michelson interferometry. However, with respect to imaging camera systems, this optical design suffers from intrinsic optical path differences (OPD) for sources off the optical axis, which restrict the interferometrically usable field of view (FOV) to typically less than 1 $arcsec^2$.

Homothetic mapping overcomes this limitation by strictly preserving the geometry of the input pupil of the twin-telescope up to the exit pupil of the beam combiner. Following this principle, no intrinsic OPD is present at the combined focus and the interferometric FOV can reach the FOV of the individual telescopes (i.e. several $arcmin^2$) (Cobb 2000 [6]). While homothetic mapping could in general be realised at any interferometric telescope, this can be accomplished easiest and best at the LBT because of its special Fizeau design (see Section 1).

2 LINC-NIRVANA

Located at one of the interferometric foci of the Large Binocular Telescope LINC-NIRVANA resembles an unique wide-field imaging camera for the near-infrared wavelength domain providing an unprecedented combination of high angular resolution ($\approx 9\ mas$), wide field of view ($\approx 100\ arcsec^2$ - both values given for J-band) and large collecting area ($\approx 100\ m^2$) (cf. Herbst et al. 2003 [10]). To manage the technical complexity and financial effort of LINC-NIRVANA the camera system is currently developed and will later on be operated by an international consortium consisting of the Max-Planck-Institute for Astronomy (PI institute), the I. Physikalische Institut of the University of Cologne, the Osservatorio Astrofisico di Arcetri and the Max-Planck-Institute for Radioastronomy.

With a scheduled start of astronomical operations in Summer 2006 LINC-NIRVANA will be the first interferometric camera at any large scale optical telescope profiting from the advantages provided by homothetic mapping.

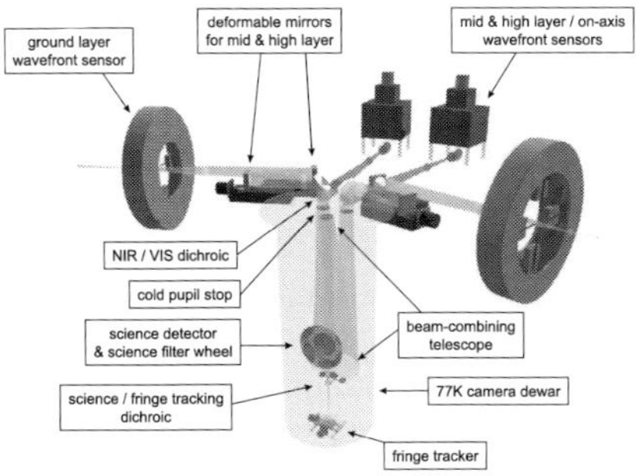

Fig. 3. The sketch above illustrates the basic sub-components of LINC-NIRVANA. All systems outside of the camera dewar (shaded in light grey) are operated at ambient temperatures.

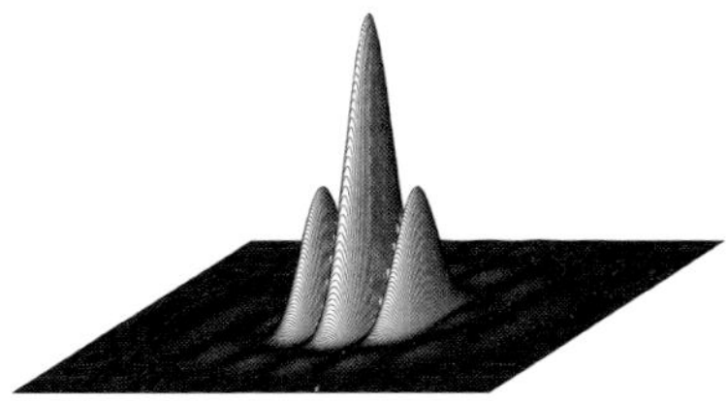

Fig. 4. This figure shows the result of a numerical simulation of the point-spread-function of LINC-NIRVANA. The interferometric baseline is aligned along the horizontal axis of the image.

To illustrate the optical design of the camera Fig. 3 sketches the fundamental sub-components of LINC-NIRVANA. With respect to its complexity a simplified division can be draw between the ambient temperature subsystems in the upper part of the image, and the cryogenic beam-combiner optics, the near-infrared science detector, and the near-infrared Fringe and Flexure Tracker inside the 77 K camera dewar.

The ambient temperature components form two sets of Multi Conjugate Adaptive Optics (MCAO) (Berkefeld et al. 2001 [12], Diolaiti et al. 2001 [13]) for the correction of high order atmospheric aberrations at each of the two telescopes. In their final implementation the MCAO systems of LINC-NIRVANA are able to analyse the wavefronts of up to 20 reference stars for up to three atmospheric layers simultaneously. Compared to conventional Adaptive Optics (AO) which use only one reference star, MCAO overcomes the limitation of the isoplanatic patch of the atmosphere, thus extending the angular size of the diffraction limited FOV and allowing a larger angular separation of the reference sources and the science object. Furthermore, since the optical signals of several reference sources are superimposed, the individual objects may be fainter than for the case of conventional AO. The combination of the last two arguments results in a significantly improved sky-coverage of MCAO systems in comparison to single star AO.

After the correction of the high order aberrations of the two separated incoming wavefronts, the ambient temperature optics form a scaled model of the input pupil of the telescope at the cold pupil at the top of the camera dewar. Within the dewar the beams of the two telescopes are then combined onto a common focal plane, forming the interferometric image of the observed objects.

However, in order to achieve time-stable conditions for interferometry, two different detector arrays are needed at the combined focus. In addition to the science detector for long integration times, a second imaging device specialised on fast readout rates is necessary as a last component of the Adaptive Optics of LINC-NIRVANA. This second array, named Fringe and Flexure Tracker (FFT), corrects the lowest order of atmospheric aberrations, the Piston aberration, a relative phase offset between the wavefronts coming in from the two telescopes (Straubmeier et al. 2003 [11]). This aberration is not detectable by the two MCAO systems of LINC-NIRVANA, since these systems operate independently of each other on the separate beams of the two telescopes and therefore are blind towards wavefront defects between one beam and the other. Located at the combined focus, the FFT analyses the positions of the interferometric fringes with respect to the centres of the Airy distributions of the two individual telescopes at frame rates of several hundred images per second. These positions give a direct measure of the pistonic aberration and allow for the generation of a corrective signal which is passed back to the MCAO systems of LINC-NIRVANA.

Based on the detailed geometry of the LBT Fig. 4 shows the result of a numerical simulation of the interferometric point spread funtion (PSF) of LINC-NIRVANA. Due to the non axial-symmetric sampling of the uv plane the PSF of a twin interferometer like the LBT is significantly different than the well known Airy distribution of a circular telescope.

References

1. O. von der Lühe, F. Derie, B. Koehler et al., *Proceedings from SPIE Conference*, Vol. 2871, p. 498-503 (1997)
2. A. Glindemann & S. Lévêque, *Proceedings of the ESO Symposium held at Antofagasta, Chile, 1-4 March 1999* p. 468 (2000)
3. A. J. Booth, M. M. Colavita, M. Shao et al., *Proceedings from ASP Conference*, Vol. 194, p. 256 (1999)
4. P. Swanson, M. Colavita, A. Boden et al., *AAS Meeting 191*, #09.12 (1997)
5. T. Herbst, P. Bizenberger, M. Ollivier et al., *Proceedings of a Workshop held July 24-29, 2000, at Ringberg Castle, Bavaria*, p. 193 (2001)
6. M. Cobb, *AAS Meeting 197*, #15.08, Vol. 32, p. 1429 (2000)
7. U. Graser & Ch. Leinert, *Annual Scientific Meeting of the Astronomische Gesellschaft at Heidelberg*, September 14-19, (1998)
8. K.-H. Hofmann, R. G. Petrov, F. Malbet et al., *Annual Scientific Meeting of the Astronomische Gesellschaft at Heidelberg*, September 14-19, (1998)
9. M. Colavita, A. Boden, S. Crawford et al., *Proceedings from SPIE Conference*, Vol. 3350, p. 776-784, (1998)
10. T. Herbst, R. Ragazzoni, D. Andersen et al., *Proceedings from SPIE Conference*, Vol. 4838, p. 456-465 (2003)

11. C. Straubmeier, A. Eckart, T. Bertram et al., *Proceedings from SPIE Conference*, Vol. 4838, p. 1271-1281 (2003)
12. Berkefeld, T., Glindemann, A. & Hippler, S., *Experimental Astronomy*, Vol. 11, Issue 1, p. 1-21 (2001)
13. Diolaiti, E., Ragazzoni, R. & Tordi, M., *Astronomy and Astrophysics*, Vol. 372, p.710-718 (2001)

The Future of Far-IR/Submillimeter Astrophysics with Single Dish Telescopes

C.M. Bradford and J. Zmuidzinas

Division of Physics, Mathematics and Astronomy, Caltech, Pasadena, CA, 91125, bradford@submm.caltech.edu, jonas@submm.caltech.edu

1 Introduction

Although far-IR – mm-wave astronomy has now been developing for several decades, access to this portion of the spectrum remains difficult and limited. Relative to their optical-wavelength counterparts, the submillimeter observatories are both small in units of wavelength, and hot in units of photon energy, which imposes limits on their angular resolution and sensitivity. Nevertheless, even the modest far-IR and submillimeter observations obtained in the last two decades have revolutionized our understanding the physics and chemistry of the ISM, the formation of stars, and the cosmic history of these processes. The demonstration by COBE that about half of the radiant energy released in the universe since decoupling comes to us in the submm/far–IR has further underscored the scientific necessity of observations in this wavelength regime [8, 4, 5]. The next decade promises even greater discoveries, as new observatories and detectors are still pushing toward fundamental limits. The paper briefly outlines some of the exciting capabilities and scientific roles of new single-dish far–IR – mm-wave telescopes and their instruments.

2 Detector Technologies

Over the past decade, one of the most important developments for far–IR through mm–wave astrophysics has been the push toward large–format direct–detector arrays. There are two key reasons why large arrays are important: (1) they provide a wider instantaneous field–of–view, thereby increasing the mapping speed; (2) they also allow better rejection of the "sky noise" due to atmospheric emissivity fluctuations, which is correlated across the array, and thereby provide better per–pixel sensitivities. Put together, these two factors provide a dramatic improvement in imaging capability.

For observations at $\lambda > 200\,\mu$m, bolometers are presently the most useful devices. Over the last 20 years, the mapping speed of bolometer arrays – equal to the ratio of the number of pixels to the square of a single pixel's sensitivity – has been improving at about a factor of two per year (faster than Moore's law !). Gains have been achieved in both single pixel sensitivity and array size. The individual detector elements are now achieving background-limited performance for ground–based astronomy or CMB studies, and the number of pixels in a typical array is ≤ 400, which is about the practical limit of what can be achieved for individually–wired detectors in a sub-Kelvin focal plane.

Multiplexing technologies for the next generation of bolometer arrays are now under development by several groups for both CMB studies and submillimeter astronomy. The most ambitious example at present is the SCUBA-2 effort, slated to deploy 5000 close-packed superconducting transition-edge detectors in each of two (450 μm and 850 μm) arrays, to be read out with a SQUID-based multiplexer [6, 3]. However, while these technologies are very promising, they are in the early phases, and there remains the substantial task of fielding robust, sensitive astronomical instruments the telescope.

The transition from the mm/submm to the far-IR ($\lambda < 200\mu$m) is a shift from ground-based observations to airborne and orbital platforms. The detectors of choice thus far have been the Ge:Ga photoconductors, which have flight heritage, require only modest cryogenic infrastructure (T$\sim$1.5 K), and can be naturally employed in large-format arrays using cryogenic CMOS multiplexers. These detectors arrays are the heart of the MIPS instrument on SIRTF (32×32 at 70μm, 2×20 at 160μm) and the PACS instrument on Herschel (16×25 at 90μm & 160μm), and will offer the most sensitive observations in the far-IR to date. Nevertheless, substantial room for improvement exists even at these shorter wavelengths: array construction is still very labor–intensive, which limits the array size, especially for stressed detectors at $\lambda > 100\,\mu$m; the quantum efficiencies are still relatively low; and the temporal response of these detectors requires sophisticated calibration protocols.

Coherent detection technology has also achieved impressive gains, driven by the needs of projects such as HIFI/Herschel and ALMA. Sensitive superconducting tunnel junction (SIS) mixers now operate up to 1.25 THz (240 μm), while superconducting hot–electron bolometer (HEB) mixers push to higher frequencies still, beyond 2.5 THz (120 μm). The improvements in local oscillator technology have been even more dramatic, especially the solid–state amplifier/multiplier chains developed for HIFI/Herschel, which are now capable of continuous tuning over > 100 GHz bandwidths and provide ample power for pumping SIS mixers at 1.25 THz and HEB mixers at 1.6 THz. These developments will revolutionize high–resolution terahertz spectroscopy, especially using airborne (SOFIA) and space (Herschel).

3 New Single-Dish Facilities

A major initiative for the submillimeter/millimeter astrophysics community in the next decade is the Atacama Large Millimeter Array (ALMA). ALMA will be an inteferometric array of 64 12-meter antennas operating in the 230, 345, 460, 650, and 850 GHz atmospheric windows, and will provide a very powerful probe in these frequency bands with its unique combination of high sensitivity and very high spatial and spectral resolution. While future observatories must carefully consider their role in this context, the development

Table 1. Future ground-based single-dish observatories

Observatory	Dia.	Surf RMS	PWV (top 25%)	Refs
SCUBA2-JCMT	15 m	20μm	0.7 mm	[6]
APEX/ASTE	12 m	18μm	0.5 mm	[9, 10]
SPT	10 m	15μm	0.4 mm	[11]
LMT	50 m	70μm	0.8 mm	[12]
A-25	25 m	12μm	0.4 mm	(adopted)

of large–format cameras – one of the major technological trends over the past decade – offers powerful new capabilities for single–dish observatories. While not widely appreciated, it is in fact possible for future single-dish facilities equipped with wide–field direct–detector cameras to be competitive with ALMA in terms of absolute point–source continuum sensitivity, and orders of magnitude faster than ALMA for wide–field imaging, thereby securing a scientifically important role which is complementary to ALMA.

Consider first the achievable point-source sensitivity of an observatory limited by the terrestrial backgrounds:

$$\mathrm{NEFD_{BG}} = \frac{\mathrm{h}\nu\,[\bar{\mathrm{n}}(\bar{\mathrm{n}}+1)]^{1/2}}{\eta_{\mathrm{inst}}\,\eta_{\mathrm{tel}}\,\eta_{\mathrm{atm}}\mathrm{A_{tel}}(\mathrm{N_{pol}}\Delta\nu)^{1/2}}, \tag{1}$$

where $\bar{\mathrm{n}} = \epsilon_{\mathrm{load}}\,\eta_{\mathrm{inst}}\,(\mathrm{e}^{\mathrm{h}\nu/\mathrm{kT}} - 1)^{-1}$ is the photon mode occupation number (equivalent to the number of photons per second per Hz of detection bandwidth in a single spatial mode arriving at a detector). Although ALMA will offer a tremendous collecting area, single–dish telescopes can offer improvements in all the other factors appearing in Eq. (1). A reduction of the atmospheric opacity improves the sensitivity twice – in increasing the transmission (η_{atm}) and in reducing the background load (ϵ_{load}). The telescope efficiency η_{tel} is determined by the antenna surface accuracy according to the Ruze formula: $\eta_{\mathrm{tel}} \approx \eta_0 \exp\left[-(4\pi\epsilon_{\mathrm{rms}}/\lambda)^2\right]$. The ALMA antennas will have $\epsilon_{\mathrm{rms}} \sim 25\ \mu$m, leaving room for improvement at the high frequencies.

3.1 Direct-Detection Continuum Observations

In addition to the sensitivity improvements that are possible by improving the telescope accuracy and atmospheric conditions, direct-detectors can provide additional sensitivity gains since one can couple both polarizations and the full bandwidth provided by the atmospheric windows. The advantage can be substantial – the 650 and 850 GHz windows are each 100 GHz wide, while ALMA will provide only 4–8 GHz instantaneous band. When combined with the efficiency factors, the per pixel sensitivity for a large single-dish telescope can be comparable to that of ALMA. The thin horizontal lines at the start of the vertical arrows in Figure 1 plot the raw sensitivities of planned and

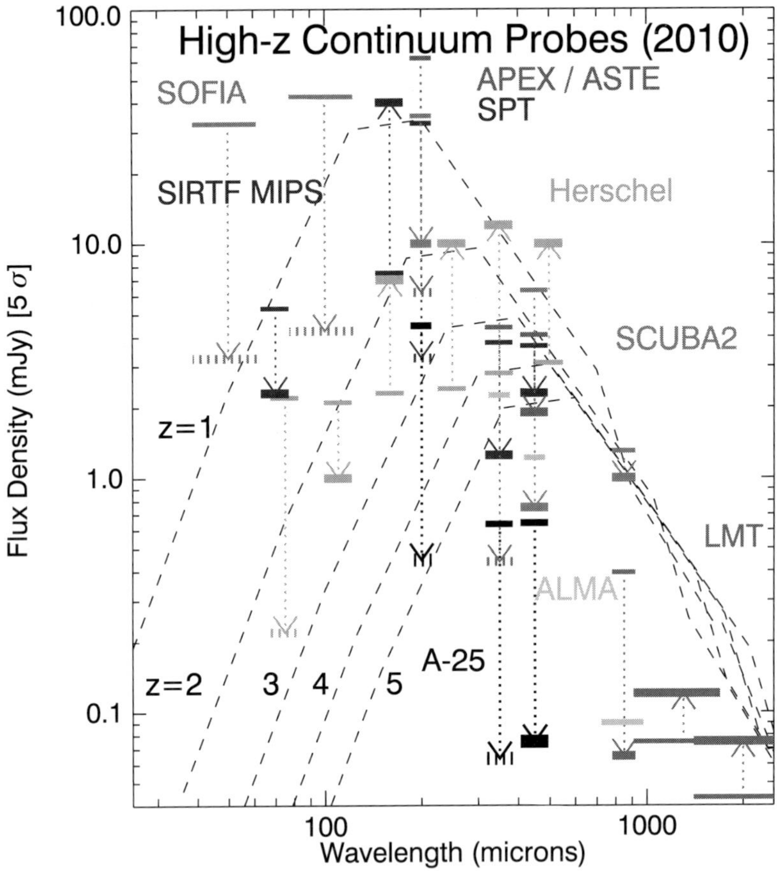

Fig. 1. Continuum capabilities in the far-IR . Raw sensitivities (5σ, 1 hour) are indicated with thin horizontal lines, taken from the references in Table 1, or calculated in the photon background limit assuming 50% instrument transmission. Estimates of the confusion limit due to extragalactic sources are indicated with thick horizontal lines at the termination of the vertical arrows. The confusion limits are taken to be 5 $\times$ the flux density at which there is 1 source per beam according to the source models of [1]. For instruments which are not confusion-limited in 100 hours, the 5σ, 100 hour sensitivity is plotted. Upward arrows imply measurements which are confusion-limited in less than 1 hour, downward arrows the converse. As a guide to the detectability of dusty galaxies, the Arp 220 SED at redshifts from 1 to 5 is overplotted in the light curves (taking $\Omega_{\mathrm{matter}} = .38$, $\Omega_\Lambda = 0.62$).

potential observatories, listed with their parameters in Table 1. In the short submillimeter windows, the 10–15 meter class telescopes at excellent sites (APEX, ASTE, SPT) offer continuum sensitivities within a factor of 2–5 of

Table 2. Estimated far-IR/submm galaxy detection rates

	200μm	450μm	850μm	1.3mm
f_{nu}, most likely [mJy]	10–1000	10–50	2–7	1–5
Platform	*Galaxy Detection Rate* [hr^{-1}]			
SIRTF MIPS	70			
Herschel PACS	100			
SPT w/ 10^3 beams		150		
APEX/ASTE (300 beams)		25		
SCUBA2		20	100	
LMT			100	700
A-25 w/ 10^3 beams	7	900	2000	
ALMA		1	60	40

ALMA. The fiducial A25 telescope, a 25–meter with $\epsilon_{\mathrm{rms}} = 12\mu$m at a very dry site would be 2–5 times *more sensitive* than ALMA in these windows.

Large Format Direct-Detector Arrays Allow Rapid Surveys

While the point source sensitivity of a large single-dish telescope can approach that of ALMA at the highest frequencies, the real advantage for single-dish telescopes is the tremendous mapping speed afforded by large-format focal plane arrays. As pointed out above, the mapping speed scales as the instantaneous solid angle and the inverse square of the sensitivity. To get a useful measure of survey speed, we convert the mapping speed to a galaxy detection rate, using power-law extrapolations to the observed source counts in a manner similar to [1]. As the estimates presented in Table 2 show, a single-dish telescope with a large detector array is much more efficient than ALMA for discovering the sources that make up the IR background. For example, at 450 μm, the most likely galaxy to be detected has a flux of a few $\times 10$ mJy. ALMA will be capable of detecting about one of these per hour, while SCUBA-2 will detect some 20 per hour, and a SCUBA-2 type array on the A-25 telescope could detect $\sim 10^3$ galaxies per hour. Naturally, the rapid mapping speed provided by arrays on single-dish telescopes is applicable to a wide range of science topics, not just distant galaxy detection.

Source Confusion Is a Limitation for the 3-Meter Class Telescopes

A key limitation of continuum surveys using single-dish telescopes is source confusion. When the number of sources in a beam approaches a substantial fraction of unity, extraction of fainter flux densities becomes difficult. Figure 1 includes estimates of this limit for the single-dish telescopes with the thick horizontal lines. We highlight some key points: 1) SIRTF and Herschel will be quickly confusion-limited at their longest wavelengths, and the resulting effective sensitivities are insufficient to probe even bright ($L = 10^{12} L_{\odot}$) sources beyond $z \sim 1$. 2) Ground-based telescopes can in principle survey bright sources to $z \sim 5$ in the 350 and 450 μm windows, but with the existing and planned 10–15-meter telescopes this will require many tens of hours of observations (of a single field) in good weather, with careful attention paid to systematics. The fiducial A-25 telescope would be a factor of 5 more sensitive than the other telescopes at these wavelengths, and yet effectively not confusion-limited because of the smaller beam. 3) At 850 μm and 1.3 mm, the observed flux density is essentially independent of redshift, and beam confusion limits a 15-meter telescope to only the bright sources, as has been found with SCUBA. In principle, the 50-meter LMT could detect sources more than 10× less luminous than its 15-meter counterpart at 850 μm, but for $\lambda > 1$mm, the confusion again prohibits the detection of fainter sources.

3.2 Far-IR–mm Spectroscopy with Single-Dish Telescopes

ALMA will be an exquisite platform for spectroscopy and will be the instrument of choice at its operational wavelengths once commissioned. Nevertheless, there are several important scientific niches for spectroscopy from single-dish telescopes that cannot be addressed with ALMA.

Full Far-IR Coverage Provided by Herschel and SOIFA

Perhaps the most important is the full far-IR spectral coverage provided by Herschel and SOFIA, which will facilitate ISM studies not possible at other wavelengths. With few exceptions, the neutral atomic and ionized gas phases are probed with spectral lines at $\lambda < 200 \mu$m, and the optical and near–IR transitions which might be important are typically obscured by dust. Key tracers of the molecular phase are also unique to the far-IR: examples include rotational transitions of the light molecular hydrides (e.g. OH, CH, HD), mid– and high–J CO lines which cool excited ($T > 100$K) molecular gas, and H_2O— a key constituent of the molecular ISM not observable from the ground. It is impossible to constrain the conditions of the dense ISM with ground-based observations alone. Much of our current understanding of the dense ISM arises is based on observations with airborne and orbital far-IR spectroscopy platforms: COBE, the KAO, ISO and SWAS. As the sensitivity of this type of observation improves by orders of magnitude in the next decade with SOFIA and Herschel, so will our understanding of the dense ISM.

Broad-Bandwidth Surveys with Direct-Detection Spectrometers

Another niche in the context of ALMA's limited bandwidth is the capability to couple a large instantaneous band from point sources for spectral surveys.

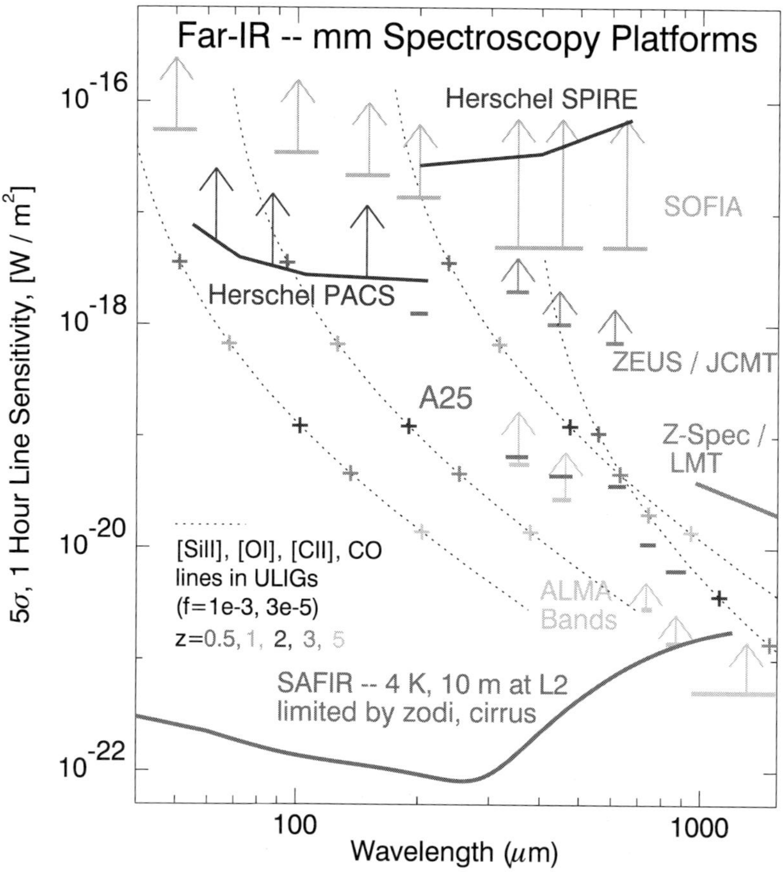

Fig. 2. Sensitivity provided by spectroscopy platforms in the far-IR and submillimeter. The upward arrows account for the effective sensitivity for line surveys, incorporating the time to cover the full atmospheric window for the ground-based instruments, or a 20% band for the SOFIA and Herschel instruments. Thin dashed curves show the intensities of diagnostic fine-structure and CO lines from high-redshift galaxies. Sensitivities for the Herschel and SOFIA instruments and ALMA are taken from the appropriate web pages, and Zeus and Z-Spec are referenced in the text. The sensitivities of SAFIR and the fiducial A25 telescope are calculated according to the photon noise limit assuming a 25% instrument transmission in a single polarization. SAFIR's background noise is due to zodiacal light and galactic cirrus, taken in regions of low intensity.

This is both a means of rapidly measuring several lines simultaneously, and potentially as a redshift measurement technique for optically-faint sources. The spectral survey rate increases as the instantaneous bandwidth of instrument, and even from the ground the instantaneous bandwidth can be an order of magnitude larger than the 4–8 GHz coupled by ALMA. It can be shown that the most sensitive approach to broad-band spectroscopy of point sources is a low-order grating spectrometer [2], and for the first time such devices are under development for the submillimeter and millimeter. We highlight ZEUS, an echelle spectrometer for the 350 μm and 450 μm windows at the JCMT [7] and Z-Spec, a new waveguide grating spectrometer which covers the full 200–300 GHz window instantaneously [1]. In their windows, these ground-based spectrometers are more sensitive than those of Herschel and SOFIA because of the aperture size, and should detect diagnostic lines in bright sources ($L \sim$ few $\times 10^{12}$ $L_\odot$) up to z$\sim$3. Furthermore, such instruments combined with the A25 telescope could improve the sensitivity by an order of magnitude – probing the more more typical ($L \sim$ few $\times$ 10^{11} $L_\odot$) galaxies.

Potential of a Cold Space Telescope

With Herschel, SOFIA, and the currently–envisioned ground-based spectrometers, we will have the capability to study the ISM conditions in local galaxies, and the very brightest galaxies out to redshifts of $\sim$ 3, but not to the earliest epochs of galaxy formation. These platforms are limited by the background noise from the warm telescope itself (and atmosphere). However, dramatic improvements in sensitivity are still possible with a space telescope cooled to $T < 10$ K. As an example, Fig. 2 includes the background-limited sensitivity for SAFIR, a 4 K, 10 meter observatory recommended in the McKee/Taylor Decadal Survey report. SAFIR will be an excellent far-IR spectroscopy platform – capable of detecting spectral lines in a Milky Way–type galaxy to redshifts of 5 or more. Note that a spectrometer can readily detect lines from several sources in the same beam, and thereby offers the potential to overcome the spatial confusion limit.

References

1. A.W. Blain: Physics Reports **369** 111 (2002)
2. C.M. Bradford et al: Proc. SPIE **4850**, 1137 (2003)
3. W.D. Duncan et al: Proc. SPIE **4855**, 19 (2003)
4. D.P. Finkbeiner, M. Davis, D.J. Schlegel: Astrophys. J. **544**, 81 (2000)
5. D.J. Fixsen, et al: Astrophys. J. **508**, 123 (1998)
6. W.S. Holland et al: Proc. SPIE **4855**, 1 (2003)
7. T. Nikola et al: Proc. SPIE **4855**, 88 (2003)
8. J.-L. Puget et al: Astrophys. J. **308**, L5 (1996)
9. http://www.mpifr-bonn.mpg.de/div/apex.html
10. http://www.no.nao.ac.jp/ lmsa/aste/aste.html
11. http://astro.uchicago.edu/spt
12. http://www.lmtgtm.org
13. http://www.safir.jpl.nasa.gov

New Astronomy Results from the Odin Satellite

Å. Hjalmarson[1] and M. Olberg[1], on behalf of the Odin Team

Centre for Astrophysics and Space Science at Chalmers University of Technology, Onsala Space Observatory, SE-439 92 Onsala, Sweden hjalmar@oso.chalmers.se, olberg@oso.chalmers.se

The Odin project, satellite and radiometers, calibration methods, as well as some highlights from the first year of Odin observations of/searches for H_2O, $H_2^{18}O$, NH_3 and O_2 in molecular clouds and comets, have been presented in the May 2003 Special Odin Issue of Astronomy & Astrophysics Letters [1].

In this paper we will discuss new results from our Odin observations of sub-millimetre wave transitions of H_2O, $H_2^{18}O$, $H_2^{17}O$, NH_3, $^{15}NH_3$ (all seen in emission and absorption), CO (5-4), ^{13}CO (5-4), $C^{18}O$ (5-4), and CI, as well as some new results from simultaneously conducted Odin searches for O_2 at 119 GHz. The first steps of spectral scans/molecular line searches towards Orion KL and Sgr B2 also will be presented.

1 A Selection of New Odin Results

We will here present and discuss in some detail a few especially interesting results from the second year of Odin observations.

1.1 Observations at 119 GHz – HC_3N, Tentative SiC in Sgr A, and Lower O_2 Limits

Whenever possible we have in parallel with our sub-millimetre wave observations also used the "fixed-tuned" HEMT receiver intended for O_2 searches. Although this receiver is no longer phase-locked, the frequency drift is very slow and the actual observing frequency can be accurately calibrated by means of the atmospheric O_2 line. An especially interesting result from these observations is displayed in Figure 1 - a spectrum observed towards Sgr A, where we can assign: i) the strongest line to the $J = 13 - 12$ transition of HC_3N (at the expected line-width and intensity), ii) a weak broader feature at the lower frequency side of HC_3N tentatively to the $^3P_2$, $J = 3 - 2$ line of SiC (in fact, the lowest energy transition of this molecule). We also note that at the high frequency end of the observed band there may be a hint of the O_2 $(1_1–1_0)$ line (at 118.750 GHz, but here shifted to a lower frequency because of a source velocity around 40 km s^{-1}, vs the assumed $v_{LSR} = 0$ km s^{-1}).

Obviously the latter tentative results will need additional Odin observations. Moreover, a secure assignment of interstellar SiC will require verifying

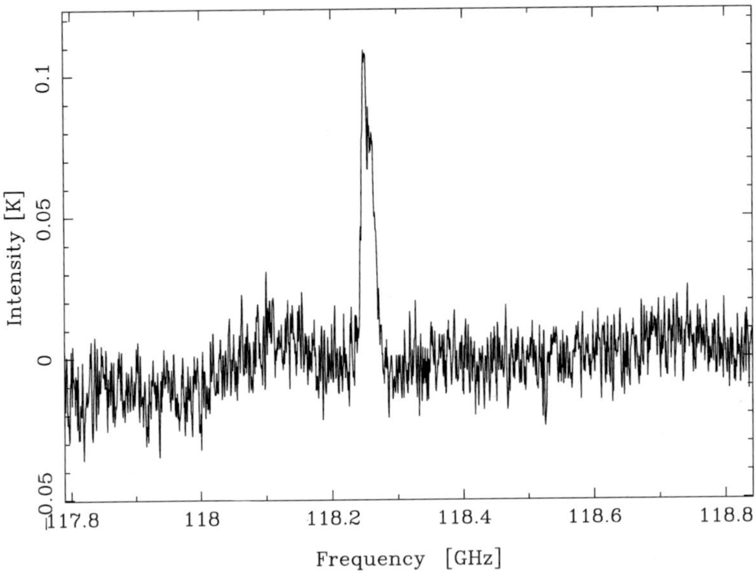

Fig. 1. Odin spectrum of the frequency range 117.8 - 118.8 GHz (assuming $v_{LSR} = 0\,\mathrm{km\,s^{-1}}$) observed in the direction of the Galactic Centre, Sgr A.

detections of other $^3P_2$ and $^3P_1$ transitions from the same molecule. SiC has been detected in the circumstellar envelope of the carbon star IRC+10216, assisted by laboratory spectroscopy ([2]). The comparatively large line-width of our tentative SiC feature is not unexpected, judging from the similarly wide SiO (v=0) lines observed in Galactic Centre clouds - presumably related to frequent cloud-cloud collisions and shocks ([6]; [5]). Similarly to the abundant SiO, the interstellar SiC may as well have a dust origin. A short communication of these new results is underway (Sandqvist, Bernath, Hjalmarson, Larsson, Pagani et al. 2004, in preparation).

We should also mention the very recent careful data reduction (by Bengt Larsson at Stockholm Observatory) for our deep O_2 (11-10) search towards the ρ Oph A molecular cloud - observations performed simultaneously with our mapping at $1'$ spacing of the 557 GHz line of H_2O and the 572 GHz line of NH_3 (Liseau et al. [1]; Liseau, Larsson et al. 2004, in preparation). The resulting O_2 abundance limit, after 300 hours of observing time, has decreased from $< 10^{-7}$ (Pagani et al. [1]), to below 2×10^{-8} (3σ). In this light the tentative SWAS detection of the 487 GHz O_2 (3_3-1_2) line in ρ Oph A [3] appears very unlikely. New Odin O_2 (1_1-1_0) search data, allowing the claim of similarly low limits (or even low level detections), now have been accumulated for a number of dark clouds/molecular clouds (Pagani et al. 2004, in preparation).

1.2 Pressure Broadened H_2O (1_{10}-1_{01}) in the Atmosphere of Mars

The rather beautiful result of Odin's mid-June 2003 observations at 557 GHz of the very pressure broadened H_2O (1_{10}-1_{01}) absorption line from Mars is shown in Figure 2. A 4 GHz wide band coverage has been synthesised by means of observations at five frequencies (offsets: 0, ±700, and ±1400 MHz). Additional observations of H_2O, $H_2^{18}O$ and CO (5-4) have been scheduled for November 2003, and analysis towards a paper is underway (Biver, Lecacheux et al. 2004, in preparation). Here the large bandwidth compared to SWAS [4] should be very useful.

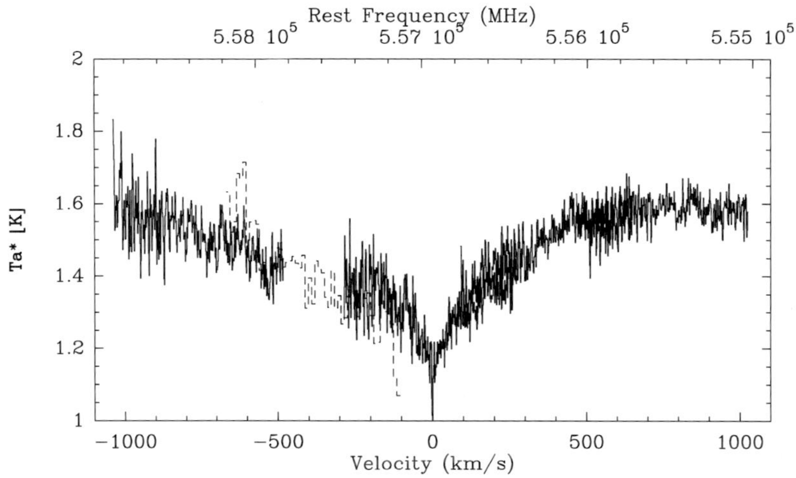

Fig. 2. Odin spectrum of the very pressure broadened H_2O (1_{10}-1_{01}) absorption line from the atmosphere of Mars. For details see text.

1.3 NH_3 (1_0-0_0) from the Surroundings of IRC+10216

A challenging result from our simultaneous Odin searches for PH and NH_3 in the carbon rich circumstellar envelope of IRC+10216 is displayed in Figure 3. While PH was not discovered, the convincing detection of the NH_3 (1_0-0_0) line at 572 GHz is very interesting. Similarly to the SWAS detection of abundant H_2O in the surroundings of the ageing star [7], the Odin NH_3 result cannot be understood by current chemical models, and hence may support their proposal of an origin in evaporation of orbiting icy bodies (such as comets), remainders from the formation of the star. Additional observations to improve our NH_3 spectrum, and to simultaneously obtain a comparison Odin H_2O spectrum, have been scheduled for November 2003, and analysis towards a paper is ongoing (Hasegawa, Justtanont, Kwok, Hans Olofsson et al. 2004, in preparation).

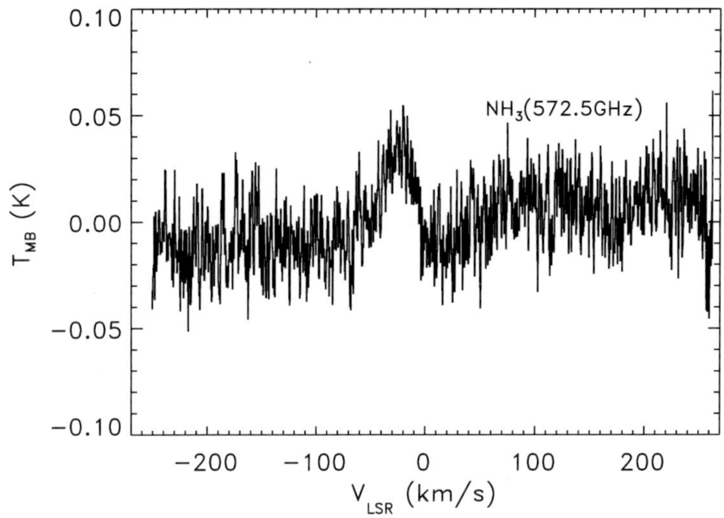

Fig. 3. Odin detection of the NH_3 (1_0-0_0) line in the surroundings of IRC+10216

1.4 H_2O (1_{10}-1_{01}) in the Circumstellar Envelope of W Hya

Long-lasting Odin observations have lead to a convincing detection of the H_2O (1_{10}-1_{01}) line of W Hya, at an intensity level three times higher than that of SWAS. Although the spectrum still is noisy, we may conclude that the line centre velocity is red-shifted compared to the stellar velocity, presumably as a result of severe "self-absorption" in the gas expanding towards us. Such H_2O self-absorptions previously have been seen in Odin observations of comets (Lecacheux et al. [1]), and in the Orion outflow (Henrik Olofsson et al. [1]). Modelling analysis towards a publication, including ISO as well as Odin data, is ongoing (Justtanont, Hans Olofsson, Bergman, Hasegawa, Kwok, Schöier-Larsen et al. 2004, in preparation).

1.5 $H_2^{17}O$ (1_{10}-1_{01}) and Other Molecular Lines in Orion KL

The spectrum resulting from our Odin observations of the $H_2^{17}O$ (1_{10}-1_{01}) towards Orion KL is shown in Figure 4. The intensity of the $H_2^{17}O$ line verifies the assumption of rather optically thin $H_2^{18}O$ (1_{10}-1_{01}) low velocity outflow emission by Olofsson et al. ([1]), which formed the basis of their water abundance determination in the inner outflow ($\approx 10^{-5}$, proposed to be a result of evaporation of icy grain mantles in this warm region). Spectral lines assigned to CH_3OCHO, $(CH_3)_2O$ and SO_2 also have been detected in this first step towards an Odin spectral scan, a topic likely to be scheduled more exten-

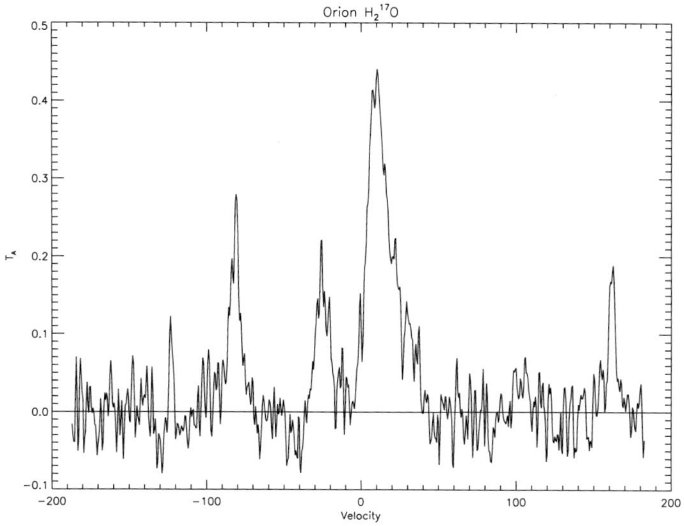

Fig. 4. Odin spectrum towards Orion KL of an 800 MHz band around 552.021 GHz

sively in the near future. A deep search for PH at 553.365 GHz as yet has been unsuccessful.

1.6 H_2O (1_{10}-1_{01}) Mapping of the DR 21/DR 21(OH) Region

The results of Odin mapping of the region around DR 21 and DR 21(OH), using two receivers simultaneously tuned to the H_2O (1_{10}-1_{01}) line, have just appeared in a Doctoral Thesis by Henrik Olofsson (2003). The very complicated spectral shapes of the water line have been interpreted in terms of broad emissions from outflow regions, intersected by strong narrow absorptions from low-excitation water in the ambient molecular cloud (at $-3\,\mathrm{km\,s^{-1}}$) and in a foreground cloud (at $+8\,\mathrm{km\,s^{-1}}$) – cf. the DR 21 spectrum in Figure 5. Odin has detected not only i) the East-West extended high-velocity outflow near DR 21 – a region of shock-enhanced H_2O and H_2 emissions – but also ii) a new EW extended high-velocity water outflow 2′N of DR 21(OH), as well as iii) an NS elongated low-velocity water outflow near DR 21 - perhaps a younger outflow, roughly perpendicular to the previously known high-velocity flow - and iv) a compact, low velocity outflow near DR 21(OH). A rather extensive paper will soon be submitted (Henrik Olofsson, Hjalmarson, Bergman, Fich et al. 2004, in preparation).

1.7 H_2O (1_{10}-1_{01}) Mapping Across the S 140 PDR

Our observations towards the S 140 molecular cloud core and bright rim have resulted in apparently very "innocent", narrow water emission lines. However,

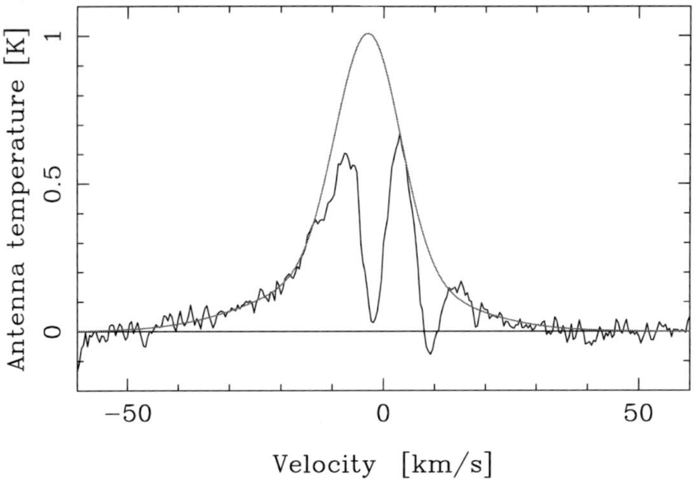

Fig. 5. H_2O (1_{10}-1_{01}) spectrum observed by Odin towards DR 21, together with a double component Gaussian model fit ($\Delta v_{FWHM} = 15$ and $40\,\mathrm{km\,s^{-1}}$)

the emission peak is very clearly red-shifted by 2–3 $\mathrm{km\,s^{-1}}$, compared to the ambient cloud velocity. A comparison with the CO, ^{13}CO, and C^{18}O, J = 3-2 lines observed by Minchin et al. [8] strongly suggests an interpretation in terms of severe self-absorption by optically thick lower-excitation water in the cloud foreground, just like that seen in the CO (3-2) line. We have now scheduled Odin mapping of the H_2O (1_{10}-1_{01}), NH_3 (1_0-0_0), CO (5-4), ^{13}CO (5-4), and O_2 (1_1-1_0) lines across the S 140 PDR, and a publication should appear rather soon (Olberg, Persson, Liljeström et al. 2004, in preparation).

1.8 NH_3 (1_0-0_0) and $^{15}NH_3$ (1_0-0_0) in Sgr B2 and Orion KL

In view of the beautiful H_2O (1_{10}-1_{01}) absorption line results already published by the SWAS team [9], we at first decided to concentrate the Odin efforts on the Galactic Centre Sgr A clouds, which resulted in perhaps even more intriguing spectra - water emission and absorption in the velocity range -200 to $+200\,\mathrm{km\,s^{-1}}$ (Sandqvist et al. [1]). However, we also performed some "pilot studies" of the H_2O (1_{10}-1_{01}), $H_2^{18}O$ (1_{10}-1_{01}), and NH_3 (1_0-0_0) bands towards Sgr B2. Our Odin NH_3 spectrum, carefully reduced by Bengt Larsson at Stockholm Observatory, is displayed in Figure 6. The spectrum not only contains a complex mixture of NH_3 (1_0-0_0) absorption and emission features but also displays a well separated weak absorption band at the $^{15}NH_3$ (1_0-0_0) frequency. Since Odin previously has detected the emission lines of NH_3 (1_0-0_0) (Larsson et al. [1]) and $^{15}NH_3$ (1_0-0_0) (still unpublished) in Orion KL, we consider identification of the $^{15}NH_3$ feature in Sgr B2 rather secure. Now a considerable amount of Sgr A and Sgr B2 Odin data for the H_2O (1_{10}-1_{01}), $H_2^{18}O$ (1_{10}-1_{01}), $H_2^{17}O$ (1_{10}-1_{01}), NH_3 (1_0-0_0), CO (5-4), ^{13}CO (5-4),

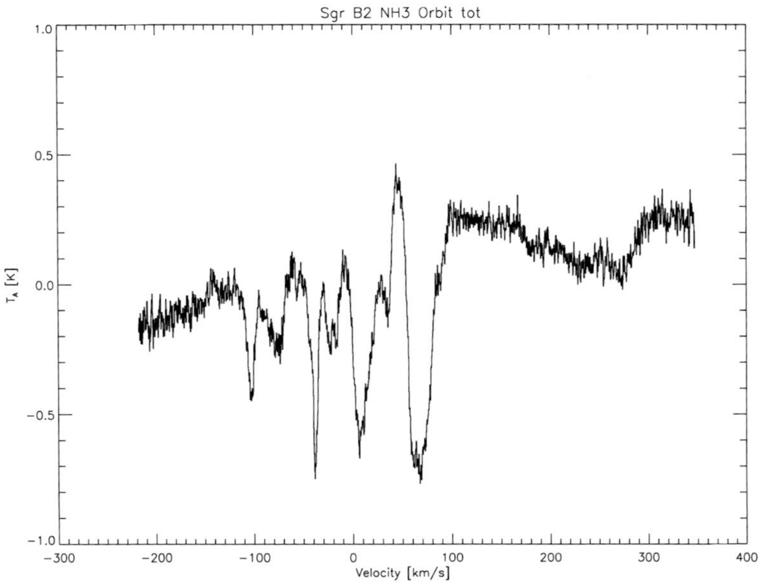

Fig. 6. NH_3 (1_0-0_0) and $^{15}NH_3$ (1_0-0_0) absorption spectra observed by Odin towards Sgr B2

$C^{18}O$ (5-4), CI, and O_2 (1_1-1_0, and 3_3-1_2) bands is awaiting final data reduction and analysis. A deep search for PH at 553.365 GHz towards Sgr B2 also is awaiting data reduction. In terms of abundance level, this search against the Sgr B2 background continuum should be highly sensitive.

2 Odin Outlook

At the time of the Zermatt meeting (September 2003) Odin had been in orbit for 30 months and no signs of ageing have appeared - rather the opposite, since the NH_3/CO receiver with its not phase-locked but very slowly drifting LO frequency, recently decided to "heal itself". Therefore it seems not too optimistic to believe in further rewarding observations during the coming years, which indeed also should be useful for guidance of future high-sensitivity observations by the Herschel Space Observatory.

Acknowledgements

The Odin team[1] is grateful for generous financial support from the Space Agencies and Research Councils in Canada, Finland, France and Sweden. The Odin Data Centre at Onsala Space Observatory also is supported by the Swedish National Facility for Radio Astronomy.

References

1. *Odin 2003 references can be found in the Odin Special Issue of* A&A Letters, **402**, L21-L81 (2003)
2. J.Cernicharo, C.A. Gottlieb, M. Guélin et al: ApJ, **341**, L25 (1989)
3. P.F. Goldsmith, D. Li, E.A. Bergin et al: ApJ, **576**, 814 (2002)
4. M.A. Gurwell, E.A. Bergin, G.J. Melnick et al: ApJ, **539**, L143 (2000)
5. S. Hüttemeister, G. Dahmen, R. Mauersberger et al: A&A, **334**, 646 (1998)
6. J. Martín-Pintado, P. de Vicente, A. Fuente, P. Planesas: ApJ, 482, L45 (1997)
7. G.J. Melnick, D.A. Neufeld, K.E.S. Ford et al: Nature, **412**, 160 (2001)
8. N.R. Minchin, G.J. White, R. Padman: A&A, **277**, 595 (1993)
9. D.A. Neufeld, M.L.N. Ashby, E.A. Bergin et al: ApJ, **539**, L111 (2000)
10. A.O.H. Olofsson: Satellite and ground based observations of massive star forming regions, Doctoral Thesis at Chalmers University of Technology, Göteborg, Sweden (2003)

[1] Odin is a Swedish-led satellite project funded jointly by the Swedish National Space Board (SNSB), the Canadian Space Agency (CSA), the National Technology Agency of Finland (Tekes), and the Centre National d'Études Spatiales (CNES, France). The Swedish Space Corporation (SSC) has been the prime industrial contractor, and is also responsible for the satellite operation from its Odin Mission Control Centre at SSC in Solna and its Odin Control Centre at ESRANGE near Kiruna in northern Sweden. See also the SNSB Odin web page: `http://www.snsb.se/eng_odin_intro.shtml`

From Molecular Cores to Planet-Forming Disks with SIRTF

N.J. Evans II[1] and the c2d Team[2]

[1] The University of Texas at Austin, Department of Astronomy, 1 University Station C1400, Austin, Texas 78712–0259 nje@astro.as.utexas.edu

[2] see the c2d website for the full list of team members http://peggysue.as.utexas.edu/SIRTF/

Summary. The *SIRTF* mission and the Legacy programs will provide coherent data bases for extra-galactic and Galactic science that will rapidly become available to researchers through a public archive. The capabilities of *SIRTF* and the six legacy programs are described briefly. Then the cores to disks (c2d) program is described in more detail. The c2d program will use all three *SIRTF* instruments (IRAC, MIPS, and IRS) to observe sources from molecular cores to protoplanetary disks, with a wide range of cloud masses, stellar masses, and star-forming environments. The *SIRTF* data will stimulate many follow-up studies, both with *SIRTF* and with other instruments.

1 Why *SIRTF*?

Much of the radiant energy in the Universe lies in the infrared region, and infrared observations are well suited to the study of distant starburst galaxies and star formation, where dust controls the flow of energy. Observational studies of galaxies and star formation have generally suffered from one or more of the following problems: biased samples, inadequate sensitivity, inadequate spatial resolution, or incomplete spectral data.

The Space Infrared Telescope Facility (*SIRTF*) [8] will provide greatly improved capabilities in the infrared region. *SIRTF* is the last of the series of four great observatories that began with the Hubble Space Telescope and continued with the Compton Gamma Ray Observatory and the Chandra X-ray Observatory. *SIRTF* covers the wavelength region from 3.6 to 160 μm with background-limited performance. Its 85-cm diameter beryllium mirror is cooled below 5.5 K, and the cryogen lifetime should be between 2.5 and 5 years. It was launched on August 25, 2003 into an earth-trailing solar orbit.

The instrument complement includes two imaging array instruments and a spectrometer. The *InfraRed Array Camera* or IRAC, covers 3.6 to 8 μm [7] in four bands; the *Multiband Imaging Photometer for SIRTF* or MIPS, covers 24 to 160 μm [5] in 3 bands; and the *InfraRed Spectrometer*, or IRS, supplies spectroscopy from 5.3 to 40 μm with resolving power $R = 60 - 120$ and from 10 to 37 μm with $R = 600$ [9]. The field of view of the imagers is $5'$ by $5'$ except at 160 μm, where the field of view is $0.5'$ by $5'$. MIPS also has an $R = 15$ spectrophotometric mode between 50 and 100 μm.

2 The Legacy Programs

Observing time on *SIRTF* is divided between Guaranteed Time Observers (GTOs), Legacy programs, and General Observer (GO) programs. The legacy program aims to ensure that coherent data bases of general and lasting value will be obtained during the limited mission lifetime. To enhance opportunities for follow-up with *SIRTF*, the legacy data have no proprietary period; after the archive has opened, about 8 months after launch, data will become available to the public and the legacy teams simultaneously. The legacy teams will provide improved data and, in many cases, ancillary data from other wavelengths and tools for analysis. Legacy programs have typically hundreds of hours of observing time and, insofar as is possible, will be observed early in the mission.

There are three extra-galactic legacy programs and three Galactic legacy programs. The Great Observatories Origins Deep Survey (GOODS) will survey 300 square arcminutes with IRAC and MIPS (24 μm only) overlapping the Hubble and Chandra deep fields. New imaging with Hubble's Advanced Camera for Surveys has been obtained. Extreme starburst galaxies may be detected to $z = 6$ [4]. The *SIRTF* Wide-area Infrared Extragalactic Survey (SWIRE) will map 100 square degrees at high Galactic latitudes with IRAC and MIPS, detecting dusty star-forming galaxies and AGNs to $z \sim 3$ [12]. The *SIRTF* Nearby Galaxies Survey (SINGS) will study 75 nearby galaxies using all the *SIRTF* instruments, providing a template for understanding more distant galaxies [10].

Turning to the Galactic legacy projects, the *SIRTF* Galactic Plane Survey (GLIMPSE) will survey 240 square degrees of the inner Galactic plane with IRAC [1]. The c2d project (described in more detail below) studies nearby low mass star formation and disk evolution [6]. The study of disk evolution is continued to later stages by the Formation and Evolution of Planetary Systems (FEPS) team [14]. The FEPS team will obtain images and spectroscopy of 300 stars with ages from 3 Myr to 1 Gyr.

3 The Cores to Disks Legacy Program

The Cores to Disks (c2d) team includes 11 co-investigators, 34 associates, and 20 affiliates. Associates will work directly on *SIRTF* data; affiliates are adding data at other wavelengths. The goal is to obtain as complete a data base as possible for nearby ($d < 350$ pc) low-mass star formation. The data base will be used to follow the evolution of molecular cloud material from starless cores to stars surrounded by planet-forming disks. The c2d team will observe stars with ages up to 3-10 Myr and will obtain spectroscopy of objects with a range of ages to trace the evolution in the nature of the dust and gas.

Using 400 hours and all three *SIRTF* instruments, the c2d program will provide the most complete data base of infrared observations for low-mass star formation. Ancillary and complementary data will be collected at wavelengths from X-ray to radio for these same regions. We will map 5 large nearby molecular clouds and 156 compact molecular cores with both IRAC and MIPS (275 hours). Photometry will be obtained for 190 stars with IRAC and MIPS (50 hours). Spectroscopy will be obtained with IRS for at least 170 targets (75 hours).

3.1 A Survey of Nearby Molecular Clouds and Cores

The *SIRTF* mission offers an unprecedented opportunity to determine the stellar content of the nearest star-forming molecular clouds, the distributions of their youngest stars and substellar objects, and the properties of their circumstellar envelopes and disks. Our program will address the questions in the following paragraph.

In the large clouds, we will study the distribution of the youngest stars and substellar objects to understand the conditions needed for their formation. Do isolated cores with associated stars preferentially harbor single stars or small stellar groups? Can brown dwarfs form in the way that ordinary low-mass stars do, or are they ejected from small stellar groups? If very young brown dwarfs are found throughout the cloud, surrounded by envelopes, an extension of low-mass star formation processes into the substellar regime would be viable. Deep IRAC imaging of dense cores will reveal their detailed structure by mapping the extinction of background stars. This structure will be compared with that derived from mapping emission from the dust continuum. The incidence of circumstellar disks in complexes and in isolated cores will elucidate the role of the environment in disk formation. Our less biased sample will allow more robust measures of the lifetimes of various stages in the star formation process. Rare objects and systems in rapid phases of evolution may be present in a large sample. We may, for example, discover the short-lived phase in which the first, molecular core forms, an as yet unobserved stage of star formation [2], [13].

We chose a sample of clouds and cores that will sample a wide range of conditions in order to separate the effects of evolution from other variables. We target five large complexes known to be forming stars in isolation, in groups, and in clusters, and 156 small, isolated cores – 110 starless and 46 with associated *IRAS* sources. The criteria for determining whether a core has an associated "star," meaning a central luminosity source at any evolutionary stage, are similar to those defined by [11].

The molecular cloud complexes and isolated cores will be observed with both IRAC and MIPS at all wavelengths from 3.6 to 160 μm, but we expect the 160 μm detector to be saturated toward these regions. The 3σ limits (based on pre-launch estimates) for the scan maps yield a limit on L_{bol} from 3.6 to 70 μm of about 10^{-3} $L_{\odot}$ at 350 pc. The IRAC bands would detect a 1 Myr old, 5 M_{Jup} brown dwarf at 350 pc, based on models [3].

Through a variety of collaborations, we are extending the data base to both longer and shorter wavelengths. The large clouds and isolated cores are being mapped in the continnum at $\lambda \sim 1$ mm, and in spectral lines. Selected regions have been observed with millimeter-wave interferometers. Further complementary data on the northern clouds in molecular lines and dust extinction and emission are being obtained by the COMPLETE[3] team, who also plan to make the data public. In addition, the large clouds and many northern cores are being imaged at shorter (R, i, and z) wavelengths to limiting magnitudes of 24.5, 22, and 22, respectively.

3.2 The Evolution of Disks up to 10 Myr

Circumstellar disks are the likely sites of planet formation. While disks around T Tauri stars have been studied extensively, much less is known about both very early stages in the formation of disks and their later evolution after accretion has ended. The c2d program on cores and clouds can detect forming disks at very early stages (disks as small as 1 M_{Earth} can be detected through $A_V = 100$ mag). We will also target "debris disks," which are found in association with nearby stars at ages up to about 1 Gyr. We can detect such disks with very small amounts of dust ($\sim 0.1 M_{Moon}$). The timescales for dispersal of disks in such systems are largely unknown.

We will target weak-line T Tauri stars (wTTs) to study disk evolution up to a few Myr. We will address the following key scientific questions. What is the frequency of debris disks in the low-mass stellar population at ages of less than 5 Myr? When does the transition from primordial to debris disk occur? Assaying the full range of disk properties at a given age will help to distinguish evolutionary effects from other factors. Do most disks develop inner holes or gaps during the dissipation process? We will detect evidence for gaps using the spectral energy distribution; the longer wavelengths accessible by *SIRTF* can reveal gaps in the main planet-building zone.

We restricted our sample to wTTs stars that lie within 5° of the nearest large molecular clouds that we are mapping. We also required evidence of youth in the form of X-ray emission or strong Li I 6707 Å absorption. These criteria should allow us to extend the evolutionary stages studied in the clouds to objects with ages up to about 10 Myr. We have worked with the FEPS team to ensure a smooth transition to their program. We will detect excess emisison by comparison to *SIRTF* standard stars and model photospheres. The survey will readily detect disks like that around β Pic, even if they have evacuated inner holes with radii as large as 30 AU. Disks an order of magnitude more tenuous will still be detected if their inner edge is at a radius of 5 AU. We have obtained spectra of the stars in the visible to determine stellar properties and observations with adaptive optics to search for multiplicity.

[3] http://cfa-www.harvard.edu/~ agoodman/research8.html

3.3 The Evolution of the Building Blocks of Planets

Our spectroscopic program with the IRS instrument will provide crucial information on the evolution in the nature of the materials that build planets, the dust, ice, and gas in circumstellar envelopes and disks. The mid-infrared wavelength range has a wealth of diagnostic features for these three components. *ISO* revealed the potential of such observations for more massive stars, and *SIRTF* will allow spectroscopic studies of solar-type stars. In the 75 hour c2d IRS program, high signal-to-noise spectra will be obtained over the full 5–40 μm range [high resolution ($R \approx 600$) over the 10–37 μm range] for all phases of star- and planet-formation up to ages of $\sim$5 Myr for at least 170 sources. The MIPS-SED mode at 50–100 μm will also be used in the second year of the program to characterize the longer wavelength silicate and ice features of a disk sub-sample.

The IRS spectra can be used to address the following questions. How do spectroscopic diagnostics evolve through the different stages of early stellar evolution? How does the chemical composition of dust and ices change from molecular clouds to planetary bodies? Mid-infrared spectroscopy of disks as the envelope is clearing will be particularly interesting. In the later evolutionary stages, changes in silicate features become the main diagnostics. The spectra will form a powerful data base to compare with spectra of Kuiper-Belt objects, comets and asteroids obtained in other *SIRTF* programs and to clarify the links between interstellar, circumstellar and solar-system material. How does the size distribution of dust grains evolve in circumstellar environments? These observations will permit independent estimates of the dust coagulation and gas dissipation time scales, processes of great importance for planet formation. What is the spectral evolution of substellar objects? The young brown dwarfs and super-Jupiters discovered in the IRAC and MIPS surveys ($\sim$ 100 expected) will be targets for follow-up observations.

We will reserve roughly half the IRS time for follow-up observations of objects found in the mapping programs. The remaining half is scheduled for observations of known embedded and pre-main-sequence stars. We can study objects down to a typical source luminosity of 0.1 $L_\odot$, allowing study of masses down to 0.1 $M_\odot$ at an age of 1 Myr. We plan to achieve a signal to noise of 50–100 on the continuum. This level will allow us to study the thermal history of the envelope through the 15 μm CO_2 bending mode profiles and to search for gas phase emission and absorption features in all phases of star formation. To obtain data with higher spectral resolving power in atmospheric windows, we are conducting flux-limited surveys in the L- and M-band windows using the VLT-ISAAC and Keck NIRSPEC instruments. A subset of IRS targets with known infrared ice absorption features is being characterized, using the CSO and OVRO facilities. The c2d-IRS team will enhance the IRS pipeline data delivered by the SSC in several ways. The most important improvement will be in the defringing of the IRS spectra, because laboratory experiments indeed show the presence of fringes.

4 Summary

The c2d program will provide a legacy for future research on star and planet formation. We hope to provide a data base for unbiased statistical studies of the formation of stars and substellar objects. Data from other wavelength regimes will add to the picture, and modeling tools will assist researchers in using the data base. Follow-up studies of these samples with *SIRTF* itself and with future missions is a natural outcome. The *SIRTF* and ancillary data will be available to the broader community from the *SIRTF Science Center* (SSC), via their Infrared Sky Archive (IRSA). Complementary data products will be made available, as far as possible, through either IRSA or public web sites. Further information on the program, including the source lists, can be found at the c2d website: `http://peggysue.as.utexas.edu/SIRTF/`.

This material is based upon work supported by the National Aeronautics and Space Administration under Contract No. 1224608 issued by the Jet Propulsion Laboratory. The Leiden SIRTF legacy team is supported by a Spinoza grant from the Netherlands Foundation for Scientific Research (NWO) and by a grant from the Netherlands Research School for Astronomy (NOVA).

References

1. Benjamin, R. A. et al. 2003, Pub. Astr. Soc. Pac., 115, 953
2. Boss, A. P. & Yorke, H. W. 1995, Astrophys. J. (Letters), 439, L55
3. Burrows, A. et al. 1997, Astrophys. J., 491, 856
4. Dickinson, M., Giavalisco, M., & The Goods Team 2003, The Mass of Galaxies at Low and High Redshift. Proceedings of the ESO Workshop held in Venice, Italy, 24-26 October 2001, p. 324., 324
5. Englebracht, C. W., Young, E. T., Rieke, G. H., Rivlis, G., Beeman, J. W., & Haller, E. E. 2000, Exp. Astron., 10, 403
6. Evans, N. J. et al. 2003, Pub. Astr. Soc. Pac., 115, 965
7. Fazio, G. G., et al. 1998, Proc. SPIE, 3354, 1024
8. Gallagher, D. B., Irace, W. R., & Werner, M. W. 2003, Proc. SPIE, 4850, 17
9. Houck, J., van Cleve, J., Brandl, B., Charmandaris, V., Devost, D., & Uchida, K. 2000, ISO beyond the peaks: The 2nd ISO workshop on analytical spectroscopy. Eds. A. Salama, M.F.Kessler, K. Leech & B. Schulz. ESA-SP 456., 456, 357 (2000)
10. Kennicutt, R. C. et al. 2003, Pub. Astr. Soc. Pac., 115, 928
11. Lee, C. W. & Myers, P. C. 1999, Astrophys. J. Suppl., 123, 233
12. Lonsdale, C. J. et al. 2003, Pub. Astr. Soc. Pac., 115, 897
13. Masunaga, H., Miyama, S. M., & Inutsuka, S. 1998 Ap.J., 495, 346
14. Meyer, M. R. et al. 2002, The Origins of Stars and Planets: The VLT View. Proceedings of the ESO Workshop held in Garching, Germany, 24-27 April 2001, p. 463., 463

Observation Programs of the *ASTRO-F* Survey

H. Shibai

Graduate School of Science, Nagoya University/The Institute of Space and Astronautical Science (ISAS) shibai@phys.nagoya-u.ac.jp

The next Japanese infrared mission in space, *ASTRO-F*, is to be launched in 2005. The observation programs related to the interstellar medium are overviewed. *ASTRO-F* will provide unique and important scientific results in this field.

1 *ASTRO-F* Mission

ASTRO-F is the first Japanese satellite ([1]) dedicated to infrared astronomy. The wavelength coverage is from 2 to 200 μm. The project was started in 1997, succeeding a small, survey-type telescope, the IRTS (Infrared Telescope in Space [2]). It follows twenty years after the *IRAS* survey [3]. A new infrared survey with higher quality is believed to be undoubtedly important as a fundamental database for astronomical research.

The *ASTRO-F* satellite weights approximately 970 kg, the maximum weight for a Sun-synchronous orbit when launched by the Japanese launch vehicle, M-V rocket. The telescope is cooled to 5.8 K to reduce the instrumental thermal emission. The effective aperture is 670 mm in diameter. The primary and secondary mirrors are fabricated using a newly developed, lightweight, SiC mirror technique. The image quality of the telescope is better than 1 arcsecond.

ASTRO-F is a hybrid-type cryostat satellite incorporating a mechanical cooler as well as cryogen. Because of this hybrid cooling system, the cryogen volume and the size of the vacuum vessel can be reduced, and therefore, a large aperture telescope fits into a small cryostat. The telescope and the focal plane instruments are cooled to 2 K by super-fluid helium (180 liter tank). By employing a mechanical cooling system (Stirling cycle cooler), most of the heat load on the cold part is compensated. The expected cryogen life is nominally one and a half years. Even after the helium cryogen has run out, near-infrared observations using the InSb array can be continued, using only the mechanical cooling system.

2 Scientific Capability

ASTRO-F has two focal plane instruments, FIS (Far-Infrared Surveyor) and IRC (Infrared Camera). In the far-infrared wavelength region, *ASTRO-F*

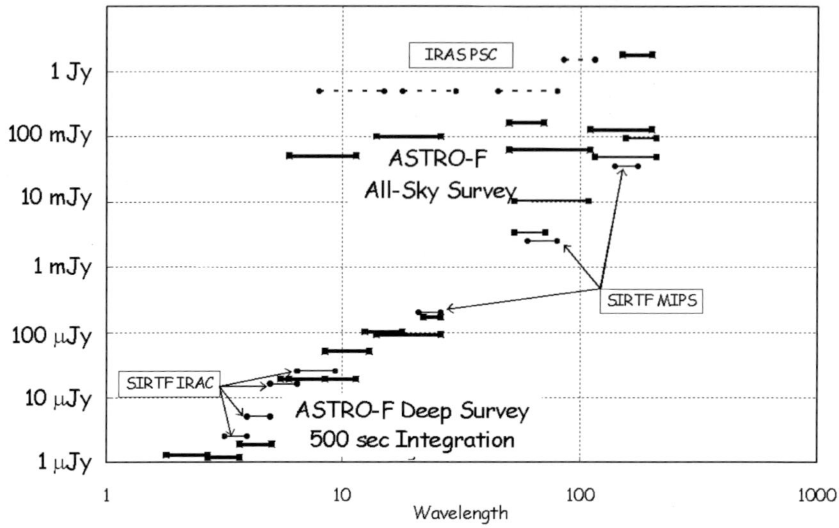

Fig. 1. Point Source Sensitivity of *ASTRO-F* (5 σ) compared to *IRAS* PSC limit [3] and SIRTF (5 σ, 200 sec integration) [6])

will execute an all-sky survey like the *IRAS* survey with a sensitivity of 100 mJy (5 σ) and several times better spatial resolution (30-50 arcsecond). Moreover, the spectral bands are added at longer than 100 micron, where most of the important objects have their peaks. The final point source catalog is expected to include distant ultra-luminous galaxies, starburst galaxies, proto-planetary disks, and other sources. The expected number of detected point sources, mainly external galaxies, is more than ten million. The data of the all-sky survey are reduced and analyzed for archive by a dedicated team formed not only by Japanese astronomers but also by Korean, UK, and Dutch astronomers. Several data products, *ASTRO-F* Flux Catalog for *IRAS* Point Sources, Point Source Catalog at High Latitudes and at Low Latitudes, and All-Sky Images, are planned to be released as soon as possible.

In addition to the all-sky survey, the Infrared Camera (IRC) will conduct a deep wide-sky survey of pre-selected parts of the sky at 2 to 25 μm. IRC has six photometric bands and low-resolution grism spectroscopy. The expected sensitivity for point sources is almost identical to that of SIRTF, and the number of spectral bands is higher. The IRC also has the low resolution spectroscopic capability. The field of view is 10 $\times$ 10 arcminutes.

3 Observing Programs

The currently planned observing programs are briefly summarized in [4]. FIS of the *ASTRO-F* is expected to contribute to almost all aspects of current astronomical research. As for the topics of interstellar matter in our Galaxy/Nearby galaxies, far-infrared spectroscopy is a powerful tool. ISO/LWS[5] revealed many important aspects of the interstellar gas and dust by measuring fine structure lines and rotational transition lines in the far-infrared. The *ASTRO-F* has the Fourier spectrometer with the array detectors that provides us far more (approx. 1000 times) efficient observations thanks to its multiplicity. Spectroscopic mapping will be made for many nearby galaxies at many spectral lines. The low-resolution spectroscopy of UIR bands of the diffuse interstellar medium is also an important topic of *ASTRO-F*. Investigation of the variation of the UIR features will reveal the variation in the composition of the interstellar medium in the Galaxy as well as in nearby galaxies.

In the near- and mid-infrared region, IRC of *ASTRO-F* has both photometric and spectroscopic capabilities with very high sensitivities and wider field of view. Very early stages of dust-enshrouded, forming stars can be detected not only for the Taurus molecular clouds but also for other distant clouds. Weak-line T-Tauri stars can be detected in the Taurus molecular cloud where *IRAS* detected only Classical T-Tauri stars. The circumstellar disk evolution of pre-main-sequence stars of medium mass will be investigated. T-Tauri stars in more distant clouds can be detected and studied statistically. *IRAS* detected large numbers of FIR sources whose energy is supplied by pre-main-sequence stars in dense molecular clouds. Many of them are associated with 'molecular cores', containing high density gas ($10^4 - 10^6 cm^{-3}$). Such *IRAS* point sources indicate newly-born stars in their parental molecular cores. *ASTRO-F* will be able to detect much fainter FIR sources powered by fainter pre-main-sequence stars, and the results may reveal the true timing of firing the stars. Near- and mid-infrared deep surveys will detect many low mass dwarfs. Low resolution spectroscopy of possible brown dwarfs in the near-infrared region will provide a more clear identification.

The Large Magellanic Cloud (LMC) is located very close to the south ecliptic pole and is accessible during two months per half year by *ASTRO-F*. Therefore, almost 20 square degrees of the LMC will be surveyed with deep pointing observations. It is expected to obtain complete NIR, MIR, and FIR maps of the LMC. The Small Magellanic Cloud (SMC) will be surveyed in a similar manner. This comprehensive survey will reveal the nature of star-formation, cluster formation, and galaxy evolution in this unique galaxy with rather low heavy-element abundance.

4 International Collaboration

ASTRO-F has a severe constraint in its pointing direction, since the telescope must be directed to within ± 1 degree from the large circle perpendicular to the direction of the Sun. Therefore, the operation planning is not so versatile. Because of the small freedom of the operation, most of the observing time will be used for the continuous all-sky survey and for pointing observations made by the *ASTRO-F* project team. The observation programming is done by a dedicated team. However, 30 % of the pointing time will be opened to the Japanese community (20 %) and to the European community (10 %). The call for open-time proposals will be announced in 2004.

5 Summary

ASTRO-F will be a unique and powerful telescope for the research of the interetellar medium complementary to the observatory-type large telescopes, such as, *SIRTF*[6], *SOFIA*[7], *Herschel*[8], *JWST*[9], and *SPICA*[10]. *ASTRO-F* will be launched in August, 2005.

References

1. H. Shibai: ASTRO-F Mission. In: *IR Space Telescopes and Instruments*, Proc. SPIE 4850, ed by J. C. Mather, pp 162–169 (2003)
2. H. Murakami, M. M. Freund, K. Ganga et al: Pub. Astron. Soc. Jp. **48**, L41–L46 (1996)
3. G. Neugebauer, H. J. Habing, R. van Duinen et al: Astrophys. J. (Letters) **278**, L1–L6 (1984)
4. H. Shibai: Adv. Sp. Res. **30 (9)**, 2089–2097 (2002)
5. M. F. Kessler, J. A. Steinz, M. E. Anderegg, et al: Astron. Astrophys. **315**, L27–L31 (1996)
6. M. W. Werner, D.B. Gallagher, W.R. Irace : Adv. Sp. Res. **30 (9)**, 2149–2150 (2002)
7. E. E. Becklin, L. J. Moon: Adv. Sp. Res. **30 (9)**, 2083–2088 (2002)
8. G. L. Pilbratt: Herschel Space Observatory mission overview. In: *IR Space Telescopes and Instruments*, Proc. SPIE 4850, ed by J. C. Mather, pp 586–597 (2003)
9. B. D. Seery: Next Generation Space Telescope (NGST): Hubble's scientific and technological successor. In: *IR Space Telescopes and Instruments*, Proc. SPIE 4850, ed by J. C. Mather, pp 170–178 (2003)
10. T. Nakagawa: Adv. Sp. Res. **30 (9)**, 2129–2134 (2002)

Exploration of Southern Sky with Atacama Submillimeter Telescope Experiment (ASTE)

K. Kohno[1], S. Yamamoto[2], R. Kawabe[3], H. Ezawa[3], S. Sakamoto[3], N. Ukita[3], T. Hasegawa[3], H. Matsuo[3], K. Tatematsu[3], Y. Sekimoto[3], K. Sunada[3], M. Saito[3], H. Iwashita[3], T. Takahashi[3], K. Nakanishi[3], N. Yamaguchi[3], T. Kamazaki[3], T. Sekiguchi[3], S. Ariyoshi[3,4], S. Yokogawa[3,4], M. Sugimoto[3,1], H. Toba[3,5], T. Oka[2], T. Sakai[2], K. Tanaka[2], H. Takahashi[5], T. Hayakawa[1], T. Okuda[1], K. Muraoka[1], Y. Fukui[6], T. Onishi[6], N. Mizuno[6], Y. Moriguchi[6], T. Minamidani[6], A. Mizuno[7], K. Suzuki[8] H. Ogawa[9], Y. Yonekura[9], S. Asayama[9], K. Kimura[9], L. Bronfman[10], and the ASTE Team

[1] Institute of Astronomy, The University of Tokyo kkohno@ioa.s.u-tokyo.ac.jp
[2] Department of Physics, The University of Tokyo
[3] National Astronomical Observatory of Japan
[4] The Graduate University for Advanced Studies
[5] Department of Astronomy, The University of Tokyo
[6] Department of Astrophysics, Nagoya University
[7] Solar-Terrestrial Environment Laboratory, Nagoya University
[8] The Instrument Development Center of School of Science, Nagoya University
[9] Department of Earth and Life Sciences, Osaka Prefecture University
[10] Departamento de Astronomía, Universidad de Chile

The Atacama Submillimeter Telescope Experiment (ASTE) is a joint project between Japan and Chile to install and operate a 10 m high precision telescope for exploration of the Southern sky at submillimeter wavelengths. Due to the excellent atmospheric conditions at the site, Pampa la Bola (4800 m) in the Atacama desert in northern Chile, ASTE offers an unique opportunity to make extensive submillimeter observations. We successfully started scientific observations in August 2003 at 800, 500, and 350 GHz bands.

1 The ASTE Project

ASTE has two main purposes. One is the on-site evaluation of high precision telescope and receiver systems, developed toward the ALMA project. The other is astronomical observations of the Southern sky at submm wavelengths.

The ASTE 10 m antenna was initially pre-installed at the Nobeyama Radio Observatory in Japan in February 2000, and preliminary tests were made from 100 to 350 GHz bands. In May 2001, the telescope was disassembled, and then shipped to Chile. After reassembly at San Pedro de Atacama (from January to March 2002), the ASTE 10 m antenna was transported to Pampa la Bola. Antenna evaluation commenced in April 2002.

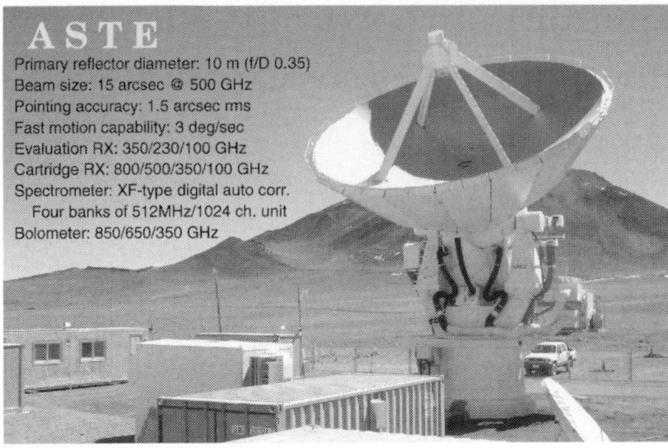

Fig. 1. ASTE observatory at Pampa la Bola (4800 m).

2 Instruments

2.1 Antenna

The ASTE antenna was designed as a prototype for the LMSA project (Ishiguro et al. 1998) employing high surface accuracy and good pointing performance with fast switching capability (Ukita et al. 2000a, 2001).

The primary reflector surface is formed by 205 remotely adjustable aluminium panels supported by CFRP backup structures and INVAR joints. Each panel was machined from a single block of aluminium achieving a surface accuracy of 5 μm (Ezawa et al. 2000). The goal is to achieve 20 μm rms surface accuracy on the primary reflector. Currently we have achieved a surface with 29 μm rms after adjustment of the panels based on preliminary holography measurements at 95 GHz. A 0.62 m diameter subreflector, mounted on the Stewart Platform, has a wobbling capability at $\sim$ 4 Hz. Global pointing accuracy was measured to be better than $1''.5$ rms using an optical pointing system. The main beam efficiencies were about 0.7, 0.6, 0.4 and 0.2 at 100, 350, 500, and 800 GHz bands, respectively, using Cartridge-type receivers (see section 2.2) placed at the 2nd Cassegrain focus. The beam efficiency at short wavelengths will be greatly improved after further panel adjustments.

2.2 Receivers

We have installed two types of heterodyne receiver systems. One is the "evaluation receivers", which covers 100, 230, and 350 GHz bands, mainly to test the performance of the antenna. For the 350 GHz band, very low receiver noise temperature ($T_{\rm RX} \sim$ 50 K or $\sim 3\ h\nu/k_{\rm B}$ in DSB) was achieved (Sekimoto et al. 2001). The other system is called "Cartridge-type receivers" (Sugimoto et al. 2003b; Yokogawa et al. 2003). Engineering models of 800, 500, and

100 GHz cartridge-type receivers were integrated into the plug-in cryogenic system employing a 3-stage GM cryocooler with temperature stabilization and mechanical vibration reduction. The system was installed at ASTE in November 2002, and we have successfully confirmed that the "ALMA-type system" works well as designed (Sugimoto et al. 2003a). Since August 2003, scientific observations have been carried out using the 800, 500, and 350 GHz receivers. $T_{\rm RX}$ (DSB) of 100 K at 492 GHz and 1000 K at 812 GHz have been achieved using Nb-based PCTJ mixers.

A three-color (850, 650, and 350 GHz) single-pixel bolometer system was installed and images of planets were obtained. The Development of a multi-pixel submm camera using SIS photon detectors (Ariyoshi et al. 2003) is now in progress for future continuum observations (Matsuo et al. 2001).

2.3 Spectrometer

An XF-type digital auto-correlator system has been installed (Sorai et al. 2000). The system consists of 4 auto-correlator units, where each unit has a bandwidth of 512 or 128 MHz with 1024 channels. The widest total frequency coverage is 2 GHz (1200 km s^{-1} at 492 GHz), and the highest frequency resolution is 0.125 MHz (0.076 km s^{-1} at 492 GHz).

2.4 Site and Infrastructure

The atmospheric and environmental conditions have been studied extensively at the site (e.g., Hirota et al. 1998; Matsuo et al. 1998; Sakamoto et al. 2000; Ukita et al. 2000b). Based on the opacity and FTS measurements, it is expected that a large fraction of the time (up to $\sim$ 50 % during the winter season) can be used for submm observations (Matsushita et al. 1999).

The on-site infrastructure to support the telescope operation includes two power generator sets (maximum capability of 200 kW), a satellite communication system ($\sim$ 64 kbps), O_2 enriched rooms for operation, and containers for storage. Currently the telescope is primarily operated from San Pedro de Atacama via satellite communication. In addition, some remote observations from Japan are made as well.

3 Initial Results from ASTE

Large area mapping of the Carina nebula region at CO(J=3–2) has been made using the 350 GHz evaluation receiver. (Fig. 2, left; Yamaguchi et al. 2003). The typical system noise temperature ($T_{\rm sys}$) was 100–200 K in DSB. It is suggested that some CO clouds are interacting with ionized fronts.

Strong CI($^3P_1 -^3 P_0$) emission from the Magellanic Clouds has been detected using the 500 GHz cartridge-type receiver (Mizuno et al. 2003; this proceedings). $T_{\rm sys}$ was about 1000 K (DSB) during these observations.

Extended CO(J=7–6) emission from the Orion A region has been observed using the 800 GHz cartridge-type receiver (Fig. 2, right; Sugimoto et al., in preparation). Thanks to the excellent atmospheric condition and low $T_{\rm sys}$ ($\sim$ 3000 K in DSB), a wide area map (more than 300 points) of CO(J=7–6) emission was obtained within just a few hours. Submm observations with ASTE (and APEX, ALMA, etc.) are indeed very promising.

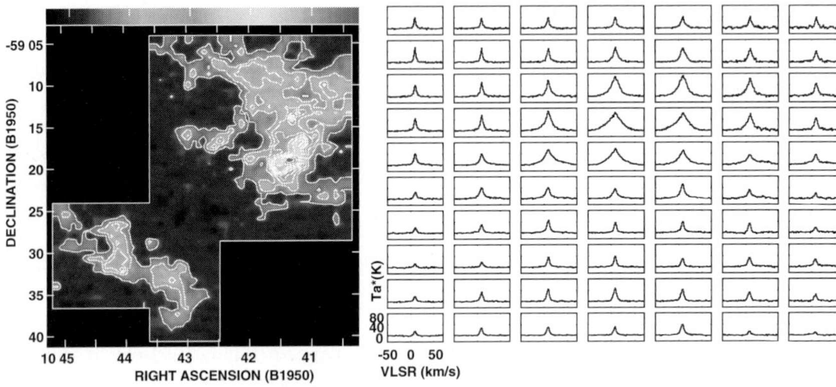

Fig. 2. Left: CO(J=3–2) map of the Carina nebula taken with the ASTE 350 GHz evaluation receiver. Right: CO(J=7–6) spectra from the Orion A region taken with the ASTE 800 GHz cartridge-type receiver.

References

1. S. Ariyoshi et al.: IEEE Trans. Applied Supercond. **13**, 1128 (2003)
2. H. Ezawa et al: Proc. SPIE, vol. 4015, 515 (2000)
3. T. Hirota et al: PASJ **50** 155 (1998)
4. M. Ishiguro et al: Proc. SPIE, vol. 3357, 244 (1998)
5. H. Matsuo et al: PASJ **50** 359 (1998)
6. H. Matsuo et al: IEEE Trans. Applied Supercond. **11**, 688 (2001)
7. S. Matsushita et al: PASJ **51** 603 (1999)
8. H. Ogawa et al: ASP Conf. Ser. vol 289, pp 23–24 (2003)
9. S. Sakamoto et al: ALMA Memo, no. 326 (2000)
10. Y. Sekimoto et al: PASJ, **53**, 951 (2001)
11. K. Sorai et al: Proc. SPIE, vol. 4015, 86 (2000)
12. M. Sugimoto et al: ALMA Memo, no. 458 (2003a)
13. M. Sugimoto et al: Cryogenics, **43**, 435 (2003b)
14. N. Ukita et al: Proc. SPIE, vol. 4015, 177 (2000a)
15. N. Ukita et al: LMSA Memo, no. 2000-9 (2000b)
16. N. Ukita et al: Publ. of Natl. Astron. Obs. Japan, **6**, 59 (2001)
17. N. Yamaguchi et al: Proc. of IAU Symp. no. 221, 336 (2003)
18. S. Yokogawa et al: PASJ, **55**, 519 (2003)

Dome C, Antarctica: The Best Accessible Sub-millimetre Site on the Planet?

P.G. Calisse[1], M.C.B. Ashley[1], M.G. Burton[1], J.S. Lawrence[1], T. Travouillon[1], J.B. Peterson[2], M.A. Phillips[1], S.J.E. Radford[3], and J.W.V. Storey[1]

[1] School of Physics, University of New South Wales, Sydney, NSW 2052 Australia
pcalisse@phys.unsw.edu.au

[2] Department of Physics, Carnegie Mellon University, PA, 15213 USA

[3] National Radio Astronomy Observatory, Tucson, AZ 85721-06625 USA

Abstract

The French-Italian station of Dome Concordia (lat. 76° S; long. 123° E; alt. 3,260 m) will be operating year-round from January 2005, offering a new outstanding opportunity for the deployment of instrumentation for sub-millimetre astronomy. In early 2003, SUMMIT, a tipper operating at 350 µm, was deployed to the site. A preliminary analysis of the winter sub-millimetre opacity data obtained is presented.

1 SUMMIT and the NRAO Sub-millimetre Tippers

The sub-millimetre tippers were originally developed at NRAO and Carnegie-Mellon University to monitor the sky opacity at sub-millimetre wavelengths at different remote sites. They have been installed at Atacama (Chile), Mauna Kea (Hawaii, US), and South Pole (Antarctica). A rugged, low power version, designed to operate unattended on the Antarctic Plateau has been developed at the University of New South Wales. It operated originally at Dome C over the summer of 2000–01, gathering the first sub-millimetre data from this site [1]. The instrument then spent two years at the South Pole, to perform a cross-calibration with the NRAO tipper permanently installed at that site. This calibration demonstrated consistency between the two instruments.

The tippers operate at a central frequency of about 350 µm (857 GHz) with a bandwidth of 108 GHz. A rotating 45° parabolic off-axis mirror allows measurement of the flux received from the atmosphere at different airmasses from 0 to 90 degrees zenith angle. Two blackbodies, maintained at different temperatures, allow an absolute calibration of the instrument. The two blackbodies are respectively maintained in thermal contact with the external environment (cold) and the inside of the supporting module AASTINO (warm), kept warm by the waste heat generated by the electrical power generator.

A radome with a transmission of 78–80% protects the system from the atmospheric agents. The beam from the sky (or from the blackbodies) is directed through a chopper, and then to a blocking filter, to an 857 GHz mesh filter and finally to a Barnes pyroelectric detector. Since February 2003 SUMMIT has been operating on the roof of the AASTINO (Automated Astrophysical Site Testing INternational Observatory), an automated, self powered module installed by UNSW at Dome C. The instrument continuously acquired data until communication with the AASTINO stopped on the 1st of July 2003.

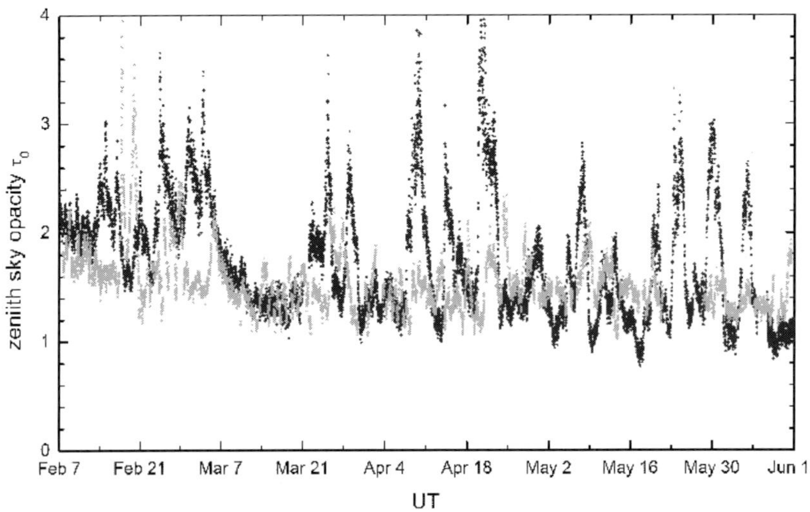

Fig. 1. Data acquired at Dome C by the SUMMIT (grey data points) compared to data acquired by the NRAO tipper at South Pole (black data points) during the same period in winter 2003

2 Opacity and Atmospheric Noise at 350 μm

Because water vapor lines cause the majority of the atmospheric attenuation at sub-millimetre wavelengths, the ideal observing site will be very high and very dry. In addition, the requirement for long integration times implies a site with stable atmospheric conditions – preferably for many hours at a time. Finally, interferometry demands good atmospheric phase stability for as long as possible and over spatial scales of a kilometre or more.

The sky opacity variability in the sub-millimetre regions depends principally on fluctuations in the precipitable water vapor content of the atmosphere. The residual zenith opacity, estimated to be about 0.7–0.8 at the elevation of Dome C for broadband instruments like the tippers, is described as dry-air opacity. However, observations carried out at South Pole at 450 m with the Viper sub-millimetre telescope, suggest that the atmospheric noise at this wavelength is not completely correlated with the zenith opacity, but can vary over a wide range – more than a factor of 10 – even when the sky opacity reaches very low values.

Dome C, Antarctica, is potentially an outstanding sub-millimetre site. With a pressure altitude approaching 4,000 m in winter, significantly lower precipitable water vapor than even the South Pole, and a remarkably stable troposphere with very low wind speeds at all altitudes, it is possibly the best accessible sub-millimetre site in the world.

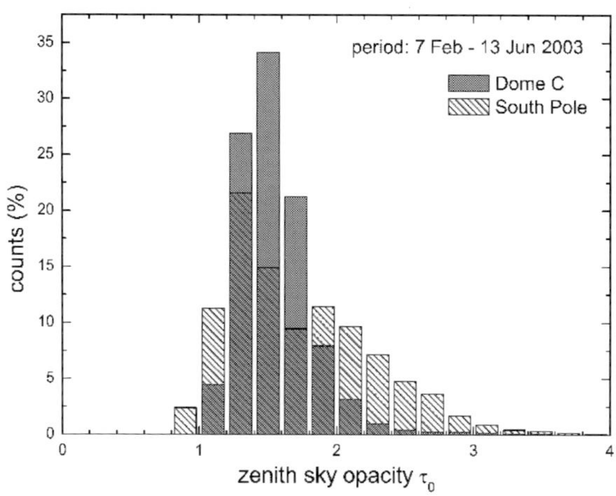

Fig. 2. Histogram of opacity data as obtained by the previous figure data

3 Data Analysis and Results

The first data obtained during summer 2000-01 at Dome C were encouraging, and show a significantly lower overall opacity than the South Pole [2, 3]. A preliminary analysis of data acquired at Dome Concordia during winter 2003 (see Figure 1) shows a more stable atmosphere than the South Pole and indeed any other site tested so far. Only in a few cases does the zenith sky opacity exceed 2 (see Figure 2).

However, because of the possible formation of a thin layer of ice on the inner surface of the SUMMIT radome, data in Figure 1 and 2 may contain an opacity offset of about 0.2 and the real opacity at the site could be, on average, even better. Simulations have shown that any such ice layer should not affect the shape of the distribution at low ($\tau = 2$) opacity values.

Acknowledgments

This work has been supported by the Australian Research Council and the Australian Antarctic Division. We thank the ENEA-PNRA (Italian Antarctic Program), IPEV (French Antarctic Program), NSF and the US Antarctic Program for their invaluable support during the 2000–2001 Antarctic campaigns and beyond.

References

1. J.B. Peterson, S.J.E. Radford, P.A.R. Ade, R.A. Chamberlin, M.J. O'Kelly, K.M. Peterson, E. Schartman: Publ. Astron. Soc. Pac. **115**, 383 (2003)
2. P.G. Calisse, M.C.B. Ashley, M.G. Burton, M.A. Phillips, J.W.V. Storey, S.J.E. Radford, J.B. Peterson: 2004, Publ. Astron. Soc. Aust. in press (2003)
3. P.G. Calisse: Publ. Astron. Soc. Aust. submitted (2004)

Part VII

Magnetic Fields in the ISM

What Drives Star Formation?

R.M. Crutcher

Department of Astronomy, University of Illinois, 1002 W. Green Street, Urbana, IL 61801, USA crutcher@uiuc.edu

1 Introduction

Understanding star formation is one of the outstanding challenges of modern astrophysics. However, in spite of significant progress in recent years, there remain unanswered fundamental questions about the basic physics of star formation. In particular, what drives the star formation process? The prevailing view has been that self-gravitating clouds are supported against collapse by magnetic fields, with ambipolar diffusion reducing support in cores and hence driving star formation, e.g., [8]. The other extreme is that molecular clouds are intermittent phenomena in an interstellar medium dominated by turbulence, and the problem of cloud support for long time periods is irrelevant, e.g., [7]. Clouds form and disperse by the operation of compressible turbulence, with clumps sometimes becoming gravitationally bound. Turbulence then dissipates rapidly, and the cores collapse to form stars. Hence, there are two competing models for driving the star formation process. The issue of what drives star formation is far from settled, on either observational or theoretical grounds. In this contribution we discuss how observations of magnetic fields in molecular clouds can distinguish between and test these models.

2 Model Predictions and Observational Tests

The ambipolar diffusion and turbulence models for driving star formation have different predictions for magnetic field morphology and strength, which form the basis for tests of the two models using observations of magnetic fields. Of course, it is clear that there are both magnetic fields and turbulence in real clouds. In order to sharpen the distinctions between the two models, here we will consider only the extreme cases of turbulence models with negligibly weak magnetic fields and magnetic support/ambipolar diffusion models without turbulence. The three main tests of the two models based on observations of magnetic fields involve: (1) the morphology of $\mathbf{B}$, (2) the scaling of $|\mathbf{B}|$ with density and internal velocity dispersion, and (3) the mass to magnetic flux ratio, M/Φ.

(1) Morphology. In the magnetic support model, the dominant $\mathbf{B}$ means field lines should be smooth, without irregular structure. Clouds will be thin disks or oblate spheroids, since thermal pressure provides the only support along field lines. Therefore, the field lines should be parallel to the minor axes of clouds. Finally, an original morphology with parallel magnetic field lines will be transformed into an hourglass morphology since it is the tension of the bent field lines that provides support. In the turbulent model, **B** will be too weak to resist twisting by the dominant turbulence, and field lines will not be smooth but chaotic, with small-scale irregular structure.

(2) Scaling. Ambipolar diffusion will leave $|\mathbf{B}|$ in cloud envelopes virtually unchanged, so at lower densities no strong correlation between $|\mathbf{B}|$ and density ρ is predicted. As M/Φ increases in the core and becomes supercritical, the subsequent collapse will increase $|\mathbf{B}|$, such that for a $|\mathbf{B}| \propto \rho^\kappa$ parametrization, $\kappa \to 0.5$; $\kappa = 0.5$ is a prediction of flux freezing. In turbulent models, in the low-density case before gravity becomes important no relation is expected between the magnetic field strength and other parameters such as density. Once gravity becomes important, a rough equipartition of energies is achieved. If we assume virial equilibrium between gravity and turbulence ($3GM^2/5R = 3M\sigma^2/2$), then $\rho R^2 \propto \sigma^2$. Flux freezing ($M \propto \Phi$) gives $\rho R \propto |\mathbf{B}|$, so $|\mathbf{B}| \propto \sigma\rho^{1/2}$ is predicted. Hence, although both models predict $\kappa \approx 0.5$ for cores, if turbulence provides significant support an additional scaling of $|\mathbf{B}|$ with the turbulent velocity dispersion σ is predicted.

(3) Mass to Flux Ratio. The ratio of the mass in a magnetic flux tube to the magnitude of the magnetic flux is a crucial parameter for the magnetic support/ambipolar diffusion model. The critical value for the mass in a uniform disk that can be supported by magnetic flux Φ is $M_{Bcrit} = \Phi/2\pi\sqrt{G}$ [9]; the precise value of the numerical coefficient is slightly model dependant. It is convenient to state M/Φ in units of the critical value, and to define $\lambda \equiv (M/\Phi)_{actual}/(M/\Phi)_{crit}$. Inferring λ from observations is possible if the column density N and the magnetic field strength $|\mathbf{B}|$ are measured:

$$\lambda = \frac{(M/\Phi)_{observed}}{(M/\Phi)_{crit}} = \frac{mNA/|\mathbf{B}|A}{1/2\pi\sqrt{G}} = 7.6 \times 10^{-21} \frac{N(H_2)}{|\mathbf{B}|} \tag{1}$$

where $m = 2.8m_H$ allowing for He, A is the cloud area of the observations, $N(H_2)$ is in cm^{-2}, and $|\mathbf{B}|$ is in μG. In the ambipolar diffusion model clouds are initially subcritical, $\lambda < 1$. Ambipolar diffusion is fastest in shielded, high-density cores, so cores become supercritical, and rapid collapse ensues. The envelope continues to be supported by the magnetic field. Hence, the prediction is that λ must be < 1 in cloud envelopes, while in collapsing cores λ becomes slightly > 1. Hence, this model tightly constrains λ. On the other hand, the turbulent model imposes no direct constraints on λ, although strong magnetic fields would resist the formation of gravitationally bound clouds by compressible turbulence. Also, if magnetic support is to be insufficient to prevent collapse of self-gravitating clumps that are formed by compressible turbulence, the field must be supercritical, $\lambda > 1$. λ may take any value > 1, although of course for the turbulence model with very weak magnetic fields that we are considering, clouds will be highly supercritical, $\lambda \gg 1$.

3 Observational Techniques

The morphology of **B** can be mapped via linear polarization observations of dust continuum or spectral-line emission, and a statistical estimate of the plane-of-sky component B_{pos} of **B** may be obtained from the dispersion in polarization position angles via the Chandrasekhar-Fermi technique [3]. However, only the Zeeman effect directly yields magnetic field strengths in molecular clouds. Stokes V spectra reveal the direction and magnitude of the line-of-sight component B_{los}; with current sensitivities, the plane-of-sky component of **B** cannot be measured, e.g., [4].

For a large number of clouds whose magnetic fields are randomly oriented with respect to the observed line of sight, it is possible to correct statistically for the fact that only $B_{los} = |\mathbf{B}| \cos\theta$ or $B_{pos} = |\mathbf{B}| \sin\theta$ is measured, where θ is the angle between the line of sight and **B**. If **B** is strong, clouds will have a disk morphology with **B** along the minor axis. To properly measure λ, one needs $|\mathbf{B}|$ and N along a flux tube, i.e., parallel to the minor axis. As noted by Crutcher [5], the path length through a disk will be too long by $1/\cos\theta$ and N will be overestimated, while B_{los} (B_{pos}) underestimates $|\mathbf{B}|$ by $\cos\theta$ ($\sin\theta$). For B_{los},

$$\overline{M/\Phi} = \int_0^{\pi/2} (M/\Phi)_{obs} \cos^2\theta \sin\theta d\theta = \frac{1}{3}(M/\Phi)_{obs}. \quad (2)$$

Similarly, for B_{pos} the same factor of 1/3 is obtained (when $\cos\theta \sin^2\theta$ is integrated). We will discuss here $\lambda_C = \frac{1}{3}\lambda_{obs}$ in order to test the magnetic support model, since a disk morphology is the prediction of that model. The turbulence model with very weak **B** predicts $\lambda_{obs} \gg 1$.

4 Observational Results

Figure 1 shows one example of the observational results – dust polarization mapping of the starless core L183 [6]. The observed polarization position angles have been rotated by 90° so the line segments are in the direction of B_{pos}. The magnetic field is fairly regular, but the dispersion in position angles of 14° (corrected for measurement error) is significant. The angle between the projected minor axis of the core and $\overline{B_{pos}}$ is $\sim 30°$. The Chandrasekhar-Fermi technique yields $B_{pos} \approx 80\ \mu G$ and $\lambda_C \approx 0.9$ for the dense core. OH Zeeman data [4] give $B_{los} < 16\ \mu$G in the envelope of this core. Space precludes reviewing all observational results, but all are included in applying the tests.

5 Test Results

(1) Morphology. Maps of dust and spectral-line linear polarization and of the Zeeman effect generally show a regular field morphology, and an hourglass

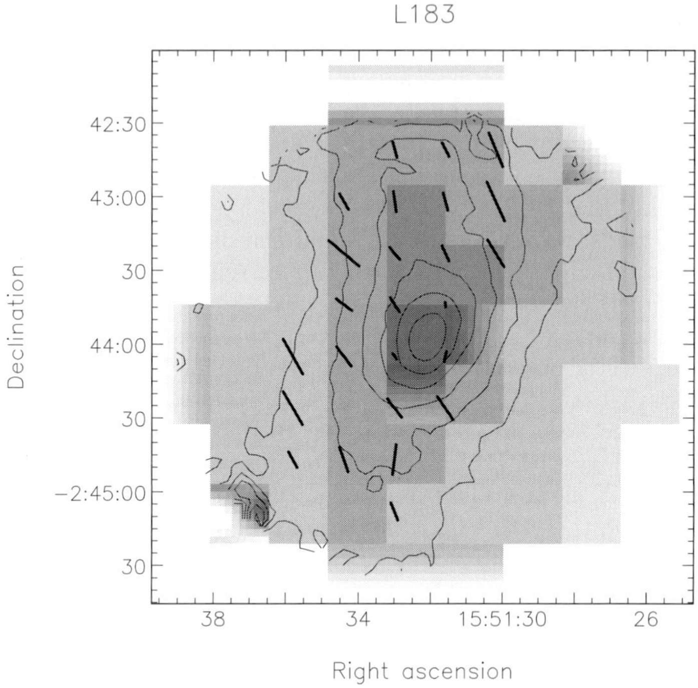

Fig. 1. Dust polarization map of the starless core L183. Grey-scale and thin contours show the dust emission at 850 μm; thick line segments show the direction of the magnetic field projected on the sky.

morphology is sometimes seen. A regular field dominating a random field and an hourglass morphology toward cores are predictions of the strong magnetic field model. However, **B** projected onto the sky is not observed to be parallel to the minor axes of starless cores as predicted by magnetic support. Finally, even though fairly small, the dispersion in polarization position angles is often greater than observational errors, implying that turbulence is producing an irregular component to **B**.

(2) Scaling. At low densities $n \sim 0.1 - 100$ cm^{-3}, it has been clear for some time that there is no correlation of $|\mathbf{B}|$ with ρ [10]. At higher densities one finds $|\mathbf{B}| \propto \rho^{\kappa}, \kappa = 0.47 \pm 0.08$ [5]. Basu [1] showed that a better correlation was obtained by fitting $|\mathbf{B}|$ *vs.* $\sigma\sqrt{\rho}$. Both the turbulence and magnetic support models predict little or no correlation between $|\mathbf{B}|$ and ρ at low densities, as observed. At higher densities, $\kappa \approx 0.5$ is the signature of flux freezing together with gravitational contraction, and does not discriminate between the two models. However, $|\mathbf{B}| \propto \sigma$ is a prediction of a significant turbulent component to the support of clouds.

(3) Mass to Flux Ratio. Figure 2 shows all of the λ_C available from Zeeman detections and sensitive limits for molecular clouds [5, 2, 11], plus the

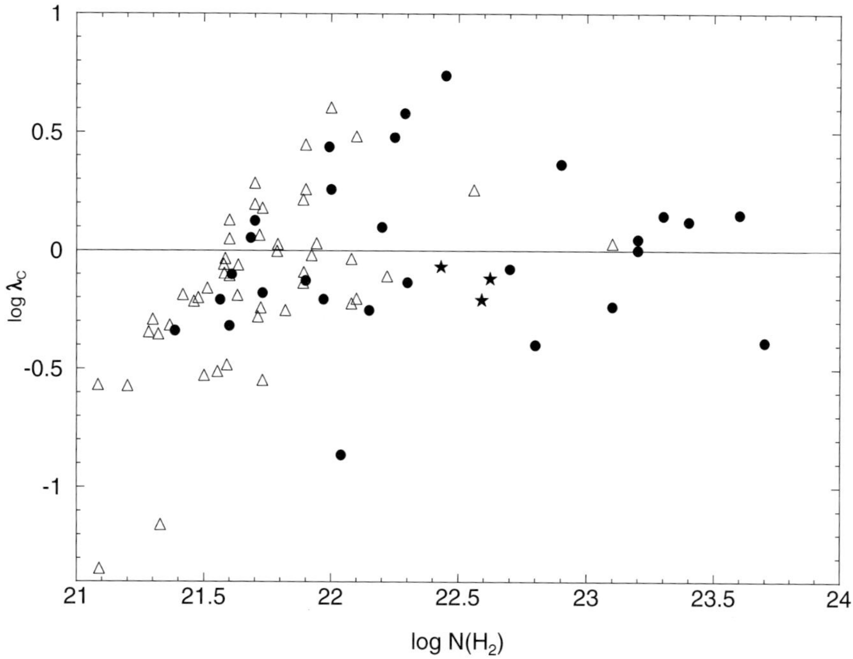

Fig. 2. λ_C are the observed mass to magnetic flux ratios, divided by 3 to correct for projection bias, in units of the critical values. $\lambda_C > 1$ is supercritical, $\lambda_C < 1$ is subcritical. Dots are for Zeeman data with $B_{los} > 3\sigma(B_{los})$, stars are for Chandrasekhar-Fermi estimates of B_{pos}, and triangles are lower limits plotted at $B_{los} = 3\sigma(B_{los})$. Although the statistical correction of 1/3 for geometrical bias has been applied to each point, so that statistically this plot should be valid, for any individual point the true λ could be higher or lower than the plotted λ_C. A large part of the scatter in figure 2 may therefore be due to geometrical projection effects.

Chandrasekhar-Fermi results for three prestellar cores [6]. The observations are distributed roughly equally above and below the $\lambda_C = 1$ line that divides subcritical and supercritical M/Φ ratios for disk geometries. Therefore, the data suggest that $\overline{\lambda} \approx 1$; that is, the typical mass to magnetic flux ratio is approximately critical. There is a slight indication that for large column densities, $\overline{\lambda}$ may be supercritical, and for small column densities, subcritical.

6 Conclusions

The generally regular morphology of polarization maps supports a strong magnetic field model. Yet the small-scale scatter in polarization position angles predicted by turbulence is seen, although this scatter is generally relatively small. As projected onto the sky, **B** is not parallel to the minor axes of cores, as predicted by the magnetic support model. The observation that $|\mathbf{B}| \propto \sigma$ is a prediction of a turbulent component of the support of clouds. The observations of molecular cloud cores are consistent with $\lambda \approx 1$. This is certainly inconsistent with the extreme turbulent model with negligible $\mathbf{B}$, for which $\lambda \gg 1$. Although $\lambda \approx 1$ is predicted to occur in cores as a result of ambipolar diffusion when the original mass-to-flux ratio was subcritical, the observations have not found direct evidence of subcritical clouds and especially of subcritical envelopes. Whether λ is slightly greater than or slightly less than 1 is crucial for whether magnetic fields provide support that is removed by the action of ambipolar diffusion. The present data are not sufficiently accurate to discriminate between these cases since λ in molecular clouds is observed to be (frustratingly) close to 1.

It therefore seems clear that the data remain inconclusive in deciding between the two extreme models for what drives star formation. However, a strong conclusion does come from the observations: *both* turbulence and strong magnetic fields are important in the physics of molecular clouds. There does not seem to be a single driver of star formation.

Tests of the role of magnetic fields are of necessity statistical in nature and must involve data covering a wide range of physical parameters. Needed are additional high sensitivity surveys of magnetic field strengths and morphologies, together with other cloud properties, in order to assess the relative importance of magnetic fields in molecular cores and envelopes.

Acknowledgement: This work was partially supported by NSF grant AST 02-05810. I thank Tom Troland for an enjoyable and productive collaboration on the study of interstellar magnetic fields.

References

1. S. Basu: ApJ **540**, L103 (2000)
2. T.L. Bourke, P.C. Myers, G. Robinson, & A.R. Hyland: ApJ **554**, 916 (2001)
3. S. Chandrasekhar & E. Fermi: ApJ **118**, 113 (1953)
4. R.M. Crutcher, T.H. Troland, A.A. Goodman, C. Heiles, I. Kazès, & P.C. Myers: ApJ **407**, 175 (1993)
5. R.M. Crutcher: ApJ **520**, 706 (1999)
6. R.M. Crutcher, D. Nutter, D. Ward-Thompson, & J.M. Kirk: ApJ, in press
7. B.G. Elmegreen: ApJ **530**, 277 (2000)
8. T.Ch. Mouschovias & G.E. Ciolek: Magnetic Fields and Star Formation: A Theory Reaching Adulthood. In: *The Origin of Stars and Planetary Systems*, ed by C. J. Lada & N. D. Kylafis (Kluwer, Dordrecht 1999) pp 305–340
9. T. Nakano & T. Nakamura: PASJ **30**, 681 (1978)
10. T.H. Troland & C. Heiles: ApJ **301**, 339 (1986)
11. T.H. Troland & R.M. Crutcher: in preparation

Tracing the Magnetic Field in Molecular Clouds

M. Houde[1], T.G. Phillips[2], R.H. Hildebrand[3], C.D. Dowell[2,4], P. Bastien[5], R. Peng[1], J.E. Vaillancourt[6], J.L. Dotson[7], and H. Yoshida[1]

[1] Caltech Submillimeter Observatory, 111 Nowelo Street, Hilo, HI 96720, USA
houde@submm.caltech.edu
[2] California Institute of Technology, Pasadena, CA 91125, USA
[3] Department of Astronomy and Astrophysics, Enrico Fermi Institute, and Department of Physics, University of Chicago, Chicago, IL 60637, USA
[4] Jet Propulsion Laboratory, 4800 Oak Grove Drive, Pasadena, CA 91109, USA
[5] Département de Physique, Université de Montréal, Montréal, Québec H3C 3J7, Canada
[6] Physics Department, University of Wisconsin, 1150 University Avenue, Madison, WI 53706, USA
[7] NASA Ames Research Center, Moffet Field, CA 94035, USA

1 Introduction

The orientation of the magnetic field vector at any given point in a molecular cloud can be specified with two parameters: α, its inclination angle to the line of sight, and β, its projection angle on the plane of the sky (eastward from north). Until recently, of these two angles, only β could be measured. At submillimeter wavelengths, this can be accomplished, for example, with polarization measurements of the continuum radiation emanating from elongated dust grains that are aligned by the local magnetic field [2]. The angle β is thus obtained from the angle of the polarization vector since the projection of the magnetic field vector in the plane of the sky is oriented at right angles to the polarization vector [5].

In this paper, we will discuss how it is possible to determine the inclination angle α through the combination of spectroscopy and polarimetry data. We will show that this is made possible by the realization [6, 7] that the magnetic field leaves its imprint on the line profiles of molecular ions. Indeed, the ion-to-neutral line width ratio, calculated from the line profiles of coexistent ion and neutral molecular species [6], is the fundamental parameter that, when combined with polarimetry data, holds the key to the determination of the inclination angle of the magnetic field.

We will summarize the technique and the first results obtained for the M17 molecular cloud and Orion A complex using HCO^+ and HCN spectroscopic data and extensive 350 μm continuum and polarimetry maps obtained with the Hertz polarimeter [3] and the SHARC II camera [4] at the Caltech Submillimeter Observatory (CSO) and published elsewhere [8, 9].

2 The Effect of the Magnetic Field on the Line Profiles of Molecular Ions

Houde et al. [6, 7] have discussed the effect that the magnetic field can have on molecular ions in the weakly ionized and turbulent regions of molecular clouds. They have shown that, because of the fact that an ion is trapped and undergoes gyromagnetic motions when in the presence of a sufficiently strong magnetic field, its linear motions are restricted to directions that are parallel to the orientation of the local field. It cannot, therefore, fully take part in the same turbulent regime that coexistent neutral species are subjected to. One direct consequence is the realization that *the width of the line profiles of molecular ions should, in general, be less than that of coexistent neutral molecular species* in cases where the orientation of the magnetic field is not, on average, aligned with the flows that characterize the turbulent motions of the neutral component of the gas. This fact has been verified experimentally in numerous molecular clouds through the observations and comparisons of coexistent neutral and ion molecular species spectra (see [6, 7] and Table 1 of [8]).

3 The Evaluation of the Inclination Angle α

Since, as was stated before, the motions of molecular ions are restricted to directions parallel to that of the magnetic field (neglecting gyrations), it should not be surprising if the width of their line profiles were to be a function of the orientation of the field to the line of sight. It can, in fact, be shown that this is so, at least when considering reasonably general models of neutral flow configurations (see Figure 2 of Houde et al. [8] and the discussion that pertains to it). Moreover, it is also the case that the ion-to-neutral line width ratio, as calculated from the line profiles of coexistent ion and neutral molecular species in the context of these models, is a well-behaved function of the inclination angle. It is maximum at approximately unity when the field is aligned with the line of sight ($\alpha = 0$) and decreases monotonically to a minimum ($\simeq 0.26$ for HCO^+) as α increases to a value of $\pi/2$ when the magnetic field lies in the plane of the sky. There is, however, an ambiguity in selecting which model of neutral flows is best suited to a given region. It is necessary to discriminate between the different alternatives in order to succeed in calculating α from the ion-to-neutral line width ratio.

This is accomplished buy using another observable parameter that is also a function of the orientation of the magnetic field relative to the line of sight. This parameter is the polarization level P that is measured, for example, from the continuum emission from dust at submillimeter wavelengths. More precisely, the polarization level P, detected by an observer studying a given region where the field is oriented with a viewing angle α, can be expressed as

$$P = P_{max} \sin^2(\alpha) , \tag{1}$$

where P_{max} is the maximum polarization level that can be detected, i.e. when the field lies in the plane of the sky ($\alpha = \pi/2$). Equation (1) can be used in conjunction with the relevant equation for the ion-to-neutral line width ratio in order to express the latter as a function of the normalized polarization level P/P_{max}. By carefully taking into account the well-known depolarization effect, characterized by a systematic decrease in the polarization level with increasing continuum flux, and assigning a suitable value for P_{max}, it then becomes possible to use polarization data to constrain the neutral flow model and, in effect, calibrate the ion-to-neutral line width ratio.

3.1 The M17 Molecular Cloud and the Orion A Complex

The previously summarized technique was used to measure the orientation of the magnetic field in the M17 molecular cloud [8] and the Orion A complex [9]. These results were acquired using 350 μm polarimetry and spectroscopic (HCN and HCO^+ in the $J \to 4-3$ transition) maps obtained using the Hertz polarimeter [3] and the CSO 300-400 GHz receiver, respectively. We also benefited from an extensive 350 μm SHARC II map [4] for the Orion A analysis.

A detailed view of the orientation of the magnetic field in M17 is shown in Figure 1 (see the caption) where α is seen not to exceed $\approx 60°$. This value was combined with Zeeman measurements [1] by Houde et al. [8] to calculate a magnitude of ≈ -750 μG for the magnetic field in this molecular cloud.

Average values for α and β were calculated for Orion A and are shown in Table 1. From these numbers, we can see that the orientation of the magnetic field is relatively stable in OMC-2 and OMC-3, while it exhibits a significant re-orientation in OMC-1. We can then infer that the field seems to be relatively unaffected by the presence of the concentrations of lower mass that characterize the OMC-2 and OMC-3 regions, while it appears to be strongly perturbed in and around the massive OMC-1 complex.

Table 1. Mean orientation of the magnetic field in parts of Orion A.

Source	α	β
OMC-3 MMS 6	73°	46°
OMC-3 MMS 9	74°	39°
OMC-2 FIR 4	80°	31°
OMC-1	60°	115°

The Caltech Submillimeter Observatory is funded by the NSF through contract AST 9980846 and the observations made with Hertz were supported by NSF Grants AST 9987441 and AST 0204886.

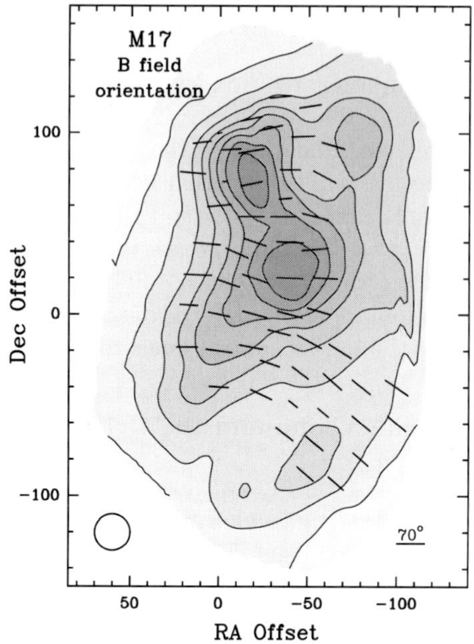

Fig. 1. Orientation of the magnetic field in M17. The orientation of the projection of the magnetic field in the plane of the sky is shown by the vectors and the viewing angle is given by the length of the vectors (using the scale shown in the bottom right corner). The contours and the grey scale delineate the total continuum flux. The beam width ($\simeq 20''$) is shown in the lower left corner.

References

1. Brogan, C. L., Troland, T. H.: ApJ, **560**, 821 (2001)
2. L. M. Davis, and J. L. Greenstein: ApJ, **114**, 206 (1952)
3. C. D. Dowell, R. H. Hildebrand, D. A. Schleuning, J. E. Vaillancourt, J. L. Dotson, G. Novak, T. Renbarger, M. Houde: ApJ, **504**, 588 (1998)
4. C. D. Dowell, C. A. Allen, S. Babu, M. M. Freund, M. Gardner, J. Groseth, M. Jhabvala, A. Kovacs, D. C. Lis, S. H. Moseley Jr., T. G. Phillips, R. Silverberg, G. Voellmer, H. Yoshida: in Proceedings of SPIE Vol. 4855 *Millimeter and Submillimeter Detectors for Astronomy*, eds. T. G. Phillips and J. Zmuidzinas, (SPIE, Bellingham, WA, 2003) pp. 73–87
5. R. H. Hildebrand: QJRAS, **29**, 327 (1988)
6. M. Houde, P. Bastien, R. Peng, T. G. Phillips, and H. Yoshida: ApJ **536**, 857 (2000)
7. M. Houde, R. Peng, T. G. Phillips, P. Bastien, and H. Yoshida: ApJ **537**, 245 (2000)
8. M. Houde, P. Bastien, J. L. Dotson, C. D. Dowell, R. H. Hildebrand, R. Peng, T. G. Phillips, J. E. Vaillancourt, and H. Yoshida,: ApJ, **569**, 803, (2002)
9. M. Houde, C. D. Dowell, R. H. Hildebrand, J. L. Dotson, J. E. Vaillancourt, T. G. Phillips, R. Peng, P. Bastien: submitted to ApJ, (2003)

Grain Alignment in Molecular Clouds

A. Lazarian

University of Wisconsin-Madison lazarian@astro.wisc.edu

Abstract. One of the most informative techniques of studying magnetic fields in molecular clouds is based on the use of starlight polarization and polarized emission arising from aligned dust. How reliable the interpretation of the polarization maps in terms of magnetic fields is the issue that the grain alignment theory addresses. I briefly review basic physical processes involved in grain alignment.

1 Why Do We Care?

The fact that the grain got aligned has been known for more than half a century. Nevertheless, it has been always a puzzle why grains get aligned in interstellar medium. Very soon after the discovery of grain alignment by Hall (1949) and Hiltner (1949) it became clear that the alignment happens in respect to magnetic field. Since that time grain alignment stopped to be the issue of pure scientific curiosity, but became an important missing link of connecting polarimetry observations with the all-important interstellar magnetic fields[1].

The history of grain alignment ideas is excited (see review by Lazarian 2003) and here we do not have space here to dwell upon it. Within this very short review we will discuss the modern understanding of grain alignment processes applicable to molecular clouds. Last decade has been marked by a substantial progress in understanding new physics associated with grain alignment. The theory has become predictive, which enables researchers to interpret observational data with more confidence.

Recent theory reviews of the grain alignment theory include Roberge (1996), Lazarian (2000, 2003). Progress in testing theory is covered in Hildebrand (2000), while particular aspects of grain dynamics are discussed in Lazarian & Yan (2003). The presentation in Lazarian (2003) goes beyond molecular cloud environment and deals with the possibility of alignment in circumstellar regions, interplanetary medium, coma of comets etc. The interested reader may use these reviews to guide her in the vast and exciting original literature on grain alignment.

[1] Additional interest to grain alignment arises from recent attempts to separate the polarized CMB radiation from the polarized foregrounds (see Lazarian & Prunet 2002 for a review).

2 How Do Grains Rotate?

Dynamics of grains in molecular clouds is pretty involved (see Fig. 1). First of all, grains rotate. The rotation can arise from chaotic gaseous bombardment of grain surface and be Brownian, or it can arise from systematic torques discovered by Purcell (1975, 1979). The most efficient among those are torques arising from H_2 formation over grain surface. One can visualize those torques imagining a grain with tiny rocket nozzles ejecting nascent high velocity hydrogen molecules (see Fig. 1). Grains are known to enable atoms of hydrogen to form molecules. The reactions are believed to take place over particular catalytic sites on grain surface. Those catalytic sites act as "Purcell rockets". Even when the surroundings of dust grains is mostly molecular, grains can rotate suprathermally, i.e. with kinetic energies much larger that kT_{gas}, due to the variation of the accommodation coefficient. Indeed, if the temperatures of gas and dust are different, those variations allow parts of the grain to bounce back impinging gaseous atoms with different efficiencies. It is easy to understand that this also results in systematic torques. In addition, Purcell (1979) identified electron ejection as yet another process that can drive grain to very large angular velocities.

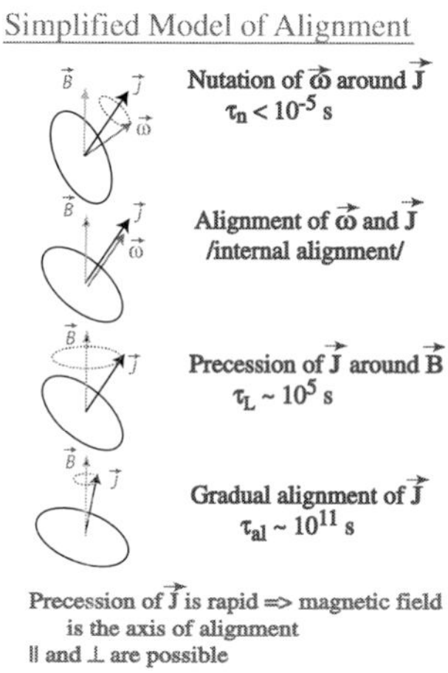

Fig. 1. Grain alignment implies several alignment processes acting simultaneously and spanning many time scales (shown for 10^{-5} cm grain in cold interstellar gas). The rotational dynamics of a grain is rather complex. The internal alignment introduced by Purcell (1979) was thought to be slower than precession until Lazarian & Draine (1999b, henceforth LD99b) showed that it happens 10^6 times faster when relaxation through induced by nuclear spins is accounted for (approximately 10^4 s for the 10^{-5} cm grains)

A very different process of grain spin-up can be found in a very important, but not timely appreciated work by Dolginov & Mytrophanov (1976). These authors considered differential scattering of photons of right and left circular polarization by an irregular dust grain. As the size of the irregularities gets comparable with the wavelength, it is natural that interaction of a grain with photons will depend on the photon polarization. Unpolarized light can be presented as a superposition of equal number of left and right circularly polarized photons. Therefore it is clear that the interaction with photons of a particular polarization would deposit angular momentum to the grain. The authors concluded that for typical diffuse ISM conditions this process should induce grain rotation at suprathermal velocities. However, while Purcell's torques became a textbook stuff, radiative torques had to wait 20 years before they were reintroduced to the field (Draine 1996, Draine & Weingartner 1996, 1997).

The minimal rotational velocity of grain is the velocity of their Brownian motion. This can be characterized by a temperature that is somewhere between that of the gas and the dust. In the case of PAHs or very small grains emitting copious microwave radiation as they rotate (Draine & Lazarian 1998) the effective rotational temperature may be subthermal (see discussion in Lazarian & Yan 2003).

It was realized by Martin (1971) that rotating charged grains will develop magnetic moment and the interaction of this moment with the interstellar magnetic field will result in grain precession. The characteristic time for the precession was found to be comparable with t_{gas}. However, soon a process that renders much larger magnetic moment was discovered (Dolginov & Mytrophanov 1976). This process is the Barnett effect, which is converse of the Einstein-de Haas effect. If in Einstein-de Haas effect a paramagnetic body starts rotating during remagnetizations as its flipping electrons transfer the angular momentum (associated with their spins) to the lattice, in the Barnett effect the rotating body shares its angular momentum with the electron subsystem causing magnetization. The magnetization is directed along the grain angular velocity and the value of the Barnett-induced magnetic moment is $\mu \approx 10^{-19}\omega_{(5)}$ erg gauss^{-1} (where $\omega_{(5)} \equiv \omega/10^5 \mathrm{s}^{-1}$). Therefore the Larmor precession has a period $t_{Lar} \approx 3 \times 10^6 B_{(5)}^{-1}$ s.

Nevertheless, suprathermal, thermal and subthermal grain rotation are just components of complex grain dynamics. As any solid body, interstellar grains can rotate about 3 different body axes. As the result they tumble while rotating. This effect was attracting attention of the early researchers (see Jones & Spitzer 1967) till Purcell (1979) identified internal relaxation within grains as the process that can suppress grain rotation about all axes, but the axis corresponding to the grain maximal moment of inertial (henceforth axis of maximal inertia). Indeed, consider a spheroidal grain, which kinetic energy can be presented as (see Lazarian & Roberge 1997) $E(\theta) = \frac{J^2}{I_{max}}\left(1 + \sin^2\theta(h-1)\right)$, where θ is the angle between the axis of major inertia and grain angular momentum. In the absence of external torques grain angular momentum is preserved. The minimum of grain energy corresponds therefore to $\theta = 0$, or grain rotating exactly about the axis of maximal inertia. As internal dissipation decreases kinetic energy, it sounds natural that $\theta = 0$ is the expected state of grain subjected to fast internal dissipation.

Purcell (1979) introduced a new process of internal dissipation which he termed "Barnett relaxation". This process may be easily understood. We know that a freely rotating grain preserves the direction of **J**, while angular velocity precesses about **J** and in grain body axes. We learned earlier that the Barnett effect results in the magnetization vector parallel to $\boldsymbol{\Omega}$. As a result, the Barnett magnetization will precess in body axes and cause paramagnetic relaxation. The "Barnett equivalent magnetic field", i.e. the equivalent external magnetic field that would cause the same magnetization of the grain material, is $H_{BE} = 5.6 \times 10^{-3}\omega_{(5)}$ G, which is much larger than the interstellar magnetic field. Therefore the Barnett relaxation happens on the scale $t_{Bar} \approx 4 \times 10^7 \omega_{(5)}^{-2}$ sec, i.e. essentially instantly compared to the time that it takes to damp grain rotation for typical molecular cloud conditions.

Even stronger relaxation process has been identified recently by Lazarian & Draine (1999a). They termed it "nuclear relaxation". This is an analog of Barnett relaxation effect that deals with nuclei. Similarly to unpaired electrons nuclei tend to get oriented in a rotating body. However the nuclear analog of "Barnett equivalent" magnetic field is much larger and Lazarian & Draine (1999a) concluded that the nuclear relaxation can be a million times faster than the Barnett relaxation.

Why would the actual relaxation rate matter? The rate of internal relaxation couples grain rotational and vibrational degrees of freedom. LD99b showed that this will result in grain "thermal flipping". Such a flipping would average out Purcell's torques and result in grain being "thermally trapped" in spite of the presence of uncompensated torques. Whether grain gets "thermally trapped" depends on its size (with the grains less than a critical size a_c rotating thermally). While Barnett and inelastic relaxation (see also Lazarian & Efroimsky 1999) results in a_c equal or less than 10^{-5} cm, the nuclear internal relaxation provides $a_c \sim 10^{-4}$ cm. This means that most grains rotate thermally in the presence of Purcell's torques. The exception to this thermallization are radiative torques that are not fixed in grain coordinates. Such torques can spin-up dust in spite of thermal flipping.

3 What Does Align Grains?

Paramagnetic Alignment

Davis-Greenstein (1951) mechanism (henceforth D-G mechanism) is based on the paramagnetic dissipation that is experienced by a rotating grain. Paramagnetic materials contain unpaired electrons which get oriented by the interstellar magnetic field **B**. The orientation of spins causes grain magnetization and the latter varies as the vector of magnetization rotates in grain body coordinates. This causes paramagnetic loses at the expense of grain rotation energy. Note, that if the grain rotational velocity $\boldsymbol{\Omega}$ is parallel to **B**, the grain magnetization does not change with time and therefore no dissipation takes place. Thus the paramagnetic dissipation acts to decrease the component of $\boldsymbol{\Omega}$ perpendicular to **B** and one may expect that eventually grains will tend to rotate with $\boldsymbol{\Omega}\|\mathbf{B}$ provided that the time of relaxation t_{D-G} is much shorter than t_{gas}, the time of randomization through chaotic gaseous bombardment. In practice, the last condition is difficult to satisfy. For 10^{-5} cm grains in the diffuse interstellar medium t_{D-G} is of the order of $7\times 10^{13}a_{(-5)}^2 B_{(5)}^{-2}$s , while t_{gas} is $3\times 10^{12} n_{(20)} T_{(2)}^{-1/2} a_{(-5)}$ s (see table 2 in Lazarian & Draine 1997) if magnetic field is 5×10^{-6} G and temperature and density of gas are 100 K and 20 cm^{-3}, respectively. However, at the time when it was introduced ,in view of uncertainties in interstellar parameters, the D-G mechanism looked plausible.

The first detailed analytical treatment of the problem of D-G alignment was given by Jones & Spitzer (1967) who described the alignment of **J** using a Fokker-Planck equation. This approach allowed them to account for magnetization fluctuations within grain material and thus provided a more accurate picture of **J** alignment. The first numerical treatment of D-G alignment was presented by Purcell (1969). By that time it became clear that the D-G mechanism is too weak to explain the observed grain alignment. However, Jones & Spitzer (1967) noticed that if interstellar grains contain superparamagnetic, ferro- or ferrimagnetic (henceforth SFM) inclusions[2], the t_{D-G} may be reduced by orders of magnitude. Since 10% of atoms in interstellar dust are iron the formation of magnetic clusters in grains was not far fetched (see Martin 1995). However, detailed calculations in Lazarian (1997), Roberge & Lazarian (1999) showed that the alignment achievable cannot account for observed polarization coming from molecular clouds provided that dust grains rotate thermally. This is the consequence of thermal fluctuations within grain material. These internal magnetic fluctuations randomize grains orientation in respect to magnetic field if grain body temperature is close to the rotational temperature.

Purcell (1979) pointed out that fast rotating grains are immune to both gaseous and internal magnetic randomization. Thermal trapping limits the range of grain sizes for which Purcell's torques can be efficient (Lazarian & Draine 1999ab). Grains with the sizes larger than the wavelength size can be spun up by the incoming starlight, however (see Draine & Weingartner 1996).

[2] The evidence for such inclusions was found much later through the study of interstellar dust particles captured in the atmosphere (Bradley 1994).

Mechanical Alignment

Gold (1951) mechanism is a process of mechanical alignment of grains. Consider a needle-like grain interacting with a stream of atoms. Assuming that collisions are inelastic, it is easy to see that every bombarding atom deposits angular momentum $\delta \mathbf{J} = m_{atom}\mathbf{r} \times \mathbf{v}_{atom}$ with the grain, which is directed perpendicular to both the needle axis $\mathbf{r}$ and the velocity of atoms $\mathbf{v}_{atom}$. It is obvious that the resulting grain angular momenta will be in the plane perpendicular to the direction of the stream. It is also easy to see that this type of alignment will be efficient only if the flow is supersonic[3]. Thus the main issue with the Gold mechanism is to provide supersonic drift of gas and grains. Gold originally proposed collisions between clouds as the means of enabling this drift, but later papers (Davis 1955) showed that the process could only align grains over limited patches of interstellar space, and thus the process cannot account for the ubiquitous grain alignment in diffuse medium.

Suprathermal rotation introduced in Purcell (1979) persuaded researchers that mechanical alignment is marginal. Indeed, fast rotation makes it difficult for gaseous bombardment to align grains. However, two new developments must be kept in mind. First of all, a number of papers proved that mechanical alignment of suprathermally rotating grains is possible (Lazarian 1995, Lazarian & Efroimsky 1996, Efroimsky 2002). Moreover, recent work on grain dynamics (Lazarian & Yan 2002, Yan & Lazarian 2003) proved that MHD turbulence can render grains with supersonic velocities. While we do not believe that mechanical alignment is the dominant process, it should be kept in mind in analyzing observations (see Rao et al. 1998).

Alignment via Radiative Torques

Anisotropic starlight radiation can both spin the grains and align them. This was first realized by Dolginov & Mytrophanov (1976), but this work came before its time. The researchers did not have reliable means to study dynamics of grains and the impact of their work was marginal. Before Bruce Draine realized that the torques can be treated with the versatile discrete dipole approximation (DDA) code (Draine & Flatau 1994) the radiative torque alignment was very speculative. For instance, earlier on difficulties associated with the analytical approach to the problem were discussed in Lazarian (1995a). However, very soon after that Draine (1996) modified the DDA code to calculate the torques acting on grains of arbitrary shape. His work revolutionized the field! The magnitude of torques were found to be substantial and present for grains of various irregular shape (Draine 1996, Draine & Weingartner 1996). After that it became impossible to ignore these torques.

[3] Otherwise grains will see atoms coming not from one direction, but from a wide cone of directions (see Lazarian 1997a) and the efficiency of alignment will decrease.

One of the problem of the earlier treatment was that in the presence of anisotropic radiation the torques will change as the grain aligns and this may result in a spin-down. Moreover, anisotropic flux of radiation will deposit angular momentum which is likely to overwhelm rather weak paramagnetic torques. These sort of questions were addressed by Draine & Weingartner (1997) and it was found that for most of the tried grain shapes the torques tend to align **J** along magnetic field. The reason for that is yet unclear and some caution is needed as the existing treatment ignores the dynamics of crossovers which is very important for the alignment of suprathermally rotating grains. A recent work by Weingartner & Draine (2003) treats flipping of grains in the presence of monochromatic radiation.

3.1 What Is Future Work?

Observational testing of alignment is extremely important. Both the dependences of the polarization degree versus wavelength that follow Serkowski law (Serkowski 1973) and studies of changes of polarization degree with the wavelength done in Far Infrared (see Hildebrand 2000) are consistent with theoretical predictions (see discussion in Lazarian, Goodman & Myers 1997). According to Lazarian (2003) the study of grain alignment at the diffuse/dense cloud interface by Whittet et al. (2001) is suggestive that grains there are being aligned by radiative torques.

Radiative torques look as the most attractive mechanism to align grains in molecular clouds. However more theoretical work is required to understand why grains subjected to anisotropic radiation get preferentially aligned with their long axes perpendicular to magnetic field.

Acknowledgments. I thank H. Yan for suggestions and help. The research is supported by the NASA grant 0830 300 N665 736.

References

1. Bradley, J.P. 1994, Sicence, 265, 925
2. Davis, L. 1955, Vistas in Astronomy, ed. A. Beer, 1, 336–
3. Davis, L. & Greenstein, J.L., 1951, ApJ, 114, 206–240
4. Dolginov A.Z. & Mytrophanov, I.G. 1976, Ap&SS, 43, 291–317
5. Draine, B.T. 1996, in Polarimetry of the Interstellar Medium, eds Roberge W.G. and Whittet, D.C.B., A.S.P. 97. 16–25
6. Draine, B.T., & Flatau, P.J. 1994, J. Opt. Soc. Am. A., 11, 1491
7. Draine, B.T. & Lazarian A. 1998a, ApJ, 494, L19–L22
8. Draine, B.T. & Lazarian A. 1998b, ApJ, 508, 157–179
9. Draine, B.T. & Lazarian A. 1999, ApJ, 512, 740–754
10. Draine, B.T. & Weingartner, J.C. 1996, ApJ, 470, 551–565
11. Draine, B.T. & Weingartner, J.C 1997, ApJ, 480, 633–646
12. Efroimsky, M. 2002, ApJ, 575, 886–899
13. Gold, T. 1951, Nature, 169, 322–323

14. Hall, J.S. 1949, Science, 109, 166–168
15. Hildebrand, R.H. 2000, PASP, 112, 1215–1235
16. Hiltner, W.A. 1949, ApJ, 109, 471
17. Jones, R.V., & Spitzer, L., Jr, 1967, ApJ, 147, 943–964
18. Lazarian, A. 1995, ApJ, , 453, 229–237
19. Lazarian, A. 1997a, ApJ, 483, 296–308
20. Lazarian, A. 1997b, MNRAS, 288, 609–617
21. Lazarian, A. 2000, in "Cosmic Evolution and Galaxy Formation", ASP v. 215, eds. Jose Franco, Elena Terlevich, Omar Lopez-Cruz, p. 69–79, astro-ph/0003314
22. Lazarian, A. 2003, Journal of Quantitative Spectroscopy and Radiative Transfer, 79, 881
23. Lazarian, A., Goodman, A.A. & Myers, P.C. 1997, ApJ, 490, 273–280
24. Lazarian, A., & Efroimsky, M. 1996, ApJ, 466, 274–281
25. Lazarian, A., & Efroimsky, M. 1999, MNRAS, 303, 673-684
26. Lazarian, A., & Draine, B.T., 1997, ApJ, 487, 248–258
27. Lazarian, A., & Draine, B.T., 1999a, ApJ, 520, L67–L70
28. Lazarian, A., & Draine, B.T., 1999b, ApJ, 516, L37–L40
29. Lazarian, A., & Prunet, S. 2002, in proc. of the AIP conf. "Astrophysical Polarized Backgrouds", eds S. Cecchini, S. Cortiglioni, R. Sault, and C. Sbarra, pp. 32–44
30. Lazarian, A., Yan, H. 2002, ApJ, 566, L105–L108
31. Lazarian, A. & Yan, H. 2003, to appear in the proceedings of Astrophysics Dust, astro-ph
32. Lazarian, A., & Roberge, W.G., 1997a ApJ, 484, 230–237
33. Lazarian, A., & Roberge, W.G., 1997b, MNRAS, 287, 941–946
34. Martin, P.G. 1971, MNRAS, 153, 279–286
35. Martin, P.G. 1971, 1995, ApJ, 445, L63–L66
36. Purcell, E.M. 1969, On the Alignment of Interstellar Dust, Physica, 41, 100–127
37. Purcell, E.M. 1975, in Dusty Universe, eds. G.B. Field & A.G.W. Cameron, New York, Neal Watson, p. 155–165
38. Purcell, E.M. 1979, ApJ, 231, 404–416
39. Purcell, E.M., & Spitzer, L., Jr 1971, ApJ, 167, 31–62
40. Rao, R, Crutcher, R.M., Plambeck, R.L., Wright, M.C.H. 1998, ApJ, 502, L75–L79
41. Roberge, W.G. 1996 in Polarimetry of the Interstellar Medium, eds, Roberge W.G. and Whittet, D.C.B., A.S.P. Vol. 97, pp. 401–416
42. Roberge, W.G., & Lazarian, A. 1999, MNRAS, 305, 615–630
43. Serkowski, K. 1973, in IAU Symp. 52, Interstellar Dust and Related Topics, ed. J.M. Greenberg & H.C. van de Hulst (Dordrecht: Kluwer), 145–160
44. Whittet, D.C.B., Gerakines, P.A., Hough, J.H. & Snenoy 2001, ApJ, 547, 872–884
45. Weingartner, J.C., & Draine, B.T. 2003, ApJ, 589, 289
46. Yan, H. & Lazarian, A. 2003a, ApJ, 592, 33L

Part VIII

Physical Conditions and Processes in the ISM

Turbulent Structure of the Interstellar Medium

M.-M. Mac Low

American Museum of Natural History, 79th Street and Central Park West, New York, NY, 10024-5192, USA `mordecai@amnh.org`

Summary. How does turbulence contribute to the formation and structure of the dense interstellar medium (ISM)? Molecular clouds are dense, high-pressure objects. It is usually argued that gravitational confinement causes the high pressures, and that the clouds are (magneto)hydrostatic objects supported by a balance between magnetic and turbulent pressures and gravity. However, magnetic pressures appear too weak, and MHD turbulent support not only requires driving, but also results in continuing gravitational collapse, as has now been demonstrated in simulations reaching 512^3 zones. Models of supernova-driven, magnetized turbulence readily form transient, high-pressure, dense regions that may form molecular clouds. They are contained not by self-gravity, but by turbulent ram pressures from the larger flow. Apparent virialization may actually be a geometrical effect. Turbulent clouds are unlikely to be in hydrostatic equilibrium, instead either collapsing or expanding, although they may appear well-fit by projected equilibrium Bonnor-Ebert spheres. Collapsing clouds probably form stars efficiently, while expanding ones can still form stars by turbulent compression, but rather inefficiently.

1 Equilibrium

The very high Reynolds numbers of the interstellar medium (ISM) ensure that it is turbulent and not laminar. How can we model a turbulent flow that is intrinsically non-steady and chaotic? Most models of the structure of the ISM rely on some notion of equilibrium: (magneto)hydrostatic equilibrium, virial equilibrium, energy equipartition, or at least statistical equilibrium (or stationarity). In this talk I use the structure of the dense ISM as a case study in the problems with these different assumptions. In this case, making such assumptions leads to the conclusion that molecular clouds are gravitationally bound, which may not generally be the case.

2 Standard Argument for Gravitationally Bound Clouds

The standard argument assumes that observed clouds are in virial equilibrium [27, 26, 24]. This was justified by apparent lifetimes estimated at 30 Myr by Blitz & Shu [9], which is many times the free-fall time $\tau_{\rm ff}$, and the seeming consistency of virial and actual mass given size-mass and size-linewidth relations. Cloud parameters can then be estimated from the assumption of virial equilibrium, and (magneto)hydrostatic models constructed by detailed balance of gravity against turbulent and magnetic pressure support.

The full Eulerian virial theorem [25] for a volume of space is a time-dependent equation for the moment of inertia of the mass in the volume

$$(1/2)\ddot{I} = 2(\mathcal{E}_{th} + \mathcal{E}_{kin} - \mathcal{T}_{th} - \mathcal{T}_{kin}) + \mathcal{M} + \mathcal{W} - (1/2)\dot{\Psi}, \tag{1}$$

where $\mathcal{E}_{th}$, $\mathcal{E}_{kin}$, $\mathcal{M}$, and $\mathcal{W}$ are thermal, kinetic, magnetic and gravitational energy, $\mathcal{T}_{th}$ and $\mathcal{T}_{kin}$ are thermal and ram pressure on the surface of the volume, and $\dot{\Psi}$ is the change in flux of the moment of inertia through the region.

This equation is usually simplified by neglecting the time-dependent and surface terms, and assuming clouds with homogeneous density and pressure. Below I will address how well justified these assumptions are, but if they are made, the equation can be recast to give external pressure [24]

$$P_{ext} = \frac{1}{4\pi}\left(-\alpha\frac{GM^2}{R^4} + \beta\frac{\Phi^2}{R^4} + 3\frac{c_s^2 M}{R^3} + \frac{\sigma^2 M}{R^3}\right), \tag{2}$$

where M, R, σ, and Φ are the mass, radius, velocity dispersion, and enclosed magnetic flux of a cloud, G is the gravitational constant, c_s is the sound speed, and α and β are geometrical constants of order unity. If gravity and magnetic pressure are ignored, we recover the generalized Boyle's law $P_{ext}V = (c_s^2 + \sigma^2)M$. If, on the other hand, external, thermal, and magnetic pressure are all ignored, the equilibrium is reduced to a balance between gravity and turbulence:

$$\alpha GM/R = \sigma^2 \Rightarrow M_{vir} = \sigma^2 R/(\alpha G), \tag{3}$$

which gives the usual definition for the "virial mass" M_{vir}. This is certainly appropriate for isolated stellar clusters, where all the other terms are indeed negligible, but may be rather more problematic for gas clouds embedded in a turbulent flow.

Clouds will rarely or never have exact equality between gravity and internal pressure (thermal, magnetic, and turbulent). McKee [24] writes the *time-averaged* virial theorem in terms of the mean internal pressure $\bar{P} = P_{ext} - \mathcal{W}(1 - \mathcal{M}/|\mathcal{W}|)/3V$. The total energy of the cloud (noting that $\mathcal{W} < 0$) is [24]

$$E = (3V/2)[P_{ext} + \mathcal{W}(1 - \mathcal{M}/|\mathcal{W}|)/3V]. \tag{4}$$

If magnetic fields are unimportant, the cloud is gravitationally bound if $P_{ext}V < -\mathcal{W}/3$, that is if the mean weight in virial equilibrium exceeds the surface pressure.

If we assume the ISM to be in pressure equilibrium, then we can take the ISM pressure to be the external pressure. Boulares & Cox [10] argue for an ISM pressure, neglecting a uniform cosmic ray component, of 1.8×10^4 K cm^{-3}, while Lallement (1998) suggests 5200–6100 K cm^{-3}. The gravitational energy of a cloud can be expressed in terms of its observable mean surface density Σ or average hydrogen column density $\bar{N}_H$ as [24]

$$-(\mathcal{W}/3V) = (3\pi a/20)G\Sigma \simeq (1.39 \times 10^5 \text{ K cm}^{-3})(\bar{N}_H/10^{22} \text{ cm}^{-2}), \quad (5)$$

where a is a geometrical factor of order unity [8]. For the local dust-to-gas ratio, $\bar{N}_H = 10^{22}$ cm^{-2} corresponds to a visual extinction of $A_V = 7.5$ mag. Since molecular clouds typically have $A_V > 2$, $P_{ext} \ll -\mathcal{W}/3V$ and so, under the assumptions given, they are gravitationally bound.

3 Validity of Standard Assumptions

Are the assumptions that go into the standard argument still well-supported? Let's begin with the assumption that molecular clouds live for many free-fall times, and therefore have time to virialize. Blitz & Shu [9] estimated that cloud lifetimes were around 30 Myr in the Milky Way based on three main arguments: the ages of stars thought to be associated with the clouds; cloud locations downstream from dust lanes thought to be associated with cloud-forming spiral-arm shocks; and the comparison between the total molecular mass and the star-formation rate in the Galaxy. In the last several years, much shorter lifetimes have been proposed by Fukui et al. [13] based on the ages of stellar clusters associated with molecular gas; and by Ballesteros-Paredes et al. [3] based on the lack of a population of post-T Tauri stars with ages > 10 Myr closely associated with molecular gas. If molecular cloud lifetimes are only 2–3 $\tau_{\rm ff}$, they may not live long enough to reach virial equilibrium.

Larson [21] found that the size of observed molecular clouds appears to correlate with the linewidth $R \propto \sigma^2$ and with the mean density of the cloud $R \propto \rho^{-1}$. These relations can be interpreted as evidence of virial equilibrium using the definition of M_{vir} given in equation (3), by noting that if they hold,

$$M_{vir} \propto R\sigma^2 \propto \rho R^3, \quad (6)$$

so the virial mass equals the actual mass of the cloud.

Ballesteros-Paredes et al. [6] pointed out that this actually only guarantees that kinetic and gravitational energy are equal, but not that full virial equilibrium holds, because of the extra terms neglected in equation (3). Even worse, both Vázquez-Semadeni et al. [29] and Ballesteros-Paredes & Mac Low [5] found that simulated observations of turbulent models appear to reproduce the density-size relationship, but that the actual density distribution did *not*. Figure 1 demonstrates this. Is even the energy equipartition argument merely an observational artifact?

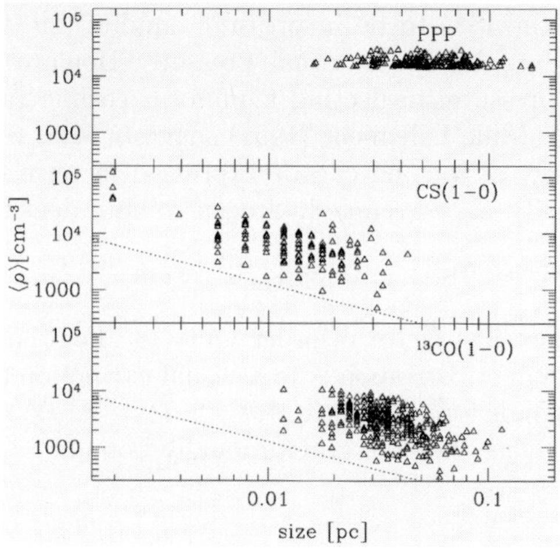

Fig. 1. Mean density-size relationship for clumps in a turbulent model measured in physical space (top) and in simulated observational coordinates for two different species (middle and bottom). The dotted line has a slope of = -1. In physical space we find no correlation, verifying the results by VBR [29], but nevertheless the simulated observations show such a correlation, as found by Larson (1981) and many others. The correlation in observational space may simply be the effect of the limited dynamic range of observations giving an effective column density cutoff.

On the other hand, the size-linewidth relationship appears in turbulent models[1] even in the complete *absence* of self-gravity [28]. This can be understood as a direct consequence of the steep velocity power spectrum expected in turbulence, with the index of the power spectrum of v^2 between -5/3 (for incompressible turbulence) and -2 (for highly compressible turbulence). Larger objects have greater velocity differences, and thus larger velocity dispersion. However, this offers no support for the hypothesis that observed velocity dispersion are primarily due to self-gravity bringing clouds to a simple equilbrium between kinetic and gravitational energy.

In summary, the approximation of virial equilibrium for molecular clouds, especially in its simplest expression of equipartition, may be misleading. Their apparent short lifetimes suggest that the transient surface terms and distortions cannot be averaged away, and may dominate the dynamics of the cloud. Ballesteros-Paredes [6] measured the transient terms in turbulent simulations, and indeed found them to be substantial.

[1] On a related subject, Brunt & Mac Low [11] examined the dispersion of velocity centroids in real and simulated observations. They found that density inhomogeneity in hypersonic turbulence cancels projection smoothing, so that observed two-dimensional velocity centroid scaling actually gives the correct three-dimensional result, by lucky coincidence. This is not true in trans- or sub-sonic turbulence.

4 Clouds in Supernova-Driven Turbulence

In order to study the structure of density enhancements in a turbulent flow more carefully, we turn to a numerical model of supernova-driven turbulence [1, 2], done with an adaptive mesh refinement code and including: heating initially balancing cooling using an equilibrium ionization cooling curve; the galactic gravitational potential (but not self-gravity); and supernovae occurring randomly at the galactic rate. In this simulation, large density enhancements form and dissipate due to the turbulent compression and cooling, without the participation of self-gravity. The left side of Figure 2 shows that pressures in some dense, cold regions reach values more than an order of magnitude higher than the average, as observed in molecular clouds. The pressure probability density function on the right side of the Figure also shows the broad distribution of pressures. The cold, high-density, high-pressure regions are not fully resolved in these models, but they have sizes, shapes, and masses consistent with giant molecular clouds. They *must* be transient, probably dispersing on an internal crossing time [30], as they cannot be in equilibrium in the absence of self-gravity.

If molecular clouds are generally transient, ram-pressure confined objects, then why do CO luminosity, dust extinction, and other observational measures of cloud mass correlate so well with apparent virial mass (see Walter in these proceedings for example)? As we noted above, a turbulent flow produces a size-linewidth relation $R \propto \sigma^2$. The usual expression for the virial mass, equation (3), can thus be rewritten as $M \propto R^2$. Any collection of

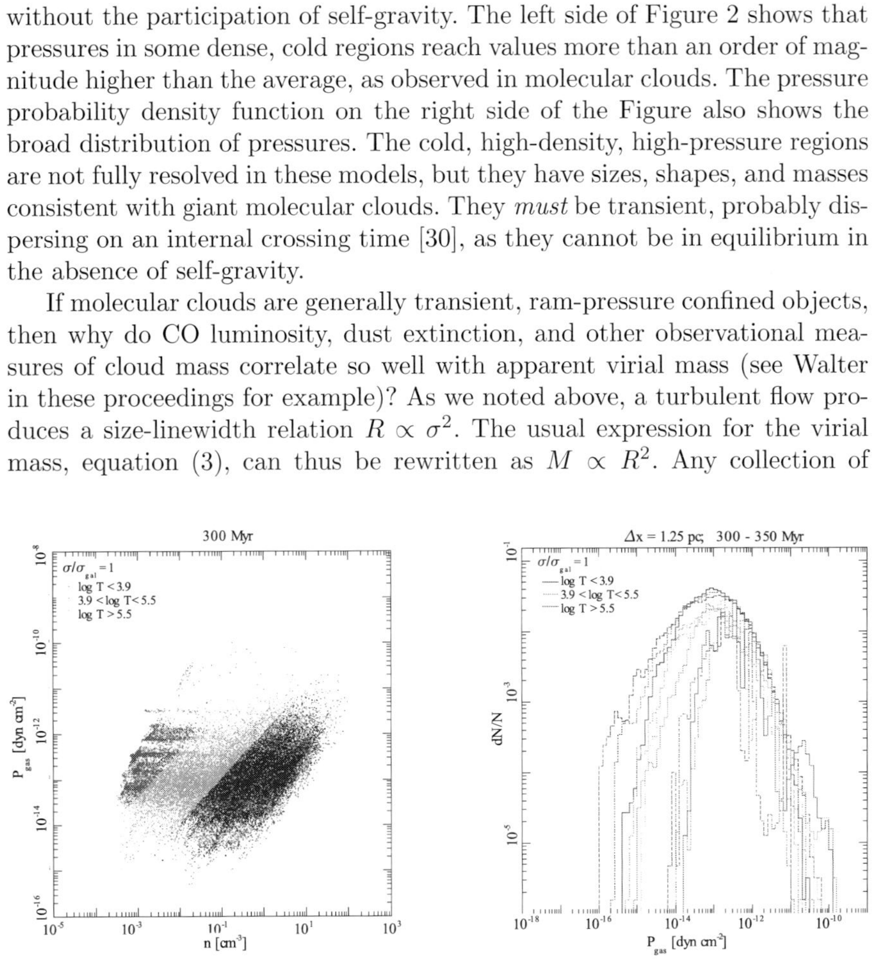

Fig. 2. *Left:* Scatter plot of pressure vs. density in the midplane of the SN-driven model showing occupation of the cold, high-pressure, high-density region associated with molecular clouds. *Right:* Volume-weighted probability distribution function of pressure in the same model. Note that the small volumes occupied by high-density, cold gas have large mass. Different lines show times separated by 25 Myr.

clouds whose mass is proportional to the square rather than the cube of their typical size will thus appear to be "virialized". Ram pressure confined objects will usually be sheetlike, rather than spheres, and will thus naturally have $M \propto R^2$ rather than $M \propto R^3$. The filamentary shapes of observed clouds is also consistent with this idea. Apparent virialization of clouds may actually just be a geometrical effect produced by a compressible turbulent flow.

5 Hydrostatic Equilibrium

The assumption of hydrostatic equilibrium between a turbulent pressure and gravity can also be misleading, because ram pressure is not isotropic. Numerical models of self-gravitating, turbulent gas reveal unexpected behavior. In the absence of continuing energy input, turbulence decays quickly [23], and collapse proceeds without substantial delay [17], whether or not fields are present [7] (so long as they cannot support magnetostatically).

Even driven turbulence strong enough that the turbulent pressure satisfies the Jeans criterion for stability against gravitational collapse does not completely prevent local collapse [19]. Adding weak magnetic fields reduces the amount of collapse, but does not prevent it entirely [14], a result that has now been confirmed at resolutions ranging from 64^3 zones to 512^3 zones by Li et al. [22], as shown in Figure 3. Collapse occurs if the mass in a region exceeds the *local* Jeans mass $m_{J,T} \propto v^3\rho^{-1/2}$. Compressible turbulence produces density enhancements with $\Delta\rho/\rho \propto v^2/c_s^2$, so even if a region is *globally* supported by supersonic turbulence, *local* regions may still become gravitationally unstable [19]. These results suggest that most observed cores are dynamically collapsing.

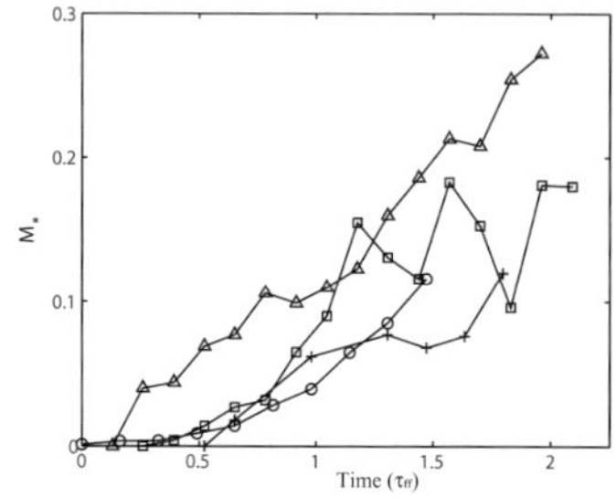

Fig. 3. Comparison of the mass accretion behavior for runs driven at $k = 1$–2 with varying resolution 64^3 (triangle), 128^3 (square), 256^3 (plus), and 512^3 (circle) [14, 22]. M_* denotes the sum of masses found in all cores, in units of box mass, determined by the modified CLUMPFIND [31]. Collapse rates vary, but collapse occurs in all cases, with the rate of collapse generally converging in the higher resolution models.

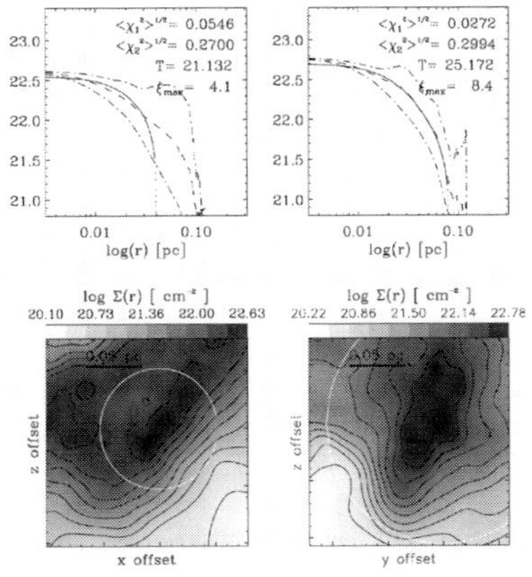

Fig. 4. Column density maps and angularly averaged radial profiles (dashed line) for xy and xz projections of a sample clump from an SPH simulation (limits of radial profiles are shown by dash-dot lines). The white circles show in each case the size of the radius used in the Bonnor-Ebert fit. Note the different morphology that the same core shows in the different projections [4].

Observations that appear to suggest cores in hydrostatic equilibrium also must be interpreted with great care. More than 50% of the cores that appear in a simulated observation of a numerical model of supersonic, hydrodynamic turbulence *without* self-gravity can be fit well by a projected Bonnor-Ebert sphere [4], as shown in Figure 4. Gravity clearly does dominate some regions in molecular clouds. However, gravitational collapse seems to proceed efficiently there, probably leading to more than half the mass forming stars [18, 19].

Large-scale gravitational instability forms regions with masses of order the thermal Jeans mass of the diffuse ISM $M_J > 10^7 M_\odot$ [15, 16]. These instabilities do not directly form single molecular clouds, but they may form regions of active star formation [12] in which giant molecular clouds form incidentally during gravitational collapse and fragmentation of the larger region.

6 Conclusions, or Questions

Let me conclude by asking the questions that I think most important. Are most or all observed clouds out of equilibrium? Are they either ram pressure confined and transient, *or* gravitationally collapsing on a free-fall time? The observational distinction between these two states may be whether associated star formation is occurring on a scale much longer than or close to the free=fall time. Is apparently bimodal star formation actually a continuum from low to high turbulent support? Finally, do clouds appear to be in virial equilibrium with $M \propto R\sigma^2$ because they are actually sheet-like objects in a turbulent flow with $M \propto R^2$ and $R \propto \sigma^2$?

Acknowledgments. I thank the organizers for partial support of my attendance. My research is partly supported by NSF grants AST99-85392, and AST03-07793, and NASA grant NAG5-10103. This work has made use of the NASA ADS Abstract Service.

References

1. M. A. Avillez: MNRAS, **315**, 479 (2000)
2. M. A. Avillez, M.-M. Mac Low: Astrophys. J. **581**, 1047 (2002)
3. J. Ballesteros-Paredes, L. Hartmann, E. Vázquez-Semadeni: Astrophys. J. **527**, 285 (1999)
4. J. Ballesteros-Paredes, R. S. Klessen, E. Vázquez-Semadeni: Astrophys. J. **592**, 188 (2003)
5. J. Ballesteros-Paredes, M.-M. Mac Low: Astrophys. J. **570**, 734 (2002)
6. J. Ballesteros-Paredes, E. Vázquez-Semadeni, J. Scalo: Astrophys. J. **515**, 286 (1999)
7. D. S. Balsara, R. M. Crutcher, A. Pouquet: Astrophys. J. **557**, 451 (2001)
8. F. Bertoldi, C. F. McKee: Astrophys. J. **395**, 140 (1992)
9. L. Blitz, F. H. Shu: Astrophys. J. **238**, 148 (1980)
10. A. Boulares, D. P. Cox: Astrophys. J. **365**, 544 (1990)
11. C. M. Brunt, M.-M. Mac Low: Astrophys. J., in revision (2003)
12. B. G. Elmegreen, Y. N. Efremov: Astrophys. J. **466**, 802 (1996)
13. Y. Fukui et al: Publ. Astron. Soc. Japan **51**, 745 (1999)
14. F. Heitsch, M.-M. Mac Low, R. S. Klessen: Astrophys. J. **547**, 280 (2001)
15. W.-T. Kim, E. C. Ostriker: Astrophys. J. **559**, 70 (2001)
16. W.-T. Kim, E. C. Ostriker, J. M. Stone: Astrophys. J. **581**, 1080 (2002)
17. R. S. Klessen, A. Burkert: Astrophys. J. Suppl. **128**, 287 (2000)
18. R. S. Klessen, A. Burkert: Astrophys. J. **549**, 386 (2001)
19. R. S. Klessen, F. Heitsch. M.-M. Mac Low: Astrophys. J. **535**, 887 (2000)
20. R. Lallement: Observations of the Local Interstellar Cloud. In: *The Local Bubble and Beyond*, ed. by D. Breitschwerdt, M. J. Freyberg, J. Trümper (Springer, Berlin 1998) pp 19–28
21. R. B. Larson: MNRAS **194**, 809 (1981)
22. P. Li, M. L. Norman, M.-M. Mac Low, F. Heitsch: Astrophys. J. in revision (2004)
23. M.-M. Mac Low: Astrophys. J. **524**, 169 (1999)
24. C. F. McKee: The Dynamical Structure and Evolution of Giant Molecular Clouds. In: *The Origin of Stars and Planetary Systems*, ed by C. J. Lada, N. D. Kylafis (Kluwer, Dordrecht 1999) pp 29–66
25. C. F. McKee, E. G. Zweibel: Astrophys. J. **399**, 551 (1992)
26. W. H. McCrea: MNRAS **117**, 562 (1957)
27. L. Mestel, L. Spitzer Jr: MNRAS **116**, 503 (1956)
28. V. Ossenkopf, M.-M. Mac Low: Astron. Astrophys. **390**, 307 (2002)
29. E. Vázquez-Semadeni, J. Ballesteros-Paredes, L. F. Rodríguez: Astrophys. J. **474**, 292 (1997)
30. E. Vázquez-Semadeni, J. Ballesteros-Paredes, R. S. Klessen: Astrophys. J. (Lett.) **585**, L131-L134 (2003)
31. J. P. Williams, E. J. de Geus, L. Blitz: Astrophys. J. **428**, 693 (1994)

Comparing Models and Observations of Turbulent Clouds

V. Ossenkopf

SRON National Institute for Space Research, P.O. Box 800, 9700 AV Groningen, the Netherlands and
1. Physikalisches Institut der Universität zu Köln, Zülpicher Straße 77, 50937 Köln, Germany ossk@ph1.uni-koeln.de

Summary. It is impossible to deduce the properties of interstellar clouds directly from astronomical observations. Statistical methods are needed to quantify the observational results and to compare them with turbulent cloud models. Different model approaches aiming at an understanding of the cloud physics are introduced. From the large variety of methods designed to quantify different properties of interstellar turbulence and to perform the comparison between models and observations three examples were selected to demonstrate that none of them provides a reliable measure for a particular physical behaviour in all circumstances but that all they have some clear distinctive power to compare models and observations.

1 Observed Molecular Cloud Structure

Observations of molecular clouds show a large variety of complex structures often described as filamentary, irregular or fractal. The confusion in these attributes already shows that there is no agreement on a unique description of the structure neither on the quantification of the structural properties. Thus reproducible and robust methods to characterize complex structures are essential as a first step for the understanding of interstellar cloud physics.

One of the properties often recognized in maps of different tracers of the interstellar medium is their self-similarity, i.e. scale invariance with respect to the actual resolution of the observations and the map size. This can be expressed in terms of a fractal dimension but the actual determination of the fractal dimension for maps of noisy data is not very reliable [12] so that more robust methods are needed. A useful approach is the determination of the power spectrum of observed maps. The scale-invariant size range is then expressed by a power law power spectrum: $P(k) \propto k^{-\beta}$

For rectangular maps, the power spectrum can be easily computed by means of a fast fourier transform (FFT) but problems arise when the measured map contains some invalid data or has a geometry which is adapted to the source an not to the needs of the FFT. Other methods have to be used then.

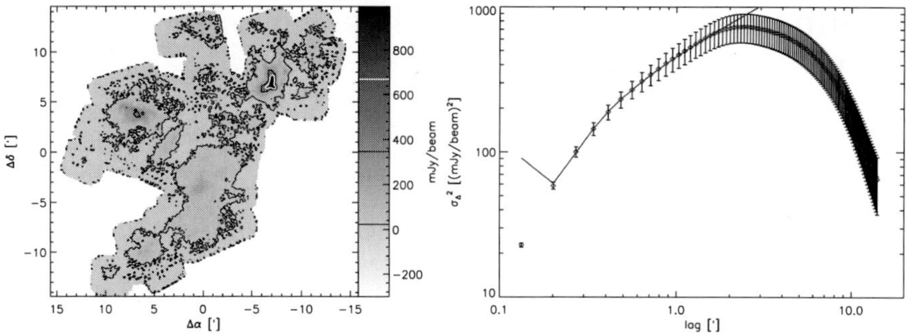

Fig. 1. 1.2 mm dust continuum map of ρ-Oph by Motte et al. [11] and the corresponding Δ-variance spectrum. The solid line shows the best fit with a self-similar scaling taking observational noise and the finite telescope beam into account.

An alternative method is the Δ-variance analysis [20, 1, 15] measuring the relative amount of structure on a given scale by filtering an observed map by a radially symmetric wavelet with a characteristic length scale l and computing the total variance in the convolved map. The spectrum of these Δ-variance values as a function of the filter size l provide a measure for the scaling behaviour of the different components contributing to an observed structure. Self-similar configurations are characterized by a power-law Δ-variance spectrum. Within the typical range of spectral indices measured in interstellar clouds the exponent of the Δ-variance spectrum can be related to the exponent of the power spectrum by $\alpha = \beta - 2$. Compared to the power spectrum the Δ-variance analysis has the big advantage that it can be performed as well for irregular maps or observations with a variable noise across the map. Fig. 1 shows such an example where the Δ-variance analysis was applied to the continuum observations of ρ Oph by Motte et al. [11]. On scales not dominated by star formation we find for all molecular clouds a significant self-similar range with $\beta = 2.5 \ldots 3.4$ [1] but here one can also identify the dominant scales in the spectrum given by the star-forming cores, as measured by Motte et al. [11].

2 Cloud Modelling

To learn from the observed cloud structure one has to create physical models of the ISM explaining all observed properties. These models should give answers to the questions:

- How were the complex cloud structures produced?
- How does the interstellar medium evolve?

2.1 Fractal Models

A first class of models is based on the phenomenological description of interstellar clouds as self-similar structures. A three-dimensional fractal is used for the density and velocity structure. Fractional Brownian motion (fBm) structures are frequently used [1, 2, 10]. They are defined by a power law power spectrum with index β and random phases and provide a perfect match to the spatial scale invariance of the observed molecular cloud maps [20]. They may represent fully developed turbulence with arbitrarily large Reynolds numbers.

Unfortunately, they also bear two severe problems. First it is not clear how one can actually learn anything on the physical behaviour of interstellar clouds from them due to their mathematical nature. Moreover, they always show a Gaussian density distribution in contrast to interstellar clouds which cannot have negative densities and show rather a log-normal distribution. Simple exponentiation would destroy their power-law scaling and the definition of a zero-cut is always arbitrary.

2.2 Hydrodynamic and Magnetohydrodynamic Models

The alternative, physically justified approach to the interstellar cloud modelling is represented by hydrodynamic or magnetohydrodynamic turbulence simulations. With the recent progress in computing power and modelling technique they start to include effects like self-gravity, various MHD modes, ambipolar diffusion, local heating and cooling, and various dissipation processes over a range of scales from the size of giant molecular clouds down to the size of protostellar cores. (see Mac Low, this volume).

With the possibility to switch different physical processes on and off they are an ideal tool to study the interplay of the processes in the structure formation. It turns out that different processes reveal themselves by different slopes in the Δ-variance spectrum so that the detection of deviations from self-similarity is an essential key to understanding the cloud physics. This may answer the questions how energy is redistributed between the different modes and mechanisms and how molecular cloud turbulence is actually driven.

The models, however, also bear two essential disadvantages. The limited numerical resolution and the unavoidable numerical viscosity limit the turbulence cascade which can be represented today to Reynolds numbers below 10^4, although the interstellar medium is characterized by Reynolds numbers above 10^7. Moreover, the definition of the boundary conditions is always a problem. With either periodic boundaries or a fixed finite box they can only represent infinitely large or very small configurations.

2.3 The Radiative Transfer Problem

With the given density and velocity structure from the models it is still not straight-forward to compute the observable properties of the clouds, like molecular line maps or profiles. Here, the full complexity of the radiative transfer problem enters. Translating densities, velocities, temperatures and abundances into the observable line/continuum intensities requires a considerable effort in computing power or sophisticated approximations.

Systematic investigations by Padoan et al. [17] and Mac Low & Ossenkopf [8, 13] have shown, that there is in general no fixed relation between the column density and any line intensity but that the Δ-variance spectra of maps from low density tracers still provide a good measure for the large-scale scaling behaviour. They show, however, that line maps from fractal models remain self-similar in all transitions in contrast to observations which show an increasing confinement of the emission when going to energetically higher molecular transitions.

3 Methods to Compare Observed and Simulated Data

Because the models do not generate an exact representation of the observed interstellar clouds the comparison of the spatial and the velocity structure between model results and observations has to be based on statistical measures. A relatively large set of independent parameters has been proposed in the literature. Characterising either the intensity distribution, the isotropic scaling behaviour, the structure in velocity space or the anisotropy of the structure, different methods have to be combined to allow a relatively complete comparison between models and observations. For a general overview see [12]. Unfortunately, our current knowledge on the cloud physics is still insufficient to determine a clear relation between a particular statistical measure and a physical mechanism, so that we have to combine many measures today to obtain conclusions on the match between a cloud model and observational data. As it is impossible to give even a rough overview over all methods here, I will introduce only a few examples which are currently discussed.

3.1 Channel Maps

A picture on the combination of the density scaling and the velocity structure can be obtained when analysing the statistical properties of channel maps. Brüll et al. (this volume) have shown that the Δ-variance spectrum of channel maps in the Galactic Ring region can be used to determine the size or distance of gravitationally collapsed cores. They allow to distinguish inter-arm material from structures in the spiral arms and the Galactic Ring.

The statistical relation between neighbouring channel maps can be quantified by means of the velocity channel analysis (VCA) introduced by Lazarian & Pogosyan [6]. For fractal clouds this method allows to deduce the spectral indices of density and velocity structure as demonstrated in Fig. 2. Theory predicts for the fractal model used here, that the channel index changes from

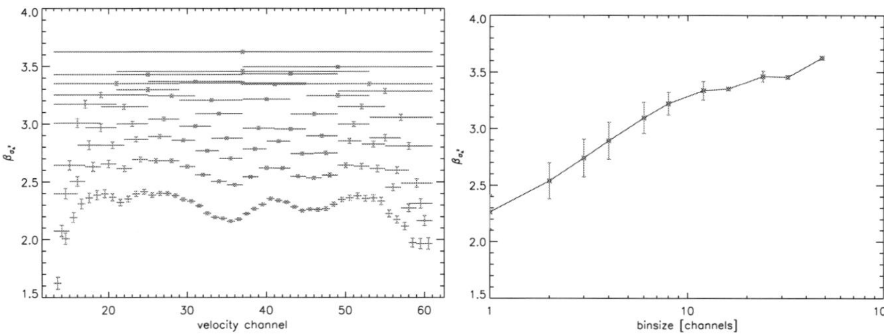

Fig. 2. Spectral indices β_{channel} of channel maps with varying width for a fractal cloud model with $\beta_n = 3.7$ and $\beta_v = 4.0$. The right panel shows the dependence of the average channel index as a function of the channel width.

$\beta_{\text{channel}} = 4.5 - \beta_v/2 = 2.5$ for narrow channels to $\beta_n = 3.7$ for broad channels. Applying the VCA to an HI map of the SMC and assuming a fractal structure Stanimirović & Lazarian [18] were able to derive the spectral indices of the density and velocity structure simultaneously.

Unfortunately, the VCA is only justified if the data represent a completely sampled fractal structure [4]. In contrast, molecular cloud observations often show indications of low-number statistics or intermittency, visible as non-Gaussian average line profiles. Numerical experiments using hydrodynamic simulations with a self-similar behaviour over a long but limited range of scales have proven that the VCA is no longer able to derive the spectral indices there. Thus the applicability of the method has to be tested for every set of observational data.

3.2 Centroid Velocity Maps

Centroids as first velocity moments are much more robust to observational effects than single channel maps so that they are often considered to be a reliable measure for the velocity structure [9]. The Δ-variance spectra of nested centroid maps from CO observations in the Polaris Flare covering three orders of magnitude in spatial scales [15] showed a continuous slope change from $\beta_{\text{centroid}} = 2.8$ at the largest scale to $\beta_{\text{centroid}} = 3.2$ at small scales. From the comparison with the results obtained in MHD computations Ossenkopf & Mac Low [14] concluded that they represent a spectral index of the underlying velocity structure which is larger by one.

In contrast, Miville-Deschênes et al. [10] found that the spectral index measured in centroid velocity maps from fBm models agrees with the spectral index of the used velocity fBm. This contradiction was noticed by Brunt & Mac Low [3], but not fully explained yet. The actual explanation of the problem arises from the variable role of projection smoothing depending on the

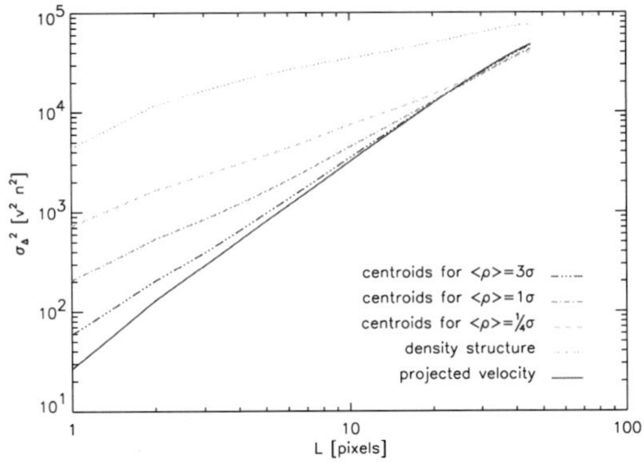

Fig. 3. Δ-variance spectra of centroid velocity maps obtained from the same fBm model when varying the average density. For comparison we have also shown the spectra of the pure density and the projected velocity structure multiplied by the variance of the other quantity to obtain the same units.

ratio between the average density and the density variation. This is demonstrated in Fig. 3 illustrating the role of the required density shift in fBm models. As discovered by Lazarian & Esquivel [7], the centroid map can be either density or velocity dominated. The figure shows that this transition from centroids reflecting the density structure to centroids reflecting the velocity structure is automatically obtained when reducing the density variation relative to the average density. In most cases we will, however find a mixture. Without an independent measure for this ratio it is impossible to distinguish both behaviours. Thus, a centroid map alone does not allow to derive any spectral index.

3.3 Clump Decomposition

Another method which is frequently applied to identify the combined density-velocity structure is the decomposition of molecular line maps into separate clumps. Various methods exists for this purpose [19, 21, 16], resulting in slightly deviating but comparable spectra of clump properties. To test the significance of the clump decomposition we have applied GAUSSCLUMPS [19] to MHD simulations. For an example model Fig. 4 displays the number of clumps as a function of their mass when determined either in the original density structure or in simulated molecular line observations of the same structure. The straight line indicates the slope which is observed in most CO observations [5]. The observed clump spectrum hardly reflects the physical distribution of clumps. When comparing individual clumps only the most massive clumps are identically identified in both data types. Moreover, the result confirms the effect of a pseudo-virialisation of the clumps in the molecular line data due to radiative transfer effects as discussed by Mac Low (this volume).

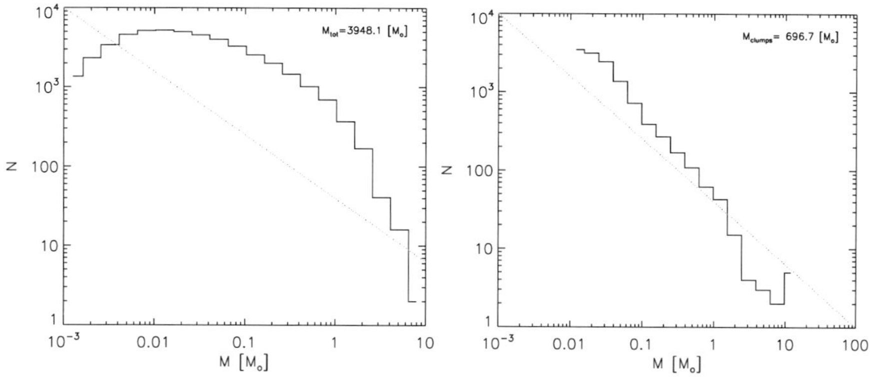

Fig. 4. Clump mass spectra computed from the original density structure and the computed ^{13}CO 1-0 map simulated with a S/N=20.

An interesting result of the clump decomposition is, however, that none of the investigated cloud models exactly reproduces the exponent measured in most observationally based spectra when computed with the same signal-to-noise ratio. The clump spectrum is able to distinguish between different models so that it is still a valuable structure parameters even if we currently do not understand its origin.

4 Summary

4.1 Results of the Comparison Between Models and Observations

None of the studied models fit all observational constraints. Fractal models show a wrong mutual relation between map statistics obtained for different tracers and either a wrong density statistics or wrong scaling laws. (Magneto-)hydrodynamic models fit a too small inertial range and provide wrong spectral indices in clump statistics and velocity channel analysis.

Nevertheless, the comparison between models and observations already allows to constrain some essential parameters governing the structure of molecular clouds. The main turbulent energy injection must occur on very large scales. Strong magnetic fields are excluded and a turbulent driving may have stopped only recently. The deviations from a self-similar scaling behaviour in star-forming regions reveals the role of self-gravity and we find first indications for the dissipation by ambipolar diffusion. The actual understanding of the molecular cloud physics is always hidden in the observation of deviations from self-similarity.

4.2 Conclusions

An iterative process of constructing cloud models, comparing their appearance in observable tracers with actual measurements, and improving the models to resemble the observational results is required to learn about the nature of interstellar turbulence. All methods for the comparison investigated so far show some distinctive power for the comparison between models and observations. However, they almost never show what they are thought to show. Currently we still do not know which method reflects a particular physical behaviour best. The use of any single method to derive cloud physics will almost necessarily fail. Beside the ongoing comparison between models and observations new classes of tools are needed to characterize e.g. the anisotropy of the structures or being sensitive to the velocity-density correlations in the clouds.

New cloud models have to provide a larger dynamic range, a self-consistent treatment of the energy balance and the the chemical structure of the clouds. Continuing this approach will help to recover the nature of interstellar turbulence finally providing a self-consistent picture for the evolution of interstellar clouds including the complex of star-formation.

References

1. Bensch F., Stutzki J., Ossenkopf V., A&A **366**, 636 (2001)
2. Brunt C.M., Heyer M.H., ApJ **566**, 276 (2002)
3. Brunt C.M., Mac Low M.-M., ApJ accepted (2003)
4. Esquivel A., Lazarian A., Pogosyan D., Cho J., MNRAS **342**, 325 (2003)
5. Kramer C., Stutzki J., Röhrig R., Corneliussen U., A&A **329**, 249 (1998)
6. Lazarian A., Pogosyan D., ApJ **537**, 720 (2000)
7. Lazarian A., Esquivel A., ApJ **592**, L37 (2003)
8. Mac Low M.-M., Ossenkopf V., A&A **353**, 339 (2000)
9. Miesch M. S., Scalo J., Bally J., ApJ **524**, 895 (1999)
10. Miville-Deschênes M.-A., Levrier F., Falgarone E., ApJ **593**, 831 (2003)
11. Motte F., André P., Neri R., A&A **336**, 150 (1998)
12. Ossenkopf V., Bensch F., Stutzki J., in: Gurzadyan V. G., Ruffini R. (eds.), *The Chaotic Universe*, World Sci. (2000), p.394
13. Ossenkopf V., A&A **391**, 295 (2002)
14. Ossenkopf V., A&A **391**, 295 (2002)
15. Ossenkopf V., Krips M., Stutzki J., A&A submitted (2002)
16. Ostriker E.C., Stone J.M., Gammie C.F., ApJ **546**, 980 (2001)
17. Padoan P., Juvela M., Bally J., Nordlund Å., ApJ **529**, 259 (2000)
18. Stanimirović S., Lazarian A., ApJ **551**, L53 (2001)
19. Stutzki J., Güsten R., ApJ **356**, 513 (1990)
20. Stutzki J., Bensch F., Heithausen A., Ossenkopf V., Zielinsky M., A&A **336**, 697 (1998)
21. Williams J.P., Blitz L., Stark A.A., ApJ **451**, 252 (1995)

Phases of the Interstellar Medium in the Milky Way

C.F. McKee[1], D. Hollenbach[2], and M.G. Wolfire[3]

[1] Departments of Physics and Astronomy, University of California, Berkeley CA 94709, USA cmckee@astro.berkeley.edu
[2] MS245-3, NASA Ames Research Center, Moffett Field CA 94035, USA hollenbach@ism.arc.nasa.gov
[3] Department of Astronomy, University of Maryland, College Park MD 20742, USA mwolfire@astro.umd.edu

Summary. The interstellar medium (ISM) is highly turbulent, with large pressure fluctuations due to energy injection by stars and by instabilities in the disk of the Galaxy. The diffuse gas exists in different states that are all at about the same pressure in a given region of the Galactic disk: Cold Neutral Medium (CNM) at $T \sim 10^2$ K, Warm Neutral Medium (WNM) and Warm Ionized Medium (WIM) at $T \sim 10^4$ K, and Hot Ionized Medium (HIM) at $T \sim 10^6$ K. This diffuse medium forms the background in which molecular clouds form. The surfaces of molecular clouds are atomic and are at the same pressure as the ambient ISM, but the bulk of the molecular gas is self-gravitating (in the Milky Way, at least) and is at a higher pressure.

The neutral atomic hydrogen (H I) in the ISM in the vicinity of the Sun is observed to be at a pressure such that the CNM and the WNM can coexist in thermal equilibrium. The condition that the H I in the disk of the Galaxy be in two-phase equilibrium can be used to *predict* the thermal pressure of the H I throughout the disk of the Milky Way. Because of interstellar turbulence, a significant fraction of the H I is at intermediate temperatures, however, where the gas is thermally unstable. The dominant heat source for the H I is photoelectric heating due to the absorption of stellar ultraviolet radiation by small grains. Fluctuations in this radiation field change the relative balance between the CNM and the WNM.

Some interstellar hydrogen is photoionized. In much of the disk of the Galaxy this ionized gas is in pressure equilibrium with the H I in the disk (the WIM). Photoionized gas within a few hundred parsecs of large OB associations (the envelopes of H II regions) absorbs a significant fraction of the ionizing radiation emitted by the associations and is at a somewhat higher pressure. Emission of ionizing radiation by old supernova remnants can account for the observations of high latitude WIM by the WHAM survey.

1 Introduction

The dense interstellar medium that is the subject of this conference is the birthplace of stars and is generally gravitationally bound. However, this dense medium forms out of atomic gas that is generally not gravitationally bound, and if we are to understand the evolution of galaxies it is essential that we understand this atomic gas. Indeed, in the Milky Way, most of the mass of the interstellar medium is atomic, not molecular: Locally, the surface density of neutral atomic gas (H I) is 5 $M_\odot$ pc^{-2} (*not* including helium) [8]; that of photoionized gas (the Warm Ionized Medium, or WIM) is about 1.0 $M_\odot$ pc^{-2} [42]; and that of the hot ionized medium (HIM) is about 0.2 $M_\odot$ pc^{-2} [52]. By contrast the local surface density of molecular gas is 1.4 $M_\odot$ pc^{-2} [4], less than a third of the atomic gas. Integrating over the entire Galaxy, the mass of H I is about 5×10^9 $M_\odot$ [53], whereas that of H_2 is about 0.8×10^9 $M_\odot$ [4].

The importance of self-gravity in the gas can be determined with the aid of the ensemble-averaged virial theorem (e.g., [32], but see MacLow in this volume for a different view), which can be expressed as

$$\bar{P} = P_s + P_G \left(1 - \frac{\mathcal{M}}{|\mathcal{W}|}\right) . \tag{1}$$

Here $\bar{P}$ is the mean total pressure (thermal plus turbulent), P_s is the surface pressure $\mathcal{M}$ is the magnetic term, and $P_G \equiv -(1/3V_{\rm cl})\mathcal{W} \equiv (1/3V_{\rm cl}) \int \rho \boldsymbol{r} \cdot \boldsymbol{g} dV$, where $V_{\rm cl}$ is the cloud volume and $\boldsymbol{g}$ the acceleration due to gravity, is proportional to the gravitational term $\mathcal{W}$. This has a simple interpretation: the mean pressure in the cloud is the sum of the surface pressure and the effective weight of the cloud; the weight can be partially counterbalanced by a magnetic field.

Consider an interstellar cloud. The thermal plus turbulent pressure at its surface is $P_s/k \simeq 2 \times 10^4$ K cm^{-3} [3]; the justification for including the turbulent pressure is given in [35]. The "gravitational pressure" $P_G \sim \frac{1}{2} G\Sigma^2 \simeq 1.4 \times 10^5 k \bar{N}^2_{\rm H,\,22}$ K cm^{-3}, where Σ is the surface density of the cloud and $\bar{N}_{\rm H,\,22}$ is the mean hydrogen column density in the cloud in units of 10^{22} cm^{-2}, corresponding to about 5 magnitudes of visual extinction. Thus, in the absence of magnetic fields, any cloud with $N_{\rm H,\,22} \gtrsim 0.5$ will have its internal pressure dominated by self-gravity. Magnetic fields can increase the required column density somewhat, but this does not qualitatively change the conclusion. Since the shielding layer required on the surface of a cloud to produce significant CO in the local FUV radiation field is about 1.4×10^{21} cm^{-2} [47], clouds in the local ISM in which a substantial fraction of the C is molecular will be gravitationally bound (see also [11]).

We can thus divide the ISM into two types of regions: those in which self-gravity is relatively unimportant ($P_s > \frac{1}{2} G\Sigma^2$) and those in which it is important ($P_s < \frac{1}{2} G\Sigma^2$). As Σ increases, self-gravity does as well, and eventually gravitational collapse ensues, leading to star formation. Here we shall focus on the regions in which self-gravity is relatively unimportant. Since the turbulent velocities in the disk of the Galaxy are approximately sonic in the warm H I (the WNM), it follows that the thermal pressure in this non-self-gravitating gas is approximately constant (see §2 for discussion of the effects of turbulence). Cold H I (the CNM) embedded in this warm gas will be at about the same thermal pressure, corresponding to an approximate two-phase medium that was first described by Field, Goldsmith & Habing [14]. Supernovae create large volumes of hot, collisionally ionized gas [7], the HIM, which is a third phase of the ISM [33]. Some of the warm gas is photoionized (the WIM), but this is not identified as a separate phase since it has about the same temperature as the WNM.

2 Neutral Atomic Hydrogen

The presence of CNM and WNM gas in the solar neighborhood is well established by H I 21 cm emission and absorption observations [30]. Locally, the phases have approximately equal mass fractions [30, 8]. The thermal pressure in the neutral gas is determined from ultraviolet absorption line studies of the fine-structure levels of C I [24, 25]. The mean thermal pressure quoted by Jenkins & Tripp [25] is $P_{\rm th}/k = 2240$ K cm^{-3}; however, corrections for a higher gas temperature appropriate for the CNM and revised atomic constants raise the thermal pressure derived from the observations to ~ 3000 K cm^{-3} [53]. A similar thermal pressure is found for neutral clouds within the Local Bubble [23].

Investigations of the thermal balance of the WNM and CNM phases in the local ISM have shown that the dominant heating source for both phases is photoelectric ejection of electrons from small grains and large molecules (polycyclic aromatic hydrocarbons; PAHs) [51, 53]. The photoelectric heating rate [2, 50] depends on the flux of stellar far-ultraviolet ($6\,{\rm eV} \leq h\nu \leq 13.6$ eV, or FUV) radiation, the PAH abundance, and the neutral PAH fraction, which in turn depends on the gas temperature, the ultraviolet flux and the free electron abundance in the gas. Free electrons in the WNM are produced by the photoionization of H and He by soft X-ray and extreme ultraviolet ($13.6\,{\rm eV} \leq h\nu \leq 10^3$ eV, or EUV) radiation emitted by stars and supernova remnants. Electrons in the CNM are produced by the photoionization of C^0 to C^+ by stellar FUV radiation. The recombination of H^+ on PAH^- is important in regulating the gas electron abundance and heating rate in the WNM [53].

The observational evidence suggests that WNM and CNM exist throughout the Galactic disk. In the inner Galaxy, H I emission (from both CNM and WNM) is seen at all velocities allowed by Galactic rotation [15]. This emission is not all WNM however, since discrete H I absorption from CNM is also seen [26, 9]. In the outer Galaxy, the presence of WNM is well established [30], whereas that of a widely distributed CNM is less so. Several high-velocity absorption components have been observed in H I [5] and Na I [43] that arise from CNM clouds at Galactic radii $R \lesssim 14$ kpc, and Kolpak et al. [26] show H I absorption to $R \lesssim 17$ kpc. Molecular clouds, which presumably form from the CNM phase, are traced to at least $R \sim 20$ kpc [54, 21]. Thus, direct observations of cold H I or molecular gas extend to at least $R \sim 20$ kpc.

For WNM and CNM gas in thermal and pressure equilibrium, Field [13] demonstrated that the gas is thermally stable where $dP_{\rm th}/dn > 0$, but is unstable to isobaric perturbations where $dP_{\rm th}/dn < 0$. Due to fine-structure line cooling of C II, the equilibrium $P_{\rm th}(n)$ curve has a characteristic shape such that two thermally stable phases may coexist in pressure equilibrium within a range of gas pressures, $P_{\rm min}$ to $P_{\rm max}$. At thermal pressures greater than $P_{\rm max}$ only the cold phase (CNM) may be present, while at thermal pressures less than $P_{\rm min}$ only the warm phase (WNM) may exist.

In light of the turbulent nature of the ISM, when does thermal instability apply? The necessary conditions for thermal instability to segregate the neutral gas into CNM and WNM phases are that the thermal pressure must dominate the turbulent pressure and that the gas has time to reach thermal equilibrium compared to the time between turbulent shocks [53, 44]. A key feature of turbulence is that the rms turbulent velocity increases with scale so that $\sigma = \sigma(1)\ell_{\rm pc}^q$, where $\sigma(1)$ is the one-dimensional turbulent velocity on the scale of 1 pc and $\ell_{\rm pc}$ is the scale in units of parsecs. Wolfire et al. [53] argued that $q \simeq 1/3$ and estimated $\sigma(1) \simeq 1.2$ km s^{-1}. The size of the region for which thermal pressure dominates, $\ell_{\rm P}$, may then be estimated by equating the thermal velocity dispersion, $\sigma_{\rm th} = 0.8(T/100\ {\rm K})^{1/2}$ km s^{-1} with the turbulent velocity dispersion $\sigma(\ell_P)$. The resulting value of $\ell_{\rm P}$ is quite uncertain; for CNM, $\ell_{\rm P} \sim 0.3$ pc, while for WNM, $\ell_{\rm P} \sim 215$ pc [53].

The second condition for a two-phase description to be valid is that the cooling time $t_{\rm cool}$ must be less than the characteristic time $t_{\rm shock}$ for a shock to induce motions of order $\sigma_{\rm th}$ or, equivalently (for an isothermal shock), to double the pressure. The time for an element of gas to return to thermal equilibrium at constant $P_{\rm th}$ after a significant perturbation is given by

$$t_{\rm cool} \simeq 7.7 \times 10^6\, T_4^{1.2} \left(\frac{P_{\rm th}/k}{3000\ {\rm K\ cm^{-3}}}\right)^{-0.8} \quad {\rm yr}, \qquad (2)$$

where $T_4 \equiv T/(10^4\,{\rm K})$ [53]. The overall accuracy is within a factor 1.5 for temperatures between $T = 55$ K and $T = 8000$ K and thermal pressures between $P_{\rm th}/k = 1000$ K cm^{-3} and $P_{\rm th}/k = 6000$ K cm^{-3}.

In order to induce motions of $\sigma_{\rm th}$, the velocity of an isothermal shock must be $\sqrt{2}\sigma_{\rm th}$. The time interval between shocks in a turbulent medium is then $t_{\rm shock} \sim \ell_{\rm P}/\sqrt{2}\sigma_{\rm th}$. The ratio $\Upsilon = t_{\rm cool}/t_{\rm shock}$ is given by

$$\Upsilon \equiv \frac{t_{\rm cool}}{t_{\rm shock}} \sim 0.3 T_4^{0.2} \left(\frac{P_{\rm th}/k}{3000\ {\rm K\ cm^{-3}}}\right)^{-0.8} \left[\frac{\sigma(1)}{1.2\ {\rm km\ s^{-1}}}\right]^3 . \qquad (3)$$

So long as $\Upsilon \lesssim 1$, a two-phase description of the H I is approximately valid on scales $\ell \lesssim \ell_P$, since the gas has time to reach thermal balance between shocks. Since ℓ_P is small for the CNM, this means that only the surface layers of a CNM cloud are likely to be in pressure equilibrium with the surrounding WNM. The Υ condition is well satisfied for the CNM ($\Upsilon \simeq 0.1$), but only marginally satisfied for the WNM ($\Upsilon \simeq 0.3$). Estimating Υ from the frequency of supernova shocks leads to a somewhat higher value for the WNM, $\Upsilon \sim 0.9$ [53]. Thus a significant fraction of WNM gas is expected to be in the thermally unstable regime. The observed fraction is $\sim 48\%$ at high Galactic latitudes [18], but the fraction in the plane is uncertain.

Numerical simulations in 1D [19, 20] and 2D [27] show that strong compression can drive stable WNM into the cold phase. Kritsuk & Norman [28] followed the development of a multiphase medium in 3D starting from non-equilibrium conditions. Simulations under turbulent ISM conditions [49, 16, 44, 48] including magnetorotational instability [41] and supernova driven turbulence [31] emphasize the dynamical ($\Upsilon \gtrsim 1$) nature of the ISM. Large ($\sim 50\%$) mass fractions of thermally unstable gas appear in low [49] and high [16] resolution (2D) models. Such large mass fractions result in part from simulated stellar sources appearing in cold peaks which drives gas to the warmer phase. Thermal instability in the presence of the magnetorotational instability maintains a two phase medium to late growth times, but with a small fraction ($\ll 50\%$) of thermally unstable gas [41].

Wolfire et al. [51, 53] have emphasized the equilibrium ($\Upsilon \lesssim 1$) nature of the ISM. They find a two-phase (WNM+CNM) equilibrium is possible in the Galactic midplane between 3 kpc and 18 kpc. Based on the mean H I density [53], and the weight of the gas layer [53, 12] thermal pressures must be greater than $P_{\rm min}$ and thus cold gas must exist in the outer Galaxy to $\sim 16-18$ kpc. Under the assumption that two phases exist, they *predict* that the thermal pressure in the Galactic plane for 3 kpc$\lesssim R \lesssim$ 18 kpc is

$$P_{\rm th}/k \simeq 1.4 \times 10^4 \exp(-R/5.5\ {\rm kpc}) \quad {\rm K\ cm}^{-3}. \tag{4}$$

3 Fluctuating FUV Fields and Neutral Phases in the ISM

The relative proportions of CNM and WNM in the Milky Way Galaxy depend not only on the thermal pressure, but on the FUV flux that controls the heating and cooling of CNM and WNM. At fixed mean interstellar gas density, high FUV fields heat CNM to WNM; low FUV fields allow WNM to cool to CNM [51, 53]. It seems likely that the dense interstellar medium (e.g., Giant Molecular Clouds) in galaxies grow from assemblages of CNM clouds. If so, the formation rate of stars in galaxies depend on the mass fraction of CNM gas. Consequently, a feedback process occurs between stars and the ISM that may control the global rate of star formation in a galaxy

[37, 38, 6]. High star formation rates lead to large numbers of OB stars that create high FUV fluxes that destroy the sources of star formation. Low rates of star formation, on the other hand, may lead to enhanced growth rates of star-forming clouds and thereby to enhanced star formation rates.

The time-averaged FUV field in the Milky Way Galaxy varies as a function of galacto-centric distance R because of the R dependence of the star formation rate, which controls the surface density of OB stars that produce FUV, and the dust grain density, which controls the mean free path for FUV photons in the ISM. Wolfire et al [53] find that the FUV mean intensity over the range 4 kpc $< R <$ 18 kpc scales as $\exp(-R/4.1$ kpc). Although the radiation field is higher at $R \sim 4$ kpc than at the solar circle, there is still considerable CNM there because of the higher mean gas densities and thermal pressures.

Even at the solar circle, $R \sim 8.5$ kpc, the FUV field varies in both time and space [40]. OB stars are generally born in associations that form inside GMCs present in the Galactic disk. The FUV field at a given point in a galaxy then depends largely on the distribution of stellar masses in relatively nearby ($\lesssim 500$ pc, limited by dust opacity) associations and on the birthrate and size distribution of these associations. The FUV luminosity of an association is an increasing function of its size, but the birthrate of associations decreases with size. The abundant low-luminosity associations together with the distant associations build an FUV background, over which is added bursts of FUV flux coming from relatively rare events (the birth of nearby large associations) that can significantly increase the FUV field for about $\sim 30 - 50$ Myr. The *time* variation of the FUV field at any given point maps directly onto the *spatial* variation of the FUV field in the galaxy at any instant of time: those regions very close to luminous associations will have high FUV fluxes, whereas large volumes of space will have much lower FUV fluxes.

At a given point in a galaxy, the time variations in the FUV flux will produce changes in the relative masses of CNM and WNM. Parravano, Hollenbach & McKee [40] have calculated the time-varying FUV field at the solar circle, and find that the time interval between epochs of high FUV flux is approximately $t \simeq 20 g_{\rm FUV}^{3/2}$ Myr, where $g_{\rm FUV}$ is the ratio of the high flux to the median flux. Using the computed time varying field, they have made preliminary models that describe the changing (in space and time) mass fractions of CNM and WNM at the solar circle, and find time and space-averaged mass fractions compatible with the observed values [39, 22]. Kritsuk & Norman [29] have performed hydrodynamical calculations of the thermal instability that demonstrate the conversion of CNM to WNM and vice versa as the FUV heating rises and falls.

4 The Warm Ionized Medium

The WIM manifests itself through optical emission lines, with intensities proportional to $\int n_e^2\, ds$, and pulsar dispersion measures, which are proportional to $\int n_e\, ds$. (Photoionized gas can also be detected via radio absorption, but this occurs primarily in the envelopes of H II regions [1], which are overpressured with respect to the general ISM. Interstellar scintillation also arises in photoionized gas, but this too likely arises in the envelopes of H II regions [46].) There are three primary sources of ionization for the WIM [33]: unshielded B stars; hot white dwarfs and central stars of planetary nebulae; and emission from old supernova remnants (SNRs). O stars (e.g., [36]) dominate the ionization in the envelopes of H II regions, but are unlikely to produce the ionization in the WIM due to self-shielding [10]. Slavin, McKee, & Hollenbach [45] have calculated the ionization due to emission from old SNRs, including the effects of non-equilibrium ionization and absorption in a cloudy ISM. Using the supernova rate per unit area of 3.8×10^{-5} kpc^{-2} yr^{-1} given in [34], they predict an Hα intensity $I^*(\mathrm{H}\alpha)\sin b = 0.53$ Rayleigh, where b is the Galactic latitude.

Reynolds and his students have pioneered the study of the WIM through observations of Hα emission, culminating in Northern Sky Survey by the Wisconsin Hα Mapper (WHAM) [17]. Their maps suggest that there are two components to the Hα emission: one component that is associated with H II regions at low latitudes and another component at high latitudes. Averaging over both components, the median observed Hα intensity is $I^*(\mathrm{H}\alpha)\sin b =$ 0.68 R. The median value dips somewhat below this average for $|b| \gtrsim 40°$. We conclude that the high latitude Hα emission is consistent with that expected from ionization by old SNRs.

Acknowledgments. The work of C.F.M. is supported by NSF grant AST00-98365; that of M.G.W. by NASA LTSA grant NAG5-9271 and that of D.J.H. by NASA RTOPs 860-10-04 and 399-25-01.

References

1. K.R. Anantharamaiah: J. Astr. Ap., **6**, 203 (1985)
2. E.L.O. Bakes, A.G.G.M. Tielens: ApJ, **427**, 822 (1994)
3. A. Boulares, D.P. Cox: ApJ, **365**, 544 (1990)
4. L. Bronfman, S. Casassus, J. May, L.-ÅNyman: A& A, **358**, 521 (2000)
5. S.W. Colgan, E.E. Salpeter, Y. Terzian: ApJ, **328**, 275 (1988)
6. E. Corbelli, E. E. Salpeter: ApJ, **450**, 32 (1995)
7. D.P. Cox, B.W. Smith: ApJ, **189**, L105 (1974)
8. J.M. Dickey, F.J. Lockman: ARAA, **28**, 215 (1990)
9. J.M. Dickey, N.M. McClure-Griffiths, B.M. Gaensler, A.J. Green: ApJ, **585**, 801
10. B.G. Elmegreen: ApJS, **205**, 405 (1976)

11. B.G. Elmegreen: In *Protostars and Planets III*, ed. E.H. Levy & J.I. Lunine. (University of Arizona Press, Tucson 1993), p. 97
12. B.G. Elmegreen, A. Parravano: ApJ, 435, L121 (1994)
13. G.B. Field: ApJ, **142**, 531 (1965)
14. G.B. Field, D.W. Goldsmith, H.J. Habing: ApJ, **155**, L149 (1969)
15. R.W. Garwood, J.M. Dickey: ApJ, **338**, 841 (1989)
16. A. Gazol, E. Vázquez-Semadeni, F. J. Sánchez-Salcedo, J. Scalo: ApJ, **557**, L121 (2001)
17. L.M. Haffner, R.J. Reynolds, S.L. Tufte, G.J. Madsen, K.P. Haehnig, J.W. Percival: ApJS, **149**, 405 (2003)
18. C. Heiles, T. H. Troland: ApJ, **586**, 1067 (2003)
19. P. Hennebelle, M. Pérault: A&A, **351**, 309 (1999)
20. P. Hennebelle, M. Pérault: A&A, **359**, 1124 (2000)
21. M.H. Heyer, J.M. Carpenter, R.L. Snell: ApJ, **551**, 852 (2001)
22. D. J. Hollenbach, A. Parravano, C. F. McKee: BAAS, **33**, 1482 (2001)
23. E.B. Jenkins: ApJ, **580**, 938 (2002)
24. E.B. Jenkins, M. Jura, M. Loewenstein: ApJ, **270**, 88 (1983)
25. E.B. Jenkins, T.M. Tripp: ApJS, **137**, 297 (2001)
26. M.A. Kolpak, J.M. Jackson, T.M. Bania, J.M. Dickey: ApJ, **578**, 868 (2002)
27. H. Koyama, S. Inutsuka: ApJ, 564, L97 (2002)
28. A.G. Kritsuk, M.L. Norman: ApJ, 569, L127 (2002)
29. A.G. Kritsuk, M.L. Norman: ApJ, 580, L51 (2002)
30. S.R. Kulkarni, C. Heiles: The Atomic Component. In: *Interstellar Processes*, ed by D. Hollenbach, H. A. Thronson, Jr. (Reidel, Dordrecht 1987) pp 87–122
31. M.-M.Mac Low, D.S. Balsara, M.A. de Avillez, J. Kim: ApJ, in revision (astro-ph/0106509) (2003)
32. C.F. McKee: In *The Origin of Stars and Planetary Systems*, ed. C.J. Lada & N.D. Kylafis. (Kluwer, Dordrecht 1999) p. 29.
33. C.F. McKee, J.P. Ostriker: ApJ, **218**, 148 (1977)
34. C.F. McKee, J.P. Williams: ApJ, **476**, 144 (1997)
35. C.F. McKee, E.G. Zweibel: ApJ, **399**, 551 (1992)
36. W.W. Miller, D.P. Cox: ApJ, **417**, 579 (1993)
37. A. Parravano: A&A, **205**, 71 (1988)
38. A. Parravano: ApJ, **347**, 812 (1989)
39. A. Parravano, D. J. Hollenbach, C. F. McKee: In *Cosmic Evolution and Galaxy Formation: Structure, Interactions, and Feedback*, ed. J. Franco, L. Terlevich, O. Lopez-Cruz, & I. Aretxaga, (ASP Publishing: San Francisco), pp 87 (2000)
40. A. Parravano, D. J. Hollenbach, C. F. McKee: ApJ, **584**, 797 (2003)
41. R.A. Piontek, E.C. Ostriker: ApJ, in press (2004)
42. R.J. Reynolds: ApJ, **372**, L17 (1991)
43. K.R. Sembach, A. C. Danks: A&A, 289, 539 (1994)
44. F.J. Sánchez-Salcedo, E. Vázquez-Semadeni, A. Gazol: ApJ, **577**, 768 (2002)
45. J.D. Slavin, C.F. McKee, D.J. Hollenbach: ApJ, **541**, 218 (2000)
46. S.R. Spangler: ApJ, **376**, 540 (1991)
47. E.F. van Dishoeck, J.H. Black: ApJ, **334**, 771 (1988)
48. E. Vázquez-Semadeni, A. Gazol, T. Passot, J. Sánchez-Salcedo: In: *Simulations of Magnetohydrodynamic Turbulence in Astrophysics*, ed T. Passot & E. Falgarone (Springer-Verlag, New York 2003), pp 213-251
49. E. Vázquez-Semadeni, A. Gazol, J. Scalo: ApJ, **540**, 271 (2000)

50. J.C. Weingartner, B.T. Draine: ApJS, **134**, 263 (2001)
51. M.G. Wolfire, D. Hollenbach, C.F. McKee, A.G.G.M. Tielens, E.L.O. Bakes: ApJ, **443**, 152 (1995)
52. M.G. Wolfire, C.F. McKee, D. Hollenbach, A.G.G.M. Tielens: ApJ, **453**, 673 (1995)
53. M.G. Wolfire, C.F. McKee, D. Hollenbach, A.G.G.M. Tielens: ApJ, **587**, 278 (2003)
54. J.G.A. Wouterloot, J. Brand: A&AS, **80**, 149 (1989)

Star Formation in Southern Bok Globules

K. Kämpgen, R. Chini, M. Nielbock, and M. Albrecht

Astronomisches Institut der Ruhr-Universität Bochum, Universitätsstraße 150, 44780 Bochum, Germany, kaempgen@astro.rub.de, chini@astro.rub.de, nielbock@astro.rub.de, albrecht@astro.rub.de

Summary. In order to investigate the formation of stars independent of violent environmental effects we surveyed a large sample of 128 small and opaque southern Bok globules at 1.2 mm continuum with SIMBA at SEST. In addition, most of these globules were and will yet be observed at 10 μm by TIMMI2 and in the far-infrared (FIR) regime by SIRTF. According to the association with IRAS sources, the sample has been divided into FIR-loud and FIR-quiet objects. Preliminary results for 36 FIR-loud sources suggest different classes of globules as witnessed by their ratio of bolometric luminosity to gas mass and – as a consequence – by the number and mass of embedded objects. At least nine FIR-quiet globules show signs of fragmentation and seem to be at the border of gravitational instability. The present study indicates that about 75% of the FIR-loud and 30% of the FIR-quiet globules do actively form stars.

1 Introduction

Majority of stars in the Galaxy are probably produced in young clusters where, in a small volume, numerous stars with similar chemical composition cover a large mass interval. Therefore, clusters provide unique opportunities for the study of the early phases of star and planet formation in a statistically meaningful way. However, our current models of star formation predominantly handle the formation of isolated stars – without the influence of nearby massive stars, their stellar winds and their hard radiation field.

In contrast, small isolated dark clouds, also known as Bok globules, are the ideal laboratory to study the formation of only a few stars in the low-mass range independent of environmental effect. Since the number of physical processes occurring within them is likely to be limited their simple structure allows them to be modeled in detail. Most Bok globules are relatively nearby (< 500 pc) with only a few exceptions being further away than 900 pc [2]. Therefore, linear scales of a few tenths of a parsec can be resolved by the current instrumentation. Their isolation turns them into objects which are mostly unaffected by the influences regulating the process of star formation in larger, more complex molecular clouds, such as large-scale motions masking the smaller scale motions of interest, energetic HII regions, and multiple star formation within one cloud which may destroy much of the evidence sought on single star forming events.

2 The BHR Sample of Southern Globules

Bourke et al. ([1], hereafter BHR) have established a list of 169 small southern molecular clouds of very high opacity and with diameters less than 10′ and thus provided a complementary sample to that of Clemens & Barvainis ([3]) in the northern sky. BHR found IRAS sources lying towards 76 of their globules. Due to the large positional uncertainty of the IRAS data, there are only ≈ 55 BHR globules with confident IRAS detections whereas for another 21 BHR globules the association with IRAS sources is questionable. The remaining 93 globules of the BHR sample do not have any IRAS counterparts. We performed a 1.2 mm continuum survey with the 37-channel bolometer SIMBA at SEST and observed 65 BHR globules from the FIR-loud sample and 63 from the FIR-quiet sample; the detection rate was 75% and 40%, respectively. In addition, 36 of the FIR-loud and 25 of the FIR-quiet globules were investigated further on by high-resolution MIR imaging with TIMMI2 at the ESO 3.6m telescope and by using archive data from 2MASS and MSX.

3 Global Properties of FIR-Loud BHR Globules

In the following we describe some global properties as they follow from the IRAS and 2MASS surveys and our own SIMBA measurements.

Bolometric luminosity. The bolometric luminosity $L_{\rm bol}$ was calculated by integrating the SED from $12\,\mu$m to 1.2 mm. In those cases where an IRAS flux is missing we regard the corresponding luminosity as an upper limit. The distance was assumed to be 500 pc whenever the distance was unknown. In this way, we obtained a luminosity distribution that ranged from 2 to $18\,{\rm L}_\odot$ with a pronounced maximum around $4\,{\rm L}_\odot$. BHR 12, 36, 92 and 23 are exceptions with luminosities of 22, 23, 62 and $1550\,{\rm L}_\odot$, respectively.

Dust temperature. For an interstellar cloud coldest dust carries the bulk of mass. In order to determine the coldest dust component, we have calculated the color temperature between $100\,\mu$m and 1.2 mm adopting the mass absorption coefficient κ_ν to be proportional to ν^2. The derived dust temperatures are $13\,{\rm K} \leq T_{\rm d} \leq 24\,{\rm K}$ with the majority of sources having $15\,{\rm K} \leq T_{\rm d} \leq 18\,{\rm K}$.

Gas mass. Using the optically thin emission of dust at 1.2 mm as a tracer for its mass, we convert the observed flux density S_ν into a mass using the relation $S_\nu = M_{\rm dust} \times \kappa_\nu \times B_\nu(T_{\rm dust})/D^2$ where B_ν is the Planck function and D the distance to the source. Adopting a gas-to-dust ratio of about 150 we derive a typical gas content of the globules of $M_{\rm gas} \sim 2 - 4\,{\rm M}_\odot$; a few exceptional objects have higher masses between 10 and $125\,M_\odot$.

Star formation efficiency. The ratio $L_{\rm bol}/M_{\rm gas}$ can be regarded as a measure for the conversion of gas into stars and thus describes the current star formation efficiency. Comparing both quantities with each other leads us to

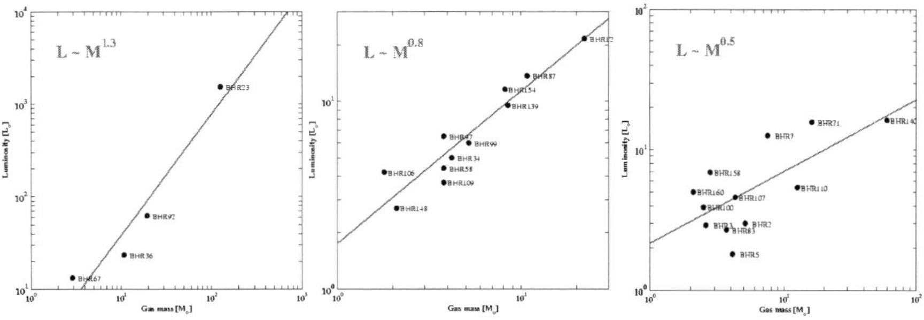

Fig. 1. The bolometric luminosity as derived from the SED as a function of gas mass as obtained from our SIMBA measurements. **Left:** Class A **Middle:** Class B **Right:** Class C

suggest that there are three classes of globules with respect to their IRAS properties:

- **Class A.** BHR 23, 36, 67 and 92 form a small group that can be characterised by a relation $L_{\rm bol} \propto M_{\rm gas}^{1.3}$, i.e. their star formation efficiency is highest among all globules. Furthermore, they have in common that they host FIR sources which were detected in all four IRAS bands. The number of embedded NIR sources ranges between 10 to 20, typically being of Class II–III. BHR 23 represents the most extreme case harbouring the earliest spectral type (B1V).
- **Class B.** There are 11 globules that form an intermediate class which is characterised by a relation $L_{\rm bol} \propto M_{\rm gas}^{0.8}$. In general, they host one or two FIR sources detected at 60 and 100 μm while fluxes at shorter IRAS wavelengths are missing. The number of associated NIR sources is below 10 which are typically of Class I–II.
- **Class C.** The remaining 12 globules that have been investigated so far host FIR sources which could only be detected at 100 μm and 1.2 mm in most cases. As such, their evolutionary stage is likely Class 0. Furthermore, their luminosity is highly uncertain and definitely underestimated. Taking, however, the derived luminosities at their face values, we obtain a relation $L_{\rm bol} \propto M_{\rm gas}^{0.5}$. The number of embedded NIR sources is generally below 5 with evolutionary stages of Class I. The most prominent example is BHR 72 hosting an embedded binary protostar.

MIR appearance. Due to the excellent spatial resolution of TIMMI 2 the 10 μm imaging of embedded IRAS sources resolves their morphology. In the case of BHR 23, for example, a single IRAS source could be resolved into a triple source. Likewise, the increased MIR sensitivity allowed us to detect another four previously unknown 10 μm sources which turns BHR 23 into an aggregate of at least seven MIR sources.

4 Global Properties of FIR-Quiet BHR Globules

One highly interesting morphological feature of the FIR-quiet globules as emerging from our SIMBA survey is that they seem to be fragmented into several emission knots. 9% are associated with one or more MIR sources as detected by the MSX satellite whereas the corresponding fraction of the FIR-loud globules is 63%. The absence of any further spectral information requires several assumptions to derive physical properties for this type of globules.

Adopting a dust temperature of 10 K, which is a typical value for externally heated dust clouds, the 1.2 mm emission may be converted into a gas mass; the corresponding masses range between 3 and 200 $M_\odot$ depending on the value of κ_ν. With the observed size of the emitting area one obtains hydrogen densities $n_H \sim 10^4 - 10^6\,\mathrm{cm}^{-3}$. Using simple arguments to estimate the gravitational stability of the fragments we calculate the Jeans-length by $L_J \sim 1.2 \times 10^{19}\sqrt{T/n_H}$. For 75% of the FIR-quiet globules showing emission at 1.2 mm the radius of the condensations is smaller than $(0.1 - 1.0)L_J$. This suggests that 30% of the observed FIR-quiet globules contain fragments that are likely to collapse.

5 Conclusions and Outlook

The results of the present study are preliminary because not all globules have been analysed completely. So far, we have reached the following conclusions:

- There are three classes of FIR-loud globules with respect to their ratio L_{bol}/M_{gas}. We find an empirical relation $L_{bol} \propto M_{gas}^{\alpha}$ with Class A having $\alpha \sim 1.3$ (10%), Class B yields $\alpha \sim 0.8$ (45%) and Class C $\alpha < 0.5$ (45%).
- The number of embedded YSOs reaches from $10 - 20$ in Class A to less than 5 in Class C.
- A single star forming event within a globule has not yet been observed.
- The FIR-quiet globules are fragmented into few clumps which seem to be close to gravitational instability.
- 75% of the FIR-loud and 30% of the FIR-quiet globules do form stars.

Currently, the ISO data archive is investigated to possibly complement the FIR data base. The global properties derived so far will be accomplished by our ongoing 10 μm survey which provides information on individual embedded objects.

References

1. T.L. Bourke, A.R. Hyland, G. Robinson: MNRAS **276**, 1052 (1995)
2. T.L. Bourke, A.R. Hyland, G. Robinson, et al.: MNRAS **276**, 1067 (1995)
3. D.P. Clemens, R. Barvainis: ApJS **68** 257 (1988)

Condensed Phases of H_2 in the ISM

D. Pfenniger

Geneva Observatory, University of Geneva

Summary. Below a temperature of 33 K molecular hydrogen can condense to liquid, and below 14 K to solid phase. These phase transitions require however a vapor pressure large enough that can be satisfied in the ISM only in particular conditions, such as during the collapse of sub-stellar mass clouds. The change from compressible gaseous to almost incompressible equation of state provides the possibility to reach stable gravitationally bound cold objects with distinct properties from stars and planets, resembling much the cometary knots ubiquitous in planetary nebulae. The process is general and the astrophysical implications numerous.

1 Overview

Many facts remain mysterious about the ISM. For example the recent deep photometry of the outer HI disk of M 31 by Cuillandre et al. (2001) challenges fundamentally the current views about the role of gas in galaxies. These authors demonstrate that stars succeed to form in the extreme outer HI disk of M 31, in regions where no CO is detected. This means that HI cools, fragments, becomes molecular, and finally forms stars. Young stars photo-dissociate the molecular invisible component and explain the mere existence of HI. Clearly the problem now is to try to understand how a very inconspicuous molecular component can exist.

2 Gas Globules Based on Liquid or Solid H_2 Cores

2.1 Gravitational Collapse and Opacity

Central to the discussion of the cold gas fate is the outcome of gravitational collapse, because a collapse is usually associated to formation of hot bodies such as stars or planets. Below we show how the opposite outcome, bodies based on a cold core, is another possibility.

In classical spherical collapse studies (e.g., Larson 1969), one learns that a spherical collapse of an initial cold ($T \approx 10\,\mathrm{K}$) ISM gas cloud increases its central density almost isothermally towards infinity in a finite time, a singular event. Isothermality results from the large transparency of the cold ISM to its own thermally generated photons, such that the cooling time remains much shorter than the free-fall time. However in the growing density collapsing core, an optical depth exceeding unity is finally reached, allowing the core to keep thermal photons longer, which leads central temperature and pressure to increase, stopping the singular density growth. All this supposes the gas equation of state remains perfect all the way.

Let us estimate the central density at which the center of a pseudo isothermal sphere $\rho(r) = \rho_0/\left[1+(r^2/r_c^2)\right]$, becomes opaque for constant opacity κ. Requiring that the optical depth $\tau_\infty = 1$ and that the pseudo isothermal sphere is in dynamical equilibrium, $\lim_{r\to\infty} GM(r)/r = 3kT/\mu m_p$, leads to a simple expression for the core radius and the critical central density above which the center begins to be opaque:

$$r_c = \frac{3\kappa kT}{8G\mu m_p}, \qquad \rho_0 = \frac{16G\mu m_p}{3\pi\kappa^2 kT} \approx \frac{2.7\cdot 10^{-8}\,(\mu/2)}{(\kappa/10^{-4})^2\,(T/10)}\,\left[\mathrm{g\,cm^{-3}}\right]. \tag{1}$$

This critical density is independent of the mass or size of the cloud. Now, for general equations of state the temperature increase in a homologous collapse is related to the density increase (see, e.g., Padmanabhan 2001 [Fig. 3.7]):

$$\frac{dT}{T} = \left(\frac{4\alpha-3}{3\delta}\right)\frac{d\rho}{\rho}, \quad \text{where} \quad \alpha \equiv \left(\frac{\partial\ln\rho}{\partial\ln P}\right)_T, \quad \delta \equiv -\left(\frac{\partial\ln\rho}{\partial\ln T}\right)_P. \tag{2}$$

So when the isothermal compressibility α decreases below 3/4, further compression leads to *cooling*, a reversal of the usual perfect gas behavior. In usual stellar or planetary conditions compressibility decreases when matter is subject to electron degeneracy, but the same effect holds for any hardly compressible matter: it cools down when compressed.

2.2 Liquid or Solid H_2

Now, molecular hydrogen, the bulk of cold gas, becomes liquid or solid, i.e., almost incompressible, below 33 K at sufficient pressure. The H_2 critical point occurs at $T_c = 33\,\mathrm{K}$, $P_c = 13\,\mathrm{bar}$, while the triple point at $T_3 = 14\,\mathrm{K}$, $P_3 = 0.07\,\mathrm{bar}$. At low T, the saturation pressure and gas density drops to interstellar values. Between 1 and 14 K the sublimation curve is well approximated by (Pfenniger & Combes 1994):

$$P_{\mathrm{solid-gas}} \approx 5.7\cdot 10^{20}\,T^{5/2}\,\mathrm{e}^{-91.75/T}\,\left[\mathrm{K\,cm^{-3}}\right]. \tag{3}$$

An analytical expression for the full H_2 equation of state $\rho(P,T)$ in cold condition has been derived in the range $[1-100]$ K, from cosmic pressures up to 1 kbar, by gathering laboratory data from the literature (Pfenniger 2004).

The H_2 gas density ρ at the critical pressure is almost the perfect gas density for the partial pressure $P_{\mathrm{solid-gas}}$. Equating this density ρ to the critically opaque density ρ_0 above provides the critical opacity κ_{crit} below which H_2 becomes incompressible *before becoming opaque to its thermal radiation*:

$$\kappa_{\mathrm{crit}} \approx 1.3 \cdot 10^{-3}\,(T/5)^{-5/4}\,\mathrm{e}^{9.18/(T/5)}\,[\mathrm{cm}^2\mathrm{g}^{-1}]\,. \tag{4}$$

Since the ISM sub-mm opacities typical of low metallicity cold clouds can be as low as $\sim 10^{-3}\,[\mathrm{cm}^2/\mathrm{g}]$ or less (see Draine 2004, [Fig. 12]), at least below $\sim 5\,\mathrm{K}$ a collapsing cloud is able to condense H_2 in the core before becoming opaque. Further compression of the condensed core, instead of heating it, cools it. The condensed core thus is able to stop a collapse without becoming hot. Somewhat warmer or more opaque initial gas can still reach H_2 condensation despite a temporary temperature increase.

We arrive thus at the surprising possibility that gravitational collapses in cold conditions lead not necessarily to a body with a hot center, such as planets or stars, but to a temperature inverted gas globule with a condensed cold core. The core is made of solid H_2 below 14 K, or liquid H_2 above, up to 33 K. Around the core an H_2-He gaseous atmosphere may extend to distances much larger than the core size, and thus makes most of the globule mass.

The typical properties of such cold gas condensations have been calculated by integrating the equilibrium equations for isothermal gravitating spheres for which the accurate H_2 equation of state is used (Pfenniger 2004). Summarizing, their main properties are:

a) The central density is determined by the solid or liquid H_2 density: $0.07 - 0.09\,\mathrm{g\,cm}^{-3}$. This is much lower, by a factor $20 - 50$, than for usual planets.

b) The maximum core radius is about $2 \cdot 10^4\,\mathrm{km}$, which corresponds to a maximum core mass similar to the Earth mass (this is the analogous of Chandrasekhar's mass for white dwarfs). At this maximum radius the gravitational and van der Waals binding energies are similar.

c) The cold globules have a relatively weak gravitational binding energy: about 0.001 eV/nucleon, which means that such globules can be rapidly evaporated by sufficient radiative or cosmic ray excitation, but the most dense parts, especially the cores, can survive several Gyr in low excitation regions. In other terms, repeated cycles of collapse/evaporation trap an increasingly fraction of the gas mass in condensed cores, until .

d) An extended atmosphere, extending up to $3 - 30\,\mathrm{AU}$ for a Jupiter mass, has rapidly an isothermal profile $\propto r^{-2}$, independently of the central core pressure. The atmosphere is expected to be transparent to sub-mm-FIR radiation, but opaque to visible/UV radiation (Combes & Pfenniger 1997).

It should be noted that the external "heating" of cold globules by penetrating radiation (FIR, cosmic rays) does not increase their apparent temperature due to the negative specific capacity of self-gravitating objects. Such "heating" expands and evaporate the atmosphere without increasing temperature, i.e., without particular emission signature.

2.3 Observational Counterparts

The best observational candidates for cold globules are the cometary knots directly observed in nearby planetary nebulae such as in the Helix Nebula (NGC 7293), but not satisfactorily explained (e.g., O'Dell et al. 2002). Their sizes ($\approx$ 100 AU), their temperature ($\sim <$ 5 K), and *their centrally concentrated profiles* (Burkert & O'Dell 1998) indicate that they are gravitationally bound around a dense center, not pressure confined clouds. They illustrate how fast cosmic gas cools and produces clumps; the condensation process leading to such gravitationally bound clouds must occur even more frequently and generally in colder conditions, such as in the outer galactic disks.

3 Conclusions

A substantial amount of cold gas provides not only an explanation for the composition of spirals, especially late types, but also how they process their gas into stars. Cold gas globules stabilized by solid/liquid H_2 cores may constitute the cold side of the star formation process. The effect is general and explains the inefficiency of usual star formation. Collapses do not need to be prevented, stars should form in higher fractions in more radiation excited, or more metal enriched (opaque) conditions, and conversely for cold globules.

The ISM fast cooling problem is removed if the outcome of collapses can be cold globules instead of stars: the ISM appears then as a very active medium constantly recycling gas from large to small scales. Cold globules are in some way the IMF low mass end. Their low binding energy allows them to play the role of a gas reservoir easily accessible via excitation. As the globule evaporation time-scale is inversely proportional to radius, at different galactic radii an equilibrium distribution of globules sizes can be reached balancing the local excitation and gas cooling rates.

References

1. Burkert, A., O'Dell, C.R. 1998, ApJ, 503, 792
2. Combes, F., & Pfenniger, D. 1997, A&A, 327, 453
3. Cuillandre, J.-Ch., Lequeux, J., Allen, et al. 2001, ApJ, 554, 190
4. Draine, B.T. 2004, in "The Cold Universe", Saas Fee Advanced Course 32, D. Pfenniger & Y. Revaz (eds.), Springer, Berlin, p. 213
5. Larson, R.B. 1969, MNRAS, 145, 271
6. O'Dell, C.R., Balick, B., Hajian, A.R., et al. 2002, AJ, 123, 3329
7. Padmanabhan, T. 2001, Theoretical Astrophysics Vol. II, Cambridge
8. Pfenniger, D. 2004, A&A, to be submitted
9. Pfenniger D., & Combes F. 1994, A&A, 285, 93 (PC94)

Collisional Excitation Processes Relevant to ISM Studies

E. Roueff[1], M.L. Dubernet[2], D.R. Flower[3], and J.T. Pottage[3]

[1] Observatoire de Paris, LUTH and UMR8102, Place J. Janssen F-92190, Meudon, France evelyne.roueff@obspm.fr
[2] LERMA - Observatoire de Paris marie-lise.dubernet@obspm.fr
[3] Physics Department, Durham University, DH1 3LE, UK david.flower@durham.ac.uk

1 Introduction

A knowledge of the rates of collisional excitation of molecules is important for both galactic and extragalactic studies in order to predict line intensities. Such processes are qualified as 'non-reactive' but 'inelastic' when the two partners of the collision keep their identity but there is an exchange between relative translational and internal energies. We shall consider mainly the theoretical aspects, but various experimental measurements may be compared with the calculations in order to establish the reliability of the theoretical data. As the critical input in the theoretical calculations is the intermolecular potential between the target molecule (which may be excited) and the perturber, it is of great value to have direct information on the potential. Van der Waals spectroscopy of the weakly bound systems is certainly the most direct probe of the intermolecular potential. Pressure broadening and double resonance experiments are other possibilities, but they involve, in addition to the intermolecular interaction, the dynamical aspects of the collision.

Section 2 outlines the theoretical context. Section 3 describes some recent results. We refer in Section 4 to the availability of a new data base for non-reactive processes, and we give in Section 5 our main conclusions and mention outstanding problems.

2 Theoretical Context

2.1 Potential Surfaces

The knowledge of appropriate intermolecular potentials is essential for scattering calculations. Inelastic collisions generally involve a single potential energy surface (PES) for the ground electronic state, but, even in this case, precise quantum chemistry calculations are very time-consuming. Angular expansions of the PES are required for the subsequent dynamical calculations, and care has to be taken when deriving such expansions. The symmetries and quasi-symmetries of the PES have to be reproduced, and this is a difficult task in the case of a multidimensional PES, as pointed out by Green [8] in the contexts of the H_2O–H_2 and NH_3–H_2 systems.

2.2 Dynamical Calculations

According to the energy range considered when calculating the cross sections, either classical or quantum mechanics should be used; the relative values of the de Broglie wavelength, $\lambda = h/(mv)$, and the dimensions of the collision partners provide a useful guide. As interstellar conditions involve low temperatures, quantum mechanics is generally more suitable. There exists a number of collision programs which are available on the web: MOLSCAT by Hutson and Green [10], MOLCOL by Flower et al. [6], and HIBRIDON by Alexander et al. [14, 1].

Various levels of approximation may be used when close coupling calculations become prohibitively expensive in computing time. The coupled states (CS) and even the infinite order sudden (IOS) approximation are appropriate at sufficiently high collision energies, where the CS approximation is generally accurate to better than a factor of 2.

Since the codes cited above are intended to be "user friendly", it might appear to be a straightforward procedure to derive collision cross-sections once the intermolecular potentials are available. However, considerable care has to be taken in the dynamical calculations themselves: the convergence of the scattering matrix elements, with respect to the number of steps and the values of the initial and final points in the propagation of the wavefunction, needs to be checked; the convergence of the cross sections with respect to the total angular momentum needs to be verified; the energy grid of the calculations has to be adapted to the presence of resonances. Finally, the range of collision energies has to be sufficient to ensure the convergence of the integral over a Maxwellian distribution, when determining rate coefficients at high temperatures.

3 Recent Results

3.1 Rotational Excitation of CO

CO has received some attention recently, and new potential surfaces have been calculated; this has encouraged different groups to determine the corresponding rate coefficients for rotational excitation by H (for rotational angular momentum $J \leq 16$: Balakrishnan et al. [2]), by He (for $J \leq 14$ up to 14: Cecchi-Pestellini et al. [3]), and by H_2 (for $J \leq 29$ for para-H_2 and $J \leq 20$ for ortho-H_2: Flower [7]). The computations were performed using the full close-coupling quantal method and either the MOLSCAT or the MOLCOL programs; the results are available on the appropriate websites. The excitation and de-excitation rate coefficients are related by detailed balance.

The collisional de-excitation rate coefficients corresponding to $\Delta J = -1$ are nearly independent of temperature between 5 and 400 K when molecular hydrogen is the perturber [7]. In addition, the rate coefficients are very similar for both para and ortho H_2. The $1 \to 0$ rate coefficient is about half that of $2 \to 1$, and the rate coefficients involving higher J with $\Delta J = -1$ are even larger but do not exceed 10^{-10} cm^3 s^{-1}.

The rate coefficients for $\Delta J = -1$ are larger than for $\Delta J = -2$ at low J. However, for values of $J > 4$, the rate coefficients for $\Delta J = -2$ transitions, induced by molecular hydrogen, become larger than the values corresponding to $\Delta J = -1$. An example is given in Figure 1, where the rate coefficients for collisional de-excitation of the $J = 7$ level of CO are shown for the various perturbers: H, He and H_2. In addition to the recent close coupling calculations, the H_2 - CO rate coefficients at high temperatures, derived by McKee et al. [15], are plotted. Their results were obtained in the IOS approximation and assuming that para-H_2 behaves similarly to He. From consideration of the reduced masses only, one would deduce that $k_{H_2} = 1.37\ k_{He}$; we see that this assumption is not justified.

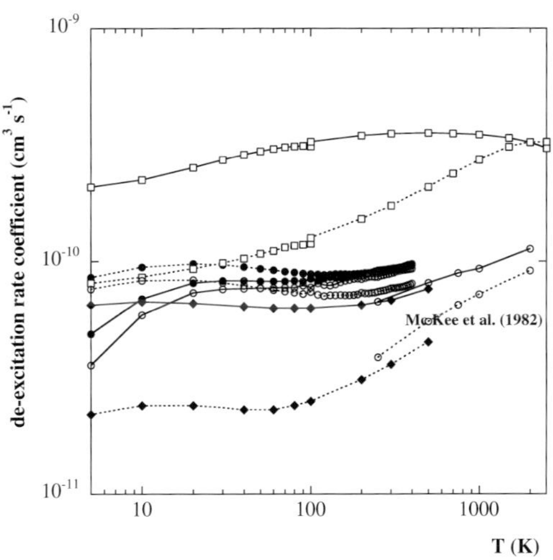

Fig. 1. Collisional de-excitation rate coefficients for CO $J = 7$; full lines refer to the $J = 7 \to 6$ transition and dotted lines to $J = 7 \to 5$. Open squares correspond to H [2], open and full circles to para- and ortho-H_2, respectively [7], and diamonds to He [3].

3.2 Rotational Excitation of H_2O

Observation of water is one of the main objectives of the HIFI instrument on Herschel, the future space infrared submillimeter telescope. Water is an important coolant of dense interstellar regions. The ortho-H_2O transition at 557 GHz has been extensively studied by the SWAS satellite, as reported by G. Melnick in this conference. Phillips et al. [18] computed the rate coefficients for the rotational excitation of H_2O by para- and ortho-H_2 in the

CS approximation, for temperatures above 20 K; the potential surface was determined on a grid of 700 different geometries [17]. This potential surface is currently being recomputed by Valiron and coworkers, using explicitly coupled cluster theory, which yields reliable results for simple closed and open shell systems (for a review, see Noga and Valiron [16]). An important goal of this work is to go beyond the rigid rotor approximation and test the effect on pure rotational excitation of the averaging over all spectator motions. Dubernet and Grosjean [4] and Grosjean, Dubernet and Ceccarelli [9] have extended the calculations of Phillips et al. [18] towards lower temperatures with the close coupling technique. Good agreement with the previous calculations is obtained at 20 K. Fig 2 displays the cross section for the $1_{10}\rightarrow 1_{01}$ de-excitation transition of ortho-H_2O, induced by para- and ortho-H_2, as a function of the collision energy.

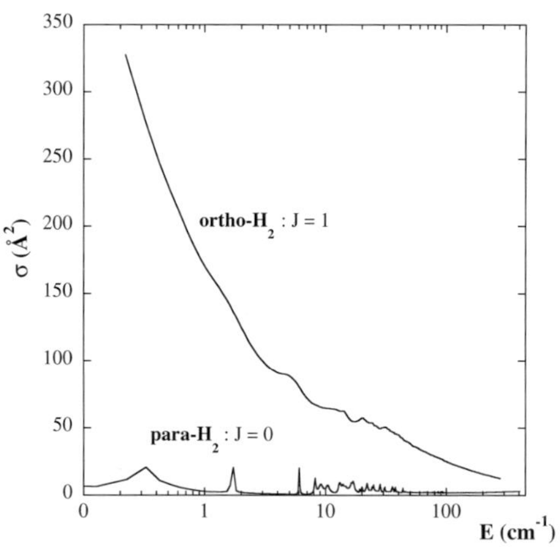

Fig. 2. Cross-section for collisional de-excitation of H_2O ($1_{10}\rightarrow 1_{01}$) by para- and ortho-H_2.

The contribution of ortho-H_2 to the collisional de-excitation is much larger than that of para-H_2. A series of resonances is apparent in the graph. Such features require that care is taken when averaging over the Maxwellian distribution to obtain the corresponding de-excitation rate coefficient. The differences between para- and ortho-H_2 arise from the strong dipole ($\mu_{H_2O} =$ 1.85 debye) - quadrupole interaction, which is present when H_2 is in its $J = 1$ rotational state.

3.3 Rotational Excitation of Methanol

Pottage, Flower and Davis [19, 20] have recently considered the collisional excitation of methanol by He in the CS approximation. The spins of the three protons of the CH_3 methyl group within CH_3OH may be aligned parallel ($I = 3/2$) or antiparallel ($I = 1/2$), corresponding to the two types of methanol (called A- and E-type, respectively), which interconvert only in proton-exchange reactions. The CH_3 and OH groups can rotate relative to each other. This internal motion is hindered by a three-fold symmetric potential with a barrier height of about 500 K. In addition to the (J, K) rotational quantum numbers, methanol has torsional states, identified by a third quantum number, ν. Rate coefficients for both A- and E-type methanol in their ground and first excited torsional states have been calculated [19, 20] for temperatures between 5 and 200 K. As a new potential surface for CH_3OH–H_2 has been computed by Davis, using many body perturbation theory, the same group is carrying out calculations of collisional excitation of methanol by para-H_2 and ortho-H_2. The results of these calculations will be used for density and temperature diagnostics, from observations of methanol lines.

4 A Data Base for Collisional Excitation Rates of Astrophysical Interest

We have described some of the recent work aimed at providing reliable collisional excitation rate coefficients for astrophysical applications. We consider that it is highly desirable to bring together such results, in preparation for the Herschel and ALMA missions. A project initiated by Marie-Lise Dubernet at the Paris Observatory aims to provide not only the rate coefficients as functions of temperature (and the corresponding fits) but also the total cross-sections as functions of energy, as well as the potential energy surfaces; the dynamical approximations will be specified and references given. Preliminary documentation is to be found at **http://basecol.obs-besancon.fr**, with special reference to the case of water - molecular hydrogen. This work is being carried out in conjunction with the Centre de Données Stellaires (CDS) in Strasbourg, with a view to the requirements of virtual observatories. It may be viewed as an extension of the data base created by the late Sheldon Green.

5 Conclusions and Future Work

We wish to emphasize the importance of advances in quantum chemistry, which now enable the (weak) interaction between a target molecule and a perturber to be determined accurately. Recent developments in studies of ultra-cold collisions have given rise to a renewal of interest in the determination of intermolecular potential energy surfaces. Tests of the Wigner law have been performed for a number of cases, including the fine structure excitation of O by He [11], incorporating reorientation and magnetic field effects [12]. A number of calculations can be undertaken with the present knowledge of intermolecular potentials, such as studying the effects of isotopic substitution. The centre of mass of the target is displaced and the symmetry properties may change considerably, as in the case of HD, which has been studied by Flower and Roueff [5, 21]. In view of the anticipated improvements in the spectral resolution of radio and submillimeter receivers, consideration should be given to the effects of hyperfine structure. We believe that it will not be possible to provide the data required to interpret the forthcoming spectroscopic observations of the molecular interstellar medium unless there is much greater investment in calculating and measuring the relevant collisional rate coefficients.

References

1. M. H. Alexander, C. Rist, D. E. Manolopoulos: J. Chem. Phys. **97**, 4836 (1992)
2. N. Balakrishnan, M. Yan, A. Dalgarno: Astrophys. J. **568**, 443 (2002)
3. C. Cecchi-Pestellini, E. Bodo, N. Balakrishnan, A. Dalgarno: Astrophys. J. **571**, 1015 (2002)
4. M.L. Dubernet, A. Grosjean: A&A **390**, 793 (2002)
5. D.R. Flower, E. Roueff: J. Phys. B **32**, 3399 (1999)
6. D.R. Flower, G. Bourhis, J.M. Launay: Comp. Phys. Com. **131**, 187 (2000), [http://ccp7.dur.ac.uk/molcol.html]
7. D.R. Flower: J. Phys. B **34**, 2731 (2001)
8. S. Green: J. Chem. Phys. **103**, 1035 (1995)
9. A. Grosjean, M.L. Dubernet, C. Ceccarelli: A&A **408**, 1197 (2003)
10. J.M. Hutson and S. Green, *MOLSCAT Version 14, distributed by Collaborative Computational Project 6* (Warrington, UK: Daresbury Laboratory) (1994), [http://www.giss.nasa.gov/molscat/]
11. R.V. Krems, A. Dalgarno: Phys. Rev. A **66**, 2702 (2002)
12. R.V. Krems, A. Dalgarno: Phys. Rev. A **68**, 3406 (2003)
13. R.P. Leavitt: J. Chem. Phys. **72**, 3472 (1980)
14. D. E. Manolopoulos, M. H. Alexander: J. Chem. Phys. **97**, 2527 (1992), [http://www.chem.umd.edu/physical/alexander/hibridon]
15. C. F. Mc Kee, J.W.V. Storey, D.M. Watson, S. Green: Astrophys.J. **259**, 647 (1982)
16. J. Noga, P. Valiron: *Computational Chemistry: Reviews of Current Trends*, vol 7, ed. J. Leszcynski (World Scientific, Singapore 2002)
17. T.R. Phillips, S. Maluendes, D. McLean, S. Green: J. Chem. Phys. **101**, 5824 (1994)
18. T.R. Phillips, S. Maluendes, S. Green: Astroph. J. Suppl. **107**, 467 (1996)
19. J.T. Pottage, D.R. Flower, S. Davis: J. Phys. B **34**, 3313 (2001)
20. J.T. Pottage, D.R. Flower, S. Davis: J. Phys. B **35**, 2541 (2002)
21. E. Roueff, D.R. Flower: MNRAS **305**, 353 (1999)

New Insights on the Thermal Dust from the Far-Infrared to the Centimeter

X. Dupac[1], J.-P. Bernard[2], N. Boudet[2], M. Giard[2], J.-M. Lamarre[3], C. Mény[2], F. Pajot[4], and I. Ristorcelli[2]

[1] European Space Agency – ESTEC, Astrophysics Division, Keplerlaan 1, 2201 AZ Noordwijk, The Netherlands, xdupac@rssd.esa.int
[2] CESR, 9 av. du Colonel Roche, BP 4346, 31028 Toulouse cedex 4, France
[3] LERMA, Obs. de Paris, 61 av. de l'Observatoire, 75014 Paris, France
[4] IAS, Campus d'Orsay, bât. 121, 15 rue Clemenceau, 91405 Orsay cedex, France

Summary. We present a compilation of PRONAOS-based results concerning the temperature dependence of the dust submillimeter spectral index, including data from Galactic cirrus, star-forming regions, dust associated to a young stellar object, and a spiral galaxy. We observe large variations of the spectral index (from 0.8 to 2.4) in a wide range of temperatures (11 to 80 K). These spectral index variations follow a hyperbolic-shaped function of the temperature, high spectral indices (1.6-2.4) being observed in cold regions (11-20 K) while low indices (0.8-1.6) are observed in warm regions (35-80 K). Three distinct effects may play a role in this temperature dependence: one is that the grain sizes change in dense environments, another is that the chemical composition of the grains is not the same in different environments, a third one is that there is an intrinsic dependence of the dust spectral index on the temperature due to quantum processes. This last effect is backed up by laboratory measurements and could be the dominant one.

We also briefly present a combined analysis of WMAP dust data together with COBE/DIRBE and COBE/FIRAS data.

1 Introduction

Precise characterization of dust emissivity properties is a major challenge of modern day astronomy. In the submillimeter domain, large grains at thermal equilibrium (e.g. [3]) dominate the dust emission. This thermal dust is characterized by a temperature and a spectral dependence of the emissivity which is usually simply modelled by a spectral index. The temperature, density and opacity of a molecular cloud are key parameters which control the structure and evolution of the clumps, and therefore, star formation. The spectral index (β) of a given dust grain population is directly linked to the internal physical mechanisms and the chemical nature of the grains.

It is generally admitted from Kramers-König relations that 1 is a lower limit for the spectral index. $\beta = 2$ is particularly invoked for isotropic crystalline grains, amorphous silicates or graphitic grains. However, it is not the case for amorphous carbon, which is thought to have a spectral index equal to 1. Spectral indices above 2 may exist, according to several laboratory measurements on grain analogs. Observations of the diffuse interstellar medium at large scales favour β around 2 (e.g. [2], [4]). In the case of molecular clouds, spectral indices are usually found to be between 1.5 and 2. However, low values (0.2-1.4) of the spectral index have been observed in circumstellar environments, as well as in molecular cloud cores.

2 The Dust Spectral Index as Seen by PRONAOS

PRONAOS (PROgramme NAtional d'Observations Submillimétriques) is a French balloon-borne submillimeter experiment [9]. Its effective wavelengths are 200, 260, 360 and 580 μm, and the angular resolutions are $2'$ in bands 1 and 2, $2.5'$ in band 3 and $3.5'$ in band 4. The data analyzed here were obtained during the second flight of PRONAOS in September 1996, at Fort Sumner, New Mexico. This experiment has observed various phases of the interstellar medium, from diffuse clouds in Polaris [1] and Taurus [10] to massive star-forming regions in Orion ([9], [5]), Messier 17 [6], Cygnus B, and the dusty envelope surrounding the young massive star GH2O 092.67+03.07 in NCS. The ρ Ophiuchi low-mass star-forming region has also been observed, as well as the edge-on spiral galaxy NGC 891 [8].

We fit a modified black body law to the spectra: $I_\nu = \epsilon_0 B_\nu(\lambda, T)\ (\lambda/\lambda_0)^{-\beta}$, where I_ν is the spectral intensity (MJy/sr), ϵ_0 is the emissivity at λ_0 of the observed dust, B_ν is the Planck function, T is the temperature and β is the spectral index. In most of the areas, we use either only PRONAOS data or PRONAOS + IRAS 100 μm data. We restrict the analysis to all fully independent ($3.5'$ side) pixels for which both relative errors on the temperature and the spectral index are less than 20%. This procedure allows to reduce the degeneracy effect between the temperature and the spectral index. Dupac et al. (2001, 2002) have shown by fitting simulated data that this artificial anticorrelation effect was small compared to the effect observed in the data.

Fig. 1 shows th observed spectral index-temperature relation. The temperature in this data set ranges from 11 to 80 K, and the spectral index also exhibits large variations from 0.8 to 2.4. We note in these plots an anticorrelation between the temperature and the spectral index, in the sense that the cold regions have high spectral indices around 2, and warmer regions have spectral indices below 1.5. In particular, no data points with $T > 35$ K and $\beta > 1.6$ can be found, nor points with $T < 20$ K and $\beta < 1.5$. This anticorrelation effect is present for all objects in which we observe a large range of temperatures, namely Orion, M17, Cygnus and ρ Ophiuchi. It is also remarkable that the few points from other regions are compatible with this general anticorrelation trend. The temperature dependence of the emissivity spectral index is well fitted by a hyperbolic approximating function.

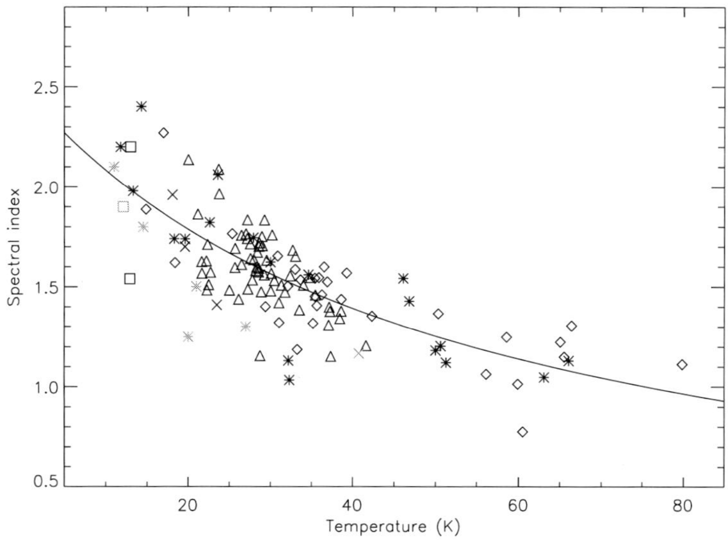

Fig. 1. Spectral index versus temperature, for fully independent pixels in Orion (black asterisks), M17 (diamonds), Cygnus (triangles), ρ Ophiuchi (grey asterisks), Polaris (black squares), Taurus (grey square), NCS (grey cross) and NGC 891 (black crosses). The full line is the result of the best hyperbolic fit: $\beta = \frac{1}{0.4+0.008T}$

Several interpretations are possible for this effect: one is that the grain sizes change in dense environments, another is that the chemical composition of the grains is not the same in different environments and that this correlates to the temperature, a third one is that there is an intrinsic dependence of the spectral index on the temperature, due to quantum processes such as two-level tunneling effects. Additional modeling, as well as additional laboratory measurements and astrophysical observations, are required in order to discriminate between these different interpretations. More details about this analysis and the possible interpretations can be found in [7].

3 The Galactic Dust Emission as Seen by COBE and WMAP

We combine data sets from COBE (DIRBE data points and FIRAS spectrum) and WMAP in order to investigate the Galactic dust spectral energy distribution on the whole sky. We convolve all data to the FIRAS 7° angular resolution, in order to obtain a consistent data set. This is done by co-adding higher resolution data from DIRBE and WMAP into 7 degree-square pixels. The resulting spectrum is shown in Fig. 2 for the Galactic center area. It is well fitted by a 21 K, β=1.86 one-dust component model. These values are well compatible with the PRONAOS temperature - spectral index relation (see Section 2). More results and conclusions from this COBE - WMAP joint analysis will soon be available.

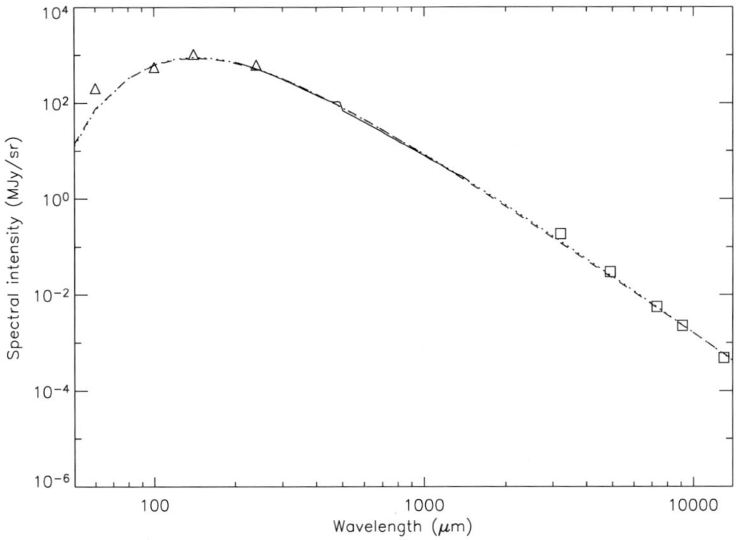

Fig. 2. Galactic dust spectrum in the Galactic center region. DIRBE data points are presented as triangles, the FIRAS spectrum is presented as a full line, and the squares are the WMAP data points. The dashed and dotted lines are the results of, respectively, the one-component and two-component fits, without the 60 μm data points.

References

1. Bernard, J.-P., Abergel, A., Ristorcelli, I., et al.: 1999, A&A, 347, 640
2. Boulanger, F., Abergel, A., Bernard, J.-P., et al.: 1996, A&A, 312, 256
3. Désert, F.-X., Boulanger, F., Puget, J.-L.: 1990, A&A, 237, 215
4. Dunne, L., Eales, S.A.: 2001, MNRAS, 327, 697
5. Dupac, X., Giard, M., Bernard, J.-P., et al.: 2001, ApJ, 553, 604
6. Dupac, X., Giard, M., Bernard, J.-P., et al.: 2002, A&A, 392, 691
7. Dupac, X., Bernard, J.-P., Boudet, N., et al.: 2003, A&A Lett., 404, L11
8. Dupac, X., del Burgo, C., Bernard, J.-P., et al.: 2003, MNRAS, 344, 105
9. Ristorcelli, I., Serra, G., Lamarre, J.-M., et al.: 1998, ApJ, 496, 267
10. Stepnik, B., Abergel, A., Bernard, J.-P., et al.: 2003, A&A, 398, 551

Photodissociation Regions

A. Sternberg

School of Physics and Astronomy, Raymond and Beverly Sackler Faculty of Exact Sciences, Tel Aviv University

1 Introduction

Photodissociation regions or photon-dominated regions (PDRs) are most broadly defined as the neutral components of the interstellar medium (ISM) where penetrating far-ultraviolet (FUV; 6-13.6 eV) radiation dominates the gas heating and thermal balance, and controls the chemical structure. Dense PDRs usually refer to high density (hydrogen densities $n_{\rm H} > 10^{-3}$) and optically thick (visual extinctions $A_V > 5$) clouds that are exposed to elevated FUV fields. Dense PDRs are sources of luminous millimeter, sub-millimeter, and far-infared line and continuum radiation, including O I 63 μm and C II 158 μm and other fine-structure emission lines, and rotational and vibrational emissions from a variety of molecules including the most abundant CO and H_2. PDRs are present at the edges of H II regions, in reflection and planetary nebulae, and perhaps most importantly in dense clumpy star-forming molecular clouds illuminated by direct and scattered stellar FUV radiation. PDRs can extend up to large cloud depths $A_V \sim 10$. As such, most of the molecular ISM may be "photon-dominated". The classical diffuse ($A_V < 1$) and translucent ($A_V < 5$) clouds may be considered examples of "low-density" PDRs.

The study of PDRs dates back to the late 70's and early 80's when advances in detector technology opened up the submillimeter and far-infrared spectral windows [1]–[6]. Comprehensive reviews of PDR theory and observations may be found in [7] and [8]. In my brief discussion I will focus on a selection of topics and some recent results. I suggest reading this article while also viewing my power-point presentation available on the conference CD.

2 Physical Processes

A beautiful example of a localized PDR is represented by the Ring Nebula [9] (see slide 4). Near-IR imaging of this source at 2.12 μm reveals prominent molecular hydrogen (H_2) vibrational emission in the 1-0 S(1) line from dense neutral gas superposed on the optical emission line tracers of the hot ionized plasma. The H_2 image clearly reveals a complex clumpy morphology, with spots of enhanced molecular emission associated with dark locations in the optical images. These are likely dense photoevaporating clumps in time-dependent PDRs [10, 11]. Vibrational H_2 emissions are readily produced in

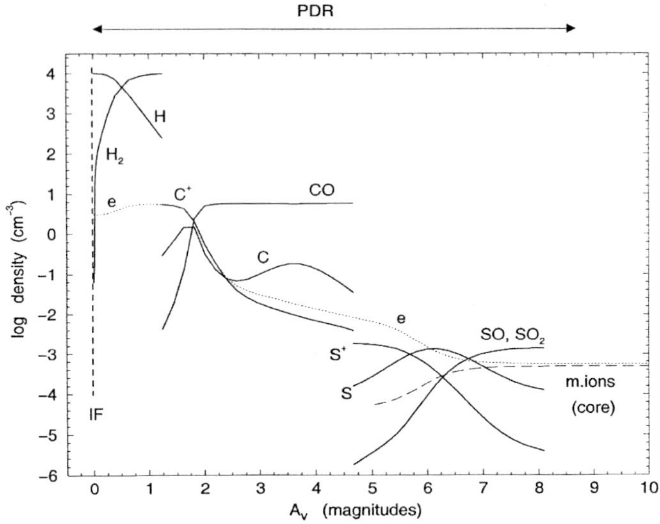

Fig. 1. Schematic structure of a PDR. FUV radiation is incident from the left.

PDRs by either radiative FUV-pumping, or by collisional excitation in hot ($> 10^3$ K) FUV heated gas [12]–[16].

Theoretical models have increased in sophistication over the years, but the basic picture is represented by one-dimensional plane parallel models (see slide 2, and Figure 1). PDRs consist of several distinct zones and transition layers, from outer photodissociated gas to inner fully molecular gas [17, 18]. The H/H_2 transition occurs closest to the cloud surface where the combined effects of dust opacity and H_2 self-shielding operate to reduce the ratio of the rates of molecular photodissociation to H_2 formation with increasing cloud depth. Molecular hydrogen forms on grain surfaces, e.g. [19], with gas phase processes becoming important in low-metallicity environments. The C^+/C/CO occurs deeper in the cloud. A thin layer of neutral atomic carbon is present between the outer parts where carbon is singly ionized, and deeper locations where all of the gas-phase carbon is locked in CO. A similar S/S^+/SO,SO_2 transition zone occurs at still larger A_V. The sizes of the various transition layers are controlled by several parameters including the cloud density or pressure, the intensity of the illuminating radiation, the grain FUV scattering properties, the H_2 formation rate coefficient, the cloud geometry and clumpiness, and the elemental gas phase abundances.

The FUV intensity is a crucial parameter. In the immediate vicinity of hot OB stars the FUV fields can be orders of magnitude more intense than at a random location in the ISM of the Galactic disk. The typical interstellar FUV field is also expected to fluctuate both temporally and spatially as individual OB associations form and disappear in the course of the Galaxy lifetime (slide 6). Ab initio computations [20] show that dust opacity limits the contributing associations out to distances of about 500 pc. The theoretical estimates are consistent with observational determinations of the mean interstellar FUV field [21, 22].

FUV energy can couple to the gas via the well studied process of grain photoelectric emission [23]–[25]. In PDRs most of the incident FUV energy is absorbed by dust particles and reradiated as quasi-thermal IR, but a small fraction of the FUV heats the gas via the photoelectric ejection of electrons from the grain surfaces The typical "work-function", for interstellar grains is $W = 6$ eV, and the expected photoelectic yield, or the fraction of electrons that are actually dislodged off the grains, is typically $Y = 0.1$. Therefore, for a mean incident FUV photon energy of 10 eV, a fraction $([10-6]/10) \times 0.1 = 0.04$ of the original photon energy is available to heat the gas. This fraction, or "heating efficiency", decreases if the grains become positively charged, as they do when the photoelectron emission rates become large relative to the grain-electron recombination rates. This argument predicts cooling emission lines (e.g. [OI] 63 μm and [CII] 158 μm fine-structure lines) with intensities of order 1% of the thermal IR dust continua, consistent with observations. Photoelectric heating is probably dominated by the smallest (< 15 Å) grains or PAHs for typical grain size distributions [24, 25].

An additional heating process that operates in dense PDRs is the collisional deexcitation of FUV-pumped H_2. The combination of this heating process with photoelectric heating, and the quenching of the cooling transitions at high density, can lead to gas temperatures in excess of 10^3 K in the outer atomic hydrogen layers in very dense PDRs.

Photoelectric heating can account for the observed intensities of the fine-structure emission lines observed in PDRs. The absolute and relative intensities of the [OI] and [CII] lines, depend on the FUV intensity and density, and the emission line ratios are useful diagnostics [26] (slide 11). For example, for fixed gas density the line strengths increase and then saturate with increasing FUV as the grains become positively charged and the heating efficiencies are diminished.

PDRs are natural sites for the production of the fine-structure emission lines from atom and ions with ionization potentials longward of the hydrogen Lyman limit. For example, long before it was first detected the C II line was recognized theoretically as a likely major cooling transition for the neutral hydrogen component of the Galactic ISM [23] - [27]. Indeed, C II fine-structure cooling plays a crucial role in the theory of the two-phase CNM/WNM structure of H I gas in the ISM [28, 29] dominating the energy losses of the cool phase.

However, the early theoretical considerations focussed primarily on relatively low density gas ($n_{\rm H} < 50$ cm^{-3}), where the fine-structure levels are subthermally populated and the C II emissivities are low (slide 9). For fixed density, the cooling efficiency decreases sharply for temperatures T below the transition energy of 92 K, and levels off at higher T, where the Boltzmann

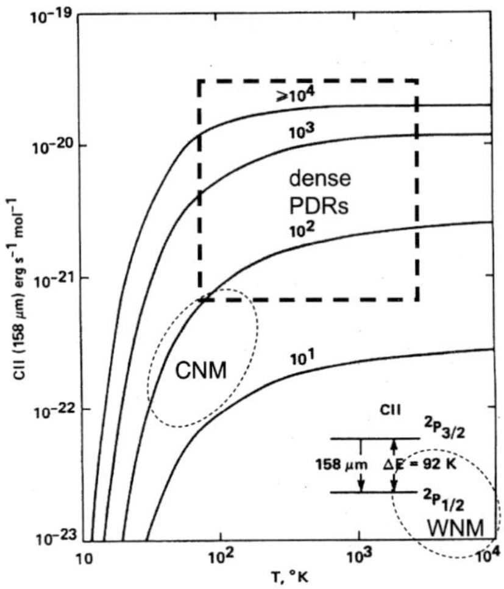

Fig. 2. C II line cooling per particle as a function of the gas temperature and hydrogen density.

factor in the collisional excitation rate coefficient approaches unity. At fixed gas temperature the efficiency increases linearly with density up to the critical hydrogen density of $\sim 10^4$ cm^{-3} above which collisional deexcitations quench the radiative emissions. For conditions characteristic of the classical WNM ($n_H \approx 0.4$ cm^{-3} $T \approx 8000$ K) and CNM ($n_H \approx 60$ cm^{-3} $T \approx 50$ K), the C II emission efficiencies are well below the maximum possible. However, for densities greater than 10^2 and up to at least 10^4 cm^{-3}, and for T above $\sim 10^2$ K, the cooling efficiencies are significantly enhanced. This range of elevated density and temperature corresponds to conditions characteristic of dense PDRs.

3 Chemistry

PDRs consist of several distinct chemical zones in which different sequences of reactions control the gas-phase production of molecules and molecular ions [18, 30]. For example, in the warm outer parts of PDRs a variety of species are produced via sequences initiated by the formation of the hydroxyl radical OH. At temperatures above a few hundred K, such as can occur in the H/H_2 transition zone [31], OH is very efficiently formed by the neutral-neutral reaction $O + H_2 \rightarrow OH + H$, and is removed by photodissociation. (In cooler gas OH is formed less efficiently via ion-molecule reactions.) In PDRs the OH radical can then readily react $OH + X \rightarrow XO + H$, where X is any of the abundant PDR species, C^+, N, O, S^+, or Si^+ (slides 14 and

15). For example, the reactions with C^+ give CO^+ which then reacts with H_2 to give HCO^+. The model computations thus predict that the abundance ratio CO^+/HCO^+ is large in PDRs, but small in shielded cores where the saturated ion HCO^+ is preferred. This molecular ion ratio is thus a useful diagnostic of PDR versus dark cloud chemistry [18].

The predictions have been nicely verified in observations of several sources including the reflection nebula NGC 7023 [32] (slide 16). This is an edge-on PDR illuminated by a Herbig B3Ve star. At the interface the FUV intensity is 2.4×10^3 (in Habing units [21]), and the density of the PDR is estimated at $n_H \approx 10^4$ cm^{-3}. In this source the CO^+/HCO^+ ratio decreases sharply with increasing depth into the cloud, from ~ 0.1 in the PDR to less than 0.001 in the shielded molecular core.

Another example of PDR chemistry is the atomic carbon produced in the $C^+/C/CO$ transition layer. Model computations [17, 18] show that the C I abundance profile is actually double peaked (see Figure 1). A first maximum occurs at visual extinctions where the C^+ recombination becomes competitive with photoionization, but where CO is still being dissociated. A second maximum occurs at higher opacities where the carbon and sulfur chemistries interact via charge transfer $S + C^+ \rightarrow S^+ + C$. This second maximum occurs at the location of the neutral atomic sulfur peak in the PDR.

However, the presence of extended C I in molecular clouds, e.g. [33], remains a nagging problem given the models which predict that atomic carbon should be confined to narrow surface layers. Extended C I may be a signature of scattered FUV radiation in a clumpy medium. Alternatively, time-dependent effects in shadowed clumps in which the cooling time is longer than the recombination time could also be playing a role [34].

The presence of PAHs may also affect the atomic carbon abundances, given that when sufficiently abundant, PAHs directly influence the electron abundances and resulting ionization fractions. For PAH abundances greater than $\sim 10^{-7}$ (relative to hydrogen) mutual neutralization $X^+ + PAH^- \rightarrow X + PAH$ between positive ions and negative PAH^- ions can compete or dominate radiative recombination $X^+ + e \rightarrow X + \nu$ as pointed out by [35]. The inclusion of PAH chemistry in PDRs leads to enhancements in the C I abundances and emissivities as neutralization occurs closer to the warmer cloud surfaces [36].

4 External Galaxies

Another nice example of PDR phenomena is the immediately recognizeable "Antennae" pair of colliding galaxies NGC 4038/4039. HST imaging shows that the system contains hundreds of compact clusters of young OB stars that are interacting with surrounding dense clouds which were likely compressed by the collision (slide 21). Far-infrared spectroscopy reveals that C II emission line is produced from the entire galaxy pair [37] but interestingly, the emission peaks at the location of the "interaction zone" located between the individual galaxy nuclei. The total luminosity in the C II line is 3.7×10^8 $L_\odot$, corresponding to about 1% of the far-IR (IRAS) luminosity. A global FUV intensity of about 500 Habing units is inferred for the interaction zone. The FUV radiation is interacting with a total gas mass of at least a few 10^7 $M_\odot$.

The C II measurements in the Antennae are limited by the low spatial resolution ($55''$) of the far-IR observations. Sub-arcsecond near-IR Keck spectroscopy [39] of the embedded mid-IR star cluster detected in ISO observations [40] reveals a spectrum containing a wealth of vibrational H_2 emission lines (slide 24). The luminosity in the 1-0 S(1) line is 9600 $L_\odot$. The relative intensities of lines in the $v = 1$ and $v = 2$ bands indicate unambiguously that the excitation mechanism is FUV-pumped fluorescence [12]–[16] (slide 25). The ortho-to-para ratio in excited states is 1.6 ± 0.07. This is nicely consistent with what is expected for FUV-pumping in optically thick absorption lines, for a ground vibrational state with an equilibrium rotational ortho-to-para ratio of 3 [41] (slides 27 and 28).

The C II to FIR luminosity ratio in the Antennae pair is consistent with expectations from PDR theory. An interesting puzzle that emerged from ISO observations is the apparent C II line deficit in ultraluminous IR galaxies (slide 22). Possible explanations for this deficit include (a) large ratios of FUV flux to gas density in the ULIRG PDRs leading to inefficient photoelectric gas heating, (b) an (unlikely) relative underabundance of OB stars, and (c) a significant absorption of FUV photons in dusty HII regions with large ionization parameters. This last possibility is favored in a recent reanalysis [38], although it remains unclear whether this idea is compatible with the overall fine-structure emission line spectra in such objects.

5 Dynamics

The recognition that the bulk of the molecular ISM in galaxies, including the Milky Way, is photon-dominated motivated the important concept of photo-ionization regulated star-formation [42]. The underlying notion is that star-formation is controlled by the rate of ambipolar diffusion flux loss in magnetically subcritical clouds [43, 44]. The ambipolar diffusion time is $t_{\rm AD} = 10^{14} x_e$ yr, where x_e is the fractional ionization. In the outer parts of PDRs $x_e \approx 10^{-4}$, where the electrons are produced by the photoionization of carbon, and C II carries the positive charge. The magnetic field is effectively trapped in such gas, and these parts of the cloud cannot collapse. In the optically thick cloud cores, cosmic-rays are the dominant source of ionization, and the fractional ionizations are much lower, with $x_e < 10^{-7}$. Molecular ions or heavy metals carry the positive charge in the cores. In the regulation scheme, the cosmic-ray cores collapse rapidly via rapid flux loss. This leads to an increased rate of star formation. Energy injection via protostellar outflows results in cloud expansion. The cloud optical depths are thereby reduced, and the mass fraction in the cosmic-ray cores decreases. This then leads to a reduced star-formation rate, cloud contraction, enhanced shielding, and a corresponding increase in the ambipolar diffusion rate and flux loss. This feedback loop yields a dynamic equilibrium at an overall star-formation rate of 4 $M_{\odot}$ yr^{-1} in the Galaxy, and also requires that star-formation be resticted to regions with high ($A_V > 4$) extinction, generally consistent with available observations.

6 Conclusion

The study of PDRs will remain important for many years to come, especially with the recent advent of the Spitzer Space Telescope, and other upcoming facilities including SOFIA, ALMA, and Herschel.

References

1. Melnick, G., Gull, G.E., & Harwit, M. 1979, ApJ, 227, L29
2. Storey, J.W.V., Watson, D.M,. & Townes, C.H., 1979, ApJ, 233, 109
3. Russell, R.W., Melnick, G., Gull, G.E., & Harwit, M. 1980, ApJ, 240, L99
4. Genzel, R., Watson, D.M., Townes, C.H., Dinerstein, H.L., Hollenbach, D., Lesterm D.F., Werner, M., & Storey, J.W.V. 1984, ApJ, 276, 551
5. Crawford, M.K., Genzel, R., Townes, C.H., & Watson, D.M. 1985, ApJ, 291, 755
6. Lugten, J.B., Watson, D.M., Crawford, M.K., & Genzel, R. 1986, ApJ, 311, L51
7. Hollenbach, D., & Tielens, A.G.G.M. 1997, ARAA, 35, 179
8. Hollenbach, D., & Tielens, A.G.G.M. 1999, Rev. Mod. Phys., 71, 173
9. Speck, A.K., Meixner, M., Jacoby, G.H., & Knezek, P.M. 2003, PASP, 115, 170
10. Natta, A., & Hollenbach, D. 1998, 337, 517
11. Goldshmidt, O., & Sternberg, A. 1995, 439, 256
12. Black, J.H., & Dalgarno, A. 1976, ApJ, 203, 132
13. Black, J.H., & van Dishoeck, E.F. 1987, ApJ, 322, 412
14. Sternberg, A. 1988, ApJ, 332, 400
15. Sternberg, A., & Dalgarno, A. 1989, ApJ, 338, 197
16. Draine, B., & Bertoldi, F. 1996, ApJ, 468, 269
17. Tielens, A.G.G.M., & Hollenbach, D. 1985, 291, 722
18. Sternberg, A., & Dalgarno, A. 1995, ApJS, 99, 565
19. Cazaux, S., & Tielens, A.G.G.M. 2003, ApJ, 577, 127
20. Parravano, A., Hollenbach, D.J., & McKee, C.F. 2003, ApJ, 584, 797
21. Habing, H.J. 1969, Bull. Astr. Inst. Netherlands, 19, 421
22. Draine, B.T. 1978, ApJS, 36, 395
23. Spitzer, L.Jr., 1948, ApJ, 107, 6

24. Bakes, E.L.O., & Tielens, A.G.G.M. 1994, ApJ, 427, 822
25. Weingartner, J.C., & Draine, B.T. 2001, ApJS, 134, 263
26. Kaufman, M.J., Wolfire, M.G., Hollenbach, D.J., & Luhman, M.L. 1999, ApJ, 527, 795
27. Dalgarno, A., McCray, R.A. 1972, ARAA, 10, 375
28. Field, G.B., Goldsmith, D.W., & Habing, H.J. 1969, ApJ, 155, L149
29. Wolfire, M.G., McKee, C.F., Hollenbach, D., Tielens, A.G.G.M. 2003, 587, 278
30. Jansen, D.J., Spaans, M., Hogerheijde, M.R., & van Dishoeck, E.F. 1995, AA, 303, 541
31. Timmermann, R., Bertoldi, F., Wright, C.M., Drapatz, S., Draine B.T., Haser, L., & Sternberg, A. 1996, AA, 315, 281
32. Fuente, A., Rodriguez-Franco, A., Garcia-Burillo, S., Martin-Pintado, J., & Black, J.H. 2003, AA, 406, 899
33. Kamegai, K., et al. 2003, ApJ, 589, 378
34. Stoerzer, H., Stutzki, J., & Sternberg, A. 1997, AA, 323, L13
35. Lepp, S., & Dalgarno, A. 1988, ApJ, 335, 769
36. Bakes, E.L.O., & Tielens, A.G.G.M. 1998, ApJ, 499, 258
37. Nikola, T., Genzel, R., Herrmann, F., Madden, S.C., Poglitsch, A., Geis, N., Townes, C.H., & Stacey, G.J. 1998, ApJ, 504, 749
38. Luhman, M.L., Satyapal, S., Fischer, J., Wolfire, M.G., Sturm, E., Dudley, C.C., Lutz, D., & Genzel, R. 2003, ApJ, 594, 758
39. Gilbert, A.M. et al. 2000, ApJ, 533, L57
40. Mirabel, I.F. et al. 1998, AA, 33, L1
41. Sternberg, A., & Neufeld, D. 1999, ApJ, 516, 371
42. McKee, C.F. 1989, ApJ, 345, 782
43. Shu, F.H., Adams, F.C., & Lizano, S. 1987, ARAA, 25, 23
44. Shu, F.H., Allen, A., Shang, H., Ostriker, E.C., & Li, Z.-Y. 1999, in The Origin of Stars and Planetary Systems, edited by C.J. Lada and N.D. Kylafis, Kluwer Academic Publishers, p. 193

Molecular Clouds in Weak FUV Fields: Observations and PDR Model

F. Bensch[1], U. Leuenhagen[2,3], J. Stutzki[3], and R. Schieder[3]

[1] Center for Astrophysics, 60 Garden Street, Cambridge, MA 02138, USA
fbensch@cfa.harvard.edu
[2] LB Kiel, Martensdamm 6, 24103 Kiel, Germany
[3] Universität zu Köln, Zülpicher Strasse 70, 50937 Köln, Germany

We present a survey of the [CI] $^3P_1 \rightarrow ^3P_0$ fine-structure transition made with the Submillimeter Wave Astronomy Satellite, SWAS, towards translucent and dark clouds in the interstellar radiation field, ISRF. The SWAS data is supplemented by ground-based observations of the three lowest rotational transitions of ^{12}CO and ^{13}CO. The comparison of the observed [CI]/CO line ratios with current photo-dominated region (PDR) models shows that the models consistently suggest a high gas volume density of $n > 10^5\,\mathrm{cm}^{-3}$ throughout the observed region. This density is uncomfortably high for a translucent molecular cloud. PDR models with the dense clumps being embedded in a low-density ($n_{\mathrm{ic}} < 10^3\,\mathrm{cm}^{-3}$) inter-clump medium with a modest H_2 column density of $N_{\mathrm{ic}}(H_2) \lesssim (0.5-1)\times 10^{20}\,\mathrm{cm}^{-2}$, in contrast, match the observed line ratios for much lower densities, $n \lesssim 10^4\,\mathrm{cm}^{-3}$.

1 Introduction

Diffuse and translucent molecular clouds are characterized by a low visual extinction of typically a few magnitudes and their chemistry is expected to be governed by the weak ISRF [1, 2]. "Translucent regions" are also found close to the surface of molecular clouds with a larger visual extinction and most of their carbon in CO. Irradiated by the ISRF, the chemistry in a layer with $0.5 \lesssim A_V \lesssim 3$ is dominated by the external FUV field, and, similar to the translucent clouds, the dissociation of CO gives rise to a detectable signal for the [CI] and [CII] fine-structure lines of neutral and ionized carbon. Here, we present [CI] and low-J CO observations of translucent regions made with the SWAS satellite and ground based radio telescopes: MCLD 123.5+24.9, a translucent cloud in the Polaris Flare, and B5, a dark cloud in the Perseus complex. We investigate whether or not the detected emission is consistent with clouds irradiated by the ISRF, using PDR models.

2 Observations

MCLD 123.5+24.9 is one of the best-studied high latitude clouds. It is a nearby cloud ($\sim$ 150 pc), allowing observations to be made with high spatial resolution (0.1 pc$\equiv$ 2$'$3). The visual extinction of $0.5 \lesssim A_v \lesssim 2.5$ [3] classifies MCLD 123.5+24.9 as a translucent cloud and a large observational database in the literature provides constraints for cloud parameters, such as the external FUV field ($\chi \approx \chi_0$) and chemical composition. This makes MCLD 123.5+24.9 an ideal test case for a translucent cloud PDR model.

The [CI] $^3P_1 \rightarrow {}^3P_0$ emission of MCLD 123.5+24.9 was mapped using the SWAS satellite. The [CI] data are supplemented with observations of low-J ^{12}CO and ^{13}CO transitions made with the FCRAO 14 m and KOSMA 3 m telescopes. The analysis of these observations is discussed in detail in [4]. Table 1 compares the results for three of the [CI]/CO line ratios to published (sub-)mm observations of translucent clouds. These line ratios are used for a comparison to PDR models in Sect. 3 below.

Table 1. $F_{[CI]}/F_{CO}$ flux density ratios for MCLD 123.5+24.9 and translucent cloud observations in the literature (F in units of erg cm^{-2} s^{-1} sr^{-1}). In addition, the ratios for 3 positions in B5 are given. The co-ordinates give the offset from RA=3^{h}44^{m}16^s Dec=+32°17$'$18$''$, J2000.

target cloud	[CI]/^{12}CO(1-0)	[CI]/^{12}CO(3-2)	[CI]/^{13}CO(1-0)
MCLD 123.5+24.9	7.0±2.2	1.1±0.4	43±22
Translucent clouds (lit.)	7.8-27	1.0-2.3	70-620
B5 (0$'$, 0$'$)	9.0±1.3	0.52±0.07	29±4
B5 (0$'$, +19$'$2)	14.8±2.5	1.05±0.17	70±12
B5 (0$'$, +25$'$5)	<4.7(3σ)	<0.91(3σ)	<271(3σ)

B5 is a dark cloud in the Perseus cloud complex exposed to the ISRF. Similarly to MCLD 123.5+24.9, B5 is relatively nearby ($d \sim 300$ pc) and has been studied in many tracers.The region harbors a dense core with a high visual extinction ($A_v > 6$) and ongoing low-mass star formation. The physical conditions and chemical composition of the diffuse, neutral halo surrounding B5 is studied in [5, 6]. SWAS [CI] observations were made along a N-S cut extending from the core across the cloud edge and into the surrounding halo. They are supplemented by mapping observations of the three lowest rotational transitions of ^{12}CO and ^{13}CO. The spectra for three selected positions along the N-S cut are shown in Fig. 1. The [CI] line intensity toward the dense core typically is $F_{[CI]} \approx 4.4$ erg cm^{-2} s^{-1} sr^{-1} and decreases along the N-S cut toward the cloud edge (Fig. 1). The emission at the cloud edge shows no limb-brightening, in contrast to the expectation for [CI] as a cloud surface tracer. No [CI] emission is detected in the cloud halo despite the high sensitivity of the observations, $dN_{CI}/dv < 2.5 \times 10^{15}$ cm^{-2} (km/s)$^{-1}$. In contrast, weak CO and ^{13}CO emission is detected in the halo, if convolved to the angular resolution of the SWAS data (4$'$3). This is surprising, because neutral carbon is expected to be significantly more abundant than CO in the low-A_v regions of the diffuse cloud halo. The line ratios derived from the observations show some variation as a function of position in the cloud (Table 1). Generally, however, these ratios are within a factor of 2-3 of the ratios derived for translucent clouds.

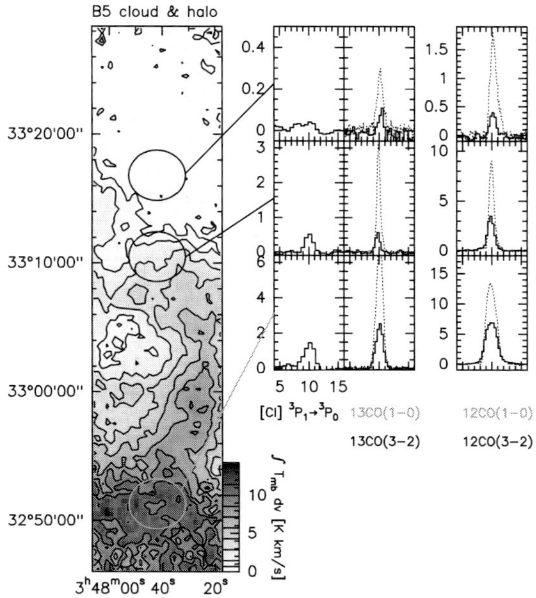

Fig. 1. [CI] and CO emission of the B5 cloud. Left: velocity integrated ^{13}CO J=1→0 map, extending from the the cloud core near the bottom end of the map across the cloud edge into the surrounding medium (0.′78 resolution). Here, we arbitrarily define the cloud edge as the 3σ contour, the lowest contour shown at $I = 0.67\,\mathrm{K\,km/s}$. Right: Sample spectra toward the cloud core, edge and the surrounding medium ($T_{\mathrm{mb}}(v)$, in [K]). All spectra have been smoothed to 4′3, the angular resolution of the SWAS observations. The transition is indicated at the bottom of each column.

3 PDR Modeling

PDR models are used to address the question whether or not the observations are consistent with clouds exposed to the ISRF. For this purpose we employ the PDR model for spherically symmetric clouds developed at the Universität zu Köln [7]. The spherical clumps have a power law density profile, $n(r) = n_s(r/r_c)^{-1.5}$, for $r_c \leq r \leq 0.2\,r_c$, and a constant density of $n(r) = 11.2\,n_s$, for $r \leq 0.2\,r_c$. The total mass of the clump is M_{cl}, and r_c, n_s denote the the clump radius and the density at the surface. The average clump density is approximately twice the clump surface density, $n_{\mathrm{cl}} \approx 2 \times n_s$. The comparison of the observations to the PDR model shows that the observed [CI]/CO line ratio consistently requires high densities, in excess of

$10^5\,\mathrm{cm}^{-3}$, *throughout* the cloud [4]. As an example, Table 2 gives the clump-averaged line ratios for a PDR model with $M_{\rm cl} = 0.1\,\mathrm{M}_\odot$, and two different densities, $n_{\rm cl} = 6 \times 10^3\,\mathrm{cm}^{-3}$ (Model 1) and $n_{\rm cl} = 6 \times 10^5\,\mathrm{cm}^{-3}$ (Model 2). This assumes that the cloud is composed of a larger number of clumps with the line emission being dominated by clumps of this size. Although high densities are not strictly ruled out by observations, it is difficult to understand which processes can maintain the pressure balance between a high density clump embedded the low density gas, in particular for translucent clouds. An alternative explanation for the observed line ratios is provided with the presence of a small column density of molecular hydrogen in a diffuse gas of low density ($n_{\rm ic} < 10^3\,\mathrm{cm}^{-3}$) surrounding the higher density clumps. The H_2 in this diffuse gas significantly reduces the photo-dissociation of CO for the embedded molecular material because of H_2 line absorption of FUV photons. As a consequence, the [CI]/CO line ratios are significantly reduced. For MCLD 123.5+24.9, the observed line ratios are consistent with a density for the line-emitting molecular cloud of $n_{\rm cl} \sim 10^4\,\mathrm{cm}^{-3}$, embedded in a diffuse inter-clump gas with a H_2 column density of $N_{\rm ic}(\mathrm{H}_2) \approx (0.5-1) \times 10^{20}$. Model 3 in Table 2 give the PDR results for a $M_{\rm cl} = 0.1\,\mathrm{M}_\odot$, $n_{\rm cl} = 6 \times 10^3\,\mathrm{cm}^{-3}$ clump embedded in diffuse gas with $N_{\rm ic}(\mathrm{H}_2) = 10^{20}\,\mathrm{cm}^{-2}$. It is interesting to note that this is consistent with the H_2 column density detected by UV absorption measurements toward the B5 cloud halo [5, 6], suggesting that the observed, low [CI]/CO line ratios indeed are due to CO shielding by H_2 in a diffuse halo surrounding the molecular cloud.

Table 2. PDR model results for selected spectral lines.

Flux density/flux density ratio	Model 1	Model 2	Model 3
[CI] $^3\mathrm{P}_1 \to {}^3\mathrm{P}_0$, in $10^{-7}\,\mathrm{erg\,s^{-1}\,cm^{-2}\,sr^{-1}}$	2.4	4.2	1.5
[CI] $^3\mathrm{P}_1 \to {}^3\mathrm{P}_0$/[CI]$^3\mathrm{P}_2 \to {}^3\mathrm{P}_1$	1.3	1.5	1.3
[CI]$^3\mathrm{P}_1 \to {}^3\mathrm{P}_0/{}^{12}$CO J=1→0	51.2	10.0	9.3
[CI]$^3\mathrm{P}_1 \to {}^3\mathrm{P}_0/{}^{12}$CO J=3→2	9.3	0.6	1.3
[CI]$^3\mathrm{P}_1 \to {}^3\mathrm{P}_0/{}^{12}$CO J=4→3	24.7	0.45	1.9
[CI]$^3\mathrm{P}_1 \to {}^3\mathrm{P}_0/{}^{13}$CO J=1→0	275	23	68
[CI]$^3\mathrm{P}_1 \to {}^3\mathrm{P}_0/{}^{13}$CO J=3→2	96	1.4	17

References

1. E.F. van Dishoeck, J.H. Black: ApJS. **62**, 109 (1986).
2. D.J. Hollenbach, T. Takahashi, A.G.G.M. Tielens: ApJ **377**, 192 (1991).
3. F. Zagury, F. Boulanger, V. Banchet: A&A **352**, 645 (1999).
4. F. Bensch, U. Leuenhagen, J. Stutzki, R. Schieder: ApJ **591**, 1013 (2003).
5. P. Wannier et al.: ApJ **510**, 291 (1999).
6. G. Kaczmarczyk: MNRAS **316**, 875 (2000).
7. H. Störzer, J. Stutzki, A. Sternberg: A&A **310**, 592 (1996).

The Influence of Grains on the Propagation and Structure of C-Type Shock Waves in Interstellar Molecular Clouds

D. Flower[1] and G. Pineau des Forêts[2]

[1] Physics Department, The University, Durham DH1 3LE, UK
david.flower@dur.ac.uk
[2] IAS, Université de Paris-Sud, F-92405 Orsay, France
guillaume.pineaudesforets@ias.u-psud.fr

1 Introduction

Grains contribute only about 1% to the mass of material in the interstellar medium (ISM). However, in dark clouds, where the charged component of the grains is believed to be a significant fraction of the total and the degree of ionization of the gas is low, the mass density of the charged grains is expected to greatly exceed that of the gaseous component. By the term 'grains', we understand both bulk materials, such as graphite or silicates, and large organic molecules (LOM), which may comprise polycyclic aromatic hydrocarbons. Grains are known to be important as both sources and sinks of molecules, notably H_2. In the context of dynamics of the ISM, the charged grains may also be important for the modes of shock propagation. In particular, the formation of steady-state C-type shock waves is constrained by the requirement that the magnetosonic speed in the charged fluid, given (at zero temperature) by

$$v_{\rm ims} = \frac{B}{(4\pi\rho_c)^{\frac{1}{2}}}, \qquad (1)$$

should exceed the shock speed, v_s. In (1), B is the magnetic induction in the preshock gas, perpendicular to the direction of propagation of the shock wave, and ρ_c is the mass density of the charged fluid. As the mass density of charged grains dominates ρ_c in dark clouds, the fraction of the grains which is charged is an important parameter.

2 The Electrical Charging of Grains in Dark Clouds

The processes which are taken into account when computing the degree of charging of grains in dark clouds are:

- electron attachment to neutral grains
- positive ion attachment to neutral grains and recombination of positive ions with negatively charged grains

– ionization of neutral grains and electron detachment from negatively charged grains by the 'internal' radiation, generated by ultraviolet fluorescence of H_2 following collisional excitation of the Lyman and Werner bands by secondary electrons, produced by cosmic ray ionization of hydrogen [3, 6]. This process is unimportant for the charge state of the LOM's but can contribute significantly to the neutralization of large grains, through electron photo-detachment.

When computing the rates of the above processes, we have been guided by the studies of Draine & Sutin [1] and Weingartner & Draine [7, 8]. However, we note that their treatment of electron attachment, based on classical mechanics, becomes questionable at the low temperatures ($T \leq 30$ K) of dark clouds, as the de Broglie wavelength of such low energy electrons is comparable with the grain dimensions.

In Fig. 1 are shown the mass densities of the charged components of the grains and the gas, both in the preshock gas and through a C-type shock wave of speed $v_s = 50$ km s^{-1}. In this model, the density of the preshock gas is $n_{\mathrm{H}} = 10^4$ cm^{-3} and the initial (transverse) magnetic field strength is $B = 100$ μG. Evident in this figure is the dominance of the contribution of charged grains to the inertia of the charged fluid. In the preshock gas, the large grains, which carry most of their mass, are predominantly neutral. This situation obtains because the small grains act as a sink for the electrons, becoming negatively charged (the "negative ions" in the Figure) and thereby reducing the number of electrons in the gas phase which is available for

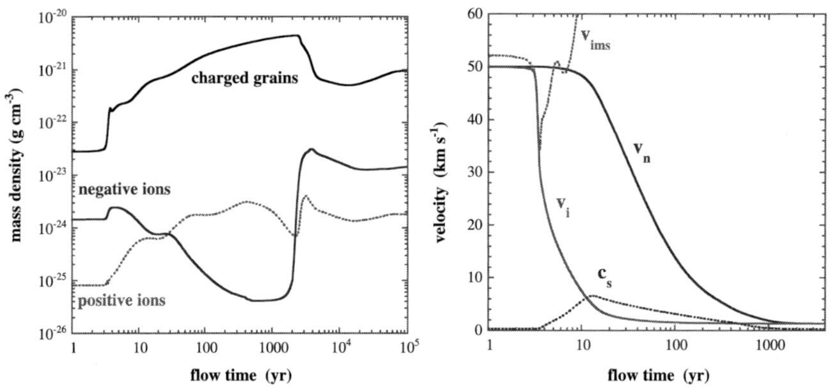

Fig. 1. Left: Contributions to the mass density of the charged fluid from large grains, small grains (large organic molecules), and positive ions, plotted against the flow time, t. The preshock gas is to the left. **Right:** The flow speeds of the charged and neutral fluids, v_i and v_n, in the shock frame, and the ion magnetosonic and the sound speeds, v_{ims} and c_s, plotted against the flow time. The shock speed is $v_s = 50$ km s^{-1}. For $t > 10$ yr, $v_{\mathrm{ims}} >> v_s$ owing to the compression of the magnetic field.

attachment to the large grains. In the model shown in Fig. 1, we have assumed a fractional number density $n_{\mathrm{LOM}}/n_{\mathrm{H}} = 10^{-6}$, which is consistent with the still uncertain estimates of the fraction of carbon which is in very small grains [4, 7].

The velocity profiles of the charged and neutral fluids, v_i and v_n, together with the local values of the sound speed, c_s, and of the ion magnetosonic speed, v_{ims}, are also shown in Fig. 1 for the same model. In this case, the ion magnetosonic speed in the preshock gas exceeds the shock speed of $v_s =$ 50 km s^{-1}, and a C-type shock obtains in steady-state. The velocity profiles of the charged and neutral fluids are expressed in the shock frame, in which the gas flows into the front of the shock wave (from the left in the Figure) at the shock speed. In this frame, the speed of the charged fluid falls rapidly as it is compressed, along with the magnetic field, and the neutral fluid subsequently recouples to the charged fluid. The recoupling time is approximately 10^3 years in this model. We note that the ion-neutral velocity difference, $v_i - v_n$ attains a maximum in excess of 40 km s^{-1}, which is sufficient to give rise to significant erosion of the charged grains through collisions with the neutral gas [5]. Also enabled by the ion-neutral drift speed is the shattering of graphite grains in collisions between their charged and neutral components [2].

3 Constraints on the Velocities of C-type Shock Waves in Dark Clouds

We have already discussed the upper limit which is imposed on the possible speeds of C-type shock waves owing to the contribution of large grains to the inertia of the charged fluid. Because the degree of charge of the large grains is dependent on fractional abundance of the LOM's, this upper limit is dependent on $n_{\mathrm{LOM}}/n_{\mathrm{H}}$. In Fig. 2 is plotted the limiting speed as a function of the gas density for $10^{-8} \leq n_{\mathrm{LOM}}/n_{\mathrm{H}} \leq 10^{-6}$. An additional constraint on the shock speed, which may be determining at high densities, is imposed by the collisional dissociation of molecular hydrogen. As H_2 is the principal coolant of the gas, its removal by collisional dissociation can result in the sound speed increasing towards the flow speed of the neutral fluid. To this sonic point in the flow there corresponds a J-type shock discontinuity, i.e., a pure C-type shock no longer obtains in steady state. The critical shock velocity, v_{crit}, for which a sonic point occurs in the flow, is also plotted in Fig. 2. It will be seen that collisional dissociation of H_2 tends to impose a more stringent constraint on the permissible shock speeds at high gas densities.

4 Conclusions

We have shown that the presence of charged grains has important consequences for the propagation of shock waves in the dark clouds of the inter-

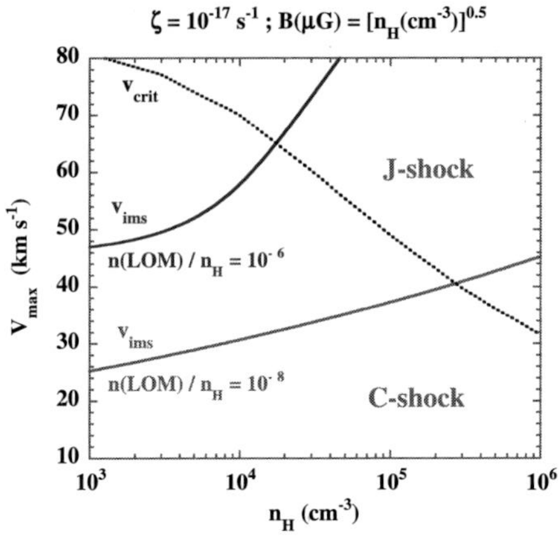

Fig. 2. The maximum speed, v_{max}, of a C-type shock wave, computed for the specified fractional abundances of small grains (LOM). The critical velocity, v_{crit}, at which a sonic point occurs in the flow, owing to collisional dissociation of H_2, is also plotted.

stellar medium. The fraction of very small grains (large organic molecules) determines the fraction of electrons in the gas phase and hence the degree of charging of the large grains, which incorporate most of their mass. The inertia of the charged large grains then determines the ion magnetosonic speed, which is an upper limit to the speeds of C-type shock waves which may obtain in steady state.

Crucial to the analysis is the rate of electron attachment to grains in dark clouds, at temperatures $T \leq 30$ K. We have noted that current theoretical treatments of this process, which are based on classical mechanics, need to be reconsidered as the de Broglie wavelength of electrons of such low kinetic energy is comparable with the grain dimensions.

References

1. Pagani, L., Bacmann, A., Motte, F. et al. 2003, A&A in press
2. Lehtinen, K. Mattila, K., Lamke, D., et al. 200 A&A 398, 571
3. Pagani, L. Pardo, J.-R., Fich, M. Motte, F. and Stepnik, B., 2002 "Infrared and Submillimeter Space Astronomy", EAS Series, Vol. 4, eds. Giard M. et al., held in Toulouse, June 2001, EdP Science, Les Ulis, p.145
4. Tine, S., Roueff, E., falgarone, E., Gerin, M., Pineau des Foret, G., A&A 356, 1039 (2000)
5. Stepnik, B., Abergel, A. Bernard J.-P. et al., A&A 398, 551 (2003)
6. Hirao, T., Amano, T., this conference (2003)

L183 (134N) Dust, Gas and Depletion

L. Pagani[1], A.J. Apponi[2], A. Bacmann[3], L. Cambrésy[4], M. Fich[5], G. Lagache[6], M.-A. Miville-Deschênes[6,7], F. Motte[8], and J.-R. Pardo[9]

[1] LERMA & UMR 8112 du CNRS, Observatoire de Paris, 61, Av. de l'Observatoire F-75014 Paris `laurent.pagani@obspm.fr`
[2] University of Arizona, Steward Observatory, 933 N. Cherry Ave., Tucson, AZ 85721
[3] ESO, Karl-Schwarzschild-Str. 2, 85748 Garching bei München, Germany
[4] Observatoire Astronomique de Strasbourg, UMR 7550, 67000 Strasbourg, France
[5] Physics Department, University of Waterloo, Waterloo, Ontario, Canada, N2L 3G1
[6] IAS, Bât. 121, Université Paris-Sud F-91435 Orsay
[7] CITA, University of Toronto, 60 St. George Street, Toronto, Ontario, Canada, M5S 3H8
[8] CEA, DSM, DAPNIA, Service d'Astrophysique, C.E. Saclay, F-91191 Gif-sur-Yvette Cedex
[9] Instituto de Estructura de la Materia, Dpto de Física Molecular CSIC, Madrid, Spain

L183, also known as L134N is a well-known starless, dark cloud. Recent observations ranging from NIR to mm allow for the first time to fully map the dust content of the cloud. The map displayed in Fig. 1 has been obtained by combining CFHT-NIR (I band star counts in the outskirts of the cloud and H-K′ reddening excess in the central regions) with ISOCAM-MIR data for the inner core itself where visual extinction is in excess of 40 mag [1].

Comparison of the ISOPHOT 200 μm emission map (see e.g. [2]) with our dust extinction map shows that the main dust features are not visible at all at 200 μm. The comparison of both sets of data (Fig. 2) clearly shows this effect. The data can be fitted by a log function above $A_V = 1.5$ (and a straight line below) which can be inverted to retrieve the temperature gradient from the border to the center of the cloud (albeit at a low angular resolution). Dust temperature as low as 7.5 K in the core is advocated. If we vary the dust profile by ± 0.5 K, the data cannot be fitted anymore. This is of course dependent upon the chosen set of dust emissivity coefficients.

From $C^{18}O$ maps obtained at Kitt Peak with the 12-m [3] and at IRAM with the 30-m [4], we can visualize the strong depletion effect on this molecule (Fig. 3). From the ratio of the two fitted slopes, we find that the CO depletion factor reaches at least 40 in the core (or 5 on the line of sight). Even N-bearing species like N_2D^+ and NH_2D seem to be depleted in the inner core. This can be seen in Fig. 4 which displays data obtained by [4] in comparison with our dust peak.

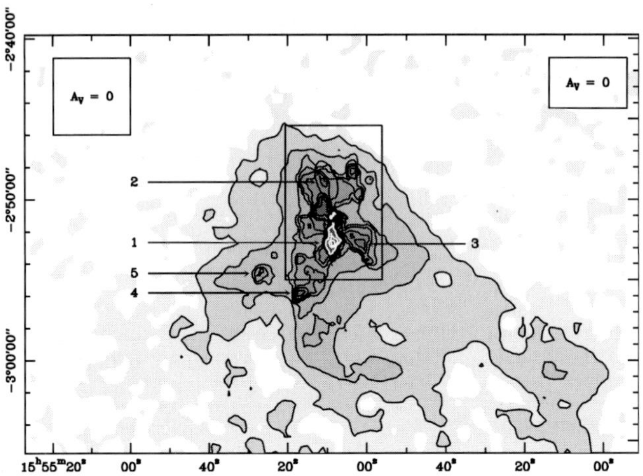

Fig. 1. Visual extinction map of L183. Levels are $A_V = 5$ to 50 by 5 (black contours) and 60 to 120 by 30 (white contours) mag. The numbers indicate the dust peaks which have a molecular counterpart. The ellipse (surrounding molecular peak #2) has no actual extinction measurements (no sources seen in H inside). Its lower visual extinction limit is 40 mag and from MAMBO measurements, the visual extinction is in the range 40-70 mag

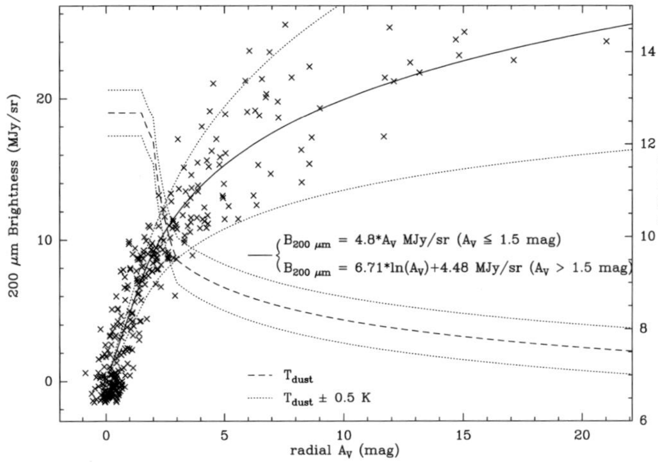

Fig. 2. The dust extinction from Fig. 1 has been correlated to the ISOPHOT 200 μm emission after convolution to the same 1.5′ resolution. The 200 μm does not trace dust beyond 7.5 mag of radial extinction (15 mag total). The best explanation is that the dust becomes too cold deep inside the cloud to emit significantly at 200 μm. The strong temperature drop at $A_V = 2$–3 is due to the change of dust emissivity from diffuse to dense which we suppose to occur at this level from [5]. The dotted lines on both sides of the fit indicate where the fit would be if we had increased or diminished the temperature by 0.5 K (corresponding dotted lines around T_{dust}) around the temperature derived from the fit inversion

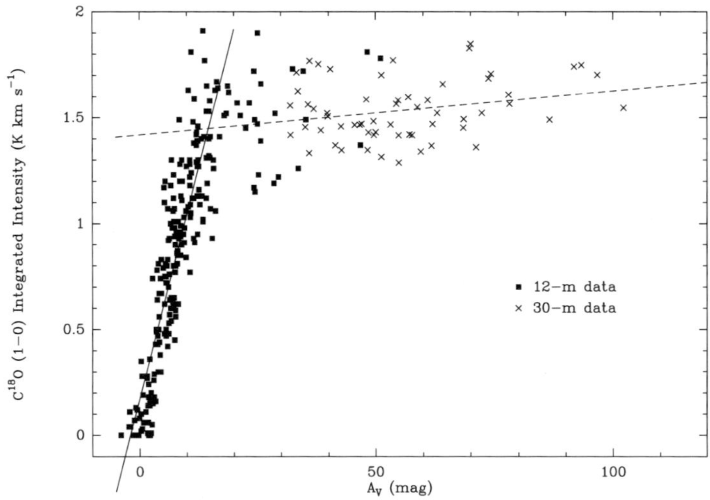

Fig. 3. $C^{18}O$ depletion. The NIR dust map from Fig. 1 has been reprojected on both the large scale (Kitt Peak 12-m) and small-scale (IRAM 30-m) maps to adjust the resolutions. $C^{18}O$ depletion is clearly visible and reaches at least 42 for the largest extinction values as measured from the ratio of the two least-square fitted slopes.

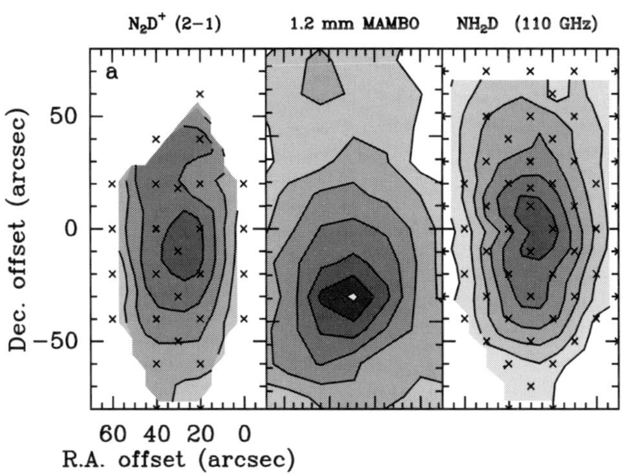

Fig. 4. Integrated intensity maps of N_2D^+ (J:2–1), and of NH_2D ($J_{KK'}$: $1_{1,1}$–$1_{0,1}$) and dust emission at 1.2 mm. Even N-bearing molecules eventually deplete in the deep interior of the dust peak. The dust emission at 1.2 mm (middle panel) is 20–30″ south of the molecular peak. The dust core is in grey contours. Map offsets are relative to 15h54m06.6s, −2°52'19.1" (J2000). Levels : 0.1 to 0.6 by 0.1 K km s^{-1}(N_2D^+), 0.1 to 0.6 by 0.1 K km s^{-1}(NH_2D). MAMBO data : 20 to 80 by 10 mJy per beam, 15″ sampling. N_2D^+ and NH_2D data from [4]

L183 is a good prototype to study depletion effects, detailed chemistry and the first steps of the star formation process ahead of the Class 0 protostar. Chemistry will indeed be better understood in this source once the excitation conditions and depletion processes will be described with some detail and taken into account in both the abundance analysis of the observed species and the chemical models themselves. This has never been done until now. Finally, to trace the very depleted core, we are left with the sole possibility to observe the very difficult line of H_2D^+ at 372 GHz (and possibly HD_2^+ of which the precise radio frequency has been announced in this conference [6], though the line, situated at 691 GHz should be even more difficult to detect).

References

1. McLaughlin D.E. & Pudritz R.E.: ApJ **469**, 194 (1996)
2. McLaughlin D.E. & Pudritz R.E.: ApJ **476**, 750 (1997)
3. Stepnik, B., Abergel, A., Bernard, J.-P., et al.: A&A **398**, 551 (2003)
4. Hirao, T. & Amano, T. : This conference (2003)
5. Tiné, S., Roueff, E., Falgarone, E., Gerin, M., Pineau des Forêts, G.: A&A **356**, 1039 (2000)
6. Hirao, T. & Amano, T. : This conference (2003)

Density Structure of the Horsehead Nebula Photo-Dissociation Region

E. Habart[1], A. Abergel[2], C.M. Walmsley[1], and D. Teyssier[3]

[1] Osservatorio Astrofisico di Arcetri, Firenze, Italy habart@arcetri.astro.it
[2] IAS, Orsay, France alain.abergel@ias.u-psud.fr
[3] SRON, Groningen teyssier@sron.rug.nl

Abstract. We present observations of dust and gas emission using ISO and ground based telescopes towards the illuminated edge of the Horsehead nebula. H_2 fluorescent emission reveals extremely sharp and bright filaments associated with the illuminated edge of the nebula which spatially coincides with the aromatic dust emission. Combination of this infrared data with millimeter observations of CO and dust continuum emission allows us to trace the penetration of the far-UV radiation field into the cloud and probe the gas density structure. From comparison with PDR model calculations, we find that a high gas density gradient at the edge of the Horsehead nebula is required to explain the observations.

1 Introduction

Photo-dissociation regions (PDRs) illuminated by bright stars are unique targets to study the interaction of stellar ultraviolet radiation with dense gas. This is important for the evolution of the cloud as well as for the star formation associated with the cloud. In this paper, we present observations of the Horsehead nebula at the western edge of the molecular cloud L1630 illuminated by the O9.5 V star σ Ori. This nearby PDR (d$\sim$400 pc) viewed edge-on offers an ideal opportunity of studying the penetration of the far-UV radiation and the density structure into a PDR. The observations presented in Sect. 2 are compared in Sect. 3 with a physical model of PDRs in order to constrain the gas density structure at the edge of the molecular cloud.

2 Observations

We show in Fig. 1 maps of the Horsehead nebula seen in different tracers. In the mid-IR image made with ISOCAM [6" of angular resolution [2]], the edge of the Horsehead nebula is delineated by a bright filament located at the edge of the dense molecular cloud as traced by its CO emission. It represents the sharpest mid-IR filament (width: 10" or 0.02 pc) detected in our Galaxy by ISOCAM and we believe that it is due to dense illuminated material seen edge-on. In order to resolve properly the PDR structure in the outer zone, we have obtained using SOFI at the NTT high angular resolution (∼1") imaging observations of the H_2 1-0 S(1) line emission. The H_2 fluorescent emission, very sensitive to both the far-UV radiation field and the gas density structure, reveals extremely sharp filaments (width: ∼5" or 0.01 pc) spatially coincident with the aromatic dust emission.

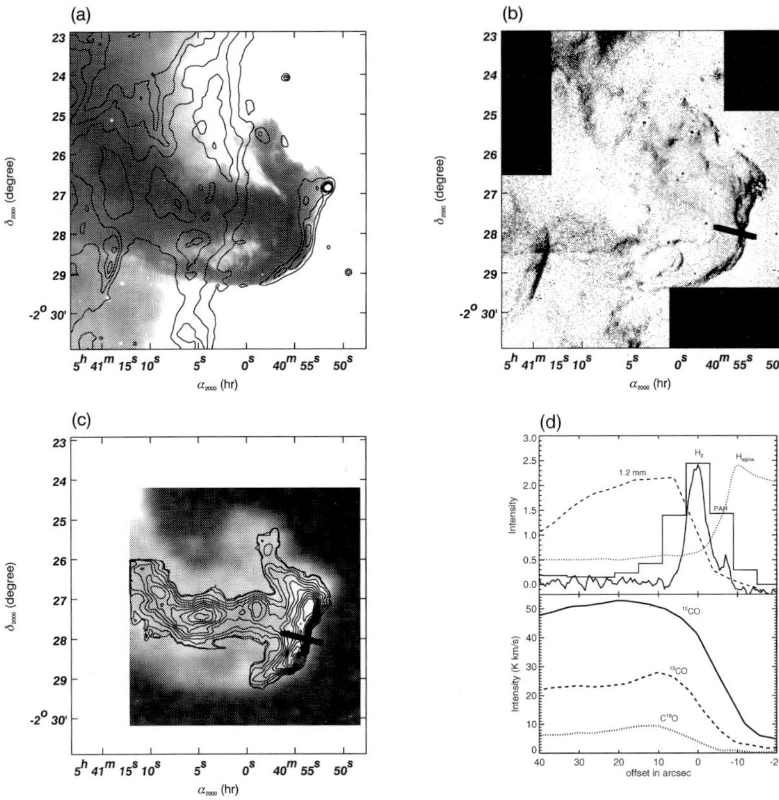

Fig. 1. (a) Map in the Hα line emission [from [5]] of the Horsehead nebula with ISOCAM contours of the LW2 emission [5-8.5 μm [2]] dominated by the aromatic hydrocarbons dust (or PAHs) emission. **(b)** SOFI/NTT map in the H_2 1-0 S(1) line emission. **(c)** Intensity map of the optically thick CO(3-2) line emission (taken at the CSO, M. Gerin, private communication) with contours of the optically thin $C^{18}O$(2-1) line emission [taken at the IRAM 30-m telescope [6]]. **(d)** Brightness profiles along the cut going through the edge of the Horsehead nebula (shown in panels (b), (c)). We show the Hα line emission (dotted lines) scaled to the H_2 1-0 S(1) line emission (in 10^{-5} erg s^{-1} cm^{-2} sr^{-1}, solid thick lines); PAH emission in the LW2 filter ($MJy/sr \times 0.14$); dust continuum emission at 1.2 mm (dashed line, mJy/beam×0.05); line emission for the $J = 2 - 1$ transition of CO, ^{13}CO, and $C^{18}O$.

In Fig. 1(d), we show the brightness profiles of the different tracers across the PDR. Although the steep edge of the cloud (~5-10") is not resolved by the millimeter observations (angular resolution : 11" at 1.3 mm), moving away from the excitation source, a layered appearance is clear: first the Hα line emission, then the H_2 emission, and finally the CO, ^{13}CO and $C^{18}O$ emission appear in succession. This spatial stratification is typical for an edge-on PDR. Outside the molecular cloud, the H_2 and CO molecules are photodissociated; the H_2 emission appears first (typically at an optical depth of $A_v \sim 1$) followed by the CO emission (at $A_v \sim 3$). Hence, the separation between the emission peaks of the different tracers measures directly the far-UV penetration size scale. Adopting a hydrogen column density, N_H, per magnitude of visual extinction, $N_H/A_v = 1.85\ 10^{21}$ cm^{-2} the observed scale size of the region (~10" per magnitude of visual extinction) implies a gas density of the order of $5\ 10^4$ cm^{-3}. However, this assumes an uniform gas density while the PDR could have a density gradient.

3 PDR Model

In this section, in order to constrain the gas density structure, we use an updated version of the stationary PDR model described in [4]. Given an incident radiation field and a gas density, this model calculates the energy balance, chemical composition, and radiative transfer in the far-UV and the cooling lines self consistently. e adopt for the far-UV intensity of the incident radiation field $G_0 \sim 100$ (in units of the Habing field) assuming that σ Ori and the Horsehead nebula are in a common plane. For the gas density, we consider a gas density profile with $n_H \sim 10^4$ and 10^5 cm^{-3} in the H_2 emitting and inner cold molecular layers, respectively. In Fig. 2, we show the model results and see that the calculated and observed emission profiles are in good agreement for this gas density profile. For a uniform gas density equal to $5\ 10^4$ cm^{-3}, the H_2 emission will peak too close to the edge while the CO emission will peak too far inside. The absolute intensities agree reasonably well with the observations with a visual extinction along the line of sight of A_v ~10, as suggested by recent 1.2 mm continuum observations [6]. Finally, note that the ratios measured between the H_2 fluorescent and PAH emission and, between the $J = 2-1$ (or $J = 3-2$) and $J = 1-0$ transitions of $C^{18}O$ (or ^{13}CO), which are essentially independent of the geometry, indicate local gas densities in the outer and inner regions, respectively, consistent with the gas density profile derived here (Habart et al. in preparation). Nevertheless, the H_2 emission profile predicted is narrower than in observations and the $C^{18}O$ emission is slighlty under-estimated and peaks more far in. Projection effects resulting from sub-structures and/or PDR inclination may be not negligible and may in part explain these discrepancies.

In summary, we require at the edge of the Horsehead nebula a high gas density gradient ($n_H \sim 10^4$ and 10^5 cm^{-3} respectively in the H_2 emitting

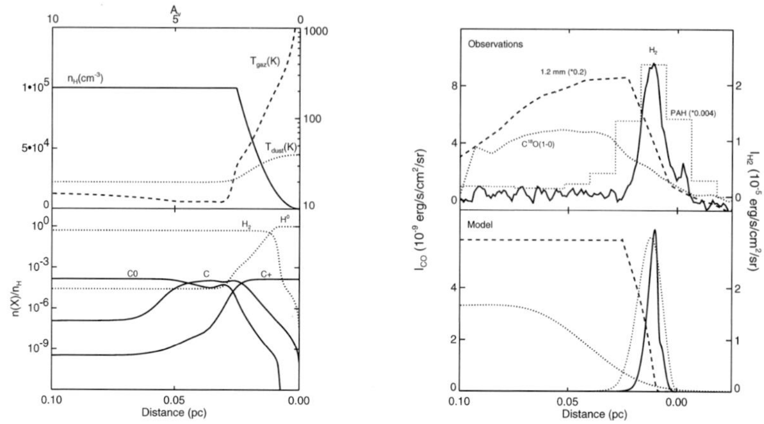

Fig. 2. Left panel : Variation of the gas density, gas temperature and chemical composition calculated by the PDR model as a function of the distance and of visual extinction from the PDR edge. **Right panel** : Comparison between the brightness profiles as observed throughout the illuminated edge of the Horsehead nebula (see cut in Fig. 1) and as predicted by the PDR model.

and inner cold molecular layers) in order to explain the observed spatial stratification of different tracers. This confirms earlier findings of such high densities on the western edge of the ρ Ophiuchi molecular cloud by [3]. This raises the question of why such high densities are found exactly at the edges of molecular clouds. The ionizing influence of the σ Ori star, which will have evaporated the more tenuous cloud material, can be a possibility.

References

1. Chièze & Pineau Des Forêts(1987)
2. Abergel, A, Teyssier, D, Bernard, J. P et al. 2003, *A&A*, 410:577.
3. Habart, E. 2001, *Ph.D thesis.* University of Paris 7.
4. Le Bourlot, J, Pineau des Forêts, G, Roueff, E et al. 1993, *A&A*, 267:L233.
5. Pound, M. W, Reipurth, B, & Bally, J. 2003, *AJ*, 125:2108.
6. Teyssier, D. 2002, *Ph.D thesis.* University of Paris 6.

Methanol: A Diagnostic Tool for High-Mass Star-Forming Regions

S. Leurini[1], P. Schilke[1], K.M. Menten[1], D.R. Flower[2], J.T. Pottage[2], and L.-H Xu[3]

[1] Max-Planck-Institut für Radioastronomie, Auf Dem Hügel 69, D-53121, Bonn sleurini@mpifr-bonn.mpg.de
[2] Physics Department, University of Durham, DH1 3LE, UK
[3] Physical Sciences Department, University of New Brunswick, Saint John, NB, Canada E2L 4L5

1 Introduction

Kinetic temperature and density are fundamental parameters for our understanding of the interstellar medium (ISM). Usually, symmetric rotors such as NH_3 are used to probe a cloud's kinetic temperature, while linear molecules, e.g. CS, probe its density. However, different spatial distributions of the tracers ("chemistry") often complicate the picture (see, e.g., [8]) as they often trace physically different and spatially non-coexisting gas components. It is thus desirable to trace all relevant physical parameters with a single molecule. Promising candidates exist among slightly asymmetric rotors, which have properties qualifying them as tracers for physical conditions. Methanol, CH_3OH, is a slightly asymmetric rotor. It is ubiquitous and associated with different regimes of star formation, from quiescent, cold (T $\sim$ 10 K), dark clouds, to "hot core" sources near high-mass (proto)stellar objects, where $[CH_3OH/H_2]$ values $\sim 10^{-7} - 10^{-6}$ are observed [3]. Up to now an extremely poor knowledge of the CH_3OH collisional rates and of their propensity rules has prevented realistic systematic studies exploiting methanol's full potential as an interstellar tracer. Recently, this situation has changed with the calculation of collisional rate coefficients by [5, 6], for collisions with helium, for both CH_3OH-A and CH_3OH-E, for levels up to $(\mathrm{J}, \mathrm{K}) = 9$.

Here we would like to focus on general aspects connected to the analysis of complex molecules' spectra, of which CH_3OH is one of the simplest examples, (for details on methanol excitation and on its probing properties see [2]).

2 Analysis Technique

The traditional approach for deriving physical parameters such as kinetic temperature and spatial density from an observed spectrum involves "by-hand" fitting of the lines with multiple components and χ^2 analysis comparing the measured quantities with statistical equilibrium calculations. However, this single line fitting procedure is extremely time-consuming and suffers from several drawbacks. With sensitive receivers nowadays available and in view of the next generation of instruments, which will provide copious amounts of data in a short time, new methods of data analysig and modelling are required.

Table 1. CH_3OH model results for G79.3P1, $\chi^2_{\nu_d}$=0.7, ν_d=22.

	best fit	3σ fit range
T_K	17 K	12–37 K
$n(H_2)$	2.5 10^5(cm^{-3})	3 10^5– 4 10^6(cm^{-3})
N(CH_3OH-A)	7.9 10^{13} (cm^{-2})	6 10^{13}–1 10^{14}(cm^{-2})
N(CH_3OH-E)	5.7 10^{13} (cm^{-2})	3 10^{13}–8 10^{13}(cm^{-2})

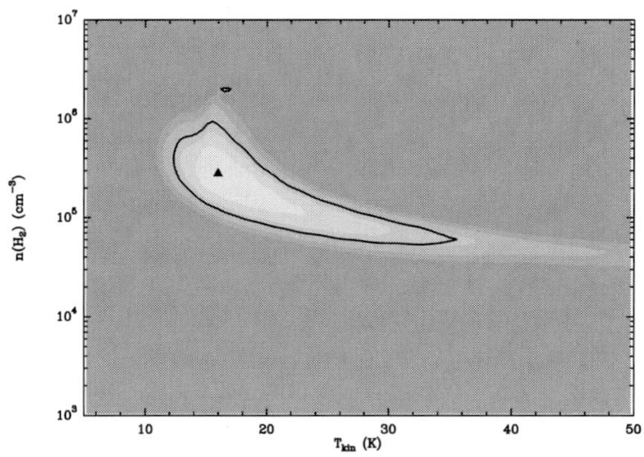

Fig. 1. χ^2 distribution in the [T_{kin},n(H_2)] plane, N(CH_3OH-A)=7.9 10^{13}, N(CH_3OH-E)=5.7 10^{13}, toward G79.3P1. A black triangle marks the minimum position, T_{kin}=17 K, n(H_2)=2.5 10^5 cm^{-3}; dashed in black the 3σ confidence surface for 22 degrees of freedom.

An innovative technique to handle the problem, proposed by [7] and recently improved by [1], is based on the simultaneous fit of the complete spectrum with a synthetic spectrum computed under Local Thermodynamic Equilibrium (LTE) conditions. Here we propose an extension to this technique using the Large Velocity Gragident (LVG) approximation, which ensures more reliable results when prominent departures from LTE are expected. The free parameters for each component are excitation temperature, molecular hydrogen density, source size, column densities for A and E states, which are treated as two independent parameters to take into account their possibly different abundance. Line width and LSR velocity are fixed parameters and assumed to be the same for all the lines in each component.

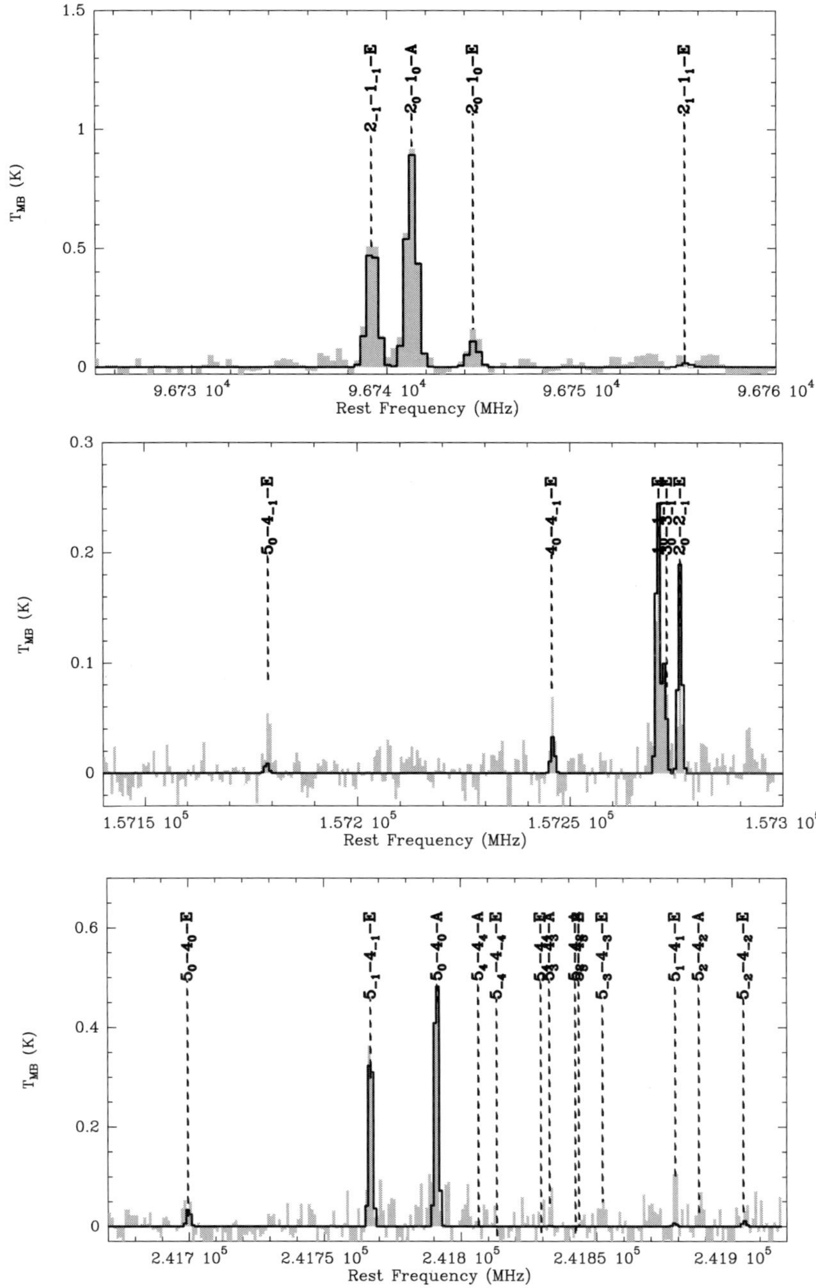

Fig. 2. CH_3OH spectra toward G79.3P1 taken with the IRAM 30m telescope; the 2mm and 1mm data have been smoothed to the 3mm beam size ot the IRAM telescope for comparison with the 3mm data. Labels show all the CH_3OH transitions. Overlaid in black the synthetic spectrum resulting from the fit.

Within a component, the optical depths for lines with a frequency separation $\nu_i - \nu_j \leq \Delta\nu_i + \Delta\nu_j$ are summed up to include local line overlap between both symmetry states. Line identification is based on the Cologne Database for Molecular Spectroscopy, (http://www.cdms.de [4]), which includes new measurements by [9].

Although the simultaneous fit of a spectrum has been demonstrated to be a powerful technique of analysis [1], it does imply several approximations. The assumption behind the analysis is that the CH_3OH distribution can be reasonably well approximated by a small number of non-interacting components, thus ignoring any structure in the source. The LVG approach adds other uncertainties to the obtained results, since it assumes only one set of physical parameters for the source and implicitly neglects any local and non-local overlap between the lines. Also, fitting CH_3OH spectra assures a reliable determination of kinetic temperature only when a large amount of lines are fitted and/or when millimeter and submillimeter data are combined, since CH_3OH shows a strong dependence on temperature mainly in the submillimeter range [2]. Another drawback is the uniqueness of the χ^2 minimization, which is not *a priori* guaranteed. Indeed the fit results occasionally depend on the input parameters: an analysis of the χ^2 distribution is therefore necessary to verify that the found minimum is global.

Having all these limitations in mind, the application of this technique to G79.3P1, an infrared-dark cloud, gives very satisfying results, (Fig. 2). Table 1 lists the best fit results and the 3σ fit range. The source size, 32″, is based on a BIMA map of CH_3OH at 3mm, (Wyrowski, priv. comm.). Fig. 1 shows the χ^2 distribution in the $[T_{kin}, n(H_2)]$ plane, with the 3σ confidence surface in black.

References

1. C. Comito: Exploring the submillimeter sky: molecular line studies at 350 μm. PhD Thesis, Rheinische Friedrich-Wilhelms Universität, Bonn (2003)
2. S. Leurini, P. Schilke, K. M. Menten et al: in preparation
3. K. M. Menten, C. M. Walmsley, C. Henkel et al: A&A **157**, 318 (1986)
4. H. S. P. Müller, S. Thorwirth, D. A Roth et al: A&A **370**, L49 (2001)
5. J.T. Pottage, D. R. Flower and S. L Davis: J. Phys. B: At. Mol. Phys. **34**, 3313 (2001)
6. J.T. Pottage, D. R. Flower and S. L Davis: J. Phys. B: At. Mol. Phys. **35**, 2541 (2002)
7. P. Schilke, T. G. Phillips and D. M. Mehringer. In: *The Physics and Chemistry of the Interstellar Medium*, ed by V. Ossenkopf, J. Stutzki, and G. Winnewisser (GCA-Verlag Herdecke, 1999) pp 330-337
8. M. Tafalla, P.C. Myers, P. Caselli, P. et al: ApJ. **569**, 815 (2002)
9. L.-H Xu and F. L. Lovas: J. Phys. Chem. Ref. Data **26**, 17 (1997)

Part IX

Astrochemistry

dedicated to Prof. G. Herzberg,
who died in 1999 at the age of 95 and would be celebrating
his 100th birthday in 2004

Astrochemistry and Observations

E.A. Bergin

University of Michigan, Ann Arbor, MI 48109 (USA) `ebergin@umich.edu`

1 Introduction

The past few years have seen dramatic improvements in our ability to extract chemical information from molecular line observations. These advances have been driven by new observational platforms and techniques. In the following I will outline the various methods and uncertainties in computing chemical information from observations across the spectrum of star formation activity (from diffuse clouds, to star-forming cores, to molecular outflows). I will show how improved chemical knowledge can be used to place stronger constraints on theoretical astrochemical models, but also open new avenues in the use of molecular emission to examine the physics of molecular clouds and star formation.

2 Diffuse Cloud Chemistry

Diffuse clouds are transition objects between the low density atomic medium and the higher density molecular phase. The allure of diffuse clouds is twofold. First they are generally, but not exclusively, studied via absorption lines. For absorption lines the conversion of equivalent width to column density is fairly straightforward and requires little information regarding source structure. In contrast, knowledge of the density and temperature structure is often needed to obtain column densities from emission lines. Thus chemical abundances derived from absorption lines can be considered more reliable. Second, these objects present rather simple laboratories to study basic physical and chemical processes.

Over the past decade studies of diffuse cloud chemistry through absorption lines has received significant attention in a series of papers by Lucas and Liszt (see Liszt & Lucas 2002 and references therein). In Figure 1 I provide a sample of this work showing detections of HCO^+ (J=1−0) and CS (J=2−1) in absorption in several line of sight clouds. To study basic chemical processes these investigations are able to observe the various species that constitute simple self-contained chemical networks. For instance, the network that creates CO in diffuse gas is believed to be: $C^+ + OH \rightarrow HCO^+$, followed by $HCO^+ + e \rightarrow CO + H$. Lucas & Liszt (2000) find that the column density of HCO^+ is capable of accounting for the observed column density of CO. However, the measured abundances of C^+ and OH are not high enough to account for observed HCO^+. Thus either our understanding of ion-molecule gas phase

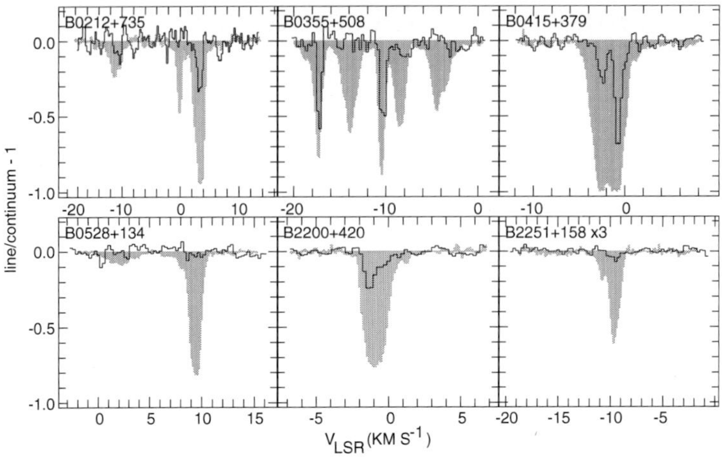

Fig. 1. CS J=2−1 (solid line) and HCO^+ J=1−0 (grey overlay) absorption spectra towards 6 quasars. Taken from Lucas & Liszt (2002).

chemistry is incorrect or diffuse clouds are not just "simple" laboratories. Indeed other more exotic non-thermal chemical processes have been invoked to overcome these failings (Zsargo & Federman 2003; Flower & Pineau des Forêts 1998).

3 Chemistry and Star Formation

The most significant recent advance in astrochemical studies of star-forming cores is the development of new techniques that move beyond line of sight average abundances, towards an examination of chemical structure as a function of depth. This is possible via new instrument capabilities which have provided images of the dust continuum emission, or information regarding the spatial distribution of dust absorption. This information has been gleaned through a variety of methods extending from 1.3mm and sub-mm continuum emission images (low-mass star-forming cores: André et al. 2000, Shirley et al. 2001; high-mass star-forming cores: van der Tak et al. 2000, Hatchell and van der Tak 2003), to images of dust absorption against the galactic mid-IR background (Bacmann et al. 2000), and near-IR extinction mapping (Alves, Lada, & Lada 2001).

Since the dust column density and mass is correlated with the H_2 column density and mass (Hildebrand 1983; Gordon 1995), these observations provide the clearest information to date on the column and volume density distribution of H_2 molecules. This significantly aids the molecular observations in two ways. (1) By indirectly confirming the location of the H_2 density and column density peak and, (2) by providing the density profile which helps to unravel the similar effects of density and abundance on excitation.

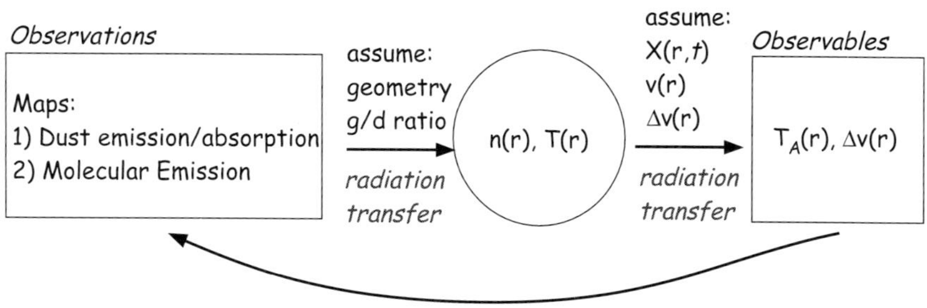

Fig. 2. Flow chart describing the iterative procedure used to determine abundance structure with depth. The abundance profile can either be assumed or predicted via a chemical model, in which case it can be time dependent.

In Figure 2 we present a flow chart that outlines this new iterative method to derive abundances. A key to this process is that spatial information is important for observations of both dust and gas. The first phase involves using dust observations to constrain core physical properties, typically the radial density profile.[1] At this point there are 2 methods that can be used to examine the abundance structure; both adopting spherical geometry.

(1) In the first method the density and temperature profile[2], and an assumed abundance profile are used as inputs to a radiative transfer model. The radiation transfer codes adopt either accelerated lambda iteration or Monte-Carlo methods. The results from this process are predictions of the emission spectra as a function of position, which can be compared directly to observations. An iterative procedure then determines the abundance profile (see van der Tak et al. 2000; Schreyer et al 2002; Tafalla et al 2002, 2003; Caselli et al 2002; Jorgensen et al. 2002; di Francesco et al. 2003; and Lee et al 2003). (2) The second method uses a chemical model (using the same density profile for the chemical and radiative transfer models) to predict the abundance profile. The abundance profile is then used as input to the radiative transfer producing observable quantities, which can be iterated by changing the parameters in the chemical model or by looking at the time-dependence (Bergin et al. 2002; Doty et al. 2002).

In the following we briefly discuss examples for pre-stellar cores and for sources which contain embedded sources.

3.1 Chemistry in Pre-stellar Cores

A representative pre-stellar core is the dark cloud B68 where the centrally concentrated H_2 distribution is traced through a map of dust visual extinction

[1] For cores with embedded stars the dust spectral energy distribution is used to estimate the radial temperature profile.

[2] Along with assumed profiles for radial velocity gradient and velocity dispersion.

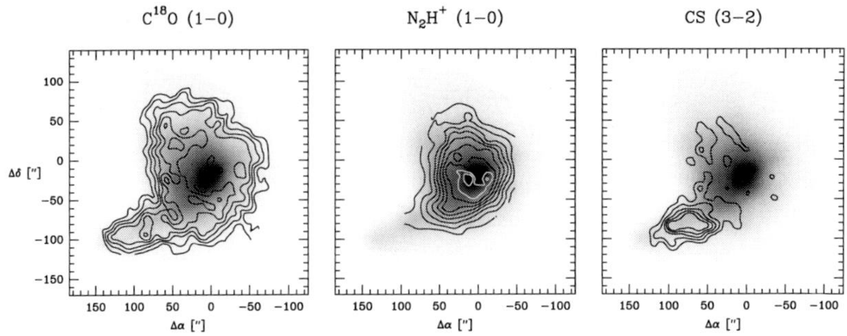

Fig. 3. The observed emission distribution of visual extinction or total column density (gray scale) in the B68 dark cloud. Contours are integrated emission maps of $C^{18}O$ (left-hand panel), N_2H^+ (middle), and CS (right-hand panel). The N_2H^+ contours are reversed in color near the central peak to show the distribution in the center over the extinction distribution. The emission distributions illustrate sequence of increasing molecular depletion from CS, to $C^{18}O$, and even N_2H^+ (Bergin et al. 2002; Lada et al. 2003).

(Alves, Lada, and Lada 2001). In Figure 3 we present the B68 A_V map along with a series of molecular emission maps. In this core we see that the N_2H^+ emission peaks in a shell partially surrounding the peak of dust extinction. Moreover, the N_2H^+ peaks inside the much larger $C^{18}O$ emission hole, which itself lies inside the CS emission depression.

Figure 4 presents the analysis of the N_2H^+ and $C^{18}O$ emission, which used a gas-grain chemical model linked to a Monte-Carlo radiative transfer code (Bergin et al. 2002). If the emission is in LTE and optically thin then the integrated emission from a source with constant abundance would appear as a straight line in this diagram. Clearly the $C^{18}O$-A_V relation deviates from a straight line at high extinctions and N_2H^+ deviates at both high and low A_V. Such structure in this relation could be due to freeze-out/depletion or high opacities (observations of a lesser abundant isotope confirm freeze-out for CO). The various solid and dashed lines show emission profiles predicted by the chemical model for various times. As time proceeds the $C^{18}O$ emission declines via freeze-out; in contrast the N_2H^+ emission increases (as a result of CO freeze-out). The dotted line is the "best fit" radial abundance profile which reproduces the emission profile shown as a solid line. For $C^{18}O$ the observations require significant freeze-out and also an abundance reduction at low extinction due to photo-dissociation. In the case of N_2H^+ an abundance reduction at low extinction is also required.[3]

B68 is just one example of a core where, using new techniques, we are inferring significant freeze-out in the early stages of star formation. Other examples can be found in the work of Tafalla et al. (2002) and references in §3. One possibility that cannot be discounted at present is that nearly

[3] This is due to freeze-out of N_2.

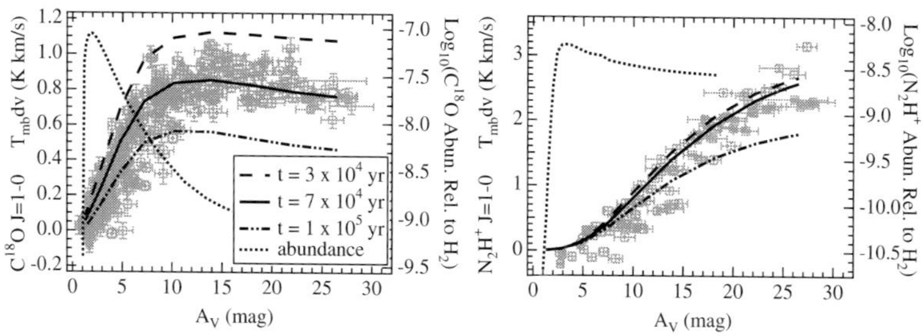

Fig. 4. Point by point comparison of $C^{18}O$ and N_2H^+ J= 1 − 0 integrated intensity with A_V the entire B68 dark cloud. In all plots the data are presented as open squares with error bars while solid curves represent the emission predicted by a model combining chemistry with a Monte-Carlo radiative transfer code. The various lines delineate the time-dependence in the chemistry. The dashed lines are the best fit molecular abundance profiles with the axis labeled to the right (abundance of N_2H^+ is multiplied by 100). Adapted from Bergin et al. (2002).

all typically used molecular tracers are absent from the gas in the innermost regions of these cores (Bergin et al. 2002); this may hinder our ability to study the earliest stages star formation. However, these sources still offer great promise as new and more precise laboratories for testing chemical theory than previous work in template sources such as TMC-1 and L134N.

3.2 Chemistry in Star-Forming Cores

The chemistry in star-forming cores overlaps with the chemical studies of hot cores. Hot cores are molecular condensations directly associated with young embedded massive stars. Stellar birth leads to the creation of sharp temperature gradients and higher temperatures lead to the release of frozen ices from grain surfaces. As a result hot cores have quite different chemical composition, extending to very complex molecular species, than typically observed in quiescent gas. While these regions are associated with massive stars it is clear that low-mass stars also can produce similar "warm" cores (Cazaux et al. 2003). The reader is referred to the following references for additional discussions (Millar 2001; van der Tak 2003).

Examples of the new iterative method to derive abundances use so-called "jump" models where abundances are enhanced in the inner core via grain surface evaporation and are lower in the outer layers. The jump transition is set by the evaporation temperature (see van der Tak 2003). Models where chemical networks are directly linked to the radiation transport code are exemplified by the work of Doty et al. (2002) and Boonman et al. (2003).

A generic result is that enhanced abundances of numerous molecules are required in the center. A key goal for hot core studies is to probe the extent of grain surface chemistry via observations of trace complex species in the gas. Grain mantles can be observed directly via vibrational modes absorbing background IR radiation, but this method lacks sensitivity for all but the most abundant ices. Thus, gas-phase observations can be used to probe the extent of grain surface chemistry beyond the production of H_2O, CH_4, and NH_3. The problem is that when molecules are released from the grain they might react in the gas (Charnley, Tielens, & Millar 1992). To draw upon a parallel to cometary observations, the products of these reactions would be ion-molecule daughter products as opposed to parent molecules that have directly evaporated from the grain surface. This is further complicated by the potential presence of high temperature gas-phase chemistry. Advances in our understanding will require continued close coupling of chemical models to observations.

4 Shock Chemistry

Another area of observational astrochemistry is the evidence for shock chemistry detected in molecular outflows. The primary example is found in the highly collimated L1157 molecular outflow, which has been subjected to a detailed chemical survey (Bachiller et al 2001). This study has shown that some species have enhanced abundances in outflow lobes (SiO, CH_3OH, HCN) while others (N_2H^+, DCO^+) are absent from the flow. Another collimated outflow, BHR71, has recently been the focus of an extensive chemical survey (shown in Figure 5). In this source the abundance trends are quite similar to L1157 in regards to the presence or absence of enhancements for most species. However, clear differences are found in the magnitude of the abundance enhancements (Bourke et al. 2004). It is not clear whether the differences are intrinsic or due to assumptions in the analysis.

Abundance enhancements are typically attributed to grain mantle core and mantle release via sputtering and/or high temperature chemistry. The effects of the latter have not been observationally explored and more attention needs to be paid to models in the literature (see series by Pineau des Forêts and collaborators). Moreover if shocks are truly liberating mantles then outflows should show high levels of deuteration as seen in hot cores. This also needs to be explored.

5 Conclusions

It is clear we are at the beginnings of a new era in astrochemistry where we will obtain more precise information regarding the abundances of molecular species, but also set limits on the abundance profile and on basic theory. In Figure 6 we present a schematic that shows a rough outline of our current understanding of the chemical structure in star forming cores. An important point here is that these zones will be different for various species (depending

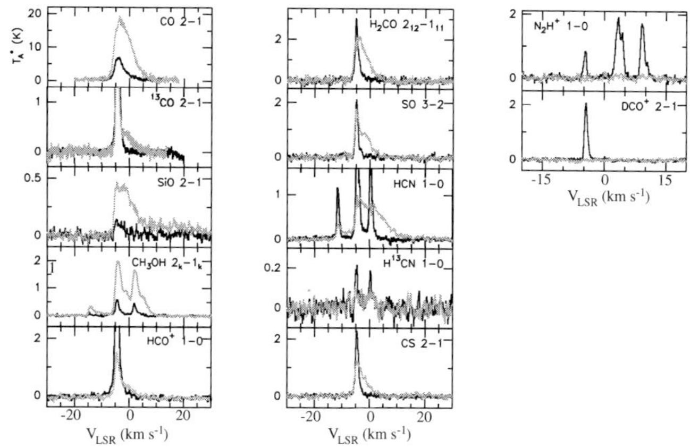

Fig. 5. Spectra of selected species taken from a chemical survey of the BHR71 outflow (Bourke et al. 2004). Dark lines indicate spectra taken towards the quiescent core while grey lines are spectra towards the peak of the red lobe of the outflow.

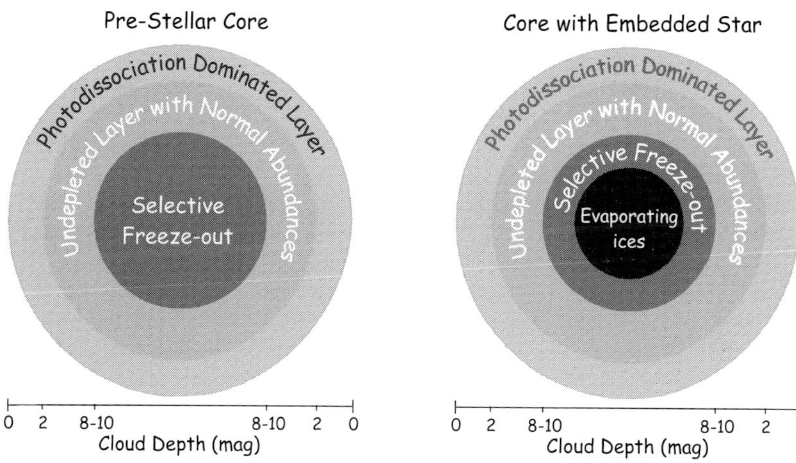

Fig. 6. Schematic of inferred chemical structure in star-forming cores. Extinction scales are calibrated to low-mass cores.

on the evaporation temperature, photo-dissociation cross-sections, depletion rate, etc.). Since molecular emission is the only way to probe gas kinematics, one can use improved chemical information as a tool. Thus, with careful choice of tracers, and knowledge of the chemistry, one can selectively probe motions from different gas along a given line of sight. This will be an important key to unraveling the star formation puzzle. In all, new observations are leading to crucial tests of chemical theory, but also presenting a completely new approach to star formation studies, bringing astrochemistry into wider application as a potent tool.

References

1. Alves, J., Lada, C. J., & Lada, E. A.: Nature, **409**, 159 (2001)
2. André, P., Ward-Thompson, D., & Barsony, M.: From Prestellar Cores to Protostars: the Initial Conditions of Star Formation. In: *Protostars and Planets IV*, ed by V. Mannings, A.P. Boss, S.S. Russell (Univ. of Arizona Press, Tucson 2000), pp 59
3. Bachiller, R. et al: A&A, **372**, 899 (2001)
4. Bacmann, A. et al: A&A, **361**, 555 (2000)
5. Bergin, E. A. et al: ApJ, **570**, L101 (2002)
6. Boonman, A. M. S. et al: A&A, **406**, 937 (2003)
7. Bourke, T. et al: in preparation (2004)
8. Caselli, P. et al: ApJ, **565**, 344 (2002)
9. Cazaux, S. et al: ApJ, **593**, L51 (2003)
10. Charnley, S. B., Tielens, A. G. G. M., & Millar, T. J.: ApJL, **399**, L71 (1992)
11. Di Francesco, J. et al: AJ, **124**, 2749 (2002)
12. Doty, S. D. et al: A&A, **389**, 446 (2002)
13. Flower, D. R. & Pineau des Forêts, G.: MNRAS, **297**, 1182 (2003)
14. Gordon, M. A.: A&A, **301**, 853 (1995)
15. Hatchell, J. & van der Tak, F. F. S.: A&A, **409**, 589 (2003)
16. Hildebrand, R. H.: QJRAS, **24**, 267 (1983)
17. Jørgensen, J. K., Schöier, F. L., & van Dishoeck, E. F.: A&A, **389**, 908 (2002)
18. Lada, C.J. et al: ApJ, **586**, 286 (2003)
19. Lee, J.-E. et al: ApJ, **583**, 789 (2002)
20. Liszt, H. & Lucas, R.: A&A, **384**, 1054 (2002)
21. Liszt, H. & Lucas, R. A&A: **355**, 333 (2000)
22. Millar, T.J.: Molecules in High-Mass Star-Forming Regions - Theory and Observation. In *Science with the Atacama Large Millimeter Array*, ed by A. Wootten (PASP Conference Proceeding, San Francisco 2001), Vol. 235, pp 45
23. Shirley, Y. L., Evans, N. J., Rawlings, J. M. C., & Gregersen, E. M.: ApJS, **131**, 249 (2000)
24. Schreyer, K. et al: A&A, **394**, 561 (2002)
25. Tafalla, M. et al: ApJ, **569**, 815 (2002)
26. Tafalla, M. et al: ApJ, in press (2003)
27. van der Tak, F. F. S. et al: ApJ, **537**, 283 (2000)
28. van der Tak, F.: Hot Molecular Cores. In *Star Formation at High Angular Resolution*, ed by M. Burton (IAU 2003), Vol. 221, 141
29. Zsargó, J. & Federman, S. R.: ApJ, **589**, 319 (2003)

Molecular Evolution in Prestellar Cores

Y. Aikawa[1], E. Herbst[2], P. Caselli[3], H. Roberts[4], and N. Ohashi[5]

[1] Dept. of Earth and Planetary Sciences, Kobe University aikawa@kobe-u.ac.jp
[2] Dept. of Physics, Astronomy and Chemistry, Ohio State University
[3] Osservatorio Astrofisco di Arcetri
[4] Dept. of Physics, Ohio State University
[5] Academia Sinica Institute of Astronomy and Astrophysics

We calculate the radial distributions of molecular abundances in various core models, and find that these distributions show reasonable agreement with observations. The enhanced NH_3/N_2H^+ ratio observed at core centers is explained by the dependence of N-chemistry on density. Since the radial distribution is determined by the balance between dynamical and chemical time scales, it can be an indicator of core contraction. Abundance variation among cores reflects the evolutionary stage or contraction time scale. High deuterium fractionation, obtained at core centers, is reproduced by our model results. The D/H ratio and N_2H^+ column density correlate positively both in our models and observational data, indicating that both values increase with evolution. Relative abundances of multiply deuterated species to normal species are calculated to lie in the range $10^{-3} - 10^{-2}$.

1 Introduction

The density distribution in starless cores is estimated from the observation of the submillimeter continuum and background stars. Comparison of the density distribution with line observations reveals molecular differentiation. CO depletion at the core center is observed in several cores, while N_2H^+ tends to show a good correlation with the continuum. Tafalla et al. (2002) solved the radiative transfer problem to estimate actual abundance distributions, and found that the N_2H^+ abundance is constant, while CO and CS are heavily depleted at the center (see also Bergin 2003 in this volume).

The spatial distribution of molecular abundances is determined by the balance between the dynamical time scale of a core and the chemical time scale of molecular species. Hence distributions can reveal not only the chemical processes but also the formation and evolution of cores. In this work, we solve a chemical network of gas-phase and modified-rate grain-surface reactions in various contracting core and static cores. Details of our models are described in Aikawa et al. (2001, 2003). Calculated spatial distributions of molecules are presented in §2. Application of our models to statistical observations and variation of molecular abundances among cores is discussed in §3. Deuterium fractionation in our models and observed cores is discussed in §4.

2 Radial Distribution of Molecular Abundances

Figure 1 (a) shows molecular evolution in a fluid parcel migrating inwards. Depletion starts in the relative order CCS, CO, NH_3 and N_2H^+. While CO is adsorbed onto grain surfaces and incorporated onto ice mantles, CS and CCS are depleted both by gas-phase reactions and the adsorption. Nitrogen species maintain a high abundance in the gas phase because their mother molecule, N_2, has a relatively low adsorption energy on the grain surface.

By calculating such molecular evolution in multiple fluid parcels, we obtain radial distributions of molecular abundances at certain evolutionary stages. Figure 1 (b) shows the distribution when the central density of the core is 3×10^6 cm^{-3}, which is similar to the value in L1544. At this stage, CCS and CO are heavily depleted at the center, while the NH_3 and N_2H^+ abundances do not vary much with radius. The distributions are in reasonable agreement with the analysis by Tafalla et al. (2002).

Let us look at the radial differentiations in more detail. Figure 2 (a) shows the estimated distribution of N_2H^+ and NH_3 in L1517B (Tafalla et al. 2003), in which the central density is $\sim 10^5$ cm^{-3}. While the N_2H^+ abundance is constant, NH_3 is more abundant at inner radii. The solid line in Fig. 2(b) represents the distribution of the ratio NH_3/N_2H^+. The central enhancement of the ratio is counterintuitive, because both species are formed from N_2 and NH_3 is expected to be adsorbed onto grains while the ion (N_2H^+) is not. We analyzed our models to resolve this puzzle. The dotted line in Fig. 2 (b) shows the computed ratio in a static core model; it is slightly enhanced at inner radii ($\leq$ 10000 AU). Although both species are formed from N_2, the formation and destruction processes are different: N_2H^+ is formed by N_2 + H_3^+ and destroyed by recombination, while NH_3 is formed by hydrogenation of N^+ and depleted mainly by adsorption onto grains. The dependence of these

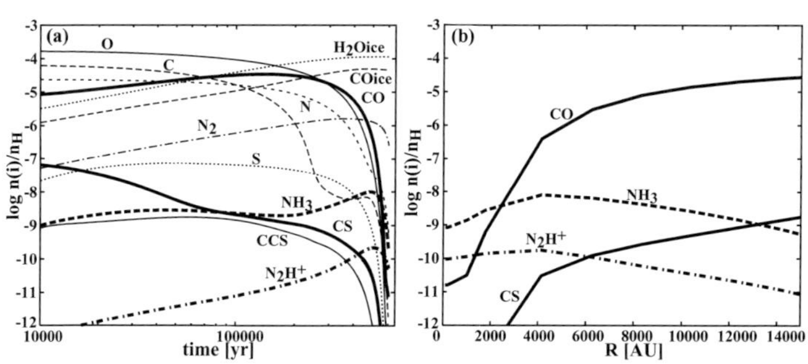

Fig. 1. (a) Molecular evolution in a fluid parcel migrating from $R = 8.2 \times 10^3$ AU to 1000 AU as the core contracts. The contraction is slower than the Larson-Penston flow (Larson 1969; Penston 1969) by a factor of 3. The gas density in the fluid parcel increases from 1.7×10^4 cm^{-3} to 5.8×10^6 cm^{-3} during the migration. (b) Molecular distribution in the core when the central density $n_H (= n(H) + 2n(H_2))$ reaches 3×10^6 cm^{-3}.

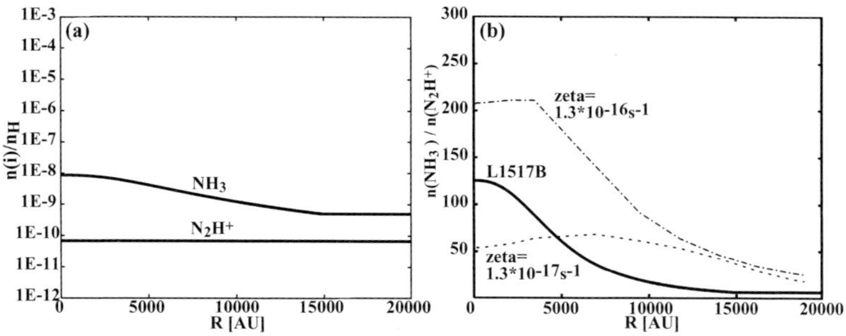

Fig. 2. (a) Radial distribution of N_2H^+ and NH_3 in L1517B. (b) Abundance ratio of NH_3/N_2H^+ in L1517B (solid line) and in models (dotted and dash-dotted lines) with the same central density ($n_H = 3 \times 10^5$ cm^{-3}) and age (3×10^5 yr) but with a different cosmic-ray ionization rate (zeta).

reactions on density causes the radial differentiation. The abundances depend differently on ionization rate and grain size as well; the central enhancement is more significant if the ionization rate is higher (dot-dashed line in Fig. 2 b), or the grain size is larger (Caselli & Aikawa in preparation).

In contrast to the nitrogen species, CO and CS suffer heavy depletion at the core center. In cold prestellar cores, the adsorption onto grains proceeds with time, and hence the molecular distribution depends on how the core contracts. The distribution in model cores is shown in Fig. 3 (a). Tafalla et al. (2002) obtained abundance distributions in L1517B (dotted lines in Fig. 3 b), which are relatively steep and similar to our static core. Recently, however, they obtained an even steeper gradient for the CO abundance from their observation with higher resolution; the data are best fitted if the abundance

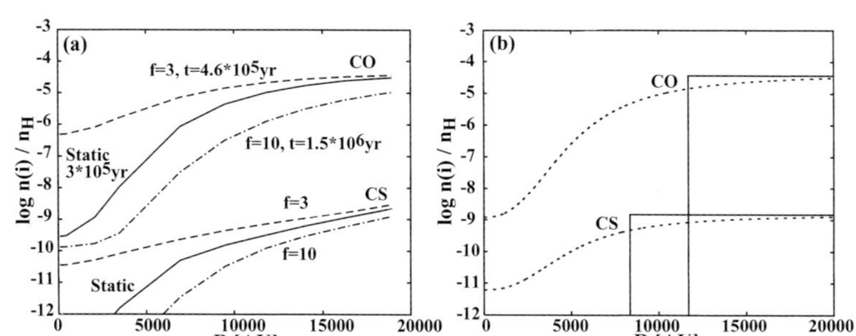

Fig. 3. (a) Radial distribution of CO and CS in a static core model (solid lines) and contracting core models with the same central density of 3×10^5 cm^{-3}. The contraction time scale is longer than that of the Larson-Penston flow by a factor of f. Note that the cloud age (t) also varies among models accordingly. (b) The distribution of CO and CS in L1517B obtained by Tafalla et al. (2002) (dotted lines) and Tafalla et al. (2003) (solid lines).

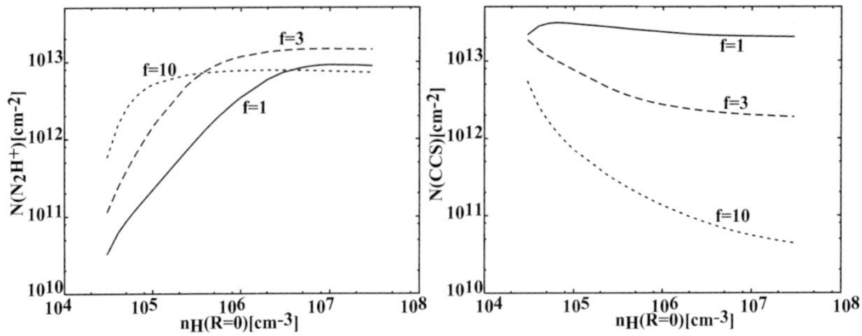

Fig. 4. Calculated column densities of N_2H^+ and CCS through the core center as a function of the central density. The core contraction is slower than the Larson-Penston flow by a factor of f.

is a step function of radius (solid lines in Fig. 3 b). Although such an estimate should be taken with caution because contamination with emission from the outer radii makes the abundance at the core center rather uncertain, the steep abundance gradient is of special interest. It is difficult to reproduce with the usual contracting or static core models. The heavy depletion at the center suggests instead that the core is old or has a long dynamical time scale, while the gas in the outer radius should be relatively young. This dichotomy in turn may suggest inhomogeneous core formation, with the outer component recently accreted from the ambient cloud. Indeed, Tafalla et al. (2003) found internal motions, which suggest such inhomogeneous or irregular core formation.

3 Variation of Molecular Column Density Among Cores

From theoretical models of collapsing cores, molecular column densities can be calculated at each evolutionary stage. Figure 4 shows N_2H^+ and CCS column densities through the core center as a function of the central density of model cores, with various contraction time scales. As contraction proceeds, N_2H^+ increases, so that this species could be a good probe of core evolution. On the other hand, CCS decreases with evolution and it also depends sensitively on collapse models.

Aikawa et al. (2003) compiled molecular column densities of prestellar cores from the literature. Cores with higher N_2H^+ column density tend to have a lower CCS column density, as predicted. In addition, cores with low N_2H^+ column density, such as L1521B and L1521E, have CCS emission peaks at the core center (Ohashi 2000; Hirota et al. 2002), while other cores such as L1544 show a central hole in the integrated CCS map. A centrally peaked CCS column density is reproduced only in the early stages of evolution in our models, strengthening the idea that the molecular column density reflects the evolutionary stage of cores.

4 Deuterium Fractionation

In low temperature (< 20 K) cores, molecular D/H ratios increase due to exothermic exchange reactions such as H_3^+ + HD → H_2D^+ + H_2. The depletion of CO, which is the main reactant for H_2D^+, further increases the ratio. Hence in the central region of evolved cores, molecular D/H ratios are expected to be much higher than the elemental ratio of 1.5×10^{-5}. The radial distribution of selected molecules in a contracting core is shown in Fig. 5 (a). The D/H ratio is enhanced at small radii. The ion H_2D^+ is of special interest; it is even more abundant than its normal counterpart, H_3^+, and is predicted to be the dominant ion at the center. Its fractional abundance of $\sim 10^{-9}$ is consistent with the observation in L1544 (Caselli et al 2003). Dissociative recombination of deuterated ions enhances the atomic D/H ratio, which is important for deuterium fractionation via grain-surface reactions.

Figure 5 (b) shows the column density ratio of DNC to HNC as a function of the N_2H^+ column density in various models and observed cores. A positive correlation between the two values can be seen, as predicted from the model.

In recent years, multi-deuterated species have been detected in several prestellar and protostellar cores. Since our model successfully reproduces the high abundance of mono-deuterated species, it is interesting to see if we can also get the proper enhancement for these species. Figure 6 shows radial distributions for gas-phase and grain-surface species in a model core. The species on grains are important for comparison with protostellar cores, since they sublimate when a protostar heats grains. In the core center ND_3 reaches an abundance of $n/n_H \sim 10^{-10}$, and the ND_3/NH_3 ratio is 10^{-2}. The ratio seems high enough to account for the observed value of $\sim 10^{-3}$ (van der Tak et al. 2002; Lis et al. 2002) even if we consider the contamination with the outer component, which possesses a low D/H ratio. The mono- and doubly-deuterated forms of methanol have abundances of 0.9 ± 0.3 and 0.2 ± 0.1

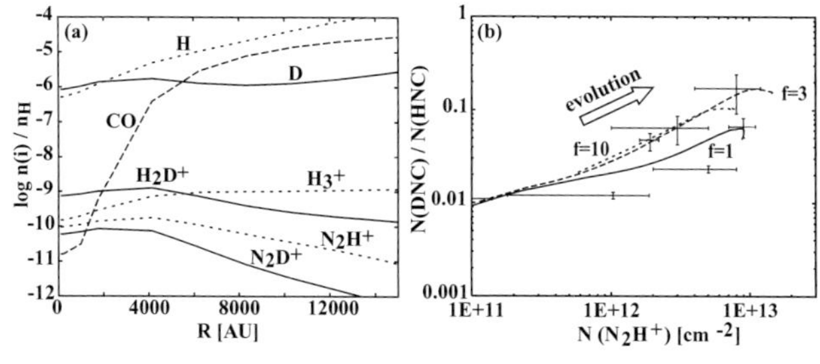

Fig. 5. (a) Calculated radial distributions of deuterated and normal species in a contracting core when the central density is $n_H = 3 \times 10^6$ cm^{-3}. The contraction is slower than the Larson-Penston flow by a factor of 3. (b) The column density ratio DNC/HNC as a function of N_2H^+ column density in contracting cores (lines) and observed cores (dots with error bars) by Caselli et al. (2002) and Hirota et al. (1998,2003).

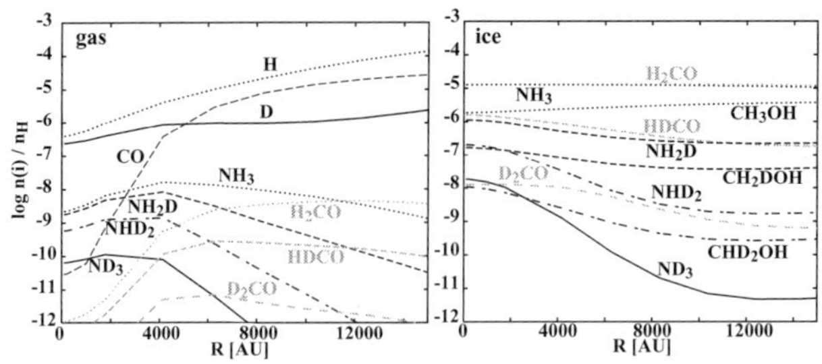

Fig. 6. Radial distribution of molecules calculated with a multi-deuterated chemical network. The core model is the same as in Fig. 5 (a). Gas-phase species are shown in the left panel, while species on the grain surface are in the right panel.

relative to the normal isotope in the protostar IRAS 16293. These values are higher than predicted in Fig. 6 (right), although better agreement is obtained in a static core model (Aikawa et al., in preparation). The predicted ratio of D_2CO/H_2CO in the gas phase is $\leq 10^{-2}$, while it is observed to be $0.01 - 0.1$ in several prestellar cores (Bacmann et al. 2003). Dependence on these ratios on physical conditions (ex. core contraction) is under investigation.

Acknowledgement. Y.A. is supported by a Grant-in-Aid for Scientific Research (14740130) and "The 21st Century COE Program of Origin and Evolution of Planetary Systems" in Ministry of Education, Culture, Sports, Science and Technology (MEXT) of Japan. E. H. thanks the NSF for its support of his research program in astrochemistry.

References

1. Y. Aikawa, N. Ohashi, S.-I. Inutsuka et al: ApJ, **552**, 639 (2001)
2. Y. Aikawa, N. Ohashi, E. Herbst: ApJ, **593**, 906 (2003)
3. A. Bacmann, B. Lefloch, C. Ceccarelli et al: ApJ, **585**, L55 (2003)
4. P. Caselli, P.J. Benson, P.C. Myers et al: ApJ, **572**, 238 (2002)
5. P. Caselli, F.F.S. van der Tak, C. Ceccarelli et al: A&A, **403**, L37 (2003)
6. T. Hirota, M. Ikeda, S. Yamamoto: ApJ, **594**, 859 (2003)
7. T. Hirota, T. Ito, S. Yamamoto: ApJ, **565**, 359, (2002)
8. T. Hirota, S. Yamamoto, H. Mikami et al: ApJ, **503**, 717 (1998)
9. R.B. Larson: MNRAS, **145**, 271 (1969)
10. D.C. Lis, E. Roueff, M. Gerin, T.G. Phillips et al: ApJ, **571**, L55 (2002)
11. N. Ohashi: In: *Astrochemistry: From Molecular Clouds to Planetary Systems*, IAU Symposium 197, ed by Y.C. Minh, & E.F. van Dishoeck (Astronomical Society of the Pacific 2000), pp 61-70
12. M.V. Penston: MNRAS, **144**, 425 (1969)
13. M. Tafalla, P.C. Myers, P. Caselli et al: ApJ, **569**, 815 (2002)
14. M. Tafalla et al. submitted to ApJ (2003)
15. F.F.S. van der Tak, P. Schilke, H.S.P. Muller et al: A&A, **388**, L53 (2002)

Deuterium Fractionation in Totally Depleted Cores

C.M. Walmsley[1], D.R. Flower[2], and G. Pineau des Forêts[3,4]

[1] INAF, Osservatorio Astrofisico di Arcetri, Largo E.Fermi 5,I-50125 Firenze,Italy walmsley@arcetri.astro.it
[2] Physics Department, The University, Durham DH1 3LE, UK David.Flower@durham.ac.uk
[3] IAS, Université de Paris-Sud, F-94205 Orsay, France gdesforets@noos.fr
[4] LUTH, Observatoire de Paris, F-92195, Meudon Cedex, France

1 Introduction

One of the important results of the past few years is that at low (∼10K) molecular cloud temperatures and at densities of greater than a few times 10^4 cm^{-3}, CO and likely most C–containing species become depleted from the gas phase. Gas–phase chemical models show that in most circumstances, CO is the source of the carbon which becomes incorporated into molecules such as HCO^+, HCN, CS etc. and thus it is reasonable that these species should be depleted under conditions not greatly different than those under which CO disappears. In fact, the observations imply that only nitrogen containing molecules such as NH_3 and N_2H^+ survive at densities of around 10^5 cm^{-3}, suggesting that the gas phase repository of N is extremely volatile. Indeed molecular nitrogen (Bergin and Langer 1997) is known to have a binding energy to certain surfaces which is smaller or of the same order as that of CO. Thus it is plausible that there are situations where mostly nitrogen hydrides remain supreme and all other heavy elements are in the form of ices on dust grain surfaces.

That being said, it seems likely that at slightly higher densities than those where NH_3 and N_2H^+ are observed, nitrogen will also freeze out onto grains. Just at what density this will happen is debatable, but freeze–out seems sufficiently likely to consider the question of what will happen in a situation where no heavy elements are available in the gas phase. There is evidence from both dust emission and absorption measurements that the hydrogen density in "prestellar cores" (i.e. gas condensations lacking a protostar which appear to be close to gravitational collapse) is as high as 10^6 cm^{-3}. It is under such conditions that one might expect something approaching "complete depletion" of heavy elements.

Such a disappearance of C, N, O from the gas phase would have certain consequences. One would be the presence of a "hole" in the distribution of the N–containing hydrides near the position of maximum density in a prestellar core. In fact some evidence of such "holes" is available in the literature. Bergin et al. (2002) see marginal evidence for an absence of N_2H^+ towards the peak of B68. Also, Belloche and André (this conference) using the Plateau de Bure interferometer see a ring of N_2H^+(1-0) emission surrounding the peak of the young protostar IRAM 04191. A schematic view of what might be the situation is given in Fig. 1 where we depict an ellipsoidal core from which the heavy elements have been depleted at the center. Surrounding this depleted nucleus is a shell where nitrogen–bearing species dominate the chemistry and an outer region with "normal" molecular abundances. This is for the moment simply a hypothesis which should be tested, but one necessary step in order to do this is to understand the conditions in the depleted center.

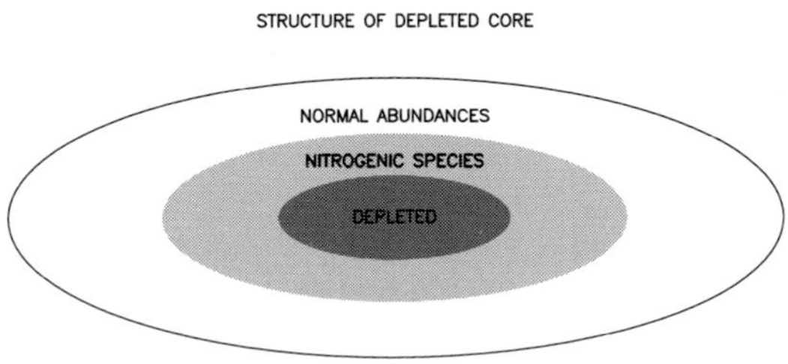

Fig. 1. A schematic view of a core close to collapse with parameters similar to L 1544. In a central "depleted" region, one can only observe hydrogenic or deuterated species such as H_2D^+. In a shell exterior to this and at somewhat lower density, nitrogen in some form (such as NH_3) survives in the gas phase but CO, CS, and other C-containing species are frozen out. In the exterior of the core, where the densities may be of order 10^4 cm^{-3}, CO is in the gas phase and abundances are "normal".

One consequence of a depleted chemistry is certainly that purely hydrogenic (or deuterated) ions come to the fore (see Roberts et al. (2003), Aikawa et al. (2003)). In the presence of species with high proton affinity (such as CO or N_2), H_3^+(and H_2D^+etc.) is rapidly destroyed but without heavy elements, the abundances of such ions will in general increase. Thus the discovery by Caselli et al. (2003) of H_2D^+ emission in the direction of the dust peak in L 1544 seems of great significance. The observed emission is compact (less than 2000 AU) and coincident with the region of highest density deduced from dust emission. This circumstantial evidence suggests that indeed a "completely depleted region" exists in the high density nucleus of L 1544 and similar dense cores. If this is the case, one expects that deuterium fractionation may be extreme. Dalgarno and Lepp(1984) pointed out many years ago that high degrees of heavy element depletion will tend to cause high degrees of deuterium fractionation and indeed this has been confirmed observationally (e.g. Caselli et al. (1999), Bacmann et al. (2003)). Accordingly, we decided that it would be useful to examine ionization equilibrium and deuterium fractionation in the extreme case of complete heavy element freeze out and with this in mind, we have carried out a study of the steady–state chemistry and ionization equilibrium in the case that heavy elements are missing from the gas phase. Here, we present some preliminary conclusions with the emphasis on deuterium enrichment. A detailed account of our results will be given in a paper presently in preparation.

2 Results

We compute the steady state abundances for the case where the chemistry is driven by cosmic ray ionization (taken to have a standard rate for $H_2$of $3{\cdot}10^{-17}$ s^{-1}) and we consider the species H, H_2, H^+, H_2^+, H_3^+ and their deuterated counterparts as well as He and He^+. Many of the rates and assumptions are based on an article of Flower and Pineau des Forêts (2003).

A difference, however, is that we consider separately the ortho and para forms of H_2, H_3^+, H_2^+, H_2D^+. A critical role in the calculations plays ion neutralisation on grain surfaces and hence the grain size distribution. Here, we have assumed in general a unique size but we have also experimented with power law size distributions. We find that the essential parameter for our calculations is the mean grain surface per H–nucleus $s_g = \langle \sigma_g\, n_g \rangle / n_{\rm H}$ which for the case of a unique grain size is inversely proportional to grain radius. Hence, our results are similar to those for a power law size distribution with the same mean surface. We have examined a range of molecular hydrogen density between 10^5 and 10^7 cm^{-3} and grain radii between 0.025 and 0.2μm (s_g between $4 \cdot 10^{-21}$ and $5 \cdot 10^{-22}$ cm^2).

One important result of our calculations is that the dominant ion of depleted cores in many circumstances is H^+ (at low densities and for grain sizes above 0.1 μm). H_3^+ and its deuterated counterparts, however, are the most abundant positive ions at higher densities and lower grain sizes. Fig. 2 gives an overview of our results for our standard model with grain radius 0.1 μm. H^+, which is destroyed by recombination on grain surfaces, has a roughly $1/n_{\rm H}$ dependence on density. H_3^+ and similar ions on the other hand are destroyed to a considerable degree by dissociative recombination with electrons which has a slower fall–off with density. This leads to the H_3^+ family becoming more abundant than protons at high densities and in fact to D_3^+ becoming the dominant ion at 10^7 cm^{-3}. This latter result is partially due to the effect that at higher densities, HD increases in abundance relative to electrons, favoring deuterium enrichment. For larger grain sizes than 0.1μm, H^+ dominates over the whole density range considered by us and for grains as small as 0.025μm, H_3^+(and its deuterated counterparts) becomes the principal ion.

From the above, one sees that it is critical to know to what extent the grain size distribution has evolved in high density cores such as L 1544. Our results depend on the correctness of the assumption that grain sizes are of order 0.1μm. How good is this assumption?

There are both theoretical (e.g. Ossenkopf and Henning 1994) and observational reasons for thinking that some grain coagulation occurs in dense

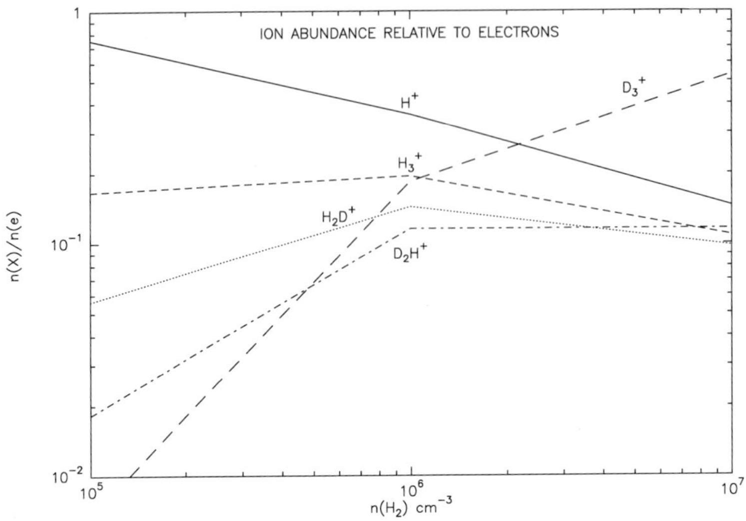

Fig. 2. Plot of ratios of abundances of H^+, H_3^+, H_2D^+, DTWHP, DTHP relative to electrons as function of H_2 density for the model with T=10 K, grain radius =0.1μm, and cosmic ray ionization rate $= 3 \cdot 10^{-17}$ s^{-1}.

interstellar cores. We expect moreover that if the heavy elements are depleted, there should also be a thick (roughly 100 Å) ice mantle accreted onto the refractory cores which are of importance in diffuse gas. It thus seems reasonable that grains smaller than 0.02μm are unlikely to exist under the conditions considered by us. On the other hand, estimates of the mm emissivity based upon comparisons of the observed millimeter continuum intensity relative to IR extinction are consistent with moderately evolved grains (e.g. Bianchi et al. 2003) with sizes not greatly different from the classical estimate of 0.1μm. We conclude that most of the grain surface area in dense cores is likely due to grains of size roughly 0.1μm.

Our results in fact allow a crude limit to be put on the grain size by attempting to fit the H_2D^+ observation of Caselli et al. (2003). Caselli et al. estimate the H_2D^+ abundance relative to H_2 to be 10^{-9} or more precisely that the o-H_2D^+ abundance is $5 \cdot 10^{-10}$. In order to fit this constraint, we need *at least* the cosmic ray ionization rate of $3 \cdot 10^{-17}$ s^{-1} of our standard model as well as a grain size less than 0.1μm(or equivalently s_g above 10^{-21} cm^2). Essentially, this is because in order to fit the constraint on o-H_2D^+, one needs both an electron abundance of at least $5 \cdot 10^{-9}$ and a relatively low proton abundance (in order that H_2D^+ is a major ion). The former condition requires a substantial cosmic ray ionization rate and the latter relatively small grains. We conclude therefore that our results add to the evidence that the dominant grain size *prior* to protostellar collapse is of order 0.1μm or somewhat less.

We have also examined the behaviour of the ortho–to–para ratios for H_3^+ and H_2D^+ in dense cores. Neither of these ratios is likely to be in LTE (see discussions by Pagani et al. (1992) and by Gerlich et al. (2002) for H_2D^+) under the conditions considered by us and both are sensitive to the fraction of H_2 in the ortho form. In steady–state, we find that the o/p ratio for H_2 should be of order 10^{-4} but our calculations show also that the timescale for reaching steady state in the H_2 o/p ratio is of order 10^5 years and thus comparable with dynamical timescales. In contrast, the timescale for reaching ionization equilibrium is short. A non steady–state o/p ratio for H_2 will cause the o/p ratios for H_3^+ and H_2D^+ to diverge from their steady state values and this induces some uncertainty e.g. in our predictions of the ortho-H_2D^+ abundance. We think it unlikely however that such problems disturb our conclusions greatly.

3 Discussion

Our calculations show that high deuterium enrichment levels can be achieved in the isotopomers of H_3^+ if all heavy elements vanish from the gas phase. This is in particular the case at high densities (above 10^6 cm^{-3}) where D_3^+ can become the most abundant ion. We also find that it is possible to explain the Caselli et al. (2003) observation of H_2D^+ towards L 1544 in terms of a model in which heavy elements are depleted in the central few thousand AU. Other possibilities cannot be excluded but we feel that complete depletion is the most plausible explanation at present.

Further progress in understanding completely depleted regions will require other tracers and, if possible, detection of absorption lines towards completely depleted regions. Measuring the o/p ratio in H_2D^+ for example would require observation of the $1_{01} - 0_{00}$ line of p-H_2D^+ at 1.37 THz and exciting the upper level of this transition (65 K above ground) probably requires temperatures above those estimated for dense cores. Observation of the $1_{10} - 1_{01}$ line of D_2H^+ at 692 GHz (upper level 33 K above ground, see Hirao and Amano this volume) seems more feasible. Absorption line measurements would be easier to interpret but require a background continuum source. This might have the implication that observing such transitions is only possible towards Class 0 sources with embedded continuum sources such as IRAM04191 and NGC1333 IRAS2.

References

1. Aikawa, Y., Ohashi, N., Herbst, E. Ap.J., **593**,906, (2003)
2. Bacmann, A., Lefloch, B., Ceccarelli, C., Steinacker, J., Castets A., Loinard, L. , Ap.J. **585**, L55 (2003)
3. Bergin, E.A., Langer, W.D. Ap.J., **486**,316 (1997)

4. Bergin, E.A., Alves, J., Huard, T., Lada, C.J. ApJ **570** , L101, (2002)
5. Bianchi, S., Gonçalves, J., Albrecht, M. et al. A&A, 399, L43 (2003)
6. Caselli, P., Walmsley, C.M., Tafalla, M., Dore, L., Myers, P.C. , Ap.J., **523**, L165 (1999)
7. Caselli, P., van der Tak F.F.S., Ceccarelli C., Bacmann, A. A&A **403** , L37 (2003)
8. Dalgarno, A., Lepp, S. Ap.J. **287**, L47 (1984)
9. Flower, D., Pineau des Forêts, G., MNRAS , **343**, 390 (2003)
10. Gerlich, D., Herbst, E., Roueff, E. Plan. Sp. Sci., **50**, 1275 (2002)
11. Ossenkopf, V., Henning, T., A&A, **291**, 943 (1994)
12. Pagani, L., Wannier, P.G., Frerking, M.A. et al. , A&A , **258**, 472 (1992)
13. Roberts, H., Herbst, E., Millar, T.J. Ap.J.,**591**,L41 (2003)

Multiply Deuterated Molecules in Star Forming Regions

C. Ceccarelli

Laboratoire d'Astrophysique de l'Observatoire de Grenoble - BP 53- F-38041, Grenoble Cedex 9, France Cecilia.Ceccarelli@obs.ujf-grenoble.fr

This contribution reviews the observations of multiply deuterated molecules in star forming regions. Just up to a few years ago, multiply deuterated molecules did not even appear in the model predictions, because they were thought to be too underabundant to be detectable. Observations in the last three/four years have demonstrated that multiply deuterated molecules not only are detectable, but provide very stringent constraints to the chemical models, and have forced major revisions.

1 Introduction

Although the cosmic abundance of deuterium is only 10^{-5} [15], it has been known since a long time that the deuterium fractionation in some molecules is enhanced to 10^{-3} – 10^{-2} in molecular clouds. The reason is the lower zero-point energy associated with the deuterated form of the molecule, which favors at low temperatures the formation of the deuterated form. For about 30 years or so singly deuterated molecules have been detected with abundances up to 10% of the main isotopes (see for example the review by [7]). Gas phase models seemed to be rather succesful in explaining the observed deuterium fractionations (e.g. [3]), and in fact have been used to derive some key parameters, like the electron abundance in cold gas [4], or to investigate the first phases of low mass star formation [23, 27].

However, in 1990 Turner [31] published the detection of the doubly deuterated form of formaldehyde in the direction of Orion. He measured a D_2CO/H_2CO ratio of around 0.003, and demonstrated that gas phase chemistry would not be able to account for the observed ratio. He argued that formaldehyde is an active grain chemistry product, and showed that the observed ratio agrees well with the predictions of the model proposed years earlier by Tielens [30]. In 1998 Ceccarelli et al. [8] measured a 25 times higher D_2CO/H_2CO ratio in the low mass protostar, IRAS16293-2422 (see also [19]). The observed ratio showed that the deuterium fractionation in formaldehyde can reach the fabulous enhancement of 8 orders of magnitude with respect to the elemental D/H ratio. The hunt for multiply deuterated molecules was open, and the models to explain all those observations had to be revised. In this contribution I will review the last three/four years of the hunt, which have seen a blooming of new detections, which in turn have led to a blooming of new chemical models to explain them.... and it is not over yet.

2 Doubly Deuterated Formaldehyde

Doubly deuterated formaldehyde was the first molecule with two deuterium atoms detected[31], and with such a large abundance that made it more than just a zoological curiousity [8]. Loinard et al. [18] derived the accurate D_2CO/H_2CO and $HDCO/H_2CO$ ratios towards IRAS16293-2422, by observing several H_2CO, $H_2^{13}CO$, HDCO and D_2CO transitions. They obtained the following ratios: $HDCO/H_2CO=0.15$ and $D_2CO/H_2CO=0.05$. Subsequently, Ceccarelli et al. [10] obtained a map of the D_2CO emission around IRAS16293-2422. D_2CO is very abundant all across the envelope of IRAS16293-2422, which size is about 60″x40″. The largest D_2CO/H_2CO abundance ratio, ~ 0.16, is observed about 20″ south of the center, and decreases to 0.03 at the borders of the envelope. This value has also been measured in the cloud L1689N, which harbors IRAS16293-2422 [11]. The relatively large D_2CO/H_2CO ratio observed both on the source and on the cloud seems to rule out that gas phase reactions are responsible for it. Rather, once again, active grain chemistry has been invoked to account for it. Note that the study of the H_2CO emission (both from several transitions and from the observed spatial profile) has allowed to reconstruct the H_2CO abundance profile in IRAS16293-2422 [9, 10], and to demonstrate that, very likely, the H_2CO and D_2CO molecules are trapped in CO-rich ices. These ices evaporate when the dust temperature reaches ~ 30 K, injecting into the gas phase also the minor components of the mantle, like formaldehyde. This explains why we observe D_2CO in the relatively cold outer envelope of IRAS16293-2422, although D_2CO is a grain surface product.

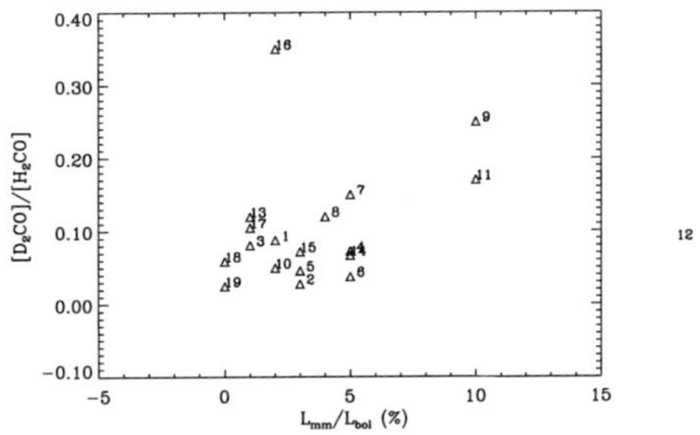

Fig. 1. D_2CO/H_2CO abundance ratio as function of the millimetric to bolometric luminosity ratio in the sample of Class 0 sources observed by Loinard et al. [18].

The first very obvious question is: *is IRAS16293-2422 a peculiar source?* To answer this question, Loinard et al carried out a survey of the D_2CO/H_2CO ratio in a sample of about two dozens protostars and found [18]:

1– D_2CO is detected in all surveyed low mass protostars, and the D_2CO/H_2CO ratio varies between 0.03 to 0.35; 2– no clear correlation is found of the D_2CO/H_2CO ratio with other quantities related to the low mass protostars, with the possible exception of the millimetric to bolometric luminosity ratio (Fig. 1); 3– D_2CO was not detected in all surveyed high mass protostars.

3 Deuteration of Methanol and Sulfide: Grain Surface Chemistry

The second very obvious question is: *is formaldehyde a peculiar molecule*? Now we know that the answer is certainly "not". Already in 2000, Roueff et al. [26] detected the doubly deuterated form of ammonia in the cold cloud L134N, and Loinard et al. [17] detected the same molecule towards a low mass protostar with a ratio NHD_2/NH_3=0.03, about six times larger than in L134N, and comparable to the D_2CO/H_2CO ratio measured in the same source [18]. Last year has also seen the detection of the triply deuterated ammonia [32, 16].

The observation of the multiply deuterated forms of methanol and sulfide, and the measure of the relevant deuteration is particularly important to constrain the grain surface chemistry models. In fact, there is ample evidence that both molecules are formed on grain surfaces, because gas phase reactions are incapable to account for the relative large abundance of these molecules in the gas phase. The most accepted view is that methanol is formed by successive hydrogenation of CO, sticked onto the grain surfaces [30, 12]. The first step of CO hydrogenation would form H_2CO, and the last CH_3OH. Therefore, the deuteration of formaldehyde and methanol follows the same route, and the singly and multiply deuterated forms of formaldehyde and methanol are predicted to be related. Parise et al. ([22, 20]) detected the singly, doubly and triply deuterated forms of methanol towards IRAS16293-2422. The comparison of these observations with the active grain surface models (see Parise on this same volume) suggests the following considerations:

a) a major fraction of methanol is in the deuterated forms, more than 30%; b) formaldehyde and methanol are approximatively in agreement with the theoretical predictions, *if the accreting gas has an atomic D/H ratio around 0.3*; c) model predictions largely overestimate the CH_3OD/CH_3OH ratio.

The D/H=0.3 is a relatively large value, when compared to the model predictions of relatively recent models. In these models, the atomic D/H ratio never reaches values larger than about 0.1, and this occurs only in extreme CO depleted conditions [25]. The detection of the doubly form of sulfide by

[33], who found D_2S/HDS=0.1, has confirmed that the accreting gas must have such large atomic D/H ratios, also in somewhat different sources.

In practice, whether the deuteration is a gas phase or active grain chemistry product, the large observed deuteration ratios cannot be reproduced if not in extreme conditions: the gas must be very dense, cold **and** severely depleted of CO. These conditions are required by gas phase chemical models, where the deuteration is set by the H_2D^+/H_3^+ ratio, for H_2D^+ transmitting the deuterium to the other molecules. The same conditions are also required by the active grain chemistry models, for the deuteration in that case is mainly set by the atomic D/H ratio of the accreting gas. Both ratios, the H_2D^+/H_3^+ and the D/H ratio, are enhanced in cold, dense and CO depleted gas. The latter is a major factor to obtain the observed extreme values. This is because CO is the main destroyer of H_3^+; if it disappears from the gas phase $H_3^+ + HD \rightarrow H_2D^+ + H_2$ becomes the main route of H_3^+ destruction, and it is unbalanced in favor of H_2D^+ in cold and dense gas.

4 Deuteration During the Pre-stellar Core Phase

The previous studies suggested that the large deuteration is not a present day product, but rather the reminiscence of the cold and dense pre-stellar core phase (e.g. [8] and all the subsequent works by the same group). As just discussed, CO depletion is a key parameter to enhance the deuterium fractionation, and pre-stellar cores are in fact the objects were the largest depletion has been measured so far [6]. In particular, Bacman et al. [1, 2] measured the CO depletion and the D_2CO/H_2CO ratio in a sample of half a dozen pre-stellar cores. The first very important result of this study is that indeed pre-stellar cores have D_2CO/H_2CO ratios similar to those observed in low mass protostars [8, 18], fully confirming the hypothesis that the large observed deuteration observed towards Class 0 sources is set during the pre-collapse phase. A second, important result is that the D_2CO/H_2CO ratio increases with the CO depletion measured on each pre-stellar core. This is a clear observational confirmation of the major role played by the CO depletion in the enhancement of the deuterium fractionation.

A further confirmation of both the role of CO depletion and the deuteration set during the pre-collapse phase came with the spectacular detection of H_2D^+ in the pre-stellar core L1544 [5]. As said, H_2D^+ is the key molecule that transmits the deuteration to the other gas phase molecules. The H_2D^+ abundance derived by these observations, $\sim 10^{-9}$, exceeds the predicted abundance by about a factor of 10. It is possible to reconcile model predictions and observations only assuming that all CNO-bearing neutral species, including atomic oxygen, are completely frozen onto the grain mantles within a radius of $\sim$ 3000 AU, which seems rather unprobable (see also the contribution by van der Tak et al. in this same volume) and not supported by the measure of the (average) CO depletion in that region, only a "modest" factor of 10. Anyway, these observations suggest that H_2D^+ is the most abundant molecular ion in the core of L1544, a fact that brought to revise the theoretical chemistry models. The very last generation of chemical models includes all the deuterated forms of H_3^+, which in turn enhance the deuteration in the other molecules, too [24]. As in previous models, the large deuterium fractionations are predicted when CO is extremely depleted. The question is whether there is enough time to build up a mantle as D-enriched as we observe, when only a few molecules are left in the gas phase.

5 The Puzzle of the Low Water–Ice Deuteration

If progresses have been made with respect to the comprehension of the deuteration observed in formaldehyde, methanol, ammonia and sulfide, it remains the big puzzle of the little deuteration observed in the solid state, i.e. on the mantles. Being the major constituent of the ices, water is the target of these studies. Despite a first claim of solid HDO detection towards a massive protostar [29], later on disclaimed [13], at present no solid HDO has been detected, with a limit $HDO/H_2O \leq 0.05$. Most of the solid state observations refer indeed to massive protostars, whose NIR continuum is bright enough to study the absorption features [13]. As shown by Loinard et al. [18], though, massive protostars do not possess the same large degree of deuteration observed in low mass protostars, and, hence, these observations were not conclusive. However, Parise et al. [21] searched for solid HDO towards four low mass protostars, two of them with measured D_2CO/H_2CO (~ 0.04). In none of the targetted sources solid HDO has been found, with a limit $HDO/H_2O \leq 0.02$. This result represents clearly a problem. It is possible that there is a selective deuteration going on which favors formaldehyde, methanol and sulfide and defavors water on the grain surfaces. But this is still to be proven. Another possibility is that gas phase and solid state observations probe different ices: "last skin" CO-rich ices the first, and the "integrated over time" H_2O-rich ices the latters. This is certainly a puzzle which needs an observational and theoretical effort to be solved.

6 Conclusions

In the last three/four years several multiply deuterated molecules have been detected with relatively large abundances. As a consequence, theoretical models of molecular fractionation have been revised, and our comprehension of the process has surely improved. It is now clear that the large deuteration observed in the gas phase molecules around low mass protostars is not a present day product, but that it is likely a reminiscence of the pre-stellar core phase, when the gas is cold, dense and depleted of CO. It is, however, less clear how exactly the deuteration occurs. Indeed, it is still debated whether gas phase reactions or grain surface chemistry are responsible for the observed deuteration, and, most important, none of the published models provides coherent predictions, capable to explain the ensemble of available observations. Very likely, both gas phase and grain surface reactions contribute, each one for different molecules/ions, but simultaneously, to the large degree of deuteration. In conclusion, although the general ideas behind the theoretical models seem to be correct, the detailed theoretical predictions are still mismatched with respect to the observations.

References

1. Bacmann, A., Lefloch, B., Ceccarelli, C., et al 2002, A&A, 389, L6
2. Bacmann, A., Lefloch, B., Ceccarelli, C., et al. 2003, ApJ, 585, L55
3. Brown, P. D. & Millar, T. J. 1989, MNRAS, 237, 661
4. Caselli, P. 2002, Planet. Space Sci., 50, 1133
5. Caselli, P., van der Tak, F., Ceccarelli, C., et al 2003, A&A, 403, L37
6. Caselli, P., Walmsley, C. M., Tafalla, M., et al 1999, ApJ, 523, L165
7. Ceccarelli, C. 2002, Planet. Space Sci., 50, 1267
8. Ceccarelli, C., Castets, A., Loinard, L., et al 1998, A&A, 338, L43
9. Ceccarelli, C., Loinard, L., Castets, A., et al 2000, A&A, 357, L9
10. Ceccarelli, C., Loinard, L., Castets, A., et al 2001, A&A, 372, 998
11. Ceccarelli, C., Vastel, C., Tielens, A., et al 2002, A&A, 381, L17
12. Charnley, S. B., Tielens, A., & Rodgers, S. D. 1997, ApJ, 482, L203
13. Dartois, E., Thi, W.-F., Geballe, T. R., et al 2003, A&A, 399, 1009
14. Jefferts, K. B., Penzias, A. A., & Wilson, R. W. 1973, ApJ, 179, L57+
15. Linsky, J. L. 1998, Space Science Reviews, 84, 285
16. Lis, D. C., Roueff, E., Gerin, M., et al 2002, ApJ, 571, L55
17. Loinard, L., Castets, A., Ceccarelli, C., et al 2001, ApJ, 552, L163
18. Loinard, L., Castets, A. et al 2002, Planet. Space Sci., 50, 1205
19. Loinard, L., Castets, A., Ceccarelli, C., et al 2000, A&A, 359, 1169
20. Parise, B., Castets, A., Herbst, E. et al 2003a, A&A, in press
21. Parise, B., Simon, T., , Caux, E. et al 2003b, A&A, in press
22. Parise, B., Ceccarelli, C., Tielens, A. et al 2002, A&A, 393, L49
23. Roberts, H., Fuller, G. A. et al 2002, Planet. Space Sci., 50, 1173
24. Roberts, H., Herbst, E., & Millar, T. J. 2003, ApJ, 591, L41
25. Roberts, H. & Millar, T. J. 2000, A&A, 361, 388
26. Roueff, E., Tiné, S., Coudert, L. H., et al 2000, A&A, 354, L63
27. Shah, R. Y. & Wootten, A. 2001, ApJ, 554, 933
28. Solomon, P. M. & Woolf, N. J. 1973, ApJ, 180, L89+
29. Teixeira, T. C., Devlin, J. P., Buch, V., et al 1999, A&A, 347, L19
30. Tielens, A. G. G. M. 1983, A&A, 119, 177
31. Turner, B. E. 1990, ApJ, 362, L29
32. van der Tak, F., Schilke, P., Müller, H., et al 2002, A&A, 388, L53
33. Vastel, C., Phillips, T. G., Ceccarelli, C., et al 2003, ApJ, 593, L97

Grain Surface Chemistry

P. Caselli[1], T. Stantcheva[2], and E. Herbst[3]

[1] INAF–Osservatorio Astrofisico di Arcetri, Largo E. Fermi 5, I–50125 Firenze, Italy caselli@arcetri.astro.it
[2] Department of Physics, The Ohio State University, Columbus, Ohio 43210, USA
[3] Departments of Physics, Astronomy, and Chemistry, The Ohio State University, Columbus, Ohio 43210, USA

1 Introduction

The formation of molecules on grain surfaces is a fundamental process for the chemical and physical evolution of molecular clouds and star forming regions ([56]; [19]). First of all, dust grains allow the formation of H_2 and thus the transition between atomic and molecular clouds ([20]; [29]; [30]; [41]; [31]; [10]; [23]). Secondly, the large abundance of water ice in cold clouds is most probably produced via hydrogenation of atomic oxygen adsorbed on grain surfaces ([54]; [25]). The large abundances of formaldehyde, methanol and complex saturated species observed in the direction of "hot cores" ([5]; [60]) as well as low mass ([11]) protostars are hard to form through gas–phase routes ([16]; [7]; [38]). Finally, recent detections of extremely large deuterium fractionation in H_2CO and CH_3OH (e.g. [12]; [37]; [14]; [36]; [39]; [2]; Ceccarelli, this symposium), as well as in H_2S ([57]), seem to require surface chemistry. Thus, coupled gas–grain chemical networks are needed to interpret the large amount of data now available and to prepare for future observations.

Unfortunately, because dust grains are discrete entities, harbouring very few reactive species, the use of simple average rate equations ([40]) cannot in general be extended to include surface chemistry ([54]; [15]). Rather, new methods are needed to include surface chemistry in gas-grain models. I will review here the efforts made to find the best way to build large–scale gas–grain models in which surface chemistry is handled "exactly", and conclude that the direct master equation method is in the long term the most promising approach. Meanwhile, "modified" rate equations can be used in chemical models of dense cloud cores (see Aikawa, this symposium). A discussion of the arguments suggesting that surface chemistry can explain the recent observations of deuterium fractionation will also be presented.

2 Surface Processes

The structure of a dust grain surface is not well characterized. It is assumed that dust grains in dense clouds grow a "dirty ice" mantle atop a carbonaceous or silicate core with radius $a \sim 0.1$ μm ([22]; [54]) and about 10^6 sites for physisorption, where the adsorption energy is < 0.2 eV. In this case, a light species such as an atomic hydrogen, after accretion, can either evaporate or travel to other sites by hopping over the barriers between sites or by quantum tunneling ([27]). The size of the barrier against diffusion is typically a fraction of the energy needed for evaporation. A chemistry based on diffusion, known as a Langmuir-Hinshelwood chemistry, can occur at very low temperatures where evaporation is not overwhelmingly rapid. Since atomic hydrogen is reactive and diffuses rapidly, it dominates the chemistry so that accreted heavy atoms and molecules can be converted to their saturated forms with an efficiency that depends on a variety of factors. If H and other accreted species are strongly bound (interaction energy $\simeq 0.5$–5 eV) they cannot move. In this case, the Eley-Rideal mechanism, in which a gas-phase species lands atop or near a stationary reactive species, can lead to reaction ([27]). This type of chemistry may occur preferentially on very small grains, where transient heating events prevent weakly bound species from remaining ([53]; [23]) and in larger grains at high temperatures ([18]). A diffusive chemistry in which migrating physisorbed species react with immobile chemisorbed species is also possible on surfaces such as graphite, which have both types of binding sites ([10]).

Further uncertainty is due to the poor understanding of non-thermal desorption mechanisms, including cosmic–ray bombardment ([33]; [24]), photodesorption via UV photons ([59]), sputtering in shocks ([47]), grain–grain collisions ([6]), energy generated by chemical reactions ([58]; [52]), and grain mantle explosions (e.g. [48]). Nevertheless, the need for surface chemistry to interpret the large amount of data now available has recently boosted the search for a correct mathematical treatment .

3 Modified Rates

The easiest approach is to modify the rate equation method so that coupled differential rate equations for gas–phase and surface species can still be utilized ([8]; [49]).

In the rate equation approach, the rate coefficients of a surface reaction between species i and j are given by ([25]):

$$K_{i,j}(s^{-1}) = \kappa(R_{diff,i} + R_{diff,j}), \tag{1}$$

where κ is the reaction probability (equal to unity if there is no activation energy), and $R_{diff,i}$ is the diffusion rate of species i over an entire grain. In the case of atomic hydrogen, $R_{diff,\mathrm{H}} \simeq 10^2$–$10^5$ s^{-1} if H sweeps over the grain via thermal hopping or quantum tunneling, respectively. This rate is orders of magnitude larger than the H accretion rate, which is $\sim 10^{-5}$ s^{-1} at typical dark cloud conditions. The use of eq. (1), in the case of a reaction without activation energy (e.g. H + O → OH), where $\kappa = 1$, implies a fractional probability for the reaction to occur between an accreting H and a species j on the grain of

$$P(\mathrm{H}, j) \simeq R_{diff,\mathrm{H}} \times t_{acc}(\mathrm{H}) = 10^7 - 10^{10} \ (!!), \tag{2}$$

before another H lands, which is unphysically large.

In the modified rate approach, the probability $P(\mathrm{H}, j)$, as well as any $P(i, j)$, is forced to not exceed unity within a physically meaningful time period: either an accretion time or an evaporation time, whichever is the smaller. In the case of H, assuming that the H evaporation rate is smaller than the H accretion rate, the modification consists of simply replacing $R_{diff,\mathrm{H}}$ with the accretion rate $R_{acc,\mathrm{H}}$ $(= 1/t_{acc}(\mathrm{H}))$ so that

$$P(\mathrm{H}, j) \simeq R_{acc,\mathrm{H}} \times t_{acc}(\mathrm{H}) = 1. \tag{3}$$

Other types of modifications are required in the case of reactions with activation energy (e.g. H + CO → HCO), where the population of the weakly reactive species (N(CO)) may become so large that the total probability of the reaction exceeds unity within a physically meaningful time. Once again, this is not allowed to happen. In addition, one needs to divide the corresponding rate coefficient by the sum of the probabilities per unit time for all reactions with H that possess activation energy (see [9] for details).

These modifications are semi–empirical and need to be tested with more appropriate stochastic methods. So far, successful tests have been carried out for simple chemical systems in a limited range of dust temperatures (5–20 K) ([49]; [9]; [51]). Comparisons between more complex chemical models based on modified rates and on stochastic methods are planned for the near future.

4 Stochastic Methods

Stochastic methods are based on the solution of the master equation, a kinetic–type equation, in which one calculates not concentrations but probabilities that a certain number of species are present. The average abundances can be calculated by suitable summations of the derived probabilities. The master equation can be solved directly, and this has been done for simple chemical systems by [4], [21], [51], [50], [42], [34]. Another way to solve the master equation is via Monte Carlo realizations ([15]).

Unlike Monte Carlo codes, the direct solution of the master equation is converted into differential equations which allow an "easy" coupling with gas phase differential rate equations. However, some appropriate approximations and simplifications are needed to avoid the integration of a huge number of differential equations, due to the coupled many–body probabilities entering the master equation ([50]; [28]). Moreover, if only reactive species (such as atomic hydrogen), which have low average abundances on grain surfaces, are included in the master equation, the number of differential equations is drastically reduced ([50]), allowing the construction of complex gas–grain chemical models (Herbst & Stantcheva, in preparation).

The direct master equation and Monte Carlo approaches have been used to check the validity of the modified rate equations for simple systems, such as that presented by [9], where the formation of singly and doubly deuterated isotopomers of formaldehyde and of singly, doubly, and multiply deuterated isotopomers of methanol on grain surfaces has been studied. This is important, because modified rates are still significantly more manageable than other methods and allow a fast exploration of the parameter space.

5 Deuterium Fractionation

One of the most challenging observational results for astrochemistry is the detection of a large deuterium fractionation in species such as H_2CO and CH_3OH (see Ceccarelli, this symposium) in protostellar sources. [39] have detected large (~ 0.2) CHD_2OH/CH_3OH abundance ratios in IRAS 16293-2422 and found that the total abundance of all deuterated methanol isotopomers exceeds that of CH_3OH. These observational results cannot be reproduced by available pure gas-phase models, even if the freeze-out of gaseous species is taken into account. Note, however, that gas-phase models can reproduce abundances of other highly deuterated species, e.g. ammonia ([44]; [45]; [46]; [35]; [1]; [26]).

We present a simple model here that is the same as in [9], but in which N and S are also included. Atomic nitrogen accretes onto dust grains and reacts with H, D and O to produce NH_3 and all its deuterated forms as well as NO, HNO and DNO. Atomic sulphur, once accreted onto grain surfaces, can form H_2S, HDS, D_2S, and NS. In analogy with the other accreting species, the gas–phase abundances of N and S are given by the steady-state values in the [32] new standard models. All surface species are assumed to evaporate into the gas once the temperature rises due to the onset of star formation.

Fig. 1 shows the abundance ratios of some singly and doubly deuterated species as a function of the D/H abundance ratio in the gas phase. The dust temperature has been assumed to be 10 K and the gas number density $n(H_2)$ to be 5×10^4 cm^{-3}. As first reported by [17], we note that the abundance ratios for doubly deuterated species rise at a slope double that of singly deuterated isotopomers in the log–log plot. Overlapped on the figure are observed values of singly and doubly deuterated methanol ([39]), doubly deuterated formaldehyde ([12]), and doubly deuterated hydrogen sulfide ([57]). The thing to note is that all the observed values are consistent with a large ($\simeq 0.1$) [D]/[H] abundance ratio, the only exception being CH_3OD, which is thought to be chemically unstable in the gas phase. These large [D]/[H] abundance ratios can be reproduced by gas–phase chemical models when freeze-out of neutral species such as CO and O is taken into account ([1]). We finally note that surface chemistry is not only able to reproduce abundance ratios, but also fractional abundances ($X(i) = n(i)/n(H_2)$) deduced from observations. For example, in the case of solid H_2S, we obtain $X(H_2S) = 3\times10^{-9}$, close to the value deduced by [57].

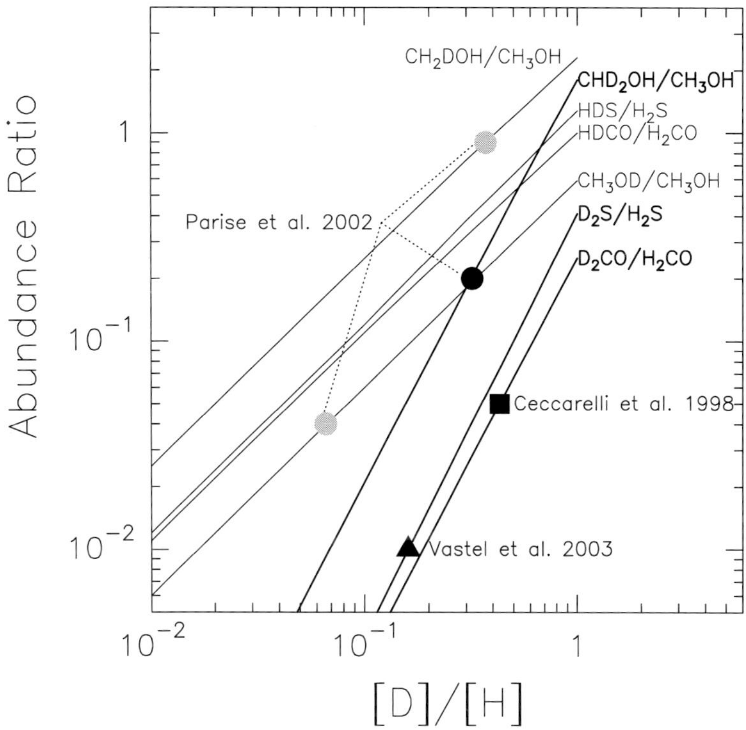

Fig. 1. Abundance ratios of singly (thin curves) and doubly (black curves) deuterated species as a function of the atomic [D]/[H] abundance ratio in the gas phase. Black (grey) symbols are observed values of doubly (singly) deuterated species toward star forming regions.

Interestingly, the doubly deuterated formaldehyde and the multiply deuterated methanol isotopomers exhibit a narrow range of temperatures ($10 \leq T_{dust} \leq 12$ K) within which they can reach the large abundances observed in star-forming regions. The drop at high T_{dust} is likely because doubly and triply deuterated species require at least one reaction with activation energy involving D. Such reactions have a probability much lower that those involving H (D is heavier than H and, consequently, is a less efficient tunneler); thus, they compete less favorably with evaporation as temperature increases ([9]).

If $T_{\rm dust}$ is lower than, say, 10 K, the H evaporation rate is so low that hydrogenation is always more efficient than deuteration. High densities ($n(H_2) >$ a few $\times 10^4$ cm^{-3}) are required to reproduce the observed deuterium fractionation. Interestingly, these values of temperature and density are typical of the regions where observations of deuterated species have been carried out. At densities $\geq 5\times10^5$ cm^{-3} and dust temperatures $\geq$ 15 K, the hydrogenation (and deuteration) of CO become negligible. Such high densities mean that the [H]/[CO] and [H]/[O] abundance ratios become very low. If the dust temperature increases above 15 K, the evaporation rates of H and D become so large that no hydrogenation and deuteration process happens on the surface.

We can conclude that, unlike ammonia, the observed D_2CO and doubly (and triply) deuterated methanol are likely to be formed on grain surfaces when the cloud density is around a few times 10^5 cm^{-3} and the dust temperature is about 10–12 K. They are released into the gas phase when the newly born protostar interacts with the surrounding medium either by heating the dust or driving shocks through powerful outflowing winds ([56]). However, one should note that (*i*) the emission of D_2CO is extended over a relatively large area ([13]), and (*ii*) large (~ 0.05) D_2CO/H_2CO abundance ratios have also been found in the direction of starless cores ([2]), where the nature of desorption mechanisms is not clear. Given that these regions are affected by CO depletion (e.g. [2]), we expect grain mantles to be mainly "apolar" and, consequently, the adsorbed species to have lower binding energies (see e.g. [55]; [3]). In this case, we speculate that the formation energy of surface species may be large enough to allow the return in the gas–phase, thus explaining the presence of multiply deuterated methanol and formaldehyde in quiescent environments.

References

1. Y. Aikawa, N. Ohashi, E. Herbst: ApJ **593**, 906 (2003)
2. A. Bacmann, B. Lefloch, C. Ceccarelli, J. Steinacker, A. Castets, L. Loinard: ApJ **585**, L55 (2003)
3. E.A. Bergin, W.D. Langer: ApJ **486**, 316 (1997)
4. O. Biham, I. Furman, V. Pirronello, G. Vidali: ApJ **553**, 595 (2001)
5. G.A. Blake, E.C. Sutton, C.R. Masson, T.G. Phillips: ApJ **315**, 621 (1987)
6. P. Caselli, T.W. Hartquist, O. Havnes: A&A **322**, 296 (1997)
7. P. Caselli, T.I. Hasegawa, E. Herbst: ApJ **408**, 548 (1993)
8. P. Caselli, T.I. Hasegawa, E. Herbst: ApJ **495**, 309 (1998)
9. P. Caselli, T. Stantcheva, O. Shalabiea, V.I. Shematovich, E. Herbst: P&SS **50**, 1257 (2002)
10. S. Cazaux, A.G.G.M. Tielens: ApJ **575**, L29 (2002)
11. S. Cazaux, A.G.G.M. Tielens, C. Ceccarelli, A. Castets, V. Wakelam, E. Caux, B. Parise, D. Teyssier: ApJ **593**, L51 (2003)
12. C. Ceccarelli, A. Castets, L. Loinard, E. Caux, A.G.G.M. Tielens: A&A **338**, L43 (1998)

13. C. Ceccarelli, L. Loinard, A. Castets, A.G.G.M. Tielens, E. Caux, B. Lefloch, C. Vastel: A&A **372**, 998 (1998)
14. C. Ceccarelli, C. Vastel, A.G.G.M. Tielens, A. Castets, A.C.A. Boogert, L. Loinard, E. Caux: A&A **381**, L17 (2002)
15. S.B. Charnley: ApJ **509**, L121 (1998)
16. S.B. Charnley, A.G.G.M. Tielens, T.J. Millar: ApJ **399**, L71 (1992)
17. S.B. Charnley, A.G.G.M. Tielens, S.D. Rodgers: ApJ **482**, L203 (1997)
18. W.W. Duley: MNRAS **279**, 591 (1996)
19. P. Ehrenfreund, S.B. Charnley: ARA&A **38**, 427 (2000)
20. R.J. Gould, E.E. Salpeter: ApJ **138**, 393 (1963)
21. N.J.B. Green, T. Toniazzo, M.J. Pilling, D.P. Ruffle, N. Bell, T.W. Hartquist: A&A **375**, 1111 (2001)
22. J.M. Greenberg, S.-S. Hong: The chemical composition and distribution of interstellar grains. In: *Galactic radio astronomy*, IAU Symposium, No. 60, ed by F.J. Kerr, S.C. Simons III (Dordrecht, D. Reidel Publishing Co. 1974) pp 155-177
23. E. Habart, F. Boulanger, L. Verstraete, C.M. Walmsley, G. Pineau des Forêts: A&A, in press (2003)
24. T.W. Hartquist, D.A. Williams: MNRAS **247**, 343 (1990)
25. T.I. Hasegawa, E. Herbst, C.M. Leung: ApJS **82**, 167 (1992)
26. J. Hatchell: A&A **403**, L25 (2003)
27. E. Herbst: Models of Gas-Grain Chemistry in Star forming Regions. In: *Astrochemistry: From Molecular Clouds to Planetary Systems*, IAU Symposium 197, ed by Y. C. Minh, E.F. van Dishoeck (Astronomical Society of the Pacific 2000) pp 147–159
28. E. Herbst, V.I. Shematovich: Ap&SS **285**, 725 (2003)
29. D. Hollenbach, E.E. Salpeter: J. Chem. Phys. **53**, 79 (1970)
30. D. Hollenbach, E.E. Salpeter: ApJ **163**, 155 (1971)
31. N. Katz, I. Furman, O. Biham, V. Pirronello, G. Vidali: ApJ **522**, 305 (1999)
32. H.-H. Lee, R.P.A. Bettens, E. Herbst: A&AS **119**, 111 (1996)
33. A. Léger, M. Jura, A. Omont: A&A **144**, 147 (1985)
34. A. Lipshtat, O. Biham: A&A **400**, 585 (2003)
35. D.C. Lis, E. Roueff, M. Gerin, T.G. Phillips, L.H. Coudert, F.F.S. van der Tak, P. Schilke,: ApJ, **571**, L55 (2002)
36. L. Loinard, A. Castets, C. Ceccarelli, et al: P&SS **50**, 1205 (2002)
37. L. Loinard, A. Castets, C. Ceccarelli, E. Caux, A.G.G.M. Tielens: ApJ **552**, L163 (2001)
38. T.J. Millar, G.H. MacDonald, A. Gibb: A&A **325**, 1163 (1997)
39. B. Parise, C. Ceccarelli, A.G.G.M. Tielens, et al: A&A **393**, L49 (2002)
40. J.B. Pickles, D.A. Williams: Ap&SS **52**, 443 (1977)
41. V. Pirronello, O. Biham, C. Liu, L. Shen, G. Vidali: ApJ **483**, L131 (1997)
42. J.G.L. Rae, N. Bell, T.W. Hartquist, M.J. Pilling, D.P. Ruffle: A&A **383**, 738 (2002)
43. H. Roberts, E. Herbst, T.J. Millar: ApJ **591**, L41 (2003)
44. H. Roberts, T.J. Millar: A&A **361**, 388 (2000a)
45. H. Roberts, T.J. Millar: A&A **364**, 780 (2000b)
46. S.D. Rodgers, S.B. Charnley: ApJ **553**, 613 (2001)
47. P. Schilke, C.M. Walmsley, G. Pineau des Forêts, D.R. Flower: A&A **321**, 293 (1997)

48. W.A. Schutte, J.M. Greenberg: A&A **244**, 190 (1991)
49. T. Stantcheva, P. Caselli, E. Herbst: A&A **375**, 673 (2001)
50. T. Stantcheva, E. Herbst: MNRAS **340**, 983 (2003)
51. T. Stantcheva, V.I. Shematovich, E. Herbst: A&A **391**, 1069 (2002)
52. J. Takahashi, D.A. Williams: MNRAS **314**, 273 (2000)
53. A.G.G.M. Tielens: PAHs and Astrochemistry. In: *Dust and Chemistry in Astronomy*, ed by T.J. Millar, D.A. Williams (London: IOP Publishing 1993) pp 103
54. A.G.G.M. Tielens, W. Hagen: A&A **114**, 245 (1982)
55. A.G.G.M. Tielens, A.T. Tokunaga, T.R. Geballe, F. Baas: ApJ **381**, 181 (1991)
56. E.F. van Dishoeck, G.A. Blake: ARA&A **36**, 317 (1998)
57. C. Vastel, T.G. Phillips, C. Ceccarelli, J. Pearson: ApJ **593**, L97 (2003)
58. K. Willacy, T.J. Millar: MNRAS **298**, 562 (1998)
59. D.A. Williams, S.D. Taylor: QJRAS **37**, 565 (1996)
60. F. Wyrowski, P. Schilke, C.M. Walmsley, K.M. Menten: ApJ **514**, 43 (1999)

Deuterated Species in Star-Forming Regions

D.C. Lis[1], M. Gerin[2], E. Roueff[3], and T.G. Phillips[1]

[1] California Institute of Technology, MC 320-47, Pasadena, CA 91125, USA
dcl@submm.caltech.edu
[2] LERMA, FRE 2460 du CNRS, Observatoire de Paris and Ecole Normale Supérieure, 24 Rue Lhomond, 75231 Paris cedex 05, France
[3] LUTH and FRE 2462 du CNRS, Observatoire de Paris, Section de Meudon, Place Jules Janssen, 92195 Meudon, France

1 Introduction

A number of multiply deuterated molecular species have been recently detected in the ISM [1]–[6] with abundances far exceeding earlier theoretical expectations. Detailed formation mechanisms for these species have not yet been satisfactorily established, but recent computations of Roberts et al. [7] show that the inclusion of HD_2^+ and D_3^+ in the gas-phase chemical network significantly enhances the fractionation of ionic and neutral species in regions of high density and heavy depletion. These models also predict the very high atomic D/H ratios necessary for grain-surface chemistry to reproduce the high formaldehyde and methanol fractionation seen in star forming regions. We argue that slow C-type shocks associated with molecular outflows may produce conditions favorable for deuterium fractionation reactions in star forming regions by *(i)* increasing the gas volume density and thus shortening the depletion and gas-phase reaction time-scales or *(ii)* evaporating grain mantle ices. In the case of triply deuterated ammonia, an extensive observational search carried out at the CSO has resulted in five detections, including only two strong detections (Barnard 1 [3] and L1689N [8]). As discussed below, in both these regions there are indications of shock interactions between molecular outflows and the ambient gas.

2 L1689N

In the best studied case of L1689N [9], where bright, spatially extended ND_3 emission has been detected [8], all molecular tracers peak near the location of IRAS 16293, a proto-binary system driving two molecular outflows. However secondary peaks are present to the east of the IRAS source and are associated with various interaction regions between the molecular outflows and the ambient cloud. The region of enhanced emission of (multiply) deuterated species is located within a blue lobe of the molecular outflow emanating from the IRAS source, near a secondary dust continuum peak. There is kinematical evidence of an interaction at this location in the form of a velocity shift between deuterated and non-deuterated tracers.

We suggested [9] that abundances of deuterated species in L1689N are strongly enhanced in the molecular gas that is impinged upon and compressed by the outflow. The shock velocity, derived from high-J CO data, is only 8–10 $km\,s^{-1}$, insufficient to destroy dust grains, but sufficient to evaporate grain mantles. At least two possible scenarios explaining the high deuteration levels observed in this region can be envisioned. *(i)* High deuteration levels are present in the grain mantle material that is released into the gas phase by the C-type shock. In this case, CO would also be released into the gas phase, resulting in a quick destruction of at least deuterated ions. This could thus be a rather short-lived phase in the evolution of a dense core, overrun by an outflow, which may explain why there are relatively few sources of this kind. *(ii)* The shock-compressed, dense gas cools efficiently to low temperatures, which sets the initial conditions for gas-phase chemistry to be re-established, at a much higher density compared to the pre-shock gas. The increased density shortens the depletion time-scale and produces conditions favorable for deuterium fractionation reactions. The gas-phase chemistry may thus subsequently drive up the abundances of deuterated molecular species.

An extensive survey of other nearby star-forming regions in the DCO^+ (3–2), CO (2–1), and 350 μm dust continuum emission is currently under way at the CSO. The source sample includes B1, NGC1333, Serpens, L1630, L1251, and HH1–2. Results for B1 and NGC1333 are briefly discussed below.

3 Barnard 1 and NGC1333

Distribution of the DCO^+(3–2) emission in B1 is shown in Figure 1 (gray-scale and white contours). The main DCO^+ peak is associated with the B1-b continuum source, where ND_3 was first detected [3] and the secondary peak to the north is associated with B1-c. The solid and dashed black contours outline the distribution of high-velocity CO emission. At least three molecular outflows can be identified in the region. An E–W, collimated outflow is centered on the B1-c. Another outflow is located west of B1-b. The blue lobe of this outflow coincides with IRAS 03301+3057 that *does not* appear to be its driving source. The third, roughly N–S outflow appears to be centered east of B1-c. The red lobe of the latter outflow encompasses the DCO^+ peak associated with B1-b. The morphology of the B1 region is thus generally similar to that in L1689N, with multiple outflows, one of which encompasses the region of the enhanced emission from deuterated molecular tracers.

Figure 2 shows the distribution of the DCO^+(3–2) emission in NGC1333 (gray-scale and white contours) with overlayed contours of high-velocity CO emission. A bright DCO^+ ridge is located NE of IRAS 4A, where ND_3 was first detected [4]. In addition, an equally bright DCO^+ peak is found $\sim 2'$ to the north, which does not coincide with any compact submillimeter or near-IR sources. This secondary DCO^+ peak is located on-axis of the famous molecular outflow associated with the HH7–11 sources, centered near SSV13

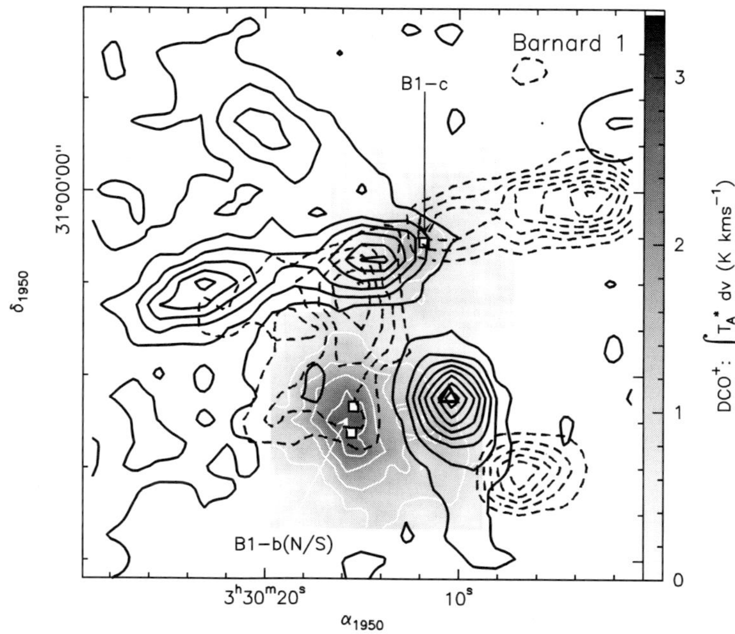

Fig. 1. Distribution of the DCO^+ (3–2) emission in B1 (grayscale and white contours) with overlayed contours of high-velocity CO (2–1) emission. Velocity ranges for the blue-shifted and red-shifted emission (solid and dashed contours, respectively) are –4 to 4 $km\,s^{-1}$ and 9 to 18 $km\,s^{-1}$, respectively. Contour levels for DCO^+ are 30 to 90% of the peak, with an interval of 20%. For CO, contour levels are 4 to 18 $K\,km\,s^{-1}$ with an interval of 2 $K\,km\,s^{-1}$. The triangle marks the location of the IRAS source 03301+3057.

and the DCO^+ source is again encompassed by the blue-shifted outflow lobe. Relatively weak ND_3 emission has also been detected at this location [8].

Similar association between the location of DCO^+ peaks and high-velocity outflow lobes is found in the remaining sources observed with the CSO.

4 Summary

There is good morphological evidence that deuterated species in *star-forming regions* appear intimately associated with high-velocity outflow lobes. Given the number of sources studied, it is unlikely to be a coincidence (e.g. a result of projection) and must indicate a real physical association. However, it has not yet been firmly established that the DCO^+ peaks are associated with the post-shock gas. In L1689N, there is clear kinematical evidence of an interaction, but the evidence is somewhat more circumstantial in the other sources. The density enhancement in the post-shock gas combined with molecular depletion, may be sufficient to account for the observed abundances of some

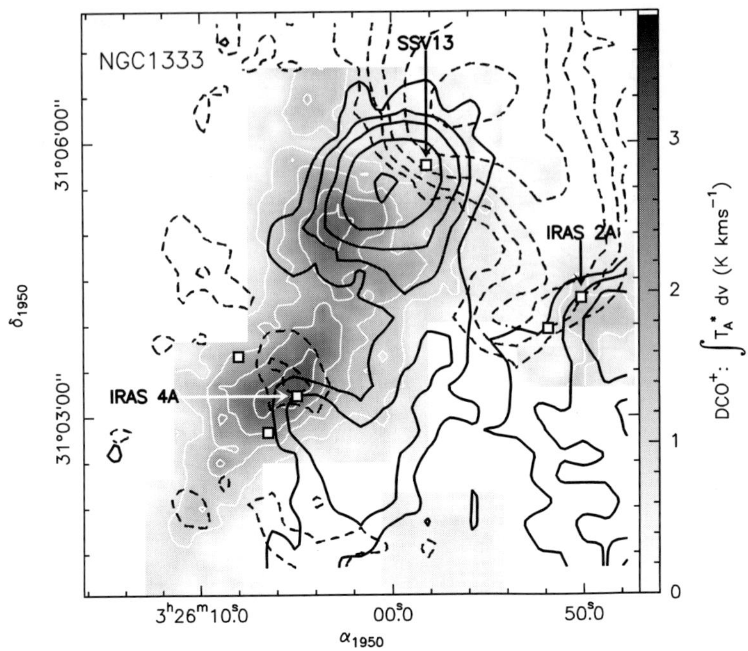

Fig. 2. Distribution of the DCO^+ (3–2) emission in NGC1333 (grayscale and white contours) with overlayed contours of high-velocity CO (2–1) emission. Velocity ranges for the blue-shifted and red-shifted emission (solid and dashed contours, respectively) are –10 to 3.5 $km\,s^{-1}$ and 11.5 to 25 $km\,s^{-1}$, respectively. Contour levels for DCO^+ are 30 to 90% of the peak, with an interval of 20%. For CO, contour levels are 4, 8, 15, 30, and 60 $K\,km\,s^{-1}$. Squares mark the locations of embedded continuum sources.

multiply deuterated species, such as ammonia. However, for other molecules, such as methanol, grain surface chemistry likely plays a role. Another interesting observational result of our study is that the DCO^+ peaks are typically *not coincident* with the embedded dust cores.

References

1. C. Ceccarelli, A. Castets, L. Loinard, et al.: A&A **338**, L43 (1998)
2. E. Roueff, S. Tiné, L.H. Coudert, et al.: A&A **354**, L63 (2000)
3. D.C. Lis, E. Roueff, M. Gerin, et al.: ApJ **571**, L55 (2002)
4. F.F.S. van der Tak, P. Schilke, H.S.P. Müller, et al.: A&A **388**, L53 (2002)
5. B. Parise, C. Ceccarelli, A.G.G.M. Tielens, et al.: A&A **393**, L49 (2002)
6. C. Vastel, T.G. Phillips, C. Ceccarelli, & J. Pearson: ApJ **593**, L97 (2003)
7. H. Roberts, E. Herbst, & T.J. Millar: ApJ **591**, L41 (2003)
8. E. Roueff, M. Gerin, F.F.S van der Tak, et al.: in preparation (2003)
9. D.C. Lis, M. Gerin, T.G. Phillips, & F. Motte: ApJ **569**, 322 (2002)

Gas-Grain Interactions and Chemical Influences on Star Forming Regions

H.J. Fraser

Raymond and Beverly Sackler Laboratory for Astrophysics at Leiden Observatory, P.O. Box 9513, 2300 RA Leiden, The Netherlands `fraser@strw.leidenuniv.nl`

Laboratory experiments have been conducted on binary ice mixtures and sequentially-deposited layers of CO with CO_2, CH_4, and CH_3OH. Spectra were recorded at 0.5 cm^{-1} resolution as the ices were heated, from 15 to 100 K. The results show that layered and mixed ices exhibit different spectroscopic traits and physical behavior. The astrophysical implications of these data are discussed, with particular reference to the recent detections of solid CO in star forming regions.

1 Introduction

In star forming regions (SFRs), ices are formed by efficient accretion of atoms and molecules from the interstellar gas onto grains. Many molecular species, including some of the simplest molecules are generated via solid state reactions at these surfaces [1]. These reaction products may remain trapped on the grain or desorb, increasing the chemical complexity of the surrounding gas. A wealth of work remains to fully comprehend the processes that influence the molecular abundances in the gas and solid phases. Consequently, the emphasis in solid-state laboratory astrophysics is now moving from a regime whereby laboratory spectra are simply matched to astronomical spectra, to a regime where experiments are designed to understand the behavior of interstellar ice analogues under controlled conditions. The remainder of this paper highlights some laboratory results on solid CO that have been used to evaluate recent observations of low and medium mass young stellar objects (YSOs).

2 Background: Observations of Solid CO Towards SFRs

Detections of solid CO are inferred by the presence of a broad absorption feature at around 2139 cm^{-1}, and have been made towards dense interstellar regions, the galactic center and YSOs, e.g. [2, 3, 4, 5, 6, 7, 8]. Recently, Pontoppidan et al. detected solid CO on lines of sight towards 39 low and medium mass YSOs [9]. They used a phenomenological fit to decompose the main CO band into only three components at 2136.5 cm^{-1}, 2139.9cm^{-1}, and 2143.7 cm^{-1}. Only the intensities of the three bands needed to be altered between different lines of sight to obtain overall fits that were significantly better than those obtained with the 'mix and match' approach, where laboratory spectra from a plethora of mixed ices at different temperatures are summed and fitted to the observed line profile [2]. Briefly, an analysis of these results indicates that between 60 and 90 % of the CO is present in its 'pure' form. Based on recent experimental data [11, 12, 13], Pontoppidan et al. show that not more than 16 % of the CO can be mixed within a H_2O environment. Laboratory astrophysics is able to address a number of questions that arise from these findings, most importantly,

- how can identical CO environments exist along all lines of sight?
- what possible physical or chemical factors can influence *only* the intensities of the fitted components?

3 Experimental

The spectroscopy and behavior of CO ices was investigated in a high vacuum (HV) chamber, capable of reaching base pressures of 1x10^{-7} mbar, using 0.5 cm^{-1} resolution Fourier Transform Infrared Transmission Spectroscopy, described in detail elsewhere [17]. Ices were deposited at a flow rate of 2x10^{17} molecules cm^{-2} s^{-1}, onto a cold CsI window, using equivalent doses of CO at 15 K. Infrared spectra were then recorded, every 3-5 K, from 15 K until 100 K, or until all the material had desorbed.

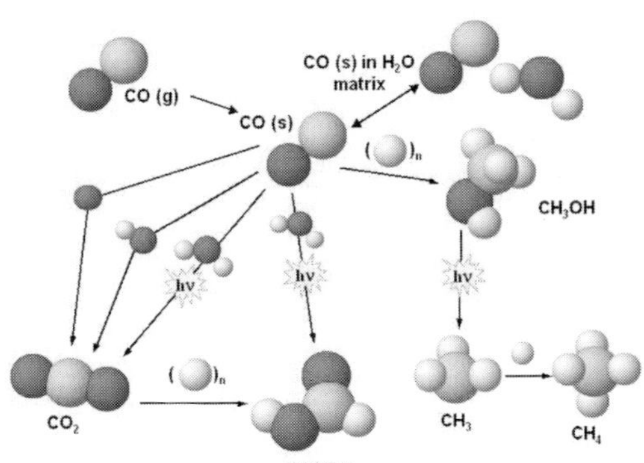

Fig. 1. Postulated solid-state chemistry of CO.

In the solid state, CO is a key source of carbon and is postulated to partake in a chemistry that involves thermally induced surface reactions between atoms/radicals and CO, or may be electron or photon mediated (see Fig. 1). On lines of sight towards low or high mass YSOs, CO is typically present at between 5 and 20 % of the H_2O abundance [3, 9]: the abundances of 'product' molecules from the CO chemistry vary significantly, but typically

Table 1. Abundances of common ice components, relative to H_2O, in different SFRs, (adapted from [18]). Species studied in this investigation are in bold type.

Constituent	Dark Cloud Elias 16	Low-Mass YSO Elias 29	High-Mass YSO W33a
H_2O	100	100	100
CO	25	5.6	9
CO_2	18	22	14
CH_4	1-2	1.6	2
CH_3OH	3	20	22
H_2CO	2-6	-	1.7-7
HCOOH	3	-	0.4-2

are also between 1 and 20 %, relative to H_2O, (see Table 1). Consequently, the ices investigated here were all studied in 1:1 ratios, either as mixed ices, or in layers. The results discussed here focus on CO with CO_2, CH_3OH, and CH_4, since H_2O-CO ice systems have been studied extensively elsewhere [11, 12] and H_2CO is prohibitively difficult to work with as it polymerizes at room temperature.

4 Results

An overview of all spectra is given in Fig. 2. The mixed and layered ices can be immediately identified from each other. By comparing the spectra on the diagonal, it can be seen that none of the pure ices studied, except CO, have solid state features in the 2135 - 2155 cm^{-1} wavelength region. No temperature dependance is observed in any of the spectra, until one or both ice constituents begin to desorb, contrary to previously reported data [20], and attributable to the fact that this work was conducted at higher resolution. A more detailed description is given in [15]: for comparison with observational spectra, the raw data files can be found at,

`http://www.strw.leidenuniv.nl/~fraser/spectra_database/`

In the layered ices, the CO peak has the same line position, a similar FWHM and overall line-shape profile as the pure CO spectrum, indicating that the layers are not miscible. Over 90 % of the CO layer desorbs between 30 and 35 K. In the CO over-layers (Fig. 2, left-hand row), the remaining CO diffuses down into the second ice layer, becoming trapped there, and desorbing only when the ice itself desorbs. This trapping effect is most pronounced in CH_3OH and CO_2, where a significant difference exists between the volatility of the two ice components. Beyond 35 K, the layered spectra more closely resemble the mixed ices. When CO is deposited underneath the second ice component it is capable of diffusing through the CH_3OH and CO_2

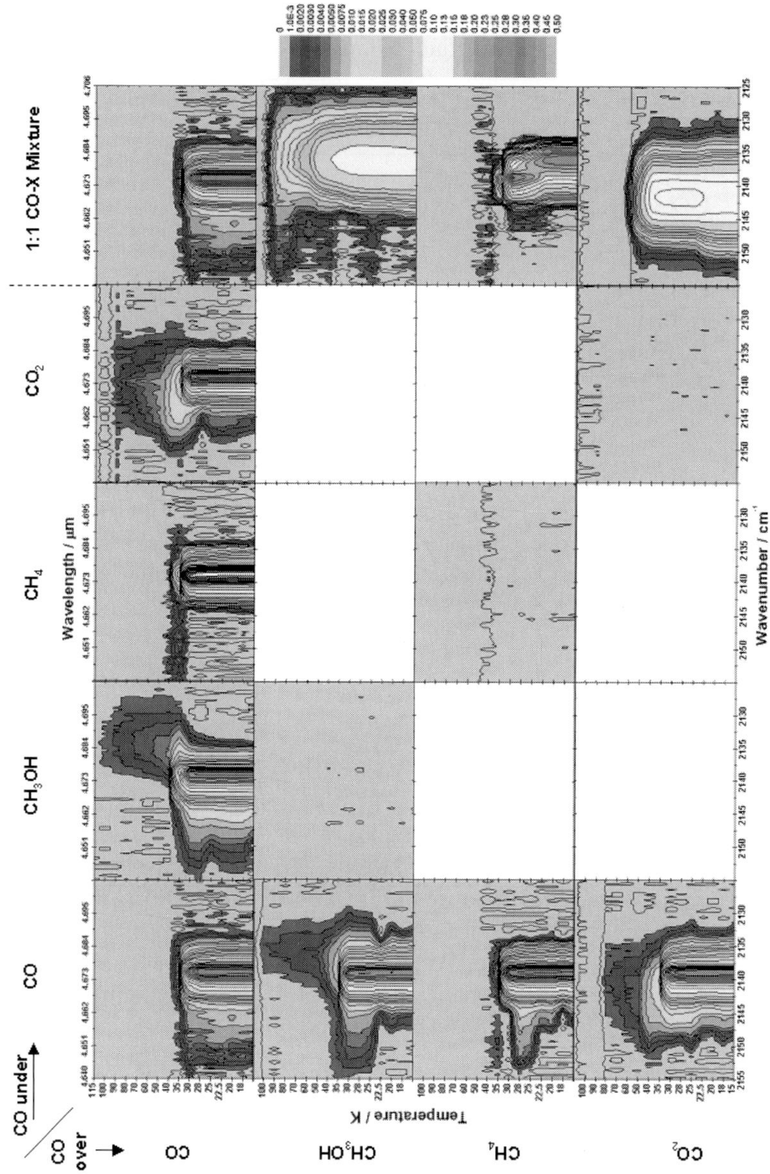

Fig. 2. Overview of all recorded CO spectra, in the 2139 cm^{-1} region, presented in a matrix of temperature-frequency data. Spectra on the diagonal represent the 'pure' ices, i.e. control spectra. On the upper row, layered spectra of CO over ice X are shown: layered spectra of CO under ice X are shown in the first column. The last column shows the mixed ice spectra. In all cases equal exposures of CO are present on the sample substrate at 15 K, equivalent to an ice thickness of approx. 840 nm

over-layers: small fractions of CO are also trapped beyond 35 K, but again over 90 % of the CO desorbs. The exception is the CH_4 over-layer system, where CO remains trapped on the surface until the CH_4 itself begins to desorb at around 40 K, suggesting that CO is unable to diffuse through CH_4 ices [14].

The spectra of the mixed ice systems bear little resemblance to the layered systems: the line profiles are not centered at the same wavelength, nor does the majority of CO desorb by 35 K. The CH_4-CO mixed ice undergoes a phase change at 22K, evinced by the simultaneous disappearance and growth of two spectroscopic bands, and an isosbestic point between the two, as discussed elsewhere [14]. In the CH_3OH-CO system, the peak position is shifted to the red, as the CO bond is weakened slightly, due to hydrogen bonding interactions between CO and OH groups of the CH_3OH. The overall line profile is broadened, reflecting the range of hydrogen bonding interactions within the ice. The majority of the CO remains trapped in the CH_3OH until 50 - 60 K, when it desorbs as the CH_3OH-ice undergoes a phase change. Around 20 % of the CO remains trapped in this new CH_3OH phase until it undergoes a second phase change around 90 K. Conversely the peak position in the mixed CO_2-CO ice system is blue-shifted in comparison to the pure CO ice position, indicating a slight strengthening of the CO bond, probably due to back-donation of electrons from the CO_2. The overall profile is broadened, again reflecting the range of CO_2-CO interactions within the ice.

5 Discussion and Astrophysical Implications

From the spectroscopy presented in Sect. 4 it appears that CO over-layers offer a very reasonable explanation for why Pontoppidan et al. observed that over 90 % of the CO resides in 'pure' CO layers. Irrespective of the underlying substrate, the CO spectra appear identical and do not vary with temperature, until the CO itself desorbs. This would explain the dominance of the 2139.9 cm^{-1} band along many lines of sight towards SFRs. It has also been shown that where CO resides in hydrogen-bonding dominated environments, a broader, red shifted component emerges in the CO spectra: this is most easily observed in mixed ices, or in ices where the layers have become mixed, due to diffusion of the CO into or through other ice constituents. Not only H_2O, but other ice constituents, e.g. CH_3OH and CH_4, are capable of such hydrogen bonding interactions. It is therefore perhaps unsurprising that the phenomenological fits identified that a maximum of two or three distinct CO ice environments exist along all lines of sight: this simply reflects the physical chemistry of CO in a variety of ice matrices. Moving from a concept where we assume that all interstellar ices are mixed, to a concept where segregated layers exist has significant implications for future astrochemical modelling and laboratory experiments on interstellar ices. The layered-ice approach has already been employed to interpret observations from a variety of SFRs, e.g. [9, 21, 22]. It is most likely that the differences in the relative intensities of each component required to fit the observational spectra in the phenomenological fits, are associated with factors such as molecular abundance, layer thickness, and the degree to which layers have become mixed, rather than chemical environment. Further investigations of these factors are currently ongoing, under ultra-high vacuum conditions, e.g. [16].

Acknowledgements

I would like to thank my co-workers involved in the work detailed here. This work was supported by NOVA, the Dutch Research School for Astronomy, and a SPINOZA Grant, awarded to Ewine van Dishoeck, to whom I am very grateful for her guidance and support.

References

1. Herbst, E., *Models Of Gas-Grain Chemistry In Star Forming Regions* in *Astrochemistry: From Molecular Clouds to Planetary Systems, Proceedings of IAU Symposium 197*, edited by Y. C. Minh and E. F. van Dishoeck, 147,(2000)
2. A.G.G.M Tielens, A.T. Tokunaga, T.R. Geballe et al, ApJ, **381**, 181, (1991)
3. J.E. Chiar, A.J. Adamson, T.H. Kerr et al, ApJ, **426**, 240, (1994)
4. J.E. Chiar, P.A. Gerakines, D.C.B. Whittet et al, ApJ, **498**, 716, (1998)
5. D.C.B. Whittet, P.A. Gerakines, A.G.G.M. Tielens et al, ApJ, **498**, L159, (1998)
6. A.C.A. Boogert, G.A. Blake, A.G.G.M Tielens, ApJ, **568**, 761, (2002)
7. A.C.A. Boogert, G.A. Blake, A.G.G.M Tielens, ApJ, **577**, 271, (2002)
8. W.F. Thi, K.M. Pontoppidan, E.F. van Dishoeck et al, A&A, **394**, L27, (2002)
9. K.M. Pontoppidan, H.J. Fraser, W.F. Thi et al, A&A, **408**, 981, (2003)
10. H.-C. Chang, H.H. Richardson, G.E. Ewing, J. Chem. Phys., **89**, 7561, (1998)
11. M.P. Collings, J.W. Dever, H.J. Fraser et al, ApJ, **583**, 1058, (2003)
12. M.P. Collings, J.W. Dever, H.J. Fraser et al, Astrophys. & Space Sci., **285**, 633, (2003)
13. H.J. Fraser, M.P. Collings, M.R.S. McCoustra et al, MNRAS, **327**, 1165, (2001)
14. W.Z. Alsindi, D.O. Gardner, E.F. van Dishoeck et al, Chem. Phys. Lett., **378**, 178, (2003)
15. H.J. Fraser, W.Z. Alsindi, S.E. Bisschop et al, A&A, submitted, (2003)
16. H.J.Fraser, E.F. van Dishoeck, Adv. Space. Res., in press, (2003)
17. P.A. Gerrakines, W.M. Schutte, J.M. Greenberg et al, A&A, **296**, 810, (1995)
18. J. Crovisier, Faraday Discuss., **109**, 437, (1998)
19. G.E. Ewing, G.C. Pimentel, J. Chem. Phys., **35**, 925, (1961)
20. P. Ehrenfreund, A.C.A. Boogert, P.A. Gerrakines et al, Faraday Discuss., **109**,463, (1998)
21. J.K. Jorgensen, F.L. Schoier, E.F. van Dishoeck, A&A, **389**, 908, (2002)
22. F.L. Schoier, J.K. Jorgensen , E.F. van Dishoeck et al, A&A, **390**, 1001, (2002)

Spectroscopy of Interstellar PAHs

A.G.G.M. Tielens[1] and E. Peeters[2]

[1] Kapteyn Astronomical Institute, PO Box 800, 9700 AV Groningen, The Netherlands
[2] Space Sciences Division, NASA Ames Research Center, MS 245-6, CA 94035-1000, USA

Summary. ISO has shown that the mid–IR spectra of almost all sources are dominated by narrow emission features collectively known as the UIR bands. The observations show that the UIR spectrum is incredibly rich with a large number of weak features, subfeatures, and shoulders; many of which were previously unknown. The peak positions and profiles of these bands show variations from source to source, reflecting the local physical conditions. The infrared spectral characteristics of a large set of Polycyclic Aromatic Hydrocarbon molecules have been investigated in the laboratory using a variety of experimental techniques. These techniques are reviewed and their advantages and disadvantages are described. Based upon this large experimental dataset, the observed spectra are analyzed and interpreted. It is concluded that the interstellar UIR bands are carried by a family of PAH molecules, that a fair fraction are charged, and that they may be N-substituted, clustered, or have adsorbed metal atoms.

1 Introduction

The Short Wavelength Spectrometer on board of the Infrared Space Observatory has provided us with a complete 2.4-45 μm spectra at moderate resolution of a large sample of objects. Analysis of these spectra have revealed a dazzling array of spectral detail. At the same time, a heroic laboratory effort using a wide range of techniques has detailed the intrinsic spectroscopic properties of Polycyclic Aromatic Hydrocarbon molecules. Together this has yielded new and surprising insights in characteristics of interstellar PAHs. It is very appropriate, therefore, to review this progress at a session devoted to Gerhard Herzberg, the grandfather of molecular astro-spectroscopy.

2 The Incredibly Rich UIR Spectrum

The SWS spectra of bright sources have revealed the incredible richness of the interstellar and circumstellar UIR spectrum. Besides the well-known UIR bands at 3.3, 6.2, 7.7, 8.6, and 11.3 μm, there is a wealth of weaker features: 3.4, 5.25, 5.65, 6.9, 10.5, 12.7, 16.4 μm to mention some of them (Fig. 2; [15],

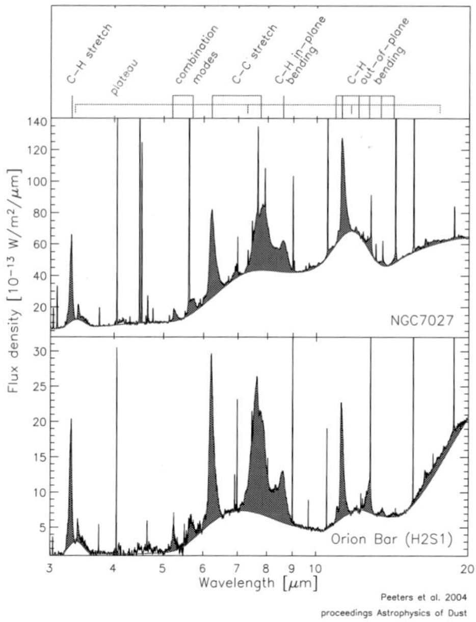

Fig. 1. The 3–20 μm emission spectra of the Orion Bar – a well known photodissociation region associated with the Orion HII region – and the planetary nebula, NGC 7027 ([15], [31]), illustrating the incredible richness of the mid-infrared emission features.

[31]). Moreover, many of the prominent bands show satellite bands (eg., 6.0, 11.0 μm), show shoulders (e.g., 7.2 μm), or break up into subfeatures (eg., 7.6 and 7.8 μm). In addition, there are broad plateaus underlying all the major features. The spectral richness of the 7-9 μm region is particularly evident in a high resolution spectrum of the reflection nebula, NGC 7023 which reveals some 10 subfeatures ([27]). The 11-15 μm region is an equally rich illustration of this spectral extravagance ([15]).

The IR emission features have been detected in a wide variety of sources: planetary nebulae, post-AGB objects, Herbig AeBe stars, HII regions, reflection nebulae, the galactic cirrus, surfaces of dark clouds, galactic nuclei, starburst galaxies, ultraluminous galaxies ([40], [41], [27], [31], [32], [15], [8], [39], [14], [38], [26], [20], [34], [25]). It is clear that the carrier is ubiquitous, abundant, and able to withstand a wide range of physical conditions.

3 The Intrinsic Spectral Characteristics of PAHs

Much progress has been made over the last decade in determining the intrinsic spectral properties of PAH molecules due to the dedicated efforts of a number of laboratory groups using a variety of complementary experimental

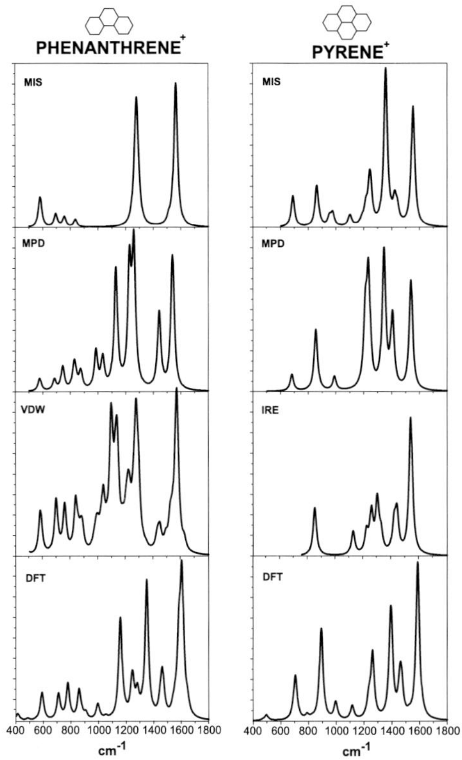

Fig. 2. The 5–20 μm spectra of the cations of phenanthrene and pyrene measured using a variety of experimental techniques. Fur a uniform presentation of the results, the measured features are represented by bands at the measured frequencies and with the appropriate strength convolved with a gaussian of 30 cm^{-1} width to mimick the interstellar spectra. MIS = Matrix Isolation Spectrscopy; MPD = MultiPhoton Dissociation; VDW = van der Waals cluster; IRE = InfraRed Emission; DFT = Density Functional Theory.

techniques. These studies have benefitted much from the guidance of theoretical calculations using density functional theory. Each of these techniques has its advantages and disadvantages which we will summarize here. For more details, see [30] and elsewhere in this volume. The focus in these studies has been on measuring the infrared characteristics of PAH cations because it had been speculated early on that ionization might have important effects on the spectra [1] and this was underpinned by theoretical calculations [13],[24]. Experimental studies lagged for a while because of the difficulty in producing a large enough concentration of charged species in a small volume. Figure 2 compares the spectra obtained for two small PAN cations using these different techniques.

Matrix isolation spectroscopy (MIS) is a well established technique which isolates very reactive or transient species in an inert matrix (eg., Ne, Ar). In these studies, neutral parent species are mixed with the noble gas species and deposited on a cold finger at 10 K or less. Irradiation of this mixture by an FUV light produces cations and their absorption spectrum can be measured in the IR, visible, and UV using standard spectrometers. However, the ionization efficiency is not high and confusion with bands due to the neutral parent molecule is always a problem. In addition, the photolysis may produce other species (eg., dehydrogenated radicals). Assignments of bands to the cationic species can be facilitated by further experiments using mixtures doped with well-known electron acceptors (eg., CCl_4). Also, small PAHs have no (electronic) absorption bands in the visible. Hence, correlation of the growth of visible with IR bands with UV dose can be helpful in this respect. There is always a problem in studying weak cationic bands in a spectral region with strong neutral bands (eg., the C-H stretching region). Also, the matrix will introduce small shifts in peak position. Furthermore, the width of the bands is always dominated by matrix interactions in MIS experiments. Extensive studies are required to assess these effects. The advantage of this technique is the relative ease with which different species can be studied and, over the last 15 years, a large sample of PAHs has been studied [37], [16], [17].

Multiphoton dissociation techniques (MPD) produce ions in the gas phase through photo-ionization with a FUV laser. These ions are stored in an ion trap, which can be manipulated to reject other species so that confusion is of no concern. However, because of space charge considerations, the total number of ions is limited, hampering the use of conventional spectroscopic techniques to measure the absorption properties of the ion. Instead, an intense, IR-tunable, pulsed, free-electron laser is used. When the wavelength of the laser is in resonance with an absorption band, the trapped species can rapidly absorb a large number of photons and, when it is highly excited, dissociation will result. The number of fragments produced and initial cations remaining can be readily measured using time of flight ion-mass spectrometry. Mass spectrometry is a very sensitive technique and even small amounts can be reliably measured. Frequency scanning of the IR laser will then result in the equivalent of an IR spectrum. While this technique has turned out to be actually very straightforward, it does require access to a free electron IR laser and only a few are world-wide available. Besides the specialized equipment, this technique has also some experimental drawbacks. In particular, the IR photon energy (0.05 eV) is much less than the typical internal energy required for dissociation ($\sim$ 30 eV). Hence, many, many photons have to be absorbed before dissociation occurs. The peak position (eg., redshift) and band strength may be adversely affected by anharmonicity effects, which are inherent to highly excited species; indeed large anharmonicities may even shift the absorption band outside of the pumping frequency before dissociation occurs. Furthermore, the spectral resolution (5 cm^{-1}) of this technique is limited by the very short pulse duration ($\sim$ 1 ps) of the laser. This technique also has a number of clear advantages: the species are in the gas phase not in a matrix, specific ions are readily isolated, it is extremely sensitive, it is relatively simple, and – turning a vice into a virtue – it also allows a study of the anharmonicity of a mode. At present, only the Dutch laser, FELIX, has been used for such studies. Nevertheless, the group at Nieuwegein has now measured a large number of small PAH cations [28], [29], [30].

The dissociation of van der Waals clusters (VDW) is a variant of this technique. In this case, a cluster of the neutral PAH with a loosely bound, noble gas atom (eg., Ar) is formed in the supersonic expansion of a molecular beam. The neutral cluster is then ionized with a double resonant ionization scheme which leaves the ionic cluster just above the ionization potential but internally cold. Absorption of a single IR photon will dissociate the cluster and the resulting bare PAH cation can be "counted" using a time of flight mass spectrometer. Again, this technique requires a tunable intense IR laser for an absorption spectrum and hence limits the spectral resolution to some 5 cm^{-1}. Also, the preparation of the species is molecule specific. The shift induced by the complexed noble atom has to be established but turns out to be small (1 cm^{-1}) compared to the spectral resolution. Some other subtle spectral effects can also occur. So, this technique is more involved than the previous two. However, it does overcome the problem of the anharmonicity associated with the MPD studies. At present, only the IR absorption spectrum of the naphthalene cation has been measured this way [33].

The most direct test of the interstellar PAH hypothesis is to measure the infrared emission spectrum of a highly excited PAH species in the gas phase. For neutral PAHs, this has been done in the 3 μm region following excitation by UV absorption or heating in an oven [11], [10], [23]. Because of the high background, this method becomes very challenging at longer wavelengths. Nevertheless, for a few neutral PAHs, this technique has been extended to longer wavelengths using a cryogenically cooled IR spectrometer capable of counting single IR photons [35], [12]. While this technique is alread very involved for neutral PAHs, studying the IR spectra of PAH cations is particularly daunting because of space charge limitations. However, the Saykally group in Berkeley has now succeeded in measuring the IR spectra of a few PAH cations in the gas phase which were ionized (and excited) by electron impacts [21], [22]. The advantage of this technique is that it measures precisely what is needed: the IR emission spectrum of a highly vibrationally excited PAH cation in the gas phase! A feat which cannot be lauded enough. The disadvantage is that this technique is very difficult and is limited to relatively small species. Also, the internal energy of the emitting species is difficult to quantify (and the exact peak position will depend on the internal energy). Mostly, of course, this technique requires the use of a unique IR spectrometer which is presently (and unwisely) not funded for astronomical studies. Hence, only the pyrene cation and some of its derivatives have been measured.

The interpretation and analysis of these experimental studies have been guided by extensive theoretical calculations using density functional theory [24], [6], [7]. The easy and speed of the calculation has to be balanced with the accuracy of the results. Extensive calculations have been made to assess various systematic effects such as those associated with the basis set. The effects of others are more difficult to evaluate; eg., the presence of low lying electronic states may introduce erroneous results. For neutrals, the theoretical results have been validated by extensive comparison with laboratory MIS studies [24]. After correction of the frequencies for the basissets (a factor 0.96), this comparison is very encouraging. All bands compare well in peak position (4 cm^{-1}) and relative intensity (15%) with the exception of the C-H stretching mode which is systematically calculated to be too intense by a factor 2. For ions the comparison seems to be less favorable (14 cm^{-1}, 50%; [3]). This technique has some very clear advantages. A large number of similar species can easily be calculated and systematic trends investigated; eg., the effects of symmetry, substitution, etc. Also, while experimental studies are hampered by the low vapor pressure of large PAHs, with current computer capabilities, it is already feasible to calculate species containing 100 C-atoms and Moore's law is on our side. It is also possible to study higher charge states which have currently not yet been studied in the laboratory but are likely to be astronomically highly relevant. Finally, this method allows us to investigate the detailed characteristics of the modes (eg., which atoms are involved in the vibrations).

The infrared spectra measured using these different techniques are compared in figure 2 for the cations of the small PAHs, phenanthrene and pyrene. The results are very encouraging: all spectra show bands at very similar frequencies and with similar relative intensities. In particular, all cations studied show strong CC modes in the 5-9 μm region relative to the CH modes at shorter or longer wavelengths. This should be contrasted to the spectra of neutral PAHs where the CH modes are invariably much stronger than the CC modes. When examined in detail, however, there are also some differences, likely related to the experimental difficulties noted above. It is clear that a critical evaluation of these differences and their origin is required to further the field. It is therefore very disappointing that the most direct technique, IR emission, is no longer supported by NASA laboratory astrophysics program.

4 Interstellar PAH Spectroscopy

The combination of a large set of high S/N spectra of interstellar IR emission features and laboratory/theoretical spectra of a large sample of PAHs has provided us with new insights in the characteristics of interstellar PAHs. Here, we will discuss the influence of charge and the CC-modes.

While the peak positions of the interstellar IR features are very characteristic for PAHs, available laboratory spectra of neutral PAHs do not produce a

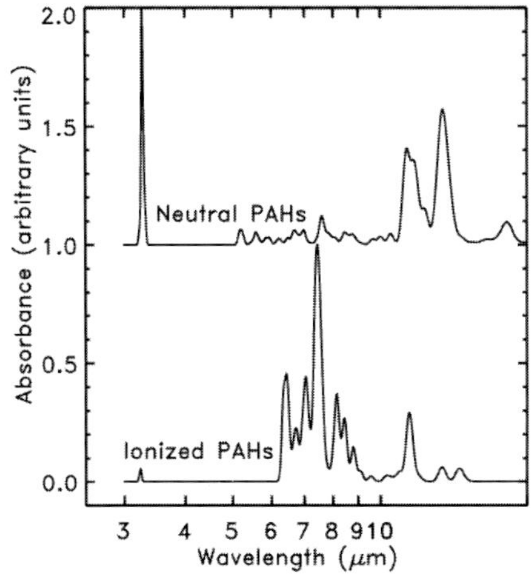

Fig. 3. The summed spectra of a sample of PAHs in neutral (top) and in cationic form (bottom). Note the large difference in the relative strength of the CH and CC modes between these two charge states. Figure taken from [31].

good fit to the data in terms of relative strength of the features. In particular, the CH out-of-plane bending modes in the 11-15 μm region are intrinsically much too strong relative to the CC stretching modes in the 5–8 μm region. Extensive laboratory and theoretical studies have unambiguously shown that charge has an important influence on the intrinsic spectral properties of PAHs (Fig. 3; cf., section 3). A properly selected combination of neutral and cationic PAHs can then provide a good fit to the overall interstellar spectrum ([31]).

A number of points should be made here. First, theoretical studies show that PAHs in the size range 50-100 C-atoms can be doubly or even triply charged under interstellar conditions. At present, the IR characteristics of such cations have only been investigated through quantum mechanical calculations. These show that multiple cationic PAHs have similar spectral properties as singly charged PAHs. Clearly, an experimental study of the IR characteristics of highly charged PAHs is important. Second, the ionization state of PAHs depends on the ratio of the ionization rate to the recombination rate and, hence, the ionization parameter, G_o/n_e, with G_o the intensity of the FUV radiation field and n_e the electron density [3], [5]. Hence, regions with very different conditions can still have very similar mid-IR emission spectra. In that respect, we note that studies of PDRs have shown that n_e scales approximately linearly with G_o. While for HII regions, this may just reflect the Strömgren condition, this proportionality extends to reflection nebulae as well [42]. Third, the observed 11.2/6.2 μm ratio is a measure of the degree of ionization of the emitting PAHs and – if properly calibrated – the IR emission features become a tool for determining the physical conditions (G_o, n_e) of the emission regions [4]. The observed 11.2/6.2 μ ratio varies by about a factor 5 in the ISO spectra [15], reinforcing the point that variations in the G_o/n_e ratio are only a factor < 10, despite the wide range of objects studied.

While the overall spectral resemblance between the observed and laboratory spectra are very encouraging (Peeters et al., 2002), there are subtle differences. These differences have the potential to provide deep insight in the properties of interstellar PAHs. The 10-15μm region is one example of a spectral region which differs in the details. The pattern of the CH modes in the 11-15 μm region is very revealing of the molecular edge structure of the emitting PAHs [16] and the inferences that can be drawn from the observed interstellar spectra have been discussed by [15]. Here we will focus on another example, the 6.2 μm band.

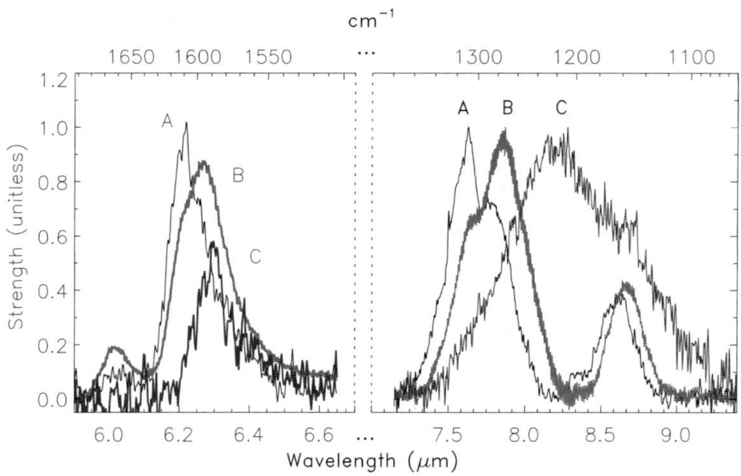

Fig. 4. The observed 6.2 and 7.7 μm emission features can be divided in three different classes, which differ in the peak position and profiles. Figure taken from [31]

An extensive study of the interstellar 6-9μm bands has revealed variations in the peak position and profile of the major IR features at 6.2 and 7.7 μm (Fig. 4; [31]). Specifically, the observed 6.2 μm band can be divided in three classes – A, B, C – with a peak position which varies from 6.2 (A) to 6.3 (C) μm. Class B, with an intermediate peak position, may actually be a combination of these other two classes. Similarly, the 7.7μm feature consists of two components – at 7.6 and 7.8 μm, respectively – whose relative strength varies from a strong 7.6 (A) to strong 7.8 (B). In a third class, C, the peak occurs at 8.2 μm. While this classification was made independently for the 6.2 and 7.7 μm features, it turns out that the sources classified as A (or B or C) for the 6.2 μm band are generally A (or B or C) for the 7.7 μm band. Hence, whatever process dominates these variations, all CC modes are affected at the same time. Moreover, these variations correlate with the astronomical characteristics of the source: HII regions and reflection nebulae are in class A, while isolated Herbig AeBe star, a few post-AGB objects and most PNe belong to class B. Class C consists of a few extreme post-AGB objects. Thus, the composition of the interstellar PAH family carrying the IR emission features varies from source to source reflecting the local physical conditions [31].

The large data set of measured or calculated IR properties of PAHs allows detailed investigation of the factors which influence the peak position of the CC modes. A systematic shift of the 6.2 μm band to shorter wavelengths with PAH size had been noticed for small PAHs (N_c < 30 C-atoms; [17]). However, this systematic trend does not persist for larger PAHs and the measured or calculated peak position levels off at 6.3 μm. Indeed, within the whole data set, no PAH has its strongest CC mode as short at 6.2 μm. Hence, while PAHs can provide a good fit to the observed peak position of the 6.3 μm band of class C, they fail for the 6.2 μm band – particularly when it is realized that the emission by a highly vibrationally excited PAH will lead to a redshift of the band [19]. Like size, nitrogen substitution in the ring also leads to a blueshift of the main CC mode in this spectral region. This shift becomes progressively larger when the N is placed deeper and deeper into the PAH skeleton and class A interstellar 6.2 μm bands can be readily explained by nitrogen-substituted PAHs [31]. It should be emphasized that the band is a CC mode – modes involving the N occur at longer wavelengths. The introduction of N in the PAH skeleton fixates the charge distribution of the PAH and activates modes which are present but IR-inactive in the "pure" PAH [18]. There may be other factors that influence the charge distribution of PAH cations in a similar way – clustering of PAHs, adsorption of metal atoms – and such PAHs still need to be investigated. At present, it is not well understood how N could become substituted deep inside the PAH skeleton in class-A sources; These are sources associated with interstellar medium material such as HII regions and reflection nebulae [31]). In these environments, the most important chemical process is likely photochemistry which would be expected to introduce variations in the molecular edge structure rather than skeletal changes [2]. Indeed, if N-substitution underlies the observed variations in the 6.2 μm band, then this suggests that PAH chemistry in the ISM is very involved and likely there are chemical routes towards PAH formation in the ISM itself as well. Finally, we emphasize that while N-substitution can explain the observed variations in the 6.2 μm band, the concomitant variations in the 7.7 μm band remain unexplained.

References

1. Allamandola, L.J., Tielens, A.G.G.M., Barker, J.R., 1985, ApJ, 290, L25
2. Allamandola, L.J., Tielens, A.G.G.M., Barker, J.R., 1989, ApJS, 71, 733

3. Bakes, E.L.O., Tielens, A.G.G.M., Bauschlicher, C., 2001, ApJ., 556, 501
4. Bakes, E.L.O., Bauschlicher, C., Tielens, A.G.G.M., 2004, in preparation
5. Bakes, E.L.O., Tielens, A.G.G.M., Bauschlicher, C., Hudgins, D., Allamandola, L.J., 2001, ApJ, 560, 261
6. Bauschlicher, C.W., Bakes, E.L.O., 2000, Chem Phys, 4, 24
7. Bauschlicher, C.W., 2002, ApJ, 564, 782
8. Boulanger, F., et al., 1996, A & A, 315, L325
9. Brechignac, P., Pino, T., 1999, A & A, 343, L49
10. Brenner, J., Barker, J.R., 1992, ApJ, 388, L39
11. Cherchneff, I, Barker, J.R., 1989, ApJ, 341, L21
12. Cook, D.J., Schlemmer, S., Balucani, N., Wagner, D.R., Harrison, J.A., Steiner, B., Saykally, R.J., 1998, J Phys Chem, 102, 1467
13. de Frees, D.J., Miller, M.D., Talbi, D., Pauzat, F., Ellinger, Y., 1993, ApJ, 408, 530
14. Helou, G., Lu, N.Y., Werner, M.W., Malhotra, S., Silbermann, N., 2000, ApJ, 532, L21
15. Hony, S., et al., 2001, A & A, 370, 1030
16. Hudgins, D., Allamandola, L.J., 1999a, ApJ, 516, L41
17. Hudgins, D., Allamandola, L.J., 1999b, ApJ, 513, L69
18. Hudgins, D., et al., 2003, in preparation
19. Joblin, C., Boissel, P., Leger, A., D'Hendecourt, L., Defourneau, D.,1995, A& A, 299, 835
20. Kahanpää, J., Mattila, K., Lehntinen, K., Leinert, C., Lemke, D, 2003, A & A, 405, 999
21. Kim, H.-S., Wagner, D.R., Saykally, R.J., 2001, Phys Rev Lett, 86, 5691
22. Kim, H.-S., Saykally, R.J., 2002, ApJS, 143, 455
23. Kurtz, J., 1992, A & A, 255, L1
24. Langhoff, S.R., 1996, J Phys Chem, 100, 2819
25. Lutz, D., et al., 1998, ApJ, 505, L103
26. Mattila, K., et al., 1996, A & A, 315, L353
27. Moutou, C., Sellgren, K., Verstraete, L., Léger, A., 1999, A & A, 347, 949
28. Oomens, J., van Roij, A.J.A, Meijer, G., von Helden, G., 2000, ApJ, 542, 404
29. Oomens, Sartakov, B., Tielens, A.G.G.M., Meijer, G., von Helden, G., 2001, ApJ, 560, L99
30. Oomens, J., Tielens, A.G.G.M., Sartakov, B., von Helden, G., Meijer, G., 2003, ApJ, 591, 568
31. Peeters, E., et al., 2002, A & A, 390, 1089
32. Peeters, E., et al., 2004, ApJ, submitted
33. Piest, H., von Helden, G., Meijers, G., 1999, ApJ, 520, L75
34. Rigopoulou, D., Spoon, H.W.W., Genzel, R., Lutz, D., Moorwood, A.F.M., Tran, Q.D., 1999, AJ, 118, 2625
35. Schlemmer, S., Cook, D.J., Harrison, J.A., Wurfel, B., Chapman, W., Saykally, R.J., 1994, Science, 265, 1686
36. Sellgren, K., 1984, ApJ, 277, 623
37. Szczepanski, J., Vala, M., 1993, ApJ, 414, 646
38. van Diedenhoven, B., et al., 2004, ApJ, submitted
39. van Kerckhoven, C., et al., 2000, A & A, 357, 1013
40. Verstraete, L., et al., 1996, A & A, 315, L337
41. Verstraete, L., et al., 2001, A & A, 372, 981
42. Young Owl, R.C., 2002, ApJ, 578, 885

Growth of Large Molecules and Small Grains

T.J. Millar

Department of Physics, UMIST, PO Box 88, Manchester M60 1QD
Tom.Millar@umist.ac.uk

1 Introduction

To date, around 125 molecules have been detected in interstellar and circumstellar regions - see Table 1. Of these molecules, some 23 are unique to circumstellar envelopes (CSEs). Several of these may be present in the interstellar medium, but are undetectable there by present techniques. Interstellar clouds can be divided into two broad classes depending on their chemical composition: (i) sources rich in organic carbon chains, that is dark clouds and carbon-rich CSEs, and (ii) sources rich in saturated organics, that is giant molecular clouds (GMCs) and hot molecular cores (HMCs). The former contain molecules such as the cyanopolyynes, $HC_{2n+1}N$, n =1–5, C_nH, n = 2–8, C_nS, n = 2-3, and the cumulenes, H_2CCC, H_2CCCC, and $H_2CCCCCC$. The latter have very few chain molecules but have large abundances of both small, H_2O, NH_3, H_2S, CH_4, and large, methanol CH_3OH, ethanol CH_3CH_2OH, dimethyl ether CH_3OCH_3, acetone CH_3COCH_3, glycolaldehyde CH_2CHOH and ethylene glycol $HOCH_2CH_2OH$, species containing several hydrogen atoms.

In addition to these species, detected in the main through high resolution rotational and/or vibrational-rotational transitions in the radio, millimeter and infrared regions, there are a group of solid-state emission and absorption bands detected in the infrared. The emission features are strong and are thought to be due to fluorescent emission from a family of molecules, the polycyclic aromatic hydrocarbons (PAHs) [12]. They are normally detected in carbon-rich environments bathed in UV photons. In diffuse interstellar clouds, few molecules are present in any abundance in the gas phase, but a major molecular component may be hidden in the species responsible for the 200 or so diffuse interstellar absorption bands (DIBs). As with the IR bands, there is, as yet, no firm identification of any particular feature with any specific molecule although detection of sub-structure in some of the stronger bands indicates that large molecules are responsible [15, 11]. The fractional abundance derived for the absorber of the 600 nm DIB observed toward ζ Oph is about 3×10^{-7} to be compared to the observed value for C_3 of 2×10^{-9} and the upper limits of 4×10^{-9} and 2×10^{-10} which have been set for the fractional abundances of C_4 and C_5 [7]. Together with the rapid photodissociation time-scales for these species in diffuse clouds, this implies that the molecules causing the DIBs must be relatively photo-resistant, or at least able to repair any photodamage on a rapid time-scale, and that their

Table 1. Observed interstellar and circumstellar molecules: December 2003.

Number of atoms							
2	3	4	5	6	7	8	≥ 9
Interstellar inorganic molecules (23)							
H_2	H_2O	NH_3					
OH	H_2S	H_3O^+					
SO	SO_2						
SiO	HNO						
SiS	NH_2						
NH	N_2H^+						
NO	N_2O						
NS	H_3^+						
PN							
HCl							
SO^+							
HF							
FeO							
Interstellar organic molecules (79)							
CH^+	HCN	H_2CO	HC_3N	CH_3OH	HC_5N	$HCOOCH_3$	HC_7N
CH	HNC	H_2CS	C_4H	CH_3CN	CH_3CCH	CH_3C_3N	CH_3OCH_3
CN	HCO	HNCO	CH_2NH	CH_3NC	CH_3NH_2	CH_3COOH	CH_3CH_2OH
CO	OCS	HNCS	CH_2CO	CH_3SH	CH_3CHO	H_2C_6	CH_3CH_2CN
CS	HCO^+	c-C_3H	NH_2CN	NH_2CHO	CH_2CHCN	CH_2OHCHO	CH_3C_4H
C_2	HOC^+	l-C_3H	HOCHO	HC_2CHO	C_6H		CH_3C_5N
CO^+	HCS^+	C_3N	c-C_3H_2	C_5H	c-C_2H_4O		CH_3COCH_3
	C_2H	C_3O	CH_2CN	H_2CCCC	CH_2CHOH		HC_9N
	C_2O	C_3S	H_2CCC	HC_3NH^+			$HC_{11}N$
	C_2S	H_2CN	HCCNC				$OHCH_2CH_2OH$
	CH_2	CH_3	HNCCC				NH_2CH_2COOH
	CO_2	C_2H_2	CH_4				
	C_3	$HOCO^+$	H_2COH^+				
		$HCNH^+$					
Circumstellar molecules (23)							
CP	SiCN	HCCN	C_5	C_2H_4		C_7H	C_8H
SiC	c-SiC_2	c-SiC_3	SiH_4	C_5N			
SiN	NaCN		SiC_4				
NaCl	MgCN						
AlCl	MgNC						
KCl	AlNC						
AlF							
SH							

synthesis does not rely on the smaller molecules. This has been confirmed by the detailed calculations of Bettens & Herbst [1, 2] who show that fractional abundances of large molecules are far too small to account for the DIBs using *in situ* formation.

2 Carbon-Chain Chemistry

The formation of complex molecules in dark clouds is more efficient than that in diffuse clouds because of the shorter collision times and because the UV photon field is essentially absent. The fundamental reaction which drives organic chemistry is the proton exchange:

$$\mathrm{H_3^+ + C \longrightarrow CH^+ + H_2}$$

Subsequent reactions with H_2 lead to the formation of protonated methane, CH_5^+ which, in dense clouds, proton transfers with abundant CO to produce methane - at typical fractional ionisations, dissociative recombination of CH_5^+ is negligible. Reactions of methane, with C^+ and methyl ions leads to formation of C–C species:

$$\begin{aligned}
\mathrm{C^+ + CH_4} &\longrightarrow \mathrm{C_2H_2^+ + H_2}\\
\mathrm{C^+ + CH_4} &\longrightarrow \mathrm{C_2H_3^+ + H}\\
\mathrm{CH_3^+ + CH_4} &\longrightarrow \mathrm{C_2H_5^+ + H_2}
\end{aligned}$$

Insertion reactions involving carbon atoms and binary reactions between hydrocarbons are also important in building the linear carbon-chain backbone:

$$\begin{aligned}
\mathrm{C_2H_2^+ + C_2H} &\longrightarrow \mathrm{C_4H_2^+ + H}\\
\mathrm{C + C_mH_n^+} &\longrightarrow \mathrm{C_{m+1}H_{n-1}^+ + H}
\end{aligned}$$

and so on. Most carbon-chain molecules contain few hydrogen atoms, despite the overwhelming abundance of H_2, because reactions between hydrocarbon ions and H_2 are generally endothermic and do not proceed at low temperatures.

Heterogeneous species are formed by reactions such as:

$$\mathrm{CN + C_2H_2 \longrightarrow HC_3N + H}$$

This chemistry is, in its detailed reactions, very similar to that in the outer CSEs of carbon-rich AGB stars (see Sect. 4). The difference is that in the traversal of the CSE, parent species, which are in the form of stable neutrals such as CO, HCN, C_2H_2, are converted to much larger species through a photon-driven chemistry. However, as these large molecules drift outward, they are eventually broken down by the incident interstellar UV radiation field. The end point is that atoms and atomic ions are returned to the interstellar medium together with the dust grains that form in these objects. In interstellar clouds the direction of the path is reversed - atoms and atomic ions form molecules - but many of the important reactions and synthetic pathways are similar.

The methyl ion is important in forming larger molecules in interstellar clouds through radiative association reactions, a process which is only efficient at the very low pressures of the interstellar medium. Such reactions have reaction rate coefficients which possess inverse temperature dependences and are particularly efficient at low temperature. Examples include:

$$\mathrm{CH_3^+ + H_2O \longrightarrow CH_3OH_2^+ + h\nu}$$
$$\mathrm{CH_3^+ + HCN \longrightarrow CH_3CNH^+ + h\nu}$$
$$\mathrm{CH_3^+ + CH_3OH \longrightarrow CH_3OCH_4^+ + h\nu}$$

which, following dissociative recombination, form methanol, CH_3OH, methyl cyanide, CH_3CN, and dimethyl ether, CH_3OCH_3, respectively.

Such reactions, when augmented with a small number of neutral-neutral reactions involving atoms and radicals, appear to account reasonably well for the observed abundances of carbon-chain and related molecules observed in cold dark clouds such as TMC-1. For example, through a careful study of ^{13}C isotopomers, Takano et al. [13] have shown that the primary synthesis of HC_3N is via the reaction of CN with C_2H_2, a reaction known from laboratory experiments to be fast at low temperatures.

3 Hot Molecular Cores

The abundances of highly saturated molecules, together with deuterium fractionation much in excess of that expected from the kinetic temperature, implies that the chemistry in hot cores has been substantially affected by the evaporation of ice mantles. Whether the molecules observed are the direct result of evaporation or of post-evaporation chemistry which occurs on a time-scale of about 10^4 yr remains unclear, and is important in understanding the process of high mass star formation. If all molecules detected in HMCs are released from grain surfaces then we can conclude that surface chemistry is complex (and not likely to be adequately modelled in the near future) although we should be able to determine the temperature at which surface chemistry occurs. On the other hand, if we can detect molecules which are solely formed in post-evaporation chemistry, then we should be able to use observations of relative molecular abundances to determine the chemical ages of hot cores. Fig. 1 shows the evolution of the column densities of complex species following evaporation of methanol and ethanol (and other molecules) in a hot core model [10]. Peak abundances of daughter molecules are reached at about 10^5 yr and then decrease as parents are destroyed and the chemistry resets itself to the physical conditions appropriate to those in the hot core.

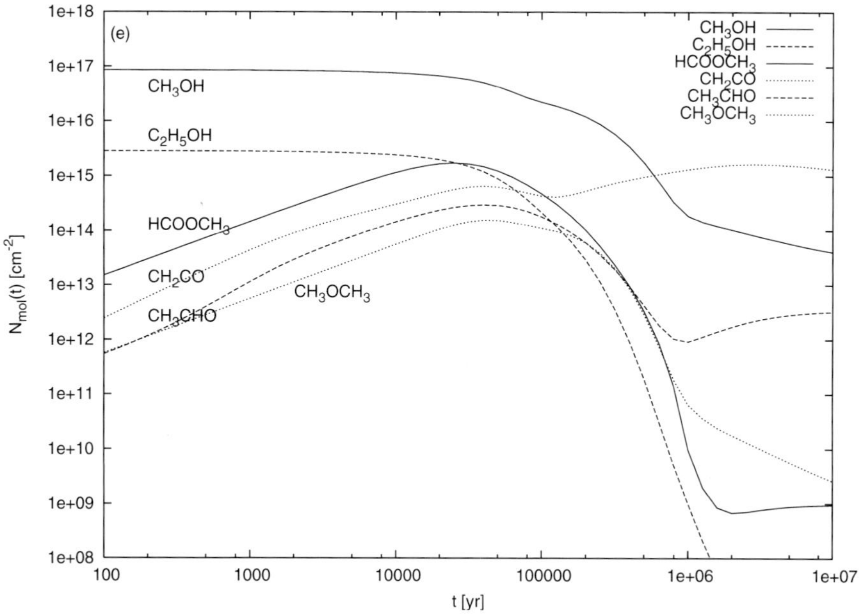

Fig. 1. The time evolution of molecular column densities in a hot core model of G34.3+0.15 [10].

One sees that daughter/parent molecular ratios vary appreciably with time in such a model.

Recent chemical models of HMCs have concentrated on post-evaporation chemistry and have incorporated better physics and a minimum set of parent (that is, evaporated) species to drive the chemistry. Viti and Williams [14] considered the effects of a gradual heating of the hot core gas to reFlect the fact that the central star evolves over time while Nomura and Millar [10] used submillimetre observations of the dust to determine radial density and temperature profiles. These were then used, with the appropriate binding energies, to determine the radial distances from the central star at which different molecules evaporate. Such an approach leads to a highly differentiated chemical distribution as shown in Fig. 2.

In some instances, daughter products predicted from earlier 'one-point' models do not appear as the parents necessary for their synthesis do not exist in the same spatial region. Of course such results depend on the unproven assumption that mantle molecules evaporate according to their individual binding energies. An extreme approach, given the very large abundance of water in ices, is to assume that all mantle species evaporate at the same point as does water. The right panel of Fig. 2 shows the resultant spatial distributions which are seen to be much different, particularly at large radial

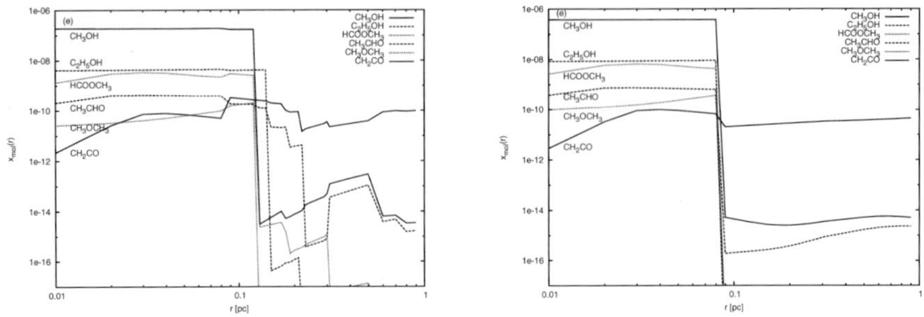

Fig. 2. The radial distribution of fractional abundances after 10^4 yr in a hot core model in which parent molecules evaporate at (left) radii which depend on their binding energy in water ice, and (right) at the same radius as water [10].

distances, from those in the left panel. Clearly interferometric studies should be able to detect such spatial differences.

4 Circumstellar Environments

Circumstellar envelopes, particularly those around late-type carbon-rich stars, provide an ideal environment for the formation of large molecules and small grains. Such CSEs can be divided into a number of physical regions in which physics and chemistry differ. Close to the stellar surface, at distances of less than 10^{13} cm, temperatures around 1300 K and H_2 densities larger than 10^{10} cm^{-3}, LTE chemistry dominates and stable molecules are formed, primarily CO, HCN, C_2H_2, etc. Stellar pulsations lead to the production of periodic shock waves which propagate outward through the envelope from a radius of about 2 10^{13} cm. These shocks help create conditions conducive to dust nucleation and growth. Material, both gas and dust, formed in the inner region is expelled by radiation pressure out into the interstellar medium. In doing so, it enters a flow with temperature and density profiles which decrease with radial distance from the star. At around 10^{16} cm, depending on the mass-loss rate and expansion velocity, the extinction falls to a low enough value that interstellar photons begin to photodissociate and photoionise these parent species, producing ions and radicals. In this outer region of the CSE, a rich chemistry develops and many carbon-chain molecules are formed. Finally, beyond about 3 10^{17} cm the UV degrades all molecules to atoms and ions and it is these that are returned, with the dust, to the interstellar medium.

The formation of PAH molecules in periodically shocked gas has been considered by Cau [4]. In any one pulsational cycle there is a competition between the addition of C_2H_2, which builds rings, and H-atom abstraction, which decreases complexity. An example of ring formation can be given through a discussion of benzene formation, for which several routes have been identified.

1. 1-buyen-3-ynyl (C_4H_3U or n-C_4H_3) to phenyl,

$$C_4H_3U + C_2H_2 \longrightarrow C_6H_5\,,$$

   where U means that the molecule has a vinyl radical located on a terminal carbon atom;
2. 1,3-butadienyl (C_4H_5U or n-C_4H_5) to benzene,

$$C_4H_5U + C_2H_2 \longrightarrow C_6H_6 + H\,,$$

3. the reaction of propargyl radicals,

$$C_3H_3 + C_3H_3 \longrightarrow C_6H_5 + H\,,$$

   which has been measured in the laboratory; and
4. formation via benzyne (C_6H_4),

$$\begin{aligned} C_4H_3U + C_2H_2 &\longrightarrow C_6H_4 + H\,, \\ C_6H_4 + H &\longrightarrow C_6H_5\,. \end{aligned}$$

Once benzene has been formed, two-ring formation occurs via either the reaction of benzene with phenyl

$$C_6H_6 + C_6H_5 \longrightarrow C_6H_5C_6H_5 + H$$

or

$$\begin{aligned} C_6H_5 + C_2H_2 &\longrightarrow C_8H_6 + H \\ C_8H_6 + H &\longrightarrow C_8H_5 + H_2 \\ C_8H_5 + C_2H_2 &\longrightarrow C_{10}H_7 \end{aligned}$$

The growth of an even larger number of rings can be envisaged as an extension of these routes, for example, using acetylene:

$$\begin{aligned} A_i(r) + C_2H_2 &\longrightarrow A_iC_2H + H\,, \\ A_iC_2H + H &\longrightarrow A_iC_2H^* + H_2\,, \\ A_iC_2H^* + C_2H_2 &\longrightarrow A_{i+1}(r)\,. \end{aligned}$$

where i is the number of rings in the molecule - one needs two reactions with C_2H_2 to add a ring.

Cau considered the formation of molecules containing up to seven benzene rings and found production to be fast. He defined the *yield* of the reaction to be the number of carbon atoms in aromatics divided by the (initial) number of carbon atoms in (parent) acetylene and found a yield of 0.003, that is about 10% of all the carbon needed in carbon dust of radius 500 nm [4]. Although this yield is too small to account for *all* of the carbon in carbonaceous grains in CSEs, Cau noted that it is sufficient to account for the carbon in the nanocrystalline cores found in pre-solar grains. That is, although PAH formation is not efficient enough that all carbon in carbon dust has been processed through a PAH molecule, all grain cores could be produced from PAHs. Subsequent growth of these nanocrystals could involve the accretion of acetylene and other hydrocarbons.

The most detailed models of photochemistry in the outer carbon-rich CSE have been made by Millar et al. [9] who considered species containing up to 23 carbon atoms. The ion-neutral and neutral-neutral chemistry is very similar to that in TMC-1 - with the exception that there is no free reactive oxygen in carbon-rich envelopes. Millar et al. showed that the photo-destruction of parents such as C_2H_2 and HCN could drive efficient growth of carbon chains with peak fractional abundances occuring at a radius of about $3\ 10^{16}$ cm for parameters appropriate to IRC+10216. Subsequently, Brown and Millar [3] included the effects of enhanced density shells, as observed by Mauron and Huggins [8], and found much improved agreement between models and observation, particularly in the spatial extent of the molecular shells.

5 Conclusions

Gas-phase chemistry is able to build large molecules in a number of different environments. However, for molecules containing 10 or more atoms, it is difficult, given inefficiencies in the chemistry, to synthesis fractional abundances greater than 10^{-11}–10^{-10} in molecular clouds. Thus, there appears to be no feasible way to grow *in situ* the molecules which are thought to be responsible for the DIBs or IR bands.

Large abundances of complex species can occur in hot cores if large abundances of precursor molecules such as water and methanol are synthesised on grains. The formation of larger species, such as the higher alcohols and gylcine [6] on grains remains an open question.

Carbon-chain molecules and dust grains are made efficiently in carbon-rich AGB stars. Kinetic modelling of the growth of small PAH molecules shows that while not all the carbon in carbonaceous dust has been through a PAH phase, the nucleation centres of all grains could have been produced as the end-point of PAH growth. Finally, although space has not allowed a discussion here, proto-planetary nebulae appear to contain large molecules [5]. Initial calculations by Woods et al. [16] show that they can be built from simple species providing that the chemistry occurs in a long-lived disk - otherwise geometrical dilution in the fast outflow prevents growth. However, other scenarios, such as molecule formation in shocked gas, have yet to be investigated in such objects.

Acknowledgements. Astrophysics at UMIST is supported by a grant from PPARC. I am very grateful to the conference organisers for supporting my attendance at this meeting.

References

1. R. P. A. Bettens, E. Herbst: ApJ, **468**, 686 (1996)
2. R. P. A. Bettens, E. Herbst: ApJ, **478**, 585 (1997)
3. J. M. Brown, T. J. Millar: MNRAS, **339**, 1041 (2003)
4. P. Cau: A&A, **392**, 203 (2002)
5. J. Cernicharo, A. M. Heras, J. R. Pardo et al: ApJ, **546**, L123 (2001)
6. S. B. Charnley, M. E. Kress, A. G. G. M. Tielens, T. J. Millar: ApJ, **448**, 232 (1995)
7. J. P. Maier, G. A. H. Walker, D. A. Bohlender: ApJ, **566**, 332 (2002)
8. N. Mauron, P. J. Huggins: A&A, **359**, 707 (2000)
9. T. J. Millar, E. Herbst, R. P. A. Bettens, MNRAS, **316**, 195 (2000)
10. H. Nomura, T. J. Millar: A&A, in press
11. P. J. Sarre, J. R. Miles, T. H. Kerr et al: MNRAS, **277**, L41 (1995)
12. K. Sellgren: ApJ., **277**, 623 (1984)
13. S. Takano, A. Masuda, Y. Hirahara, H. Suzuki, M. Ohishi: ApJ, **329**, 1156 (1998)
14. S. Viti, D. A. Williams: MNRAS, **305**, 755 (1999)
15. G. A. H. Walker, A. S. Webster, D. A. Bohlender, J. Krelowski: ApJ, **561**, 272 (2001)
16. P. M. Woods, T. J. Millar, E. Herbst, A. A. Zijlstra: A&A, **402**, 189 (2003)

PAHs in the Interstellar Carbon Cycle: The Experimental View of PIRENEA

C. Joblin

Centre d'Etude Spatiale des Rayonnements, CNRS-UPS-OMP, BP 4346, 31028 Toulouse Cedex 04, France – christine.joblin@cesr.fr

1 Astrophysical Context

The identification of interstellar polycyclic aromatic hydrocarbons (PAHs) has motivated a lot of theoretical and laboratory work. These species are so far the best candidates to account for the Aromatic Infrared emission Bands (AIBs) between 3.3 and 12.7 μm. They could contain 10 to 20% of total carbon making them an important component of interstellar matter. However, trying to match the AIBs with the calculated or measured spectra does not appear to be a successful method to identify individual species. More insight into their chemical structure could be gained by considering how these species are formed and how they evolve due to environmental conditions (UV irradiation, gas and dust interactions). This implies considering photodissociation processes and chemical reactions. In these studies, one has to bear in mind the physical conditions of interstellar clouds with low densities, typically less than $10^6\,\mathrm{cm}^{-3}$ and cold temperatures, a few tens of K for PAHs between the absorption of two UV photons. Processes that occur on long time scales such as fragmentation in competition with infrared cooling and radiative association are expected to play an important role whereas such processes are difficult to access in laboratory conditions. The PIRENEA experimental set-up was developed to address these studies, the specifications (high isolation and low temperatures) being met by combining the trapping capabilities of an ion cyclotron resonance (ICR) cell with cryogenic shielding.

Experiments performed with PIRENEA are motivated by observational results. Three recent ones appear of special interest:

-i- Small hydrocarbons such as C_2H, c-C_3H_2, C_4H and C_6H have been detected with radiotelescopes in photodissociation regions (D. Teyssier et al., these proceedings).

-ii- Analysis of spectro-imagery data from the Infrared Space Observatory shows the presence of a population of Very Small Grains closely connected to PAHs, possibly PAH clusters which photoevaporate when processed by UV photons (Rapacioli et al., this conference)

-iii- Recent results obtained with the ODIN sub-millimetre wave spectroscopy satellite give upper limits for the abundance of O_2 in molecular clouds that cannot be accounted for by chemical models [1].

Items **-i-** and **-ii-** raise the question of the formation and destruction of interstellar PAHs. Item **-iii-** suggests to explore alternative mechanisms for the destruction of O_2. Can PAHs play a role?

2 Photodissociation

Most of the experiments were conducted on the coronene cation $^{12}C_{24}H_{12}^+$. The cations produced by laser ablation are isolated in the trap and then heated by the UV-visible radiation of a Xe arc lamp, allowing the dissociation threshold to be reached in competition with IR cooling (time scale of seconds). The major results obtained are that photodissociation proceeds by sequential loss of H down to the bare carbon cluster C_{24}^+. Species containing an odd number of hydrogen atoms are less stable than those containing an even number : $C_{24}H_{2n}^+$ and $C_{24}H_{2n+1}^+$ species photodissociate with respectively 10 and 5 eV of mean internal energy [2]. The C_2H_2 channel was found to be negligible in the photodissociation of the compact PAH $C_{24}H_{12}^+$, whereas it is known to be a major channel in the fragmentation of small species n≤14 [3]. The destruction of the carbon skeleton was observed only for C_{24}^+ leading to a series of carbon clusters [19].

3 Chemical Attack by O_2

Reactivity of $C_{24}H_p^+$ (p=12,..,0) was studied with molecules such as H_2, O_2 and H_2O. No reactivity was observed with H_2 whereas $[C_{24}H_p - H_2O]^+$ complexes are formed when injecting H_2O. The case of reactivity with O_2 is discussed here in more detail. One of the advantages of ICR cells is the associated Fourier transform mass spectrometry analysis which is non-destructive, sensitive and provides very high mass resolution. Also, the technique allows to selectively eject some species out of the cell by exciting them excessively into their cyclotron motion. This procedure is used to isolate from the dissociation cascade the species $C_{24}H_p^+$ with a given p. Then, O_2 gas is injected and the reactivity with $C_{24}H_p^+$ is observed as a function of the injection time. During injection, average typical pressure in the vacuum chamber is $\sim 10^{-7}$ mbar which is low enough to avoid significant contribution from three-body collisions. The reactivity of $C_{24}H_p^+$ with O_2 was not observed for p even and led to $C_{23}H_pO^+$ (+ ejection of CO) for p odd (Fig. 1). This implies a role of radical sites; hydrogens are arranged in pairs at the periphery of coronene (cf. Fig. 2). When pairs of H are lost, an energetically favorable rearrangement for the molecule is to form triple bonds at the periphery. For an odd number of H, there exists an unpaired electron forming a radical site which is chemically very active. Information on the structure of the reaction product is obtained by further irradiation with the Xe lamp. In less than 2 seconds of irradiation, the following steps are observed: $[C_{23}H_{2n+1}O]^+ + h\nu \rightarrow C_{22}H_{2n+1}^+ + CO$.

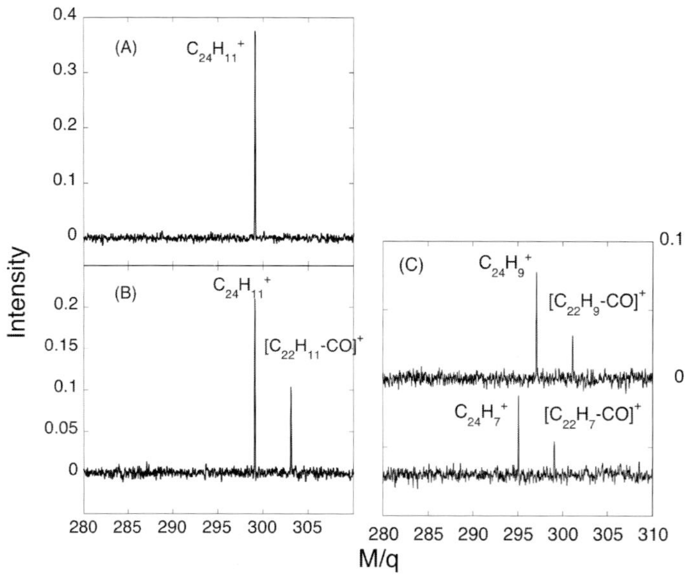

Fig. 1. (A) Isolation of $^{12}C_{24}H_{11}^+$ obtained by photodissocia tion of $^{12}C_{24}H_{12}^+$ in the PIRENEA ICR cell. (B) Reaction of $C_{24}H_{11}^+$ with O_2 (10 s at P$\sim 2\,10^{-7}$ mbar) forming $[C_{22}H_{11} - CO]^+$. (C) Same as (B) but with $C_{24}H_9^+$ and $C_{24}H_7^+$.

This occurs very quickly and could suggest that CO was present as a sidegroup. However, a preferred structure is expected to be with a five-membered ring as shown in Fig. 2 (J.C. Tabet, private communication). Follow the reactions $C_{22}H_{2n+1}^+ + h\nu \rightarrow C_{22}H_{2n}^+ + H$, species containing an even number of H being more stable, and then the condensation reaction $C_{22}H_{2n}^+ + h\nu \rightarrow C_{20}H_{2n}^+ + C_2$. The final product of the photodissociation process is $C_{20}H_{10}^+$ when starting with $C_{24}H_{11}^+$ and $C_{20}H_{p-1}^+$ for $C_{24}H_p^+$ with $p < 11$. The final molecules are very likely to be corannulene (Fig. 2) and its dehydrogenated forms. This is an interesting result since corannulene is a curved PAH which is often invoked as a precursor of fullerenes.

4 Discussion and Perspectives

The use of the PIRENEA set-up is illustrated here to study the photodissociation and chemical reactivity of PAHs in conditions that approach those of interstellar space. In the case of the compact PAH $C_{24}H_{12}^+$, it was observed that photodissociation of the carbon skeleton occurs mainly after complete dehydrogenation. However, ring opening can happen via chemical attack. This is the case for instance when O_2 meets a radical site on dehydrogenated coronene. The reaction appears to be without a barrier and ejects CO. Under

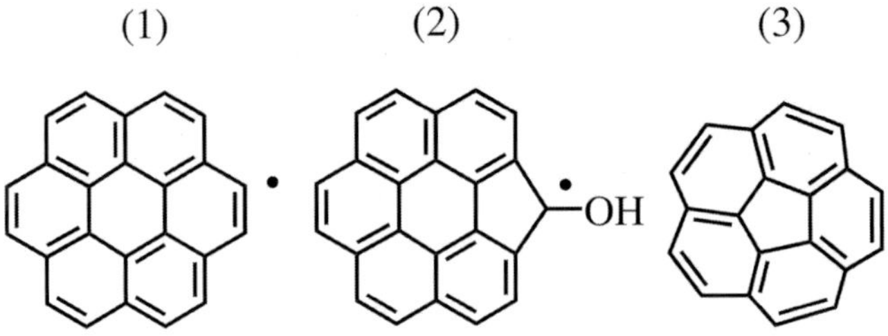

Fig. 2. Structural formulae of (1) $C_{24}H_{11}^+$: coronene without one H, (2) $[C_{23}H_{11}O]^+$: product of reaction of $C_{24}H_{11}^+$ with O_2 and (3) $C_{20}H_{10}^+$: corannulene produced by UV irradiation of (2). For simplicity, hydrogen atoms at the periphery are not represented, nor the charge of the cations. The dot represents an unpaired electron creating a radical site.

UV irradiation the product rearranges toward a more stable form which happens to be corannulene. The presence of a five-membered ring at the center gives this molecule its curved shape. These studies have several astrophysical implications. First, new types of PAH-like molecules are generated and they are good interstellar candidates to be studied. Second, the ease of reactivity of O_2 with PAH radicals is a process that has to be considered as a possible loss channel for O_2 in interstellar clouds. Clearly more quantitative work is required. Kinetic curves have been drawn and values for the chemical rate extracted but these values have to be calibrated. Then the implication of this new reaction route will be estimated in chemical models of interstellar clouds.

Acknowledgments

M. Armengaud and P. Frabel are acknowledged for their technical involvement in the PIRENEA set-up, the Programme National de Physique et Chimie du Milieu Interstellaire and the Région Midi-Pyrénées for their financial support.

References

1. Pagani, L., Olofsson, A.O.H., Bergman, P., et al., 2003, A&A 402, L77
2. Joblin, C., Toublanc, D., Pech, C., Boissel, P., & Armengaud, M., Frabel, P., 2003, JCP, submitted
3. Jochims, H. W., Rühl, E., Baumgärtel, H., Tobita, S., and Leach, S., 1994, ApJ, 420, 307
4. Joblin, C., In : Société Française d'Astronomie et d'Astrophysique, Scientific Highlights 2003, ed by F. Combes, D. Barret and T. Contini (EDP Sciences, les Ulis, 2003) pp.

Small Carbon Chains and Rings in Photo-Dominated Regions

D. Teyssier[1,3], J. Pety[2,3], M. Gerin[3], D. Fosse[3], A. Abergel[4], E. Roueff[5], and C. Joblin[6]

[1] SRON, P.O.Box 800, 9700 AV Groningen, The Netherlands, teyssier@sron.rug.nl
[2] IRAM, 300 rue de la Piscine, Domaine Universitaire, 38406 Saint Martin d'Hères, France
[3] LERMA, UMR 8540, CNRS/Obs. de Paris, 24 rue Lhomond, 75231 Paris Cedex 5, France
[4] IAS, UMR-8617, Université Paris-Sud, bâtiment 121, 91405 Orsay, France
[5] LUTH, Bat. 18, Place J. Janssen, 92195 Meudon, France
[6] CESR, BP4346, 9, Av. du Colonel Roche, 31028 Toulouse Cedex 04, France

1 Introduction

Carbon is the fourth most abundant element in the interstellar medium (ISM), and also the most versatile for building molecules. Carbon chemistry can therefore be considered as the core of interstellar chemistry. Of the nearly 130 molecules now observed in various sources, about 75% have at least one carbon atom, while one fourth are hydrocarbons. These molecules, initially reported in circumstellar shells (e.g. [8], [1]), are now observed in many other media, ranging from molecular dark clouds (up to C_8H, [4]) to the diffuse gas (C_2, C_3, C_4 – [13], [2] –, C_2H and c-C_3H_2– [10]). A natural question therefore arises: if carbon chains are present in the diffuse ISM, what does happen in the Photon-Dominated Regions (hereafter PDRs)? As in diffuse clouds, the chemical processes are dominated by the radiation but the gas is denser. Knowledge about the distribution of carbon chains and rings in PDRs is however yet scarce and mostly limited to the works of [6] and [7], who reported observations of C_2H at some positions in NGC7023, the Orion Bar and NGC7027. In this work we extend the study of hydrocarbons in PDRs through an extensive inventory of carbon chains and rings of up to six carbon atoms.

2 Observations

Carbon chains and rings were observed at the IRAM 30-m in a sample of 5 PDRs, namely the Horsehead nebula, the ρ Oph L1688 cloud, IC63, NGC7023 and the Orion Bar. The radiation fields experienced by these PDRs respectively range from $G_0 = 100$ to some 10^4 (in units of Habing field). Fig. 1 illustrates some of the maps obtained in transitions of C_2H, c-C_3H_2 and

$C^{18}O$. Cuts across the considered PDRs are shown in Fig. 2. In addition, some weaker lines were probed are positions of interest. They include the linear forms of C_3H, C_3H_2 and C_4H_2, c-C_3H, as well as C_6H.

3 Results

3.1 Spatial Distribution

In all sources, the hydrocarbon emission is observed until the border of the PDR (e.g. Horsehead, IC63, see also [5]), or even extends towards the illuminating stars in regions where the molecular ridge emission traced by $C^{18}O$ has significantly decayed (Fig. 1 and 2). This is surprising as they are expected to be quickly destroyed by UV radiation. On large scale, there is a general spatial correlation between the hydrocarbon distribution and the emission of the aromatic infrared bands (AIBs). This is less true for the Orion Bar but extended emission coincident with the PAH line at 3.3 μm is nevertheless observed. CCH, c-C_3H_2 and C_4H are ubiquitously found in all targets. CCH and c-C_3H_2 correlate remarkably well, a behavior apparently widespread in the diffuse ISM ([10]). The weaker chains and rings are on the other hand only barely detected, with the exception of the Horsehead nebula where most heavier species have been successively observed at a position coincident with the AIB emission peak. At this position we report the first detection of C_6H in a PDR.

3.2 Abundances

The molecular abundances with respect to H_2 were computed under LTE or LVG assumptions, depending on the species. The hydrocarbon abundances appear reasonably constant across the different objects and positions. This constancy confirms the general trend already reported in the diffuse gas ([10]). Although error bars remain high (essentially due to uncertainties on $N(H_2)$), the observed values result in average abundances of $[C_2H] = 1.9 \pm 0.4 \times 10^{-8}$, $[c\text{-}C_3H_2] = 1.4 \pm 0.2 \times 10^{-9}$ and $[C_4H] = 1.1 \pm 0.2 \times 10^{-9}$. Abundance ratios between molecules show a somewhat similar trend and seem to be hierarchically distributed. On average, the following ratios are found: $[c\text{-}C_3H_2]/[C_4H] \sim 1$ and $[C_2H]/[c\text{-}C_3H_2] \sim 15$. The latter result is of the same order as the value reported in the diffuse gas (ratio of 28) by [10]. Mean values for the targets are compiled in Tab. 1, in combination with representative values reported in diffuse clouds ([10]) and the TMC-1 dark cloud ([4]).

3.3 Link to the AIB Carriers

As mentioned above, the hydrocarbon emission seems to correlate well with the AIB emission, or, in some other cases, hydrocarbons exist in areas where

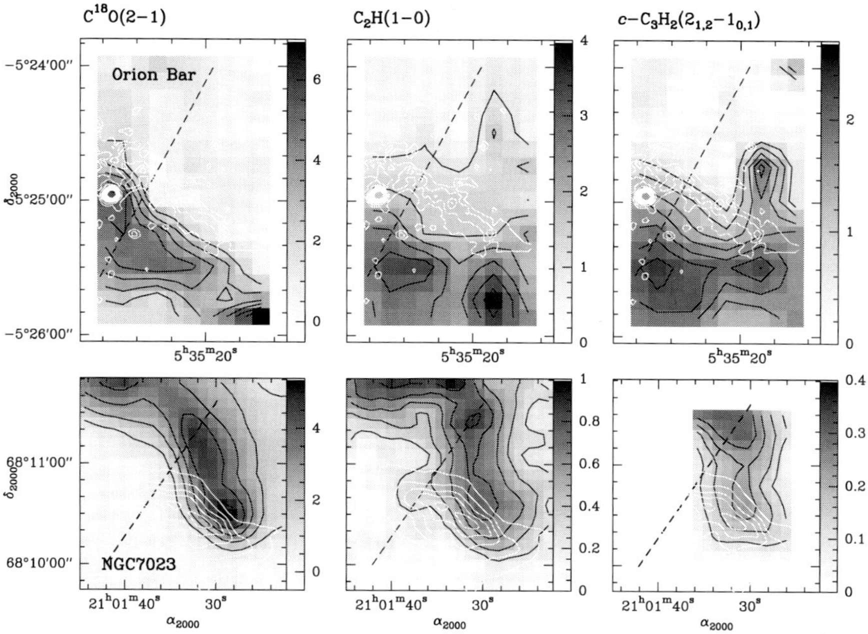

Fig. 1. Integrated intensity maps (K.km/s, T_A^* scale) of $C^{18}O(2–1)$, C_2H and c-C_3H_2 in the Orion Bar and NGC7023. White contours represent the PAH emission at 3.3 μm (Orion Bar, [14]) and 6.2 μm (NGC7023, [3]). Dashed lines indicate the positions of the cuts displayed in Fig. 2.

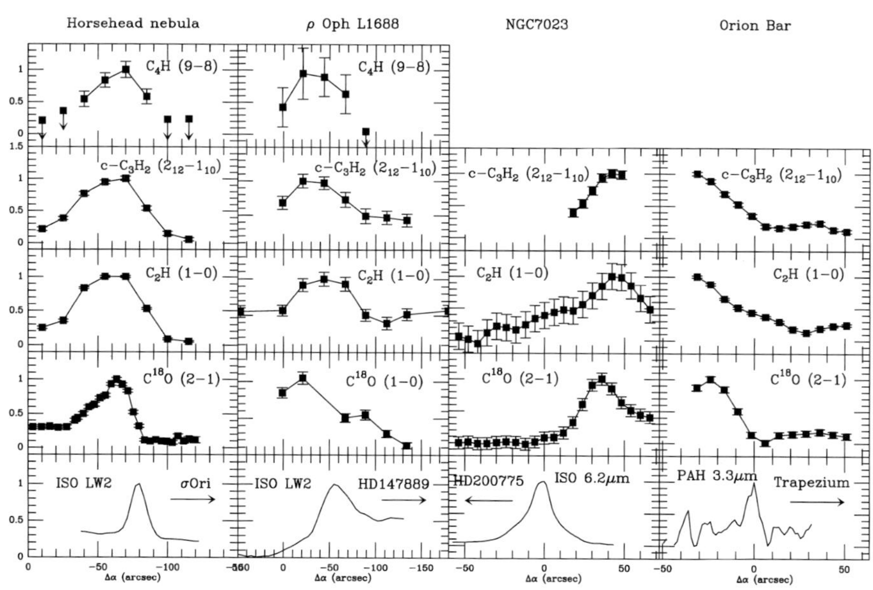

Fig. 2. Normalized intensity cuts across the PDR sample for various species.

Table 1. Mean molecular abundances relative to H_2 for PDRs, TMC-1 (Cyanopolyyne peak, CP) and diffuse clouds.

Molecule	PDR	TMC-1 (CP)	Diffuse clouds
C_2H	1.9e-8	7e-8	2.9e-8
c-C_3H_2	1.4e-9	3e-8	1.4e-9
l-C_3H_2	<3.4e-11	1e-10	5e-10
HC_3N	6e-11	8e-9	–
C_4H	1.1e-9	2e-8	–
C_6H	1.1e-10	2e-8	–
$[C_2H]/[c\text{-}C_3H_2]$	15 ± 4	2.3	28 ± 8
$[c\text{-}C_3H_2]/[C_4H]$	1	1.5	–

$C^{18}O$ is already efficiently photo-dissociated. This result suggests a chemical link to the AIB carriers, confirmed by recent observations of C_2H, c-C_3H_2 and C_4H performed on the Horsehead nebula by [12]. We propose that the photo-erosion of UV-irradiated carbonaceous components could feed the medium with small carbon clusters or molecules, thus overcoming their destruction rate by the FUV at the cloud interface. This production route is suggested also by laboratory experiments ([9]), and could be invoked to explain the discrepancy between the observed abundances and the results of pure gas-phase chemistry models of PDRs.

Acknowledgment

We are very grateful to J. Cernicharo for providing and helping us with his LVG code upgraded with several of the species presented in this paper.

References

1. Cernicharo J., Kahane C., Gómez-González J., Guélin M., 1986, A&A 167, L5
2. Cernicharo J., Goicoechea J.R., Bénilan Y., 2002, ApJ 580, L157
3. Cesarsky D., Lequeux J., Abergel A., et al., 1996, A&A 315, L305
4. Dickens J.E., Langer W.D., Velusamy T., 2001, ApJ 558, 693
5. Fossé D., Teyssier D., Gerin M., et al., in *SFChem 2002*, Waterloo, Ontario, Canada. Eds C.L. Curry & M. Fich.
6. Fuente A., Martín-Pintado J., Cernicharo J., Bachiller R., 1993, A&A 276, 473
7. Fuente A., Rodrîguez-Franco A., García-Burillo S., et al., 2003, A&A 406, 899
8. Guélin M., Green S., Thaddeus P., 1978, ApJ 224, L27
9. Joblin C., this conference
10. Lucas R., Liszt H.S., 2000, A&A 358, 1069
11. Oka, T., Thorburn J.A., McCall B.J., et al., 2003, ApJ 582, 823
12. Pety J., Teyssier D., Gerin M., et al., in preparation
13. Roueff E., Felenbok P., Black J.H., Gry C., 2002, A&A 384, 629
14. Tielens A.G.G.M., Meixner M.M., v. d. Werf P.P., et al., 1993, Science 262, 86

Dust Chemistry in the Laboratory

V. Pirronello[1], G. Manicó[1], J. Roser[2], and G. Vidali[2]

[1] DMFCI, Universitá di Catania, Viale Doria 6, 95125 Catania, Sicily, Italy
pirronello@dmfci.unict.it, gmanico@dmfci.unict.it
[2] Physics Department, Syracuse University, Syracuse, NY13244, USA
gvidali@mailbox.syr.edu, jroser@mailbox.syr.edu

1 Introduction

The presence of dust grains in space is of fundamental importance for the emergence of chemical complexity in the Interstellar Medium (ISM) because grains may catalyze chemical reactions on their surfaces enabling the formation of species that cannot be produced easily in the gas phase, such as H_2 [1, 2, 3, 4, 5] and CO_2 [6].

Two main mechanisms exist: the Eley-Rideal (E-R) mechanism and the Langmuir-Hinshelwood (L-H) mechanism. In the E-R mechanism the formation of the products occurs when either an atom or a molecule, coming from the gas phase, hits directly an already adsorbed atom or molecule and reacts with it. Because the efficiency of this process is linear with the surface density (coverage) of absorbed species and such a coverage is usually very small on interstellar grains, this mechanism has not been considered to be important in the ISM.

In the L-H mechanism the reaction takes place between two partners that are already adsorbed on the surface of the solids. If at least one of the partners is mobile they may encounter and react.

An important variant of the E-R mechanism is the so called Hot Atom mechanism that is active during the landing of an atom coming from the gas phase, i.e. when the atom in the attempt to accommodate on the surfaces suffers multiple jumps and collisions with it and in this way while exploring numerous sites has an enhanced probability of encountering an already adsorbed partner and of reacting with it.

In this paper we present experimental results of studies of catalytic processes occurring on dust grains analogues in simulated interstellar medium conditions. In the next section we will briefly describe our experimental apparatus and we then present selected results of the study of the formation of molecular hydrogen and carbon dioxide on dust grain analogues. Each presentation of data is followed by an illustration of how the data can be used to understand processes occurring in the actual ISM.

2 Experimental Apparatus and Methods

The apparatus, housed in the Physics Department of Syracuse University, and the experimental methods have been described in great detail [1, 8, 9]. Here we recall that measurements are done in a Ultra-High Vacuum (UHV) stainless steel chamber. The pumping system of the main chamber consists of a cryopump, a turbomolecular pump and an ion pump; the operating pressure is in the low 10^{-10} torr range. The catalytic surface is mounted on a liquid helium continuous flow cryostat placed in the center of the UHV chamber; the sample temperature can be varied between 5 K and the room value by a careful adjustment of the liquid helium flow and the power to a heater located behind the sample. The sample temperature is measured by an iron-gold/chromel thermocouple and/or a calibrated silicon diode placed in contact with it. Two triple differentially pumped atomic beam lines are aimed at the surface of the sample and are used to send reactants (atoms, molecules, radicals) on the surface. Each of the two beam lines is equipped with a radio-frequency cavity in which molecular species are dissociated. In order to simulate as closely as possible the kinetic energy of gas phase species in the ISM, the beams can be cooled to a kinetic temperature $\sim$ 200 K by passing them through a liquid nitrogen cooled aluminum nozzle. The velocity distribution of the beam can be measured with a slotted wheel mounted on a high speed water-cooled, in-vacuo motor. Dissociation rates are typically in the 75 to 90 % range, and are constant throughout a run. The estimated fluxes are as low as 10^{12} atoms $cm^{-2}s^{-1}$.

Products of reactions are collected by a quadrupole mass spectrometer mounted on a rotatable flange. This detector is also used to measure the flux of incoming atoms and hence the dose that is sent on the sample.

3 Hydrogen Recombination on Grain Analogues

H_2 is the most important molecule in space. It does not form efficiently enough when two neutral H atoms collide in the gas phase in order to make up for the photo-dissociation rate in diffuse clouds. It was therefore proposed [10] that interstellar dust grains might act as catalysts.

In the ISM, as far as the atom-surface interaction is concerned, there are three main classes of dust surfaces to consider: silicates, carbonaceous particles, and ice. We measured the recombination efficiency, that is, the probability that two H atoms impinging on a dust grain form a molecule, as a function of the grains temperature on: a silicate slab [1, 2], an amorphous carbon sample [3], and an amorphous water ice layer [4, 5]. To increase the signal/noise ratio, and to avoid spurious effects due to the deposition on the surface of the undissociated fraction of the hydrogen beam, we sent on the sample H and D atoms and measured the amount of HD produced. This signal is divided by one half of the total number of H and D atoms sent and by the probability of forming HD vs. H_2 and vs. D_2. The HD signal is measured both during irradiation when H and D are sent onto the surface, to detect molecules that are promptly formed and released, and also at the end of irradiation, when the sample temperature is quickly ($\sim$ 0.06 K/sec) raised, a procedure called Temperature Programmed Desorption (TPD). TPD is performed to detect molecules that remained stuck on the surface (if formed during irradiation) or that are formed during the temperature increase due to the increased mobility. This procedure allows to accelerate the process of molecule formation and ejection that otherwise would last too long.

During H atom irradiation the efficiency of recombination is rather low, between .05 and .1. The values of the efficiency measured using the TPD technique are as high as 0.35 for olivine, 0.5 for amorphous carbon at the lowest "irradiation temperatures" (the temperature of the surface during irradiation with atoms) of 5-7 K and almost .5 on low density amorphous water ice, but they decrease at increasing irradiation temperatures.

For irradiations performed at the lowest possible temperature, the HD TPD signal has two distinctive signatures depending on the H,D surface density (coverage) . The best data set for this purpose is the one from olivine because, being polycrystalline, its surface morphology is more regular and flatter than the others. At high coverage there is first order reaction kinetics, i.e., the desorption rate is directly proportional to the number of molecules on the surface. This occurs when at the beginning of the TPD there are already formed molecules on the surface. When the coverage is low the desorption is of the "second order", i.e. the desorption rate is proportional to the square of the number of species on the surface. This means that when the TPD begins, there are no already formed molecules but only atoms. The increase in temperature set H and D atoms in motion that eventually meet, form HD and are then released in the gas phase.

From the second order desorption observed at sufficiently low coverage we deduce that adatom mobility depends on surface temperature and that tunneling (which is temperature independent) does not assure by itself high enough mobility (at least on realistic surfaces). Thermal hopping is a possibility, but another one is that of thermally assisted tunneling, a composite mechanism in which atoms in the ground state of the atom-surface potential are first thermally raised inside the well to higher energy levels and only then (when the activation energy barriers are lower and narrower) are able to tunnel to a neighboring site.

If tunneling is not efficient, the presence of already formed molecules at the beginning of TPDs when a first order desorption is observed at high coverage can be understood as the formation of molecules taking place during the irradiation phase by the Hot Atom mechanism, i.e., when atoms landing from the gas phase during the accommodation process suffer multiple hits with the surface, lose energy and have the chance to react with already adsorbed partners.

We investigated also the energetics of the molecular formation process. The formation of molecular hydrogen releases 4.48 eV of energy; it is therefore important to know how this energy is partitioned between the molecule and the solid. If the kinetic energy of the ejected molecule is substantial, then this energy can contribute to the heating of the cloud. The rovibrational energy, on the other hand, can be re-radiated into the infrared where the cloud is transparent, thus reducing the energy budget of the cloud itself. Using the time-of-flight technique we performed the first measurement of the kinetic energy of HD formed on the surface of high density amorphous water ice [5].

Our results can be fitted using a flight time distribution corresponding to a Maxwell-Boltzmann velocity distribution; our analysis shows that HD comes out with a kinetic temperature that is approximately the temperature of the surface at the time of the desorption, T$\sim$ 24 K. This result can be explained by the fact that HD forms on the highly porous high density amorphous water ice; thus, before escaping, HD suffers several collisions that equilibrate its kinetic temperature with that of the ice. An experiment done at lower H and D exposure and/or on less porous surfaces might yield a very different result. Such experiments are under way.

We used the Rate Equation approach to describe the formation of molecular hydrogen in our experiments. However, in order to take properly into account the stochastic nature of the process, especially in ISM conditions and on small grains, the Master Equation approach [11] has to be used.

Let N_H be the population of hydrogen atoms on a single interstellar grain and R_{H_2} the rate of H_2 molecules produced and immediately released in the gas phase. Temporal variation of the population of H adatoms on the grain surface and the subsequent rate of H_2 formation are given by the two equations: $\dot{N}_H = FAS - pN_H - \alpha {N_H}^2$ and $R_{\mathrm{H}_2} = \frac{1}{2}\,\alpha {N_H}^2$, where $F = n_{\mathrm{H}} v_{\mathrm{H}}$ is the flux of H atoms; p the probability of desorption in the unit of time of hydrogen as atoms, i.e. it is equal to the inverse of their residence time on the surface, so that $p = 1/t_{\mathrm{H}}$. The first term represents a source term of H atoms landing from the gas phase; the second term is the loss of H adatoms due to their evaporation (still as atoms) and the third term represents the rate at which H adatoms are lost due to the recombination into H_2 and its immediate release in the gas phase.

The steady state population of H adatoms on the grain, when $\dot{N}_H = 0$, is given by $N_H = [-p + (p^2 + 4\alpha FAS)^{1/2}]/(2\alpha)$ and the production rate of H_2 from a single grain is

$$R_{\mathrm{H}_2} = \frac{p^2 - p(p^2 + 4\alpha FAS)^{1/2} + 2\alpha FAS}{4\alpha}. \tag{1}$$

This expression represents the "exact" solution of the problem of H recombination on grains under the given conditions and yields the production rate of molecular hydrogen per unit time and volume once it is multiplied by the grain number density n_g. Two limiting cases are particularly interesting. The first case is when the adatom evaporation rate (as atoms) is negligible compared to their recombination rate on the surface. This is the limit of $p^2 \ll 2\alpha FAS$. In this case it is easily obtained the expression deduced by Hollenbach et al. [12] in their seminal paper. In the other limit, $p^2 \gg \alpha FAS$, when evaporation of hydrogen atoms is not negligible, the expression proposed by Pirronello et al. [2] based on the analysis of the experimental results summarized previously is recovered. The same two limiting cases are very instructive for the interpretation of the first and second order desorption observed experimentally.

To evaluate the recombination rate in interstellar applications the numerical values of α and t_H are necessary and they have been obtained by Katz et al. [13] by solving numerically the same equations in the non steady state case. They found that a good fit of desorption spectra was possible only by dropping the assumption that all H_2 molecules are immediately released in the gas phase upon formation and allowing a fraction μ of the molecules that are formed to remain on the surface. The parameters of the simulation are μ and the energy barriers for desorption and mobility of H and H_2. The results obtained by Katz et al. [13] allow quantitative applications of the experimental measurements to astrophysically relevant situations and are summarized in Table 1.

Table 1. Diffusion and desorption barriers in meV

Material	E_{diff}	$E_{H_{des}}$	$E_{H2_{des}}$	μ
Olivine	24.7	32.1	27.1	.33
Amorphous Carbon	44.0	56.7	46.7	.42

The strong temperature dependence in the evaporation rate of physisorbed H atoms (the only one considered in our experiments) at temperatures higher than 20-25 K limits the efficiency of molecular hydrogen formation by the investigated mechanisms to dense and diffuse clouds. For those interstellar environments where dust temperature is high, like in Photon Dominated Regions, it is imperative to consider the involvement of chemisorption sites where H atoms may dwell to high temperature of the solid and render again efficient the L-H and the Hot Atom mechanism.

4 Carbon Dioxide Formation

An example of a process in which the surface is actively involved is that of the formation of carbon dioxide by the association of CO molecules and O atoms on a water ice surface [6]. CO_2 has been observed by ISO (Infrared Space Observatory) in a variety of environments [14, 15]. Its high abundance in the solid phase [16], and its very low abundance in the gas-phase, suggest that CO_2 is formed almost exclusively in the solid phase and that it might be a sensitive probe of gas-grain chemistry. Several experiments showed that CO_2 formation occurs in ices containing CO pure or mixed with H_2O under UV or particle irradiation [17, 18, 19, 20]. However, ISO observations indicate that CO_2 is abundant also in quiescent dark clouds, such as toward Elias 16 [21], where UV and particle fluxes are negligible. Therefore there must be a mechanism that is able to produce CO_2 even in the absence of ionizing radiation as foreseen by Tielens & Hagen [22].

Fournier et al. [23] and Grim & d'Hendecourt [24] irradiated with UV photons mixtures of CO and either N_2O or O_2 at low temperatures in order to produce atomic oxygen, reaching opposite conclusions: activationless formation and no formation of carbon dioxide. Roser et al. [6] carried out experiments on carbon dioxide formation in the same ultra-high vacuum (UHV) apparatus described above. The ^{18}O beam was produced by radio-frequency dissociation of $^{18}O_2$. Isotopically labeled $^{13}C^{16}O$ was introduced into the UHV chamber concurrently with O atom irradiation of the sample in order to obtain an intimate mixture in the solid phase.

CO_2 formation was investigated over a range of CO/O abundance ratio from 5.6 to 21 and with a CO-ice thickness of $\sim$ 100 layers. The dilution of O in CO has been chosen to minimize the probability of undesired interaction between two oxygen atoms or between an oxygen atom and the undissociated O_2. We tuned the detector to mass 47 to measure the amount of $^{13}C^{16}O^{18}O$ formed by the interaction of $^{13}C^{16}O$-ice and ^{18}O atoms.

In a first set of experiments, after each deposition at about 5 $\sim$ K, a TPD was carried out. In a second set of experiments a water-ice amorphous cap of $\sim$ 100 layers was placed on top of the O and CO ice. In the experiments done without the water-ice cap no detectable amount of CO_2 was produced. CO and O sublimed completely showing the existence of a reaction energy barrier, perhaps only slightly higher than the desorption one.

In the second set of experiments, done with the water-ice cap, the signal of $^{13}C^{16}O^{18}O$ was detected when the substrate temperature was high enough for the sublimation of the water-ice layer (above 150 K).

The formation of CO_2 most probably occurs during the migration of CO and O through the amorphous water-ice layer, but CO_2 molecules (larger than CO and O), once formed, remain trapped in the water-ice until the whole ice sublimes, an effect that had already been observed previously by Pirronello et al. [25]. In the range of relative abundances investigated, the yield of carbon dioxide is roughly proportional to the oxygen dose. The fact that in these experiments the O abundance constitutes the rate limiting factor in the formation of CO_2 is in agreement with observations [14, 26].

To gain insights in the kinetics of CO_2 formation we built a simple model that takes into account the migration and escape of CO and O from the water ice during warm-up. To model the process it was assumed that the reaction $CO + O \longrightarrow CO_2$ takes place during the diffusion of both CO and O in the amorphous water-ice cap and that such a diffusion can be regarded as a sequence of desorptions from and adsorptions on sites in the inner walls of cavities in the amorphous ice cap. In this model the barriers to obtain carbon dioxide molecules are simply the desorption energy barrier and the formation barrier, if any. For CO the desorption barrier from an adsorption site in a cavity is roughly equal to 1740 K, which is the measured surface binding energy for CO on water-ice [27]. The desorption barrier for O atoms has been evaluated to be around 800 K [22]. Integrating the set of differential equations that describe the mobility of CO and O and the recombination probability when they encounter in the same adsorption site we obtained a value of $\sim$ 290 K for the reaction energy barrier.

The presence of the water-ice cap certainly allows to maintain CO and O in the solid to higher temperatures but also, as already noticed by Fournier et al. [23], the icy matrix allows the otherwise forbidden transition to occur. Using the activation energy barrier, E_a=290 K, it is possible to estimate the timescale to produce a carbon dioxide molecule when an O atom from the gas phase lands on a site of the ice mantle occupied by a CO molecule or vice versa. Such a timescale is of the order of a second at 10 K and is already negligible at grain temperatures of 15 K - that is, close to the temperature values of the ice mixture in quiescent clouds toward Elias16 [21] and in Sgr A [14]. Such timescales are short enough to allow a large fraction of solid CO to be converted into CO_2 in mantles on grains of quiescent clouds, without the intervention of energetic agents in agreement with Gerakines et al. [14].

5 Summary

We have described the importance of surface reactions to form key species that are difficult to be produced by gas phase reactions in interstellar conditions. We focused on the formation of molecular hydrogen and carbon dioxide and presented both experimental surface science methods to gain insight in the processes that actually occur and theoretical methods to apply the results obtained in the laboratory in astrophysically relevant conditions.

6 Acknowledgments

V.P. was supported by the Italian Ministry for University and Scientific Research (MIUR) through grant 21043088 and by the Agenzia Spaziale Italiana (ASI). G.V. was supported by NASA through grants NAG5-4998, NAG5-6822 and NAG5-9093.

References

1. V. Pirronello, C. Liu, L. Shen, G. Vidali: ApJ **475**, L69 (1997a)
2. V. Pirronello, O. Biham, C. Liu, L. Shen, G. Vidali, G.: ApJ **483**, L131 (1997b)

3. V. Pirronello, C. Liu, J.E. Roser, and G. Vidali: A&A **344**, 681 (1999)
4. G. Manicó, G. Ragunì, V. Pirronello, J.E. Roser, G. Vidali: ApJ **548**, L253 (2001)
5. J.E. Roser, G. Manicó, V. Pirronello, G. Vidali: ApJ **581**, 276 (2002)
6. J.E. Roser, G. Vidali, G. Manicò, V. Pirronello: ApJ **555**, L61 (2001)
7. O. Biham, I. Furman, N. Katz, V. Pirronello, G. Vidali: MNRAS **296**, 869 (1998)
8. G. Vidali, V. Pirronello, C. Liu, L.Y. Shen: Astrophys. Lett. Comm. **35**, 423 (1998)
9. O. Biham, V. Pirronello, G. Vidali: Chemical Reactions on Solid Surfaces of Astrophysical Interest. In: *Solid State Astrochemistry*, ed by V. Pirronello et al., (Kluwer, Dordrecht 2003) pp 211-250
10. R.J. Gould, E.E. Salpeter: ApJ **138**, 393 (1963)
11. O. Biham, I. Furman, V. Pirronello, and G. Vidali: ApJ **553**, 595 (2001)
12. D. Hollenbach, M.W. Werner, and E.E. Salpeter: ApJ **163**, 165 (1971)
13. N. Katz, I. Furman, O. Biham, V. Pirronello, and G. Vidali: ApJ **522**, 305 (1999)
14. P.A. Gerakines et al.: ApJ **522**, 357 (1999)
15. A.C.A. Boogert, et al.: A&A **353**, 349 (2000)
16. E.F. van Dishoeck, et al.: A&A **315**, L349 (1996)
17. L.B. d'Hendecourt, L.J. Allamandola, R.J.A. Grim, and J.M. Greenberg: A&A **158**, 119 (1986)
18. R.J.A. Grim, J.M. Greenberg, M.S. de Groot, F. Baas, W.A. Schutte, and B. Schmitt: A&A Supp. Ser. **78**, 161 (1989)
19. P.A. Gerakines, W.A. Schutte, P. Ehrenfreund: A&A **312**, 289 (1996)
20. M.H. Moore, R. Khanna, and B. Donn: J. Geophys. Res. **96**, 17541 (1991)
21. D.C.B. Whittet et al.: ApJ **498**, L159 (1998)
22. A.G.G.M. Tielens and W. Hagen: A&A **114**, 245 (1982)
23. J. Fournier, J. Deson, C. Vermeil, and G.C. Pimentel: J. Chem. Phys. **70**, 5726 (1979)
24. R.J.A. Grim, and L.B. d'Hendecourt: A&A **167**, 161 (1986)
25. V. Pirronello, W.L. Brown, L.J. Lanzerotti, at al.: ApJ **262**, 636 (1982)
26. J.E. Chiar: Orig. Life Evol. Biosph. **27**, 79 (1997)
27. S.A. Sandford, and L.J. Allamandola: Icarus **76**, 201 (1988)

Far–IR Spectroscopy Towards Sagittarius B2

J.R. Goicoechea[1] and J. Cernicharo[2]

[1] Departamento de Astrofísica Molecular e Infrarroja, IEM/CSIC, Serrano 121, 28006 Madrid, Spain. javier@damir.iem.csic.es
[2] Departamento de Astrofísica Molecular e Infrarroja, IEM/CSIC, Serrano 121, 28006 Madrid, Spain. cerni@damir.iem.csic.es

The far–IR is a unique wavelength range for Astrophysical studies, however, it can only be fully sampled from space platforms. The fundamental rotational transitions of light molecules, the high–J transitions of polyatomic species, the bending modes of non-polar molecules, several atomic fine structure lines and many frequencies blocked by the earth atmosphere can only be observed between 50 and 200 μm (6.0 and 1.5 THz). In this contribution we present the far–IR spectrum of Sgr B2 at a resolution of ~35 km s^{-1}, the *"Rosetta stone"* of ISO's spectra. We also discuss the perspectives of the far–IR Astronomy in the context of the future telescopes under development.

1 Introduction

Due to the atmospheric opacity, the far–IR domain has been the last window used in Astronomy. For the first time, the *Infrared Space Observatory* (ISO) has opened this spectral frequency range through *molecular spectroscopy*. The sensitivity of the instrumentation on board this platform has no comparison with the few previous space missions or airborne observations carried before the launch of ISO. Almost all the operative range of ISO in the far–IR was not explored before. The far–IR spectrum of the most significant galactic sources was unknown. In particular, the main radiation emitters, the molecules, were unidentified. Far–IR observations are specially suitable to the study of the *warm gas* in molecular clouds. Among these sources, Sgr B2 in the Galactic Center, is a paradigmatic object for our knowledge of the *chemical complexity* of the Galaxy.

The molecular species and the atomic fine structure lines that can be detected in the far–IR domain are essential for the study of the physical and chemical conditions of the interstellar medium. The bulk of these species can only be observed from space platforms. As a example, the water vapor abundance can determine if stars will be formed during the gravitational collapse of a molecular cloud. Another example are the non–polar carbon chains. These species can be the *"skeletons"* from which the large carbon molecules responsible for a great part of the IR emission in the Galaxy can be formed. Due to the lack of permanent electric dipole, these species do not have a rotational spectrum to be observed from radio telescopes.

In this contribution we present the main results of our detailed study of the *Long–wavelength spectrometer* (LWS) spectrum of Sgr B2(M) between 43 μm (7.0 THz) and 197 μm (1.5 THz). Both with the grating ($\lambda/\Delta\lambda\sim200$) and with the Fabry–Perot (FP; $\lambda/\Delta\lambda\sim10000$).

2 The Far–IR Spectrum of Sgr B2(M)

Figure 1 shows the grating and FP observations of Sgr B2(M). The far–IR spectrum of Sgr B2 is dominated by the high thermal emission of the dust ($\sim$28000 Jy at $\sim$80 μm [$\sim$3.75 THz]) and by the molecular/atomic lines.

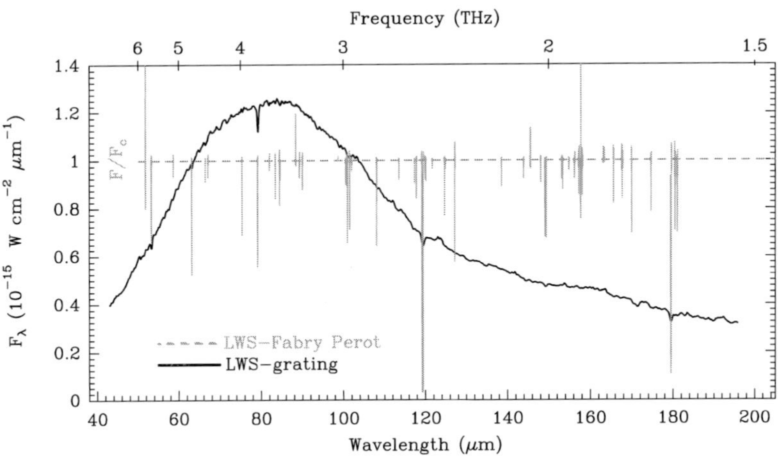

Fig. 1. LWS/grating spectrum of Sgr B2(M) (α=17^{h}44^{m}10.61^s, δ=-28°22$'$30.0$''$ [J1950]) between 43 and 197 μm and main detections with the LWS/FP. Ordinate is valid for the grating (in W cm^{-2} μm^{-1}) and for the FP observations (in F/F$_c$).

Figure 2 shows the most abundant species that can be detected with ISO in the far–IR (see Goicoechea et al. 2004) and gives insights of which could be detected in other molecular clouds of the interstellar medium (ISM).

The molecular richness in the outer layers of Sgr B2 is probed by the FP detections towards Sgr B2(M), where more than 70 lines from 15 molecular and atomic species are observed at high signal to noise ratio.

The spectral lines that appear in the far–IR and that have been observed in Sgr B2 can be classified in:

- **Rotational lines** of light *O–bearing* molecules such as H_2O, $H_2^{18}O$, OH, ^{18}OH, and H_3O^+, *N–bearing* molecules such as NH, NH_2 and NH_3 and other diatomic species such as CH, HD or HF.
- **Bending modes** of non-polar species such as the C_n carbon chains.
- **Atomic fine structure lines** of [O I], [O III], [C II], [N II] and [N III].

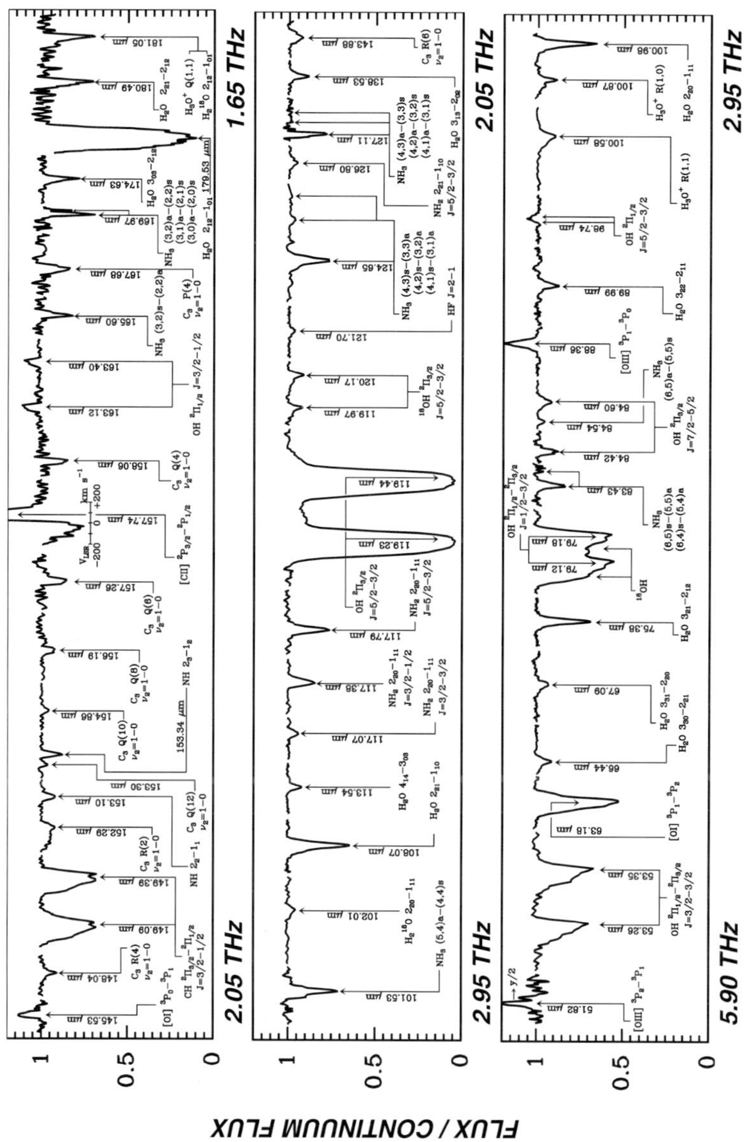

Fig. 2. Far–IR features in the spectrum of Sgr B2(M) taken with the ISO/LWS Fabry-Perot. Note the wavelength discontinuity of the spectrum after each line.

The far–IR observations have offered the opportunity of studying the chemistry of *O–* and *N–bearing* species through the $H_3O^+/H_2O/OH/O$ (Goicoechea & Cernicharo 2001, 2002) and $NH_3/NH_2/NH$ observations. In addition, the abundance of non–polar molecules such as C_3 have been determined for the first time (Cernicharo et al. 2000). Finally, the influence of the UV radiation field in the external layers of Sgr B2 (the environment of the cloud) has been shown through the analysis of the atomic fine structure lines

(Goicoechea et al 2004). All these new scientific goals can be obtained only through space observations in the far–IR spectral range.

3 Future Perspectives

The spatial scales of the physical and kinetical phenomena within molecular clouds and the amount of molecular species that could be present, claim for a much better sensitivity and spectral/spatial resolution in the far–IR. In the next years, most of these handicaps will be overcome by *ALMA* (∼2010) for interferometric observations with $\lambda > 350$ μm. On the other hand, the *Herschel Space Telescope* (∼2007) will offer much better spectral resolution than ISO for $\lambda > 157$ μm. However, the spatial resolution will be similar to that of current millimeter single–dish telescopes.

In any case, the full spectral window between 50 and 200 μm (6.0 and 1.5 THz) is not going to be sampled by any planned space mission in the next decade. Taking into account all the background learned from the ISO/LWS observations (which is only the top of the iceberg), the request of high spectral and spatial resolution is mandatory in order to understand and get deeper into the unique phenomena occurring in the far–IR domain. The high spectral resolution (< 1 km s^{-1}) can be achieved by a single-dish heterodyne instrument, while the high spatial resolution can only be achieved with a large single dish telescope with direct/heterodyne detection techniques or with a heterodyne space interferometer.

Acknowledgements

We thank Spanish DGES and PNIE for funding support under grant PANAYA 2000–1784 and ESP2001-4516. We also thank Nemesio J. Rodríguez-Fernández for stimulating discussions about Sgr B2 and the Galactic Center.

References

1. Cernicharo, J., Goicoechea, J.R., & Caux. E. 2000, *ApJ*, 534, L199
2. Goicoechea, J.R., & Cernicharo, J. 2001, *ApJ*, 554, L213
3. Goicoechea, J.R., & Cernicharo, J. 2002, *ApJ*, 576, L77
4. Goicoechea, J.R., Rodríguez-Fernández, N.J., & Cernicharo, J. 2004, *ApJ*, in press. astro-ph/0309353

H_3^+ Line Survey Towards the Galactic Center

M. Goto[1,2], B.J. McCall[3], T.R. Geballe[4], T. Usuda[2], and T. Oka[5]

[1] IfA, University of Hawaii mgoto@duke.ifa.hawaii.edu
[2] Subaru Telescope mgoto@duke.ifa.hawaii.edu
[3] University of California, Berkeley bjmccall@astron.berkeley.edu
[4] Gemini Observatory tgeballe@gemini.edu
[5] University of Chicago t-oka@uchicago.edu

1 H_3^+ Survey Towards the Galactic Center

The H_3^+ in the interstellar medium is a unique probe of physical and chemical environments in dense and diffuse clouds. As the observational study in many lines of sight is becoming matured, we have as many problems newly found. The overabundance of H_3^+ in diffuse clouds is one of them [3, 4]. We undertook the absorption line survey of H_3^+ toward the Galactic center sources to shed light on it [1].

The Galactic center sources are ideal line of sight to start the survey since they suffer heavy visual extinction of A_V=25–40, the highest in the Galaxy among those obscured by diffuse clouds. The intervening clouds have different radial velocities, and the nature of them have been studied in previous spectroscopic observations from near-infrared to radio wavelengths. We observed two infrared sources, GCS 3-2 in the Quintuplet cluster and GC IRS 3 near Sgr A*. Both have intrinsically featureless spectra, and sufficiently luminous to provide good continuum fluxes in the 3 μm region.

The observations were made in June 2001 using the Infrared Camera and Spectrograph (IRCS; Tokunaga et al. 1998; Kobayashi et al. 2000) at the 8.2 m Subaru Telescope. Subaru IRCS is equipped with an echelle and a cross-dispersing gratings. The high resolution (R=20,000) spectroscopy with wide wavelength coverage is unique characteristic of the instrument. It allows us to record one third of 3.2–4.0 μm with a single exposure. This makes the system very suitable for any line-survey projects, in particular for H_3^+ vibrational transition lines that scatter all over the 3 μm region.

2 Result and Discussion

2.1 Velocity Components

All transitions within our spectral coverage starting from the ground states $(J, K) = (1, 0)$ and $(1, 1)$ are successfully detected (Figure 1). At least four velocity components are resolved in both GCS 3-2 and GC IRS 3 (Figure 2).

The bluest component of GC IRS 3 at -140 km s^{-1} is attributed to the expanding molecular ring [7]. The expanding molecular ring is a chain of the

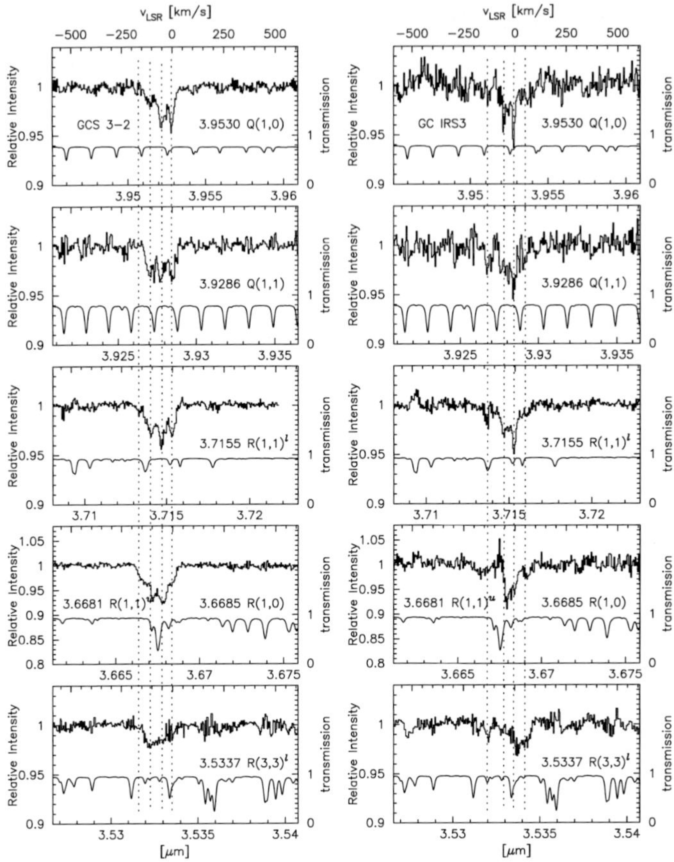

Fig. 1. H_3^+ absorption lines toward GCS 3-2 (right) and GC IRS 3 (left).

molecular clouds that rotates around the nucleus at 200 pc away from the Galactic center, slowly receding from it. The same component appears at -110 km s^{-1} in the line of sight toward GCS 3-2.

The two absorption minima at -60 km s^{-1} and 0 km s^{-1} toward GC IRS 3 closely match with those of H_2CO in the same line of sight. The HI spectrum toward Sgr A shows self-absorption at the same velocities. It indicates that those H_3^+ absorption to GC IRS 3 are originated from diffuse interstellar clouds. We attributed the -60 km s^{-1} component to the clouds in the 3 kpc arm because the radial velocity is consistent. The signature of the 3 kpc arm also appears at similar velocity in absorption in GCS 3-2.

The 0 km s^{-1} components are usually attributed to the local clouds within a few kps from the solar system. However, in contrast to the absorption features at 0 km s^{-1} ubiquitously seen in the HI 21 cm spectroscopy, the H_3^+ absorption at 0 km s^{-1} with this intensity has not been observed in many lines of sight. There could be a contribution from the low-velocity clouds close to the Galactic center region. Some of such clouds have been reported by [9] to be within about 50 pc of Sgr A.

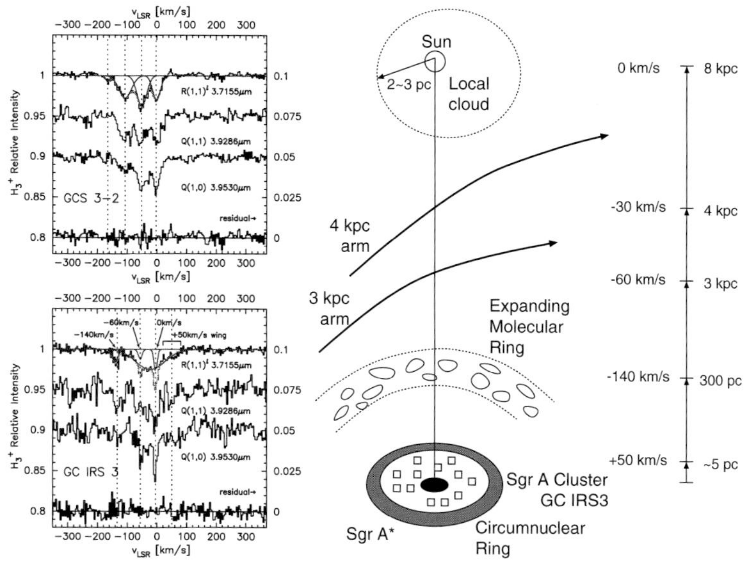

Fig. 2. Velocity components and their attribution.

2.2 Metastable H_3^+

The highlight of the present observation is the first detection of the absorption line starting from a metastable state $(J, K) = (3, 3)$. The only absorption line we detected, and is not starting from the ground state (1, 1) or (1, 0), is $R(3, 3)^l$ at 3.534 μm. The rotationally excited state (3, 3) is special because spontaneous emission from this level to lower levels is forbidden by the ortho–para selection rule. [6] predicted (3, 3) level of interstellar H_3^+ is excessively populated. Other transition lines starting from $J > 1$ levels, $R(2, 1)^u$, $R(2, 2)^l$, $Q(2, 1)^l$, and $Q(3, 0)$, are also covered by our observation, but end up with negative detection. This clearly demonstrates the metastability of (3, 3) level, and theoretical prediction of [6].

However, the problem is that the kinematic temperature of the surrounding molecular hydrogen should be very hot in order to collisionally excite the H_3^+ to the (3, 3) level [5]. The (3, 3) level is higher than the lowest (1, 1) level by 361 K. The $R(3, 3)$ and $R(1, 1)$ have almost similar absorption depths and transition dipole moments. The kinetic temperature of hydrogen should be over 200 K to populate (3, 3) in the same extent with (1, 1). However, large population of clouds with so much warm molecular gas has not been recognized to date.

The nature of the warm cloud is still in mystery, however, a hint of it could be found in the comparison of the line profiles of H_3^+ and CO (Figure 3). The line profile of H_3^+ starting from (1, 1) and metastable state (3, 3) are apparently different. The velocity components of CO ν=2–0 and H_3^+ (1, 1) show reasonable agreement, and most likely two molecules are physically associated. However, a quantitative difference may be noticeable. The H_3^+ (1, 1) has broad and strong pedestal component, while the CO trace goes back to the continuum level in between the absorption lines. The last panel in Figure 3 is the differential spectrum of H_3^+(3, 3) and H_3^+(1, 1) overlaid with CO ν=2–0. The match between the two spectra is convincing. Our working hypothesis is that the CO ν=2–0 traces cold (10 K) and dense molecular cloud, while H_3^+(3, 3) traces hot (200 K) diffuse cloud, *exclusively*. Only H_3^+ (1, 1) absorption occurs in both of them, producing sharp absorption lines superposed on the broad pedestal absorption. The hypothesis provides *ad-hoc* accounts for the differences in the line profiles, however, we need more observations with higher spectral resolution to prove the presence of interstellar clouds as warm as about 200 K.

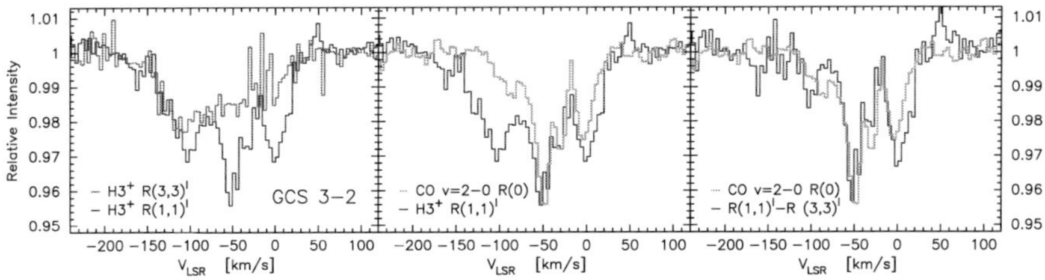

Fig. 3. Comparison of absorption line profiles.

References

1. Goto, M., McCall, B. J., Geballe, T. R., Usuda, T., Kobayashi, N., Terada, H., & Oka, T. 2002, PASJ, **54**, 951
2. Kobayashi, N., et al. 2000, Proc. SPIE, **4008**, 1056
3. McCall, B. J., et al. 2002, ApJ, **567**, 391
4. McCall, B. J., et al. 2003, Nature, **22**, 500
5. Oka, T., & Epp 2003, in this proceedings.
6. Pan, F.-S., & Oka, T. 1986, ApJ, **305**, 518
7. Scoville, N. Z. 1972, ApJ, **175**, L127
8. Tokunaga, A. T., et al. 1998, Proc. SPIE, **3354**, 512
9. Whiteoak, J. B., & Gardner, F. F. 1978, Proc. Astron. Soc. Australia. **3**, 266

A Far UV Spectroscopic Study of the HD 34078 Sightline

F. Le Petit[1], P. Boissé[2], E. Rollinde[3], E. Roueff[4], G. Pineau des Forêts[5], C. Gry[6], and P. Felenbok[7]

[1] LUTH – Observatory of Paris Franck.LePetit@obspm.fr
[2] ENS/LERMA – Observatory of Paris Boisse@lra.ens.fr
[3] Institut d'Astrophysique de Paris Rollinde@iap.fr
[4] LUTH – Observatory of Paris Evelyne.Roueff@obspm.fr
[5] IAS – Université d'Orsay – France Guillaume.PineaudesForets@obspm.fr
[6] ESA – Vilspa – Spain cgry@iso.vilspa.esa.es
[7] LESIA – Obseravtory of Paris Paul.Felenbok@obspm.fr

1 Introduction

Five FUSE observations towards the runaway star HD 34078 have been performed between January 2000 and October 2002 to study small scale structure in the spatial distribution of H_2. The spectra turned out to be exceptionally rich since for the first time H_2 is observed from its ground state up to its J = 11 level. Other species (HD, C I, CO, Ar I, ...) are clearly detected. We also observed this line of sight (LOS) at the CFHT in order to search for OH.

2 Description of the Line of Sight

HD 34078 (AE Aurigae) is a O9.5 Ve star of visual magnitude $m_V = 5.95$ located at 500 pc from the Earth. It is a runaway star with a transverse velocity of 100 km·s^{-1}. The color excess has been estimated to 0.52, corresponding to $A_V = 1.8$ ($R_V = 3.42$, Patriarchi et al. [8]). Many molecules have been previously detected on this LOS. H and CO have been detected by McLachlan and Nandy [4], CH and CH^+ by Rollinde et al. [9], CN, C_2 by Federman [2] and C_3 by Oka et al. [7].

3 Data Analysis and Results

3.1 FUSE Observations

Part of the first FUSE spectrum is presented on Fig.1. The S/N ratio is 30 per 15 mÅ pixel. CH observations at high resolution show only one major molecular component (Rollinde et al. [9]). Column densities have been measured using MIDAS and the fitting program *Owens* developed by M. Lemoine (IAP). Detailed information concerning the analysis method can be found in Boissé et al. [1]. 9 giving a b value between 4 and 6 km s^{-1}. Since b(CH) = 3 km s^{-1}, we adopted this value for J = 3 and increased b up to 5 km s^{-1} with J (up to J = 9). Due to the imprecise knowledge of b we take into account in the errorbars an incertitude of ±1 km s^{-1} and give them at 3σ. For levels higher than J = 9, it is possible to find in the spectrum opticaly thin lines and so to deduce column densities from the equivalent width.

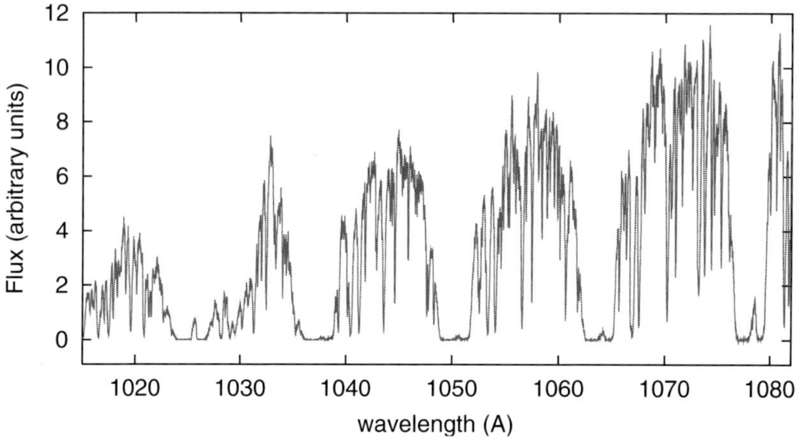

Fig. 1. January 2000 FUSE spectrum towards HD 34078, detector LiF1A. Regions with no signal correspond to absorption of H_2 in levels J = 0, 1, 2. Most of the other lines correspond to excited H_2.

The implied total column density of H_2 is: $N(H_2) = 6.4\ (\pm 0.4) \times 10^{20}$ cm^{-2}. Column densities for individual levels are summarized in the excitation diagram (Fig. 2). The temperature derived from J = 0 and 1 levels is $T_{01}(H_2)$ = 77 K.

Due to the numerous lines in the spectrum and to the resulting overlap, the column density of HD J = 0 is difficult to determine. The only two suitable lines give two inconsistent values. On the other hand, the 7-0 R(1) HD line at 1022.62 Å is clearly detected: N(HD, J=1) = 5.4×10^{13} cm^{-2}. This line has been previously observed by Wright and Morton [10] towards ζ Oph.

Strong H I Lyman β absorption is present in our FUSE spectra. The derived column density is 2.7×10^{21} cm^{-2}, a value significantly higher than the one derived from an IUE spectrum taken in 1979, N(H I) = 2.0×10^{21} cm^{-2} (value obtained from a re-analysis of IUE spectrum and slightly higher than the one given by Mc Lachlan and Nandy [4]).

Lines of C I in its three fine structure levels appear clearly in the spectrum above 1100 Å. Corresponding column densities are 9.4, 5.8 and 2.2×10^{15} cm^{-2}.

3.2 CFHT Observations

CFHT observations have been performed towards HD 34078 with the Coudé f/4 spectrometer Gecko with a nominal resolution of 120 000. The three expected optically thin lines of OH at 3072, 3075, 3081 Å are detected and give N(OH) = $3.5(\pm 2.1)\times 10^{13}$ cm^{-2}.

4 Modelling and Discussion

Observations towards HD 34078 can be summarized as: 1) column densities and $T_{01}(H_2)$ are typical of what can be found in standard transluscent LOS, 2) H_2 is unusually excited on this LOS. This suggests the contribution of two components: 1) a standard diffuse component, not affected by the star and containing most of the molecular species, 2) some H_2, UV-pumped by the HD 34078 radiation field. According to Mac Low et al. [5], H_2 should exist in a layer of the bow shock surrounding runaway stars such as HD 34078 (comparable to a PDR).

Using our numerical model of interstellar chemistry (Le Bourlot et al. 1993, Nehmé et al. in preparation), we modelled these two components (Trans. cloud: n_H = 700 cm^{-3} in agreement with C I excitation, χ = ISRF, T = $T_{01}(H_2)$; PDR: $n_H = 10^4$ cm^{-3}, $\chi = 10^4\times$ ISRF). An additional MHD C-shock (v = 25 km s^{-1}, pre-shock density = 20 cm^{-3}, B = 7μG) is also required to reproduce N(CH^+). It does not contribute significantly to the H_2 excitation. Results are in good agreement with observations (comparison between computed and observed column densities can be found in Boissé et al. [1]). The computed H_2 excitation diagram is presented in Fig. 2.

This modelling work shows that most of the molecules are produced in the transluscent component and that illumination by HD 34078 UV radiation of dense gas surrounding the O star satisfactorily reproduces the H_2 excitation above J = 2 (considering the limited number of free parameters). However, some observations suggest that the diffuse cloud and PDR component are not unrelated; this needs to be clarified.

5 Small Scale Structure

The 5 FUSE spectra spanning over a time of 2.7 yr have been compared. The largest separation probed is about 45 AU.

• Comparison of H_2 lines among successive spectra.
No N(H_2) changes above 5% are detected from the comparison of damped line profiles. Then, the medium probed does not present ubiquitous H_2 small scale structure such as the one seen in H_2CO, OH or in the atomic phase through H I 21 cm absorption. Moreover, since column densities of H_2 in excited levels (assumed to trace the bow shock) do not present fluctuations

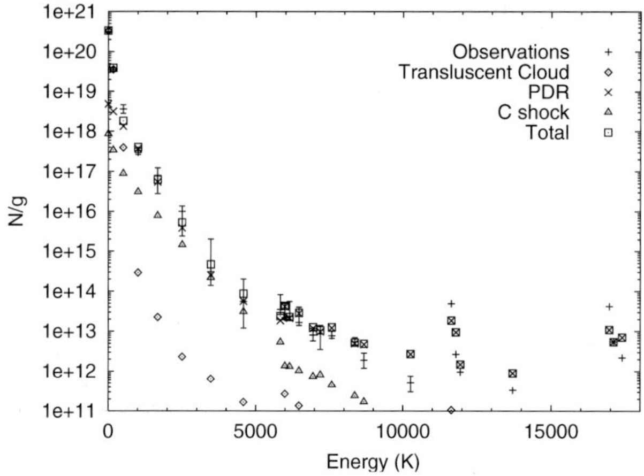

Fig. 2. H_2 excitation diagram towards HD 34078. Observations without errorbars correspond to upper limits. The PDR component correspond to the H_2 in the bow shock. Total corresponds to the addition of the two components plus the C-shock.

either, one can conclude that the medium around HD 34078 is relatively smooth and that no instability develops at the shocked boundary.

• **Comparison of Lyman β absorptions.**
No variations of N(HI) higher than 10% are detected between the five successive FUSE spectra. On the other hand, the FUSE value (log(N(HI)) = 21.43) is larger than IUE value (21.30). This corresponds to an increase of N(HI) of 1.8% per year between 1979 and 2002 (for an assumed constant increase rate). Such a variation is comparable to the variation of N(CH) on this line of sight (Rollinde et al. [9]).

References

1. P. Boissé, F. Le Petit, E. Rollinde, et al., to be submitted, A&A
2. S. R. Federman, C. J. Strom, D. L. Lambert,et al: ApJ, **424**, 772 (1994)
3. J. Le Bourlot, G. Pineau des Forêts, E. Roueff, D. Flower: A&A **267**, 233, (1993)
4. A. Mc Lachlan and K. Nandy: MNRAS, **207**, 355 (1984)
5. M. Mac Low, D. van Buren, D. Wood and E. Churchwell E: ApJ **369**, 395 (1991)
6. C. Nehmé, F. Le Petit, P. Hily-Blant et al., A&A, in preparation
7. T. Oka, J. A. Thorburn, B. J. McCall et al.: A&A, **372**, 644., (2003)
8. A. Patriarchi, L. Morbidelli, M. Perinotto and G. Barbaro: A&A **372**, 644 (2001)
9. E. Rollinde, P. Boissé, S. R.Federman, K. Pan: A&A, **401**, 215 (2003)
10. E. L. Wright and D. C. Morton: ApJ **227**, 483 (1979)

Infrared Spectroscopy of Ionic PAHs and Related Compounds

J. Oomens[1], G. von Helden[2], and G. Meijer[1,2]

[1] FOM Institute for Plasma Physics "Rijnhuizen", Edisonbaan 14, NL-3439MN Nieuwegein, The Netherlands joso@rijnh.nl

[2] Fritz-Haber-Institut der Max-Planck-Gesellschaft, Faradayweg 4-6, D-14195 Berlin, Germany

Despite the fact that the interstellar Unidentified InfraRed emission bands (UIRs) are now widely accepted to be due to emission from (ionic) gas-phase polycyclic aromatic hydrocarbons (PAHs)[1, 2], close inspection of the interstellar spectra reveal many unexplained issues[3]. As to the carriers of the bands, various possible (PAH-related) systems have been suggested besides PAHs[4] and many of them await laboratory verification. In addition, detailed spectral information on highly excited PAH species[5] is very limited.

In recent years, we have developed a method to investigate the infrared spectroscopy of relatively large molecular ions in the gas phase, which has been applied to a variety of ionic PAHs up to the size of coronene[6]. The method employs infrared multiple photon dissociation (IRMPD[7]) spectroscopy of the ions stored in an ion trap[8], using the intense narrow-band infrared radiation from a free electron laser (FEL)[9]. Here we show how this method may help to understand details of the spectra of ionic PAHs and hence of the UIR spectrum. In particular, information is obtained on anharmonic shifts of highly excited PAHs and infrared spectra of PAH derivatives are investigated.

1 Vibrational Anharmonicity

Originating from highly vibrationally excited species, the UIR spectrum is affected by the anharmonicity of vibrational potentials[10, 11, 12]. Nonetheless, very little information on vibrational anharmonicity of PAH species, neither experimental nor theoretical, has been reported to date. IRMPD spectroscopy[7] involves the absorption of many, in our experiments typically hundreds of infrared photons. Hence, these spectra show the effects of anharmonicity, mainly in the form of red-shifted and broadened absorption bands. We have recently reported a detailed comparison of IRMPD spectra to linear absorption spectra[13, 14] to obtain an estimate of the anharmonic parameters of ionic PAHs[12].

Redshifts averaged over various bands of seven ionic PAHs have been determined for the three main mid-infrared spectral regions[12]. The redshift is approximately proportional to the band frequency, *i.e.* the shift in the CC

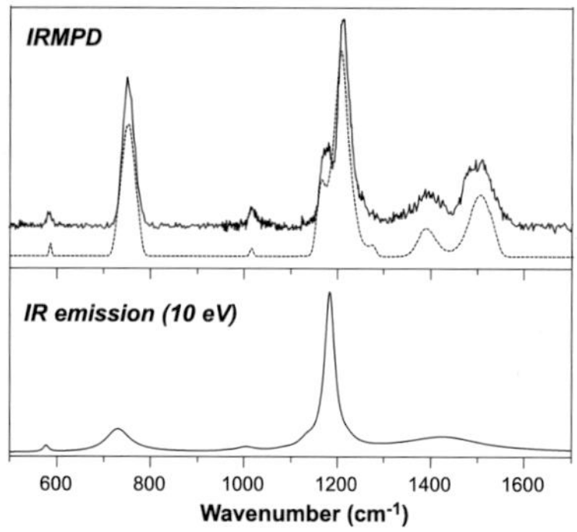

Fig. 1. Experimental IRMPD spectrum of the naphthalene cation (solid) along with a model fit of the data (dashed), using linear absorption data[14] as input. The lower panel shows the calculated emission spectrum using the anharmonicity parameters obtained from this fit.

stretching region is roughly twice that in the CH out-of-plane bending region. Secondly, substantially larger (relative) scattering is observed in the in-plane CH bending (7-9 μm) region than in the CH out-of-plane bending (11-13 μm) and CC stretching (6 μm) regions. Interestingly, also the band positions in the CH in-plane bending region show more scattering, whereas in the other two regions, they line up fairly well for the different PAH ions. Moreover, it may be noted that the UIR spectrum shows much broader features in the 7–9 μm range than in the other two spectral ranges.

It should be noted that multiple photon absorption, as employed in our experiments, and infrared emission, as occurring in the interstellar medium, are two fundamentally different processes. For instance, the overlap between the laser bandwidth and the molecular absorption band, of which the latter redshifts as the internal energy increases, determines the efficiency of the IRMPD process, whereas no such limitations apply to the emission process. Hence, the average band shifts mentioned above cannot be used (directly) to estimate redshifts in an emission spectrum.

Therefore, the IRMPD spectrum of the naphthalene cation has been modelled numerically[7, 12]. The model spectrum, which uses the linear absorption spectrum[14] as input, is fitted to the experimental IRMPD spectrum (Fig. 1) thereby yielding the mode specific anharmonicity parameters. These parameters are then used to compute an emission spectrum[11] of the naphthalene ion. A detailed account is given in [12].

2 PAH-Related Species

Virtually all experimental data on ionic PAHs available to date concern the cations of "regular", *i.e.* chemically stable and often commercially available,

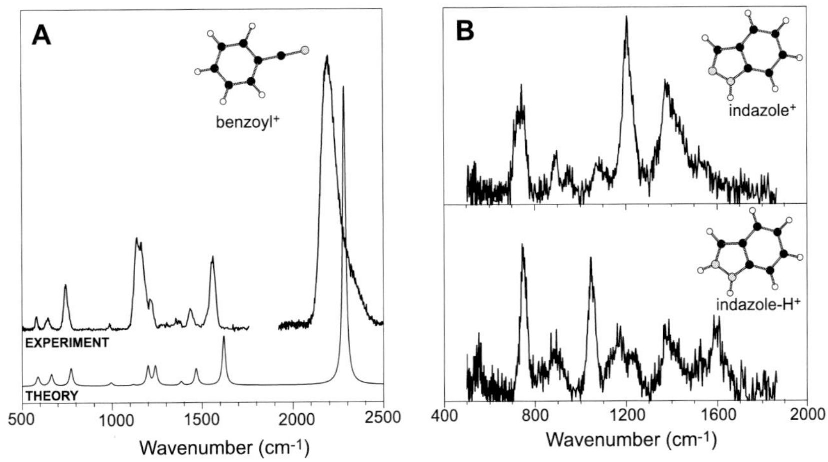

Fig. 2. IRMPD spectrum of the benzoyl cation compared to a density functional theory calculation (**A**) as well as spectra of cationic and protonated indazole (**B**). These spectra illustrate the possibilities of ion trap based IRMPD spectroscopy for the study of fragmented and protonated aromatic systems.

PAH molecules. However, depending on parameters such as UV flux and hydrogen density, it is likely that many more "exotic" PAH-like species occur in the interstellar medium. As examples of such species one could think of dehydrogenated, fragmented and hydrogenated PAHs. Contrary to the regular PAH radical cations, these species may possess closed-shell electronic configurations, which likely increases their stability. Indeed, several such systems have recently been suggested as possible UIR carriers[4] but none have been spectroscopically studied in the laboratory thus far.

With the use of mass-spectrometric techniques, it is possible to tackle the infrared spectroscopy of various of these PAH-related systems. Particularly, ion trap based techniques allow to isolate species of a specific mass out of a heterogeneous ensemble of *in-situ* generated ions. This strategy, which employs the mass-selective instability properties of the ion trap[8], has been used to record the IRMPD spectrum of the benzoyl cation (C_6H_5CO), one of the photo-fragments of benzoic acid[15]. The IRMPD spectrum of this closed-shell fragment ion is shown in Fig. 2.

Protonated aromatic systems[16] are another class of astrophysically relevant species that can be studied in an ion trap, as is exemplified by the spectrum of protonated indazole (indazolium). Ion-molecule reactions between neutral indazole ($C_7H_6N_2$) vapor and ionic fragments in the trap produce indazolium. Its spectrum is subsequently recorded simultaneously with the spectrum of the bare indazole cation, as shown in Fig. 2. The two spectra are clearly very different. Details will be given in a forthcoming publication.

The observed interstellar depletion of transition metals and the hypothesized occurrence of their adducts with aromatic hydrocarbons[17] render these systems interesting subjects for spectroscopic investigation. Various experimental approaches to study such systems have become available recently[18, 19, 20]. Among them, ion trap based techniques have the advantage of mass-selectivity prior to IR spectroscopic interrogation. Using a Fourier Transform mass spectrometer has the additional advantage of superior mass resolution and it provides the possibility to produce a wealth of metal-organics via ion-molecule reactions[20].

Acknowledgment. This work is part of the research program of FOM, which is financially supported by the Nederlandse Organisatie voor Wetenschappelijk Onderzoek (NWO).

References

1. A. Léger, J.L. Puget: A&A **137**, L5 (1984)
2. L.J. Allamandola, A.G.G.M. Tielens, J.R. Barker: ApJ **290**, L25 (1985)
3. D.A. Beintema et al., A&A **315**, L369 (1996); S. Hony et al., A&A **370**, 1030 (2001); E. Peeters et al., A&A **390**, 1089 (2002)
4. P. Ehrenfreund, S.B. Charnley, Ann. Rev. Astron. Astrophys. **38**, 427 (2000); L.W. Beegle et al., Spectrochim. Acta A **57**, 737 (2001); D.M. Hudgins et al., Spectrochim. Acta A **57**, 907 (2001); E.L.O. Bakes et al., ApJ **560**, 261 (2001); V. Le Page et al., ApJ **584**, 316 (2003)
5. C. Joblin, et al.: A&A **299**, 835 (1995)
6. J. Oomens, A.J.A. van Roij, G. Meijer, G. von Helden: ApJ **542**, 404 (2000); J. Oomens, B.G. Sartakov, A.G.G.M. Tielens, G. Meijer, G. von Helden: ApJ **560**, L99 (2001)
7. E. Grant, P. Schulz, A. Sudbo, Y. Shen, Y.T. Lee: PRL **40**, 115 (1978)
8. See e.g. R.E. March: J. Mass Spectrom. **32**, 351 (1997)
9. See e.g. W.B. Colson, et al. Physics Today **55**, 35 (2002)
10. J.R. Barker, L.J. Allamandola, A.G.G.M. Tielens: ApJ **315**, L61 (1987); F.C. Witteborn, S.A. Sandford, J.D. Bregman, L.J. Allamandola, M. Cohen, D.H. Wooden, A.L. Graps: ApJ **341**, 270 (1989)
11. C. Pech, C. Joblin, P. Boissel: A&A **388**, 639 (2002)
12. J. Oomens, A.G.G.M. Tielens, B.G. Sartakov, G. von Helden, G. Meijer: ApJ **591**, 968 (2003)
13. J. Szczepanski, M. Vala: ApJ **414**, 646 (1992); D.M. Hudgins, L.J. Allamandola: J. Phys. Chem. **99**, 3033 (1995)
14. J.A. Piest, G. von Helden, G. Meijer: ApJ **520**, L75 (1999)
15. J. Oomens, et al.: Chem. Phys. Lett. **367**, 576 (2003)
16. N. Solcà, O. Dopfer: J. Am. Chem. Soc. **125**, 1421 (2002)
17. G. Serra et al. : A&A **260**, 489 (1992); A. Klotz et al.: A&A **304**, 520 (1995)
18. M.A. Duncan: Int. Rev. Phys. Chem. **22**, 407 (2003)
19. D. van Heijnsbergen, G. von Helden, G. Meijer, P. Maitre, M.A. Duncan: J. Am. Chem. Soc. **124**, 1562 (2002)
20. J. Lemaire et al.: Phys. Rev. Lett. **89**, 273002 (2002); J. Oomens, D.T. Moore, G. von Helden, G. Meijer, R.C. Dunbar: submitted

Chemistry and Kinematics of the Pre-Stellar Core L1544: Constraints from H_2D^+

F. van der Tak[1], P. Caselli[2], C.M. Walmsley[2], C. Ceccarelli[3], A. Bacmann[4], and A. Crapsi[2]

[1] Max-Planck-Institut für Radioastronomie, Auf dem Hügel 69, 53111 Bonn, Germany vdtak@mpifr-bonn.mpg.de
[2] Osservatorio Astrofisico di Arcetri, Largo E. Fermi 5, 50125 Firenze, Italy
[3] Laboratoire d'Astrophysique de l'Observatoire de Grenoble, BP 53, 38041 Grenoble Cedex, France
[4] European Southern Observatory, Karl-Schwarzschild-Straße 2, 85748 Garching bei München, Germany

This paper explores the sensitivity of line profiles of H_2D^+, HCO^+ and N_2H^+, observed towards the center of L 1544, to various kinematic and chemical parameters. The total width of the H_2D^+ line can be matched by a static model and by models invoking ambipolar diffusion and gravitational collapse. The derived turbulent line width is b=0.15 km s^{-1} for the static case and $\lesssim$0.05 km s^{-1} for the collapse case. However, line profiles of $HC^{18}O^+$ and N_2H^+ rule out the static solution. The double-peaked H_2D^+ line shape requires either infall speeds in the center that are much higher than predicted by ambipolar diffusion models, or a shell-type distribution of H_2D^+, as is the case for HCO^+ and N_2H^+. At an offset of $\approx$20″ from the dust peak, the H_2D^+ abundance drops by a factor of $\approx$5.

1 Introduction

Deuterium-bearing molecules are important as probes of the very cold phases of molecular clouds prior to star formation. The H_2D^+ ion is especially important as tracer of H_3^+, the primary ion in dense molecular clouds, which does not have a dipole moment and hence no pure rotational lines. In addition, at low temperatures ($\lesssim$10 K), H_2D^+ has the ability to channel D atoms from their main reservoir, HD, into heavier species. This process leads to abundance ratios of DCO^+/HCO^+ and N_2D^+/N_2H^+ of $\sim 10^{-3} - 10^{-1}$ observed towards dense cores, much larger than the elemental D/H ratio of $\sim 10^{-5}$. Recent observations of multiply deuterated H_2CO, CH_3OH, H_2S and NH_3 (see Ceccarelli, this volume) suggest that under extreme conditions, a significant fraction of D may be transferred to heavy molecules. We wish to quantify the role of H_2D^+ in this process, and compare with other processes such as grain surface reactions (Caselli, this volume).

The ground-state 1_{01}–0_{00} transition of para-H_2D^+ at 1370 GHz will be a prime target for GREAT on SOFIA. However, with the upper energy level

65 K above ground, this line will only be excited in relatively warm ($\gtrsim$ 20 K) regions, where chemical fractionation is ineffective. For colder sources, the 1_{10}–1_{11} ground-state transition of ortho-H_2D^+ at 372 GHz is more suitable, which can be observed from the ground under good conditions. At the low temperatures ($\lesssim$ 10 K) and high densities ($> 10^5$ cm^{-3}) of pre-stellar cores, reactive collisions with ortho–H_2 keep the ortho-para ratio of H_2D^+ at $\sim$1, orders of magnitude above the LTE value [1, 2].

Until 2002, only two detections of H_2D^+ had been obtained, which indicated abundances of $\sim 10^{-11}$–10^{-12} towards Class 0 objects [3, 4]. In October 2002, we [5] observed strong H_2D^+ emission towards the pre-stellar core L 1544, and derived an abundance of $\sim 1\times10^{-9}$ in the central $\approx 20''$. Such a high abundance suggests that in this region, all CNO-bearing species are depleted onto dust grains, a situation explored in more detail by Walmsley (this volume). Data taken in June 2003 at the CSO indicate that the same phenomenon takes place in at least five other pre-stellar cores. More observations are scheduled for December 2003. These data will be presented in a forthcoming paper. Here we investigate the line profile of H_2D^+ in L 1544, and derive its abundance outside the central region.

2 Kinematics

The line profile of H_2D^+ towards L 1544 appears double-peaked, although the signal-to-noise ratio is not high (Fig. 1). Fitting two Gaussians to the profile yields results very similar to those for the profiles of $HC^{18}O^+$, $D^{13}CO^+$ and N_2H^+, observed by [6]: two thermal components separated by $\approx$0.26 km s^{-1} (Table 1). We therefore investigate whether the kinematic models that fit the HCO^+ and N_2H^+ data also reproduce the H_2D^+ line profile.

Table 1. Centroids and widths of the two velocity components. Numbers in brackets denote uncertainties in units of the last decimal.

Line	V_{LSR} km s^{-1}	ΔV_{obs} km s^{-1}	ΔV_T^a km s^{-1}
H_2D^+ (1_{10}–1_{11})	7.06(3)	0.22(5)	0.28–0.34
	7.34(2)	0.25(6)	
$HC^{18}O^+$ (1–0)	7.04(1)	0.18(3)	0.10–0.12
	7.28(1)	0.23(3)	
$D^{13}CO^+$ (2–1)	7.08(2)	0.20(4)	0.10–0.12
	7.35(4)	0.20(8)	
N_2H^+ (1–0, F_1F=10–11)	7.08(1)	0.19(1)	0.11–0.13
	7.33(1)	0.20(2)	

^{a} Thermal line width at T_{kin}=7 and 10 K.

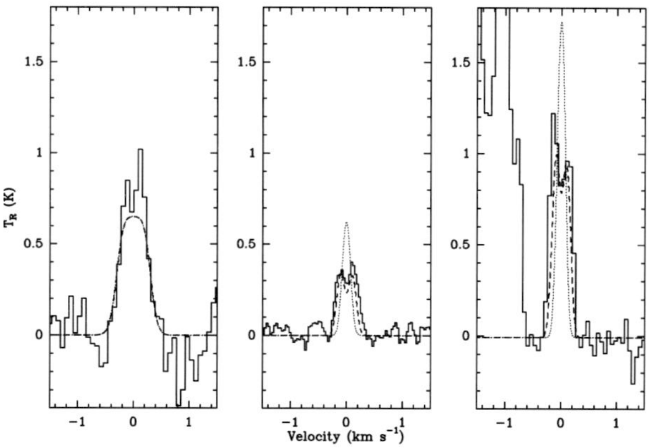

Fig. 1. Observed line profiles of H_2D^+ (left), $HC^{18}O^+$ (middle) and N_2H^+ (right), and synthetic profiles for the static (dotted) and infall (dashed) models.

The line profile of H_2D^+ was modeled using a Monte Carlo radiative transfer program [7][5]. Figure 2 shows the adopted temperature and density structure of L1544, taken from [8]. See [5] for details of the excitation model; at $T_{\rm kin} \lesssim 10$ K, ortho-H_2D^+ is essentially a two-level system, so that our results are not sensitive to the collision rates of non-radiative transitions between high-lying levels. For H_2D^+ we adopt an abundance of 1×10^{-9} in the central 20″ [5]. For HCO^+, DCO^+, N_2H^+ and N_2D^+, we used the abundance profiles from [9], Model 3, and assumed zero abundance inside r=2500 AU. We explored static models, and models with velocity fields from the ambipolar diffusion models 't3' and 't5' of [10] (see [6] for details). For the turbulent broadening, Doppler parameters b between 0.05 and 0.25 km s^{-1} were tried. Smaller values of b are overwhelmed by thermal broadening; larger ones do not fit the data.

We find that the total width of the H_2D^+ line can be matched using either velocity field. While the best-fit static model has b=0.15, b=0.05 gives the best fits with the infall velocity fields. However, the $HC^{18}O^+$ and N_2H^+ J=1-0 observations from [6] rule out the static model, which does not give a double-peaked line shape. Using the infall velocity fields, the data are matched with b=0.05, consistent with the H_2D^+ results (Fig. 1).

None of our adopted velocity fields reproduces the double-peaked H_2D^+ line shape that the observations indicate. One possibility is that the infall speeds of ≈0.1 km s^{-1} continue further inwards than in the models by [10]. Alternatively, the distribution of H_2D^+ may have a central hole, not because of adsorption onto dust grains (as for CO and N_2, the precursors of HCO^+ and N_2H^+), but due to conversion into D_2H^+ and D_3^+.

[5] http://www.mpifr-bonn.mpg.de/staff/fvandertak/ratran/

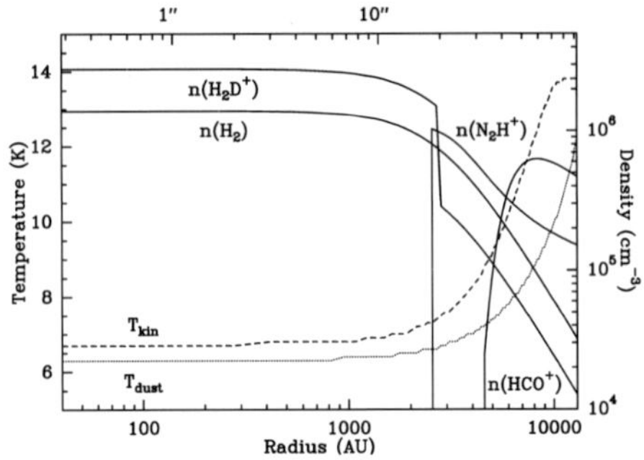

Fig. 2. Radial profiles of dust and gas temperature, and the densities of H_2, H_2D^+ ($\times 2 \times 10^9$), HCO^+ ($\times 10^{12}$) and N_2H^+ ($\times 10^{10}$) in our model of L 1544.

3 Abundance Profile of H_2D^+

The abundance of H_2D^+ away from the dust peak of L 1544 was estimated to be a factor of two lower than toward the dust peak [5]. However, this abundance may be an overestimate because some fraction of the emission at the 20″ offset positions is pickup from the central core. We have run Monte Carlo models of the H_2D^+ emission, using the same temperature and density structure as before, and dropping the H_2D^+ abundance at a 20″ radius by factors of 2–10 from its central value of 1×10^{-9}. The H_2D^+ intensity at the 20″ offset position is best matched if the abundance drops by a factor of ≈ 5 at this radius. This result is independent of the velocity field. Models where this factor is 3 or 10 produce clearly worse matches to the data.

In a future paper, we will follow our results up with two-dimensional models and an exploration of different velocity fields. We will also model line profiles at offset positions.

References

1. Pagani, L., Salez, M., & Wannier, P. G.: A&A, 258, 479 (1992)
2. Gerlich, D., Herbst, E., & Roueff, E.: Plan. Sp. Science, 50, 1275 (2002)
3. Stark, R., van der Tak, F. F. S., & van Dishoeck, E. F.: ApJ, 521, L67 (1999)
4. Stark, R., Sandell, G., Beck, S., et al.: ApJ, submitted (2003)
5. Caselli, P., van der Tak, F. F. S., Ceccarelli, C., & Bacmann, A.: A&A, 403, L37 (2003)
6. Caselli, P., Walmsley, C. M., Zucconi, A., Tafalla, M., Dore, L., & Myers, P. C.: ApJ, 565, 331 (2002)
7. Hogerheijde, M.R., & van der Tak, F. F. S.: A&A, 362, 697 (2000)
8. Galli, D., Walmsley, M., & Gonçalves, J.: A&A, 394, 275 (2002)
9. Caselli, P., Walmsley, C. M., Zucconi, A., Tafalla, M., Dore, L., & Myers, P. C.: ApJ, 565, 344 (2002)
10. Ciolek, G. E. & Basu, S.: ApJ, 529, 925 (2000)

Part X

Star Formation and Late Stages of Stellar Evolution

The Ionization State of Protoplanetary Disks: The Chemical View

Th. Henning[1], D. Semenov[1], and D. Wiebe[2]

[1] Max-Planck-Institute for Astronomy, Königstuhl 17, D-69117 Heidelberg, Germany `henning@mpia.de`, `semenov@mpia.de`
[2] Institute of Astronomy, Pyatnitskaya 48, Moscow 119017, Russia `dwiebe@inasan.ru`

Summary. We describe how a reduction technique can be applied to a chemical model of a typical protoplanetary disk in order to determine the primary chemical processes, relevant to the evolution of the disk ionization degree. It is shown that the number of involved species and reactions significantly varies throughout the disk, serving as a direct indicator for the chemical complexity. Additionally, we discuss where in the disk the chemical equilibrium approach can be used to compute the fractional ionization.

1 Introduction

During the last decade an efficient mechanism for the angular momentum transport in protoplanetary disks has been proposed, where anomalous viscosity is produced by MHD turbulence driven by the magnetorotational instability (MRI; [1]). The value of the ionization degree in these objects is a key factor in determining the efficiency of MRI and thus accretion in general. A typical young protoplanetary disk covers a wide range of physical conditions, like surface density, which results in the well-known layered structure: (1) an "active" surface subject to high-energy stellar radiation and thus being unstable to MRI, and (2) a "quiescent" midplane ("dead" zone) where the ionization degree is very low (e.g., [2]). Though the location of the border between these two disk regions is sensitive to the chemical composition, in MHD studies of protoplanetary disks a chemical equilibrium is often assumed with

$$x_{\rm e}({\rm eq}) = \sqrt{\frac{\zeta}{\beta n_{\rm H}}}, \tag{1}$$

where ζ is the ionization rate, β is a recombination coefficient between an electron and a typical ion, and $n_{\rm H}$ is the hydrogen number density. Obviously, such a simple approach may not be applicable everywhere in the disk due to complexity of the involved chemical and physical processes. To verify where and when this approach is reasonably accurate, the disk fractional ionization must be pre-computed with a chemical model. On the other hand, it is intuitively clear that it is not necessarily required to take into account

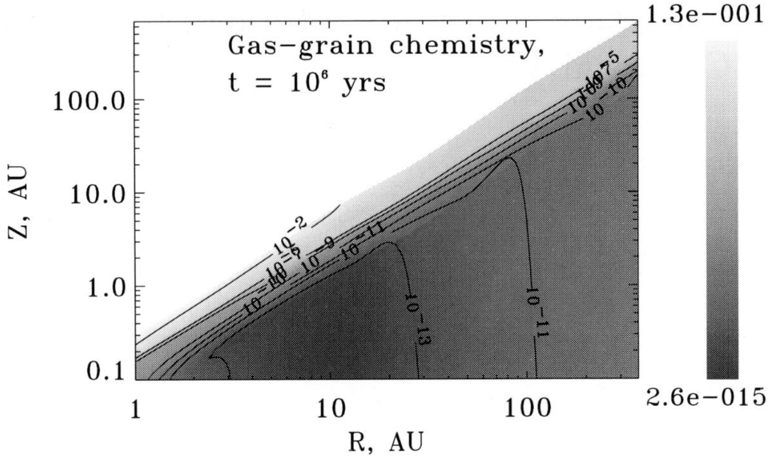

Fig. 1. Shown is the ionization structure of the disk at 1 Myr.

thousands of reactions among hundreds of species included in a typical astrochemical database to determine the ionization structure. Thus, a rigorous mathematical method is needed to distinguish between those species and reactions which are indeed responsible for the evolution of x_e in various disk locations. In this contribution we present the ionization structure of a protoplanetary disk, summarize chemical processes which dominate the ionization degree in various disk regions, and discuss the applicability of the chemical equilibrium approach (see also [3]).

2 Model

As a representative model of the disk, we adopt a steady-state flared disk model of [4] with a central T Tauri star which radiates high X-ray and UV fluxes. Cosmic rays and interstellar radiation UV field are considered to be other ionizing factors. The disk ranges from 1 to about 400 AU in radial direction and $\sim$ 0.01–700 AU in vertical direction, which implies a broad range of physical parameters. With this model in hand, we calculate the time-dependent evolution of the disk fractional ionization over 1 Myr, using a gas-grain chemical model based on the UMIST 95 database [5] and the set of surface reactions of [6]. The photodesorption of mantle species is taken into account along with thermal and cosmic ray-induced desorption mechanisms. As initial abundances, we utilize those calculated for a dark cloud model.

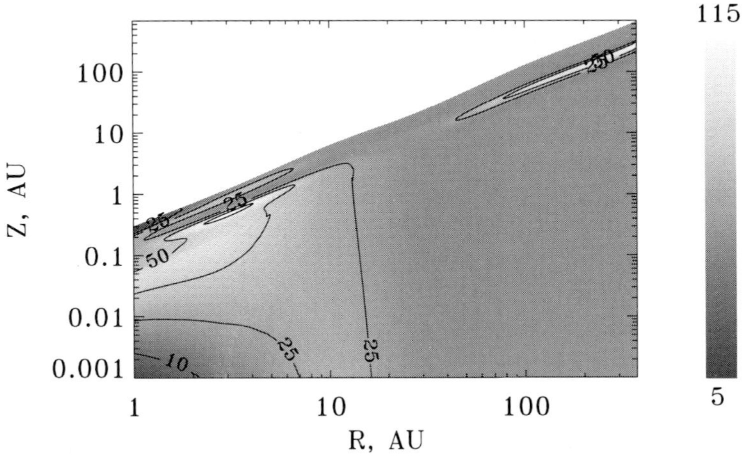

Fig. 2. The distribution of the number of species in the reduced networks governing fractional ionization through the disk is presented.

3 Results

The disk ionization degree at 1 Myr calculated with the adopted physical and chemical model is shown in Fig. 1. It can be clearly seen that the disk can be divided into three distinct regions. The midplane is opaque to both X-rays and UV photons but not to cosmic rays, resulting in a low degree of ionization, $x_e < 10^{-12}$. Above the midplane, x_e grows in response to the increasing X-ray intensity. We define the lower boundary of the *intermediate layer* as $z_1 < 0.05\,r^{1.42}$ AU. Further away from the disk interior the ionization by UV photons starts to play a role. This height is taken as an upper border for the intermediate layer and the lower boundary of the disk surface ($z_2 < 0.17\,r^{1.27}$ AU). With the reduction technique described in [7], we analyze which chemical processes govern the time-dependent evolution of the ionization degree in these three disk regions.

We found that in the midplane and surface layer it is sufficient to consider a few of species and several tens of reactions in order to guarantee an accuracy better than 50 per cent in the resulting x_e. In the inner warm midplane region these are the ionization of molecular hydrogen by cosmic rays, followed by formation of H_3^+ ions which rapidly react with metals and electrons. The metal ions are slowly neutralized by recombination with electrons and thus become the dominant charged species. In the outer midplane region, metals are irreversibly depleted onto dust grains and therefore ion-molecule reactions of H_3^+ with simple neutrals like CO and N_2 are important. Finally, these molecular ions recombine with dust grains and electrons. The dominant ions at 1 Myr are HCO^+, H_3^+, and N_2H^+. In the surface layer the life-cycle of the electrons includes the photoionization of carbon atoms and simple radicals, like OH and CO, as well as X-ray ionization of atomic and molecular

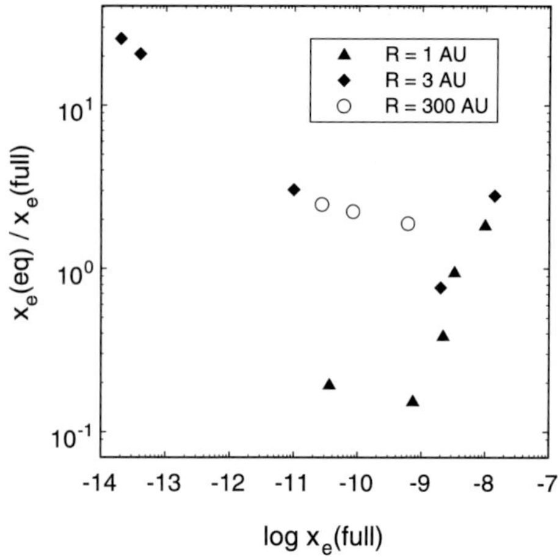

Fig. 3. A comparison of the equilibrium and time-dependent ionization fractions at 1 million years for different disk locations is shown.

hydrogen. Then hydrogen ions and H_3^+ react with oxygen, water, hydroxyl, and CO which results in the formation of HCO^+, H_3O^+, and OH^+. These molecular ions further react with either their neutral analogs or electrons, thus closing the chemical cycle. The chemical equilibrium of the fractional ionization in these regions is achieved after 100 – 1000 years.

This is in contrast to the case of the intermediate layer, where the ionization equilibrium is not reached during the entire considered evolutionary time. The evolution of the ionization degree in this disk region is regulated by the following chemical processes. Initially abundant HCO^+ and H_3^+ recombine very rapidly, while metal ions are neutralized and stick to dust grains. Then, few equilibrium states are reached with NH_4^+, H_3CO^+, and HCO^+ as the dominant ions, where their abundances in this sequence increase. After the initial decline, the evolution of the fractional ionization shows a step-like behaviour. To reproduce the behaviour with an accuracy of less than 20 per cent, one needs a reduced network that includes a few tens of species involved in cycles of synthesis and destruction of the above ions.

However, at $3 < R < 30$ AU the situation is more complicated because the density is high and the temperature is low enough to allow efficient gas-grain interactions. The simple step-like evolutionary trend does not work anymore since the incorporation of carbon atoms into long carbon chains becomes important. This process results in a simultaneous decrease of CH_4 abundance, which regulates the amount of ionized sulfur in the gas-phase. Therefore, S^+ rapidly becomes a dominant ion at $t \sim 3 \cdot 10^3 - 3 \cdot 10^5$ years. However, oxidation slowly destroys long carbon chains and the methane abundance in the gas-phase grows again which restores the CH_4–S^+ equilibrium cycle. At $t > 3 \cdot 10^5$ years the abundance of H_3^+ determines the disk fractional ionization. Due to complexity of the involved chemical processes, the reduced network must incorporate almost all species with up to 6 carbon atoms and some *surface reactions*. As far as we know, this is the first indication that the surface chemistry may have a great impact on the evolution of fractional ionization in protoplanetary disks.

In Fig. 2 we show the distribution of the number of species involved in chemical reactions, relevant for the evolution of x_e at 10^6 years.

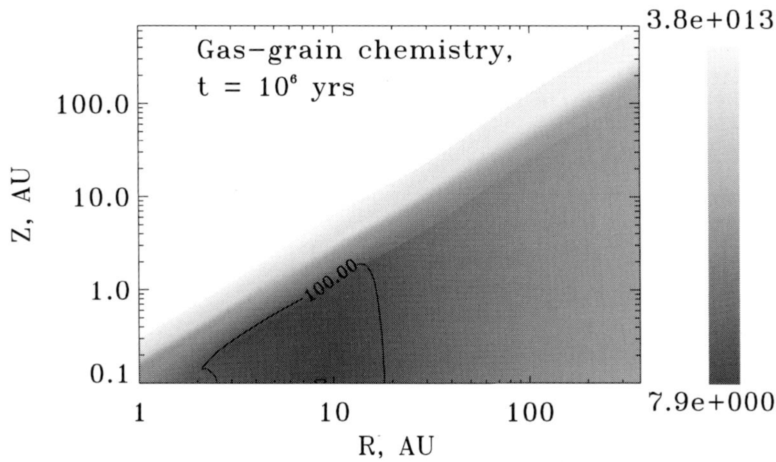

Fig. 4. The distribution of the magnetic Reynolds numbers within the disk shown. The solid line denotes $Re_M^{\mathrm{crit}} = 100$.

4 Discussion

In Fig. 3 we compare the disk fractional ionization, computed from Eq. (1), to the ionization degree calculated with the full chemical network. As it can be seen, this ratio exceeds 10 at low values of the ionization degree, which is easy to explain. If x_e is small then ion-grain interactions should be taken into account. At a moderate ionization degree level ($10^{-10}-10^{-8}$), the equilibrium x_e differs from the "exact" value by a factor of a few at $t \sim 1$ Myr and by a larger factor at earlier times.

The intermediate layer shows a very different behaviour. As mentioned above, the equilibrium state is not reached there and therefore $x_e(\mathrm{eq})$ does not reflect the disk ionization degree for most of the computational time.

In the surface layer the fractional ionization is controlled by both X-rays and UV radiation which implies that an important change should be made to the equation (1). There, the evolution of x_e is tightly connected with the evolution of C^+. Since the corresponding recombination coefficient of ionized carbon is much lower than usually assumed in MHD calculations, teh quantity β in Eq. (1) should be replaced by the appropriate value. The ionization equilibrium in the surface layer is reached in less than 100 years.

Thus, we conclude that the equilibrium approach is applicable most of the time and in most parts of the disk, but it should be used with care.

As a by-product of our study, we find the location and size of the so-called "dead" zone within the disk, which is the region where MRI cannot be operational and thus efficient accretion is not possible. For that we use a quantity that characterizes the coupling between matter and magnetic field, namely magnetic Reynolds number Re_M (e.g., [8]). The MHD turbulence can be sustained only if this quantity exceeds a poorly-defined critical value, $1 < Re_M^{\rm crit} < 100$. The distribution of the Re_M values within the disk is presented in Fig. 4 From this plot it becomes obvious that for $Re_M^{\rm crit} = 1$ there is a "dead" zone, while for $Re_M^{\rm crit} = 100$ the "dead" zone will occupy a region defined by $2 < r < 20$ AU and $z < 0.06\, r^{1.27}$.

DS was supported by the *Deutsche Forshungsgemeinschaft*, DFG project "Research Group Laboratory Astrophysics" (He 1935/17-2). DW acknowledges support from the grant NSh-162.2003.2.

References

1. S.A. Balbus & J.F. Hawley: ApJ **376**, 214, (1991)
2. Ch. Gammie: ApJ **457**, 355, (1996)
3. D.A. Semenov, D.S. Wiebe, & Th. Henning: A&A, in press, (2004)
4. P. D'Alessio, N. Calvet, L. Hartmann, S. Lizano, & J. Canto: ApJ **527**, 893, (1999)
5. T.J. Millar, P.R.A. Fraquhar, & K. Willacy: A&AS **121**, 139, (1997)
6. T.I. Hasegawa, E. Herbst, & C.M. Leung: ApJS **82**, 167, (1992)
7. D.S. Wiebe, D.A. Semenov, & Th. Henning: A&A **399**, 197, (2003)
8. S. Fromang, C. Terquem, & S.A. Balbus: MNRAS **329**, 18, (2002)

Modeling Molecular-Line Emission from Circumstellar Disks

M.R. Hogerheijde

Sterrewacht Leiden, P.O. Box 9513, 2300 RA, Leiden, The Netherlands
michiel@strw.leidenuniv.nl

1 Introduction

Most young stars are surrounded by accretion disks [1], which are likely progenitors of planetary systems. The study of the physical structure, chemical composition, and evolution of these disks is therefore clearly a topic of great interest. The thermal emission of dust grains in disks has provided a wealth of information on the disks' properties. At near-infrared wavelengths, the thermal emission is dominated by a small amount ($\sim 10^{-5}$ $M_\odot$) of warm material close to the star, and the emission is often optically thick. These wavelengths serve as excellent probes of the presence or absence of a disk. At the longer (sub) millimeter wavelengths, the emission probes the cold material at larger radii (> 1 AU), where the emission is usually mostly optically thin. These wavelengths are good probes of the total mass of a disk.

Dust, however, only makes up 1% of the total mass of the disk, if we adopt the standard interstellar gas-to-dust mass ratio. Ignoring the difficult to observe H_2 molecule (but see, e.g., [2, 3]), circumstellar disks are now commonly observed through lines of molecules such as CO, HCO^+, HCN, CN, *etcetera*. Although it is found that the molecular abundances are often strongly depleted through freezing out onto dust grains (§2) and the lines correspondingly weaker, molecular-line observations offer a powerful array of diagnostics to study disks: from the relative intensities of transitions, the molecular excitation can be derived and the density and temperature of the gas constrained; from the line strengths, the molecular abundances are found and the chemical composition inferred; and from the resolved line profiles, the velocity field in the disk is directly probed. This contribution discusses the methods with which this wealth of information can be extracted from molecular lines, and presents examples of recent research that illustrates this.

2 The Physical and Chemical Structure of Disks

Within the framework of standard accretion disks [4, 5], recent theoretical models specifically address their vertical structure [6, 7]. In a system where the central object dominates the mass and the luminosity, the disk is expected to assume a flared configuration, where the hydrostatic scale height increases with radius [8, 9]. This flared surface intercepts more stellar light, which raises the temperature of the surface and, through reprocessing, of the disk interior (e.g., [7]). Refinements to these models include the effects of dust sublimation and self-shadowing of the outer disk by inner, flared regions [10, 11, 12, 13].

The resulting descriptions of the density and temperature as function of radius and height then serve as input to calculate the abundances of various molecules (e.g., [14, 15, 16, 17, 18, 19, 20, 21, 22, 23]; and Henning, this volume). Three processes affect the molecular gas-phase abundances (reviewed in [24]): (1) freeze-out of molecules onto dust grains in the dense and cold midplane, reducing the gas-phase abundances; (2) evaporation of these ice mantles at intermediate vertical heights and near the star, where the temperatures exceed the evaporation values (~ 22 K for CO, ~ 90 K for H_2O) and where species return to the gas; and (3) dissociation and ionisation by (inter-) stellar ultraviolet photons in the disk surface where there is insufficient shielding by dust, H_2, and CO, and where a chemistry akin to a photon-dominated region is set up (see Sternberg, this volume). Radial inflow and vertical convection further influence the disk's chemical structure.

3 Radiative Transfer and Molecular Excitation

With a model in hand for the physical and chemical structure of the disk (density, temperature, systematic and turbulent velocity fields, and gas-phase abundances – all as functions of radius R and height z), the emergent emission (or absorption) lines can be calculated. An iterative approach is usually chosen: Starting with an educated guess for the level populations throughout the disk (or at a sufficient number of grid points), the absorption and emission coefficients are evaluated and the radiative transfer through the disk is solved (along a sufficient number of photon paths). This yields the intensity of the radiation field throughout the disk, with which the equations of statistical equilibrium can be solved and an update to the level populations found. This procedure is repeated until the solution converges.

Several factors complicate this prescription in circumstellar disks. Disks have a large range in density and temperature, with molecular excitation ranging from thermodynamic equilibrium to sub-thermal. Molecular abundances can vary by several orders of magnitude within the disk. And opacities, both in continuum and in lines, can be large. These factors require that the input model contains sufficient detail to cover the variations in excitation and abundance, and necessitate careful assessment of the convergence of the solution: large opacities can strongly slow down the convergence, because at each iteration step, changes in the population only propagate over short distances (corresponding to $\tau \sim 1$).

Increased computer speed and memory has made realistic model calculations tractable. In 1999 in a workshop at the Lorentz Centre in Leiden, several of the leading codes for molecular excitation and radiative transfer were (successfully!) tested against each other[1]. The reliability of the obtained solutions depends on the quality of the molecular collision rates that are used; several projects are under way to provide accurate rates for astrophysically relevant species (see Roueff, this volume; and Schöier et al. in prep.[2]).

Slow convergence of iterative methods in the presence of large opacities is not only an inconvenience for the impatient, but may lead to incorrect results. Convergence is often judged on the difference between subsequent iterations, with the implicit assumption that this will be small only when the solution is close to the correct value. The small differences between subsequent solutions due to opacity is easily misinterpreted as convergence. In the modeling of stellar atmospheres these problems are well understood, and have lead to the development of acceleration techniques such as Accelerated Lambda Iteration [25] (ALI; Lambda Iteration refers to iterative solution methods). Acceleration techniques have been included in Monte Carlo approaches to solve the radiative transfer [26][3]. This combines the reliability of accelerated convergence with flexibility in addressing arbitrary geometries.

4 Examples

Resolved Disk Emission of LkCa 15. Using the millimeter array of the Owens Valley Radio Observatory (OVRO), the emission of the 450 AU radius, 0.18 $M_\odot$ disk around the young star LkCa 15 has been resolved [27, 28]. ^{12}CO J=2–1 emission peaks at the source, with velocities consistent with Keplerian rotation of an inclined ($i \approx 60°$) disk around a 0.8 $M_\odot$ star. In stark contrast, the emission of CN 1–0 and HCN 1–0 is broken up in two peaks, symmetrically displaced from the star by $\sim 2'' \approx 300$ AU.

The appearance of HCN and CN can be reproduced by a simple radial 'step function' for the abundance, with CN and HCN only present outside 300 AU [29]. This can be explained by chemistry in a flared disk, where only outside $\sim$ 300 AU densities are low enough to allow ultraviolet radiation to penetrate and set up a PDR with increased abundances of HCN and CN. Self-consistent calculations are currently under way (Kessler et al., in prep.).

Two-dimensional Transport of Ultraviolet Radiation. To accurately account for the penetration of ultraviolet radiation from the central star into the flared disk, the full details of the two-dimensional transfer must be included [30]. Compared to one-dimensional calculations that only include vertical transport, the predominantly forward-scattering nature of the grains causes the ultraviolet field to penetrate deeper. A larger fraction of the disk is subject to photo-chemistry, with increased abundances of radicals like CN and C_2H.

[1] For more information, see `http://www.strw.leidenuniv.nl/~radtrans`.

[2] Available through `http://www.strw.leidenuniv.nl/~moldata` and `http://basecol.obs-besancon.fr`.

[3] Available at `http://www.strw.leidenuniv.nl/~michiel/ratran.html`.

An interesting application is the effect of the accretion-generated ultraviolet radiation field on the HCN and CN chemistry. Two-dimensional calculations show that the CN/HCN abundance ratio increases by a factor of 10 if an ultraviolet excess similar to that of TW Hya is included on top of a 4000 K stellar blackbody [30]. The ratio of the CN 3–2 line emission over HCN 4–3 increases by a similar amount, albeit confined to $\sim 1''$ region from the star at a typical distance of 140 pc for the object. Instruments like the Smithsonian Submillimeter Array (SMA) and the Atacama Large Millimeter Array (ALMA) will be able to trace such regions.

Far-infrared ^{12}CO Line Emission from Superheated Disk Surfaces. The object Elias 29 is a T Tauri star with a face-on disk of 0.012 $M_\odot$ that is obscured behind several layers of interstellar clouds [31]. It shows ^{13}CO 6–5 and ^{12}CO lines up to at least 20–19, indicating the presence of ~ 0.001 $M_\odot$ of warm (> 200 K) gas [32]. This reservoir of warm gas may be explained by material heated in outflow shocks, but is also comparable to the amount of material expected in the superheated surface layer of the flared disk around Elias 29.

Standard flared disk models [7] cannot produce sufficient far-infrared ^{12}CO line emission: although the amount of warm gas is sufficient, its density is too low to excite these lines [32]. In the standard flared-disk model, perfect mixing between gas and dust is assumed. If, however, the grains have settled toward the midplane, the superheated layer extends further into the denser disk interior. Calculations indicate that ^{12}CO far-infrared line strengths in this case can even exceed the observed values, Although more detailed modeling is required, including the fate of small dust grains that will not decouple from the gas, this scenario suggests that the superheated layer is a viable source of far-infrared CO emission.

Rotation and Infall in the Disk Around L1489 IRS. Submillimeter-continuum and millimeter aperture-synthesis observations show that the object L1489 IRS is surrounded by a flattened structure of ~ 2000 AU radius [33] and ~ 0.02 $M_\odot$ mass [34]. While this mass is entirely typical for disks around T Tauri stars, its size exceeds by a factor of 2–3 the largest known gas disks. Through modeling of the HCO^+ 1–0 and 3–2 interferometer images, the velocity field in the disk can be obtained [33]. In addition to a Keplerian component, suggesting a central mass of 0.65 $M_\odot$, inward motions amounting to $\sim 10\%$ of the total velocity vector are found. If such motions continue inward to the star (the interferometer data only sample scales > 700 AU), the disk lifetime is $\sim 2 \times 10^4$ yr. This is much shorter than the estimated duration of the embedded (few times 10^5 yr) and T Tauri (few times 10^6 yr) phases. It suggests that L1489 IRS represents a short-lived transitional stage, where the collapsing cloud core is settling onto a rotationally supported disk.

The inward motions can be followed to within 0.1 AU from the star through ^{12}CO fundamental ro-vibrational absorption lines against the stellar continuum at 4.7 μm [35]. The line profiles of the individual components of the P- and R- branches show prominent wings extending to +100 km s^{-1}. Such velocities are expected to occur within 0.1 AU from the star, based on the velocity model derived from the HCO^+ measurements above. The presence of absorption line originating from rotational levels of J=6 and above indicate temperature exceeding 100–200 K, which also are only expected to be populated close to the star. Taken together, this shows that the inward motions are present from 2000 AU as traced by the HCO^+ data to within 0.1 AU probed by the CO absorption.

There are several objections against the interpretation that the entire disk around L1489 IRS is contracting. The implied mass accretion rate exceeds observational limits [36] by more than an order of magnitude. And the theoretical models for accretion flow predict subsonic motions, while the observed speeds are supersonic. A possible solution is that only a thin, unstable surface layer participates in the inflow, while the bulk of the disk is Keplerian. Another solution posits the existence of two regions of inflow: one at large radii (probed by the HCO^+ data) where the cloud's material spirals onto the outer disk, and one at small radii (probed by the CO line wings) where material streams from the disk onto the star. The fact that both regions can be fit with the same velocity model does not require them to be physically connected, because the motions occur in the gravitational potential of the same star. Further study of the L1489 IRS aims at distinguishing these scenarios (Hogerheijde et al., in preparation).

5 Conclusion

In this contribution I have shown that there now exist increasingly realistic two-dimensional and three-dimensional models of the physical and chemical structure of disks. Radiative transfer and molecular excitation tools have been developed that can calculate the emergent line spectrum on the basis of these disks models. On the observational side, current data are beginning to probe the structure of disks, although with no more than a few resolution elements across the objects. The first results show interesting structural differences between various probe molecules, which current chemical models can explain.

In the next few years, increasing detail is to be expected in disk models, as the increased speed of computers will bring more extensive modeling within reach. With the arrival of new observational facilities in the submillimeter (SMA, CARMA, ALMA) and the infrared (SIRTF, SOFIA, Herschel), molecular-line modeling will form an essential link in comparing models with data. By the end of this decade, our understanding of protoplanetary disks will likely have progressed significantly.

MRH wishes to thank the organisers of the conference and the Leids Kerkhoven-Bosscha Fonds for their generous financial support.

References

1. S. V. W. Beckwith and A. I. Sargent: Nature **383**, 139 (1996)
2. W. F. Thi, E. F. van Dishoeck, G. A. Blake, G. J. van Zadelhoff, J. Horn, E. E. Becklin, V. Mannings, A. I. Sargent, M. E. van den Ancker, A. Natta, and J. Kessler: ApJ **561**, 1074 (2001)
3. M. J. Richter, D. T. Jaffe, G. A. Blake, and J. H. Lacy: ApJL **572**, L161 (2002)
4. D. Lynden-Bell and J.E. Pringle: MNRAS **168**, 603 (1974)
5. J. E. Pringle: ARA&A **19**, 137 (1981)
6. P. D'Alessio, J. Canto, N. Calvet, and S. Lizano: ApJ **500**, 411 (1998)
7. E. I. Chiang and P. Goldreich: ApJ **490**, 368 (1997)
8. N. I. Shakura and R. A. Sunyaev: A&A **24**, 337 (1973)
9. S. J. Kenyon and L. Hartmann: ApJ **323**, 714 (1987)
10. C. P. Dullemond: A&A **361**, L17 (2000)
11. C. P. Dullemond, C. Dominik, and A. Natta: ApJ **560**, 957 (2001)
12. C. P. Dullemond, G. J. van Zadelhoff, and A. Natta: A&A **389**, 464 (2002)
13. C. P. Dullemond and A. Natta: A&A **405**, 597 (2003)
14. Y. Aikawa, S. M. Miyama, T. Nakano, and T. Umebayashi: ApJ **467**, 684 (1996)
15. Y. Aikawa, T. Umebayashi, T. Nakano, and S. M. Miyama: ApJL **486**, L51 (1997)
16. Y. Aikawa, T. Umebayashi, T. Nakano, and S. M. Miyama: ApJ **519**, 705 (1999)
17. Y. Aikawa and E. Herbst: A&A **351**, 233 (1999)
18. Y. Aikawa and E. Herbst: ApJ **526**, 314 (1999)
19. Y. Aikawa and E. Herbst: A&A **371**, 1107 (2001)
20. Y. Aikawa, G. J. van Zadelhoff, E. F. van Dishoeck, and E. Herbst: A&A **386**, 622 (2002)
21. K. Willacy and W. D. Langer: ApJ **544**, 903 (2000)
22. A. J. Markwick, M. Ilgner, T. J. Millar, and T. Henning: A&A **385**, 632 (2002)
23. E. Bergin, N. Calvet, P. D'Alessio, and G. J. Herczeg: ApJL **591**, L159 (2003)
24. E. F. van Dishoeck and G. A. Blake: ARA&A **36**, 317 (1998)
25. G. B. Rybicki and D. G. Hummer: A&A **262**, 209 (1992)
26. M. R. Hogerheijde and F. F. S. van der Tak: A&A **362**, 697 (2000)
27. C. Qi: Ph.D. Thesis, Calif. Inst. of Tech., Pasadena (2001)
28. G. Duvert, S. Guilloteau, F. Ménard, M. Simon, and A. Dutrey: A&A **355**, 165 (2000)
29. J. E. Kessler, G. A. Blake, C. Qi, and J. Brown: BAAS **34**, 1319 (2002)
30. G.-J. van Zadelhoff, Y. Aikawa, M. R. Hogerheijde, and E. F. van Dishoeck: A&A **397**, 789 (2003)
31. A. C. A. Boogert, M. R. Hogerheijde, C. Ceccarelli, A. G. G. M. Tielens, E. F. van Dishoeck, G. A. Blake, W. B. Latter, and F. Motte: ApJ **570**, 708 (2002)
32. C. Ceccarelli, A. C. A. Boogert, A. G. G. M. Tielens, E. Caux, M. R. Hogerheijde, and B. Parise: A&A **395**, 863 (2002)
33. M. R. Hogerheijde: ApJ **553**, 618 (2001)
34. M. R. Hogerheijde and G. Sandell: ApJ **534**, 880 2000.
35. A. C. A. Boogert, M. R. Hogerheijde, and G. A. Blake: ApJ **568**, 761 (2002)
36. J. Muzerolle, L. Hartmann, and N. Calvet: AJ **116**, 2965 (1998)

NGC 7538 S – A High-Mass Protostar with a Rotating Disk

G. Sandell[1], M. Wright[2], and J.R. Forster[2]

[1] USRA-SOFIA, NASA Ames Research Center, MS 144-2,Moffett Field, CA 94035, U.S.A. gsandell@mail.arc.nasa.gov
[2] RAL, University of California, Berkeley 601 Campbell Hall, Berkeley, CA 94720, U.S.A wright@astron.berkeley.edu, rick@hcro.berkeley.edu

We present new 1.4 mm imaging with the BIMA[3] array of the high mass protostar NGC 7538 S in continuum and several molecular transitions. We resolve the continuum emission at 1.4 mm into an elliptical source with a size of $\sim 8'' \times 3''$ centered on the OH/H_2O maser position. Both DCN and the dust continuum show the elliptical dust disk as well as fainter, more extended emission from the surrounding cloud core. The DCN J = 3 → 2 maps confirm the velocity gradient seen in $H^{13}CN$ J = 1 → 0 and show an even larger velocity gradient than seen $H^{13}CN$, although near the center of the accretion disk some of the emission appears to originate from the outflow. ^{13}CO J = 2 → 1 is heavily self absorbed at the cloud velocity, but the line wings show the outflow with a similar extent to that seen in HCO^+ J = 1 → 0. CH_3CN J = 11 → 10 shows a more centrally condensed ($\leq 5''$) and hotter source coinciding with the center of the accretion disk. The emitting region for CH_3CN gets progressively smaller for higher K transitions, which shows that the temperature increases towards the center of the accretion disk. The mass of the continuum "disk" is $\sim$ 85 $M_\odot$ and the luminosity is $\sim 10^4$ $L_\odot$ at an assumed distance of 2.8 kpc, while the enclosed mass estimated from DCN gives $\sim$ 80 $M_\odot$.

1 Introduction

We still know very little about high–mass protostars. It is assumed that high–mass protostars are surrounded by accretion disks and that they drive molecular outflows, similarly to those seen in low mass protostars. However, since high-mass stars evolve much more rapidly than low–mass stars, there are hardly any examples of true high-mass protostars. Many high–mass stars power outflows [1], but whether they also have accretion disks is less clear. High-mass star forming regions are more distant and high-mass stars generally form in clusters or small groups making it difficult to isolate and map disks around high–mass protostars, because of confusion and poor spatial

[3] The BIMA array is operated by the Universities of California (Berkeley), Illinois, and Maryland with support from the National Science Foundation.

resolution. Although there are now more than 30 low-mass Class 0 sources (i.e. true protostars), there are only a few high-mass protostellar candidates that appear to share the same characteristics as their low mass counterparts. Molinari et al. [2] identified IRAS 23385 + 6053 as a high–mass Class 0 object driving a molecular outflow, but this protostar is at a distance of 4.9 kpc. They did not have enough resolution to confirm a rotating disk. IRAS 20126 + 4104 [3] has been the best example of a young high-mass (proto)star driving an outflow and being surrounded by a rotating disk. Recently, however, Sandell and Sievers [4] identified NGC 7538 S as a high–mass protostar. Follow up observations by Sandell et al. [5] showed that it drives a compact bipolar outflow and that it has a rotating massive disk orthogonal to the outflow. In this paper we present new BIMA observations on this massive protostellar disk and the dense cloud that surrounds it.

2 Background

The molecular cloud southwest of the H II region NGC 7538 is a well known site of high-mass star formation [6]. However, the OH and H_2O maser source 80″ south of IRS 1, NGC 7538 S, is even younger and more extreme than any of the other known high–mass (proto)stars in this region. It is very deeply embedded in a cold massive cloud core and is undetected in the NIR and MIR. It has only weak free-free emission, which appears to originate in an ionized jet. Sandell et al. [5] observed NGC 7538 S with the BIMA array at 3 mm and show that this source is a high mass protostar, which powers a luminous young ($\leq$ 2,000 yr) bipolar outflow. Imaging in the optically thin $H^{13}CN$ J = 1 $\rightarrow$ 0 line shows that the maser source is surrounded by a rotating disk seen nearly edge on and perpendicular to the outflow. The symmetry axis of the outflow is centered on the OH/H_2O maser and the dynamical center of the disk. The rotation curve derived from the $H^{13}CN$ map gives an enclosed mass for the disk of $\sim$ 40 $M_\odot$, while line and continuum observations predict a mass of $\geq$ 80 $M_\odot$. The surrounding cloud core has a mass of $\sim$ 1000 $M_\odot$ within a radius of 0.15 pc.

3 New Results

During the 2002/2003 winter season we observed NGC 7538 S at 1 mm with two frequency settings. One targeting H_2CO $3_{2,2} \rightarrow 2_{2,1}$ and $3_{0,3} \rightarrow 2_{0,2}$ and including SO $5_5 \rightarrow 4_4$; the other one primarily DCN J = 3 $\rightarrow$ 2 and CH_3CN J = 12 $\rightarrow$ 11 (K = 0 to 5), but also including ^{13}CO J = 3 $\rightarrow$ 2. Both frequency settings had line–free bands for getting a good continuum map of the protostellar disk. Fig. 1 shows selected spectra towards the center of the protostar. Most, if not all of the observed molecules show some high velocity blue–shifted emission towards the center and in the vicinity of the ionized jet. Emission in H_2CO $3_{0,3} \rightarrow 2_{0,2}$ and DCN are dominated by the disk and both molecules show that the disk is rotating with a Keplerian–like rotation curve. The deep DCN image shows an even larger velocity gradient than Sandell et al. [5] measured in $H^{13}CN$ J $= 1 \rightarrow 0$. If one assumes that the gas is still gravitationally bound, the rotation curve measured from DCN gives an enclosed mass of $\sim$ 80 $M_\odot$, which is in good agreement with the mass derived from optically thin molecules assuming standard abundances. CH_3CN J $= 11 \rightarrow 10$ shows a more centrally condensed ($\leq 5''$) and hotter source ($T_k > 100$ K) coinciding with the center of the accretion disk. The emitting region for CH_3CN gets progressively smaller for higher K transitions, which shows that the temperature increases towards the center of the accretion disk. CH_3CN also shows a striking blue wing (see e.g. Fig 1), which becomes more dominant at high K transitions, suggesting that the high velocity gas is very hot near the center of the accretion disk.

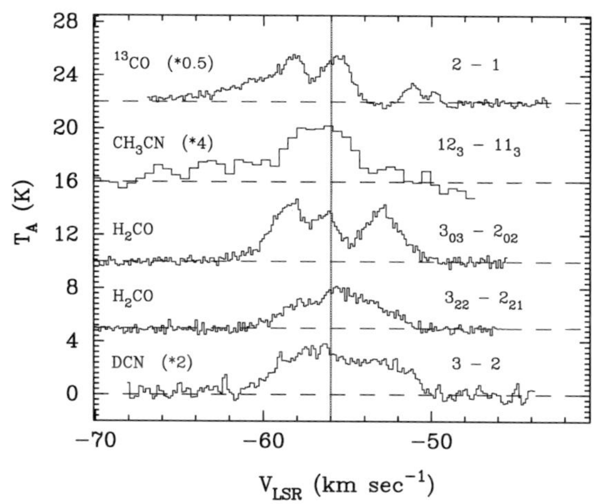

Fig. 1. Selected 1.4 mm spectra towards the center of the protostellar disk. Note the strong self-absorption in H_2CO $3_{03} \rightarrow 2_{02}$ and ^{13}CO J $= 1 \rightarrow 0$, suggesting infall. Most lines show some blue–shifted high velocity gas. All spectra except H_2CO have been scaled; the scaling factor is given in parenthesis.

Continuum observations at 220 GHz (Fig 2) with $\sim 3.5''$ resolution show not only the disk, but also the cold surrounding cloud core and looks almost identical to the integrated DCN map. A two component Gaussian fit gives a size of $7.6'' \times 2.8''$ at a position angle (p.a.) of 57° and a total flux density of 0.8 Jy.

Assuming Hildebrand [7] dust opacity, a dust emissivity of 1.5, a gas–to–dust ratio of 100 and a dust temperature of 35 K gives a total disk mass of $\sim$ 85 $M_\odot$, in good agreement with the mass derived from molecular lines and with the dynamical mass. A Gaussian fit to the integrated DCN emission

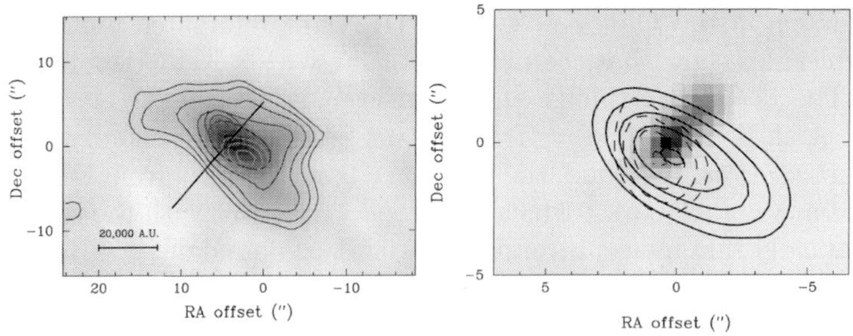

Fig. 2. left: Greyscale image of the 1.4 mm continuum map overlaid with contours; the position of the protostar is at +3.7″, +0.5″. The overlaid grey contours show the blue– (northeast) and red–shifted (southwest) emission of DCN ±2 km s^{-1} from the systemic velocity of the disk and integrated over 1 km s^{-1}. The symmetry axis and extent of the outflow is indicated by the solid line. **right:** A close up view of the continuum disk centered at 0″, 0″, with the emission from the surrounding cloud suppressed. The continuum disk is plotted with linear contours and overlaid on a 6 cm VLA grey scale map showing an ionized jet emerging from the protostar. The dotted contours show the extent of the CH_3CN K=3 emission.

gives a nearly identical size (8.5″ × 3.4″ at a p.a. of 51°). Although the disk does not appear perfectly symmetric around the dynamical center of the disk (Fig 2), we have not found any evidence for more than one active star formation center in the NGC 7538 S cloud core. The DCN map shows a secondary peak 6″ to the northwest of the protostar (at $V_{lsr} \sim -59$ km s^{-1}), which is also evident in the continuum map. However, whether this is a separate protostellar condensation or due to interaction with the blue–shifted outflow is still unclear.

All our observations show that this is an extremely massive young rotating disk, which is in the process of forming one or several high–mass stars. Such a disk cannot be stable, but is probably supported by the high pressure from the dense cloud core surrounding it.

References

1. H. Beuther, P. Schilke, T. K. Sridharan et al: Astron. Astrophys. **383**, 892 (2002)
2. S. Molinari, L. Testi, J. Brand et al: Astrophys. J. **505**, L39 (1998)
3. Q. Zhang, T. R. Hunter, T. K. Shridharan: Astrophys. J. **505**, L151 (1998)
4. G. Sandell, A. Sievers: Astrophys. J., in press (2004)
5. G. Sandell, M. Wright, J. R. Forster: Astrophys. J. **590**, L45 (2003)
6. M. Werner, et al: Mon. Not. R. Astron. Soc. **50**, 101, (1978)
7. R. Hildebrand: J. Roy. Astron. Soc. **24**, 267 (1983)

The Circumstellar Environments of High Mass Protostellar Objects

G.A. Fuller[1], S.J. Williams[1], and T.K. Sridharan[2]

[1] Physics Department, UMIST, PO Box 88, Manchester M60 1QD, UK
G.Fuller@umist.ac.uk, sjw@janus4.phy.umist.ac.uk
[2] Center for Astrphysics, 60 Garden Street, Cambridge, MA 02138, USA
tksridha@cfa.harvard.edu

1 Introduction

We have recently been studying a sample of isolated, infrared bright but radio quiet high mass protostellar objects (HMPOs) initially identified by [7]. These sources are promising isolated protostellar sources which have similar dust temperatures and masses to UCHII regions, they match the IRAS criteria for UCHII regions([11]), but have not formed their own UCHII region and are not near other UCHII regions. The sample comprises 69 candidate HPMOs which have luminosities from $10^3 L_\odot$ to $10^5 L_\odot$. This sample has also been studyed by [8], [1] and [2].

2 Dust Emission

SCUBA 850μ m and 450μ m images of the sources obtained at the James Clerk Maxwell Telescope (JCMT) in Hawaii show that there are on average 1.6 submillimetre sources associated with each target. These sources are clustered with an inferred spatial distribution which varies at $r^{-0.75}$ where r is the separation of the companions. Adopting the cold component temperatures estimated for each source by [8] the clump masses range from $\sim 1 M_\odot$ up to $10^3 M_\odot$ with a median value of $100 M_\odot$. The distribution of clump mass above about $100 M_\odot$ has a slope, s, $N(M > m) \propto M^s$ of -1.3, which is close to the slope of the IMF suggesting that if these cores form stars with a mass distribution like the IMF, star formation efficiency is independent of the clump mass ([9]).

Maps of the spectral index from 450μ m to 850μ m show that the temperature structure of the majority of clumps is consistent with the clumps being centrally heated. However a number of sources show spectral index gradients suggesting external heating or changing grain properties across the cores. The dust in these sources appears quite different from that associated with UCHII regions. Toward these HMPO sources the dust has an opacity which varies as the frequency to the 0.9 power, considerably flatter than for UCHII regions ([12]). Such a shallow dependance of dust opacity on frequency is typically interpreted as indicating the presence of large dust grains. This

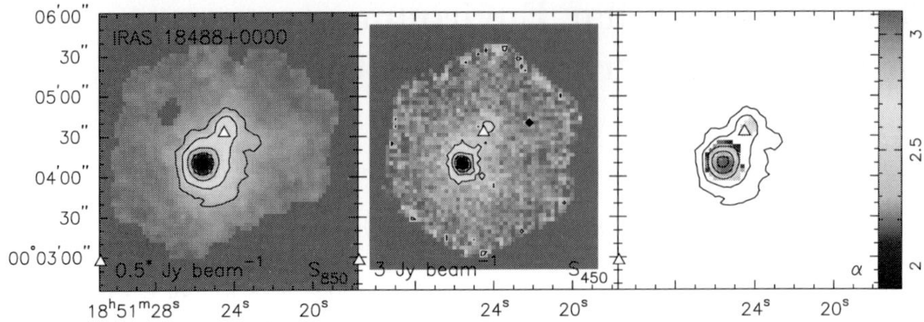

Fig. 1. SCUBA images of IRAS18488+0000 at 850μ m (left) and 450μ m (centre). The righthand panel shows the 450μ m to 850μ m spectral index map. The triangular markers show the location of the 8μ m MSX point sources in the region.

suggests that the majority of the dust in these sources may not yet have been heated sufficiently to fully evaporate the grain mantles which accumulated on the cold grains prior to the formation of massive stars.

For a subset of 38 objects for which there was only one submillimetre source associated with each IRAS source, we have modelled the far-infrared and submillimeter continuum emission ([10]) using the one dimensional radiative transfer code DUSTY ([4]). Each source was modelled as a central luminosity source surrounded by a spherically symmetric dust shell described by the temperature at the inner edge of the shell, $T_{\rm inner}$, which for a given luminosity central source determines the radius of the inner edge of the shell, a shell thickness, Y, and the density profile within the shell. We parameterized the density within the shell, $\rho(r)$, as $\rho(r) \propto r^{-p}$. The values of p, $T_{\rm inner}$ and Y where then varied to find the simultaneous best fit to both the source SED and image. Figure 2 shows an example of the best fit for one of the sources. Only two sources could not be satisfactory fitted with these kinds of models.

The best fit value of p was found to range from 0.5 to 2.5 with an average of 1.7 ± 0.5, consistent with values found by [1] for 1.2mm observations of this sample and [6] for a different sample of similar high mass sources. Interestingly if the sources are subdivided into the three groups identified [1] as a possible evolutionary sequence, there is a suggestion of an evolutionary trend in the power-law index. For sources which show neither resolved radio continuum emission nor evidence of hot gas through emission from CH_3CN and/or CH_3OH, presumably the youngest sources, the mean value of p is 2.0 ± 0.3. For sources which do have have CH_3CN/CH_3OH emission but not radio emission the mean value of p is 1.8 ± 0.1 while for the supposed oldest sources, which have both radio continuum emission and CH_3CN/CH_3OH emission p has a mean value of 1.5 ± 0.1. However since the number of sources which can currently be assigned to these groups is small, 2, 9 and 8 respectively, the existence of an evolutionary trend must be regarded as tenative and in need of further investigation.

3 Evidence of Infall?

An important aspect of understanding star formation is identifying and studying the accretion flow which is adding mass to a forming star. In an infalling envelope around a young star, the differential extinction suffered by the redshifted and blueshifted emission of an optically thick line leads to an asymmetric line profile, with stronger blueshifted than redshifted emission (e.g. [14]). The peak positions of 68 850μ m sources identified from our SCUBA observations have been surveyed using the IRAM 30m telescope and the JCMT to search for such blue asymmetric line profiles ([3]). Each position has been observed in up to five transitions, N_2H^+ J=1-0, HCO^+ J=1-0, 3-2, 4-3 and $H^{13}CO^+$ J=3-2. Fits to the hyperfine structure of the N_2H^+ have been used to measure the velocity dispersion of the material and its systemic velocity. For sources observed in both $H^{13}CO^+$ J=3-2 and N_2H^+, the line velocities and line widths are in good agreement.

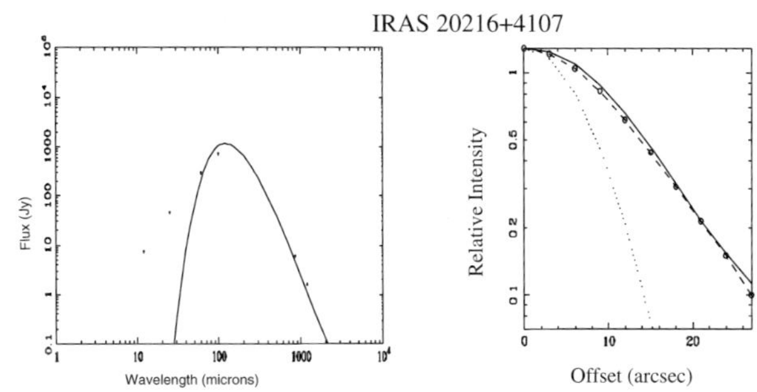

Fig. 2. DUSTY model fits to the spectral energy distribution (left) and azimuthal average of the 850μ m emission (right) of IRAS 20216+4107. The points and dashed curve show the observations and the solid curve the best fit model. In the righthand panel the dotted line shows the beam profile. The best fit model has $p = 2.5$.

With respect to the N_2H^+ velocity, 45 sources in the sample show at least one significantly asymmetric line profile with $|\delta v| > 0.25$ where δv is the asymmetry parameter defined by [5]. These sources show a predominance of blue asymmetric profiles, with values for the blue excess ([5]) of up to 30%. For example considering the 26 sources observed in HCO^+ $1-0$ with confirmed distances of 6 kpc or less, there are 9 sources with blue asymmetric profiles but only one with a red asymmetric profile. The probability of an excess of blue profiles this large arising by chance is less that 1% ([3]). The presence of such a large fraction of blue asymmetric profiles is suggestive of the presence of infall of material around these sources. A similar conclusion has recently been reached for a different sample of sources using different tracers by [13]. Our observations suggest infall velocities in the range of 0.1 km/s to 1 km/s and mass infall rates of $4-40\times10^{-4}M_\odot$/yr. Although much work still remains

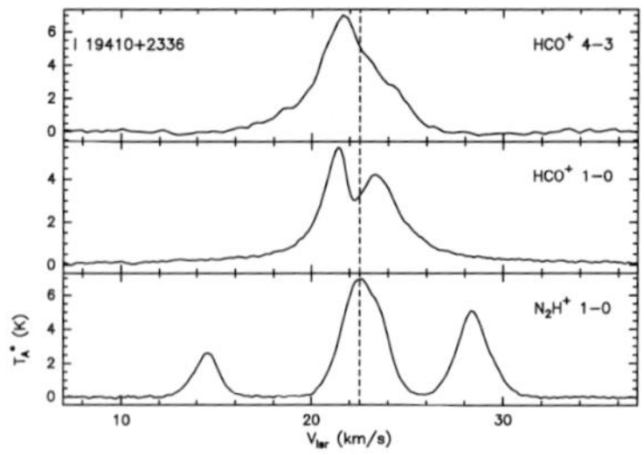

Fig. 3. Spectra of N_2H^+ J= 1 − 0, HCO^+ J= 1 − 0 and J= 4 − 3 towards IRAS19410+2336. The vertical line marks the source velocity from a fit to the N_2H^+. Both the HCO^+ transitions show significant blue asymmetries with respect to this velocity.

be done to conclusively demonstrate whether these blue profiles are tracing infalling material, these observations open up the intriguing possibility of studying the accretion flows around forming high mass protostellar objects.

References

1. H. Beuther, P. Schilke, K. M. Menten, F. Motte, T. K. Sridharan, F. Wyrowski: ApJ, **566**, 945 (2002)
2. H. Beuther, P. Schilke, T. K. Sridharan, K. Menten, C. M. Walmsley, F. Wyrowski: A&A, **383**, 892 (2002)
3. G.A. Fuller, S.J. Williams, T. K. Sridharan: submitted to A&A
4. Z. Ivezić, M. Nenkova, M., Elitzur: User Manual for DUSTY, University of Kentucky Internal Report (1999)
5. D. Mardones, P.C.Myers, M. Tafalla, D.J. Wilner, R. Bachiller, G. Garay: ApJ, **489**, 719 (1997)
6. K.E. Mueller, Y.L. Shirley, N.J. Evans, Jacobson, H.R.: ApJS, **143**, 469 (2002)
7. B. Ramesh,& T. K. Sridharan: MNRAS, **284**, 1001 (1997)
8. T. K. Sridharan, H. Beuther, P. Schilke, K. M. Menten, F. Wyrowski: ApJ, **566**, 931 (2002)
9. S.J. Williams, G.A. Fuller, T.K. Sridharan: A&A in press (2003)
10. S.J. Williams, G.A. Fuller, T.K. Sridharan: submitted to A&A
11. D.O.S. Wood & E. Churchwell: ApJ, **340**, 265 (1989)
12. T.R. Hunter, PhD thesis, Caltech (1997)
13. J. Wu, N.J. Evans: ApJL, **592**, 79 (2003)
14. S. Zhou: ApJ, **394**, 204 (1992)

Star Formation: Theory and Modelling

S. Inutsuka

Department of Physics, Kyoto University, Kyoto, 606-8502, Japan
inutsuka@tap.scphys.kyoto-u.ac.jp

Summary. Star formation (SF) is a complex combination of various processes that include the effects of self-gravity, magnetic field, radiation, chemical reactions, etc. Theoretical modelling of SF requires division of the processes into separate pieces. In terms of basic physics, we review our current theoretical understanding of those processes: thermal evolution of gas, gravitational fragmentation to form molecular cloud cores with and without magnetic field, radiation hydrodynamical evolution of a low-mass protostar, and the evolution of a circumstellar disc.

1 Thermal Budget

Star-forming process is a sequence of hydrodynamical processes that include the effects of magnetic field. The basic element of hydrodynamics is the effective equation of state, that depends on radiative interaction in interstellar medium (ISM). In the low density regime, where the column density is smaller than $\sim 2\times 10^{21}\mathrm{cm}^{-1}$, temperature T of the ISM of given density can be determined by equating the optically thin radiative cooling and heating. The solid curve in Figure 1a shows the equilibrium temperature of ISM as a function of number density. The solid curve in Figure 1b shows the pressure of the ISM. The stable branch of radiative equilirium state is called "cold neutral medium" (CNM) and corresponds to $T < 10^2 K$. The dashed curve in Figure 1a,b shows the evolutionary track of the fluid element compressed by a shock and go through the thermally unstable region indicated by shade [1]. In this way, CNM will experience thermal instability which affects the dynamics if it is suddenly heated by some mechanism. Note that this process has not

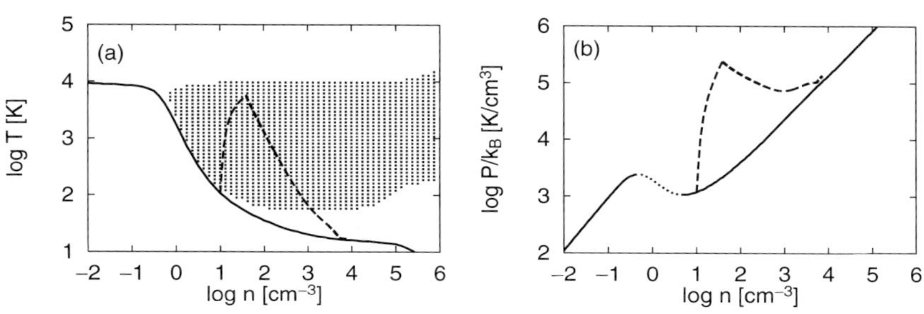

Fig. 1. A phase diagram of ISM and an evolutionary path of shocked CNM [1].

been fully analysed in recent numerical simulations of supersonic turbulence in ISM.

The column densities of typical molecular clouds are larger than $\sim 2 \times 10^{21}\mathrm{cm}^{-1}$, thus, temperature of gas becomes even smaller owing to smaller heating rate. When the density of the gas is sufficiently large so that the gas and dust grains are thermally well coupled, the gas becomes isothermal. Thermal balance of dust grains can be described by the following equation:

$$4\pi\kappa\rho I + \Gamma_{\mathrm{g}} = 4\pi\kappa\rho\sigma_{\mathrm{SB}}T^4 \equiv \Lambda_{\mathrm{thin}}, \tag{1}$$

where κ, I, Γ_{g} denote opacity, the intensity of interstellar radiation field, the heating rate due to collision with gas, respectively. With typical parameters for ISM, the above equation gives $T \sim 10K$. The strong temperature sensitivity of the cooling rate makes the dust grains, and hence, gas to be nearly isothermal. In a simple gravitational collapse process, this isothermality persist until density becomes very large so that compressional heating dominates the cooling (see §3.3). Note that this simple property is not expected in turbulent molecular clouds where the supersonic motions create shocks and rapidly heat the gas and thermaly decouple gas from grains.

2 Modes of Star Formation

It is convenient to distinguish the modes of SF as "Spontaneous Star Formation" or "Triggered Star Formation."

2.1 Spontaneous Star Formation

In spontaneous mode of SF, the molecular cloud are supported by magnetic field, and its quasi-static evolution is driven by ambipolar diffusion until the cloud becomes *supercritical* and starts dynamical collapse (e.g., [2]). Thus the timescale of the evolution is much larger than the free-fall timescale, which tends to lower the averaged SF rate in the Galaxy. The leakage of magetic field occurs more rapidly in denser regions, so that the density profile becomes centrally condensed before the dynamical collapse sets in. This centrally peaked profile is not favored in the context of binary formation [3].

2.2 Triggered Star Formation

In triggered mode of SF, the initial condition of the gravitationally unstable cloud is provided by some external triggering mechanism. The timescale of this dynamical evolution is determined by the initial free-fall time of the cloud. Although massive stars formed in GMCs have significant effect on the destruction of the parental clouds, low-mass-star-forming regions do not seem to have an efficient mechanism for halting SF and lowering SF efficiency. The

SF efficiency in the triggered mode of SF should be studied in more detail in light of observed low SF rates ($\sim M_{\odot}/\mathrm{yr}$) in the Galaxy.

Another question about this mode is the nature of triggering mechanisms: where are the triggers? In the well-known picture of ISM any cloud in the Galactic thin disk is swept up by supernova remnants once in $\sim 10^6$ years [4], which is sufficiently frequent in the context of dynamical evolution of ISM and SF. To solve the problem on the nature of the triggering mechanisms, we should study the effect of the old phase of the supernova remnants on ISM and SF [1, 6].

2.3 Turbulent Star Formation

Recently the turbulence-driven SF is studied very actively [5]. The timescale of the dissipation of supersonic turbulence in isothermal or barotropic gas is on the order of the crossing time and is very short (c.f. [6]). Thus, if a cloud is supported by turbulence against self-gravity, the evolution timescale is on the order of free-fall time. Therefore this SF mode can be considered, at present, a sub-category of the triggered SF.

The critical difference between these hypothetical SF modes are the strengths of the magnetic fields in the molecular clouds. The observational determination of the field strength in each star forming region would clarify validity of each scenario of SF, but remains difficult [7]. From a theoretical viewpoint, the difference of these modes of SF should originate in the formation process of molecular clouds, which will be the target of theoretical research in the future.

3 Gravitational Fragmentation

Observed molecular clouds have masses much larger than the (apparent) Jeans mass in each cloud, and often show remarkable filamentary structures. One of the most extensively studied star-forming regions is Taurus dark cloud, where the infrared polarization measurements show the direction of well organized magnetic field lines that is perpendicular to the remarkable filaments. This kind of information on the morphology has some suggestion on the mode of SF. On larger scale, Taurus dark cloud is located in a circular hole found in the map ($0 < V_{\mathrm{LSR}} < 2\mathrm{km/s}$) of HI 21cm line emission [8]. In general, HI holes may be formed by the expansion of supernova remnants or the effect of strong stellar winds. The observational study on the relation of the HI hole and the origin of Taurus dark cloud will be interesting future work.

3.1 Formation of Filamentary Structure

In the following, we briefly describe how filamentary structures form and evolve in the context of Triggered SF, in which a molecular cloud is, at first,

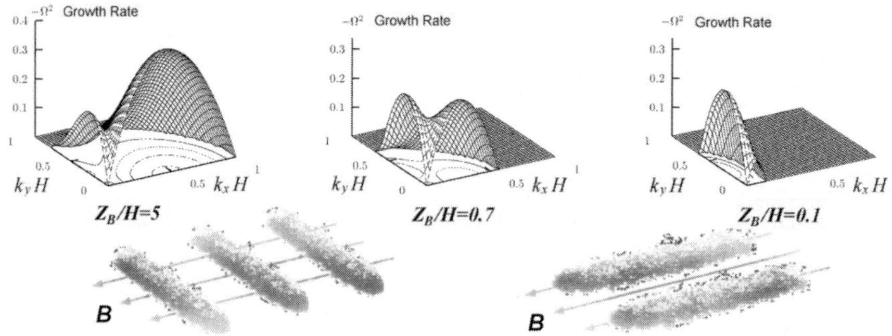

Fig. 2. Fragmentation of magnetized sheet-like clouds [9]. The normal vector of the sheet is in the z-direction and H denotes the semi-thickness. The unperturbed magnetic field lines are in the x-direction.

compressed one-dimensionally or swept up into a sheet-like configuration (for a review, see [12]). We assume that the magnetic field lines are also passively compressed and the resultant direction of the magnetic field lines are in the plane of the sheet, because typical magnetic pressure in ISM is much smaller than, e.g., the ram pressure of a supernova remnant. The modes of gravitational fragmentation of the magnetized sheet-like cloud can be summarized in Fig.2 where we plot the dispersion relation for the linear perturbation, and schematically show the resultant structures. The character of the fragmentation depends on the surface density of the sheet-like cloud [9].

For the layer with a thickness larger than the pressure scale height H $(= C_{\rm s}[2\pi G\rho_c]^{-1/2}$ where $C_{\rm s}$ is the isothermal sound speed and ρ_c is the mid-plane density), perturbations whose wavevector is *parallel* to the magnetic fields grow faster than those perpendicular to the fields. Therefore the layer fragments into filaments, and the direction of longitudinal axis of each filament is perpendicular to the magnetic field lines. The further evolution is explained in the next subsections.

On the other hand, the layer with a thickness much smaller than H becomes more unstable for perturbations *perpendicular* to the magnetic fields. In this case it fragments into filaments, and the direction of longitudinal axis of each filament is *parallel* to the magnetic field lines. The compressional motion is negligible and the resulting filaments are stable against radial collapse. Such quasi-equilibrium filaments may possibly break up into pieces again mainly through the motion along the axis direction of filaments, depending on the strength of the magnetic field. The resulting nearly spherical clumps, however, will not collapse into stars because the mass of each clump tends to be less than that of Bonner-Ebert sphere. Thus, this case is not directly related to active star formation.

In this way, the fragmentation direction is determined only by the thickness (and hence, the surface density) of the sheet-like cloud, and not by the strength of the magnetic field. Simple analytical explanation for the reason for the difference of the fragmentation processes can be found in [12].

3.2 Evolution of Filaments

In the previous subsection, the dynamical fragmentation of the sheet-like cloud is investigated in terms of the instability of the *equilibrium* sheet. This is justified because one-dimensional compression always provides quasi-equilibrium sheet-like configuration (see below).

It is important to distinguish a remarkable difference between a planar collapse and a cylindrical collapse. Consider a one-dimensional planar collapse in which the half-thickness, $z_{\rm b}$, of a sheet-like cloud is decreasing. The forces per unit mass due to pressure and gravity are as follows:

$$F_P = -\frac{1}{\rho}\frac{\partial P}{\partial z} \sim \frac{C_{\rm s}^2}{z_{\rm b}}, \quad F_G = -\frac{\partial \Phi}{\partial z} \sim -2\pi G \Sigma, \tag{2}$$

Where Σ is the column density. Therefore a planar collapse is inevitably halted by pressure forces when the thickness becomes sufficiently small. On the other hand, during a cylindrical collapse in which the radius or the scale height, R, of the cylinder is decreasing, the two forces take the following forms:

$$F_P = -\frac{1}{\rho}\frac{\partial P}{\partial r} \sim \frac{C_{\rm s}^2}{R}, \quad F_G = -\frac{\partial \Phi}{\partial r} \sim -\frac{2GM_{\rm line}}{R}, \tag{3}$$

where $M_{\rm line}$ is the mass per unit length (line-mass), $M_{\rm line} = \int_0^R 2\pi\rho r dr$. In the isothermal case, which is very important in the theory of star-formation, the above two forces have the same dependence to the radius. Therefore, radial collapse cannot be halted by the increasing pressure unless the gravity is sufficiently small from the beginning. The critical line-mass corresponding to an isothermal cylindrical equilibrium is $2C_{\rm s}^2/G$ which depends only on the temperature. Hence, a filamentary isothermal gas cloud cannot be in equilibrium for any radial density distribution and eventually collapses unless its line-mass is smaller than the above value. This collapse of isothermal filaments with various line-masses was described in terms of the one-parameter family of self-similar solutions [10]. In general, the fragmentation of a sheet-like cloud produces filaments whose line-mass is larger than the critical value. Therefore within actual filaments, equilibrium is overly idealized state. This property of the filamentary configuration is not altered by the inclusion of the weak magnetic field that is perpendicular to the axis of the filament. The magnetic flux density and the magnetic force per unit mass scales as

$$B \sim B_0 \frac{R_0}{R}, \quad F_B = -\frac{1}{8\pi\rho}\frac{\partial B^2}{\partial r} \sim \frac{B_0^2}{8\pi\rho_0 R}. \tag{4}$$

This scaling is again the same as the scaling of the gravity (eq.[3]). Therefore, if the gravity is initially greater than the magnetic force, the gravity continues to dominate the magnetic force during the radial collapse.

We must study massive collapsing filaments to determine the evolution of actual interstellar molecular clouds. The numerical calculations of the fragmentation of massive collapsing filaments show that the fragmention is not effective during the rapid radial collapse, unless perturbations with very large amplitude are imposed initially [11]. Therefore there is no characteristic density with which the fragmentation occurs during the isothermal phase. If the radial collapse is stopped by the mechanism for increasing the gas temperature or C_s^2 in eq.(2), the filament will fragment. The question of how and when the isothermal evolution terminates was explored in [14].

3.3 Characteristic Mass for Fragmentation

In general, molecular clouds are under thermal balance described by $\Gamma_{\rm g} + \Gamma_{\rm ext} = \Lambda_{\rm th}$, where $\Gamma_{\rm g}$, $\Gamma_{\rm ext}$, and $\Lambda_{\rm th}$ are the compressional heating rate of gas, the heating rate due to the external sources such as cosmic rays and photons in ISM, and the radiative cooling rate, respectively. The isothermal collapse proceeds during $\Gamma_{\rm g} \ll \Lambda_{\rm th}$ and thus $\Gamma_{\rm ext} \approx \Lambda_{\rm th}$, but isothermality is broken down when the compression of gas becomes so effective (i.e., $\Gamma_{\rm g} > \Lambda_{\rm th}$) as to heat up the cloud sufficiently against the cooling. The critical density for the violation of isothermality can be obtained by equating $\Gamma_{\rm g}$ and $\Lambda_{\rm th}$. Whether we should use optically thin cooling or optically thick diffusion cooling depends on the initial temprature of the parental cloud. Figure 3 shows the result where the horizontal axis denotes the temperature of the collapsing cloud. The observed temperature of Taurus Dark Cloud is about $10K$. Thus, we see in the Figure 3 that the violation of isothermality occurs when the optical depth, τ, of the cloud is much smaller than unity, i.e., when the density is much smaller than the density at which $\tau \sim 1$. The expected minimum mass of the fragmentation can be calculated as $M_{\rm min} \sim$

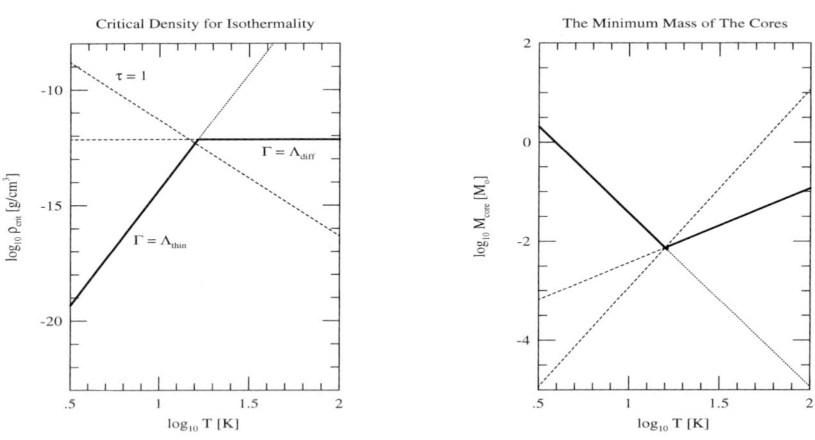

Fig. 3. The characteristic mass for fragmentation [14].

$\lambda_J M_{\rm line}$, where λ_J is the wavelength (Jeans length) of the most unstable mode for fragmentation calculated with the critical density. Note that the objects with larger masses can be formed depending on the initial spectrum of perturbations [15].

4 Formation and Evolution of Protostars

The further evolution of the collapsing cores and the formation and evolution of protostars are radiation-hydrodynamical processes. Although we need to model these processes by solving radiative transfer equation and hydrodynamics simultaneously in multi-dimensions, the most sophisticated theoretical modellings are axisymmetric and are based on the (flux-limited) diffusion approximation [13].

Here we focus on the theoretical models that account for the detailed radiative transfer processes in *spherical* symmetry [16]. The classical results based on gray-approximation were confirmed by the recent work on this problem that solved the frequency-dependent RHD equations by "Variable Eddington Factor Method" that provided the time evolution of the apparent spectrum of radiation field (Spectral Energy Distribution) in addition to the detailed dynamical evolution of the protostar [17]. Molecular line emmision profiles of various important species were also calculated in self-consistent dynamical models [18].

Figure 4 shows the time evolution of the central temperature (as a function of the central density) of the collapsing cloud. Once the compressional heating dominates the radiative cooling, the central temperature increases gradually from the low value ($\sim 10K$) of ISM. The first slope of the increasing temperature corresponds to the ratio of specific heats $\gamma = 5/3$: $T(\rho) \propto \rho^{2/3}$ for $10^1K < T < 10^2K$. This monatomic gas property is due to the fact that the rotational degree of freedom of the molecular hydrogen is not excited in this low temperaure regime: e.g., $E(J = 2 - 0)/k_{\rm B} = 512K$. When the temperature becomes larger than $\sim 10^2K$, the slope becomes that

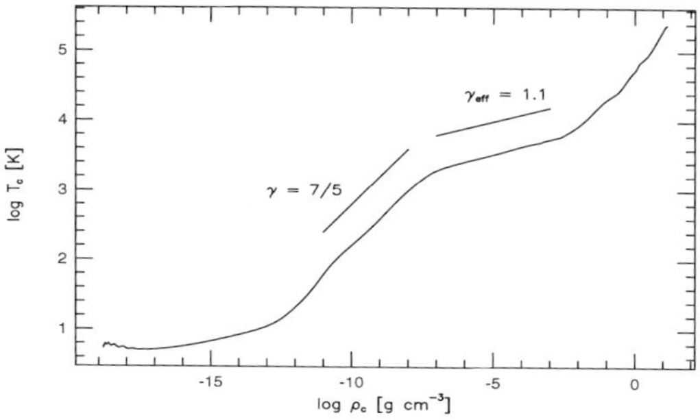

Fig. 4. The temperature evolution in the center of the collapsing core [17].

of $\gamma = 7/5$ of the diatomic molecule. The effective ratio of specific heats is larger than the critical value that make the gas pressure support against the spherical self-gravitational force: $\gamma > \gamma_{\rm crit} \equiv 4/3$. Thus, the collapsing velocity is decelerated and forms a shock at the surface of quasi-adiabatic object, which is called "the first core". Its radius is about 1 AU. It consists mainly of H_2. The increase of density and temperature inside the first core is slow but monotonic. When the temperature becomes $> 10^3 K$, the dessociation of H_2 starts. The binding energy of H_2 is about 4.5 eV that is much larger than the thermal energy per hydrogen molecule in this temperature regime. Therefore the dessociation of H_2 acts as an efficient cooling of the gas, which makes the effective $\gamma < 4/3$, and triggers the second dynamical collapse. In this second collapse phase, the collapsing velocity becomes very large and engulfs the first core. As a result, the first core lasts only for $\sim 10^3$ years. In the course of the second collapse, the central density attains the stellar value $\rho \sim 1$g/cc, and the second adiabatic core "protostar" forms. The time evolution of the SED obtained from the self-consistent RHD calculation can be found in [17].

Importance of First Cores

The formation of the 'short-lived' first core has important theoretical implications for the other processes that are not accounted for in the spherically symmetric modelling. This is because this tentative pressure support helps the formation of rapidly rotating disk-like structure around the center if the gravitational collapse start in the cloud with angular momentum.

Formation of Binaries. It is well-known that most of stars are formed in binary or multiple system. Theorists have done extensive works on the mechanisms responsible for the formation of multiple systems (see, e.g., the review in [3]). Parts of the results can be summarized as follows: the gravitational fragmentation in a collapsing molecular cloud core is not expected during the early isothermal phase of evolution, unless the mass of the core is much larger than the initial Jeans mass. This property is demonstrated in numerical simulations and supported by semi-analytic calculations in the case of the initially uniform cloud with solid rotation [20]. Initial central concentration in the density profile makes the fragmentation even more difficult. Theorists are now studying the effect of turbulent motions that are expected to enhance the fragmentation [21, 22].

When we follow the evolution to higher density regime where gas becomes adiabatic, a disk-like structure forms and enables another mode of binary formation even without (initially input) internal turbulent motion, i.e., the fragmentation of the disk around the central star. For example, the recent calculations based on the piecewise polytropic equation of state show that the central portion of the collapsing core becomes adiabatic and forms a disc-like structure around the central object, which subsequently fragments into "satellite" objects. [23]. The study of the effect of realistic radiative transfer on these modes of fragmentation remains important and interesting future work. These phenomena correspond to the fragmentation of/around the first core, which might account for the formation of binaries with separations $\gtrsim 1$AU. > 1AU. Possible formation of binaries with even shorter separations occurring in the disk around the second core is also proposed [24].

Triggers for MHD outflows. Another effect of the formation of the first core can be found in the study of MHD evolution of the self-gravitationally collapsing core. The magnetically supercritical core whose rotation axis is parallel to the mean direction of magnetic field lines produces self-similarly collapsing state as long as the equation of state is isothermal. Once the first core is formed owing to the change of the effective equation of state, a rapidly rotating disk-like structure is formed and its rapid rotation winds up the field lines and creates significant amount of toroidal component of magnetic field lines. This enhanced toroidal magnetic field produces a bipolar outflow driven by magnetic pressure [25]. Thus, the formation of the first core plays a critical role in launching the molecular outflows. If the similar process would happen again around the second core, the ejected outflow would have very high velocity and might be related to the observed optical jets or high-velocity neutral winds. Prospective observational detection of the first core will not only prove the validity of RHD theory, but also constrain the theoretical models for the various related phenomena.

5 Evolution of Disks

The further evolution of the protostar is through the gas accretion from the envelope and through the discs.

5.1 Early Phase

Even if the disc survives the gravitational fragmentation into companion stars, the early phase of the circumstellar disc should be characteized by its massiveness and its marginally stability to the local gravitational instability. In this phase, the global "m=2" or a-few-armed spiral mode plays the most dominant role in promoting the rapid gas accretion onto the central object. The gas infall from the envelope onto the surface of the disc enlarges the surface density of the disc and lowers the value of Toomre's $Q = \kappa C_s/(\pi G \Sigma)$, thereby it tends to amplify the spiral modes in the disc. The increased gravitational torque by the amplified modes promotes the gas accretion through the disc, and thus in turn, lowers the the surface density of the disc. The self-regulation processes in the self-gravitating discs are studied extensively without including the gas infall or mass supply to the disc [26]. The importance of the thermal processes is also emphasized [27]: the stability depends on temperature ($Q \propto \sqrt{T}$). Obviously, theoretical study of the gas infall from the envelope of the collapsing core onto the disc and the self-gravitationally driven accretion to the central star requires three-dimensional modelling of RHD with a large dynamic range of space, which remains an interesting but difficult future work.

5.2 Late Phase

The surface density and the total mass of the disc should decrease soon after the gas infall from the envelope terminates. The dynamical evolution of the disc in this late phase should resort to a mechanism not based on self-gravity. The most promising mechanism is provided by "magnetorotational instability" (MRI) [28].

In the last decade, our understanding of MRI was improved considerably with respect to, e.g., ambipolar diffusion [29], Ohmic dissipation [30], and Hall effect [31]. An interesting "Fluctuation-Dissipation Relation" in discs is also confirmed [32]. The saturation level of the Maxwell stress of the magnetic field in MRI-driven MHD turbulence determines the effective torque, and hence, the mass accretion rate in the disc. Thus, we need to understand the physical mechanism of the saturation process of the MRI-driven turbulence. In the case where the magnetic Reynolds number $Re_M > 1$, the most unstable wavelength of the MRI is given by $\lambda_{\rm MRI} \sim v_{\rm A}/\Omega \propto v_{\rm A}$ where $v_{\rm A}$ denotes the Alfvén wave speed. If the magnetic field is amplified by the growth of the MRI, $\lambda_{\rm MRI}$ shifts to the longer wavelength. Therefore the energy spectrum natually shows the inverse cascade, and tends to form the channel flow. When the field strength in this large scale channel flow becomes sufficiently large, the magnetic reconnection happens and destroys the channel flow. This cycle of phenomena continues recurrently [32]. The study of the non-linear phase of the MRI-driven turbulence in discs demonstrated the importance of the rapid process of the magnetic reconnection.

To determine the dynamical evolution of discs in this phase, we should also determine the ionization degree and the effective resistivity in the discs. A realistic modelling of the physical parameters requires detailed calculation of the ionization equilibrium with chemical reactions in gas phase and on the surface of dust grains [33]. This line of research will be studied actively and linked to the researches of planet formation in the discs.

References

1. Koyama, H. & Inutsuka, S. 2000, ApJ 532, 980
2. Shu, F. H., Adams, F. C., & Lizano, S. 1987, ARAA, 25, 23
3. Bodenheimer, P., Burkert, A., Klein, R. I., & Boss, A. P. 2000: Protostars and Planets IV, eds. Mannings, V., Boss, A.P., & Russell, S. S. (Tucson: Univ. of Arizona Press), 675
4. McKee, C. F., & Ostriker, J. P. 1977, ApJ, 218, 148

5. Mac Low, M.-M. & Klessen, R. : Rev. Mod. Phys. 2003, in press
6. Koyama, H. & Inutsuka, S. 2002, ApJ 564, L97
7. Crutcher, R. M. 2003, in this volume
8. Hartmann, D. & Burton, W. B.: *Atlas of Galactic Neutral Hydrogen* (Cambridge UP; New York 1997); Morino et al (2003) poster presented in the conference
9. Nagai, T., Inutsuka, S. & Miyama, S. M. 1998, ApJ, 506, 306
10. Inutsuka, S. & Miyama, S. M. 1992, ApJ, 388, 392
11. Inutsuka, S. & Miyama, S. M. 1997, ApJ, 480, 681
12. Inutsuka, S. & Tsuribe, T. *IAU Symp. 200, The Formation of Binary Stars*, eds. H. Zinnecker & R. Mathieu (San Francisco: ASP 2001) 391
13. Bodenheimer, P., Yorke, H. W., Rozyczka, M., Tohline, J. E. 1990, ApJ 355, 651; Yorke, H. W. & Bodenheimer, P., 1999, ApJ 525, 330; Yorke, H. W., Bodenheimer, P., & Laughlin, G. 1993, ApJ 411, 274
14. Masunaga, H. & Inutsuka, S. 1999, ApJ, 510, 822
15. Inutsuka, S. 2001, ApJ 559, L149, 2001
16. Larson, R. B. 1969, MNRAS, 145, 271; Narita, S., Nakano, T., & Hayashi, C.: Prog. Theor. Phys., 43, 942 (1970); Stahler, S. W., Shu, F. H., & Taam, R. E. 1980, ApJ, 241, 637; Winkler, K.-H. A., & Newman, M. J. 1980, ApJ, 236, 201; 1980, ApJ, 238, 311; Masunaga, H., Miyama, S.M. & Inutsuka, S. 1998, ApJ, 495, 346
17. Masunaga, H. & Inutsuka, S. 2000, ApJ, 531, 350
18. Masunaga, H. & Inutsuka, S. 2000, ApJ, 536, 406
19. Boss, A. P., Fisher, R. T., Klein, R. I., & McKee, C. F. 2000, ApJ, 528, 325
20. Tsuribe, T. & Inutsuka, S. 1999, ApJ, 523, L155; 1999, ApJ, 526, 307
21. Klein, R.I., Fisher, R., & McKee, C. F.: *IAU Symp. 200, The Formation of Binary Stars*, eds. H. Zinnecker & R. Mathieu (San Francisco: ASP 2001) 361
22. Bate, M. R., Bonnell, I. A., & Bromm, V. 2002, MN, 336, 705
23. Matsumoto, T. & Hanawa, T. 2003, ApJ 595, 913
24. Bonnell, I. A. & Bate, M. R. 1994, MNRAS 271, 999
25. Tomisaka, K. 2002, ApJ 575, 306
26. Laughlin, Korchagin, & Adams 1997, ApJ 477, 410; 1998, ApJ 504, 945
27. Pickett, B. et al. 2000, ApJ 529, 1034
28. Balbus, S. & Hawley, J. 1991, ApJ 376, 214
29. Blaes, O. & Balbus, S. 1994, ApJ 421, 163; Hawley, J. F. & Stone J. M. 1998, ApJ, 501, 758
30. Jin, L. 1996, ApJ, 457, 798; Sano, T. Miyama, S. M. & Inutsuka, S. 1998, ApJ 506, L57; Fleming, T. & Stone, J. 2000, ApJ 530, 464; 2003, ApJ 585, 908
31. Balbus, S. & Terquem, C. 2001, ApJ 552, 235; Sano, T. & Stone, J. 2002, ApJ 570, 314; 2002, ApJ 577, 534
32. Sano, T. & Inutsuka, S., ApJ 561, L179, 2001
33. Sano, T. et al. 2000, ApJ 543, 486; Fromang, S. et al. 2002, MN 329, 18

Shock-Excited Rotational H_2 Lines in Class 0 Sources as Probes of Magnetic Field and Chemistry in the Dense Interstellar Medium

S. Cabrit[1], B. Lefloch[2], J. Cernicharo[3], G. Pineau des Forêts[4,5], J. Lebourlot[5], and D. Flower[6]

[1] LERMA, Observatoire de Paris, F-75014 Paris sylvie.cabrit@obspm.fr
[2] LAOG, Observatoire de Grenoble, BP 53, F-38041 Grenoble Cedex 9
[3] CSIC, Instituto de Estructura de la Materia, Serrano 123, E-28006 Madrid
[4] IAS, Université de Paris-Sud, F-92405 Orsay
[5] LUTH, Observatoire de Paris, F-92195 Meudon
[6] Physics Department, The University, Durham DH1 3LE, UK

1 Summary: The Diagnostic Potential of v=0-0 H_2 Lines

Strong and collimated outflow activity is among the most spectacular characteristic of Class 0 sources, and actually led to discover the first examples of this class of objects (e.g. [1]). The impact of the bipolar protostellar wind with dense ambient gas creates strong shocks in which molecular hydrogen is expected (and observed) to be a major coolant [2], together with CO and H_2O [3]. An advantage of H_2 over the latter species is that it is not affected by freezing/desorption processes on grains, which are clearly very important in cold and dense protostellar cores but still difficult to model accurately (see related contributions in this volume).

We will concentrate here on the pure rotational lines of H_2, which are a powerful discriminant between hydrodynamical "jump" (J-type) shocks where kinetic energy is dissipated by molecular viscosity, and magneto-hydrodynamical "continuous" (C-type) shocks, where energy is dissipated by friction between the decoupled charged and neutral fluids (see [4]). Using recent observations of Class 0 sources in rotational H_2 lines obtained with the ISO satellite, in particular towards HH 2, we will illustrate how comparison with detailed MHD shock models can yield useful constraints not only on the type of shock (C or J) and on the shock speed, but also on physical and chemical conditions in the dense interstellar medium around the protostar (transverse magnetic field, H_2 ortho-para ratio, and abundance of small grains).

2 The Case of HH 2

2.1 General Characteristics

The Herbig-Haro (HH) object HH 2 in Orion, and its symmetrical counterpart HH 1, are the optically brightest HH objects in the sky, with a high-excitation spectrum indicative of shock speeds exceeding 100 km s^{-1}. Their well-defined bowshock morphology, and their oppositely directed proper motions of up to 200-400 km s^{-1} [5] pointing away from an embedded VLA source [6], make them a prototypical example of a bipolar outflow system from a young star [7]. Therefore, they represent an excellent laboratory to investigate the interaction of supersonic jets with the dense interstellar medium.

Several molecular tracers of this interaction have been imaged, including a bipolar outflow in CO(2-1) [8], localized emission peaks in HCO^+ [9], and ro-vibrational H_2 emission in the close vicinity of optical knots [10],[11]. Observations in 20"-80" beams with ISOSWS and ISOLWS reveal both warm molecular gas (rotational H_2, high-J CO) indicative of slow non-dissociative shocks in a dense medium, as well as atomic lines of [Ne II], [Si II] indicative of fast J-shocks in a lower density medium [12].

2.2 ISOCAM Spectro-Imaging of HH 2 in Atomic and H_2 Lines

Higher angular resolution observations of HH 2 with ISOCAM-CVF at 6"/pixel have recently enabled the spatial relationship between the molecular and high-excitation atomic components to be clarified. Spectra and color images can be seen in [13]. It is found that [Ne II] and [Ne III] lines are restricted to the optical knot of highest excitation and proper motion: HH 2 H. Hence, they probably trace the *Mach disk* in the driving jet. Pure rotational H_2 lines, in contrast, avoid this knot and peak towards the periphery of HH 2, as expected if they trace ambient gas swept-up by the jet *bowshock*. The detection of tracers with widely different excitation conditions and preshock densities in HH 2 noted by [12] can thus be naturally explained.

Figure 1a shows that the 0-0 lines S(3) to S(7) come from shocked gas of excitation temperature $T_J \simeq 1000 - 1500$ K, with a low, non-equilibrium ortho-para ratio (hereafter OTP) of 1-2. Spatial variations in T_J and OTP are related to the optical morphology of the HH object in Fig. 1b. In the following, we use the example of HH 2 to show how comparison of v=0-0 H_2 observations with shock models can strongly constrain the shock physics and the chemical conditions in the preshock medium.

3 Comparison of H_2 v=0-0 Lines with Shock Models

3.1 Constraints on the Preshock Transverse Magnetic Field

Figure 2 plots predicted rotational H_2 fluxes and excitation for stationary, planar MHD molecular shock models in a dark medium with various *transverse* magnetic field strengths B, parametrized by $b = B(\mu G)/\sqrt{n_H/\mathrm{cm}^{-3}}$. For $b = 0.1$, the speed limit between C-shocks and J-shocks occurs at $V_s \simeq$ 5 km s^{-1} (see Fig. 4 below) and therefore only J-shocks are plotted here (V_s = 5-25 km s^{-1}; models from [14]). For b=1-10, C-shocks survive up to much

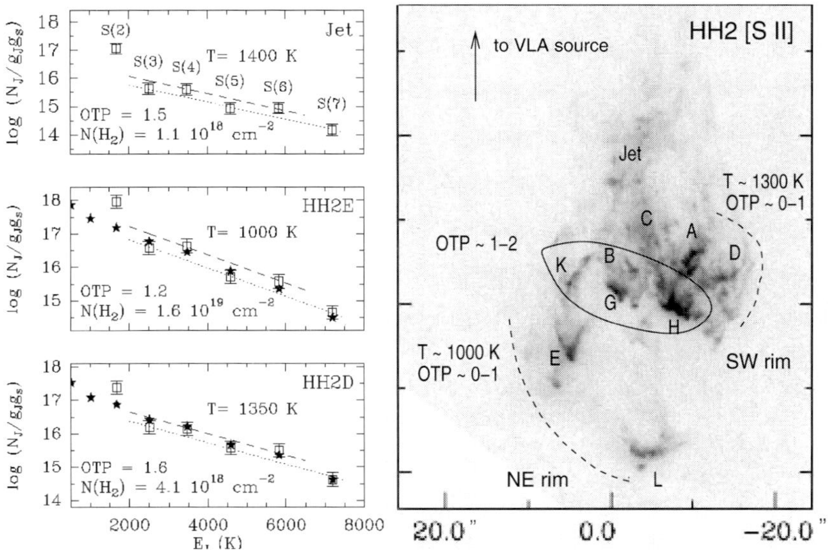

Fig. 1. a(left) H_2 v=0 population diagrams, excitation temperature (T) and ortho-para ratio (OTP) at several positions in HH 2 derived from ISOCAM-CVF observations (open squares). Filled stars plot C-shock models. From [13]. **b(right)** Spatial trends in T and initial OTP towards optical knots, on an HST image from [7].

higher speeds of 40-100 km s^{-1} (models taken from [15]). The sensitivity of the 0-0 S(5) line flux with b is clearly apparent in the Figure.

Filled symbols in Fig. 2 show ISO observations of H_2 in outflows from several Class 0 sources: HH 2 from [13]; HH 54 and L1157 from [2]; Cep A from [16]; Orion Peak 1 from [17]. It can be seen that, with very few exceptions, C-type shocks with values of b ranging from 0.3 to 10 are required to explain the strength of the 0-0 S(5) line, assuming a filling-factor ≤ 1. This is consistent with the range of b inferred from Zeeman splitting observations in dense molecular clouds (Crutcher, this volume). On the other hand, the preshock density n_H is not well constrained.

3.2 Constraints on Shock Speed and Preshock H_2 Ortho-Para Ratio

Once a value of b is adopted, constraints on the shock speed and the initial H_2 ortho-para ratio (OTP) can be obtained from comparison with C-shock models, as illustrated in Fig. 3 in the case of HH 2 (a constant value of $b = 1$ is adopted here for simplicity, although the curved shape of the bowshock could conceivably lead to variations in transverse B with position).

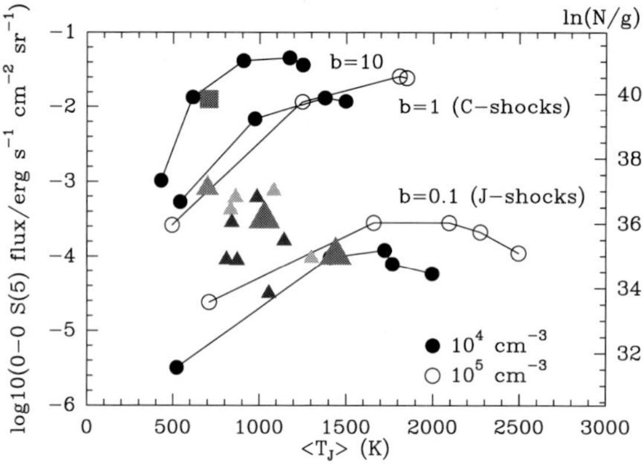

Fig. 2. Effect of B-field on the 0-0 S(5) line intensity, as a function of excitation temperature in v=0 levels. The right-hand scale shows the corresponding ln(N/g) in the upper level (v=0,J=7). Curves show stationary MHD shocks with various values of the *transverse* preshock magnetic field parameter $b = B(\mu G)/\sqrt{n_H/\mathrm{cm}^{-3}}$, and a preshock density 10^4 cm^{-3}(filled circles) or 10^5 cm^{-3}(open circles). Shock speed increases from left to right. The initial OTP is 1. Color symbols show ISO observations of shock-excited H_2 in bipolar flows from various Class 0 sources (see text). C-shocks with minimum b values $\simeq$ 0.3 to 10 appear typically required.

We shall first consider measurements along the rims of HH 2 (see Fig. 2b), which correspond to the lower set of points in Fig. 3. They tend to lie along the same theoretical curve, meaning that their observed variations in OTP and T could be explained by a change in shock speed, with the same initial b, n_H and OTP. Adopting the high $n_H = 10^5$ cm^{-3} found for a quiescent HCO^+ clump ahead of HH 2 [18], we infer an initial OTP of 1 and shock speeds ranging from 15 (North-East rim) to 25 km s^{-1}(South-West rim). For a lower preshock density $n_H = 10^4$ cm^{-3}, the initial OTP would be close to 0 and shock speeds would range from 30 (NE rim) to 40 km s^{-1} (SW rim). In either case, the OTP in the preshock gas is low (0-1), which is quite surprising given the evidence for previous outflow episodes in HH 1-2. Some pristine cold clumps must have remained ahead of the HH 2 bowshock. As for shock speeds, they agree well with optical proper motions in the NE rim (30-40 km s^{-1} in knots E and L; [7]), but not with those measured in the SW rim ($\simeq$ 120 km s^{-1}; [7]). The mixture of fast jet gas and slow shocked ambient medium in the SW rim could possibly result from the off-axis impact of the current Mach disk HH 2 H on this rim.

If we now turn to the knots close to the center of HH 2 (ellipse in Fig. 2b), which correspond to the upper set of points in Fig. 3, we see that they fall on a different model curve. A higher initial OTP (by $\simeq$ 1 for the same n_H) or a lower preshock density (e.g. $n_H = 10^4$ cm^{-3} instead of $n_H = 10^5$ cm^{-3}, for OTP = 1) are indicated. It could be that these knots propagate into a "wake"

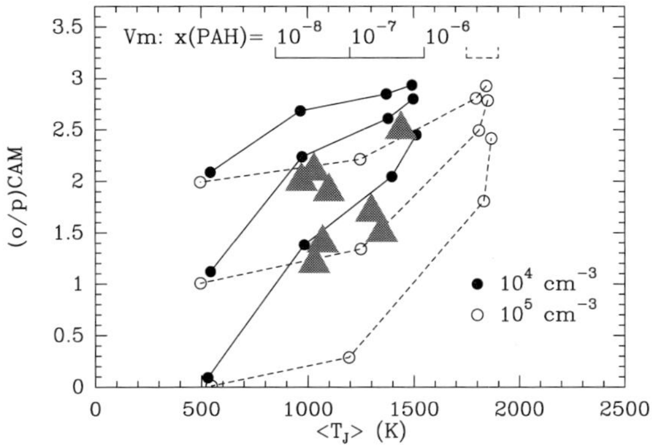

Fig. 3. Constraints on shock speed and preshock H_2 ortho-para ratio (OTP), using the observed ortho-para ratio as a function of excitation temperature in v=0 levels. The curves show stationary MHD shock models with $b = 1$ and initial OTP = 0, 1, 2 for preshock densities of 10^4 cm^{-3} (filled circles, V_s = 20-50 km s^{-1}) and 10^5 cm^{-3} (open circles, V_s = 10-40 km s^{-1}). Triangles show ISOCAM observations in HH 2 knots, from [13]. Maximum excitation temperatures set by the magnetosonic limit are drawn at the top for each n_H and a range of PAH abundances (see Section 3.3).

of freshly-shocked ambient gas left behind the bowshock rim. Interestingly, the above C-shock models systematically underpredict the 0-0 S(2) line in HH 2 (cf. starred symbols in Fig. 2a). Comparison with CO(2-1) [8] and 0-0 S(1) [12] yields $T \simeq 400$ K for this excess component [13], suggesting slow C-shocks (10-20 km s^{-1}) in molecular bullets or highly oblique bow sections.

3.3 Constraints on the Preshock PAH Abundance

As discussed by Flower & Pineau des Forêts (this volume and [19]), the maximum speed for C-shocks depends not only on the magnetic field strength, but also on the abundance of very small grains (hereafter "PAH"), which determines the fraction of large grains that are initially charged, and thus the inertia and magnetosonic speed of the charged fluid. Both effects are illustrated in Figure 4 for $n_H = 10^4$ cm^{-3}. It can be seen that relatively high PAH abundances $\simeq 2 \times 10^{-8} - 2 \times 10^{-7}$ are required to allow the C-shock speeds of 30-40 km s^{-1} inferred in HH 2 for $n_H = 10^4$ cm^{-3} and $b = 1$. Instead, if the preshock density in HH 2 is closer to $n_H = 10^5$ cm^{-3}, the magnetosonic limit is ≥ 38 km s^{-1} for PAH abundances $\geq 10^{-8}$ (see [19]), and the corresponding C-shock speeds of 15-25 km s^{-1} do not set as stringent a constraint. Interestingly, our ISOCAM-CVF observations do reveal clear PAH signatures at 11.3μm in the vicinity of HH 2. Detailed modelling is under way to infer the corresponding abundance (Lefloch et al., in preparation). This determination may also have indirect bearings on the origin of H_2 rovibrational lines in HH 2. Their high vibrational temperatures $\simeq$ 2000 K are

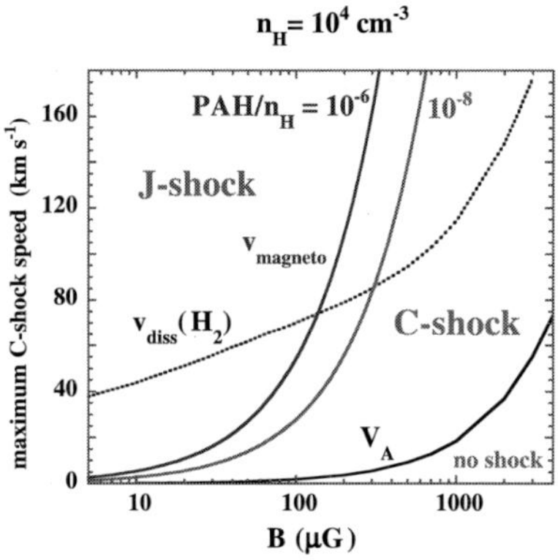

Fig. 4. Speed limit between C-shocks and J-shocks as a function of *transverse* magnetic field strength B for a preshock density $n_H = 10^4$ cm^{-3}. The limit set by H_2 dissociation [15] is shown as a dotted curve; the limit set by the magnetosonic speed of charged grains [19] is shown by thick curves for two abundances of very small grains (PAH). The neutral Alfven speed V_A, below which no shock occurs, is also shown.

currently attributed to fast C-shocks in unresolved "mini-bowshocks" [11]. Another possibility might be that ro-vib lines trace dissociative J-shocks in localized regions where the shock speed exceeds the magnetosonic limit.

References

1. R. Bachiller, P. André, S. Cabrit: A&A **241**, L43 (1991)
2. S. Cabrit et al. In *The Universe as seen by ISO* (ESA SP-247, 1999) pp 449–452
3. T. Giannini, B. Nisini, D. Lorenzetti ApJ **555**, 40 (2001)
4. D.R. Flower, et al. MNRAS **341**, 70 (2003)
5. G.H. Herbig, B.F. Jones: AJ **86**, 1232 (1981)
6. S.H. Pravdo et al.: ApJ **293**, L35 (1985)
7. J. Bally et al.: AJ **123**, 2627 (2002)
8. A. Moro-Martín, et al. : ApJ **520**, L111 (1999)
9. W.R.F. Dent, R.S. Furuya, C.J. Davis: MNRAS **339**, 633 (2003)
10. A. Noriega-Crespo, P. Garnavich: AJ **108**, 1432 (1994)
11. J. Eisloeffel, M.D. Smith, C.J. Davis: A&A **359**, 1147 (2000)
12. S. Molinari, A. Noriega-Crespo: AJ **123**, 2010 (2002)
13. B. Lefloch, et al. : ApJ **590**, L41 (2003)
14. D. Wilgenbus et al.: A&A **356**, 1010 (2000)
15. J. Le Bourlot, G. Pineau des Forêts, D. Flower, S. Cabrit MNRAS **332**, 985 (2002). Grid for o/p=3 at `http://ccp7.dur.ac.uk/newderm.html`
16. C.M. Wright et al: A&A **315**, L301 (1996)
17. D. Rosenthal, F. Bertoldi, S. Drapatz: A&A **356**, 705 (2000)
18. J.M. Girart et al.: A&A **388**, 1004 (2002)
19. D. Flower, G. Pineau des Forêts: MNRAS **343**, 390 (2003)

The Role of Pre-planetary Nebulae in the Shaping Asymmetric Planetary Nebulae

J. Alcolea

Observatorio Astronómico Nacional (Spain) J.Alcolea@oan.es

It is firmly established since more than 20 years, that Planetary Nebulae (PNe) are the result of the evolution of the Circumstellar Envelopes (CEs) around intermediate-mass stars in the Asymptotic Giant Branch (AGB), once these stars leave the AGB and become much hotter [16, 18]. However we still have problems in explaining the origin and formation of non spherical PNe, also known as asymmetric PNe, since yet they must be the remnants of the isotropic AGB CEs; how this metamorphosis can happen still is a matter of debate. In order to solve the mystery, in the past 15 yr, a growing attention has been focused on the so called pre-Planetary Nebulae (pPNe). These intermediate stage sources are nebulae around post-AGB stars not hot enough to ionize them. This is not a minor problem since in the last few years we have realized that most PNe and almost all pPNe are not spherically symmetric but strongly bipolar and/or elongated.

Today we are far from knowing all the details of this wonderful transformation, but some points have been cleared up. There is building evidence that by the end of the AGB the mass loss mechanism changes, from being slow, isotropic, and radiation driven, to much faster, highly-bipolar, and of unknown origin; the interaction of these two winds would be responsible for the shaping of pPNe and ultimately of non spherical PNe. In this contribution I will review on some of the latest results in the field of pPNe studies, focusing on those particularly relevant for understanding the morphology and kinematics of these early post-AGB sources.

1 Mass Loss Along and Beyond the AGB Phase

In the Main Sequence, low and intermediate mass stars live on their H fuel at their cores. When this is exhausted, the star moves up along the Giant Branch, until He starts its ignition in the core. Once the core He is exhausted too, the star finally ascends through the Asymptotic Giant Branch (AGB). At this point, due to the low mass of the star, the Carbon and Oxygen rich core becomes degenerate and the evolution of the star can not proceed any further; the star is very luminous and red, and its power is supplied by shells of H and He that burns alternatively in its envelope. While on the AGB, stars become strong radial pulsators with periods of about one year. On each of these pulsations the size of the star can change by as much as 50%, and the outermost layers of the atmosphere alternatively rise and sink at speeds up to $10 - 20\,\mathrm{km\,s^{-1}}$.

A dense extended atmosphere develops around the star, as the combined result of the low escape velocity, strong pulsation, and cool surface. Those layers are rich in molecules, that in the case of refractory elements condense into dust grains, which absorb and reflect the stellar photons, being pushed away from the star. As the grains move out, thanks to this radiation pressure, they collide with the molecules and the gas is carried along too. This results in the formation of a circumstellar envelope (CE), rich in molecules and dust, which gently expands into the Interstellar Medium isotropically, with a constant velocity V_∞ of a few $\mathrm{km\,s^{-1}}$. This expanding circumstellar envelope represents a mass loss, that in the latest stages of the AGB (at the super-wind phase) can be as large as several $10^{-5}\,\mathrm{M_\odot\,yr^{-1}}$. In fact this mass loss rate is about the maximum that can be sustained by radiation pressure. The momentum provided by the stellar photons per unit of time can not surpass the limit $10L_\star c$ (where c is the speed of light), even considering multiple scattering, and should equal the momentum loose by the CE per unit of time, which is simply $\dot{M}V_\infty$. For normal conditions, this translates into an upper limit for the mass loss rate of about $10^{-4}\,\mathrm{M_\odot\,yr^{-1}}$ [6].

Obviously, the star can not endure such an enormous mass loss rate for a long time: after a few ten thousand years the H/He-burning shells become literally exposed, and the star turns extremely hot and tiny becoming a blue dwarf. This transformation occurs very rapidly, lasting no more than a few thousand years. [Later on, the star will slowly cool down and end its days as a white dwarf.] At the same time, the circumstellar envelope is also dramatically altered. Since the copious mass loss has stopped, the CE becomes thiner and thiner as it expands, being first photo-dissociated and then photo-ionized by the hard UV radiation emanating from the hot central star (once it becomes hotter than 30,000 K), finally leading to the formation of a Planetary Nebula (PN). The formation of the PN is also due to the interaction of the slow wind from the AGB phase (the former CE) with a more tenuous but faster and hotter one that is released by the central star at this stage. This Interacting Stellar Wind model was proposed by Kwok in 1978 [12] to explain the morphology of spherical PNe.

Traditionally for explaining non-spherically symmetric PNe it has been assumed that during the super-wind phase, at the end of the AGB, the mass loss is not isotropic but somehow enhanced along certain equatorial plane. The interaction of this non-isotropic AGB wind with the isotropic wind in the PN phase would naturally result in the formation of asymmetric PNe: this modification of the ISW model is known as the GISW (Generalized ISW) model [14]. However there is a major problem: AGB envelopes are systematically spherically symmetric as one would expect from an isotropic loose of mass. This has been proved by means of observations of the CO emission [15], OH and SiO masers [10], and the detection of arcs of dust scattered light around AGB and post-AGB sources [2].

2 pPNe: a Couple of Well Studied Cases

The symmetry break-up should occur then either at the very end of the AGB or on the post-AGB phase. To better study this process it makes sense to pay some attention to the very few sources (about 200) we should expect to be in the intermediate phase between the AGB CEs and the PNe, i.e. the pre-Planetary Nebulae (pPNe). The properties of these nebulae were poorly known until very detailed studies of few of them were performed in the last half of the past decade. I will quickly review in two of the best studied cases before describing their properties in a more statistical way.

2.1 M 1–92: Minkowski's Footprint

M 1–92 is a pPNe with a beautiful bipolar image in the optical [5]. From interferometric observations of their CO emission [4], we know that the two lobes are due to the stellar light scattered by the dust located on the walls of two cavities, which are divided by a thick equatorial structure. The kinematics of the nebula is dominated by a large constant velocity gradient along the symmetry axis, resulting in expansion velocities of $70\,\mathrm{km\,s^{-1}}$ at both tips of the nebula. On the contrary, in the equatorial direction, only radial expansion with a constant velocity of $8\,\mathrm{km\,s^{-1}}$ is detected: i.e. a velocity field very much like those in AGB envelopes. The kinematic age of the fast molecular outflow is only 900 yr.

This kinematics and shape are thought to be the result of the interaction of some post-AGB ejecta, very collimated, with the formerly spherical CE, resulting in the stretching and acceleration of the nebula along its present symmetry axis. The presence of large constant velocity gradients is also a characteristic in many PNe. This sort of Hubble-like law can be explained either by an acceleration event which lasted much less that the kinematic age of the outflow (< 100 yr. in the case of M 1–92), or by a momentum-driven (i.e. isothermal) shock between two winds both with densities $1/r^2$ [2]. The linear momentum carried by the fast molecular outflow is so extremely large, that this totally rules out photon pressure on grains as the mechanism responsible for the ejecta that reshaped this nebula 900 yr ago.

2.2 OH 231.8+4.2: The Rotten Egg Nebula

OH 231.8+4.2 is also a beautiful bipolar nebula which has been studied deeply and in great detail. In addition to the thick equatorial disk and the two emptied lobes (as in M 1–92), the very massive ($\sim 1\,M_\odot$) molecular component of the nebula shows a pair of long tails at the end of the lobes [1]. Both the lobes and the tails are expanding at high velocity, showing a very well defined Hubble-like law, corresponding to a kinematic age of 770 yr. Because of the large collimation and velocity gradient of the tails, at least in this object, a sudden acceleration event (lasting probably less than 100 yr) is the most plausible explanation for the observed velocity field.

HST images also precisely trace this massive molecular component by the dust scattered light, but in addition a series of shock fronts are seen in Hα and some forbidden atomic lines [7], which are due to the (adiabatic) shocks between the molecular high velocity jets and the very old shell of gas lost along the AGB before the super-wind phase.

The linear momentum carried out by the fast expanding gas is again much larger than what can be provided by photon pressure on grains, even if we consider that the interaction could last 770 yr. Yet, if photon pressure on grains was responsible for this bipolar outflow, to have expansion velocities larger than $400\,\mathrm{km\,s^{-1}}$ (as it is observed at the end of the south molecular jet) the drag velocity between dust and gas might be as large as $100\,\mathrm{km\,s^{-1}}$. On the contrary a perfect spatial match between these two components is observed [1, 7].

3 The Statistical Approach

In the previous section I have presented two examples of very well studied pPNe. However one may wonder whether these cases are representative of the entire pPNe population, or even if this population is homogeneous or on the contrary we can define different types of pPNe. At this respect I would like to comment on two statistical works, one on the morphology and one on the dynamics of pPNe, which I found very relevant for understanding the role of the pPN phase in the shaping of asymmetrical PNe.

3.1 Is There Any Spherical pPNe?

The first work is that by Ueta et al. [20]. These authors imaged 27 pPNe in the optical using the HST-WFPC2. Of these sources, six did not show any nebulosity around, so their morphological study could not be performed. For the remaining 21 targets, the authors conclude that they split in two different populations, having about the same number of members, which they name SOLE (11 sources) and DUPLEX (10 sources). SOLE sources are characterized by clearly bimodal SEDs, consisting in the star and dust shell emissions. Accordingly, their images in the optical are dominated by the central source, that is surrounded by a much fainter nebulosity which is always found to be elongated: they resemble elliptical PNe but with a very bright central source. On the contrary, the members of the DUPLEX group do not show the stellar component in their SEDs, and accordingly the star is not seen in the optical images, that clearly show in all cases a bipolar morphology, i.e. two reflecting lobes divided by an obscuring equatorial disk.

The authors propose that both types of sources can be explained by a single nebular structure, in which the mass distribution depends on the polar angle (the angle with respect to the symmetry axis). They assume a pair of conical cavities at the two poles of the nebula, the main difference between the two groups being the overall opacity of the pPN [13]. For DUPLEXs, the opacity is large, so the star is not directly seen (unless the nebula is pole-on), and the scattered light only comes from the walls of the emptied cones. On the contrary, for SOLEs the opacity is low, and the star can always be directly seen (regardless of the orientation of the nebula), and the reflected light emanates from all the nebula, the elongation being due to the lower extinction of the light along the polar directions. The origin of the polar cones would be in both cases due to an anisotropic mass loss during the super-wind phase.

I would agree with that interpretation except for this last point, simply because such an asymmetric mass loss has never been observed in the AGB phase. However I would have no objection if these cavities are formed during the post-AGB phase, as the kinematics observed in M 1–92 & OH 231.8+4.2 and many other pPNe is telling us. If this is the case I would conclude that all pPNe are bipolar, in the sense that all of them have been stretched and accelerated preferentially along the symmetry axis, as a result of some early post-AGB mass loss. It is clear that, at least for DUPLEX sources this process also results in the formation of two cavities (the two lobes seen in optical images) at both sides of the equatorial plane of the nebula (we have many well studied examples). However this may be also true for SOLE sources, or at least it is true for their prototype IRAS 17436+5003. I have performed Plateu de Bure maps of the CO emission of this nebula, concluding that the kinematics and morphology in this case is not very different from those of M 1–92. Therefore, for the sake of simplicity, I would dare to say that all pPNe are bipolar also in the sense that their morphology is dominated by a somewhat denser equatorial structure dividing the two cavities at both sides of the symmetry axis.

3.2 The Prevalence of High-Velocity Winds in pPNe and Its Consequences

The second statistical study I would like to comment is that performed by Bujarrabal et al. [6]. The aim of this work was to find out up to what extent pPNe emitting in CO show high-velocity molecular emission, and whether these outflows always show the momentum excess problem (w.r.t. the maximum that can be obtained from radiation pressure on grains) often found in well studied pPNe. The mass, axial momentum and kinetic energy of these outflows can be computed just from single dish spectra of CO, provided that one knows how to convert line intensities into masses. In order to do that single dish spectra of ^{12}CO and ^{13}CO J=2–1 and 1–0 were obtained for a sample of 37 objects. After removing high-mass (hyper-giants) and low-mass ($< 1\,M_\odot$) objects, high-velocity molecular emission was found in 30 (94%) cases of the remaining 32 pPNe in the sample. Since we know that all pPNe are morphologically bipolar [20], this result suggests that this bipolarity is due in all cases to some post-AGB high-velocity ejecta.

Due to the poor quality of the spectra in some cases, the authors were able to computed the axial momentum in these outflows, $P_{\rm jet}$, in only 22 cases. When comparing $P_{\rm jet}$ with the (radial averaged) maximum linear momentum that radiation pressure on grains is capable of provide per unit of time, $10\,L_\star\,c$, it is found that in 21 cases (95%) radiation pressure requires times larger than a reasonable upper limit for the acceleration time of these outflows ($\sim$200 yr), the only exception being the Red Rectangle. Even if we assume that the outflows are not due to a sudden acceleration, but to a continuous interaction of two $1/r^2$-winds, for about half of the sources the required time is larger than the kinematic age for the outflow. Note that this result is very robust since the ratio $P_{\rm jet}/10\,L_\star\,c$ does not depend on the distance to the source (usually a very badly known parameter).

This result means that the bipolar ejecta responsible for shaping of the pPNe can not be powered by radiation pressure on grains. Therefore, it seems that at the very point at which the star leaves the AGB, the mass loss changes from isotropic to bipolar, resulting in the stretching and acceleration of the nebula along its present symmetry axis. But at the same time the mass loss mechanism also changes from being due to radiation pressure to some other mechanism not known yet. This symmetry break-up, responsible for the shaping of asymmetric PNe, occurs very early in the post-AGB phase: later on, when the star becomes hot enough, the hot and fast wind released from the central source will collide with the pPN resulting in the final shape of the PN.

4 Do Orbiting Disks Exist Around Post-AGB Sources?

When it comes to think in other mechanisms capable of powering these post-AGB bipolar ejecta, it is impossible to forget the universal link in Astrophysics between bipolarity (the release bipolar jests) and angular momentum (rotation). This link holds for bipolar outflows from young stellar objects, symbiotic stars, AGNs, pulsars, and even some type of GRBs. Note that not in all these scenarios accretion is needed to power the outflows, but certainly lots of angular momentum are available in all cases.

In fact it is well accepted that to explain the existence of bipolar PNe and pPNe rotation must play a fundamental role. Discrepancies arise when trying to identify the source of the angular momentum. Some authors favor the hypothesis that only multiple systems will be able to produce bipolar ejecta [19]. However, if bipolarity is a characteristic of all pPNe, as it seems to be the case, we will need multiple system in all post-AGB sources. Other authors speculate that the rotation of the star, mainly the core, will be enough to provide the required amounts of angular momentum: this, of course, promises to be a more universal mechanism [3].

Regardless of the source of angular momentum, if asymmetrical pPNe and PNe and due to the bipolar-rotation connection, of should expect to have equatorial rotating structures, at least in young pPNe (where the dense gas is still close to the star and the axial acceleration is still going on or has stopped very recently). The detection of any rotating structure in pPNe has been very elusive until very recently. Equatorial structures in pPNe are normally found in radial expansion not in rotation, since they are the least disturbed remnants of the former AGB shell. This is not surprising since these are huge structures and simply because of angular momentum conservation, the rotation velocity must be very small in comparison with the expansion velocity, which makes its detection virtually impossible.

Finally, very recently, equatorial rotating structures close to the post-AGB sources have been detected in two pPNe. The first one is the detection of a ring of SiO masers around QX Pup, the Mira variable that it is assumed to be the central star of the pPN OH 231.8+4.2 [17]. These authors observed the SiO masers at 7 mm using the VLBA. They found that the maser spot locations and velocities are compatible with a model in which the emission arises from a rotating and in-falling disk, of 12 AU in diameter, perpendicular to the symmetry of the nebula. Second epoch VLBA observations SiO masers at 3.5 mm seem to confirm these results [9]. However, I should mention that some authors think that QX Pup, the Mira variable powering the SiO masers, it may not be the central post-AGB source of the pPN, since apparently it is displaced from the axis of symmetry of the nebula by at least 1000 AU [11].

The second case is the already mentioned Red Rectangle [8]. Plateu de Bure maps of the CO emission at 2.6 and 1.3 mm show that the molecular gas in this bipolar pPN is confined to a thin equatorial disk of $5''$ in diameter: no wonder why this source shows no momentum excess problem, presently it has no CO outflow ! Although the spatial resolution of these observations is poor in comparison with the disk size ($3.0''\times1.5''$), the position vs. velocity diagram along the equatorial direction clearly shows the $1/\sqrt{r}$-function shape characteristic of Keplerian rotation (the CO velocity wings in this source trace not an outflow but the fast rotating inner parts of the disk). The observations are very well fitted by a model in which a disk of $0.04\,M_\odot$ orbits a $0.9\,M_\odot$ central mass, rotating at maximum velocity of $7\,\mathrm{km\,s^{-1}}$ at its inner boundary. A very low expansion of $0.4\,\mathrm{km\,s^{-1}}$ is also required in the outer parts of the disk, to explain the strong self-absorption in the blue-side of the optically thick J=2–1 CO spectrum.

In spite of the large importance of this first unambiguous detection of an orbiting structure around a post-AGB star, the relevance of this discovery in the statistical sense still needs to be investigated. On the one hand, it could happen that such structures are not present in all pPNe. The central source of the Red Rectangle is a close binary system [21]: the detected disk is in fact circum-binary. On the other hand it is also possible that structures like are very common in pPNe, the reasons because they have been not detected till now being a) the much stronger emission from the rest of the envelope (not present in the Red Rectangle), and b) the much larger distance of most pPNe (1–3 kpc) in comparison with this relatively nearby source (350–700 pc).

References

1. J. Alcolea, V. Bujarrabal, C. Sánchez-Contreras, R. Neri, J. Zweigle: A&A **373**, 932 (2001)
2. B. Balick, A. Frank: ARA&A **40**, 439 (2002)
3. E.G. Blackman, A. Frank, J.A. Markiel, J.H. Thomas, H.M. van Horn: Nature **409**, 485 (2001)
4. V. Bujarrabal, J. Alcolea, R. Neri: ApJ **504**, 915 (1998)
5. V. Bujarrabal, J. Alcolea, R. Sahai, J. Zamorano, A.A. Zijlstra: A&A **331**, 361 (1998)
6. V. Bujarrabal, A. Castro-Carrizo, J. Alcolea, C. Sánchez-Contreras: A&A **377**, 868 (2001)
7. V. Bujarrabal, J. Alcolea, C. Sánchez-Contreras, R. Sahai: A&A **389**, 271 (2002)
8. V. Bujarrabal, R. Neri, J. Alcolea, C. Kahane: A&A **409**, 573 (2003)
9. J.F. Desmurs, C. Sánchez-Contreras, V. Bujarrabal, J. Alcolea, F. Colomer: in prep.
10. H.J. Habing: A&A Rev. **7**, 97 (1996)
11. Y. Gómez, L.F. Rodríguez: ApJ **557**, L109 (2001)
12. S. Kwok, C.R. Purton, P.M. Fitzgerald: ApJ **219**, L125 (1978)
13. M. Meixner, T. Ueta, M. Bobrowsky, A. Speck: ApJ **571**, 936 (2002)
14. G. Mellema, A. Frank: MNRAS **273**, 401 (1995)
15. R. Neri C. Kahane, R. Lucas, V. Bujarrabal, C. Loup: A&ASS **130**, 1 (1998)
16. B. Paczyński: Acta. Astr. **21**, 417 (1971)
17. C. Sánchez-Contreras, J.F. Desmurs, V. Bujarrabal, J. Alcolea, F. Colomer: A&A **385**, L1 (2002)
18. D. Schönberner: A&A **79**, 108 (1979)
19. N. Soker, S. Rappaport: ApJ **538**, 241 (2000)
20. T. Ueta, M. Meixner, M. Bobrowsky: ApJ **528**, 861 (2000)
21. H. Van Winckel, C. Waelkens, L.B.F.M. Waters: A&A **293**, L25 (1995)

The Mass Function of Newly Formed Stars

L.A. Hillenbrand

California Institute of Technology; MS 105-24; Pasadena, CA 91125
lah@astro.caltech.edu

1 Introduction

The topic of the stellar "original mass function" has a nearly 50 year history, dating to the publication in 1955 of Salpeter's seminal paper. In this review I discuss the many more recent results that have emerged on the initial mass function (IMF), as it is now called, from studies over the last decade of resolved populations in star forming regions and young open clusters.

The spiral structure of our Galaxy can be traced over many kilo parsecs near the plane ($b < 5°$) via indicators of imminent, ongoing, or recent star formation such as molecular gas, HII regions, and OB stars. Locally, however, many of these same diagnostics can be found far above the Galactic plane. The ring of local star formation, historically termed the Gould Belt (Gould 1879), is inclined by $\sim 20°$ with respect to the plane, and extends as close as 150 pc at its nearest and 750 pc at its furthest. Towards the outer part of the Galaxy along this ring are the Perseus, Orion, and Monoceros clouds, crossing the plane in Cygnus/Cepheus, and towards the inner Galaxy the Ophiuchus, Scorpius, Centaurus, Lupus, and Chamaeleon clouds again crossing the plane in Vela. Notably the famed Taurus-Auriga star forming region is not part of the Gould's Belt but rather located in the vicinity of the inferred center of the expanding ring, near the PerOB3 groups and dispersed Cas-Tau association. The Gould's Belt phenomenon recently has been repopularized based on the nearby young (<100 Myr) field stellar population uncovered by the Hipparcos and ROSAT all-sky surveys, as well as from new appreciation of the spatial distribution of young neutron stars, pulsars, and gamma ray sources. The fortuitous geometry of local star formation enables us to study stellar populations relatively free of the foreground / background contamination and severe crowding and confusion which dominate investigations of more distant Galactic plane star formation.

Looking in detail at an individual star-forming molecular cloud one finds that the stellar distribution is not random across the face of the cloud, but rather that young stars tend to congregate into small groups or aggregates and denser clusters. In Taurus, for example, there are distinct regions of near-simultaneous aggregated star formation in addition to similarly aged stars sparsely spread across the cloud (e.g. Hartmann et al 2001). In Orion, which is a more massive molecular complex by a factor of $\sim$10, the clusters (such as the Orion Nebula Cluster and NGC 2024) are denser but a similar dispersed population is still present (e.g. Strom, Strom & Merrill 1993; Carpenter 2000). Summarizing the global clustering characteristics of the local regions of star formation, cluster densities range from unbound aggregates or groupings of tens of stars over a 0.1–0.5 pc radius ($< 10^2$stars/pc^3) to possibly bound clusters containing hundreds to thousands of stars over a 0.5–5 pc radius (10^3–10^5stars/pc^3). Furthermore, higher mass stars are associated with denser though not necessarily more populous clusters (Hillenbrand 1995; Testi et al. 1997). High-mass and low-mass stars do form in the same places and at the same times; drastically different star formation mechanisms, e.g. for the production of massive stars, apparently are not required. Given the wide dynamic range in mass of stars and indeed sub-stellar mass objects in star forming regions, such areas are natural targets for studies of the IMF.

Investigation of the detailed form of the stellar/sub-stellar IMF in young clusters is motivated by the fact that the distribution of masses is one of the most basic characteristics of a stellar system. For a single region or cluster, a successful theory of star formation should predict the salient features of the IMF such as minimum / maximum masses and the location / width of any peak in the distribution. A related predisposition concerns the universality of the IMF. Molecular cloud fragmentation and star formation processes could lead to an IMF that depends on variables such as time, location, metallicity, local stellar density, cloud density or temperature, the pre-existence of massive stars, or whether star formation is stochastic or triggered (e.g. via external cloud compression). The detailed shape of the IMF is also important for several utilitarian reasons, particularly if the form is a universal one, including deduction of the past star formation rate in the Galaxy (increasing, decreasing, or constant in time), interpretation of integrated light from distant regions in our Galaxy or other galaxies, and constraining the fraction of baryonic mass tied up in low luminosity objects.

In advance of summarizing the large body of work on the IMF in star forming regions and young open clusters it is useful to call attention to pioneers of field star IMF studies, including Salpeter (1955), Miller & Scalo (1979), Scalo (1986), Rana (1987), Kroupa, Tout & Gilmore (1993), Kroupa (2001). The field IMF near the stellar/sub-stellar boundary has been explored by Reid et al. (1999) and Burgasser et al. (2001). High mass populations largely in clusters are summarized by Massey (1995). These papers include many analysis steps in transforming observational data to masses. First are the conversions of photometric and spectroscopic data to absolute magnitudes or luminosities and effective temperatures, and then the corrections for incompleteness, star formation histories, stellar evolution, and in some cases effects such as binarity. Useful are recent review articles on the IMF including those by lead authors Kroupa, Meyer, Elmegreen, Scalo, and Larson.

What we know from these field star studies is that the distribution of masses rises as $dN/dM \propto M^{-\alpha}$ with the exponent α changing from $\sim$2.3 at the highest masses (nearly identical to the original Salpeter result which probed masses 0.3–16 $M_\odot$) to $\sim$0.3 near and across the sub-stellar boundary.

While the slope does flatten towards lower masses, it should be emphasized that the IMF extends in a continuous manner from the stellar domain well into to the brown dwarf regime. As found both in the field and in star forming regions, there is nothing special about the hydrogen-burning limit in terms of the production rates of gravitationally bound objects from the ISM.

The various field star IMF's that have been proposed do have different implications for the average stellar mass and the mass fraction in low- / intermediate- / high-mass objects. For example, considering a stellar population which extends from 0.1–120 $M_\odot$, a Salpeter (1955) IMF predicts that 96% of the stars and 60% of the mass are found at masses $< 1\ M_\odot$, with a mean stellar mass of 0.35 $M_\odot$. The log-normal form of the Miller-Scalo (1979) IMF, on the other hand, predicts that 78% of the stars and 31% of the mass are found at masses $< 1 M_\odot$ with a mean stellar mass of 0.89 $M_\odot$, while the multiple power law Kroupa (2001) prediction is that 90% of the stars and 45% of the mass are found at masses $< 1 M_\odot$, with a mean stellar mass of 0.80 $M_\odot$. Extending the lower mass limit of the population through the brown dwarf domain to 0.01 $M_\odot$ implies for the Kroupa IMF that 37% of objects and 4% of the mass are found in brown dwarfs, and that the mean mass per object (star / brown dwarf) drops to 0.38 $M_\odot$. An additional way in which to think of the IMF as a probability distribution is to ask how many stars are expected in a cluster before a star of given mass is present. Again using the Kroupa multiple power-law slope one finds that 16 stars are needed before a 1 $M_\odot$ star is expected, 245 before the first 8 $M_\odot$ star is expected, and 8274 before a 120 $M_\odot$ object is expected. Drastically different numbers result from the various proposed IMF's and the implications are not inconsequential for calculations of quantities such as the number of ionizing photons or the number of proto-supernovae in a cluster.

2 Advantages and Disadvantages of Studying the IMF in Star Forming Regions and Young Open Clusters

In contrast to the field star studies, investigations of youthful environments have the advantage that all stars formed in the vicinity are still present; none have evolved into stellar remnants and none have escaped the gravitational grasp of the molecular cloud and/or dense stellar cluster. Thus what is measured is not only the present day mass function but the initial mass function as well, with no corrections needed. The IMF can thus be measured over a large dynamic range, ideally from $>100\ M_\odot$ down to $<0.08\ M_\odot$ in a single region. Importantly, young low mass stars and brown dwarfs are several (3–5) orders of magnitude more luminous at 1–10 Myr of age than they are on the main sequence, compressing the mass-luminosity relationship. Further, forming sub-stellar populations are much warmer than those in the solar neighborhood field which have been cooling for many Gyr. At 1 Myr of age any object with spectral type later than M6–M7 (i.e. $T < 3000$ K) is a brown dwarf ($<0.08\ M_\odot \approx 80\ M_{Jupiter}$) and by mid-L has mass $<10\ M_{Jupiter}$ according to most current pre-main sequence contraction models. Another advantage of young cluster populations for IMF studies is that individual regions are co-eval, equi-distant, and have the same metallicity. Finally, presence of molecular gas can screen against contaminating background stars, making membership easier to assume. Results from one region can be compared directly to those from another with perhaps different total mass, stellar density, or metallicity, and also to those from the solar neighborhood field which represents a mix of star formation episodes occurring across kpc of space and Gyr of time in our Galaxy.

However, there are also several disadvantages to working in youthful environments. These include the poorly calibrated evolution of the mass-luminosity relationship through the pre-main sequence stage, the presence of large and spatially variable extinction in star-forming regions, the possibility of photometric excesses due to processes associated with circumstellar (accretion) disks, the spatial crowding and nebulosity which complicate observations, and finally the lack of truly superb statistics for individual regions.

3 Methodology for IMF Study of Young Stars

With the above benefits and limitations in mind, various techniques have been developed in attempting to measure the IMF in star-forming regions and young open clusters. These include the use of luminosity functions and a model of the star formation history (e.g. Zinnecker, Rieke, Comeron, Wilking, Strom, Lada, Alves, Muench), employment of color-magnitude diagrams and a photometric de-reddening technique as well as a model of the star formation history (e.g. Meyer, Carpenter, McCaughrean, Figer), and finally the construction of Hertzsprung-Russell diagrams which require both spectroscopy and photometry (e.g. Herbig, Strom, Hillenbrand, Brown, Luhman, Preibisch, Rebull). All of these methods must assume a set of theoretical mass-luminosity relationships and translate them into the needed format for comparison to the data.

The spectroscopic techniques are the most robust, but consequently require the most observational and analytical effort. Feasibility with large telescopes (8–10 m) extends well into the sub-stellar regime for the more proximate (<500 pc) regions of recent star formation, but the technique is applicable only to the upper end of the IMF in more distant (>1–2 kpc) regions. Spectroscopy is usually performed in the V, I, J, or K bands (0.5–2 μm) and enables the photospheric temperature to be estimated through adoption of a spectral-type to effective temperature relationship. Spectral types are derived from ratios of specific atomic lines or molecular bands with the most diagnostic features generally found in the spectral regime closest to the peak wavelength of the stellar blackbody. The calibrations to effective temperature are established through fundamental techniques such as the infrared flux method, measured diameters, or for intrinsically less luminous objects via theoretical modelling of stellar atmospheres. Once a spectral type is determined an intrinsic color scale is used to compare observed to expected (given the spectral type) colors and thereby derive the color excess. This can be converted using an adopted extinction law into equivalent V-band extinction. De-reddened photometry is then combined with a bolometric correction scale to produce bolometric luminosities. Finally, the derived effective temperature and bolometric luminosity are compared to those predicted from pre-main sequence evolutionary models to infer stellar masses and ages.

It should not go without mention that the pre-main sequence evolutionary models (e.g. those by D'Antona & Mazzitelli 1997; Swenson et al. 1994; Baraffe et al. 1998; Siess et al. 2000; Palla & Stahler 2000; Yi et al. 2003) remain poorly calibrated. Variations between these predictions can be as much as factors of 2 in mass and factors of 3 in age. In most cases such models combined with quantities derived from observational data such as Hertzsprung-Russell diagrams, color-magnitude diagrams, or luminosity functions are the only means available for inferring stellar masses. However, fundamentally determined masses are those calculated from observations of dynamical systems such as eclipsing double-line spectroscopic binaries or, to lower precision, systems surrounded by circumstellar disks whose keplerian rotation can be measured. Comparisons between the (limited) available dynamical masses and pre-main sequence evolutionary tracks demonstrate that most existing sets of tracks systematically under-predict low-mass pre-main sequence masses by up to 20–30% (Hillenbrand & White 2004).

4 IMF Results in Youthful Regions

Before summarizing and synthesizing the results on stellar/sub-stellar IMFs in star-forming regions and young open clusters over the past decade, I call attention to a few recent results involving my own collaborators as examples of the sort of work being done. First at the high-mass end, Slesnick, Hillenbrand, & Massey (2002) performed a UBV photometric + BV-band spectroscopy survey of the H χ Per region, finding a typical Salpeter-like IMF for each of the two clusters ($\Gamma = -1.36 \pm 0.20$ and $\Gamma = -1.25 \pm 0.23$) over the mass range 4–40 $M_\odot$ with $\sim$40 past supernova expected. At the other end of the mass spectrum, Slesnick, Hillenbrand, & Carpenter (2004) have studied the sub-stellar domain of the Orion Nebula Cluster using JHK photometry and J/K spectra, confirming the turnover suspected by several authors based on analyses at higher masses (Hillenbrand 1997; Luhman 2000; Muench et al 2002). Depending on the behavior of the mass-luminosity-temperature relationships – which are still highly uncertain at such low masses – there may even be a flattening or potential rise towards the lowest mass brown dwarfs.

Generalizing now to the local set of star forming regions and young open clusters, it is often the case that incompleteness and overall poor statistics preclude robust measurement of the stellar/sub-stellar IMF. Assembly of existing literature demonstrates that close to 5000 stars in 24 regions can be located on the HR diagram and compared to theoretical evolutionary tracks. However, only a few regions are reasonably complete into the brown dwarf regime, namely the Orion Nebula Cluster, IC 348, and the Pleiades. These regions compare well to a log-normal type (Miller-Scalo 1979) mass function with acceptable statistical and systematic (due to methodology) variation.

To provide a meaningful assessment of the IMF, observers need to measure its shape in detail, including the dynamic range, peak, and width of the peak. Attempts at functional fits are welcome. At present, however, most authors simply force single power laws through their data and quote the resulting best-fit slope, sometimes even with an errorbar. Taken en ensemble, however, these forced power laws present a very different picture of the IMF. Figure 1 is a plot inspired by similar graphs presented by Scalo (1998) and Kroupa (2001) of the slope of the initial mass function Γ where $dN(\log M)/d\log M \propto M^{\Gamma}$ and $dN(M)/dM \propto M^{-\alpha}$ such that Γ = 1-α, shown as a function of $< M > /M_{\odot}$ where $< M >$ is the average mass over which a common slope is claimed to apply. The data are drawn from an extensive list of IMF primary references appearing over the last decade. Many different techniques are assimilated, including spectroscopic investigations, color-magnitude diagrams, and luminosity functions. Similar plots of Γ as a function of minimum mass or maximum mass, rather than the average mass, are also instructive. What is clear from these Figures is that at the high mass end (M>2–3 $M_{\odot}$) reported IMF slopes are constant as a function of mass and distributed around the original Salpeter value of $\Gamma = -1.35$. At lower masses (M<1–2 $M_{\odot}$), however, the reported slopes deviate from Salpeter in a manner such that the IMF slope continually becomes shallower / flatter towards lower masses.

There scatter in Figure 1 amounts to ~1 dex for any mean mass. Although it may be tempting to interpret this scatter as real variation in the IMF, there are several affects which may in part explain it. As stated above there are differences in methodology between authors, and while spectroscopic investigations are indeed the most robust, they are not practical for all regions; of note is that there are no systematic differences in Figure 1 between the spectroscopic investigations and those that are purely photometric (color-magnitude diagrams or luminosity functions). Another possible effect is that of systematics present in the different effective temperature or bolometric correction scales adopted by different authors, but again, offsets between authors or groups are not evident in the results as presented in the format of Figure 1. The adopted set of pre-main sequence theory is yet another way to introduce systematic differences. Finally, the statistical robustness of the results should be considered in lieu of generally small sample sizes, particularly at the high mass end where by the very nature of the IMF the poisson statistics are the poorest. Kroupa (2001) has investigated this last point by

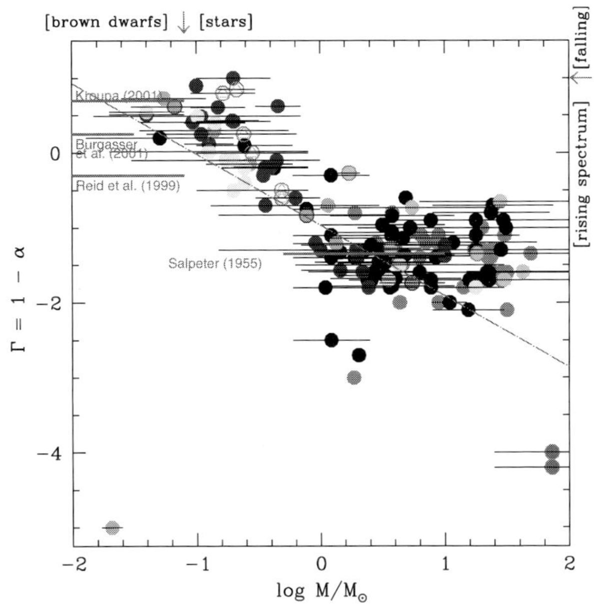

Fig. 1. Values quoted in the literature for the slope of the initial mass function plotted against mean mass (points), with the range of masses over which the single slope is claimed to apply indicated (horizontal lines). Filled circles represent star-forming regions and young open clusters while triangles are globular clusters. Continual change of the mass function slope with mean mass below about 1 $M_\odot$ demonstrates that a quadratic or log-normal function best describes the composite data at low masses. Above several $M_\odot$ the scatter is more considerable, and the quadratic or log-normal function yields to a flatter slope that averages close to the Salpeter (1955) value. The dash-dot line shows a Miller-Scalo (1979) log-normal.

simulating a version of Figure 1 using model populations differing in total numbers of stars (10^2–10^6) sampled many times. The observed scatter about the Salpeter slope can in fact be reproduced through sampling error alone. Kroupa (2001) also investigates the scatter introduced by dynamical evolution of clusters on time scales of a few Myr to tens of Myr, and concludes that some of the scatter may be attributed to the redistribution of mass within or exterior to the gravitational bounds of a cluster.

In summary, the general trends emerging from the literature are for a universal IMF with a single Salpeter-like slope above masses of 1–3 $M_\odot$ and a constantly flattening slope towards lower masses, extending smoothly and continuously from low mass stars into the sub-stellar regime. At present, there is little evidence for substantial variations in the IMF from region to region, or over time. The scatter present in Figures of IMF slope vs mass are explainable by a combination of methodology differences, theory differences, and sampling statistics. Yes, there are outliers in such plots, and the regions in question should be investigated further to determine the accuracy of the reported results. But the dominant trends remain.

5 Future Goals of IMF Studies

Having crossed the sub-stellar boundary over the past few years, the next major threshold for low mass IMF studies to reach is the theoretical limit for fragmentation out of a molecular cloud, derived by calculating the ability of a Jeans' mass / radius object contracting on the free-fall time to radiate its own gravitational energy. Thus far no low-mass cut-off in the IMF has been observed. Observationally, the Next Generation Space Telescope (now JWST) should be sensitive at the distance and age of the Orion Nebula Cluster, for example, to objects as low in mass as 1 $M_{Jupiter}$. Given an appropriate means of sorting between field star contaminants and cluster members at such faint magnitudes, the goal of determining the lowest free-forming mass object should be reached in this dense and populous star forming region.

Another major question concerns the universality of the IMF. Is the mass spectrum the same everywhere, that is, invariant with respect to time, location, size or density of the star-forming environment, metallicity, other initial conditions such as temperature or pressure? Can we relate the IMF to some physical aspect of the star formation process? Thus far no systematic variations have been detected, for example between high-mass star forming environments of varying metallicity (e.g. Massey et al. 1995ab) or between low-mass star forming environments of different stellar density (e.g. Luhman 2000), or even in "extreme" environments such as R 136 (Massey & Hunter 1998) in the Large Magellenic Cloud. That is not to say that no such trends are present. Rather, to within our ability to provide robust statistical assessment, variations have not yet been detected. A very few regions with "unusual" IMFs have been presented (e.g. the Hillenbrand et al. 1995 discussion of the bias of the BD+40 4124 region towards higher mass stars and the Briceno et al. 2002 presentation of the lack of brown dwarfs in Taurus) though whether these regions are truly unusual or merely the expected statistical outliers remains unquantified. A goal then is to more rigorously quantify the invariance of the IMF.

6 Summary

- The IMF is not a single power law but rather continually changes its slope below masses of 2–3 $M_\odot$.
- There is nothing special about the hydrogen burning limit in terms of the frequency distribution of masses above or below this boundary.

- Thus far no low-mass cut off has been observed in the IMF, with limits reaching perhaps as low as 20–30 $M_{Jupiter}$ in some star formation regions. The next fundamental boundary is that of opacity-limited fragmentation.
- The lack of strong evidence for IMF variations suggest that its functional form may be universal and therefore independent of any of the physical variables suspected to play a role in the star formation process.

References

1. I. Baraffe, G. Chabrier, F. Allard, and P.H. Hauschildt, AA, **337**, 403 (1998)
2. C. Briceno, K.L. Luhman, L. Hartmann, J.R. Stauffer, and J.D. Kirkpatrick, ApJ, **580**, 317 (2002)
3. A.J. Burgasser PhD Thesis, California Institute of Technology (2001)
4. J.M. Carpenter AJ **120**, 3139 (2000)
5. F. D'Antona, and I. Mazzitelli, in Cool stars in Clusters and Associations, ed. R. Pallavicini, & G. Micela, Mem. S. A. It., **68**, 807 (1997)
6. B.A. Gould: Uranometria Argentina, ed. P.E. Corni, Buenos Aires, p. 354 (1879)
7. L. Hartmann, J. Ballesteros-Paredes, and E.A Bergin ApJ **562**, 852 (2001)
8. L.A. Hillenbrand PhD Thesis, University of Massachusetts (1995)
9. L.A. Hillenbrand, AJ, **113**, 1733 (1997)
10. L.A. Hillenbrand and R.J. White, ApJ, in press (2004).
11. L.A. Hillenbrand, M.R. Meyer, S.E. Strom, and M.F. Skrutskie, AJ **109**, 280 (1995)
12. P. Kroupa, MNRAS **322**, 231 (2001)
13. P. Kroupa, MNRAS **322**, 231 (2001)
14. P. Kroupa, C.A. Tout, and G. Gilmore, MNRAS **262**, 545 (1993)
15. K.L. Luhman, G.H. Rieke, E.T. Young, et al., ApJ, **540**, 1016 (2000)
16. P. Massey and D.A. Hunter, ApJ **493**, 180 (1998)
17. P. Massey, K.E. Johnson, and K. Degioia-Eastwood, ApJ **454**, 151 (1995a)
18. P. Massey, C.C. Lang, K. Degioia-Eastwood, and C.D. Garmany, ApJ **438**, 188 (1995b)
19. G.E. Miller and J.M. Scalo, ApJS **41**, 513 (1979)
20. F. Palla, and S.W. Stahler, ApJ, **525**, 772 (1999)
21. N.C. Rana, AA, **184**, 104 (1987)
22. I.N. Reid, J.D. Kirkpatrick, J. Liebert, et al., ApJ, **521**, 613 (1999)
23. E.E. Salpeter ApJ, **121**, 161 (1955)
24. J.M. Scalo, Fund. Cos. Phys. **11**, 1 (1986)
25. J.M. Scalo, *The Stellar Initial Mass Function (38th Herstmonceux Conference)*, edited by G. Gilmore and D. Howell, ASP Conf. Ser. **142**, 201 (1998)
26. L. Siess, E. Dufour, and M. Forestini, AA, **358**, 593 (2000)
27. C.L. Slesnick, J.M. Carpenter, and L.A. Hillenbrand, ApJ, submitted (2004)
28. C.L. Slesnick, L.A. Hillenbrand, and P. Massey, ApJ, **576**, 880 (2002)
29. K.M. Strom, S.E. Strom, and K.M. Merrill ApJ **412**, 233 (1993)
30. F.J. Swenson, J. Faulkner, F.J. Rogers, and C.A. Iglesias ApJ, 425, 286 (1994)
31. L. Testi, F. Palla, T. Prusti, A. Natta, and S. Maltagliati, AA **320**, 159 (1997)
32. S. Yi, Y.-C. Kim, and P. Demarque, ApJS, **144**, 259 (2003)

A Square Degree Survey for Molecular Flows in TMC1

K.-F. Schuster[1], B. Lefloch[2], H. Ungerechts[3], C. Thum[1], F. Gueth[1], M. Sterzik[4], and H. Wiesemeyer[1]

[1] IRAM, 300 Rue de la Piscine, 38406 St Martin d'Heres, France
schuster@iram.fr
[2] IRAM, Avenida Divina Pastora, 7, E 18012 Granada, Spain
[3] LAOG, 414 Rue de la Piscine, 38406 St Martin d'Heres, France
[4] ESO, Alonso de Cordova 3107, Santiago, Chile

Abstract. We conducted a high-sensitivity high-resolution survey in the 230 GHz CO(2-1) line throughout a major fraction of the TMC1 complex (Heiles cloud 2) to search for molecular flows from young stellar objects. Our observations show a large number of unknown flows from known or unknown objects. The flows have a wide range of morphologies on scale sizes from 0.01 to 1 pc. A first analysis suggests an alignment among flows in parallel to magnetic fields as derived from optical polarization measurements. The area density of flows is sufficiently high to suggest an important impact on the cloud and the star formation within.

1 Introduction

Since the first detection of molecular flows from young stellar objects, the underlying physical processes have been subject of intense observational and theoretical work. A great number of problems related to the various phenomena connected to molecular flows (morphology, shocks, jets, etc.) have been modelled and are fairly well understood (see [1]). Other problems such as the entrainment of molecular gas and its relation to the parameters of the central object remain uncertain for the time being. A different but equally important question is the one of the impact of flows on the surrounding cloud and whether a statistically significant sample of flows can serve as an indicator for correlations among individual starforming events [2].

To obtain such a statistically significant and homogeneous sample, we conducted a sensitive, large-scale, high-resolution survey for molecular flows in the TMC1 region in the CO(2-1) line. The Taurus molecular cloud region has been chosen for the following reasons: With Taurus at a distance of 140 pc the spatial resolution of 140 AU ($11''$ with the IRAM 30m telescope) allows us to sample the molecular gas on parsec scale sizes down to the size of gravitationally bound envelopes. Furthermore, Taurus is relatively unaffected by high mass stars and therefore radiative effects on the cloud are limited. It is therefore conceivable that the effects of flows are unperturbed. Finally,

Taurus has been intensively studied in a large number of low-resolution, large-scale surveys (for example [6] [7]) in the mm range and also in a large number of observations in complementary wavelength regions.

2 Observations

A rectangular field of 1.5×1 degrees was observed with 6 square subfields of 30 × 30 minutes of arc. The observations were done with HERA, the 1.3mm heterodyne multibeam system at the IRAM 30m telescope [3] and in a frequency switched on-the-fly (OTF) mode. A frequency throw of 13.9 MHz was used which is a multiple of the natural standing wave pattern caused by the reflections between the SIS mixers and the telescope subreflector. Such a throw, together with the high switching frequency of 5 Hz results in a very flat spectral baseline even under medium weather conditions. The 3 × 3 array was swept either in RA or Dec with the array inclined by 18.4 degrees to the scanning direction (Fig. 1). Given the 24″ distance between beams this generates a sampling of 7.6″ perpendicular to the scanning direction; 6″ sampling was applied in the scanning direction. The autocorrelator spectrometer VESPA was used with a frequency resolution of 80 kHz (100 m/s at 230 GHz) and a bandwidth of 40 MHz per multi-beam element. Pointings and focus scans were made on planets (Mars and Venus) and the pointing errors were consistently better than 4″. The data were calibrated and reduced with the standard OTF calibration software (IRAM OTF-CAL). Baselines of order up to 3 were subtracted with a fixed window for line emission.

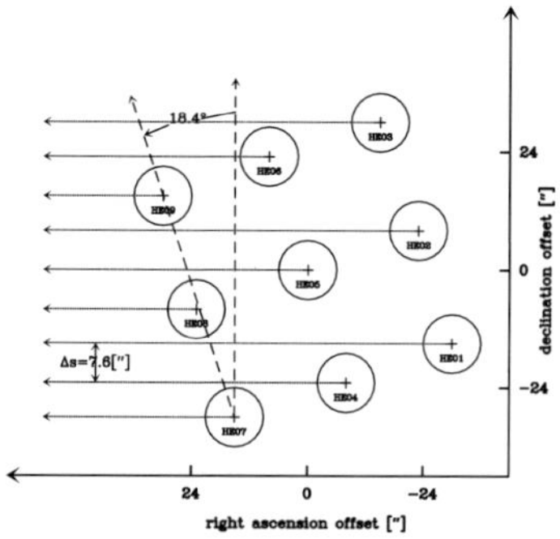

Fig. 1. Scanning scheme for the on-the-fly observations as used in the survey.

The relative calibration errors between different pixels were corrected as to reduce stripes. The correction factors were calculated by comparing the total integrated emission for the individual pixels for the individual submaps. Given the large size of the maps, this method gives very good results and only very small differences of the total emission might be introduced by fine structure in the emission. The resulting corrected maps are nearly free of gain related scanning effects.

3 Results

Within our survey we found a large number (18) of molecular flows. The detections of flows can be classified into three categories: a) four detections of known flows from known objects. b) four detections of previously unknown flows originating from known objects (4) and c) ten detections of previously unknown flows from unknown objects (10). A large additional number of tentative identifications of flows were made which need further investigation and are therefore not counted here. The problem in the identifications of these flows lies in the strong crowding of voids and streamers in the field.

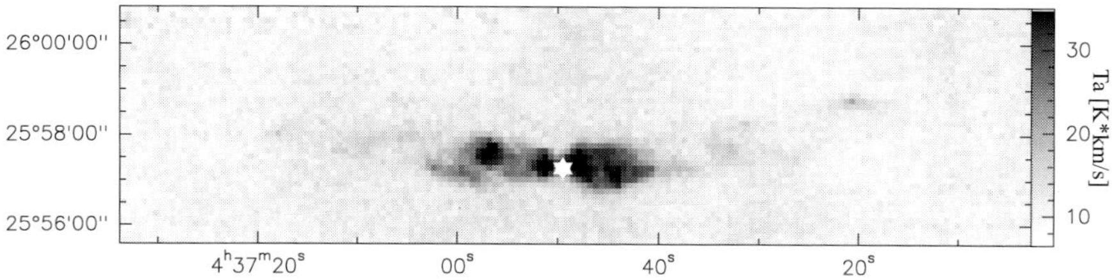

Fig. 2. Integrated CO(2-1) line emission from a parsec-scale molecular flow originating from the class 0 source L1527 (coord. 1950).

With the increased sensitivity of the present survey, an important number of previously known flows turn out to be considerably larger than in earlier observations [4] [5] with sizes on parsec scales (TMC1, L1527 for example, see Fig. 2). These large sizes have implications for the ages of the driving sources and the flows. Our finding reconfirms that parsec-scale molecular flows are common also in Taurus. The relative density of such objects implies that flows play an important role not only in the disruption of starforming cores but also in the dispersion of entire dark clouds. These dispersal effects might compete locally with with other effects (shock compression) that trigger starformation.

The preliminary analysis of our data reveal an important number of flows with unknown driving sources. In particular we detect a number of highly collimated and relatively short flows which may originate from very young sources.

The ensemble of detected flows shows a clear tendency for alignment along a position angle of 45° which is in agreement with the large scale magnetic field as derived from optical polarization measurements of background stars [8].

The survey also a reveals a previously unknown large-scale cloud component in the southern part of the field. This component has a very well

Fig. 3. The 1.5 × 1 degree field in the 4.1–4.4 km/s velocity range. The center of the field corresponds to RA 04h37m06s Dec 25°30′00″ (1950). The field shows a newly found cloud component with a northern rim running in a large angle to the filaments of the Heiles 2 cloud. The previously unknown, parsec-scale northern flow lobe of TMC1 is also clearly visible in the upper left part of the map.

defined rim which is dynamically and spatially distinct from the filaments of the Heiles 2 cloud (Fig. 3). This finding suggests that global dynamics in this region might be more complex than previously thought.

References

1. IAUS 182, Herbig-Haro Flows and the Birth of Low Mass Stars, Edt B. Reipurth and C. Bertout, 1997.
2. S. E. Strom, K. M. Strom 1987, IAUS 115, 255.
3. K.-F. Schuster, C. Boucher, W. Brunswig, M. Carter, J.Y. Chenu, B. Foullieux, A. Greve, D. John, B. Lazareff, S. Navarro, A. Perrigouard, J.-L. Pollet, A. Sievers, C. Thum and H. Wiesemeyer 2003, submitted to A&A
4. S. Zhou, N.J. Evans, & Y. Wang 1996, ApJ 466, 296.
5. C.J. Chandler, S. Terebey, M. Barsony, T.J. Moore, T.N. Gautier 1996, ApJ 471, 308.
6. H. Ungerechts & P. Thaddeus 1987, ApJ SS, 63:645–660.
7. T. Onishi, A. Mizuno, A. Kawamura, H. Ogawa & Y. Fukui 1998, ApJ 502, 296.
8. G. Duchêne, F. Ménard 2003, Proc. of "Open Issues in Local Star Formation and Early Stellar Evolution", Ouro Preto (Brazil)

Submillimetre Continuum Mapping: Finding Clues to the Physics of Star Formation

D. Johnstone

National Research Council Canada, Herzberg Institute of Astrophysics, 5071 West Saanich Road, Victoria, BC, V9E 2E7, Canada
doug.johnstone@nrc-crnc.gc.ca

1 Introduction

Our knowledge of the small-scale submillimetre emission within star-forming regions has grown considerably since the last Zermatt Symposium in 1998. At that time only a few locations of star formation had been mapped, most notably in Ophiuchus [1] and Orion A [2]. Since the advent of sensitive submillimetre bolometer arrays on a large number of telescopes (e.g. CSO, IRAM, JCMT), 1000 square-arcminute maps have been routinely produced, imaging nearby star-forming cores in their entirety. These observations are sensitive to sub-stellar mass condensations and have confirmed the earlier results that the masses of dense clumps span the range of stellar masses and that the cumulative mass function is steeper than the mass function of molecular clouds and cloud fragments [1, 3, 4, 5]. Much theoretical and observational work at present, is centered around the interpretation of these phenomena.

2 Observational Considerations

Despite the relative simplicity with which submillimetre continuum observations are obtained, reduction of the data is not entirely straightforward. Most detectors still use a chopping procedure to remove the submillimetre bright and quickly varying atmosphere of the Earth. Producing reliable maps of star-forming regions requires careful consideration of the information contained in the chopped data. While features that are smaller or similar in size to the chop throws used are easily reconstructed, larger features are often poorly retained. Indeed, algorithms which attempt to reproduce these large spatial modes also enhance the noise associated with incomplete removal of the bright sky, producing low amplitude, long spatial wavelength artifacts in the reconstructed maps [6, 7]. Thus, within the large submillimetre images, only the small-scale (i.e. up to a few arcminutes) structure is reliable. In addition, bright features in the maps tend to have negative bowls around them due to the missing long spatial wavelength information, complicating measurements of structure near strong sources.

Once reconstructed maps are produced, interpretation of the submillimetre emission is mostly straightforward. The observed emission is almost always due to thermally radiating dust, typically at temperatures of $T_d = 8 - 40\,\mathrm{K}$. Additionally, the emission is usually optically thin. Thus, the continuum emission is a reasonable tracer of the column density of dust, and by extrapolation gas, within the map. Constraints can be placed on the dust properties, particularly the dust temperature and emissivity profile, through observations at multiple wavelengths [8, 9] and spectral energy distribution (SED) fitting [10]. Often, however, the dust temperature is not reliably determined and must be estimated from either the luminosity of the embedded source (e.g. measured by IRAS) or the strength of the nearby interstellar radiation field. At submillimetre wavelengths, the dust emission is approximately linear with temperature, except for in the coldest regions, and thus precise knowledge of the temperature is not crucial. The emissivity of dust at submillimetre wavelengths is much more uncertain, with noticeable dependence on the dust composition, shape, and size [11]. Assuming, however, that the dust is at a near constant temperature and has uniform composition, relative column densities and masses may be measured within a map and statistical measures of the small-scale properties may be obtained.

3 Structure Within Star-Forming Regions

A great deal of structure is observed in the submillimetre maps of star-forming regions. Figure 1 shows the Integral Shaped Filament (ISF) in Orion A, observed at 450μm and $850\,\mu$m [2]. An extended ridge runs north-south, connecting each of the well known star-forming cores OMC1, OMC2, and OMC3. Near OMC1, additional filamentary fingers can be seen radiating from the core and likely tracing the boundaries of protostellar jets and outflows. A dust shell is visible just to the north-east of OMC1, surrounding the HII region M43. The ratio of the 450μm to $850\,\mu$m emission reveals that the dust becomes colder where the submillimetre emission is brightest, such as along the spine of the filament. This result is also observed for isolated pre-protostellar cores[12]. The high optical depths within these regions prevents the penetration of the interstellar radiation field, leading to a reduced equilibrium temperature. Care must be taken in interpreting these results, however, as evolution of dust properties with density or contamination of the observed continuum band by molecular line emission may complicate the analysis [11, 13].

The extended ISF has a radial column density profile which drops off too slowly with radius to be fit by a purely hydrostatic isothermal cylinder [2, 14]. A number of massive filaments have been observed with density profiles similar to that of the ISF [15, 16, 17]. Fiege and Pudritz [18] provide reasonable fits to these profiles by including a helical magnetic field embedded within the filament. The helical magnetic model in also in agreement with submillimetre polarimetry of the ISF [19]. Recently, the first filament which may be fit by an isothermal cylinder model has been identified [8] using

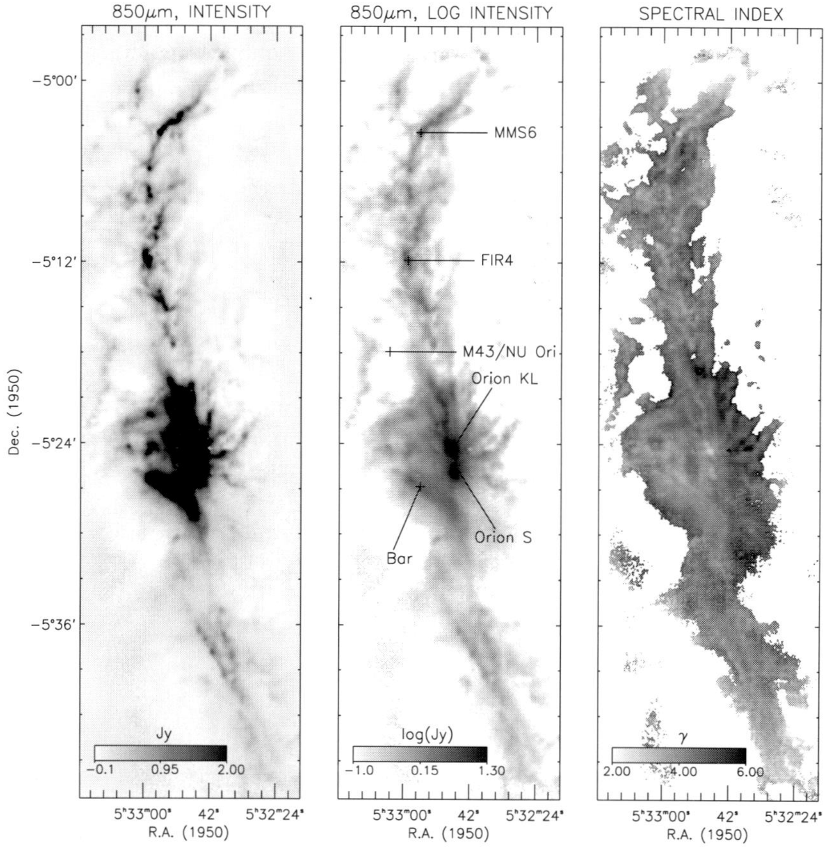

Fig. 1. The 850μm emission and 450μm to 850μm spectral index distribution in the Integral Shaped Filament (ISF) at the northern end of the Orion A molecular cloud. The Orion Nebula is located directly in front of the main concentration of emission from the OMC1 core at the center of the image. Note the filaments extending radially away from this core. **(a).** The 850μm image showing the observed flux from -0.1 to 2 Jy/beam with a linear transfer function. **(b).** The 850μm image showing the observed flux from 100 mJy/beam to 20 Jy/beam with a logarithmic transfer function. **(c).** The 450μm to 850μm spectral index (γ) in the range 2 to 6 with a linear transfer function.[2]

submillimetre continuum imaging. Understanding the equilibrium physics of the observed filamentary structures should shed light onto their formation mechanisms and ultimately the physical processes at play within molecular clouds.

Imaging of lower-mass nearby star-forming regions, such as L1551 in Taurus, in submillimetre continuum emission also reveals structure (Fig. 2). The locations of protostellar sources are illuminated, revealing the enshrouding material out of which the stars are forming, (e.g. IRS 5 and HL Tau) and circumstellar disk material (HH 30). As well, evidence for sculpting motions within the molecular cloud are observed in the enhanced column densities seen at the extremes of outflows, such as the L1551 flow emanating from the image centre. Finally, extended, diffuse condensations are observed, coincident with known optical extinction, such as the dark bay around HH 265 to the northwest of L1551.

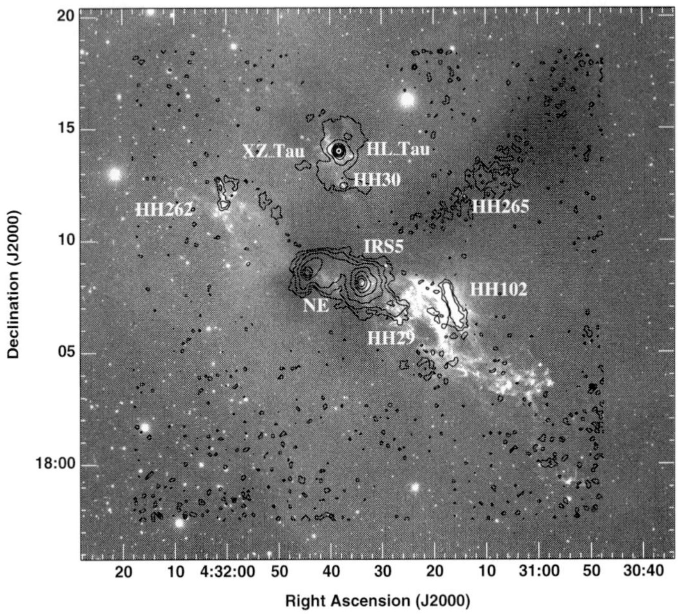

Fig. 2. Contours of 850μm emission superposed on a SII image of the L1151 region in Taurus.[20]

4 Physical Properties of the Small-Scale Structure

The most important result of the large-area submillimetre continuum mapping of star-forming regions has been the detection of small-scale condensations. Many of these are without corresponding IR emission and are thus candidate pre-protostellar objects [1, 3, 21, 4]. The individual clumps are either unresolved or barely resolved, yielding sizes of order $R_c \sim 3-10\times10^3$ AU and have total masses in the range $M \sim 0.1 - 10\,M_\odot$, assuming typical dust properties. Figure 3 plots the distribution of clumps mass versus size for the Ophiuchus core [3]. A majority of the clumps lie well within the *stable* regime, where thermal pressure within the clump provides adequate support against gravitational forces. Only the few, outlying points at the top of the figure lie clearly in the unstable regime.

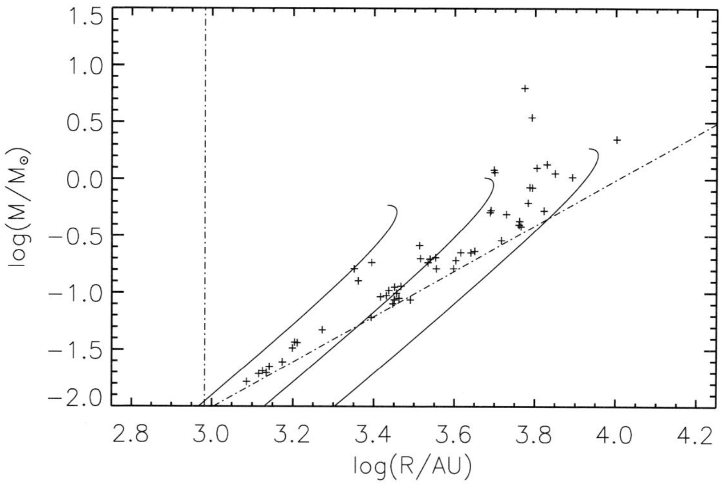

Fig. 3. Derived mass and size for each of the identified clumps in Ophiuchus. Also plotted are the resolution limit and the minimum mass that a clump must have to be visible above the background (dashed lines). The three solid curves from left to right denote the mass-radius relation for Bonnor-Ebert spheres with 20 K internal temperatures and external pressures $P/k = 1, 3.3, 10 \times 10^6\,\mathrm{K\,cm^{-3}}$. Note that only those clumps lying above the turn-over of the solid curves are unstable to collapse.[3]

Figure 3 was obtained under the assumption that the dust temperature within each clump is 20 K. Lowering the assumed dust temperature increases the mass within each clump, bringing them closer to instability. To better analyse clump stability, the emission profile of each clump can be measured. As a gas cloud becomes gravitationally dominated, the central density rises significantly and the surface density profile becomes peaked. Thus, the observed concentration of a clump is a measure of the importance of gravity within each submillimetre source. In both Ophiuchus [3] and Orion B [4] the majority of clumps have concentrations consistent with support dominated by dominated by thermal pressure. Only a handful of sources appear to be unstable to collapse.

Once the concentration is measured, each source may be compared with equilibrium isothermal spheres subjected to a finite boundary pressure (Bonnor-Ebert spheres [22, 23, 3, 4]) to deduce the physical conditions of the clump (mass, temperature, and bounding pressure) from the observational constraints (brightness, size, concentration). Figure 4 shows the results of such an analysis in Ophiuchus [3]. The derived temperatures of the clumps lie in the range 10 – 20 K while the bounding pressures are $P/k \sim 10^{6-7}\,\mathrm{K\,cm^{-3}}$. The deduced bounding pressures for the clumps is two to three orders of magnitude greater than the Galactic ISM value but agrees well with calculations of the pressure deep within molecular clouds [3, 4].

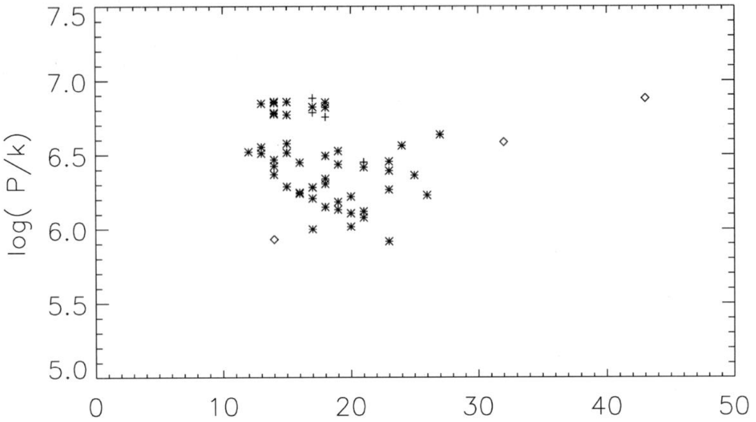

Fig. 4. Results of determining the bounding pressure and dust temperature of clumps in Ophiuchus using the assumption that they are well represented by Bonnor-Ebert spheres with measured concentrations.[3]

Significant caveats apply to these measured clump properties. Most clumps are only moderately resolved, complicating the measure of concentration. Additionally, the observational concentrations have been compared to static pre-collapse structures. Accreting protostellar objects, particularly during the Class 0 stage when the infalling envelope depletes the central region of the clump, may have low concentrations similar to those observed. Thus, while the lack of high concentration sources constrains the number of unstable pre-protostellar objects, additional tests are required to determine whether the less concentrated majority of sources are stable or not.

The mass function of the clumps is independent of the model for clump structure, assuming that all clumps have a similar temperature. Figure 5 plots the results for Ophiuchus [3], showing that the cumulative number function of clumps is steep, i.e. $N_c \propto M^{-1.5}$ or $M^{-1.0}$. This result is in agreement with other submillimetre observations of Ophiuchus [1], Serpens [24], and Orion B [4, 21] and reveals that the distribution of clump masses is similar to the Salpeter initial mass function of stars, i.e. $N_* \propto M^{-1.35}$. The distribution of molecular clouds and molecular cloud cores is much flatter [5] than this relation, i.e. $N_{mc} \propto M^{-0.5}$. While most of the mass in clouds resides in the few large objects, most of the mass in the submillimetre clumps, and in stars, resides in the numerous small objects.

Small-scale clumps have masses on par with stellar masses and have a steep cumulative mass function. Whether the similar mass function provides evidence for a direct link between the clumps and stars remains controversial. When modeled as hydrostatic structures, the physical properties of the clumps are found to favour stability versus collapse, requiring extreme increases in either bounding pressure or clump mass to become unstable. Al-

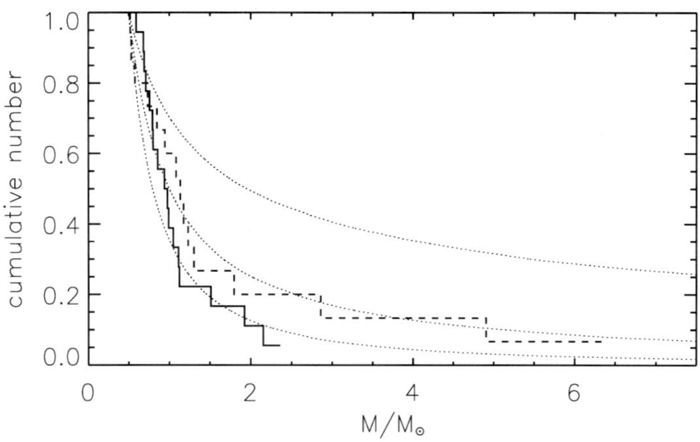

Fig. 5. Cumulative number function for clumps with mass greater than $M = 0.5, M_{\odot}$. The dashed line converts the measured integrated flux for each clump directly to mass using a constant temperature $T_d = 20\,\mathrm{K}$. The thick line converts the flux to mass using the derived temperature from fitting the clumps to Bonnor-Ebert spheres. The overlaid dotted curves, from bottom to top, show the expected relation if the clump number function satisfies a power law with $M^{-1.5}$, $M^{-1.0}$, and $M^{-0.5}$ respectively.[3]

ternative models exist, however. For example, turbulent cloud calculations predict short-lived small-scale clumping which may mimic the appearance of stable hydrostatic sources [25]. These wave interfaces are only transient phenomena and are unlikely to collapse. More information on both the individual clumps and the environments in which the clumps reside are required before definitive answers are found.

5 The Future of Sub-millimetre Mapping

Large-area submillimetre mapping of molecular clouds is now routine. To date most of the observations have been biased toward known locations of star formation within each cloud. It is unclear whether the distribution of clumps and clump masses will change significantly as lower column density regions are observed within each cloud. Circumstantial evidence is accumulating, however, that shows the structure in the submillimetre emission diminishes. The submillimetre sources in Orion B cluster within the high density CS cores, avoiding regions of low column density [4].

Theoretical arguments suggest that submillimetre bright clumps are likely biased toward large optical depth, high-pressure cores within each molecular cloud [26]. These cores should have lower ionization fractions then the rest of the cloud, allowing for significant damping of magnetically-driven turbulence and the formation of small, dense, thermalized structures. Even if thermalized clumps form throughout the cloud, however, it is only in the high pressure regions that they will be small and dense.

The first unbiased submillimetre maps of entire molecular clouds are just now being obtained, mostly in conjunction with the SIRTF Legacy program c2d [27]. Preliminary results on a four square degree field in Ophiuchus suggest a lack of bright small-scale structure anywhere except the previously-observed core [28]. When completed these surveys will make significant progress in our understanding of the most massive thermal clumps in molecular cores. Much deeper maps will be necessary, however, to probe the complete stellar mass range in low pressure regions of the molecular cloud. The arrival of SCUBA2 (about 1000x faster than SCUBA) at the JCMT in 2006 should revolutionize our understanding of submillimetre structure in the entire molecular cloud, just in time for Zermatt 2008!

References

1. F. Motte, P. André, R. Neri: A&A **336**, 150 (1998)
2. D. Johnstone, J. Bally: ApJ **510** L49 (1999)
3. D. Johnstone, C.D. Wilson, G. Moriarty-Schieven et al: ApJ **545**, 327 (2000)
4. D. Johnstone, M. Fich, G.F. Mitchel et al: ApJ **559**, 307 (2001)
5. C. Kramer, J. Stutzki, R. Röhrig et al: A&A **329**, 249 (1998)
6. D.T. Emerson: In: *Multi-Feed Systems for Radio Telescopes*, ASP Conf. Ser. 75, ed by D.T. Emerson, J.M. Payne (ASP, San Francisco 1995) p 309
7. D. Johnstone, C.D. Wilson, G. Moriarty-Schieven et al: ApJS **131**, 505 (2000)
8. D. Johnstone, J.D. Fiege, R.O. Redman et al: ApJ **588**, L37 (2003)
9. C. Kramer, J. Richter, B. Mookerjea et al: A&A **399**, 1073 (2003)
10. R. Chini, B. Reipurth, D. Ward-Thompson, et al: ApJ, **474**, L135 (1997)
11. V. Ossenkopf, Th. Henning: A&A **291**, 943 (1994)
12. N.J. Evans, J.M.C. Rawlings, Y.L. Shirley et al: ApJ, **557**, 193 (2001)
13. D. Johnstone, A.M.S. Boonman, E.F. van Dishoeck: A&A, in press (2003)
14. J. Ostriker: ApJ **140**, 1056 (1964)
15. J. Alves, C.J. Lada, E.A. Lada, et al: ApJ **506**, 292 (1998)
16. C.J. Lada, J. Alves, E.A. Lada: ApJ, **512**, 250 (1999)
17. B. Stepnik et al: A&A, **398**, 551 (2003)
18. J.D. Fiege, R.E. Pudritz: ApJ, **544**, 830 (2000)
19. B. C. Matthews, J.D. Fiege, G. Moriarty-Schieven: ApJ, **569**, 304 (2001)
20. G. Moriarty-Schieven, D. Johnstone, J. Bally: in preparation.
21. F. Motte, P. André, D. Ward-Thompson et al: A&A, **372**, L41 (2001)
22. W.B. Bonnor: MNRAS, **116**, 351 (1956)
23. R. Ebert: Z. Astrophys., **37**, 217 (1955)
24. L. Testi, A.I. Sargent: ApJ, **508**, L91 (1998)
25. J. Ballesteros-Paredes, R. Klessen, E. Vázquez-Semadeni: ApJ, **592**, 199 (2003)
26. C.F. McKee: ApJ, **345**, 782 (1989)
27. N.J. Evans, L.E. Allen, G.A. Blake et al: PASP, **115**, 965 (2003)
28. D. Johnstone, J. Di Francesco, A. Goodman et al: in preparation.

Survey for High-Mass Protostars in Cygnus X

S. Bontemps[1], F. Motte[2], N. Schneider[3], and P. Schilke[4]

[1] Observatoire de Bordeaux, BP89, 33270 Floirac, France
bontemps@obs.u-bordeaux1.fr
[2] SAp, CEA/Saclay, 91191 Gif-sur-Yvette, France
motte@discovery.saclay.cea.fr
[3] Observatoire de Bordeaux, BP89, 33270 Floirac, France
schneider@obs.u-bordeaux1.fr
[4] MPIfR, Auf dem Hügel 69, 53121 Bonn, Germany
schilke@mpifr-bonn.mpg.de

1 Introduction

The process of the formation of high-mass stars is still largely unknown and poorly constrained by observations. Both theoretical and observational approaches are paved with serious difficulties. The models face the same limitations as for the low-mass regime but in an even more extreme case in terms of short timescales, large infall rates and large central masses (e.g. Stahler et al. [11]). The rare occurence of massive stars renders observations and surveys for proto-OB stars difficult. The search for and the studies of collapsing high-mass protostellar objects (HMPOs) are therefore of prime importance (e.g. Kurtz et al. [5]).

Since HMPOs are expected to be deeply embedded inside infalling envelopes, and to be as luminous as OB stars (10^3 to $10^6\,L_\odot$), these rare objects are usually believed to be among the brightest IR sources of the sky and have therefore been actively surveyed using the IRAS catalog (e.g. Wood & Churchwell [12]). Compelling evidences for the existence of a "cold" phase of HMPOs which would be missed by IRAS-based samples have recently been driven mainly by dust millimeter continuum imaging of high-mass star formation sites (e.g. Sandell [10]; Molinari et al. [6]; Motte et al. [9]; Garay et al. [4]). It demonstrates that like for the low-mass regime (e.g. André et al. [1]), IR samples of protostellar objects suffer from severe biases toward bright IR sources, and are not suitable to homogeneously follow the evolutionary path from pre-stellar condensations to young embedded stars. In contrast, since the dust continuum emission is optically thin in the millimeter range and therefore directly traces the circumstellar masses, surveys based on millimeter continuum imaging are much less biased.

As part of a long-term and global study of the star formation sites we recently started in Cygnus X, we present here some first results of a complete imaging survey of the complex in continuum at 1.2 mm using MAMBO at the IRAM 30 m telescope. It is part of a comprehensive observational project in Cygnus X which also comprises a large-scale $^{13}CO(3–2)/(2–1)$ and $N_2H^+(1–$

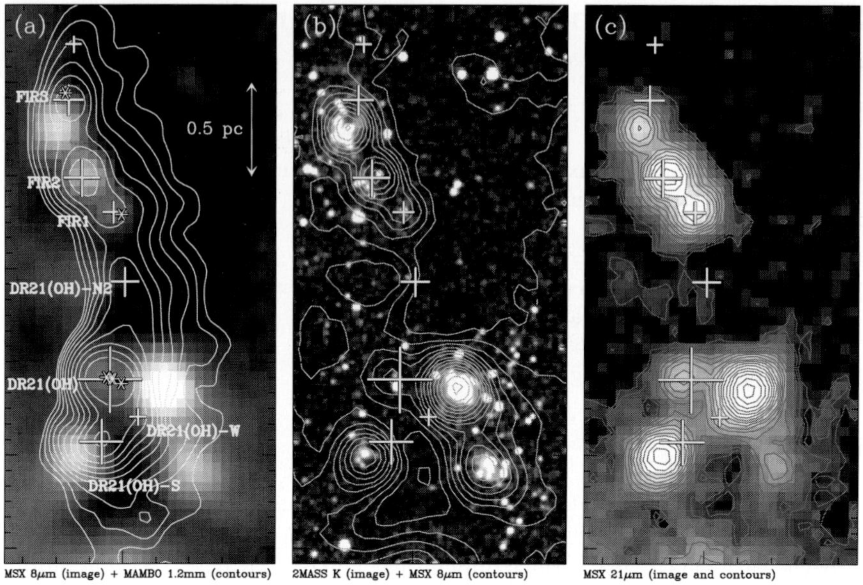

Fig. 1. (a) $3' \times 6'$ MAMBO 1.2 mm contour map around DR21(OH) overlayed on MSX 8 μm. Crosses: millimeter "protostars" detected in the region (sizes of symbols scale with envelope masses). Source names from Chandler et al. [3]. Star symbols: water masers from Braz & Epchtein [2]. **(b)** and **(c)** near-IR (2MASS) versus mid-IR (MSX) emission. The mid-IR sources coincide well with the nearby young star clusters but only marginally with the protostars.

0) survey with KOSMA and the FCRAO, as well as a full analysis of IR and radio databases (2MASS, MSX, IRAS, HI) toward this region.

2 Observations

The millimeter continuum and heterodyne observations were both obtained at the IRAM 30 m telescope. The continuum imaging has been achieved during the 2000/2001 winter (0.7 sq. deg. with the MAMBO 37 channel array) and the 2002/2003 winter (2.3 sq. deg. with the MAMBO-II 117 channel array). With MAMBO-II a fast mapping technique using a $8''$/sec scanning speed and a $46''$ step between the scans has been used leading to a high imaging speed of $\sim$ 0.1 squ. deg. per hour, and reaching a final rms of $\sim$ 15 mJy (see Motte, Bontemps et al. in prep). The molecular lines were observed in Sept. 2003 using 2 different settings of the 4-receiver configuration: [SiO(2–1), CH_3CN(4–3), H_2CO 218 GHz, CH_3OH 241 GHz] and [N_2H^+(1–0), CS(2–1), CS(5–4), HCO^+(3–2)].

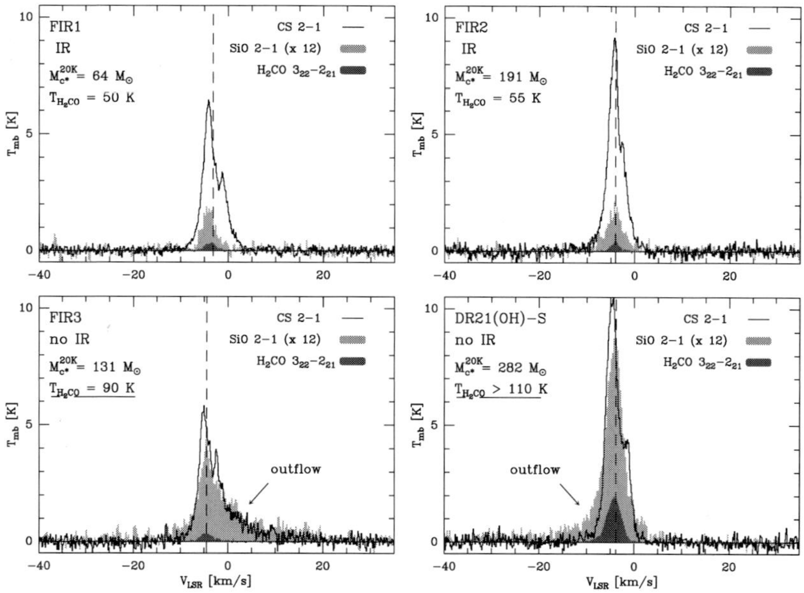

Fig. 2. IRAM 30 m spectra toward the 4 most massive "millimeter" protostars after DR21(OH) from Fig.1. FIR1 and FIR2 may have weak outflows but probably do not contain any significant hot core (low H_2CO temperatures), while the 2 IR-quiet sources, DR21(OH)-S and FIR3, show evidences for outflows and hot cores. The N_2H^+ rest velocities are shown as dashed vertical lines. T_{H_2CO} is the kinetic temperature derived from the $3_{22} - 2_{21}/3_{03} - 2_{02}$ line ratio.

3 A Population of IR-Quiet HMPOs in Cygnus X

Using the source detection algorithm described by Motte et al. ([7], [9]), a total of 125 compact millimeter sources have been extracted. The gas masses of these millimeter cores derived from their millimeter fluxes (with T_{dust} = 20 K, d = 1.7 kpc and a typical dust emissivity, e.g. Motte & André [8]) range from a few to several hundreds of $M_\odot$, and reach a total gas mass for all sources of almost 6000 $M_\odot$. The most massive cores are in the high-mass range, and since the average densities are typically a few 10^5 to a few $10^6\,cm^{-3}$, these cores are excellent candidates to host massive protostellar precursors (i.e. HMPOs), either pre-stellar or protostellar in nature.

We have cross-correlated this sample of "millimeter" HMPOs with near-IR and mid-IR data from 2MASS and MSX, and it was surprising to find that a large fraction of the millimeter HMPOs does not coincide with any bright IR source. Figure 1 shows the DR21(OH) region well illustrating this point for the whole Cygnus X survey (3 sq. deg. in total). Among the eight millimeter sources in this region, only two coincide with relatively bright MSX 8 and 21 μm sources, one coincides with a weak mid-IR source (DR21(OH)-S), and five are not emitting in the mid-IR or are pretty weak sources (typically at a level of less than 1–2 Jy at 21 μm).

The molecular line observations of these HMPO candidates, and especially of the IR-quiet objects, clearly show that they are actual protostars since they show outflows and hot core tracers (see Figure 2).

4 Conclusions

The nearby Cygnus X molecular complex is a rich high-mass star forming regions providing the first unbiased sample of massive HMPOs in a single complex. A significant population of IR-quiet HMPOs is found and will be discussed in more detail in forthcoming papers. The clear indications of outflows and sometimes of hot cores demonstrate the protostellar nature of these objects. Their properties appear very similar to the Class 0 low-mass protostars, and it is tempting to assimilate these IR-quiet HMPOs to massive Class 0s.

The proximity of the Cygnus X complex compared to most galaxy-wide works found in the literature allowed us to resolve acceptably well the star-forming sites. It shows that bright IR sources are more often associated with young embedded clusters (i.e. with the OB young stars inside them) than with the nearby HMPOs. One should therefore be very cautious in associating bright IR sources with HMPOs when the spatial resolution is not large enough to resolve individual cores of typical size 0.1 pc.

References

1. P. André, D. Ward-Thompson, M. Barsony. In: *Protostars and Planets IV*, ed by Mannings, Boss, & Russell (Univ. of Arizona Press, Tucson 2000) p 59
2. M.A. Braz, N. Epchtein: A&A Sup. Ser. **54**, 167 (1983)
3. C. Chandler, W.K. Gear, R. Chini: MNRAS **260**, 337 (1993)
4. G. Garay, K. Brooks, D. Mardones, R. Norris, M. Burton: ApJ **579**, 678 (2002)
5. S. Kurtz, R. Cesaroni, E. Churchwell, et al. In: *Protostars and Planets IV*, ed by Mannings, Boss, & Russell (Univ. of Arizona Press, Tucson 2000) p 299
6. S. Molinari, L. Testi, J. Brand, R. Cesaroni, F. Palla: ApJ **505**, L39 (1998)
7. F. Motte, P. André, R. Neri: A&A **336**, 150 (1998)
8. F. Motte, P. André: A&A **365**, 440 (2001)
9. F. Motte, P. Schilke, D.C. Lis: ApJ **582**, 277 (2003)
10. G. Sandell: A&A **358**, 242 (2000)
11. S.W. Stahler, F. Palla, P.T.P. Ho. In: *Protostars and Planets IV*, ed by Mannings, Boss, & Russell (Univ. of Arizona Press, Tucson 2000) pp 327
12. D.O.S. Wood, E. Churchwell: ApJ **340**, 265 (1989)

Induced Massive Star Formation in Dense Molecular Clouds

Cometary Globules in HII Regions

B. Lefloch[1], J. Cernicharo[2], L. Deharveng[3], and A. Zavagno[3]

[1] Laboratoire d'Astrophysique, Observatoire de Grenoble, BP53X, 38041 Grenoble cedex 9, France, lefloch@obs.ujf-grenoble.fr
[2] DAMIR, IEM, C. Serrano, 121, 28006 Madrid, Spain, cerni@damir.iem.csic.es
[3] Observatoire de Marseille, France, lise.deharveng@oamp.fr
[4] Observatoire de Marseille, France, annie.zavagno@oamp.fr

1 Introduction

Many observations suggest that HII regions could play an important role in spreading star formation throughout the Galaxy. As reviewed in [6], most of the embedded young stellar clusters in the solar neighbourhood are adjacent to HII regions excited by more evolved stars : Orion, Perseus OB2, the Rosette nebula, W3-5, M17, to name but a few. [13] estimate that 10%–30% of stars in mass in the Galaxy are formed under the influence of adjacent HII regions.

Various scenarii of star formation triggered in the expansion of an HII region have been proposed but detailed comparison with the observations, especially in the early protostellar phase, is still missing however. In order to better characterize the triggering mechanisms and the induced protostellar population, we have started a systematic search for induced star formation in Cometary Globules (CGs), i.e. the dense condensations located in the molecular shell (the "PDR" region) around HII regions. Several theoretical and observational studies ([1, 7, 8]) have shown the important role played by the Ly-continuum radiation of the nebula in the evolution of these condensations, which is now well understood. Since the first evidences that CGs *do* form stars ([12]) it has become clear that these are very efficient stellar nurseries. Systematic studies ([14, 15, 4]) show that star formation is rather common in CGs and that they tend to form more massive and luminous (proto)stars, with luminosities $\geq 10^3\ L_\odot$. [14] estimate that more than 15% of the massive stars form in CGs. It is therefore not so surprising that most of the "true" high-/intermediate-mass protostars have been discovered in the vicinity of HII regions (see contribution by Sandell).

Our search was based on two criteria : 1) a morphology of the ionized nebula nebula and the Photon Dominated Region as simple as possible 2) the presence of a young stellar cluster at the border of the nebula. We discuss below the cases of Sh104 (Sect. 2) and the Trifid nebula (Sect. 3), which illustrate two different modes of induced star formation.

2 The "Collect and Collapse Case": Sh104

Sh104 is the prototype region of massive star formation induced by the fragmentation of the molecular layer surrounding an expanding HII region. In this scenario, first proposed by [5], material of the ambient cloud accumulates in a dense layer between the ionization front and the shock front preceding it, as the HII region expands with supersonic velocity. Instabilities may develop when the layer decelerates; two characteristic timescales come into play. The short timescale ($\sim$ internal sound crossing time) leads to the formation of low-mass stars. The layer may also fragment and collapse gravitationally on the longer timescale $\tau_l \sim 0.5\,(G\rho_0)^{-1/2}$ (1–3 Myr in gas of density $\rho_0 = 10^3$–$10^2\,\mathrm{cm}^{-3}$, respectively). Collapse then occurs along the layer's length and the fragments are massive enough to form massive stars or clusters.

Sh104 is an optically visible HII region whose main exciting star is an O6V, at $\sim 4.0\,\mathrm{kpc}$. Both the size (8 pc) and the low electron density ($n_e \sim 1 - 10\,\mathrm{cm}^{-3}$) indicate the region is at an advanced evolutionary stage. An UC HII region lies at its eastern border; it coincides with the IRAS source 20160+3636 and a young near-IR stellar cluster. The large luminosity of the cluster ($L \sim 3 \times 10^4\,L_\odot$) implies the presence of at least one massive star.

The low-density material, mapped in the ^{12}CO(2-1) line, (see Fig. 1 and [3]) is distributed along a ring, which entirely surrounds the HII region. The bulk of the emission arises in a narrow velocity range ($4\,\mathrm{km\,s^{-1}}$) and does

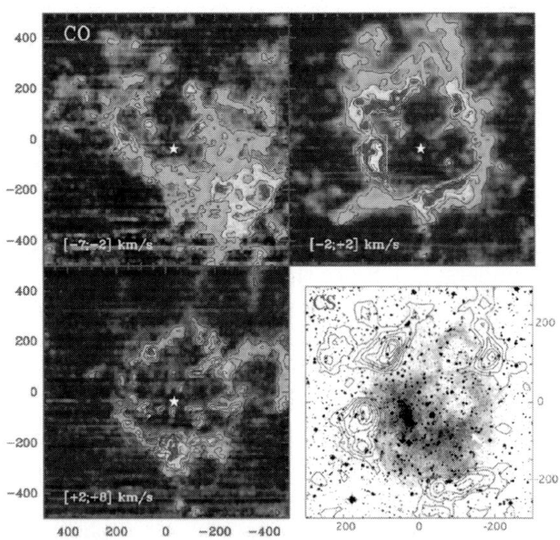

Fig. 1. Sh104 : the first three panels show the ^{12}CO(2-1) velocity integrated emission. The white star marks the position of the exciting star. Lower right panel : distribution of the CS(2-1) flux integrated between -3 and $+5\,\mathrm{km\,s^{-1}}$. The contours are superposed on the DSS2-red frame of Sh104.

not extend very far away from the ring . The dense molecular gas, as traced by CS (Fig. 1), is distributed in four main fragments *regularly* spaced. The properties of the fragments (size, density, mass) are all very similar. The eastern fragment is divided in dense massive subunits or "cores" (70–100 $M_\odot$), separated by one Jeans length. The young stellar cluster appears embedded in one of the fragments and lies halfway between two cores. The gravitational free-fall timescale in these cores is $\sim 7 \times 10^4$ yr, whereas the age of the region is probably ~ 1 Myr. This is most probably the origin of the observed cluster.

The regular spacing of the fragments and the location of the stellar cluster in one of these fragments provides strong evidence in favor of the "collect and collapse model". It excludes that the stellar cluster formed from a pre-existing clump a) that collapsed spontaneously and was subsequently "immersed" in the compressed layer, or b) whose collapse was induced by collision with the compressed layer. The cluster probably represents the second generation of stars, which formed from the fragmentation of the swept-up layer due to the gravitational instability which develops on a timescale τ_l .

It is possible to reproduce the fragment properties (Whitworth 2004, in prep) by assuming that the accumulation of material in the layer occurs in a cloud of *finite* mass and stops at some point (unlike the simple case usually considered, like in [16]). Our CO observations show indeed that the low-density gas distribution is almost confined in the neutral shell (apart from a higher velocity component with low gas column density). The location of the fragments around the rim of the nebula is naturally accounted for if the parental cloud was initially slightly aspherical.

3 Shock-Induced Star Formation in the Trifid Nebula

The Trifid nebula (M20) illustrates a case of shock interaction with the ambient cloud that leads to the formation of another generation of stars. This HII region is excited by an O 7.5 III star (HD 164492 A), which provides the same number of ionizing photons as in f Sh104. M20 is however much closer (1.7 kpc), more compact (its diameter is ~ 3.5 pc) and ten times denser Either the time-radius relation for HII regions or the analysis of the photoionized globule TC2 in the South of M20 ([11]) indicate an early dynamical age of $\simeq 0.3$ Myr. This is consistent with the detection of photoionized disks (proplyds) around companions from the stellar cluster associated with HD 164492 A ([10]). A multiwavelength study has been carried out using the ISO satellite and the largest millimeter and submillimeter ground-based telescopes (Fig. 2).

3.1 Protostellar Content

The first result is the discovery of several massive protostellar cores in the dense shell surrounding the nebula ([2]). Apart from tiny gas clumps which

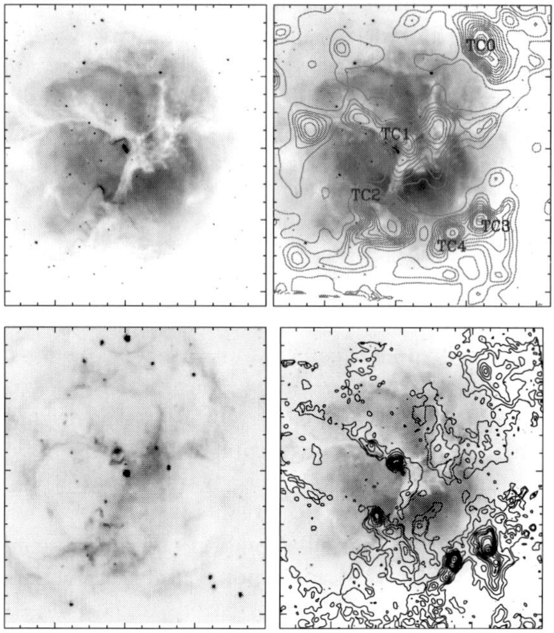

Fig. 2. M20 : (up. left) in the optical ($H\alpha$); (low left) at $11\mu m$ seen with ISOCAM; (up. right) in $HCO^+ J = 1 \rightarrow 0$ seen with the SEST; (low. right) at 1.3mm with the IRAM 30m telescope.

appear to survive inside the ionized phases, the emission of the dense molecular gas, as traced with HCO^+, is distributed over a thin shell surrounding the HII region. The emission from the dust lanes (the "arms" of the Trifid, seen in absorption in the optical in Fig. 2) shows that the nebula is expanding at about 5–8 $\mathrm{km\,s^{-1}}$. Five condensations (dubbed TC0-4) have been detected in the dense shell (TC1 lies in the Eastern dust lane). There is no evidence of ongoing star formation towards TC0. Flux isocontours rather suggest the condensation is just being reached by the ionization front of M20. Mapping of the thermal dust emission at 1.3mm has revealed strong peaked emission towards TC1-4. The dust cores are compact ($\sim$ 0.1–0.2 pc) and massive (20 to 90 $M_\odot$), with luminosities in the range 500–2000 $L_\odot$. Four of these cores (TC1-4) exhibits signs of protostellar activity, either Herbig-Haro jet (TC2, [11]) or bipolar outflows (TC1, 3, 4; [9]). Only TC4 was detected with ISOCAM in the mid-IR, implying at least some of the sources to be "true" (Class 0) protostars. The parameters of the cores suggest TC1-4 to be intermediate-mass objects protostars. A detailed molecular line analysis of the TC3 and TC4 cores and their outflows confirms this result ([9]). The molecular line emission in these cores is similar to that detected in the Orion protostellar cores, once scaled to the same distance.

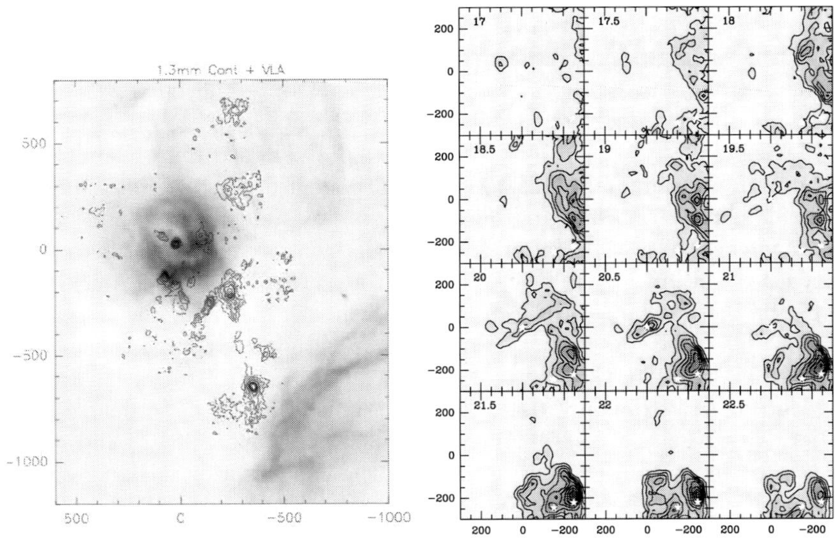

Fig. 3. (*left*) 1.3mm dust emission map (contours) superposed on a 20cm VLA image of the M20 region. (*right*) Velocity-channel map of the $C^{18}OJ = 1-0$ emission on the Western side of the Trifid. White stars mark the position of HD 164492 A, TC3-4.

3.2 Interaction with the Ambient Medium

The mapping of the molecular gas and dust emission over a large field around M20 ($20' \times 30'$) shows the nebula to be immersed in a complex web of filaments that extend over a scale of several parsecs (Fig. 3). Two additional condensations have been discovered : a cold clump (TC0n) $\sim 400''$ North of TC0, and a massive young protostar (TC5) South of TC3, at the border of the W28 supernova remnant. These four condensations are regularly spaced, distant by ~ 3.5 pc from each other. The Northern condensations appear definitely less massive and exhibit a shallower emissivity distribution. TC0n is gravitationnally unbound, which shows indirect evidence for the role of magnetic field in supporting the fragment.

The condensations are therefore fragments of a filament 12 pc long oriented North-South at the border of M20. The $C^{18}O$ velocity channel map shows that the emission from TC3 and TC4 is anticorrelated with the emission from the rest of the filament; there is a perfect match between the core and the "hole" left in the filament, which are separated by about $3\,\mathrm{km\,s^{-1}}$ (compare panels at $v = 18\,\mathrm{km\,s^{-1}}$ and $v = 21$–$22\,\mathrm{km\,s^{-1}}$). This is a direct evidence that TC3 and TC4 formed in material of the filament that has been shocked. Incidentally, extended SiO emission was detected in the fragment. Dense molecular gas tracers such as CS, SiO, $H^{13}CO^{+}$ peak exactly at the same velocity towards TC3 and TC4, despite their different optical depth, hence probing different layers of the core. It is therefore consistent with star formation having occurred *after* the shock impact. In the opposite case, one would expect optically thin tracers to keep track of the initial velocity distribution in the inner core regions. Similar results are found towards TC5.

To conclude, several young (Class 0-I) protostars of intermediate mass have been discovered around M20. They are identified as fragments of large-scale filaments 10 pc long and most likely pre-existed the birth of M20. The only evidences of star formation along the filament are found in the fragments which have been impacted by M20 and the W28 SNR in the South (TC3, 4, 5). From the analysis of the velocity field of different molecular tracers we conclude that star formation occured *after* the shock impact, hence was induced by the latter.

More observations are needed to determine the magnetic field configuration with respect to the filament (parallel ? perpendicular ?) and its role in the fragmentation and the support of the condensations. Higher-angular resolution are also needed to study the content of the individual protostellar cores and the shock impact on source multiplicity and the core accretion rate.

References

1. Bertoldi, F., ApJ, **346**, 735, 1989
2. Cernicharo, J., Lefloch, B., Cox, P. et al. , Nature (1998)
3. Deharveng, L., Lefloch, B., Zavagno, A., A&A, **408**, L25 (2003)
4. Dobashi, K., Yonekura, Y., Matsumoto, T., et al., PASJ, **53**, 85, (2001)
5. Elmegreen, B.G., Lada, C.J., ApJ, **214**, 725 (1977)
6. B.G. Elmegreen, Y.N., Efremov, R.E., Pudritz, H., Zinnecker. In: *Protostar & Planets IV*, ed V.G. Mannings, A.P. Boss & S.S. Russell (Tucson, Univ. Arizon Press 2000)
7. Lefloch, B., Lazareff, B., A&A, **289**, 559, (1994)
8. Lefloch, B., Lazareff, B., A&A, **301**, 522, (1995)
9. Lefloch, B., Cernicharo, J., ApJ, **545**, 340, (2000)
10. Lefloch, B., Cernicharo, J., Cesarsky, D., et al. A&A, **368**, L13, (2001)
11. Lefloch, B., Cernicharo, J., Rodriguez, L.F., et al., **581**, 335 (2002)
12. Reipurth, B., A&A, **117**, 183 (1983)
13. Yamaguchi, R., Saito, H., Mizuno, N., et al., PASJ, **51**, 791 (1999)
14. Sugitani, K., Fukui, Y., Ogura, K., ApJS, **77**, 59, (1991)
15. Sugitani, K., Ogura, K., ApJS, **92**, 163, (1994)
16. Whitworth, A.P., Bhattal, A.S., Chapman, S.J., et al., MNRAS, **268**, 291 (1994)

Disappearance of N_2H^+ from the Gas Phase in the Class 0 Protostar IRAM 04191

A. Belloche[1,2] and P. André[3]

[1] Max-Planck-Institut für Radioastronomie, Auf dem Hügel 69, D-53121 Bonn, Germany, belloche@mpifr-bonn.mpg.de

[2] LRA/LERMA, Ecole Normale Supérieure, 24 rue Lhomond, F-75231 Paris Cedex 05, France

[3] Service d'Astrophysique, CEA/Saclay, F-91191 Gif-sur-Yvette, France, pandre@cea.fr

Abstract. We present a high-resolution spectroscopic study of the envelope of the young Class 0 protostar IRAM 04191+1522 in Taurus. N_2H^+(1-0) observations with the Plateau de Bure Interferometer and the IRAM 30m telescope demonstrate that the molecular ion N_2H^+ disappears from the gas phase in the inner envelope ($r < 1600$ AU, $n_{H_2} > 4-7 \times 10^5$ cm^{-3}). This may result from N_2 depletion on polar ice mantles and enhanced grain chemistry.

1 The Class 0 Protostar IRAM 04191

With an age $t \sim 1-3 \times 10^4$ yr measured from the beginning of the accretion phase, IRAM 04191+1522 is the youngest Class 0 protostar known so far in the Taurus molecular cloud (d = 140 pc). It has a prominent envelope (1.5 $M_\odot$) and an associated powerful bipolar outflow [3]. Belloche et al. [4] showed that the envelope is undergoing both extended infall motions and fast, differential rotation. They proposed that the rapidly rotating inner part of the envelope ($r <$ 3500 AU) corresponds to a magnetically supercritical core decoupling from an environment still supported by magnetic fields and strongly affected by magnetic braking.

We recently carried out new observations in the N_2H^+(1-0) line with the Plateau de Bure Interferometer (PdBI) to probe the inner structure of the IRAM 04191 envelope. Here, we present and discuss the results of these high-resolution ($\sim 5''$) observations of the small-scale distribution of N_2H^+.

2 Small-Scale N_2H^+(1-0) Emission

Fig. 1a shows the N_2H^+(1-0) integrated intensity map obtained with PdBI, which filters out the extended emission observed with the IRAM 30m telescope. The (naturally-weighted) synthesized half-power beamwidth is $5.8'' \times 4''$. Two emission peaks are seen approximately along the direction perpendicular to the outflow axis, on either side of a gap very close to the central

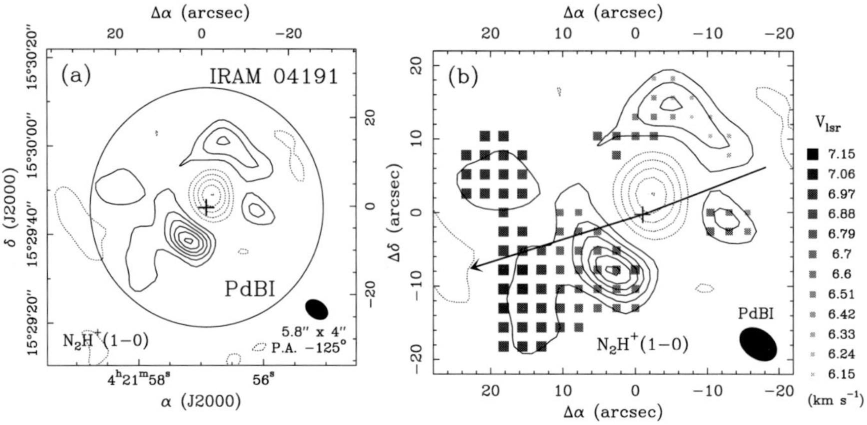

Fig. 1. (a) N_2H^+(1-0) integrated intensity map of the IRAM 04191 envelope obtained with PdBI. The contour step is 0.1 Jy beam^{-1} km s^{-1} (i.e. 3 times the rms noise). The big circle represents the primary beam. The emission gap very close to the central position (*black cross*) suggests a strong decrease of the N_2H^+ abundance in the inner part of the envelope. **(b)** Map of the LSR velocity (*squares*) as derived from Gaussian hfs fits to the N_2H^+(1-0) spectra. The map of integrated intensity is overlaid as contours. We measure a mean velocity gradient of $\sim$ 26 km s^{-1} pc^{-1} across the envelope (*black arrow*), whose direction at PA $\sim$ 107° is consistent with the rotation seen on larger scale.

position. This suggests a strong decrease of the N_2H^+ abundance toward the center. Besides, the two emission peaks have very different centroid velocities (see Fig. 1b): we measure a strong velocity gradient of $\sim$ 26 km s^{-1} pc^{-1} along the axis with PA = 107°, i.e. approximately the same axis along which a rotational velocity gradient is detected on larger scale [4]. The gradient measured with PdBI is 8 times larger than the mean velocity gradient observed with the 30m telescope toward the core, which confirms the presence of fast, differential rotation in the IRAM 04191 envelope.

3 N_2H^+ Disappearance from the Gas Phase

Since the interferometer filters out the extended emission, we added short spacings obtained with the 30m telescope [4] to the PdBI data. As shown in the N_2H^+(1-0) integrated intensity map (see Fig. 2a), the emission is resolved in a ringlike structure with two local peaks, surrounding a gap close to the center. Gaussian hfs fits to the 7-component N_2H^+(1-0) spectra yield a mean total opacity $\tau_{\rm tot} = 10 \pm 4$ in the central part ($\theta < 10''$), corresponding to $\tau = 2.6 \pm 1.0$ for the strongest component and $\tau = 1.1 \pm 0.4$ for the isolated component. However, the opacity map does not show any enhancement toward the center and the gap close to the center does not have the highest opac-

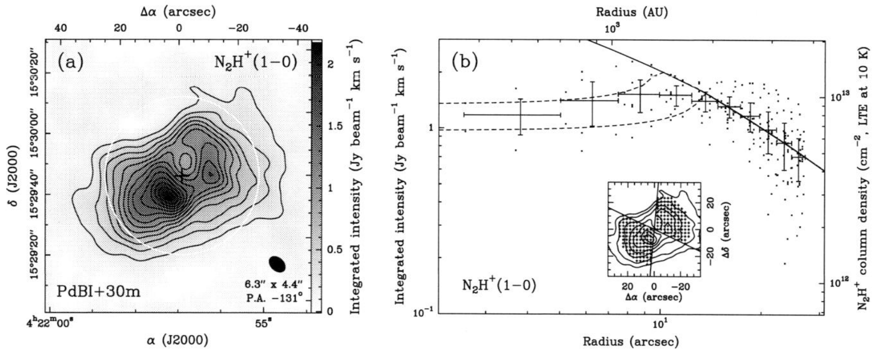

Fig. 2. (a) N_2H^+(1-0) integrated intensity map of the IRAM 04191 envelope combining the PdBI data and the short spacings obtained at the 30m telescope. The contour step is 0.16 Jy beam^{-1} km s^{-1} (i.e. 7 times the rms noise). The white circle represents the PdBI primary beam. **(b)** Radial profile of the N_2H^+(1-0) integrated intensity corrected for the primary beam attenuation. The dots with error bars represent the average intensity in concentric shells limited to the positions shown in the insert. The solid curve indicates the H_2 column density profile [10, 4] scaled to the 9th shell for comparison. The dashed curves show the effect of a hole of N_2H^+(1-0) emission with radius r = 1350 and 1850 AU, respectively. They suggest a strong decrease of the N_2H^+ abundance in the inner part of the envelope ($r < 1600$ AU).

ity. Besides, the intensity maps integrated over each ΔF_1 group of N_2H^+(1–0) hyperfine components separately show the same pattern. We are thus confident that the observed emission gap does not result from an opacity effect.

Assuming a constant excitation temperature for N_2H^+(1-0), we compare in Fig. 2b the circularly-averaged profile of the N_2H^+(1-0) integrated intensity, corrected for PdBI primary beam attenuation, with the H_2 column density profile derived from the dust emission [10, 4]. The N_2H^+(1-0) profile departs from the H_2 column density profile for $\theta \leq 13''$. A good model fit is obtained when one assumes a hole of N_2H^+(1-0) emission in the central part of the envelope ($r <$ 1350-1850 AU, see dashed curves). This crude comparison thus strongly suggests that the N_2H^+ ion disappears from the gas phase above a density of $n_{H_2} \sim 4-7\times10^5$ cm^{-3} in the IRAM 04191 envelope.

4 Discussion

Very little evidence for N_2H^+ abundance drop has been reported so far in dense cores (see however the case of B68 [6]). The N_2H^+(1-0) emission usually follows the dust emission very well in prestellar cores and has long been thought to be one of the best molecular tracers of dense gas [12, 7]. From a theoretical point of view, all recent chemical models predict a centrally-peaked N_2H^+ column density profile during the prestellar collapse phase [5, 1, 2, 9, 11]. In particular, the models calculated by Aikawa et al. [2] including gas-phase reactions, gas-dust interactions and diffusive grain-surface reactions predict an increase of the N_2H^+ abundance during the core collapse up to $n_{H_2} \sim 1 \times 10^7$ cm^{-3} or more. Since only a few 10^4 yr have elapsed and the density at $r \sim 1600$ AU cannot have changed much between point mass formation (the last stage of this model) and the present stage of IRAM 04191, the drop observed in the N_2H^+ abundance above $n_{H_2} \sim 4 - 7 \times 10^5$ cm^{-3} is strongly at variance with current theoretical predictions.

Our result implies either a stronger dependence of the N_2H^+ abundance on density (i.e. an earlier drop in abundance during the prestellar phase) or an abundance drop on a timescale as short as a few 10^4 yr. Since N_2 is the precursor of N_2H^+, the drop in N_2H^+ abundance may result from stronger N_2 depletion onto grain surfaces. This could happen if N_2 were depleted on H_2O ice mantles characterized by a higher binding energy (see [6], Model C of [1] and Fig. 6 of [11]) and/or if the chemistry transforming N_2 in less volatile species on grain surfaces were enhanced. It is also possible that N_2H^+ destroyers such as CO have returned to the gas phase in the inner envelope, following heating by the central protostar or shock by the outflow. However, the dust temperature is below 10 K at $r \sim 1000$ AU and our $C^{18}O$(2-1) observations [4] at a resolution of 800 AU in radius show depletion toward the envelope center by a factor of 3 in column density when compared to the 1.3mm continuum emission. This latter explanation seems thus less likely.

A better tracer than N_2H^+ will thus have to be found to probe the velocity structure of cold protostellar envelopes at high resolution with the next generation of (sub)millimeter instruments such as ALMA. H_2D^+ may be such a good tracer [6, 8].

Acknowledgements. We would like to thank Y. Aikawa, E. Bergin, H. Fraser, M. Gerin, D. Johnstone, G. Pineau des Forêts and M. Walmsley for enlightening discussions on chemistry, and F. Gueth, R. Neri and J. Pety for their help with the reduction of our PdBI observations.

References

1. Y. Aikawa, N. Ohashi, S.I. Inutsuka, et al.: ApJ, **552**, 639 (2001)
2. Y. Aikawa, N. Ohashi, & E. Herbst: ApJ, **593**, 906 (2003)
3. P. André, F. Motte, & A. Bacmann: ApJ, **513**, L57 (1999)
4. A. Belloche, P. André, D. Despois, & S. Blinder: A&A, **393**, 927 (2002)
5. E.A. Bergin, & W.D. Langer: ApJ, **486**, 316 (1997)
6. E.A. Bergin, J. Alves, T.L. Huard, & C.J. Lada: ApJ, **570**, L101 (2002)
7. P. Caselli, P.J. Benson, P.C. Myers, & M. Tafalla: ApJ, **572**, 238 (2002)
8. P. Caselli et al: A&A, **403**, L37 (2003)
9. Z.Y. Li, V.I. Shematovich, D.S. Wiebe, & B.M. Shustov: ApJ, **569**, 792 (2002)
10. F. Motte, & P. André: A&A, **365**, 440 (2001)
11. V.I. Shematovich, D.S. Wiebe, B.M. Shustov, & Z.Y. Li: ApJ, **588**, 894 (2003)
12. M. Tafalla, P.C. Myers, P. Caselli, et al.: ApJ, **569**, 815 (2002)

The Massive Star Forming Region RCW 38 – from X-Rays to Molecular Gas

T.L. Bourke[1], S.J. Wolk[1], M. Vigil[2], and J. Alves[3]

[1] Harvard-Smithsonian Center for Astrophysics tbourke@cfa.harvard.edu
[2] Massachusetts Institute of Technology
[3] European Southern Observatory

1 Introduction

The HII region RCW 38 is one of the nearest (1.7 kpc) embedded ($A_V \sim 10$) cluster forming regions containing an O star (O5.5 – IRS2; [1] [2] [3]). However the region has received only scant attention, no doubt due to its southern location. At 10 μm [1] [2] the brightest feature lies to the NW of IRS2, called IRS1, and the dust in the region interior to IRS1, centred on IRS2, is depleted by a factor of 100 w.r.t. to the interstellar value [2] [4]. Low resolution 1-mm continuum observations [5] suggest A_V up to 800, implying that the cluster lies near the front surface of a dense molecular cloud, and so represents an embedded blister HII region. Previous near-infrared (NIR) observations have revealed the embedded cluster associated with this region [6] [7] but none had the spatial resolution or sensitivity required to obtain a complete picture.

2 Observations

Here we present new NIR, X-ray, radio continuum, and molecular line observations of RCW 38. The NIR data were obtained during ISAAC commisioning on the VLT (http://www.eso.org/outreach/press-rel/pr-1998/pr-19-98.html; ESO Messenger #94), the X-ray data are from a 100 ks Chandra observation ([8]; http://chandra.harvard.edu/photo/2002/rcw38/index.html), the radio continuum from ATCA observations at 1.6 and 4.8 GHz [3], and the molecular line data (CS 2-1/5-4, ^{13}CO 1-0/2-1, C^{18}O 2-1) are from SEST and Mopra [9] [10].

3 Results

The results are present in more detail in [3], [8], [9], and [11]. In this paper we use figures to illustrate the relationship between the different components of RCW 38. Note that all these figures can be found in colour in the poster version contained in the CD included in these proceedings.

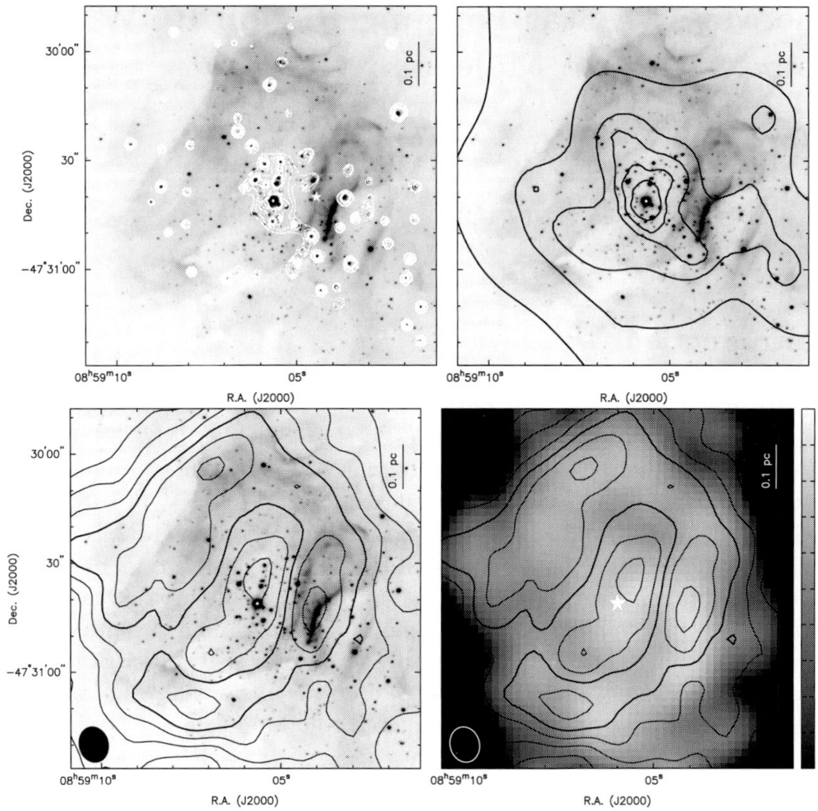

Fig. 1. The underlying image in the first 3 panels is the VLT K-band image. The central overexposed star is IRS2 (O5.5). Extensive nebulosity is present in a non-symmetric ring-like structure around IRS2. To the W of IRS2 are two bright ridges: the inner, weaker ridge coincides with the brightest 10 μm ridge and IRS1 [1] [2], marked with a white star. Numerous NIR cluster members are seen. The NIR colour-colour diagram indicates that more than 50% of the sources possess an infrared excess, and the KLF indicates an age of < 1 Myr [11]. The distribution of these sources in the environment of an O star is under investigation [11]. In the **top left panel** the contours show the X-ray emission from Chandra with resolution $\sim 1''$. Only the brightest sources are highlighted. A large number of the IR sources have X-ray counterparts. At low emission levels almost all the IR sources with an IR excess show X-ray emission above the background. Notably IRS1 shows no X-ray emission. The source immediately to the W of IRS1 on the bright ridge is a hard X-ray source (e.g., deeply embedded), and has no clear NIR counterpart – a possible Class 0 source (a handful of other sources exist). In the **top right panel** the point sources have been carefully removed from the X-ray data (see [8] for details) showing the extended X-ray emission in the region. In the inner few tenths of a parsec around IRS2 the emission is best fitted with a power-law profile, suggesting non-thermal emission, as might be expected from a SNR. No other evidence for a SNR in this young region exists. In the outer regions the profile is thermal.

The radio flux due to this non-thermal component is predicted to be only ~ 2 mJy at 1 GHz [8]. The expected dominant X-ray emission is thermal (wind-driven) and the presence of non-thermal emission is currently unexplained. Recent NIR AO imaging with the VLT [12] resolves IRS2 into a binary. The X-ray spectrum of IRS2 shows prominant emission lines and is well fitted by a thermal spectrum. The **bottom left panel** shows the 1.6 GHz radio emission (beam at lower left) in contours on the K-band image. The heavier contour represents the 65% level of the peak flux. The 4.8 GHz data show a similar structure but with poorer (u, v) coverage. The 1.6 GHz emission shows a ring-like structure and is similar to the NIR emission. Note by comparison with the top right panel that the diffuse X-ray gas is interior to the 1.6 GHz emission. The emission levels are measured in Jy, and so any low-level mJy emission due to the non-thermal emission cannot be distinguished. The radio spectral index map is shown in the **bottom right panel**, with contours of 1.6 GHz emission, clearly shows that the dominant radio emission in the region is thermal (the white star marks the position of IRS2). The spectral index is steeper than that expected for optically thin emission, suggesting that the turnover frequency occurs between 1.6 and 4.8 GHz, as is typical for HII regions.

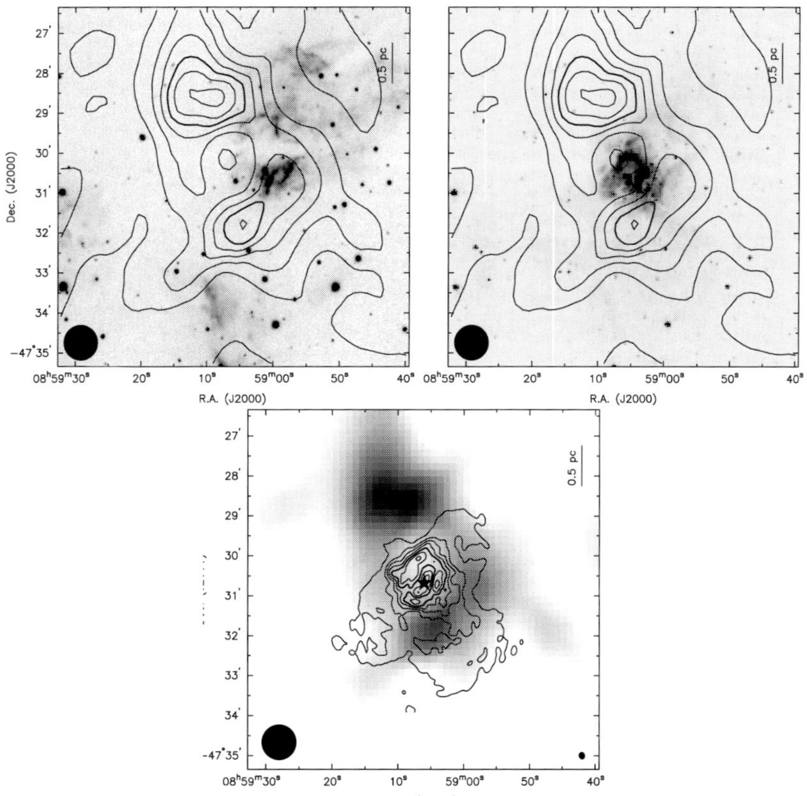

Fig. 2. In the **top left panel** the underlying image is SII (ANU RSAA 40″), while in the **top right panel** it is a K-band mosaic constructed from 2MASS QuickLook

Images. Both images are overlaid with contours of integrated CS 2-1 emission (SEST – beam size shown at lower left). The heavier line indicates 65% of the peak CS 2-1 emission. IRS2 is visible near the centre of the SII image. Comparison of the SII and K-band images shows that the HII region appears to be breaking out of the nearside of the cloud in the NW. As seen in the colour VLT image (ISAAC commissioning, CD poster), the extinction in the W is lower, supporting this view. In the **bottom panel** the 1.6 GHz emission (beam size is shown at lower right) is overlaid on the integrated CS 2-1 emission. The CS 2-1 neatly bounds the 1.6 GHz emission, but the 1.6 GHz emission is extended in the NW-SE direction, in agreement with the SII image in the NW. In the SW extended emission is also seen, suggesting the 1.6 GHz emission is escaping around the molecular gas. This is not the case in the NE. The CS emission does not seem to support the view that the HII region lies at the near side of the molecular cloud, but appears to be embedded within. Probably the previous 1-mm continuum observations did not have sufficient resolution to show such structure. Our recently obtained SEST SIMBA data (1.2 mm) indicates a lack of emission in the central region around IRS2, in agreement with the data presented here and the analysis of [2] who infer a gas-to-dust ratio 100 times lower than interstellar to explain their MIR data.

References

1. Frogel, G.A., & Persson, S.E., 1974, ApJ, **192**, 351
2. Smith, C.M., Bourke, T.L., et al., 1999, MNRAS, textbf303, 367
3. Vigil, M., Bourke, T.L., Wolk, S.J., in prep
4. Furniss, I., Jennings, R.E., & Moorwood, A.F.M., 1975, ApJ, **202**, 400
5. Cheung, L.H., Frogel, J.A., Hauser, M.G., & Gezari, D.Y., 1980, ApJ, **240**, 74
6. Storey, J.W.V., & Bailey, J, 1982, PASAu, **4**, 429
7. Ligori, S., Moretti, A., Robberto, M., Guarnieri, M.D., Zinnecker, H. 1994, MmSAI, **65**, 815
8. Wolk, S.J., Bourke, T.L., Smith, R.K., Spitzbart, B., & Alves, J., 2002, ApJ, **580**, L161
9. Bourke, T.L., Vigil, M., et al., in prep
10. Bourke, T.L., Myers, P.C., Robinson, G., Hyland, A.R., 2001, ApJ, **554**, 916
11. Wolk, S.J., Spitzbart, B., Bourke, T.L., Alves, J., Vigil, M., in prep
12. Megeath, S.T., Bourke, T.L., & Alves, J., in prep

Observation of Multiply-Deuterated Methanol in the Low-Mass Protostar IRAS16293–2422

B. Parise

Centre d'Etude Spatiale des Rayonnements
9 avenue du Colonel Roche - BP 4346 - 31028 Toulouse Cedex 04 - France
parise@cesr.fr

Summary. We present observations of singly, doubly and triply-deuterated metha- nol towards the low-mass protostar IRAS16293−2422. Comparison of the fractionation ratios derived for these species is a unique tool to understand the chemical processes that lead to their formation. The present observations are consistent with formation of methanol on grain surfaces if the atomic D/H ratio is 0.1 to 0.2.

1 Introduction

Deuterated molecules are puzzling since, despite the relatively low elemental abundance of deuterium in space (D/H $\sim 1.5\times10^{-5}$, [5]), extremely large abundances of multiply-deuterated molecules have been observed towards low-mass protostars and dark clouds (e.g. $D_2CO/H_2CO \sim 10$ % in IRAS 16293–2422 [2, 3, 6], $ND_3/NH_3 \sim 10^{-3}$ in NGC1333-IRAS4A [13]).

Comparison of the abundance of several isotopomers of the same species is a useful tool to understand the formation processes of molecules. In particular, such studies may help disentangle between gas-phase or grain-surface chemistry. We report here the observation of singly (both CH_2DOH and CH_3OD), doubly and triply-deuterated methanol towards the low-mass protostar IRAS16293–2422. The present observations are consistent with formation of methanol on grain surfaces by successive hydrogenations of CO, if the atomic D/H ratio is 0.1 to 0.2.

2 Observations

The observations were performed using the IRAM 30m telescope (Pico Veleta, Spain) in May 2002 [7] and April 2003 [8]. We detected 23 CH_2DOH lines, 6 CH_3OD lines, and respectively 15 and 12 lines for CHD_2OH and CD_3OH. The telescope was pointed at the coordinates $\alpha(2000)=16^h32^m22.6^s$ and $\delta(2000)=-24^o28'33.0''$.

Examples of observed spectra for CD_3OH are shown in Fig. 1.

Two $^{13}CH_3OH$ lines were observed simultaneously to the CD_3OH lines. In addition, we analysed 330 GHz $^{13}CH_3OH$ lines observed at the JCMT in January 2000.

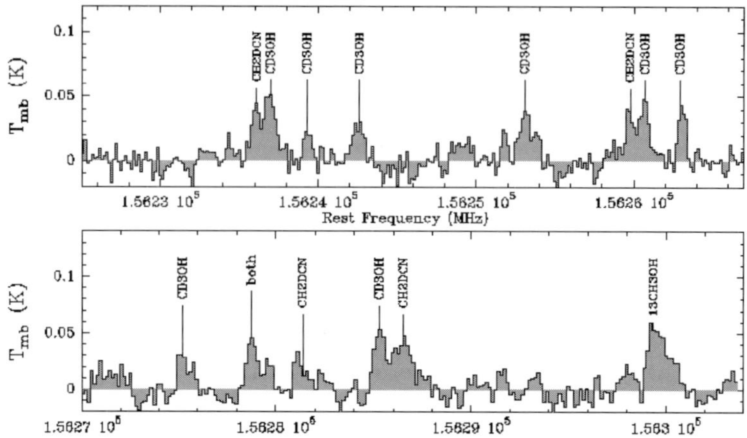

Fig. 1. Some CD_3OH detected lines. The intensities are reported in main-beam brightness temperature.

3 Derivation of the Column Densities

The column density for each molecule was derived using the population diagrams presented in Fig. 2. For clarity, the points for $^{13}CH_3OH$, CH_3OD, CHD_2OH and CD_3OH have been shifted by +3, -2, -4 and -4 respectively. For CH_3OD and CD_3OH, the derived rotational temperatures are substantially different from that for the other isotopomers. This may come from the lack of data at high energy for CH_3OD. We then derived the column density by fixing the rotational temperature to that derived with other isotopomers.

The CH_3OH column density was computed using the $^{12}CO/^{13}CO$ ratio of 70 derived by [1]. The derived column densities and fractionation ratios for each molecule are listed in Table 1.

Table 1. Derived column densities and fractionation ratios relative to CH_3OH for deuterated methanols. [a]fixed temperature (see text).

Molecule	T_{rot}(K)	N (cm^{-2})	fractionation
CD_3OH	85 ± 28	$(1.4 \pm 0.9)\times10^{14}$	1.4 ± 1.4 %
	47 ± 7^a	$(7.8 \pm 2.3)\times10^{13}$	0.8 ± 0.6 %
CHD_2OH	47 ± 7	$(6.0 \pm 2.2)\times10^{14}$	6 ± 5 %
CH_2DOH	48 ± 3	$(3.0 \pm 0.6)\times10^{15}$	30 ± 20 %
CH_3OD	20 ± 4	$(1.5 \pm 0.7)\times10^{14}$	2 ± 1 %
	47 ± 7^a	$(2.8 \pm 0.3)\times10^{14}$	3 ± 2 %

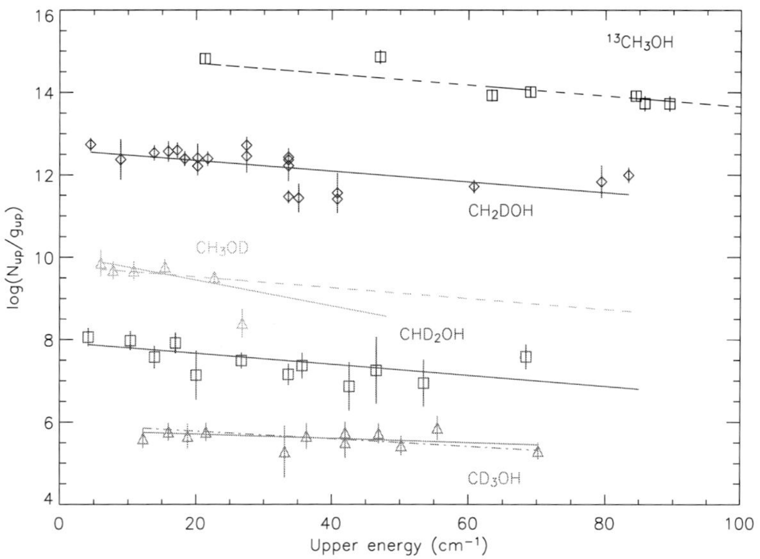

Fig. 2. Rotational diagrams for the 5 species. Note that the points for $^{13}CH_3OH$, CH_3OD, CHD_2OH and CD_3OH have been shifted by +3, -2, -4 and -4 respectively.

4 Formation on the Grains?

Since gas-phase chemistry models are unable to reproduce the large absolute abundances of methanol in the envelope of low-mass protostars, methanol is believed to be formed on the surface of dust grains, probably from successive hydrogenations (or deuterations) of CO [12] [4]. In this scheme, the fractionations of the different isotopomers should be linked and depend only on the relative accretion rates for H and D atoms [10].

We compare the deuterium fractionations observed to those predicted by the grain chemistry model described in [11] at 10K. Note that at 10K, the results of this model are similar to a random distribution of deuterium atoms [10]. Observed fractionation ratios have been overlaid on each curve, allowing the derivation of the required atomic D/H ratio in the gas-phase.

The CD_3OH, CHD_2OH and CH_2DOH observations are consistent with a formation on grain surfaces with an atomic D/H abundance ratio of 0.1–0.2. Such a high atomic fractionation ratio in the gas phase is predicted by the recent gas-phase model of [9], which involves not only H_2D^+ but also D_2H^+ and D_3^+ as precursors for deuterium fractionation. Nevertheless, this high atomic D/H ratio only appears after much of the CO has depleted and it is not clear yet if the large absolute abundance of deuterated methanol can be formed in this late stage. Modelling coupling both gas and grain chemistry is therefore required to fully address this question.

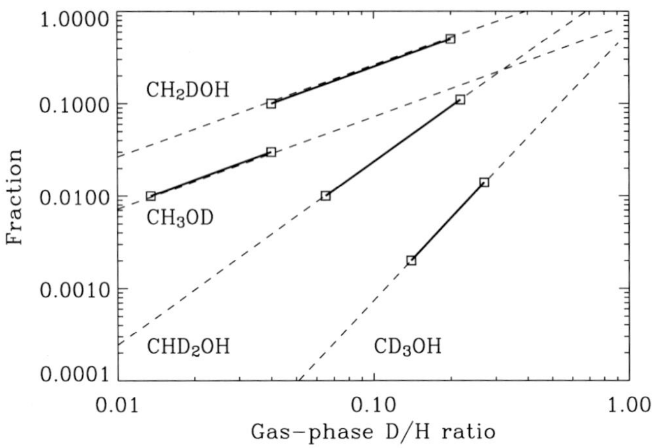

Fig. 3. Deuterium fractionation ratio for each isotopomer versus the abundance ratio of deuterium to hydrogen atoms in the gas phase. Dashed lines: model results [11]. Thick lines: observations of IRAS 16293.

As can be seen in Fig. 3, CH_3OD appears to be under-deuterated when compared with the grain chemical predictions. It is possible that the CH_3OD fractionation may be affected in the warm gas; e.g., this isotopomer may be preferentially converted into CH_3OH when released in the gas-phase by protonation reactions followed by dissociative recombination with an electron [4] [7].

References

1. A.C.A. Boogert, G.A. Blake & A.G.G.M. Tielens: ApJ **577**, 271 (2002)
2. C. Ceccarelli, A. Castets, L. Loinard et al: A&A **338**, L43 (1998)
3. C. Ceccarelli, L. Loinard, A.Castets et al: A&A **372**, 998 (2001)
4. S.B. Charnley, A.G.G.M. Tielens & S.D. Rodgers: Apj **482**, L203 (2002)
5. J.L. Linsky: Space Sci. Rev. **84**, 285 (1998)
6. L. Loinard, A. Castets, C. Ceccarelli et al: A&A **359**, 1169 (2000)
7. B. Parise, C. Ceccarelli, A.G.G.M. Tielens et al: A&A **393**, L49 (2002)
8. B. Parise et al: submitted (2003)
9. H. Roberts, E. Herbst & T.J. Millar: Apj **591**, 41 (2003)
10. S.D. Rodgers & S.B. Charnley: Planet. Space Sci. **50**, 1125 (2002)
11. T. Stantcheva & E. Herbst: MNRAS **340**, 983
12. A.G.G.M. Tielens & W. Hagen: A&A **114**, 254 (1982)
13. F.F.S. van der Tak, P. Schilke, H.S.P. Müller et al: A&A **388**, L53 (2002)

UYSO 1 – An Extremely Young Massive Stellar Object

J. Forbrich[1], K. Schreyer[2], B. Posselt[3], R. Klein[4], and Th. Henning[5]

[1] Max-Planck-Institut für Radioastronomie, Auf dem Hügel 69, D-53113 Bonn, Germany forbrich@mpifr-bonn.mpg.de
[2] Astrophysikalisches Institut und Universitäts-Sternwarte, Schillergäßchen 2–3, D-07745 Jena, Germany martin@astro.uni-jena.de
[3] Astrophysikalisches Institut und Universitäts-Sternwarte, Schillergäßchen 2–3, D-07745 Jena, Germany posselt@astro.uni-jena.de
[4] Max-Planck-Institut für extraterrestrische Physik, Giessenbachstraße, D-85748 Garching, Germany rklein@mpe.mpg.de
[5] Max-Planck-Institut für Astronomie, Königstuhl 17, D-69117 Heidelberg, Germany henning@mpia.de

In the course of a comprehensive mm/submm survey of massive star-forming regions, a particularly interesting object has been found in the surroundings of the bright FIR source IRAS 07029-1215 , in a distance of 1 kpc. The object – named UYSO 1 (*Unidentified Young Stellar Object 1*) – is cold ($T = 40$ K), it has a massive envelope, and it is associated with an energetic molecular outflow. No infrared point source has been detected at its position for $\lambda \leq 20$ μm. Therefore, it is a very good candidate for a member of the long searched for group of massive protostars.

1 Introduction

Whether the formation of massive stars significantly differs from the formation of low mass stars is not yet fully understood. While disk accretion seems possible under special circumstances (e.g. [9]), coalescence of smaller objects into one massive star is an alternative model (e.g. [2]). Observations of molecular outflows in regions of massive star formation indicate, however, that the accretion model is indeed applicable. Even though little is known about the initial conditions of [4], young massive protostars are expected not to be detectable at NIR/MIR wavelengths, but mainly in the submm continuum.

Using SCUBA and IRAM bolometers, we investigated a sample of 47 luminous IRAS sources, searching for massive protostellar objects in their close vicinities [13, 10]. Near IRAS 07029-1215, we discovered a deeply embedded object in a distance of about 1 kpc [14], powering a high-velocity bipolar CO outflow. The IRAS source is located in the bright H II region S 297, it has an IRAS luminosity of 1700 $L_\odot$ [7].

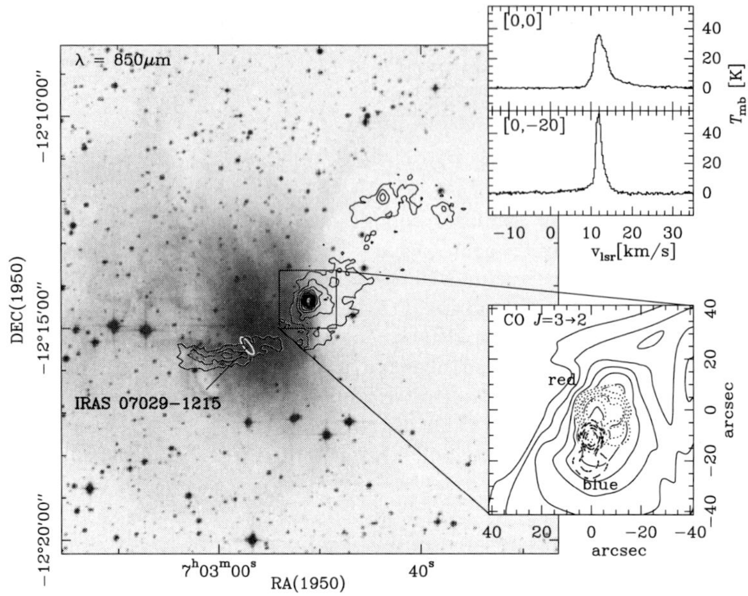

Fig. 1. Left: UYSO 1, seen in $\lambda = 850$ μm continuum emission (contours, background: POSS). Upper Right: Two examples of the CO($J = 3 \rightarrow 2$) line wings. Lower Right: Velocity-integrated emission of the CO line and its wings.

2 Observations

Directly following the SCUBA continuum observations at $\lambda = 450$ μm and $\lambda = 850$ μm taken in October 1999, a map in CO($J = 3 \rightarrow 2$) was obtained using the B3 facility receiver). Additional observations in CS($J = 2 \rightarrow 1$), CS($J = 5 \rightarrow 4$), SiO($J = 2 \rightarrow 1$) and especially $H_2CO(J = 3_{03} - 2_{02})/(J = 3_{22} - 2_{21})$ were carried out at the IRAM 30m telescope in March 2003.

3 Results

The surroundings of IRAS 07029-1215 are shown in Fig. 1 as a POSS image. The SCUBA 850 μm data, overlaid as contours, show UYSO 1 as a compact emission peak at $\alpha_{(B1950.0)} = 07^h02^m51^s, \delta_{(B1950.0)} = -12°14'26''$, between S 297 and a dark cloud. Two additional dusty filaments are visible at 850 μm, one of them containing the IRAS source.

UYSO 1 was not detected at near- and mid-infrared wavelengths (λ =2.2.. 20 μm) by MSX and 2MASS.

The CO map in Fig. 1 (lower right) is a close-up view of UYSO 1. The line wings extending from -20 km s^{-1} at [0",-20"] to +40 km s^{-1} at [0",0"] can be clearly distinguished from well-determined baselines. These line wings are due to a high-velocity bipolar outflow originating from UYSO 1. The outflow is less clearly visible in the CS data, however lines on respective positions have corresponding shapes. The spatial resolution of the observations as of now is insufficient to conclude whether the bipolar outflow is composed of only one or more components [1].

4 Analysis

4.1 Temperature and Mass Estimates

A temperature estimate was obtained from the $H_2CO(J = 3_{03} \rightarrow 2_{02})/(J = 3_{22} \rightarrow 2_{21})$ line ratio, following [11], resulting in T=40±5 K for a relatively wide density range.

Following [8], we derive hydrogen column densities from the CO $J = 3 \rightarrow 2$ transition. This leads to $M_{\rm gas} = 40\ M_\odot$. Another estimate for the total mass has been calculated by [13] from an analysis of the continuum emission at $\lambda = 850\ \mu$m, resulting in $M = 15\ M_\odot$. The discrepancy can at least partly be explained by a smaller region taken into account in the latter case as well as the dust continuum emission tracing only the high-density central regions while the CO emission is collected along the line of sight throughout the low-density halo. Thus, the envelope mass estimates based on line and dust emission are compatible when keeping in mind their respective limitations and range between 15 and 40 $M_\odot$. The virial mass is much higher ($M_{\rm vir} = 180\ M_\odot$), which is certainly caused by an overestimation of line width due to optical thickness as well as turbulent motion, in addition to the cloud not being in virial equilibrium.

4.2 Outflow Properties

The CO line wings were analyzed to their respective 10% contour levels, corresponding to a 2 σ detection for the redshifted line wing and a 4 σ detection for the blueshifted line wing. The masses contained inside the two outflow lobes were calculated by integration of H_2 column densities, resulting in $M_{\rm blue} = 1.8\ M_\odot$ and $M_{\rm red} = 3.6\ M_\odot$, the total outflow mass thus being $M_{\rm out} = 5.4\ M_\odot$. An analysis of the dynamical timescale, given for an inclination of $i = 57.3°$ (the most probable value, cf. [3]), leads to $t_d(i = 57.3°) = 1700$ yrs. We note that this is not a good age indicator. The corresponding mass outflow rate is $\dot{M}(57.3°) = 3.3 \times 10^{-3}\ M_\odot$/yr. Only a massive central object is able to uphold such an outflow rate.

4.3 Spectral Energy Distribution

Given the measured continuum fluxes at 450 μm and 850 μm, flux estimates from the IRAS atlas images and NIR/MIR upper limits from MSX and 2MASS, a tentative SED of UYSO 1 could be derived. Integration leads to a luminosity estimate of $< 1900\ L_\odot$. A modified blackbody accounting for the dust optical depth was fitted to the data with $T = 45$ K and $\beta = 2$. The continuum emission was then modeled using the radiative transfer code developed by [12] with standard ISM parameters and in 1D mode. A model with an envelope mass of $M_{env} = 30\ M_\odot$ and an assumed single central object in the B2.5 class fits the data well.

5 Conclusions

The emerging picture of UYSO 1 is that of an early B star surrounded by an envelope of about 30–40 $M_\odot$. The object seems to be in a particularly early evolutionary state, however already shows a bipolar outflow with a total mass of $M_{outflow} = 5.4\ M_\odot$. To our knowledge, there are only very few comparable sources that are still undetected at infrared wavelengths (see e.g. [5]). With additional high-resolution observations, we hope to learn more about this early evolutionary stage of massive star formation. For more details, see [6].

References

1. H. Beuther, P. Schilke, T.K. Sridharan, K.M. Menten, C.M. Walmsley, F. Wyrowski: A&A **383**, 892 (2002)
2. I.A. Bonnell, M.R. Bate, H. Zinnecker: MNRAS **298**, 93 (1998)
3. S. Bontemps, D. Ward-Thompson, P. André: A&A **314**, 477 (1996)
4. N.J. Evans II, Y.L. Shirley, K.E. Mueller, C. Knez: The Formation and Early Evolution of Massive Stars. In: *ASP Conf. Ser. 267, Hot Star Workshop III: The Earliest Phases of Massive Star Birth*, ed. P.A. Crowther (ASP, San Francisco 2002)
5. F. Fontani, R. Cesaroni, L. Testi, M. Walmsley: IRAS 23385+6053: An Example of Candidate Massive Protostar. In: *IAU Symposium 221: Star Formation at High Resolution*, eds. R. Jayawardhana, M. Burton, and T. Bourke (ASP IAU Publications Volume S-221, *in press*)
6. J. Forbrich, K. Schreyer, B. Posselt, R. Klein, Th. Henning: ApJ, *in press*
7. Th. Henning, R. Cesaroni, M. Walmsley, W. Pfau: A&AS **93**, 525 (1992)
8. Th. Henning, K. Schreyer, R. Launhardt, A. Burkert: A&A **353**, 211 (2000)
9. J. Jijina, F.C. Adams: ApJ **462**, 874 (1996)
10. R. Klein, B. Posselt, K. Schreyer, Th. Henning, *in prep.*
11. J.G. Mangum, A. Wootten: ApJS **89**, 123 (1993)
12. V. Manske, Th. Henning, A.B. Men'shchikov: A&A **331**, 52 (1998)
13. B. Posselt: Millimeterbeobachtungen in Sternentstehungsgebieten. Diplomarbeit, Friedrich-Schiller-Universität, Jena (2003)
14. J.G.A. Wouterloot, J. Brand: A&AS **80**, 149 (1989)

Millimeter Properties of the Protoplanetary Disk Surrounding HH 30

J. Pety[1,2], F. Gueth[1], A. Dutrey[3,4], and S. Guilloteau[1,4]

[1] IRAM, 300 rue de la Piscine, F-38406 Saint Martin d'Hères, France, pety@iram.fr
[2] LERMA, Obs. de Paris, 51 avenue de l'Observatoire, F-75014 Paris, France
[3] LAOG, Obs. de Grenoble, B.P. 53, F-38041 Grenoble Cedex 9, France
[4] Obs. de Bordeaux, France

1 Introduction

HH 30 is a well-known PMS star located in the Taurus molecular cloud, at a distance of ~140 pc. Burrows et al. (1996) [1] obtained spectacular images with the Hubble Space Telescope, revealing *i)* a flared bipolar nebula of diameter ~450 AU, separated in two parts by a dust lane and *ii)* a highly-collimated jet emanating out of the central region. Scattered light models have shown that the reflection nebulae are most likely the upper and lower surfaces of a nearly edge-on circumstellar disk in hydrostatic equilibrium, with a mass of about $10^{-3} M_\odot$ [2, and references therein]. These properties make HH 30 a textbook case for the study of the disk/jet system in PMS stars.

In the last five years, we observed using the IRAM Plateau de Bure Interferometer: *i)* ^{12}CO (J=2–1) emission, which reveals the structure of the north-east outflow lobe: a conical lobe emanating out of the innermost part of the circumstellar disk, and surrounding the optical jet (See Fig. 1.a); *ii)* ^{13}CO (J=2–1) emission, which reveals the Keplerian disk avoiding confusion from the outflow (See Fig. 1.b); *iii)* Continuum at 3.4, 2.7 and 1.3 mm, which show the dust disk structure (See Fig. 1.c). Typical resolutions are 0.7" at 1.3 mm and 1.8" at 3.4 mm.

While the optical and near-IR observations trace the dust emission (scattered regime), mm observations can be used to trace the thermal emission of the dust, but also the gas component, thus giving a direct access to the disk kinematics. We thus present here preliminary results of a detailed study of the thermal and kinematics properties of the HH 30 protoplanetary disk.

2 ^{13}CO (J=2–1) Observations

The ^{13}CO (J=2–1) emission in HH 30 was observed in the CD configuration in December 2000 and February 2001. Fig. 1.b shows 3–sigma contours of the blue-shifted and red-shifted integrated emission. Moreover, our measurements of the continuum fluxes (~ 4 mJy at 3 mm and ~ 18 mJy at 1 mm) agree with previous ones made at OVRO [3].

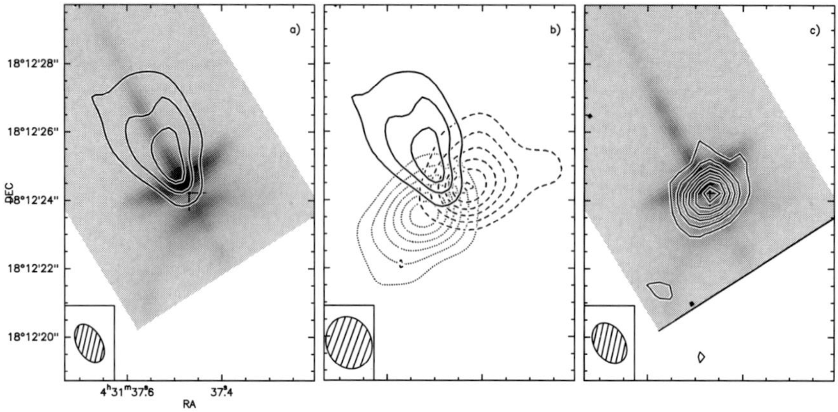

Fig. 1. a) Contours of the ^{12}CO (J=2–1) emission of HH 30 are plotted over a composite HST image (670 and 787.7 nm, see [1]). Emission is only integrated at extreme velocities (≤ 4 km s^{-1} and ≥ 11 km s^{-1}) to avoid contamination by the disk emission. Contours are at the 3σ level (48 mJy/Beam) with a spatial resolution of $1.23 \times 0.75''$ PA 31° (compared to a pixel size of 0.1″ for the HST image). b) Superimposition of 3σ–level contours of *i)* the outflow emission as traced by the ^{12}CO(2–1) emission integrated at extreme velocities (cf. left panel) and *ii)* the blue-shifted (dotted line) and red-shifted (dashed line) emission of the disk as traced by the ^{13}CO(2–1) line. The spatial resolution is $1.57 \times 1.35''$ PA 25° and $3\sigma = 29$ mJy/Beam. c) Superimposition of the 3σ–level contours (1.4 mJy/ beam) of the continuum emission at 1.3 mm over the same composite HST image. The spatial resolution is $1.29 \times 0.91''$ PA 33°. The thermal emission is clearly tracing the dust disk.

3 Fitting Procedure

A standard disk model [4] was used to fit the *uv*-plane visibilities. This model takes into account emission from both the ^{13}CO (J=2–1) line and the dust continuum. It assumes an axi-symmetrical geometry, local thermodynamic equilibrium (to solve the radiative transfer), hydrostatic equilibrium and the absence of a vertical gradient of temperature. Turbulence is phenomenologically introduced by adding a turbulent width dV to the thermal width. Spatial evolutions of temperature, density and velocity are assumed to be power law normalized at $r_{100} = 100$ AU ($T = T_{100}\,(r/r_{100})^{-q}$, $n = n_{100}\,(r/r_{100})^{-p}$ and $V = V_{100}\,(r/r_{100})^{-v}$). A value of 0.5 for the velocity scaling exponent v implies that the disk rotation is Keplerian. Finally, fitting is done directly in the *uv* plane to avoid the delicate stage of deconvolution.

4 Detailed Results

Table 1 display the results of the fitting procedure. The main physical results are: *i)* The disk outer radius is ~ 420 AU; *ii)* The disk rotation is Keplerian;

Table 1. Detailed results of multi-dimensional fits of the ^{13}CO (J=2–1) emission. The quoted uncertainties are the 1σ formal uncertainties from the χ^2 fit. However, these uncertainties do not follow Gaussian statistics. The error bars must thus correctly be interpreted! Moreover, if most parameters are very well defined (*e.g.* $R_{\rm out}$), some are poorly constrained (*e.g.* only a maximum value of q can be deduced). The CO abundance in the disk X(^{13}CO) is computed from the total disk mass, itself measured from the continuum data.

	^{13}CO & continuum data	
Assumed Distance	D (pc) =	140
Systemic velocity	$V_{\rm LSR}$ (km.s^{-1}) =	7.30 ±0.04
Orientation	PA (°) =	32 ±2
Inclination	i (°) =	80 ±5
Outer radius	$R_{\rm out}$ (AU) =	$420 \pm^{50}_{20}$
Turbulent linewidth	Δv (km.s^{-1}) =	0.40 ±0.04
Abundance & H_2 Density law: $n(r) = n_{100}\left(\frac{r}{100\text{AU}}\right)^{-s}$		
^{13}CO reference	$X^{13}_{\rm TMC1}$ =	$1\ 10^{-6}$
^{13}CO abundance	X(^{13}CO) =	$1.7\ 10^{-7} \pm 0.2\ 10^{-7}$
^{13}CO depletion	f(^{13}CO) =	6 ±1
Density at 100 AU	n_{100} (cm^{-3}) $\simeq$	$4.7\ 10^{7} \pm 0.9\ 10^{7}$
Density exponent	$s \simeq$	2.8 ±0.3
Temperature law: $T(r) = T_{100}\left(\frac{r}{100\text{ AU}}\right)^{-q}$		
Temperature at 100 AU	T_{100} (K) =	$28 \pm^{5}_{4}$
Temperature exponent	$q \simeq$	0 (< 0.6)
Velocity law: $V(r) = V_{100}\left(\frac{r}{100\text{ AU}}\right)^{-v}$		
Velocity at 100 AU	V_{100} (km.s^{-1}) =	-1.95 ±0.05
Velocity exponent	v =	0.48 ±0.06
Stellar mass	M_* ($M_\odot$) =	0.44 ±0.03
Surface Density law: $\Sigma(r) = \Sigma_{100}\left(\frac{r}{100\text{ AU}}\right)^{-p}$		
Surf. dens. at 100 AU	Σ_{100} (cm^{-2}) =	$2.1\ 10^{22} \pm 0.4\ 10^{22}$
	Σ_{100} (g.cm^{-2}) =	0.092 ±0.018
Surf. dens. exponent	$p \simeq$	1 – 1.5
Scale Height law: $H(r) = H_{100}\left(\frac{r}{100\text{ AU}}\right)^{-h}$		
Scale Height at 100 AU	H_{100} (AU) $\simeq$	19
Scale Height exponent	$h \simeq$	1.4

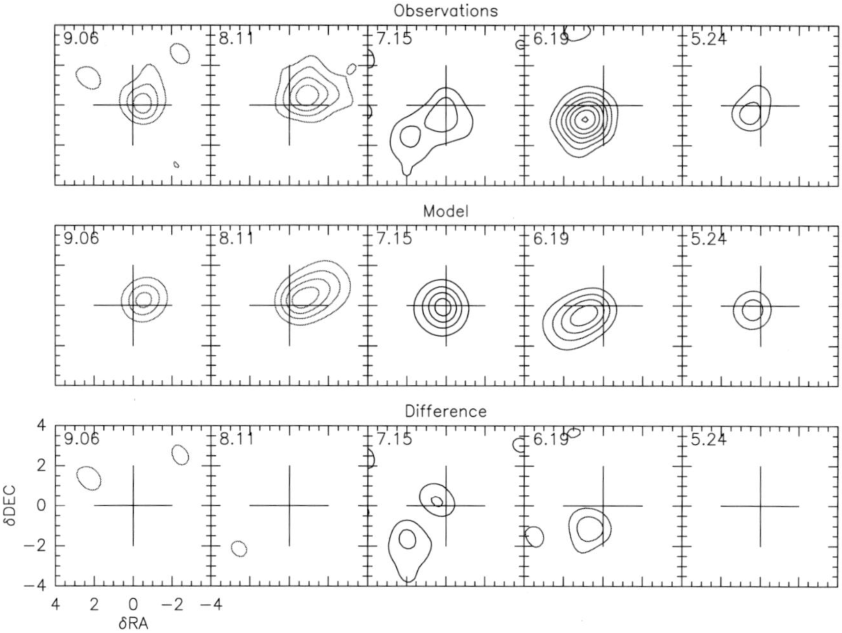

Fig. 2. Channel maps of the ^{13}CO (J=2–1) emission. Observations, best model and (observation-model) are plotted from top to bottom. All contours are equally spaced by a value of 29 mJy/beam (3 sigma). An asymmetry (due to confusion with the parent cloud) is detected at the 6σ level close to the systemic velocity (i.e. in the 6.19 and 7.15 $\mathrm{km\,s^{-1}}$ velocity channels).

iii) Using a distance of 140 pc and the magnitude of the Keplerian rotation, we deduce a mass of $0.44\,M_\odot$ for the protostar. A detailed comparison with previous optical/NIR studies will be made in a forthcoming paper.

Fig. 2 can be used to evaluate the level of confidence of the parameters of Table 1. An asymetry is clearly seen in the 6.19 and 7.15 $\mathrm{km\,s^{-1}}$ velocity channels. We assign this asymetry to confusion from the parent cloud as a significant detection of $C^{18}O$ (not shown here) is observed in the same blue shifted velocity channels. To avoid contamination of the derived parameters, only velocity channels devoid of confusion were used in the fitting procedure.

References

1. Burrows, C. J. and Stapelfeldt et al., K. R., 1996, Apj **473**, 437
2. Wood, K. and Wolff, M. J. and Bjorkman, J. E. and Whitney, B., 2002, Apj **564**, 887–895
3. Stapelfeldt, K. R. and Padgett, D. L., *Science with the Atacama Large Millimeter Array (ALMA)*,(1999)
4. Guilloteau, S. and Dutrey, A., 1998, A&A**339**, 467-476

Author Index

Subject Index

Source Index

Molecule Index

SPRINGER PROCEEDINGS IN PHYSICS

50 **Magnetic Properties of Low-Dimensional Systems II**
New Developments
Editors: L.M. Falicov, F. Mejía-Lira, and J.L. Morán-López

51 **The Physics and Chemistry of Organic Superconductors**
Editors: G. Saito and S. Kagoshima

52 **Dynamics and Patterns in Complex Fluids**
New Aspects of the Physics–Chemistry Interface
Editors: A. Onuki and K. Kawasaki

53 **Computer Simulation Studies in Condensed-Matter Physics III**
Editors: D.P. Landau, K.K. Mon, and H.-B. Schüttler

54 **Polycrystalline Semiconducturs II**
Editors: J.H. Werner and H.P. Strunk

55 **Nonlinear Dynamics and Quantum Phenomena in Optical Systems**
Editors: R. Vilaseca and R. Corbalán

56 **Amorphous and Crystalline Silicon Carbide III, and Other Group IV–IV Materials**
Editors: G.L. Harris, M.G. Spencer, and C.Y. Yang

57 **Evolutionaly Trends in the Physical Sciences**
Editors: M. Suzuki and R. Kubo

58 **New Trends in Nuclear Collective Dynamics**
Editors: Y. Abe, H. Horiuchi, and K. Matsuyanagi

59 **Exotic Atoms in Condensed Matter**
Editors: G. Benedek and H. Schneuwly

60 **The Physics and Chemistry of Oxide Superconductors**
Editors: Y. Iye and H. Yasuoka

61 **Surface X-Ray and Neutron Scattering**
Editors: H. Zabel and I.K. Robinson

62 **Surface Science**
Lectures on Basic Concepts and Applications
Editors: F.A. Ponce and M. Cardona

63 **Coherent Raman Spectroscopy**
Recent Advances
Editors: G. Marowsky and V.V. Smirnov

64 **Superconducting Devices and Their Applications**
Editors: H. Koch and H. Lübbing

65 **Present and Future of High-Energy Physics**
Editors: K.-I. Aoki and M. Kobayashi

66 **The Structure and Conformation of Amphiphilic Membranes**
Editors: R. Lipowsky, D. Richter, and K. Kremer

67 **Nonlinearity with Disorder**
Editors: F. Abdullaev, A.R. Bishop, and S. Pnevmatikos

68 **Time-Resolved Vibrational Spectroscopy V**
Editor: H. Takahashi

69 **Evolution of Dynamical Structures in Complex Systems**
Editors: R. Friedrich and A. Wunderlin

70 **Computational Approaches in Condensed-Matter Physics**
Editors: S. Miyashita, M. Imada, and H. Takayama

71 **Amorphous and Crystalline Silicon Carbide IV**
Editors: C.Y. Yang, M.M. Rahman, and G.L. Harris

72 **Computer Simulation Studies in Condensed-Matter Physics IV**
Editors: D.P. Landau, K.K. Mon, and H.-B. Schüttler

73 **Surface Science**
Principles and Applications
Editors: R.F. Howe, R.N: Lamb, and K. Wandelt

74 **Time-Resolved Vibrational Spectroscopy VI**
Editors: A. Lau, F. Siebert, and W. Werncke

75 **Computer Simulation Studies in Condensed-Matter Physics V**
Editors: D.P. Landau, K.K. Mon, and H.-B. Schüttler

76 **Computer Simulation Studies in Condensed-Matter Physics VI**
Editors: D.P. Landau, K.K. Mon, and H.-B. Schüttler

77 **Quantum Optics VI**
Editors: D.F. Walls and J.D. Harvey

78 **Computer Simulation Studies in Condensed-Matter Physics VII**
Editors: D.P. Landau, K.K. Mon, and H.-B. Schüttler

Printing: Strauss GmbH, Mörlenbach
Binding: Schäffer, Grünstadt